AF358637

Molecular Targets of Oxidative Stress: Focus on the Nrf2 Signaling Pathway in Health and Disease

Molecular Targets of Oxidative Stress: Focus on the Nrf2 Signaling Pathway in Health and Disease

Editor

Marcel Bonay

Basel • Beijing • Wuhan • Barcelona • Belgrade • Novi Sad • Cluj • Manchester

Editor
Marcel Bonay
University of Paris Saclay
(UVSQ)
Versailles
France

Editorial Office
MDPI AG
Grosspeteranlage 5
4052 Basel, Switzerland

This is a reprint of articles from the Special Issue published online in the open access journal *Antioxidants* (ISSN 2076-3921) (available at: https://www.mdpi.com/journal/antioxidants/special_issues/Nrf2_Signaling_Pathway).

For citation purposes, cite each article independently as indicated on the article page online and as indicated below:

Lastname, A.A.; Lastname, B.B. Article Title. *Journal Name* **Year**, *Volume Number*, Page Range.

ISBN 978-3-7258-2227-0 (Hbk)
ISBN 978-3-7258-2228-7 (PDF)
doi.org/10.3390/books978-3-7258-2228-7

Contents

 *antioxidants*

Editorial

Molecular Targets of Oxidative Stress: Focus on the Nrf2 Signaling Pathway in Health and Disease

Marcel Bonay [1,2]

1 UVSQ, INSERM END-ICAP, Université Paris-Saclay, 78000 Versailles, France; marcel.bonay@aphp.fr; Tel.: +33-170-429-415

2 Service de Physiologie-Explorations Fonctionnelles Bi-Sites, Hôpitaux Ambroise Paré et Bicêtre, Assistance Publique-Hôpitaux de Paris, 92104 Boulogne, France

1. Introduction

Oxidative stress, known to increase the risk of multiple metabolic and chronic disorders or cancer development, is defined as an imbalance between the production of reactive oxygen species (ROS) and the capacity of antioxidants to counteract the deleterious effects of oxidants. To regulate the oxidation/reduction (redox) balance, numerous antioxidant enzymes and nonenzymatic antioxidants exist. Free radicals activate transcription factors to promote antioxidant production and mitochondrial biogenesis. One of these transcription factors, nuclear factor erythroid 2-related factor 2 (Nrf2), is a master regulator of antioxidant and anti-inflammatory responses. Indeed, Nrf2 contributes to redox balance by initiating the transcription of hundreds of genes involved in antioxidant and cytoprotective responses. Better understanding of the molecular targets of oxidative stress and their interaction with the Nrf2 signaling pathway would strengthen the relevance of their preventive or therapeutic use in health and diseases.

For this Special Issue, researchers were invited to submit original articles or review articles on different aspects of the modulation of oxidative stress in animal models or in humans. The topics included focused on biological and physiological effects of the Nrf2 signaling pathway in chronic diseases or their prevention.

2. An Overview of Published Articles

Four original articles and three reviews fell under the neuroscience theme and/or focused on neurologic disorders.

Amyotrophic lateral sclerosis is a progressive and severe disease caused by the degeneration of motor neurons; however, available treatments have only limited clinical benefits. Jiménez-Villegas et al. (contribution 1) realized targeted transcriptional profiling in leukocytes from patients with hexanucleotide expansion located in the first intron of the C9orf72 gene, a genetic alteration observed in 40% of familial amyotrophic lateral sclerosis (ALS) patients, and found an altered expression of 10 redox genes compared with healthy controls. In an in vitro model that reproduced toxic mechanisms attributed to C9orf72 pathology, Nrf2 activation by dimethyl fumarate was shown to protect motor-neuron-like hybrid cells against dipeptide repeat toxicity. This study highlights Nrf2 as a potential therapeutic target for ALS patients with C9orf72 gene expansion repeats.

In the treatment of other neurological disorders without therapeutic resources apart from mechanical ventilation, high spinal cord injuries (SCIs) induce the deafferentation of phrenic motoneurons, leading to respiratory muscle paralysis. Michel-Flutot et al. (contribution 2) assessed the antioxidant response in phrenic motoneurons involving the AMPK–Nrf2 signaling pathway following C2 spinal cord lateral hemi-section in rats, and found a reduced expression of phosphorylated AMPK and Nrf2 at one hour post injury followed by a rebound of expression at one day post injury. In the total spinal cord around phrenic motoneurons, increases in phosphorylated AMPK and Nrf2 occurred at

Citation: Bonay, M. Molecular Targets of Oxidative Stress: Focus on the Nrf2 Signaling Pathway in Health and Disease. *Antioxidants* 2024, 13, 262. https://doi.org/10.3390/antiox13030262

Received: 16 February 2024
Accepted: 18 February 2024
Published: 21 February 2024

three days post injury, showing the differential antioxidant responses between phrenic motoneurons and other cell types. The modulation of the AMPK–Nrf2 signaling pathway could improve the antioxidant response and help in spinal rewiring to these deafferented phrenic motoneurons in patients with high SCIs.

Neuroinflammation and failing redox homeostasis might be involved in the pathophysiology of neurological sequelae in long-COVID. Affecting nearly 65 million people worldwide, long-COVID is the subject of growing research [1]. Ercegovac et al. (contribution 3) investigated whether variations in antioxidant genetic profile were associated with neurological sequelae in long-COVID. Neurological examination and antioxidant genetic profile (SOD2, GPXs, and GSTs) determination, as well as genotype analysis of the key redox-sensitive transcription factors Nrf2 and ACE2, which affects binding and internalization of the coronavirus, were conducted in 167 COVID-19 patients. They found that individuals carrying the GSTP1 Val or GSTO1 Asp allele exhibit lower odds of long-COVID myalgia development, both independently and in combination. Furthermore, the combined presence of GSTP1 Ile and GSTO1 Ala alleles exhibited cumulative risk regarding long-COVID myalgia in carriers of the combined GPX1LeuLeu/GPX3CC genotype. Moreover, individuals carrying the combined GSTM1-null/GPX1LeuLeu genotype were more prone to developing long-COVID "brain fog", while probability was further enhanced if the Nrf2 A allele was also present. The fact that certain genetic variants of antioxidant enzymes affect the probability of long-COVID manifestations further highlights the involvement of genetic susceptibility, both when SARS-CoV-2 infection is initiated in the host cells, but also months later.

The P2X7 receptor (P2X7R) is a cation-permeable ATP ligand-gated ion channel; its activation is involved in neuronal excitability, neuroinflammation, and functions of astrocytes and microglia. Lee et al. (contribution 4) assessed responses to LPS in the mouse hippocampus in vivo and showed that P2X7R increases LPS-induced neuroinflammation by leading to Nrf2 degradation, aberrant glutamate–glutamine cycle activity, and impaired cystine/cysteine uptake, which inhibit glutathione biosynthesis. They suggest that the targeting of P2X7R, which would exert nitrosative stress with iNOS in a positive feedback manner, may be one of the most important therapeutic strategies against nitrosative stress under pathophysiological conditions.

Neurodegenerative diseases, such Alzheimer's disease and Parkinson's disease, represent a major health problem, with millions of people affected worldwide; these conditions are leading causes of disability [2]. In their review, Amoroso et al. (contribution 5) described the implication of oxidative stress and Nrf2 activation in neurodegenerative diseases. The potential therapeutic interests of small molecules inducing Nrf2, natural Nrf2 activators, and multitargeting Nrf2 activators were discussed.

In another review, Sani et al. (contribution 6) highlighted the role of Nrf2 in depression. Most studies were performed in murine models of depression, and indicated that the Nrf2–antioxidant pathway repaired neuroinflammation and restored behavioral test performance. Although Nrf2 might represent a potential therapeutic target, the authors emphasized that animal models of depression are difficult to translate to humans.

Genetic diversity and interindividual differences in post-translational modifications might be critical for developing oxidative-stress-related diseases, such as cancer and neurological disorders. The example of Human NAD(P)H:quinone oxidoreductase 1 (hNQO1), a multifunctional and antioxidant stress protein whose expression is controlled by the Nrf2 signaling pathway, is highlighted by Pey, A.L. (contribution 7) in an exhaustive review.

In vitro or ex vivo models were used in five studies to assess oxidative stress and the Nrf2 pathway in human clinical disorders.

Lim et al. (contribution 8) showed that a methanol extract of Ehretia tinifolia protected mouse immortalized Kupffer cells from lipopolysaccharide (LPS)-mediated oxidative stress and excessive inflammatory responses by activating antioxidant Nrf2/HO-1 and inhibiting pro-inflammatory NF-κB and MAPKs.

Deramaudt et al. (contribution 9) highlighted the capacity of repetitive magnetic stimulation (rMS) to activate the non-canonical Nrf2 pathway, modulate macrophage function, and enhance the host's defense against bacterial infection.

Tabolacci et al. (contribution 10) demonstrated how rutin, a bioflavonoid found in some vegetables and fruits, may play a potentially cytoprotective role against UVA-induced skin damage through a purely antiapoptotic mechanism.

Pattabiraman et al. (contribution 11) found that various microRNAs carried by exosomes of non-pigmentary ciliary epithelium subjected to acute and chronic oxidative stress can regulate Nrf2 and Keap1. They suggest that this regulation may influence trabecular meshwork health and functionality, aqueous humor drainage, intraocular pressure, and primary open-angle glaucoma pathophysiology.

Slaven et al. (contribution 12) suggest that the cellular response of human mesenchymal stem cells and human lung microvascular endothelial cells to chronic low-dose-rate gamma radiation, with a downregulation of Nrf2 and related genes, is significantly different from the cellular response to acute, high-dose-rate radiation. Instead, the primary cellular response to chronic low-dose-rate gamma radiation is hypoxia and iron deficiency, with increased hypoxia signaling and the increased activation of pathways regulated by iron deficiency.

Two original articles with in vivo models and four reviews are dedicated to the Nrf2 signaling pathway in metabolism, inflammation, and cancers.

Many studies have shown that exercise training improves skeletal muscle health via multiple adaptive pathways. Bhat et al. (contribution 13) assessed the effects of 3 weeks of treadmill exercise training in wild-type and iMS-Nrf2flox/flox inducible muscle-specific Nrf2 mice. Their results suggest a critical role of Nrf2 in the beneficial effects of skeletal muscle and adaptation to exercise training.

Inflammatory bowel diseases (IBDs), such as Crohn's disease and ulcerative colitis, are chronic inflammatory disorders that affect the gastrointestinal tract, and their incidence and clinical severity have increased worldwide [3]. Tian et al. (contribution 14) assessed the effects of dietary glucoraphanin supplementation on mitochondrial dysfunction and oxidative stress in an acute colitis mouse model induced by dextran sulfate sodium. Glucoraphanin supplementation protected the colonic histological structure, suppressed inflammatory cytokines, reduced macrophage infiltration in colonic tissues, activated AMP-activated protein kinase (AMPK), peroxisome proliferator-activated receptor-gamma coactivator (PGC)-1α, and nuclear factor erythroid 2-related factor 2 pathways in the colonic tissues of dextran-sulfate-sodium-treated mice. They suggest that dietary glucoraphanin provides a dietary strategy to alleviate IBD symptoms.

In their review, Yan et al. (contribution 15) discussed the mechanisms of ferroptosis and focused on the regulation of ferroptosis by Nrf2. A list of some clinical applications of targeting the Nrf2 signaling pathway in the treatment of diseases, such as cancers and neurodegenerative and ischemic diseases, was presented. Ishii et al. (contribution 16) clarified how stress activated MAP kinases and cyclin-dependent kinase 5 mediated the nuclear translocation of Nrf2 via Hsp90α-Pin1-Dynein motor transport machinery. Malignant tumors often express enhanced Pin1-Hsp90α signaling pathways; therefore, the authors proposed how these systems might represent a potential therapeutic target for tumor treatments.

In their review, Xia et al. (contribution 17) explained the complex interaction between Nrf2, oxidative stress, lipid metabolism, insulin signaling, and chronic inflammation in obesity. Finally, Sabatino, L. (contribution 18) described the main aspect of thyroid hormone signaling and depicted the role of Nrf2 in oxidant–antioxidant homeostasis in the thyroid hormones system.

3. Conclusions

Most of the studies presented in this Special Issue were dedicated to multiple metabolic and chronic disorders or cancer development. They contribute to improving understanding of the molecular targets of oxidative stress and their interaction with the Nrf2 signaling pathway in some oxidative-stress-related diseases.

Conflicts of Interest: The author declares no conflict of interest.

List of Contributions

1. Jiménez-Villegas, J.; Kirby, J.; Mata, A.; Cadenas, S.; Turner, M.R.; Malaspina, A.; Shaw, P.J.; Cuadrado, A.; Rojo, A.I. Dipeptide Repeat Pathology in C9orf72-ALS Is Associated with Redox, Mitochondrial and NRF2 Pathway Imbalance. *Antioxidants* **2022**, *11*, 1897. https://doi.org/10.3390/antiox11101897.
2. Michel-Flutot, P.; Efthimiadi, L.; Djerbal, L.; Deramaudt, T.B.; Bonay, M.; Vinit, S. AMPK-Nrf2 Signaling Pathway in Phrenic Motoneurons following Cervical Spinal Cord Injury. *Antioxidants* **2022**, *11*, 1665. https://doi.org/10.3390/antiox11091665.
3. Ercegovac, M.; Asanin, M.; Savic-Radojevic, A.; Ranin, J.; Matic, M.; Djukic, T.; Coric, V.; Jerotic, D.; Todorovic, N.; Milosevic, I.; et al. Antioxidant Genetic Profile Modifies Probability of Developing Neurological Sequelae in Long-COVID. *Antioxidants* **2022**, *11*, 954. https://doi.org/10.3390/antiox11050954.
4. Lee, D.-S.; Kim, J.-E. P2X7 Receptor Augments LPS-Induced Nitrosative Stress by Regulating Nrf2 and GSH Levels in the Mouse Hippocampus. *Antioxidants* **2022**, *11*, 778. https://doi.org/10.3390/antiox11040778.
5. Amoroso, R.; Maccallini, C.; Bellezza, I. Activators of Nrf2 to Counteract Neurodegenerative Diseases. *Antioxidants* **2023**, *12*, 778. https://doi.org/10.3390/antiox12030778.
6. Sani, G.; Margoni, S.; Brugnami, A.; Ferrara, O.M.; Bernardi, E.; Simonetti, A.; Monti, L.; Mazza, M.; Janiri, D.; Moccia, L.; et al. The Nrf2 Pathway in Depressive Disorders: A Systematic Review of Animal and Human Studies. *Antioxidants* **2023**, *12*, 817. https://doi.org/10.3390/antiox12040817.
7. Pey, A.L. Phenotypic Modulation of Cancer-Associated Antioxidant NQO1 Activity by Post-Translational Modifications and the Natural Diversity of the Human Genome. *Antioxidants* **2023**, *12*, 379. https://doi.org/10.3390/antiox12020379.
8. Lim, J.S.; Lee, S.H.; Yun, H.; Lee, D.Y.; Cho, N.; Yoo, G.; Choi, J.U.; Lee, K.Y.; Bach, T.T.; Park, S.-J.; et al. Inhibitory Effects of Ehretia tinifolia Extract on the Excessive Oxidative and Inflammatory Responses in Lipopolysaccharide-Stimulated Mouse Kupffer Cells. *Antioxidants* **2023**, *12*, 1792. https://doi.org/10.3390/antiox12101792.
9. Deramaudt, T.B.; Chehaitly, A.; Charrière, T.; Arnaud, J.; Bonay, M. High-Frequency Repetitive Magnetic Stimulation Activates Bactericidal Activity of Macrophages via Modulation of p62/Keap1/Nrf2 and p38 MAPK Pathways. *Antioxidants* **2023**, *12*, 1695. https://doi.org/10.3390/antiox12091695.
10. Tabolacci, E.; Tringali, G.; Nobile, V.; Duca, S.; Pizzoferrato, M.; Bottoni, P.; Clementi, M.E. Rutin Protects Fibroblasts from UVA Radiation through Stimulation of Nrf2 Pathway. *Antioxidants* **2023**, *12*, 820. https://doi.org/10.3390/antiox12040820.
11. Pattabiraman, P.P.; Feinstein, V.; Beit-Yannai, E. Profiling the miRNA from Exosomes of Non-Pigmented Ciliary Epithelium-Derived Identifies Key Gene Targets Relevant to Primary Open-Angle Glaucoma. *Antioxidants* **2023**, *12*, 405. https://doi.org/10.3390/antiox12020405.
12. Slaven, J.E.; Wilkerson, M.; Soltis, A.R.; Rittase, W.B.; Bradfield, D.T.; Bylicky, M.; Cary, L.; Tsioplaya, A.; Bouten, R.; Dalgard, C.; et al. Transcriptomic Profiling and Pathway Analysis of Mesenchymal Stem Cells Following Low Dose-Rate Radiation Exposure. *Antioxidants* **2023**, *12*, 241. https://doi.org/10.3390/antiox12020241.
13. Bhat, A.; Abu, R.; Jagadesan, S.; Vellichirammal, N.N.; Pendyala, V.V.; Yu, L.; Rudebush, T.L.; Guda, C.; Zucker, I.H.; Kumar, V.; et al. Quantitative Proteomics Identifies Novel Nrf2-Mediated Adaptative Signaling Pathways in Skeletal Muscle Following Exercise Training. *Antioxidants* **2023**, *12*, 151. https://doi.org/10.3390/antiox12010151.
14. Tian, Q.; Xu, Z.; Sun, Q.; Iniguez, A.B.; Du, M.; Zhu, M.-J. Broccoli-Derived Glucoraphanin Activates AMPK/PGC1α/NRF2 Pathway and Ameliorates Dextran-Sulphate-Sodium-Induced Colitis in Mice. *Antioxidants* **2022**, *11*, 2404. https://doi.org/10.3390/antiox11122404.
15. Yan, R.; Lin, B.; Jin, W.; Tang, L.; Hu, S.; Cai, R. NRF2, a Superstar of Ferroptosis. *Antioxidants* **2023**, *12*, 1739. https://doi.org/10.3390/antiox12091739.

16. Ishii, T.; Warabi, E.; Mann, G.E. Stress Activated MAP Kinases and Cyclin-Dependent Kinase 5 Mediate Nuclear Translocation of Nrf2 via Hsp90α-Pin1-Dynein Motor Transport Machinery. *Antioxidants* **2023**, *12*, 274. https://doi.org/10.3390/antiox12020274.
17. Xia, Y.; Zhai, X.; Qiu, Y.; Lu, X.; Jiao, Y. The Nrf2 in Obesity: A Friend or Foe? *Antioxidants* **2022**, *11*, 2067. https://doi.org/10.3390/antiox11102067.
18. Sabatino, L. Nrf2-Mediated Antioxidant Defense and Thyroid Hormone Signaling: A Focus on Cardioprotective Effects. *Antioxidants* **2023**, *12*, 1177. https://doi.org/10.3390/antiox12061177.

References

1. Cervia-Hasler, C.; Brüningk, S.C.; Hoch, T.; Fan, B.; Muzio, G.; Thompson, R.C.; Ceglarek, L.; Meledin, R.; Westermann, P.; Emmenegger, M.; et al. Persistent complement dysregulation with signs of thromboinflammation in active Long Covid. *Science* **2024**, *383*, eadg7942. [CrossRef] [PubMed]
2. GBD 2015 Neurological Disorders Collaborator Group. Global, regional, and national burden of neurological disorders during 1990–2015: A systematic analysis for the Global Burden of Disease Study 2015. *Lancet Neurol.* **2017**, *16*, 877–897. [CrossRef] [PubMed]
3. Ng, S.C.; Shi, H.Y.; Hamidi, N.; Underwood, F.E.; Tang, W.; Benchimol, E.I.; Panaccione, R.; Ghosh, S.; Wu, J.C.Y.; Chan, F.K.L.; et al. Worldwide incidence and prevalence of inflammatory bowel disease in the 21st century: A systematic review of population-based studies. *Lancet* **2017**, *390*, 2769–2778; Erratum in *Lancet* **2020**, *396*, e56. [CrossRef] [PubMed]

Article

P2X7 Receptor Augments LPS-Induced Nitrosative Stress by Regulating Nrf2 and GSH Levels in the Mouse Hippocampus

Duk-Shin Lee and Ji-Eun Kim *

Department of Anatomy and Neurobiology, Institute of Epilepsy Research, College of Medicine, Hallym University, Chuncheon 24252, Korea; dslee84@hallym.ac.kr
* Correspondence: jieunkim@hallym.ac.kr

Abstract: P2X7 receptor (P2X7R) regulates inducible nitric oxide synthase (iNOS) expression/activity in response to various harmful insults. Since P2X7R deletion paradoxically decreases the basal glutathione (GSH) level in the mouse hippocampus, it is likely that P2X7R may increase the demand for GSH for the maintenance of the intracellular redox state or affect other antioxidant defense systems. Therefore, the present study was designed to elucidate whether P2X7R affects nuclear factor-erythroid 2-related factor 2 (Nrf2) activity/expression and GSH synthesis under nitrosative stress in response to lipopolysaccharide (LPS)-induced neuroinflammation. In the present study, P2X7R deletion attenuated iNOS upregulation and Nrf2 degradation induced by LPS. Compatible with iNOS induction, P2X7R deletion decreased S-nitrosylated (SNO)-cysteine production under physiological and post-LPS treated conditions. P2X7R deletion also ameliorated the decreases in GSH, glutathione synthetase, GS and ASCT2 levels concomitant with the reduced S-nitrosylations of GS and ASCT2 following LPS treatment. Furthermore, LPS upregulated cystine:glutamate transporter (xCT) and glutaminase in $P2X7R^{+/+}$ mice, which were abrogated by P2X7R deletion. LPS did not affect GCLC level in both $P2X7R^{+/+}$ and $P2X7R^{-/-}$ mice. Therefore, our findings indicate that P2X7R may augment LPS-induced neuroinflammation by leading to Nrf2 degradation, aberrant glutamate-glutamine cycle and impaired cystine/cysteine uptake, which would inhibit GSH biosynthesis. Therefore, we suggest that the targeting of P2X7R, which would exert nitrosative stress with iNOS in a positive feedback manner, may be one of the important therapeutic strategies of nitrosative stress under pathophysiological conditions.

Keywords: ASCT2; glutamate-glutamine cycle; glutaminase; glutamine synthase; glutathione synthetase; iNOS; S-nitrosylated cysteine; xCT

Citation: Lee, D.-S.; Kim, J.-E. P2X7 Receptor Augments LPS-Induced Nitrosative Stress by Regulating Nrf2 and GSH Levels in the Mouse Hippocampus. *Antioxidants* **2022**, *11*, 778. https://doi.org/10.3390/antiox11040778

Academic Editor: Marcel Bonay

Received: 23 March 2022
Accepted: 12 April 2022
Published: 13 April 2022

Publisher's Note: MDPI stays neutral with regard to jurisdictional claims in published maps and institutional affiliations.

1. Introduction

P2X7 receptor (P2X7R) is one of the cation-permeable ATP ligand-gated ion channels. P2X7R activation is involved in neuronal excitability, neuroinflammation and functions of astrocytes, as well as microglia [1–4]. P2X7R also regulates the generation of reactive oxygens species (ROS) and nitric oxide (NO) in response to various harmful insults [5,6]. Indeed, P2X7R activation enhances lipopolysaccharide (LPS)-induced inducible NO synthase (iNOS) expression [7–9]. NO can alter protein structure and function and exert biological effects directly by binding to free thiol groups on cysteine residues of target proteins forming S-nitrosylated (SNO)-proteins (known as S-nitrosylation) [10]. Furthermore, we have recently reported that S-nitrosylation of P2X7R facilitates its trafficking on the cell membrane [11]. Therefore, it is plausible that P2X7R may augment nitrosative stress with iNOS in a positive feedback manner, which would play an important role in the pathogenesis of various neurological diseases related to inflammations.

Glutathione (GSH) is an endogenous tripeptide (glutamate-cysteine-glycine) antioxidant. GSH is synthesized via several steps. Glutamate cysteine ligase (GCLC) is the rate-limiting enzyme that converts glutamate and cysteine (mostly derived from cystine;

the oxidized dimer form of cysteine) to γ-glutamylcysteine. GSH synthetase (GSHS) adds glycine (derived from exogenous glycine or serine) to γ-glutamylcysteine for generating GSH in an ATP-driven reaction. Glutaminase (GLS) and glutamine synthase (GS) are also involved in GSH synthesis by regulating the glutamate-glutamine cycle [12]. In addition, some membrane transporters mediate GSH synthesis to supply neutral amino acids. Solute carrier 1 (SLC1) A4 and A5 known as ASCT1 and ASCT2, respectively, preferentially transfer the substrates alanine, serine and cysteine (term ASC) [13]. Cystine:glutamate transporter (xCT or SLC7a11) exchanges cystine for glutamate (or cysteine) with a molar ratio of 1:1 by the substrate gradients across the plasma membrane under physiological conditions [14,15]. Interestingly, P2X7R activation decreases glutamate uptake and GS activity in astrocytes, although P2X7R cannot affect the release of GSH. Furthermore, P2X7R activation regulates xCT-mediated glutamate and ASCT2-mediated D-serine releases from astrocytes [4,14–17]. Therefore, it is plausible that P2X7R activity would negatively regulate GSH levels in the brain. However, we have recently reported that P2X7R deletion paradoxically decreases the basal GSH level in the mouse hippocampus, although it does not influence GCLC, GSHS and GLS expression levels [18]. Considering P2X7R-mediated generations of ROS and NO [5,6], it is likely that P2X7R deletion may reduce the demand of GSH for maintenance of the intracellular redox state or affect other antioxidant defense systems, which has been elusive.

Nuclear factor-erythroid 2-related factor 2 (Nrf2), a redox-sensitive transcription factor, regulates antioxidant-response element (ARE)-dependent transcription and the expression of antioxidant enzymes, which contributes to a broad spectrum of cellular functions, such as redox balance, cell cycle, cell death, immunity, metabolism, selective protein degradation, development, aging and carcinogenesis [19,20]. Under physiological conditions, a cytoplasmic repressor Kelch-like erythroid cell-derived protein with CNC homology (ECH)-associated protein 1 (Keap1) binds to Nrf2, which prevents nuclear Nrf2 translocation and mediates Nrf2 polyubiquitinylation and subsequent proteasomal degradation through the cullin-3 (Cul3)-based E3 ubiquitin ligase complex [21–23]. Oxidation and S-nitrosylation of SH-groups in Keap1 lead to the liberation of Nrf2 from Keap1 binding and increase Nrf2-mediated transactivation of multiple ARE-bearing genes, such as GSH synthetic enzymes [21–23]. Since LPS administration decreases GSH levels in the brain [24], it is presumable that P2X7R-mediated nitrosative stress may affect Nrf2 activity, which regulates GSH synthesis. Indeed, P2X7R deletion prevents the diminished efficacy of N-acetylcysteine (NAC, a GSH precursor) in GSH synthesis following SIN-1 (500 μM, a generator of nitric oxide, superoxide and peroxynitrite) treatment [18]. Furthermore, glutamine regulates both GSHS and Nrf2 levels [25]. However, it is unknown whether P2X7R-mediated nitrosative stress affects Nrf2 activity/expression and GSH synthesis in response to LPS-induced neuroinflammation.

Here, we demonstrate, for the first time, that P2X7R augmented nitrosative stress by Nrf2 degradation, impaired GSH synthesis, aberrant glutamate-glutamine cycle and dysfunctions of cysteine transporter following LPS treatment. Therefore, we suggest that P2X7R may be responsible for nitrosative stress in inflammatory conditions.

2. Materials and Methods

2.1. Experimental Animals, Chemicals and LPS Treatment

We used male C57BL/6J ($P2X7R^{+/+}$, WT) and $P2X7R^{-/-}$ (KO) mice (60- to 90-day-old, 25–30 g, The Jackson Laboratory, USA) in the present study. Animals were given a commercial diet and water *ad libitum* under controlled conditions (22 °C ± 2 °C, 55% ± 5% humidity, and 12-h light/12-h dark cycle). All experimental protocols described below were approved by the Institutional Animal Care and Use Committee of Hallym University (Chuncheon, South Korea, Code number: Hallym 2018-3, approval date: 30 April 2018 and Hallym 2021-30, approval date: 17 May 2021). Every effort was made to reduce the number of animals employed and to minimize animal discomfort. All reagents were obtained from Sigma-Aldrich (St. Louis, MO, USA), except as noted. Animals were treated with LPS

(5 mg/kg i.p.). Control animals received an equal volume of normal saline instead of LPS. Three days after LPS injection, animals were used for GSH assay, immunohistochemistry, Western blot and measurements of *S*-nitrosylation.

2.2. GSH Assay

Animals were sacrificed by decapitation. Hippocampal tissues were rapidly removed and sonicated with 0.5 mL of 5% sulfosalicylic acid and centrifuged at $10,000 \times g$ for 10 min at 4 °C. The supernatant was mixed with 1 mL of dithiobis-2-nitrobenzoic acid and 1 mL EDTA in 100 mL sodium phosphate buffer, pH 7.5, and 1 mL NADPH and 200 U/mL of glutathione reductase was added. GSH standards were treated identically, and optical absorbance of samples and standards was measured at 405 nm. Values were normalized to protein content as determined with a BCA protein assay kit (Thermo Scientific, Waltham, MA, USA) [18].

2.3. Immunohistochemistry

Animals were anesthetized with urethane anesthesia (1.5 g/kg, i.p.) and perfused transcardially with 4% paraformaldehyde in 0.1 M phosphate buffer (PB, pH 7.4). Brains were post-fixed in the same fixative overnight. Brain tissues were cryoprotected by infiltration with 30% sucrose overnight. Thereafter, Brains were cryosectioned at 30 μm. Free-floating sections were washed 3 times in PBS (0.1 M, pH 7.3) and incubated with 3% bovine serum albumin in PBS for 30 min at room temperature. Later, sections were incubated with a cocktail solution containing primary antibodies (Table 1) in PBS containing 0.3% Triton X-100 overnight at room temperature. Thereafter, sections were visualized with appropriate Cy2- and Cy3-conjugated secondary antibodies. Some tissues were incubated in biotinylated IgG and avidin-peroxidate complex and developed in 3,3′-diaminobenzidine in 0.1 M Tris buffer. Immunoreaction was observed using an Axio Scope microscope (Carl Zeiss Korea, Seoul, South Korea). To establish the specificity of the immunostaining, a negative control test was carried out with preimmune serum instead of the primary antibody. All experimental procedures in this study were performed under the same conditions and in parallel. To measure fluorescent intensity, five areas/animals (300 μm^2/area) were randomly selected within the hippocampus (5 sections from each animal, $n = 7$ in each group). Thereafter, the mean intensity of each section was measured by using AxioVision Rel. 4.8 and ImageJ software. Intensity measurements were represented as the number of a 256 grayscale. The intensity of each section was standardized by setting the threshold level (mean background intensity obtained from five image inputs). Manipulation of the images was restricted to threshold and brightness adjustments to the whole image.

Table 1. Primary antibodies and lectin used in the present study.

Antigen	Host	Manufacturer (Catalog Number)	Dilution Used
ASCT2	Rabbit	Alomone labs (Jerusalem, Israel) (#ANT-082)	1:500 (WB)
GCLC	Rabbit	Abcam (Waltham, MA, USA) (#ab190685)	1:2000 (WB)
GFAP	Mouse	Millipore (Burlington, MS, USA) (#MAB3402)	1:2000 (IH)
GLS	Rabbit	Abcam (#ab93434)	1:1000 (WB)
GS	Mouse	Millipore (#MAB302)	1:1000 (WB)
GSHS	Rabbit	Abcam (#ab133592)	1:2000 (WB)
IB4		Vector (Los Altos, CA, USA) (#B-1205)	1:200 (histochemistry)
SNO-cysteine	Rabbit	Abcam (#ab94930)	1:1000 (IH)
iNOS	Rabbit	Novus Biologicals (Centennial, CO, USA) (#NB300-605)	1:100 (IH) 1:500 (WB)

Table 1. *Cont.*

Antigen	Host	Manufacturer (Catalog Number)	Dilution Used
Iba-1	Rabbit	Biocare Medical (Pacheco, CA, USA) (#CP 290)	1:500 (IH)
Nrf2	Rabbit	Abcam (#ab137550)	1:200 (IH) 1:1000 (IH)
xCT	Rabbit	Abcam (#ab175186)	1:1000 (WB)
β-actin	Mouse	Sigma (#A5316)	1:5000 (WB)

IH: Immunohistochemistry; WB: Western blot.

2.4. Western Blot

Animals were decapitated under urethane anesthesia (1.5 g/kg, i.p.). The hippocampus was rapidly dissected out and homogenized in lysis buffer. The protein concentration in the supernatant was determined using a Micro BCA Protein Assay Kit (Pierce Chemical, Dallas, TX, USA). Thereafter, Western blot was performed by the standard protocol ($n = 7$ in each group). The primary antibodies used in the present study are listed in Table 1. The bands were detected and quantified on an ImageQuant LAS4000 system (GE Healthcare Korea, Seoul, South Korea). As an internal reference, rabbit anti-β-actin primary antibody (1:5000) was used. The values of each sample were normalized with the corresponding amount of β-actin.

2.5. Measurement of S-Nitrosylation on GS and ASCT2

Modified biotin switch assay was performed with the *S*-nitrosylation Western Blot Kit (ThermoFisher) according to the manufacturer's protocol. Briefly, lysates were reacted with ascorbate in HENS buffer for specific labeling with iodoTMTzero reagents with MMT pretreatment. Protein labeling can be confirmed by Western blot using TMT antibodies. Thereafter, TMT-labeled proteins were purified by Anti-TMT Resin, eluted by TMT elution buffer, and identified by Western blot according to standard procedures. For technical controls, we omitted ascorbate for each sample. The ratio of SNO-protein to total protein was described as *S*-nitrosylation levels [11].

2.6. Data Analysis

Quantitative data are expressed as mean ± standard error of the mean. After the Shapiro–Wilk *W*-test was used to evaluate the values on normality, data were analyzed by the Student *t*-test, paired Student *t*-test, or one-way analysis of variance (ANOVA) followed by Newman–Keuls posthoc test. A $p < 0.05$ is considered to be statistically different.

3. Results

3.1. P2X7R Deletion Ameliorates Microglial Activation, but Not Reactive Astrogliosis in Response to LPS

First, we evaluated the role of P2X7R in glial responses to LPS in the mouse hippocampus in vivo. In $P2X7R^{+/+}$ mice, LPS increased glial fibrillary acidic protein (GFAP, an astroglial marker) ($F_{(1,12)} = 36.75$, $p < 0.001$, one-way ANOVA, $n = 7$, respectively) and ionized calcium-binding adapter molecule-1 (Iba-1, a microglial marker) ($F_{(1,12)} = 60.0$, $p < 0.001$, one-way ANOVA, $n = 7$, respectively) intensities indicating reactive astrogliosis and microgliosis, respectively (Figure 1A–C). P2X7R deletion did not affect GFAP and Iba-1 intensities under physiological condition (Figure 1A–C) but attenuated the increased Iba-1 ($F_{(1,12)} = 44.17$, $p < 0.001$, one-way ANOVA, $n = 7$, respectively), but not GFAP, intensity induced by LPS (Figure 1A–C). These findings indicate that P2X7R may play an important role in microglial activation rather than reactive astrogliosis following LPS treatment.

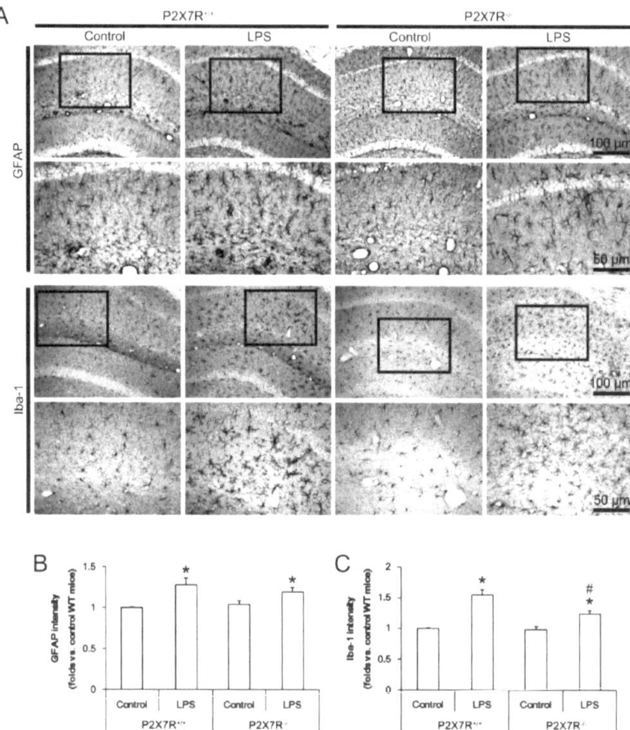

Figure 1. Effects of P2X7R deletion on microglial and astroglial responses to LPS. P2X7R deletion attenuates microglial activation, but not reactive astrogliosis induced by LPS. (**A**) Representative images for GFAP (an astroglial marker) and Iba-1 (a microglial marker) positive cells. Low panels are high magnification photos of boxes in upper panels. (**B**,**C**) Quantification of effects of P2X7R on GFAP and Iba-1 intensities following LPS treatment. Error bars indicate S.E.M. (*,# $p < 0.05$ vs. control and WT mice, $n = 7$, respectively).

3.2. P2X7R Deletion Attenuates LPS-Induced iNOS Induction in Microglia Rather Than Astrocytes

P2X7R activation increases NO production in response to LPS [7], while oxidized ATP (OxATP, a P2X7R antagonist) blocks LPS-induced NO production in vitro [26]. Therefore, we investigated the effect of P2X7R deletion on iNOS induction in responses to LPS in vivo. Under physiological conditions, no difference in iNOS protein levels between $P2X7R^{+/+}$ and $P2X7R^{-/-}$ mice (Figure 2A,B). LPS increased iNOS expression to 1.72- and 1.45-fold of control level in $P2X7R^{+/+}$ and $P2X7R^{-/-}$ mice, respectively ($F_{(3,24)} = 163.7$, $p < 0.001$, one-way ANOVA, $n = 7$, respectively; Figure 2A,B). LPS-induced iNOS induction was lower in $P2X7R^{-/-}$ mice than that in $P2X7R^{+/+}$ mice ($F_{(1,12)} = 27.3$, $p < 0.001$, one-way ANOVA, $n = 7$, respectively; Figure 2A,B). An immunohistochemical study revealed that LPS led to iNOS induction in microglia and astrocytes in $P2X7R^{+/+}$ mice (Figure 2C). LPS-induced iNOS upregulation was higher in microglia than that in astrocytes ($t_{(6)} = 12.7$, $p < 0.001$, paired Student t-test, $n = 7$, respectively; Figure 2C,D). In $P2X7R^{-/-}$ mice, LPS-induced iNOS expression in microglia was 0.28-fold of $P2X7R^{+/+}$ mice level ($F_{(1,12)} = 170.5$, $p < 0.001$, one-way ANOVA, $n = 7$, respectively; Figure 2C,D). Astroglial iNOS level in $P2X7R^{-/-}$ mice was similar to that in $P2X7R^{+/+}$ mice. Thus, iNOS level was higher in astrocytes than that in microglia unlike $P2X7R^{+/+}$ mice ($t_{(6)} = 3.27$, $p = 0.02$, paired Student t-test, $n = 7$, respectively; Figure 2C,D). These findings indicate that P2X7R may enhance iNOS induction in microglia more than astrocytes following LPS treatment.

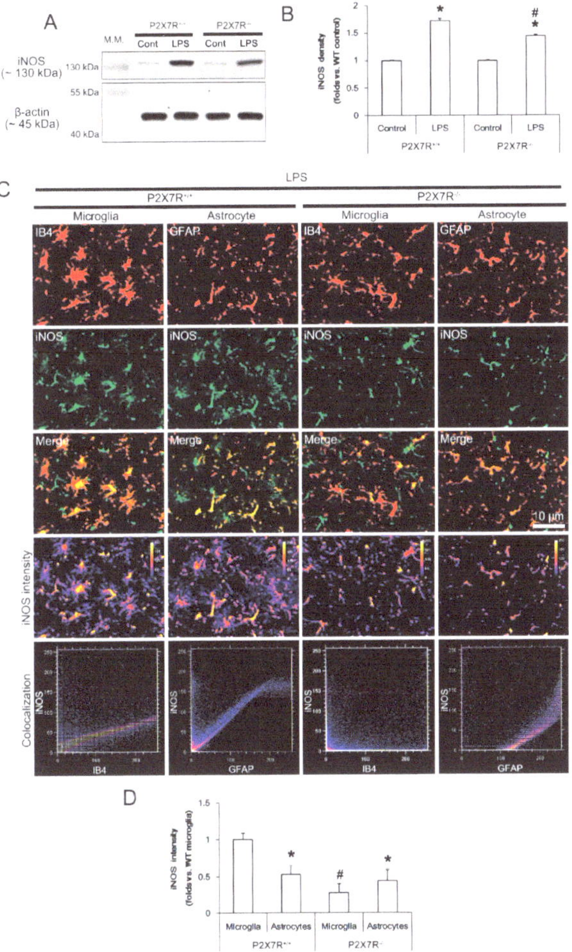

Figure 2. Effects of P2X7R deletion on LPS-induced iNOS induction in microglia and astrocytes. P2X7R deletion ameliorates iNOS induction in microglia rather than astrocytes following LPS injection. (**A**) Representative Western blot of iNOS in the whole hippocampus. (**B**) Quantification of iNOS protein level based on Western blot data. Error bars indicate S.E.M. (*,# $p < 0.05$ vs. control and WT mice, $n = 7$, respectively). (**C**) Representative photos of iNOS expression, intensity and the degree of colocalization in IB4 (a microglial marker) and GFAP (an astroglial marker) positive cells. (**D**) Quantification of iNOS induction in microglia and astrocytes. Error bars indicate S.E.M. (*,# $p < 0.05$ vs. control and WT mice, $n = 7$, respectively). Full-length gel images of Western blot data in (**A**) could be found in Supplementary Figure S1.

3.3. P2X7R Deletion Attenuates LPS-Induced SNO-Cysteine Production in Microglia and Astrocytes

To confirm the effects of P2X7R deletion on iNOS-mediated NO synthesis, we performed the immunohistochemical study using an antibody detecting SNO-cysteine in vivo. Under physiological conditions, SNO-cysteine level in the hippocampus was higher in *P2X7R*$^{+/+}$ mice than that in *P2X7R*$^{-/-}$ mice (Figure 3A). Double immunofluorescent data demonstrated that SNO-cysteine signal was mainly detected in microglia in *P2X7R*$^{+/+}$ mice, while it was weakly observed in *P2X7R*$^{-/-}$ mice (Figure 3B). Following LPS treatment, SNO-cysteine levels were increased in microglia and astrocytes in *P2X7R*$^{+/+}$ and *P2X7R*$^{-/-}$ mice (Figure 3B,C). LPS increased SNO-cysteine production to 3.7- and 1.65-fold of control level in $P2X7R^{+/+}$ and $P2X7R^{-/-}$ mice, respectively ($F_{(3,24)} = 109.1$, $p < 0.001$, one-way ANOVA, $n = 7$, respectively; Figure 3B,C), indicating that LPS-induced SNO-cysteine production was significantly lower in $P2X7R^{-/-}$ mice than that in $P2X7R^{+/+}$ mice ($F_{(1,12)} = 65.6$, $p < 0.001$, one-way ANOVA, $n = 7$, respectively; Figure 3B,C). These findings indicate that P2X7R may reinforce LPS-induced iNOS upregulation that would increase SNO-cysteine production.

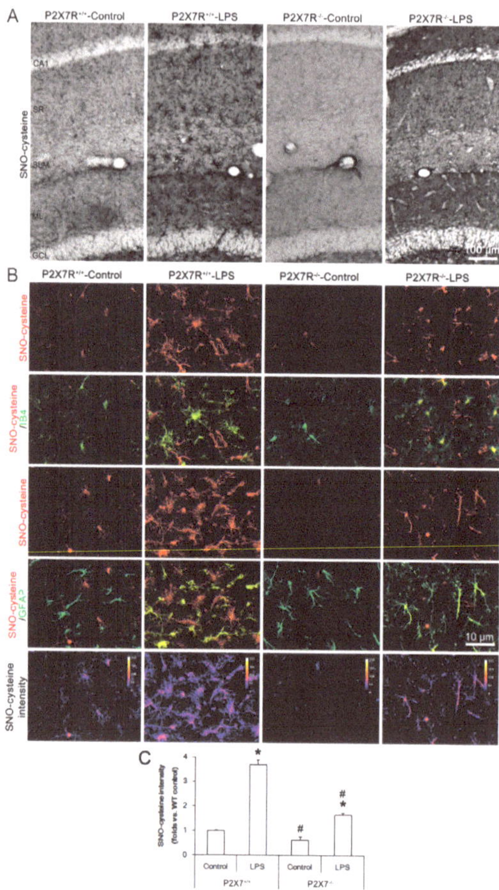

Figure 3. Effects of P2X7R deletion on LPS-induced SNO-cysteine production in microglia and astrocytes. Under physiological conditions, SNO-cysteine level in the hippocampus is higher in $P2X7R^{+/+}$ mice than that in $P2X7R^{-/-}$ mice. LPS increases SNO-cysteine production in microglia and astrocytes within the hippocampus of $P2X7R^{+/+}$ more than $P2X7R^{-/-}$ mice. In $P2X7R^{-/-}$ mice, SNO-cysteine level is lower in microglia than that in astrocytes. (**A**) Representative images for SNO-cysteine in the hippocampus. (**B**) Representative photos of SO-cysteine production in IB4 (a microglial marker) and GFAP (an astroglial marker) positive cells. (**C**) Quantification of SNO-cysteine production in the hippocampus. Error bars indicate S.E.M. (*,# $p < 0.05$ vs. control and WT mice, $n = 7$, respectively).

3.4. P2X7R Deletion Ameliorates LPS-Induced Nrf2 Downregulation

LPS elevates ROS level [27,28], which subsequently decreases Nrf2 level [29]. Furthermore, Nrf2 activation (nuclear accumulation) effectively inhibits LPS-induced iNOS upregulation [30,31]. Therefore, it is likely that P2X7R may facilitate LPS-induced iNOS expression via Nrf2 downregulation, which has been unknown. Thus, we validated Nrf2 protein level in the hippocampi of $P2X7R^{+/+}$ and $P2X7R^{-/-}$ mice following LPS treatment. Under physiological condition, there was no difference in Nrf2 protein level between

$P2X7R^{+/+}$ and $P2X7R^{-/-}$ mice ($F_{(1,12)} = 0.463$, $p = 0.51$, one-way ANOVA, $n = 7$, respectively; Figure 4A,B). LPS decreased total Nrf2 protein level to 0.74-fold of control level in $P2X7R^{+/+}$ mice ($F_{(1,12)} = 33.177$, $p < 0.001$, one-way ANOVA, $n = 7$, respectively; Figure 4A,B), but not $P2X7R^{-/-}$ mice ($F_{(1,12)} = 0.025$, $p = 0.878$, one-way ANOVA, $n = 7$, respectively; Figure 4A,B). LPS decreased total Nrf2 level and its nuclear accumulation in microglia to 0.27- and 0.35-fold of control level in $P2X7R^{+/+}$ mice, respectively, ($F_{(1,12)} = 318.7$ and 84.2, $p < 0.001$, respectively, one-way ANOVA, $n = 7$, respectively; Figure 4C–E) but not in $P2X7R^{-/-}$ mice (Figure 4C–E). LPS also reduced total Nrf2 level and its nuclear accumulation in astrocytes to 0.25- and 0.23-fold of control level in $P2X7R^{+/+}$ mice, respectively ($F_{(1,12)} = 461.8$ and 428.6, $p < 0.001$, respectively, one-way ANOVA, $n = 7$, respectively; Figure 5A–C). In $P2X7R^{-/-}$ mice, LPS declined nuclear Nrf2 accumulation in astrocytes to 0.74-fold of control level without altering total Nrf2 level ($F_{(1,12)} = 15.2$, $p = 0.002$, one-way ANOVA, $n = 7$, respectively; Figure 5A–C). These findings indicate that P2X7R may facilitate the decreases in total Nrf2 level and its nuclear accumulation in microglia and astrocytes following LPS treatment.

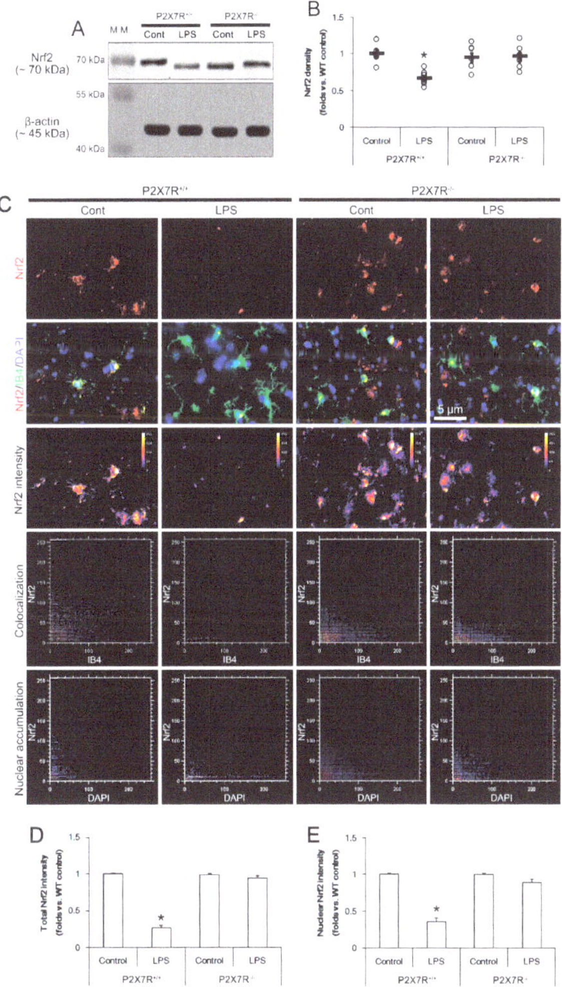

Figure 4. Effects of P2X7R deletion on LPS-induced Nrf2 downregulation in microglia. LPS decreases Nrf2 protein level in the hippocampus of $P2X7R^{+/+}$ mice, but not $P2X7R^{-/-}$ mice, since total Nrf2 level

and its nuclear accumulation are reduced in microglia. (**A**) Representative Western blot of Nrf2 in the whole hippocampus. (**B**) Quantification of iNOS protein level based on Western blot data. Open circles indicate each individual value. Horizontal and error bars indicate the mean value and S.E.M., respectively (*,# $p < 0.05$ vs. control and WT mice, $n = 7$, respectively). (**C**) Representative photos of Nrf2 expression, intensity and the degree of colocalization in IB4 (a microglial marker) positive cells and DAPI (a nuclear marker). (**D**,**E**) Quantification of total and nuclear Nrf2 intensity in microglial. Error bars indicate S.E.M. (* $p < 0.05$ vs. control and WT mice, $n = 7$, respectively). Full-length gel images of Western blot data in this figure could be found in Supplementary Figure S2.

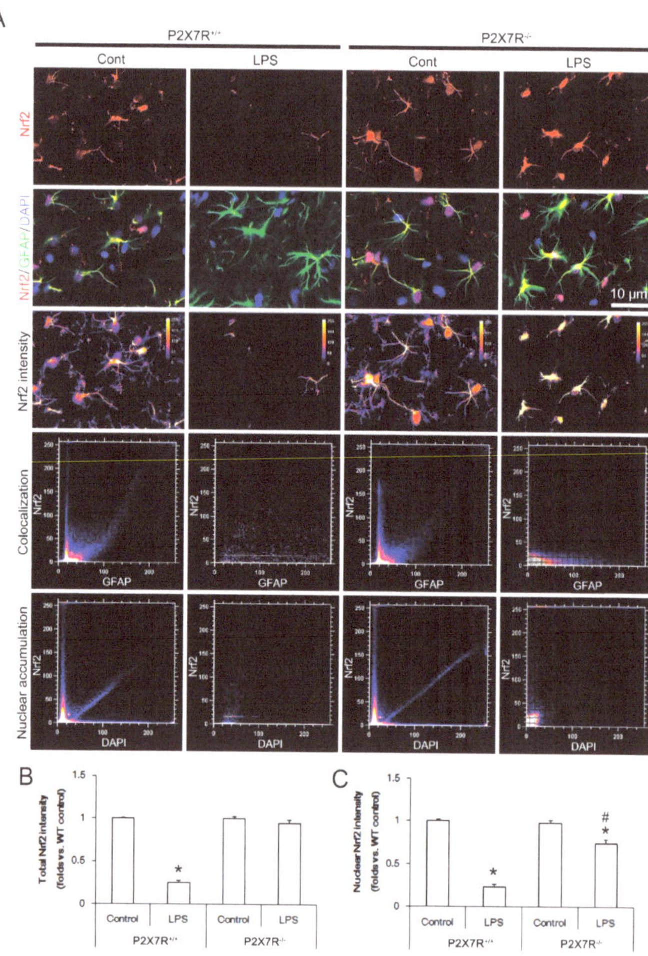

Figure 5. Effects of P2X7R deletion on LPS-induced Nrf2 downregulation in astrocytes. LPS decreases total Nrf2 protein level in astrocytes of *P2X7R⁺/⁺* mice, while it does not in astrocytes of *P2X7R⁻/⁻* mice. LPS also diminishes nuclear Nrf2 protein level in astrocytes of *P2X7R⁺/⁺* mice. LPS-induced Nrf2 downregulation in astrocytes is attenuated in *P2X7R⁻/⁻* mice. (**A**) Representative photos of Nrf2 expression, intensity and the degree of colocalization in GFAP (an astroglial marker) positive cells and DAPI (a nuclear marker). (**B**,**C**) Quantification of total and nuclear Nrf2 intensity in astrocytes. Error bars indicate S.E.M. (*,# $p < 0.05$ vs. control and WT mice, $n = 7$, respectively).

3.5. P2X7R Aggravates the Decreased GSH Concentration Induced by LPS

Recently, we have reported that P2X7R deletion reduces the total GSH level in the hippocampus by regulating the glutamate-glutamine cycle and neutral amino acid transports under physiological conditions, which may be a consequent response to the absence of P2X7R-mediated oxidative or nitrosative stresses [18]. Since Nrf2 plays a key role in the regulation of GSH synthesis [32], it is likely that P2X7R-mediated Nrf2 downregulation may influence GSH levels following LPS injection. Consistent with our previous study [18], the present data showed that total GSH level in $P2X7R^{-/-}$ mice (4.55 ± 0.02 μg/mg protein) was lower than that in $P2X7R^{+/+}$ mice (5 ± 0.14 μg/mg protein; $F_{(1,12)}$ = 21.361, $p < 0.001$; one-way ANOVA, $n = 7$, respectively; Figure 6A). LPS decreased total GSH concentration in $P2X7R^{+/+}$ mice (4.29 ± 0.02 μg/mg protein, 86% of control level; $F_{(1,12)}$ = 11.618, $p = 0.005$, $n = 7$, respectively) more than $P2X7R^{-/-}$ mice (4.31 ± 0.02 μg/mg protein, 95% of control level; $F_{(1,12)}$ = 19.835, $p < 0.001$, $n = 7$, respectively). Thus, there was no difference in total GSH level in both groups following LPS treatment (Figure 6A). Considering GSH decreases LPS-induced NO production by inhibiting iNOS expression [33], our findings indicate that LPS-induced nitrosative stress may be more severe in $P2X7R^{+/+}$ mice and lead to the higher GSH consumption than those in $P2X7R^{-/-}$ mice.

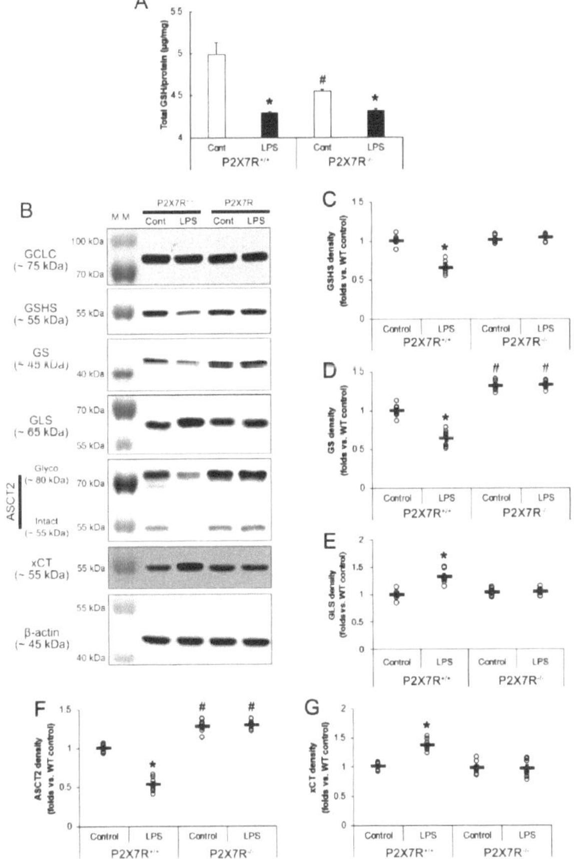

Figure 6. Effects of P2X7R deletion on GSH concentration and expressions of GCLC, GSHS, GS, GLS, ASCT2 and xCT following LPS injection. Under physiological condition, P2X7R deletion reduces GSH level in the hippocampus. However, P2X7R deletion increases GS and ASCT2 levels. LPS declines GSH concentration in $P2X7R^{+/+}$ mice more than in $P2X7R^{-/-}$ mice. LPS decreases GSHS, GS and ASCT2 levels, but increases GLS and xCT levels only in the $P2X7R^{+/+}$ mice. (**A**) Total GSH level in the hippocampus under physiological and post-LPS treated conditions. (**B**) Representative Western blot

of GCLC, GSHS, GS, GLS, ASCT2 and xCT in the whole hippocampi of that $P2X7R^{+/+}$ and $P2X7R^{-/-}$ mice. (**C–G**) Quantification of GSHS, GS, GLS, ASCT2 and xCT levels based on Western blot data. Open circles indicate each individual value. Horizontal and error bars indicate the mean value and S.E.M., respectively (*,# $p < 0.05$ vs. control and WT mice, $n = 7$, respectively). Full-length gel images of Western blot data in (**B**) could be found in Supplementary Figure S3.

3.6. P2X7R Downregulates GSHS, but Not GCLC Expression following LPS Treatment

Next, we explored if P2X7R deletion would also influence GSH production. GCLC is the rate-limiting enzyme in GSH biosynthesis [34]. Thus, we investigated whether P2X7R deletion and/or LPS affect GCLC expression in the mouse hippocampus. Under physiological conditions, GCLC expression level was similarly observed in $P2X7R^{+/+}$ and $P2X7R^{-/-}$ mice. Consistent with previous studies [35,36], LPS did not affect GCLC expression levels in both $P2X7R^{+/+}$ and $P2X7R^{-/-}$ mice ($F_{(3,24)} = 0.12$, $p = 0.95$, one-way ANOVA, $n = 7$, respectively; Figure 6B).

GSHS catalyzes γ-glutamylcysteine and glycine to GSH. The present data showed that P2X7R deletion did not affect GSHS expression under physiological conditions. Consistent with a previous study demonstrating a marked reduction in GSHS expression induced by LPS [37], $P2X7R^{+/+}$ mice showed the decreased GSHS expression (66% of control level) in the hippocampus following LPS treatment, while $P2X7R^{-/-}$ mice did not ($F_{(3,24)} = 55.67$, $p < 0.001$, one-way ANOVA, $n = 7$, respectively; Figure 6B,C). These findings indicate that P2X7R deletion may attenuate the decreased GSH concentration by maintaining GSHS expression following LPS treatment.

3.7. P2X7R Downregulates GS, but Increases GLS Expression following LPS Treatment

Glutamate and glutamine are equally used as precursors for GSH synthesis in astrocytes, which is regulated by Nrf2 [32]. GS catalyzes the conversion of glutamate and ammonia to glutamine and plays a major role in ammonia detoxification, interorgan nitrogen flux, acid–base regulation, cell proliferation and protection from apoptotic stimuli [38]. Recently, we have reported that GS expression is higher in $P2X7R^{-/-}$ mice than $P2X7R^{+/+}$ mice under physiological conditions [18]. Thus, we investigated whether LPS distinctly influences GS expression between $P2X7R^{+/+}$ and $P2X7R^{-/-}$ mice. Consistent with our previous studies [18], GS expression in $P2X7R^{-/-}$ mice was 1.31-fold higher than that in $P2X7R^{+/+}$ mice under physiological condition ($t_{(12)} = 7.871$, $p < 0.001$, Student t-test, $n = 7$, respectively; Figure 6B,D). LPS decreased GS protein level to 0.66-fold of control level in $P2X7R^{+/+}$ mice, but not $P2X7R^{-/-}$ mice ($F_{(3,24)} = 115.13$, $p < 0.001$, one-way ANOVA, $n = 7$, respectively; Figure 6B,D). These findings indicate that LPS may reduce GS protein levels, which may be abrogated by P2X7R deletion.

Astrocytes use glutamine as a precursor for GSH synthesis via the glutamate-glutamine cycle mediated by GLS and GS [39]. Since LPS activates GLS activity/expression [40], we also validated the effect of LPS on GLS protein levels in $P2X7R^{+/+}$ and $P2X7R^{-/-}$ mice. Under physiological conditions, there was no difference in GLS expression between $P2X7R^{+/+}$ and $P2X7R^{-/-}$ mice (Figure 6B,E). LPS increased GLS protein level in $P2X7R^{+/+}$, but not $P2X7R^{-/-}$ mice ($F_{(3,24)} = 19.16$, $p < 0.001$, one-way ANOVA, $n = 7$, respectively; Figure 6B,E). Together with the altered GS expression, our findings suggest that P2X7R deletion may inhibit the changed glutamate-glutamine cycle induced by LPS.

3.8. P2X7R Upregulates xCT, but Decreases ASCT2 Expression following LPS Treatment

The increased GS expression and glutamine concentration potentially facilitate glutamine efflux from astrocytes by inducing the trafficking of ASCT2 [41,42]. Since P2X7R deletion increases ASCT2 expression [18], we investigated whether P2X7R affects ASCT2 expression following LPS treatment. In the present study, ASCT2 expression in $P2X7R^{-/-}$ mice was 1.33-fold higher than that in $P2X7R^{+/+}$ mice under physiological condition ($t_{(12)} = 7.71$, $p < 0.001$, Student t-test, $n = 7$, respectively; Figure 6B,F). LPS decreased ASCT2 protein expression to 0.53-fold of control level in $P2X7R^{+/+}$ mice, but not $P2X7R^{-/-}$ mice

($F_{(3,24)} = 171.12$, $p < 0.001$, one-way ANOVA, $n = 7$, respectively; Figure 6B,F). These findings indicate that P2X7R may inhibit ASCT2-mediated cysteine uptake and glutamine efflux.

xCT also influences GSH synthesis to supply neutral amino acids [14,15]. Thus, we explored whether LPS affects xCT protein levels in $P2X7R^{+/+}$ and $P2X7R^{-/-}$ mice. Under physiological conditions, xCT expression level was similarly detected in $P2X7R^{+/+}$ and $P2X7R^{-/-}$ mice (Figure 6B,G). However, LPS increased xCT expression levels in $P2X7R^{+/+}$ (137% of control level), but not $P2X7R^{-/-}$ mice ($F_{(3,24)} = 20$, $p < 0.001$, one-way ANOVA, $n = 7$, respectively; Figure 6B,G). Considering the changed ASCT2 expression, our findings indicate that P2X7R may inhibit GSH synthesis via dysfunction of cysteine uptake as well as aberrant glutamate-glutamine cycle following LPS treatment.

3.9. P2X7R Regulates S-Nitrosylation of GS and ASCT2 under Physiological and Post-LPS Treated Condition

NO can deplete GSH levels by *S*-nitrosylation of GSH metabolic enzymes [43]. Furthermore, four cysteine residues of GS are *S*-nitrosylated by NO: cysteine 99, 183, 269 and 346 [44]. Indeed, GS activity is highly susceptible to reactive nitrogen and oxygen species, and the inhibition of NO synthesis increases GS activity in rat brain and cultured rat astrocytes [45,46]. On the other hand, NO inhibits ASCT2 transporter activity by oxidation of cysteine residues [47]. To compensate NO-mediated ASCT2 inhibition, NO upregulates ASCT2 protein via *de novo* synthesis in vitro [48]. Thus, we also explored whether P2X7R deletion affects *S*-nitrosylation of GS and ASCT2 induced by LPS. Under physiological conditions, SNO-GS level in $P2X7R^{-/-}$ mice was 0.67-fold of $P2X7R^{+/+}$ mice level ($t_{(12)} = 6.352$, $p < 0.001$, Student *t*-test, $n = 7$, respectively; Figure 7A,B). LPS increased SNO-GS level to 1.4-fold of control level in $P2X7R^{+/+}$ mice ($t_{(12)} = 6.48$, $p < 0.001$, Student *t*-test, $n = 7$, respectively; Figure 7C,D), but not $P2X7R^{-/-}$ mice ($t_{(12)} = 0.98$, $p = 0.346$, Student *t*-test, $n = 7$, respectively; Figure 7E,F). These findings indicate that P2X7R deletion may ameliorate *S*-nitrosylation of GS under physiological- and post-LPS conditions. Similar to the *S*-nitrosylation of GS, SNO-ASCT2 level in $P2X7R^{-/-}$ mice was 0.68-fold of $P2X7R^{+/+}$ mice level under physiological condition ($t_{(12)} = 6.475$, $p < 0.001$, Student *t*-test, $n = 7$, respectively; Figure 8A,B). However, LPS increased SNO-ASCT2 level to 1.32- and 1.16-fold of control level in $P2X7R^{+/+}$ ($t_{(12)} = 5.349$, $p < 0.001$, Student *t*-test, $n = 7$, respectively; Figure 8C,D) and $P2X7R^{-/-}$ mice ($t_{(12)} = 5.16$, $p < 0.001$, Student *t*-test, $n = 7$, respectively; Figure 8E,F), respectively. Thus, P2X7R deletion attenuated increased SNO-ASCT2 production induced by LPS ($F_{(1,12)} = 9.338$, $p = 0.01$, one-way ANOVA, $n = 7$, respectively; Figure 8E,F).

Considering that the increased GS expression induces ASCT2 trafficking [41,42], we also analyzed the correlations of expression/*S*-nitrosylation level between GS and ASCT2. Linear regression analysis showed a direct proportional relationship between GS and ASCT2 levels with linear correlation coefficients of 0.8519 ($t_{(19)} = 7.09$, $p < 0.001$, $n = 21$, respectively; Figure 8G) and 0.8582 ($t_{(19)} = 7.28$, $p < 0.001$, $n = 21$, respectively; Figure 8G) in $P2X7R^{+/+}$ and $P2X7R^{-/-}$ mice, respectively. The SNO-GS level also showed a direct proportional relationship with SNO-ASCT2 level in $P2X7R^{+/+}$ (linear correlation coefficients, 0.7547; $t_{(19)} = 5.01$, $p < 0.001$, $n = 21$, respectively; Figure 8G) and $P2X7R^{-/-}$ mice (linear correlation coefficients, 0.8379; $t_{(19)} = 6.69$, $p < 0.001$; Figure 8G), respectively. These findings indicate that P2X7R may be involved in GS-mediated ASCT2 regulation under physiological and post-LPS treated conditions.

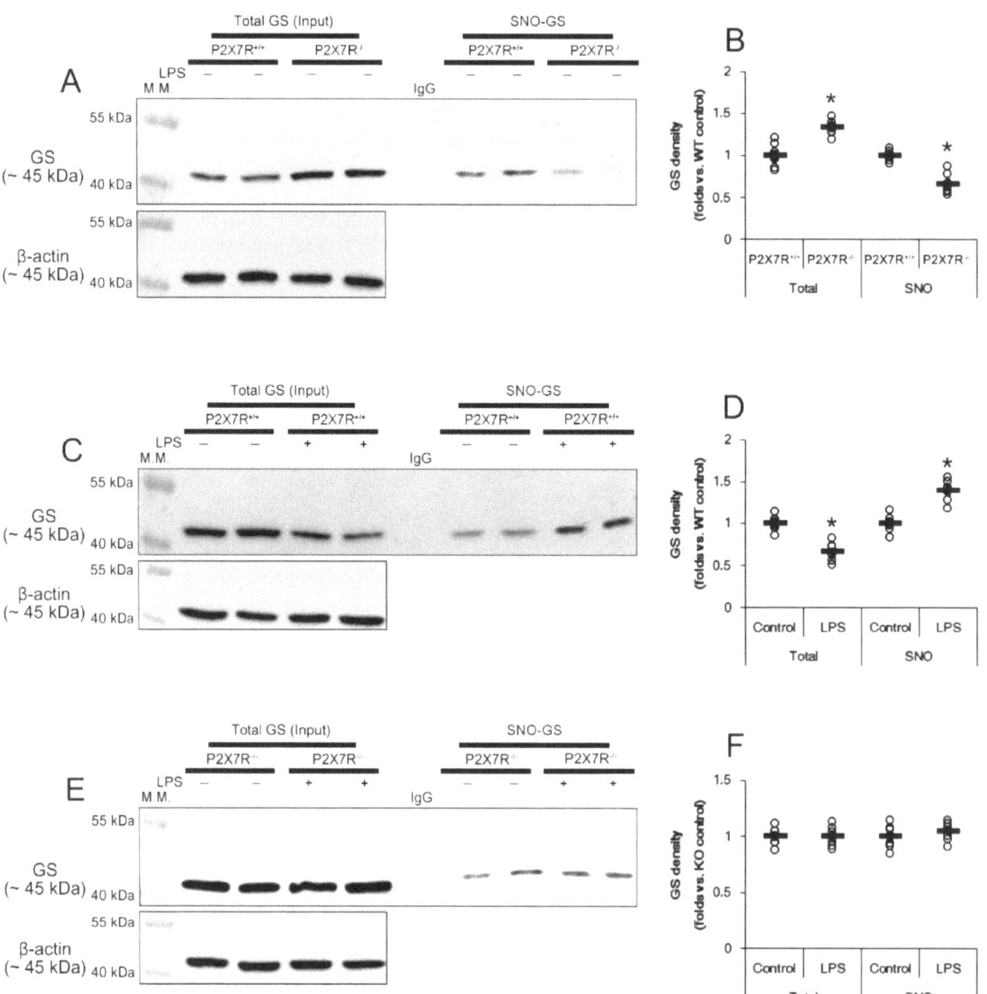

Figure 7. Effects of P2X7R deletion on S-nitrosylation of GS following LPS injection. Under physiological conditions, total GS level in $P2X7R^{-/-}$ mice is higher than that of $P2X7R^{+/+}$ mice. However, the SNO-GS level in $P2X7R^{-/-}$ mice is lower than that of $P2X7R^{+/+}$ mice. LPS decreases total GS level but increases SNO-GS level in $P2X7R^{+/+}$ mice. LPS does not affect them in $P2X7R^{-/-}$ mice. (**A**) Representative Western blot of total- and SNO-GS in the whole hippocampi of that $P2X7R^{+/+}$ and $P2X7R^{-/-}$ mice. (**B**) Quantification of the total- and SNO-GS level based on Western blot data. Open circles indicate each individual value. Horizontal and error bars indicate the mean value and S.E.M., respectively (* $p < 0.05$ vs. WT mice, $n = 7$, respectively). (**C**) Representative Western blot of total- and SNO-GS in the whole hippocampi of that $P2X7R^{+/+}$ mice following LPS treatment. (**D**) Quantification of total- and SNO-GS level based on Western blot data (* $p < 0.05$ vs. control mice, $n = 7$, respectively). (**E**) Representative Western blot of total- and SNO-GS in the whole hippocampi of that $P2X7R^{-/-}$ mice following LPS treatment. (**F**) Quantification of total- and SNO-GS level based on Western blot data. Horizontal and error bars indicate the mean value and S.E.M., respectively ($n = 7$, respectively). Full-length gel images of Western blot data in this figure could be found in Supplementary Figure S4.

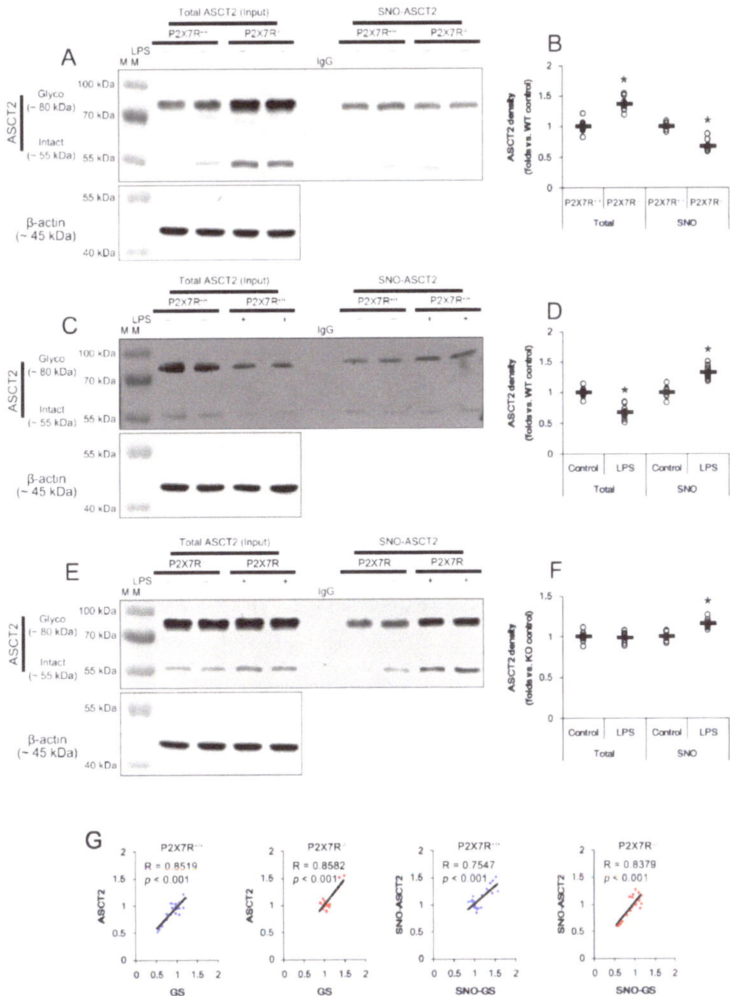

Figure 8. Effects of P2X7R deletion on ASCT2 expression and its S-nitrosylation following LPS injection. Under physiological conditions, total ASCT2 level in $P2X7R^{-/-}$ mice is higher than that of $P2X7R^{+/+}$ mice. However, SNO-ASCT2 level in $P2X7R^{-/-}$ mice is lower than that of $P2X7R^{+/+}$ mice. LPS decreases total ASCT2 level but increases SNO-ASCT2 level in $P2X7R^{+/+}$ mice. LPS also increases SNO-ASCT2 level in $P2X7R^{-/-}$ mice without affecting total ASCT2 level. Total- and SNO-ASCT2 levels show a direct proportional relationship with Total- and SNO-GS levels in $P2X7R^{+/+}$ and $P2X7R^{-/-}$ mice, respectively. (**A**) Representative Western blot of total- and SNO- ASCT2 in the whole hippocampi of that $P2X7R^{+/+}$ and $P2X7R^{-/-}$ mice. (**B**) Quantification of total- and SNO-ASCT2 level based on Western blot data. Open circles indicate each individual value. Horizontal and error bars indicate the mean value and S.E.M., respectively (* $p < 0.05$ vs. WT mice, $n = 7$, respectively). (**C**) Representative Western blot of total- and SNO-ASCT2 in the whole hippocampi of that $P2X7R^{+/+}$ mice following LPS treatment. (**D**) Quantification of total- and SNO-ASCT2 level based on Western blot data (* $p < 0.05$ vs. control mice, $n = 7$, respectively). (**E**) Representative Western blot of total- and SNO-ASCT2 in the whole hippocampi of that $P2X7R^{-/-}$ mice following LPS treatment. (**F**) Quantification of total- and SNO-ASCT2 level based on Western blot data (* $p < 0.05$ vs. control mice, $n = 7$, respectively). (**G**) Linear regression analyses of total- and SNO proteins between ASCT2 and GS in $P2X7R^{+/+}$ and $P2X7R^{-/-}$ mice. Full-length gel images of Western blot data in this figure could be found in Supplementary Figure S5.

4. Discussion

LPS is a gram-negative bacterial cell surface proteoglycan, which triggers neuroinflammation via toll-like receptor 4 (TLR4)-mediated microglial and astroglial activation in the brain [49]. After exposure to LPS, P2X7R augments iNOS expression and production of NO in microglia and astrocytes [26,50]. Therefore, P2X7R is one of the modulators of neuroinflammatory responses. In the present study, SNO-cysteine level in the hippocampus was higher in $P2X7R^{+/+}$ mice than that in $P2X7R^{-/-}$ mice under physiological conditions. Furthermore, LPS led to microglial activation and reactive astrogliosis in $P2X7R^{+/+}$ mice concomitant with increases in iNOS expression and SNO-cysteine production, which were attenuated by P2X7R deletion. These findings indicate that P2X7R may regulate nitrosative stress in the brain under physiological and inflammatory conditions. Interestingly, *S*-nitrosylation facilitates the trafficking of P2X7R, which promotes microglial activation and astroglial dysfunction following status epilepticus (a sustained seizure activity) [11]. Therefore, our findings suggest that P2X7R and iNOS may exert nitrosative stress in a positive feedback manner under inflammatory conditions.

Nrf2 plays a role in the regulation of cellular redox homeostasis [32]. Under nitrosative stress, Nrf2 sequestered by Keap1 is transported to the nucleus and promotes ARE-related gene expressions. However, excessive nitrosative stress leads to Nrf2 proteasomal degradation [51,52]. Indeed, a low dose of LPS (0.5 mg/kg) increases Nrf2 expression accompanied by the upregulated heme oxygenase-1 (HO-1, one of the downstream genes of Nrf2) at 4 h after treatment [53]. However, a high dose of LPS (1 mg/kg/day) leads to Nrf2 downregulation 6 days after administration [54]. Furthermore, a high dose of LPS (1 mg/kg) decreases Nrf2 expression and is coupled with reduced HO-1 expression and the upregulations of interleukin-6 that are regulated by Nrf2 [54]. In the present study, LPS (5 mg/kg) decreased the total Nrf2 protein level and its nuclear accumulation in microglia and astrocytes 3 days after treatment, which are ameliorated by P2X7R deletion. Considering these previous studies and the present data, it is plausible that Nrf2 upregulation may be an adaptive response against oxidative and/or nitrosative stress in response to a low dose of LPS at the early time window. In contrast, a high dose of LPS would lead to Nrf2 downregulation (or degradation), and in turn a decreased Nrf2 capacity for defending against oxidative- or nitrosative stress would contribute to LPS-induced neuroinflammation at the late time window. Since Nrf2 inhibits iNOS upregulation and attenuates the formation of SNO-proteins induced by LPS [10,30,31]; therefore, our findings suggest that P2X7R may reinforce iNOS-mediated nitrosative stress by facilitating Nrf2 degradation following LPS treatment.

Recently, we have reported that P2X7R deletion reduces the total GSH level in the hippocampus under physiological conditions, as an adaptive response to the absence of oxidative or nitrosative stresses mediated by P2X7R [18]. Consistent with this report, the present data show that the total GSH level in $P2X7R^{-/-}$ mice was lower than that in $P2X7R^{+/+}$ mice. Furthermore, LPS decreased total GSH concentration to 86% and 95% of control levels in $P2X7R^{+/+}$ and $P2X7R^{-/-}$ mice, respectively. Regarding that Nrf2 activates GSH biosynthesis [55,56] and P2X7R deletion prevented LPS-induced GSHS downregulation in the present study, our findings indicate that P2X7R may aggravate LPS-induced nitrosative stress, which would increase GSH consumption or reduce GSH synthesis.

On the other hand, Nrf2 activates GCLC and xCT which maintain intracellular GSH levels by regulating the rate-limiting steps for GSH synthesis. Furthermore, GSH depletion increases the transcription of *Nrf2* and *xCT* [55–57]. Therefore, GSH concentration and the Nrf2 system may be reciprocally regulated by each other. In the present study, P2X7R deletion did not affect Nrf2, xCT and GCLC levels under physiological conditions. Unexpectedly, LPS increased Nrf2 degradation and xCT expression, which were abrogated by P2X7R deletion. Furthermore, LPS did not affect GCLC levels in both $P2X7R^{+/+}$ and $P2X7R^{-/-}$ mice. Considering the xCT-mediated cystine-glutamate shuttle [58], it is likely that xCT upregulation with unaltered GCLC expression may be an Nrf2 independent compensatory response to GSH depletion induced by LPS. Indeed, $Nrf2^{-/-}$ mice show no genotypic difference in xCT level [59]. γ-Tocopheryl quinone (a powerful chemotherapeutic agent as an oxidative metabolite of γ-tocopherol) can increase cellular GSH levels without any considerable change in GCLC but facilitates the availability of cystine through Nrf2-independent xCT induction [60]. xCT level in astrocytes is also upregulated by intracellular GSH depletion, independent of Nrf2 [61]. However, xCT constitutes a cystine-cysteine shuttle whereby cystine uptake drives cysteine release, and extracellular cysteine provided by this shuttle is necessary for the transfer of NO equivalents [62]. Therefore, our findings provide the possibility that LPS-induced xCT upregulation may participate in the clearance of SNO-proteins against P2X7R-mediated nitrosative stress but may exacerbate GSH depletion by excessive cysteine efflux. Further studies are needed to elucidate the role of xCT upregulation under neuroinflammatory conditions.

Glutamate and glutamine are equally required for GSH biosynthesis through the glutamate-glutamine cycle that is regulated by GLS and GS [32]. LPS activates GLS accompanied by increased NO production [40,63]. The present data also demonstrate that LPS increased GLS protein level in $P2X7R^{+/+}$, but not $P2X7R^{-/-}$ mice, although P2X7R deletion did not affect GLS expression under physiological conditions. Since xCT exchanges glutamate for cystine influx [14,15], GLS upregulation may increase glutamate gradients to facilitate xCT-mediated cystine uptake for GSH synthesis. Indeed, GLS hyperactivation increases glutamate release and xCT upregulation amplifies glutamate efflux [64,65]. In contrast, LPS reduces GS expression, which enhances the release of inflammatory mediators and leads to perturbation of the redox balance [66,67]. Consistent with our previous study [18], the present study shows that P2X7R deletion increased GS expression, but reduced SNO-GS levels under physiological conditions. LPS decreased GS expression, accompanied by the increased SNO-GS level in $P2X7R^{+/+}$ mice, while it did not affect them in $P2X7R^{-/-}$ mice. Since S-nitrosylation of GS leads to its degradation by the 20S proteasome [68], our findings indicate that P2X7R may decrease GS level by accelerating S-nitrosylation-mediated GS degradation under physiological- and post-LPS conditions. Considering that GS converts glutamate to glutamine [38], this GS downregulation may contribute to an increase in intracellular glutamate concentration representing an aberrant glutamate-glutamine cycle under inflammatory conditions. Interestingly, glutamine enhances Nrf2 and GSHS activities [25]. Therefore, it is likely that GS upregulation in $P2X7R^{-/-}$ mice may also play an important role in the preservation of Nrf2 and GSHS levels following LPS treatment. Taken together, our findings suggest that P2X7R may modulate LPS-induce neuroinflammation by regulating the glutamate-glutamine cycle.

ASCT2 is a glutamine:cysteine exchanger that participates in GSH biosynthesis [41,42]. S-nitrosylation inhibits ASCT2 activity [47] and leads to ASCT2 upregulation as an adaptive response [48]. In the present study, P2X7R deletion increased ASCT2 expression, but reduced SNO-ASCT2 level under physiological conditions. LPS decreased ASCT2 expression, accompanied by the increased SNO-ASCT2 level in $P2X7R^{+/+}$ mice, while it enhanced only SNO-ASCT2 level in $P2X7R^{-/-}$ mice. The present data also reveal that GS and SNO-GS levels had direct proportional relationships to ASCT2 and SNO-ASCT2 levels, respectively, in both $P2X7R^{+/+}$ and $P2X7R^{-/-}$ mice, indicating that GS activity/expression may regulate ASCT2 expression independent of P2X7R. Compatible with the reduced GS expression; therefore, our findings suggest that P2X7R may inhibit ASCT2-mediated glutamine:cysteine exchange under neuroinflammatory conditions, which would reduce GSH biosynthesis.

In the present study, P2X7R deletion relieved, not completely inhibited, the upregulations of iNOS expression and SNO-cysteine level induced by LPS. However, SNO-GS level was unaffected by LPS in $P2X7R^{-/-}$ mice. Although we cannot provide the underlying mechanisms of this phenomenon, the possibility would considerable. In contrast to the case of SNO-GS level, the present data show the increased SNO-ASCT2 level induced by LPS in both $P2X7R^{+/+}$ and $P2X7R^{-/-}$ mice. Therefore, it is likely that NO generated from iNOS may lead to S-nitrosylation of ASCT2 rather than GS due to the distinct affinity of NO bindings between ASCT2 and GS. The affinity test for NO binding to target proteins would be useful to understand the underlying mechanisms of nitrosative stress under neuroinflammatory conditions.

5. Conclusions

In the present study, we demonstrate that P2X7R deletion (1) attenuated iNOS upregulation and Nrf2 degradation induced by LPS, (2) decreased SNO-cysteine production under physiological and post-LPS treated conditions, (3) ameliorated the LPS-induced decreases in GSH, GSHS, GS and ASCT2 levels without altering GCLC level, (4) inhibited LPS-induced xCT and GLS upregulation, and (5) reduced *S*-nitrosylations of GS and ASCT2. These findings indicate that P2X7R may augment LPS-induced neuroinflammation by inducing Nrf2 degradation, aberrant glutamate-glutamine cycle and impaired cystine/cysteine uptake, which would inhibit GSH biosynthesis. Therefore, we suggest the targeting of P2X7R, which would exert nitrosative stress with iNOS in a positive feedback manner, and may be one of the important therapeutic strategies of nitrosative stress under pathophysiological conditions.

Supplementary Materials: The following supporting information can be downloaded at: https://www.mdpi.com/article/10.3390/antiox11040778/s1, Figure S1: Full-length gel images of Western blot data in Figure 2A; Figure S2: Full-length gel images of Western blot data in Figure 4A; Figure S3: Full-length gel images of Western blot data in Figure 6B; Figure S4: Full-length gel images of Western blot data in Figure 7; Figure S5: Full-length gel images of Western blot data in Figure 8.

Author Contributions: J.-E.K. designed and supervised the project. D.-S.L. and J.-E.K. performed the experiments described in the manuscript with J.-E.K. and analyzed the data. D.-S.L. and J.-E.K. wrote the manuscript. All authors have read and agreed to the published version of the manuscript.

Funding: This study was supported by a grant of National Research Foundation of Korea (NRF) grant (No. 2021R1A2C4002003).

Institutional Review Board Statement: All experimental protocols were approved by the Institutional Animal Care and Use Committee of Hallym University (Chuncheon, South Korea, Code number: Hallym 2018-3, approval date: 30 April 2018 and Hallym 2021-30, approval date: 17 May 2021).

Informed Consent Statement: Not applicable.

Data Availability Statement: The data presented in this study are available in the article and Supplementary Materials.

Conflicts of Interest: The authors declare no conflict of interest. The funders had no role in the design of the study; in the collection, analyses, or interpretation of data; in the writing of the manuscript, or in the decision to publish the results.

References

1. Kim, J.E.; Ko, A.R.; Hyun, H.W.; Min, S.J.; Kang, T.C. P2RX7-MAPK1/2-SP1 axis inhibits MTOR independent HSPB1-mediated astroglial autophagy. *Cell Death Dis.* **2018**, *9*, 546. [CrossRef] [PubMed]
2. Kim, J.E.; Kang, T.C. The P2X7 receptor-pannexin-1 complex decreases muscarinic acetylcholine receptor-mediated seizure susceptibility in mice. *J. Clin. Investig.* **2011**, *121*, 2037–2047. [CrossRef] [PubMed]
3. Kim, J.E.; Ryu, H.J.; Yeo, S.I.; Kang, T.C. P2X7 receptor regulates leukocyte infiltrations in rat frontoparietal cortex following status epilepticus. *J. Neuroinflamm.* **2010**, *7*, 65. [CrossRef] [PubMed]
4. Lo, J.C.; Huang, W.C.; Chou, Y.C.; Tseng, C.H.; Lee, W.L.; Sun, S.H. Activation of P2X(7) receptors decreases glutamate uptake and glutamine synthetase activity in RBA-2 astrocytes via distinct mechanisms. *J. Neurochem.* **2008**, *105*, 151–164. [CrossRef] [PubMed]
5. Codocedo, J.F.; Godoy, J.A.; Poblete, M.I.; Inestrosa, N.C.; Huidobro-Toro, J.P. ATP induces NO production in hippocampal neurons by P2X(7) receptor activation independent of glutamate signaling. *PLoS ONE* **2013**, *8*, e57626. [CrossRef]
6. Ficker, C.; Rozmer, K.; Kató, E.; Andó, R.D.; Schumann, L.; Krügel, U.; Franke, H.; Sperlágh, B.; Riedel, T.; Illes, P. Astrocyte-neuron interaction in the substantia gelatinosa of the spinal cord dorsal horn via P2X7 receptor-mediated release of glutamate and reactive oxygen species. *Glia* **2014**, *62*, 1671–1686. [CrossRef]
7. Sperlágh, B.; Haskó, G.; Németh, Z.; Vizi, E.S. ATP released by LPS increases nitric oxide production in raw 264.7 macrophage cell line via P2Z/P2X7 receptors. *Neurochem. Int.* **1998**, *33*, 209–215. [CrossRef]

8. Hu, Y.; Fisette, P.L.; Denlinger, L.C.; Guadarrama, A.G.; Sommer, J.A.; Proctor, R.A.; Bertics, P.J. Purinergic receptor modulation of lipopolysaccharide signaling and inducible nitric-oxide synthase expression in RAW 264.7 macrophages. *J. Biol. Chem.* **1998**, *273*, 27170–27175. [CrossRef]

9. Choi, H.B.; Ryu, J.K.; Kim, S.U.; McLarnon, J.G. Modulation of the purinergic P2X7 receptor attenuates lipopolysaccharide-mediated microglial activation and neuronal damage in inflamed brain. *J. Neurosci.* **2007**, *27*, 4957–4968. [CrossRef]

10. Qu, Z.; Meng, F.; Zhou, H.; Li, J.; Wang, Q.; Wei, F.; Cheng, J.; Greenlief, M.C.; Lubahn, B.M.; Sun, Y.G.; et al. NitroDIGE analysis reveals inhibition of protein S-nitrosylation by epigallocatechin gallates in lipopolysaccharide-stimulated microglial cells. *J. Neuroinflamm.* **2014**, *11*, 17. [CrossRef]

11. Lee, D.S.; Kim, J.E. Protein disulfide isomerase-mediated S-nitrosylation facilitates surface expression of P2X7 receptor following status epilepticus. *J. Neuroinflamm.* **2021**, *18*, 14. [CrossRef] [PubMed]

12. Dringen, R.; Brandmann, M.; Hohnholt, M.C.; Blumrich, E.M. Glutathione-dependent detoxification processes in astrocytes. *Neurochem. Res.* **2015**, *40*, 2570–2582. [CrossRef] [PubMed]

13. Utsunomiya-Tate, N.; Endou, H.; Kanai, Y. Cloning and functional characterization of a system ASC-like Na+-dependent neutral amino acid transporter. *J. Biol. Chem.* **1996**, *271*, 14883–14890. [CrossRef] [PubMed]

14. Kalivas, P.W. The glutamate homeostasis hypothesis of addiction. *Nat. Rev. Neurosci.* **2009**, *10*, 561–572. [CrossRef] [PubMed]

15. Van Liefferinge, J.; Bentea, E.; Demuyser, T.; Albertini, G.; Follin-Arbelet, V.; Holmseth, S.; Merckx, E.; Sato, H.; Aerts, J.L.; Smolders, I.; et al. Comparative analysis of antibodies to xCT (Slc7a11): Forewarned is forearmed. *J. Comp. Neurol.* **2016**, *524*, 1015–1032. [CrossRef] [PubMed]

16. Fu, W.; Ruangkittisakul, A.; MacTavish, D.; Baker, G.B.; Ballanyi, K.; Jhamandas, J.H. Activity and metabolism-related Ca^{2+} and mitochondrial dynamics in co-cultured human fetal cortical neurons and astrocytes. *Neuroscience* **2013**, *250*, 520–535. [CrossRef]

17. Pan, H.C.; Chou, Y.C.; Sun, S.H. P2X7 R-mediated $Ca^{(2+)}$-independent d-serine release via pannexin-1 of the P2X7 R-pannexin-1 complex in astrocytes. *Glia* **2015**, *63*, 877–893. [CrossRef]

18. Park, H.; Kim, J.E. Deletion of P2X7 receptor decreases basal glutathione level by changing glutamate-glutamine cycle and neutral amino acid transporters. *Cells* **2020**, *9*, 995. [CrossRef]

19. Ishii, T.; Itoh, K.; Takahashi, S.; Sato, H.; Yanagawa, T.; Katoh, Y.; Bannai, S.; Yamamoto, M. Transcription factor Nrf2 coordinately regulates a group of oxidative stress-inducible genes in macrophages. *J. Biol Chem.* **2000**, *275*, 16023–16029. [CrossRef]

20. Shih, A.Y.; Johnson, D.A.; Wong, G.; Kraft, A.D.; Jiang, L.; Erb, H.; Johnson, J.A.; Murphy, T.H. Coordinate regulation of glutathione biosynthesis and release by Nrf2-expressing glia potently protects neurons from oxidative stress. *J. Neurosci.* **2003**, *23*, 3394–3406. [CrossRef]

21. Itoh, K.; Wakabayashi, N.; Katoh, Y.; Ishii, T.; Igarashi, K.; Engel, J.D.; Yamamoto, M. Keap1 represses nuclear activation of antioxidant responsive elements by Nrf2 through binding to the amino-terminal Neh2 domain. *Genes Dev.* **1999**, *13*, 76–86. [CrossRef] [PubMed]

22. McMahon, M.; Itoh, K.; Yamamoto, M.; Hayes, J.D. Keap1-dependent proteasomal degradation of transcription factor Nrf2 contributes to the negative regulation of antioxidant response element-driven gene expression. *J. Biol. Chem.* **2003**, *278*, 21592–21600. [CrossRef] [PubMed]

23. Kobayashi, A.; Ohta, T.; Yamamoto, M. Unique function of the Nrf2-Keap1 pathway in the inducible expression of antioxidant and detoxifying enzymes. *Methods Enzymol.* **2004**, *378*, 273–286. [PubMed]

24. Noh, H.; Jeon, J.; Seo, H. Systemic injection of LPS induces region-specific neuroinflammation and mitochondrial dysfunction in normal mouse brain. *Neurochem. Int.* **2014**, *69*, 35–40. [CrossRef]

25. Venoji, R.; Amirtharaj, G.J.; Kini, A.; Vanaparthi, S.; Venkatraman, A.; Ramachandran, A. Enteral glutamine differentially regulates Nrf 2 along the villus-crypt axis of the intestine to enhance glutathione levels. *J. Gastroenterol. Hepatol.* **2015**, *30*, 1740–1747. [CrossRef]

26. Guerra, A.N.; Fisette, P.L.; Pfeiffer, Z.A.; Quinchia-Rios, B.H.; Prabhu, U.; Aga, M.; Denlinger, L.C.; Gaudarrama, A.G.; Abozeid, S.; Sommer, J.A.; et al. Purinergic receptor regulation of LPS-induced signaling and pathophysiology. *J. Endotoxin Res.* **2003**, *9*, 256–263. [CrossRef]

27. Muhammad, T.; Ikram, M.; Ullah, R.; Rehman, S.U.; Kim, M.O. Hesperetin, a citrus flavonoid, attenuates LPS-induced neuroinflammation, apoptosis and memory impairments by modulating TLR4/NF-kappaB signaling. *Nutrients* **2019**, *11*, 648. [CrossRef]

28. Noworyta-Sokolowska, K.; Gorska, A.; Golembiowska, K. LPS-induced oxidative stress and inflammatory reaction in the rat striatum. *Pharmacol. Rep.* **2013**, *65*, 863–869. [CrossRef]

29. Shah, S.A.; Khan, M.; Jo, M.H.; Jo, M.G.; Amin, F.U.; Kim, M.O. Melatonin stimulates the SIRT1/Nrf2 signaling pathway counteracting lipopolysaccharide (LPS)-induced oxidative stress to rescue postnatal rat brain. *CNS Neurosci. Ther.* **2017**, *23*, 33–44. [CrossRef]

30. Townsend, B.E.; Johnson, R.W. Sulforaphane induces Nrf2 target genes and attenuates inflammatory gene expression in microglia from brain of young adult and aged mice. *Exp. Gerontol.* **2016**, *73*, 42–48. [CrossRef]

31. Lee, D.S.; Kwon, K.H.; Cheong, S.H. Taurine chloramine suppresses LPS-induced neuroinflammatory responses through Nrf2-mediated heme oxygenase-1 expression in mouse BV2 microglial cells. *Adv. Exp. Med. Biol.* **2017**, *975*, 131–143. [PubMed]

32. Sun, X.; Erb, H.; Murphy, T.H. Coordinate regulation of glutathione metabolism in astrocytes by Nrf2. *Biochem. Biophys. Res. Commun.* **2005**, *326*, 371–377. [CrossRef] [PubMed]

33. Erickson, A.M.; Nevarea, Z.; Gipp, J.J.; Mulcahy, R.T. Identification of a variant antioxidant response element in the promoter of the human glutamate-cysteine ligase modifier subunit gene. Revision of the ARE consensus sequence. *J. Biol. Chem.* **2002**, *277*, 30730–30737. [CrossRef] [PubMed]

34. McElroy, P.B.; Sri Hari, A.; Day, B.J.; Patel, M. Post-translational activation of glutamate cysteine ligase with dimercaprol: A novel mechanism of inhibiting neuroinflammation in vitro. *J. Biol. Chem.* **2017**, *292*, 5532–5545. [CrossRef]

35. Chen, J.; Yin, W.; Tu, Y.; Wang, S.; Yang, X.; Chen, Q.; Zhang, X.; Han, Y.; Pi, R. L-F001, a novel multifunctional ROCK inhibitor, suppresses neuroinflammation in vitro and in vivo: Involvement of NF-κB inhibition and Nrf2 pathway activation. *Eur. J. Pharmacol.* **2017**, *806*, 1–9. [CrossRef]

36. Wu, P.S.; Ding, H.Y.; Yen, J.H.; Chen, S.F.; Lee, K.H.; Wu, M.J. Anti-inflammatory activity of 8-hydroxydaidzein in LPS-stimulated BV2 microglial cells via activation of Nrf2-antioxidant and attenuation of Akt/NF-κB-inflammatory signaling pathways, as well as inhibition of COX-2 activity. *J. Agric. Food Chem.* **2018**, *66*, 5790–5801. [CrossRef]

37. Ko, K.; Yang, H.; Noureddin, M.; Iglesia-Ara, A.; Xia, M.; Wagner, C.; Luka, Z.; Mato, J.M.; Lu, S.C. Changes in S-adenosylmethionine and GSH homeostasis during endotoxemia in mice. *Lab. Investig.* **2008**, *88*, 1121–1129. [CrossRef]

38. Häberle, J.; Görg, B.; Toutain, A.; Rutsch, F.; Benoist, J.F.; Gelot, A.; Suc, A.-L.; Koch, H.G.; Schliess, F.; Häussinger, D. Inborn error of amino acid synthesis: Human glutamine synthetase deficiency. *J. Inherit. Metab. Dis.* **2006**, *29*, 352–358. [CrossRef]

39. Hayashi, M.K. Structure-function relationship of transporters in the glutamate-glutamine cycle of the central nervous system. *Int. J. Mol. Sci.* **2018**, *19*, 1177. [CrossRef]

40. Gao, G.; Zhao, S.; Xia, X.; Li, C.; Li, C.; Ji, C.; Sheng, S.; Tang, Y.; Zhu, J.; Wang, Y.; et al. Glutaminase C regulates microglial activation and pro-inflammatory exosome release: Relevance to the pathogenesis of Alzheimer's disease. *Front. Cell. Neurosci.* **2019**, *13*, 264. [CrossRef]

41. Gegelashvili, M.; Rodriguez-Kern, A.; Pirozhkova, I.; Zhang, J.; Sung, L.; Gegelashvili, G. High-affinity glutamate transporter GLAST/EAAT1 regulates cell surface expression of glutamine/neutral amino acid transporter ASCT2 in human fetal astrocytes. *Neurochem. Int.* **2006**, *48*, 611–615. [CrossRef] [PubMed]

42. Bröer, A.; Brookes, N.; Ganapathy, V.; Dimmer, K.S.; Wagner, C.A.; Lang, F.; Bröer, S. The astroglial ASCT2 amino acid transporter as a mediator of glutamine efflux. *J. Neurochem.* **1999**, *73*, 2184–2194. [PubMed]

43. Asahi, M.; Fujii, J.; Suzuki, K.; Seo, H.G.; Kuzuya, T.; Hori, M.; Tada, M.; Fujii, S.; Taniguchi, N. Inactivation of glutathione peroxidase by nitric oxide. Implication for cytotoxicity. *J. Biol. Chem.* **1995**, *270*, 21035–21039. [CrossRef] [PubMed]

44. Raju, K.; Doulias, P.T.; Evans, P.; Krizman, E.N.; Jackson, J.G.; Horyn, O.; Daikhin, Y.; Nissim, I.; Yudkoff, M.; Nissim, I.; et al. Regulation of brain glutamate metabolism by nitric oxide and S-nitrosylation. *Sci. Signal.* **2015**, *8*, ra68. [CrossRef]

45. Miñana, M.D.; Kosenko, E.; Marcaida, G.; Hermenegildo, C.; Montoliu, C.; Grisolía, S.; Felipo, V. Modulation of glutamine synthesis in cultured astrocytes by nitric oxide. *Cell. Mol. Neurobiol.* **1997**, *17*, 433–445. [CrossRef]

46. Kosenko, E.; Llansola, M.; Montoliu, C.; Monfort, P.; Rodrigo, R.; Hernandez-Viadel, M.; Erceg, S.; Sánchez-Perez, A.M.; Felipo, V. Glutamine synthetase activity and glutamine content in brain: Modulation by NMDA receptors and nitric oxide. *Neurochem. Int.* **2003**, *43*, 493–499. [CrossRef]

47. Scalise, M.; Pochini, L.; Console, L.; Pappacoda, G.; Pingitore, P.; Hedfalk, K.; Indiveri, C. Cys site-directed mutagenesis of the human SLC1A5 (ASCT2) transporter: Structure/function relationships and crucial role of Cys467 for redox sensing and glutamine transport. *Int. J. Mol. Sci.* **2018**, *19*, 648. [CrossRef]

48. Uchiyama, T.; Matsuda, Y.; Wada, M.; Takahashi, S.; Fujita, T. Functional regulation of Na$^+$-dependent neutral amino acid transporter ASCT2 by S-nitrosothiols and nitric oxide in Caco-2 cells. *FEBS Lett.* **2005**, *579*, 2499–2506. [CrossRef]

49. Barbierato, M.; Facci, L.; Argentini, C.; Marinelli, C.; Skaper, S.D.; Giusti, P. Astrocyte-microglia cooperation in the expression of a pro-inflammatory phenotype. *CNS Neurol. Disord. Drug Targets* **2013**, *12*, 608–618. [CrossRef]

50. Aga, M.; Watters, J.J.; Pfeiffer, Z.A.; Wiepz, G.J.; Sommer, J.A.; Bertics, P.J. Evidence for nucleotide receptor modulation of cross talk between MAP kinase and NF-kappa B signaling pathways in murine RAW 264.7 macrophages. *Am. J. Physiol. Cell Physiol.* **2004**, *286*, 923–930. [CrossRef]

51. Mann, G.E. Nrf2-mediated redox signalling in vascular health and disease. *Free Radic. Biol. Med.* **2014**, *75*, S1. [CrossRef] [PubMed]

52. Owuor, E.D.; Kong, A.N. Antioxidants and oxidants regulated signal transduction pathways. *Biochem. Pharmacol.* **2002**, *64*, 765–770. [CrossRef]

53. Leite, J.A.; Isaksen, T.J.; Heuck, A.; Scavone, C.; Lykke-Hartmann, K. The α_2 Na$^+$/K$^+$-ATPase isoform mediates LPS-induced neuroinflammation. *Sci. Rep.* **2020**, *10*, 14180. [CrossRef] [PubMed]

54. Ali, T.; Rahman, S.U.; Hao, Q.; Li, W.; Liu, Z.; Ali Shah, F.; Murtaza, I.; Zhang, Z.; Yang, X.; Liu, G.; et al. Melatonin prevents neuroinflammation and relieves depression by attenuating autophagy impairment through FOXO3a regulation. *J. Pineal Res.* **2020**, *69*, e12667. [CrossRef] [PubMed]

55. Steele, M.L.; Fuller, S.; Patel, M.; Kersaitis, C.; Ooi, L.; Münch, G. Effect of Nrf2 activators on release of glutathione, cysteinylglycine and homocysteine by human U373 astroglial cells. *Redox Biol.* **2013**, *1*, 441–445. [CrossRef] [PubMed]

56. Ishii, T.; Mann, G.E. Redox status in mammalian cells and stem cells during culture in vitro: Critical roles of Nrf2 and cystine transporter activity in the maintenance of redox balance. *Redox Biol.* **2014**, *2*, 786–794. [CrossRef]

57. Valdovinos-Flores, C.; Limón-Pacheco, J.H.; León-Rodríguez, R.; Petrosyan, P.; Garza-Lombó, C.; Gonsebatt, M.E. Systemic L-buthionine-S-R-sulfoximine treatment increases plasma NGF and upregulates L-cys/L-cys2 transporter and γ-glutamylcysteine ligase mRNAs through the NGF/TrkA/Akt/Nrf2 pathway in the striatum. *Front. Cell. Neurosci.* **2019**, *13*, 325. [CrossRef]

58. Taguchi, K.; Tamba, M.; Bannai, S.; Sato, H. Induction of cystine/glutamate transporter in bacterial lipopolysaccharide induced endotoxemia in mice. *J. Inflamm.* **2007**, *4*, 20. [CrossRef]
59. Pacchioni, A.M.; Vallone, J.; Melendez, R.I.; Shih, A.; Murphy, T.H.; Kalivas, P.W. Nrf2 gene deletion fails to alter psychostimulant-induced behavior or neurotoxicity. *Brain Res.* **2007**, *1127*, 26–35. [CrossRef]
60. Ogawa, Y.; Saito, Y.; Nishio, K.; Yoshida, Y.; Ashida, H.; Niki, E. Gamma-tocopheryl quinone, not alpha-tocopheryl quinone, induces adaptive response through up-regulation of cellular glutathione and cysteine availability via activation of ATF4. *Free Radic. Res.* **2008**, *42*, 674–687. [CrossRef]
61. Seib, T.M.; Patel, S.A.; Bridges, R.J. Regulation of the system x(C)-cystine/glutamate exchanger by intracellular glutathione levels in rat astrocyte primary cultures. *Glia* **2011**, *59*, 1387–1401. [CrossRef] [PubMed]
62. Zhu, J.; Li, S.; Marshall, Z.M.; Whorton, A.R. A cystine-cysteine shuttle mediated by xCT facilitates cellular responses to S-nitrosoalbumin. *Am. J. Physiol. Cell. Physiol.* **2008**, *294*, C1012–C1020. [CrossRef] [PubMed]
63. Hollinger, K.R.; Zhu, X.; Khoury, E.S.; Thomas, A.G.; Liaw, K.; Tallon, C.; Ying, W.; Prchalova, E.; Atsushi, K.; Camilo, R.; et al. Glutamine antagonist JHU-083 normalizes aberrant hippocampal glutaminase activity and improves cognition in APOE4 mice. *J. Alzheimers Dis.* **2020**, *77*, 437–447. [CrossRef]
64. Thomas, A.G.; O'Driscoll, C.M.; Bressler, J.; Kaufmann, W.; Rojas, C.J.; Slusher, B.S. Small molecule glutaminase inhibitors block glutamate release from stimulated microglia. *Biochem. Biophys. Res. Commun.* **2014**, *443*, 32–36. [CrossRef] [PubMed]
65. Fan, Z.; Wirth, A.K.; Chen, D.; Wruck, C.J.; Rauh, M.; Buchfelder, M.; Savaskan, N. Nrf2-Keap1 pathway promotes cell proliferation and diminishes ferroptosis. *Oncogenesis* **2017**, *6*, e371. [CrossRef]
66. Palmieri, E.M.; Menga, A.; Lebrun, A.; Hooper, D.C.; Butterfield, D.A.; Mazzone, M.; Castegna, A. Blockade of glutamine synthetase enhances inflammatory response in microglial cells. *Antioxid. Redox Signal.* **2017**, *26*, 351–363. [CrossRef]
67. Wang, S.; Zhang, H.; Geng, B.; Xie, Q.; Li, W.; Deng, Y.; Shi, W.; Pan, Y.; Kang, X.; Wang, J. 2-arachidonyl glycerol modulates astrocytic glutamine synthetase via p38 and ERK1/2 pathways. *J. Neuroinflamm.* **2018**, *15*, 220. [CrossRef]
68. Görg, B.; Qvartskhava, N.; Voss, P.; Grune, T.; Häussinger, D.; Schliess, F. Reversible inhibition of mammalian glutamine synthetase by tyrosine nitration. *FEBS Lett.* **2007**, *581*, 84–90. [CrossRef]

Article

Antioxidant Genetic Profile Modifies Probability of Developing Neurological Sequelae in Long-COVID

Marko Ercegovac [1,2,†], Milika Asanin [1,3,†], Ana Savic-Radojevic [1,4], Jovan Ranin [1,5], Marija Matic [1,4], Tatjana Djukic [1,4], Vesna Coric [1,4], Djurdja Jerotic [1,4], Nevena Todorovic [5], Ivana Milosevic [1,5], Goran Stevanovic [1,5], Tatjana Simic [1,4,6], Zoran Bukumiric [1,7,*] and Marija Pljesa-Ercegovac [1,4,*]

1   Faculty of Medicine, University of Belgrade, 11000 Belgrade, Serbia; ercegovacmarko@gmail.com (M.E.);
    masanin@gmail.com (M.A.); ana.savic-radojevic@med.bg.ac.rs (A.S.-R.); njranin@gmail.com (J.R.);
    marija.matic@med.bg.ac.rs (M.M.); tatjana.djukic@med.bg.ac.rs (T.D.); vesna.coric@med.bg.ac.rs (V.C.);
    djurdja.jovanovic@med.bg.ac.rs (D.J.); ivana.milosevic00@gmail.com (I.M.); goran_drste@yahoo.com (G.S.);
    tatjana.simic@med.bg.ac.rs (T.S.)
2   Clinic of Neurology, Clinical Centre of Serbia, 11000 Belgrade, Serbia
3   Clinic of Cardiology, Clinical Centre of Serbia, 11000 Belgrade, Serbia
4   Institute of Medical and Clinical Biochemistry, 11000 Belgrade, Serbia
5   Clinic of Infectious and Tropical Diseases, Clinical Centre of Serbia, 11000 Belgrade, Serbia;
    nevena.todorovic.1992@gmail.com
6   Department of Medical Sciences, Serbian Academy of Sciences and Arts, 11000 Belgrade, Serbia
7   Institute of Medical Statistics and Informatics, 11000 Belgrade, Serbia
*   Correspondence: zoran.bukumiric@med.bg.ac.rs (Z.B.); marija.pljesa-ercegovac@med.bg.ac.rs (M.P.-E.);
    Tel.: +381-11-2681964 (Z.B.); +381-11-3636249 (M.P.-E.)
†   These authors contributed equally to this work.

Citation: Ercegovac, M.; Asanin, M.; Savic-Radojevic, A.; Ranin, J.; Matic, M.; Djukic, T.; Coric, V.; Jerotic, D.; Todorovic, N.; Milosevic, I.; et al. Antioxidant Genetic Profile Modifies Probability of Developing Neurological Sequelae in Long-COVID. *Antioxidants* **2022**, *11*, 954. https://doi.org/10.3390/antiox11050954

Academic Editor: Marcel Bonay

Received: 13 April 2022
Accepted: 8 May 2022
Published: 12 May 2022

Publisher's Note: MDPI stays neutral with regard to jurisdictional claims in published maps and institutional affiliations.

Abstract: Understanding the sequelae of COVID-19 is of utmost importance. Neuroinflammation and disturbed redox homeostasis are suggested as prevailing underlying mechanisms in neurological sequelae propagation in long-COVID. We aimed to investigate whether variations in antioxidant genetic profile might be associated with neurological sequelae in long-COVID. Neurological examination and antioxidant genetic profile (SOD2, GPXs and GSTs) determination, as well as, genotype analysis of Nrf2 and ACE2, were conducted on 167 COVID-19 patients. Polymorphisms were determined by the appropriate PCR methods. Only polymorphisms in GSTP1AB and GSTO1 were independently associated with long-COVID manifestations. Indeed, individuals carrying *GSTP1 Val* or *GSTO1 Asp* allele exhibited lower odds of long-COVID myalgia development, both independently and in combination. Furthermore, the combined presence of *GSTP1 Ile* and *GSTO1 Ala* alleles exhibited cumulative risk regarding long-COVID myalgia in carriers of the combined *GPX1 LeuLeu/GPX3 CC* genotype. Moreover, individuals carrying combined *GSTM1-null/GPX1LeuLeu* genotype were more prone to developing long-COVID "brain fog", while this probability further enlarged if the *Nrf2 A* allele was also present. The fact that certain genetic variants of antioxidant enzymes, independently or in combination, affect the probability of long-COVID manifestations, further emphasizes the involvement of genetic susceptibility when SARS-CoV-2 infection is initiated in the host cells, and also months after.

Keywords: neurological manifestations; long-COVID; antioxidant enzymes; glutathione transferases; polymorphisms

1. Introduction

Although the main focus regarding the ongoing coronavirus disease 2019 (COVID-19) pandemic has been the acute clinical manifestations of the disease, at this point, it is clear that understanding the sequelae of the disease is of equal, if not of utmost importance [1–3]. Indeed, the novel term "post-COVID-19 syndrome" or "long-COVID" has been introduced for those individuals in whom COVID-19 can cause symptoms that last weeks or months after the infection has resolved. Still, there are no precise criteria for this multi-organ disorder, which includes a wide range of clinical manifestations, including pulmonary, cardiovascular, immunological, neurological, hematologic, endocrine, renal, gastrointestinal, dermatologic and/or psychiatric disorders [4–7]. Interestingly, according to the National Health System of the United Kingdom, the odds of having long-term symptoms do not seem to be linked to the severity of acute COVID-19 [8].

It has been shown that the prevalence of neurological symptoms in the acute and subacute COVID-19 is up to 90%, varying from sense of smell and taste disorders to headaches, myalgias, cognitive impairment and memory problems, even comprising encephalopathy, delirium, cerebrovascular disease, seizures and neuropathy in some cases [9–14]. Manifestations such as abnormal movements, psychomotor agitation, syncope and autonomic dysfunction seem to be less frequent in COVID-19 patients [1,3]. These neurological disorders might be explained by a myriad of ethological factors, including inflammation, cytokine storm, cerebral hypoxia and hypoperfusion [15–18]. Namely, even mild inflammation may disrupt the blood–brain barrier, possibly even initiating protein aggregation [17,18].

In the course of the acute phase of COVID-19, significant alterations in redox homeostasis also contribute to the vicious cycle of inflammation and oxidative stress [19–22]. Based on this premise, we have recently recognized the role of the antioxidant genetic profile in the susceptibility and severity of acute COVID-19. Apart from superoxide dismutases (SODs) and glutathione peroxidases (GPXs), a superfamily of glutathione transferases (GSTs) also contributes greatly to antioxidant defense. These multifunctional enzymes are involved in a number of catalytic and non-catalytic processes, however primarily recognized as phase II cellular detoxification system enzymes [23,24]. Importantly, certain GSTs, precisely GSTO1, exhibit a pro-inflammatory role by modulating the pro-inflammatory lipopolysaccharide (LPS)/Toll-like receptor (TLR-4)-induced activation of the nuclear factor kappa B (NF-B) pathway in macrophages [25,26]. Furthermore, these glutathione (GSH)-dependent enzymes demand this important thiol for their function, thus modulating GSH homeostasis. Interestingly, in the case of COVID-19, it has been shown that the severe form of the disease is triggered by conditions leading to decreased glutathione levels [27], which might also be affected by polymorphisms in GSH-dependent enzymes and their impact on glutathione utilization. Indeed, we found that the polymorphisms in *GSTP1*, *GSTM3*, *GSTO1* and *GSTO2* genes were associated with significant odds of COVID-19 development, while the combined *GSTP1* and *GSTM3* genotype even exhibited cumulative risk regarding both COVID-19 occurrence and COVID-19 severity [26,28]. Furthermore, we have shown the influence of *SOD2* and *GPX1* polymorphisms on the inflammation and coagulation parameters in COVID-19 patients [29].

In long-COVID, numerous data indicate toward an unexpectedly high prevalence of prolonged neurological symptoms among convalescent patients, which are of special interest due to their potential to be life altering [1,2,30]. Fatigue, cognitive impairment and "brain fog", myalgia, headache, mood disorders, sensorimotor deficits, as well as, smell or taste disorders are among the most frequent neurological sequels of COVID-19 [31]. It is important to note that even in patients in whom neurological symptoms were absent in the acute phase of the disease, microstructure changes, as well as, cerebral blood flow changes were observed at 3-month follow-up [32]. Although mechanisms underlying the manifestation of long-COVID are still unclear, neuroinflammation and disturbed redox homeostasis are suggested as prevailing in neurological sequelae propagation [1]. Indeed, numerous neurological diseases are characterized by disturbances in redox homeostasis, while in some of them (e.g. Parkinson's disease, Alzheimer's disease, amyotrophic lateral sclerosis, and epilepsy) polymorphisms occurring in antioxidant genes are even recognized as risk factors [33–37]. In this line, transcription factor Nrf2 (nuclear factor (erythroid-derived 2)-like2), one of the key regulators of cellular redox state that modulates the expression of a large number of genes, including antioxidant enzymes, might also be an important contributing factor [38,39]. Namely, Kelch-like ECH-associated protein 1 or Keap1/Nrf2 pathway reacts to changes in cellular levels of reactive oxygen species (ROS).

When ROS levels are increased, Keap1 undergoes allosteric changes that lead to decreased proteasomal degradation and accumulation of Nrf2. Its accumulation, followed by nuclear translocation, results in intensified transcription of Nrf2 target genes [38,39]. Interestingly, the role of Nrf2 has also been investigated in relation to the expression of genes involved in inflammasome assembly, as well as in inflammasome signaling [40,41].

Since there is limited evidence on the pathophysiological mechanisms implicated in the manifestation of long-COVID, and oxidative stress is supposed to be one of them, we aimed to investigate whether variations in genetic profile, especially in genes encoding immediate (superoxide dismutase) and first-line defense (glutathione transferases and glutathione peroxidases) antioxidant enzymes might be associated with neurological sequels in long-COVID.

2. Materials and Methods

The study was conducted on 167 COVID-19 patients (100 men and 67 women, with an average age of 55.9 ± 12.3 years) treated in the Clinical Centre of Serbia, between July 2020 and February 2021. All participants were Caucasians by ethnicity. Positive SARS-CoV2 reverse transcription (RT)-PCR test performed from nasopharyngeal and oropharyngeal swabs according to World Health Organization guidelines and using available RT-PCR protocols, age ($\geq$18 years old) and willingness to provide written informed consent were the inclusion criteria for participation in the study.

The principles of the International Conference on Harmonization (ICH) Good Clinical Practice, the "Declaration of Helsinki," and national and international ethical guidelines were followed during this study with approval obtained from the Ethics Committee of the Clinical Centre of Serbia. Clinical, demographic and epidemiological data were collected using the RedCap® based questionnaire (RedCap®. Available online: https://redcap.med.bg.ac.rs/, AntioxIdentification, accessed on 21 February 2022).

The data on medical history, signs and symptoms of the disease, comorbidities and laboratory parameters were obtained from the patients' clinical records. Follow-up neurological examination conducted 3 months after acute phase of COVID-19 included: evaluation of mental status (orientation and recent memory), examination of cranial nerves by assessment of pupils for size, symmetry and reactivity to light, primary eye position, motor (tone, signs of rigidity and spasticity) and sensitivity response, deep tendon reflexes and pathologic reflexes (Babinski sign), neck stiffness, coordination testing, nystagmus, tremor, assessment of ataxia and gait. The presence of myalgia, fatigue and "brain fog" was also investigated. Precisely, myalgia was evaluated by static and dynamic manual palpation of soft tissue and joints, as well as, the presence of muscle pain during movement. Fatigue was evaluated using the Fatigue Assessment Scale (FAS), which represents a 10-item general fatigue questionnaire. It is based on 10 statements that refer to how individuals usually feel, where individuals answer this 10-item scale using a five-point scale ranging from 1 ("never") to 2 ("sometimes"), 3 ("regularly"), 4 ("often") or 5 ("always"). A total FAS score of < 22 indicates no fatigue, while a score of $\geq$ 22 indicates fatigue [42]. Regarding "brain fog", it was evaluated based on individuals' agreement (agree or disagree) with a list of 19 brain fog descriptors (e.g. forgetful, difficulty thinking and focusing, slow, sleepy, etc.). Agreement in 10 or more descriptors was considered as presence of "brain fog" [43]. DNA isolation was performed from EDTA-anticoagulated peripheral blood obtained from the study participants using PureLink™ Genomic DNA Mini Kit (ThermoFisher Scientific, Waltham, MA, USA).

GSTM1 and *GSTT1* deletion polymorphisms were determined by multiplex PCR, using the *CYP1A1* gene as an internal control. *GSTA1* rs3957357 and *GPX1* rs1050450 polymorphisms were determined using PCR restriction fragment length polymorphism (PCR-RFLP). Primer sequences of *GSTM1, GSTT1, CYP1A1, GSTA1* and *GPX1* genes and PCR protocols details were as previously described [28,29]. *GSTP1* rs1695, *GSTP1* rs1138272, *GSTM3* rs1332018, *GSTO1* rs4925, *GSTO2* rs156697, *ACE2* rs4646116, *ACE2* rs143936283, *SOD2* rs4880 and *GPX3* rs8177412 polymorphisms were determined by real-time PCR onMastercycler ep realplex (Eppendorf, Germany), using TaqMan Drug Metabolism Genotyping assays (Life Technologies, Applied Biosystems, United States). Assays' IDs were as follows: C_3237198_20, C_1049615_20, C_3184522_30, C_11309430_30, C_3223136_1_, C_32336201_30, C_175425292_10, C_8709053_10 and C_2596717_20, respectively). Nrf2 rs6721961 polymorphism was determined by confronting 2-pair primers (CTPP) PCR method [44]. PCR products were separated on 2% agarose gel stained with SYBR® Safe DNA Gel Stain (Invitrogen, United States) and visualized on Chemidoc (Biorad, United States).

Statistical data analysis was performed using IBM SPSS Statistics 22 (SPSS Inc., Chicago, IL, United States). Results were presented as n (%) and mean $\pm$ SD. Pearson's chi-squared test or Fisher's exact test were used to test for the association between acute COVID-19 manifestations regarding the long COVID-19 sequel. We used univariate logistic regression for calculating the odds ratio (OR) and 95% confidence interval (95%CI) in order to determine the potential association between assessed genotypes and odds for the development of long-COVID. Due to unfavorable relations between the number of outcomes and potential predictors, multivariable analysis was not conducted. All p-values less than 0.05 were considered significant.

3. Results

The clinical characteristics of 167 convalescent COVID-19 patients included in this study are summarized in Table 1.

As presented, the average BMI indicates that the majority of patients were overweight and 47% were smokers sometime during their lifetime. Regarding the most frequent COVID-19 comorbidities, 35% of patients had hypertension and 11% had diabetes. During the acute phase of COVID-19 febricity (oral temperature exceeded 37.2 °C) was observed in 86% of patients, while 49% had a temperature over 38 °C (fever). Almost all patients who presented with pneumonia were hospitalized (86%) with 38% requiring O_2 support.

Out of a total number of COVID-19 patients, around 31% reported the loss of smell and taste during the acute phase of the disease. Weakness and general malaise were present in approximately 90% of the patients, while 44% presented with myalgia. Almost half of the COVID-19 patients (46%) suffered from headaches, while in 40%, the headache was associated with fever (Table 2).

Neurological manifestations evaluated during follow-up included fatigue, which was present in 80% of patients, myalgia in 28%, "brain fog" in 13%, as well as instability and paresthesia, present in 9% and 10%, respectively (Table 3). Further analysis showed that 42 COVID-19 patients had both fatigue and myalgia, 20 had both fatigue and "brain fog", while 6 that had both myalgia and "brain fog" also had fatigue.

The distribution of analyzed genotypes among COVID-19 patients is summarized in Table 4. As presented, genetic variations were assessed for 11 antioxidant enzymes, including SOD2, GPX1 and GPX3, as well as, eight members of the GSTs family. Furthermore, polymorphic expressions of the key redox-sensitive transcription factor Nrf2 and, in addition, two functional polymorphisms in ACE2 were also evaluated (Table 4). As visible, the frequencies of certain genotypes differ from those generally observed in the healthy Caucasian population, especially in the case of *GSTM1*, *GSTP1AB* and *GSTP1CD*, *GSTO1*, *GSTO2* and *GPX3*, as previously described [26,28,29].

Figure 1A–C depict the distribution analysis of acute COVID-19 clinical manifestations assessed in the population of patients with the most prominent long-COVID neurological manifestations, including fatigue, myalgia and "brain fog", which were further separately analyzed.

Long-COVID fatigue was present in 134 (80%) COVID-19 patients. The majority of them were hospitalized (88.1%), had febricity (88.8%), suffered from COVID-19 pneumonia (92.5%) and reported general malaise (95.5%) or feeling of weakness (96.3%) in the acute phase of the disease (Figure 1A). As presented, pneumonia, feeling of weakness, general malaise and febricity over 38 °C were significantly more frequent (p = 0.047, p < 0.001,

$p < 0.001$ and $p = 0.043$, respectively) in patients with long-COVID fatigue in comparison to those who lacked this manifestation.

Regarding long-COVID myalgia, it was present in 46 (28%) COVID-19 patients, among which the vast majority (84.4%) were hospitalized during the acute phase of COVID-19. These patients presented with COVID-19 pneumonia (91.3%), febricity (93.5%), weakness (97.8%) and general malaise (95.7%) (Figure 1B). When compared to COVID patients without long-COVID myalgia, headache with or without fever and myalgia were significantly more frequent in patients who developed this long-COVID manifestation ($p = 0.039$, $p = 0.014$ and $p < 0.001$, respectively) in the acute phase of the disease.

Interestingly, all 21 (13%) patients experiencing long-COVID-19 "brain fog" suffered from COVID-19 pneumonia and felt weak during the acute phase of the disease (100%). Almost all of them were hospitalized (90.5%), febrile (95.2%) and felt general malaise (95.2%) as well (Figure 1C).

In the next step, a possible association between genetic variability and the probability of developing neurological manifestations in long-COVID were analyzed in the study group, see Figure 2A–C.

Table 1. Clinical characteristics of COVID-19 patients.

	COVID-19 Patients
Age (years) [a]	55.9 ± 12.3
Gender, *n* (%)	
Male	100 (60)
Female	67 (40)
BMI (kg/m^2) [a]	29.0 ± 5.0
Smoking, *n* (%)	
Never	84 (53)
Former	57 (36)
Ever	17 (11)
Hypertension, *n* (%)	
No	109 (65)
Yes	58 (35)
Diabetes, *n* (%)	
No	149 (89)
Yes	18 (11)
Febricity, *n* (%)	
No	23 (14)
Yes	144 (86)
Febricity over 38 °C, *n* (%)	
No	85 (51)
Yes	82 (49)
Hospitalization, *n* (%)	
No	24 (14)
Yes	143 (86)
Pneumonia, *n* (%)	
No	17 (10)
Yes	150 (90)
O$_2$ support, *n* (%)	
No	103 (62)
Yes	64 (38)

[a] Values are presented as mean ± SD; BMI, body mass index.

Table 2. Neurological manifestations in acute COVID-19.

ACUTE COVID-19	COVID-19 Patients
Loss of smell, *n* (%)	
No	115 (69)
Yes	52 (31)
Loss of taste, *n* (%)	
No	119 (71)
Yes	48 (29)
Myalgia, *n* (%)	
No	94 (56)
Yes	73 (44)
Weakness, *n* (%)	
No	16 (10)
Yes	151 (90)
General malaise, *n* (%)	
No	15 (9)
Yes	152 (91)
Headache, *n* (%)	
No	91 (54)
Yes	76 (46)
Headache with fever, *n* (%)	
No	101 (60)
Yes	66 (40)

Table 3. Neurological manifestations in long-COVID.

LONG-COVID	COVID-19 Patients
Fatigue, *n* (%)	
No	33 (20)
Yes	134 (80)
Myalgia, *n* (%)	
No	121 (72)
Yes	46 (28)
"Brain fog", *n* (%)	
No	146 (87)
Yes	21 (13)
Instability, *n* (%)	
No	152 (91)
Yes	15 (9)
Paresthesia, *n* (%)	
No	150 (90)
Yes	17 (10)

Although long-COVID fatigue represents the most frequent neurological manifestation of long-COVID in our study, none of the analyzed fourteen genes was found to significantly affect the odds of its development.

On the other hand, regarding the long-COVID myalgia, we observed that the carriers of at least one *GSTP1AB Val* allele were more than twice less prone to this neurological manifestation (OR = 0.43, 95%CI 0.21–0.87). Similarly, the *GSTO1Asp* allele decreased the odds of long-COVID myalgia nearly two times (OR = 0.51, 95%CI 0.25–1.02) (Figure 2B). What is more, in individuals carrying combined *GSTP1ABValVal/GSTO1 AspAsp* genotype, the odds of having myalgia were even lower (OR = 0.24, 95%CI 0.09–0.66) when compared to carriers of combined *GSTP1IleIle/GSTO1AlaAla* genotype. On the contrary, genetic variability in glutathione peroxidases seems to increase the odds of developing long-COVID myalgia. Namely, *GPX1 Leu* and *GPX3 CC* alleles are associated with increased odds (Figure 2B), which was more than threefold in individuals carrying combined *GPX1LeuLeu/GPX3CC* genotype (OR = 3.06, 95%CI 1.07–8.77). Furthermore, when all potentially "increased-odds related genotypes" for long-COVID myalgia were analyzed together, it was shown that the carriers of combined *GSTP1ABIleIle/GSTO1AlaAla/GPX1LeuLeu/GPX3CC* genotype had 10-fold increased odds (OR = 10.5, 95%CI 1.67–66.09) of having this neurological manifestation in comparison to individuals with *GSTP1ABValVal/GSTO1AspAsp/GPX1ProPro/GPX3TT* genotype (data not shown).

Table 4. Distribution of analyzed genotypes in COVID-19 patients.

Variants	Genotype, *n* (%)	Variants	Genotype, *n* (%)
GSTM1		*GSTT1*	
active [a]	65 (39)	*active* [a]	131 (79)
null [b]	100 (61)	*null* [b]	34 (21)
GSTM3 (rs1332018)		*GSTA1 (rs3957357)*	
AA	61 (37)	*CC*	52 (37)
AC	51 (31)	*CT*	66 (46)
CC	53 (32)	*TT*	24 (17)
GSTP1 (rs1695)		*GSTP1 (rs1138272)*	
IleIle	82 (49)	*AlaAla*	129 (78)
IleVal	66 (40)	*AlaVal*	36 (21)
ValVal	18 (11)	*ValVal*	1 (1)
GSTO1 (rs4925)		*GSTO2 (rs156697)*	
AlaAla	79 (48)	*AsnAsn*	79 (47)
AlaAsp	48 (29)	*AsnAsp*	46 (28)
AspAsp	38 (23)	*AspAsp*	41 (25)
Nrf2 (rs672196)		*SOD2 (rs4880)*	
CC	124 (75)	*AlaAla*	39 (23)
CA	38 (23)	*AlaVal*	89 (54)
AA	3 (2)	*ValVal*	39 (23)
GPX1 (rs1050450)		*GPX3 (rs8177412)*	
ProPro	75 (45)	*TT*	126 (75)
ProLeu	71 (43)	*TC*	39 (24)
LeuLeu	20 (12)	*CC*	2 (1)
ACE2 (rs4646116)		*ACE2 (rs143936283)*	
LysLys	154 (94)	*GluGlu*	37 (23)
LysArg	9 (5)	*GluGly*	126 (77)
ArgArg	1 (1)	*GlyGly*	0 (0)

[a] Active, if at least one active allele present; [b] Null if no active alleles present.

One more neurological sequelae of COVID-19 seems to be affected by variations in genes encoding antioxidant proteins. Namely, *GSTM1-null* genotype and *GPX1 Leu* alleles independently increase the odds of "brain fog" manifestation in long-COVID more than twice (Figure 2C). This effect is even more potentiated when these genotypes are present in combination, reaching almost 13 fold increased odds (OR = 12.98 (95% CI 1.56–1693.46) in individuals carrying *GSTM1-null/GPX1LeuLeu* genotype in contrast to carriers of *GSTM1-active/GPX1ProPro* genotype. Similarly, Nrf2 polymorphism individually increases the odds of long-COVID "brain fog" by approximately 50%. However, when analyzed in combination with GSTM1 and GPX1 genotypes, it seems that the presence of the combined *GSTM1-null/GPX1LeuLeu/Nrf2AA* genotype yields more than 15 times increased odds (OR = 15.55, 95%CI 1.56–2102.22) for this neurological manifestation (data not shown).

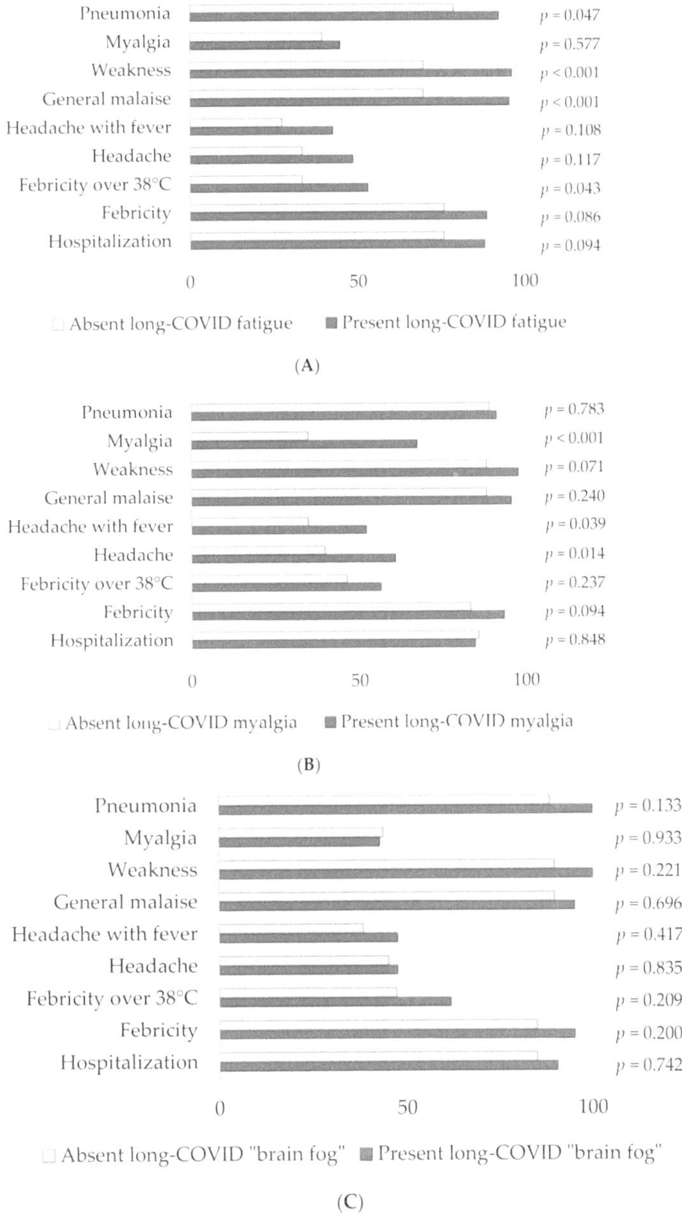

Figure 1. (**A**) The frequency of acute COVID-19 manifestations in population of patients without and with long- COVID fatigue. (**B**) The frequency of acute COVID-19 manifestations in population of patients without and with long-COVID myalgia. (**C**) The frequency of acute COVID-19 manifestations in population of patients without and with long-COVID "brain fog".

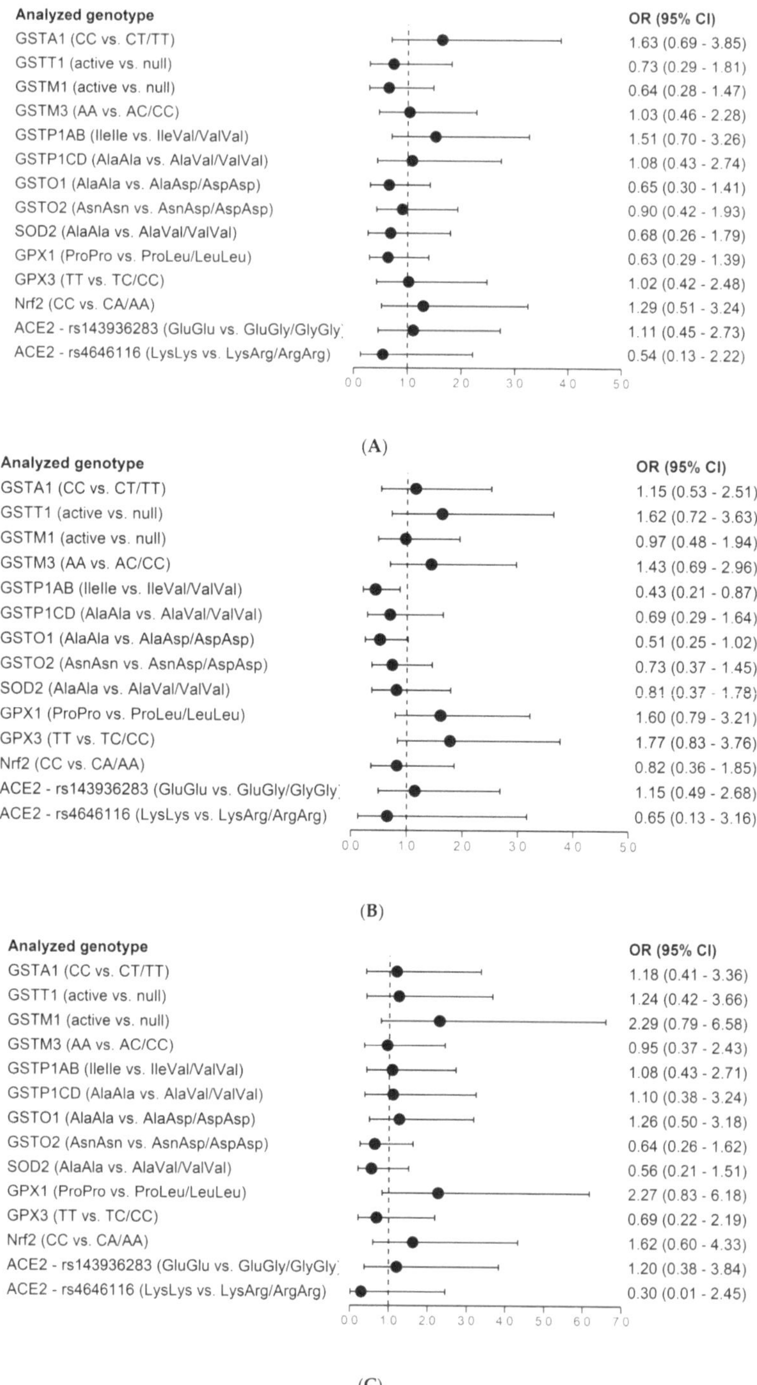

Figure 2. (**A**). The association of genetic variations and long-COVID fatigue. (**B**) The association of genetic variations and long-COVID myalgia. (**C**). The association of genetic variations and long-COVID "brain fog".

4. Discussion

Lately, both scientists' and clinicians' attention has been raised towards understanding why some people develop long-COVID. Oxidative stress, due to impaired expression of antioxidant enzymes, as well as cytoprotective proteins under the control of the antioxidative response element in the DNA, has been suggested as one of the molecular basis of long-COVID [45,46]. Furthermore, apart from the role in the maintenance of the intracellular redox state, the key redox-sensitive transcription factor Nrf2 has been associated with inflammatory response and inflammasome formation in acute COVID-19, proposing its role in long-COVID development too. Therefore, we speculated that genetic variations in antioxidant enzymes, including different members of the GST family, as well as Nrf2, might modulate individual susceptibility towards the development of long-COVID neurological manifestations.

Among the fourteen polymorphisms analyzed in this study, only GSTP1AB and GSTO1 were found to be independently associated with long-COVID manifestations. Indeed, the data obtained showed that individuals carrying GSTP1 Val or GSTO1 Asp allele exhibit lower odds of long-COVID myalgia development, which was even more potentiated in individuals carrying both of these alleles. On the other hand, the presence of GSTP1 Ile and GSTO1 Ala alleles exhibited cumulative risk regarding long-COVID myalgia in carriers of combined GPX1LeuLeu/GPX3CC genotype. This is in line with the fact that individuals carrying combined GPX1 Leu and GPX3 C alleles have an increased probability of developing this long-COVID manifestation. In addition, one particular genotype combination seems to be associated with another long-COVID neurological manifestation. Precisely, individuals carrying combined GSTM1-null/GPX1LeuLeu genotype are more prone to developing long-COVID "brain fog", while probability further enlarges if the Nrf2 A allele was also present. To our knowledge, this is one of the first investigations that addressed the association of antioxidant genetic profile and long-COVID manifestations.

Long-COVID, or post-acute sequelae of SARS-CoV-2, refers to "signs and symptoms that develop during or after an infection consistent with COVID-19, which continue for more than 12 weeks and are not explained by an alternative diagnosis" [45]. The proposition that a multitude of organs might be responsible for long-COVID [47] is in line with the fact that COVID-19 is regarded as an endothelial disease, since all organs depend on perfusion by the vascular microcirculation, precisely those capillaries that are composed of endothelium and pericytes [48–50]. Since both of these cell types express the ACE-2 protein, after binding to its receptor, SARS-CoV-2 may cause vascular injury on one side, but can also cause blood clot formation in blood vessels both in the periphery and brain [51,52]. Indeed, there is evidence of severe hypoxia in tissues surrounding endothelial cells undergoing apoptosis [53]. This multi-organ disease includes, among others, neurological manifestations that seem to be present in nearly one-third of COVID-19 patients within the first 6 months following acute COVID-19 [31]. In the case of neurological sequelae of COVID-19, apart from capillary dysfunction and hypoxia, impaired glucose metabolism, as well as neuro-inflammation and disturbed redox homeostasis are suggested as a main pathophysiological basis [1,53,54].

Regarding redox disbalance, the sequence of events once SARS-CoV-2 enters the host cells seems to be clear now. Namely, the virus alters biochemical processes and, consequently, causes the overproduction of reactive oxygen species (ROS) and oxidative stress. This further causes mitochondrial dysfunction and DNA damage, but at the same time, it affects the expression of the key redox-sensitive transcription factor Nrf2 [45,55]. Nrf2 deficiency decreases the ability of tissue to properly react to disturbances in redox homeostasis, since many genes targeted by Nrf2 encode enzymes essential in antioxidative stress response [41,56]. Since Nrf2 is also shown to down-regulate the expression of genes involved in inflammasome assembly, including NLRP3, caspase 1, IL-1, and IL-18, in that way inhibiting NLRP3 inflammasome activity [40], this further contributes to the process of viral-induced inflammation and the release of proinflammatory cytokines and other molecules that amplify the innate immune response and lead to cell death [45,46]. However, the question remains whether, and in which way, this sequence of events in acute infection explains clinical manifestations in long-COVID?

It is well established that SARS-CoV-2 binds to the host cell and internalizes using a surface ACE2 protein [57,58]. It has further been speculated that genetic polymorphism in the ACE2 gene and the disruption of ACE2 protein 3D structure could influence the virus-ACE2 interaction, consequentially affecting the viral load and patients' response to the infection in terms of disease progression and severity [57,59]. Our results suggest that the ACE2 genotype might also be associated with the odds of developing long-COVID manifestations. Namely, the missense ACE2 rs4646116 SNP analyzed in our study, shown to destabilize ACE2 protein [59], seems to decrease the odds of fatigue, myalgia and "brain fog" in long-COVID patients. In this line, certain ACE2 variants, which are regarded as COVID-19-susceptible or -protective in the acute COVID-19, might also be analyzed in terms of susceptibility to long-COVID clinical manifestations.

As suggested, once the virus is in the cell, it disrupts redox homeostasis and triggers activation of redox-sensitive Nrf2 [41,45]. Nrf2 rs6721961 single nucleotide polymorphism, analyzed in our study, is positioned in the middle of the antioxidant response element (ARE) motif and affects the binding of Nrf2 to the ARE. Therefore, individuals carrying the homozygous *AA* genotype exhibit a lower level of Nrf2 mRNA, which further leads to its lower activity [60]. Although Nrf2 polymorphism itself cannot be associated with the probability of long-COVID manifestations in our study, it has been shown that one of its target genes can. Namely, among various enzymes encoded by Nrf2 target genes and regulated by its binding to AREs are glutathione transferases, especially GSTP1 [39,61]. Apart from its role in the detoxification, chemoresistance and antioxidant defense, GSTP1 also possesses binding activity toward macromolecules, as well as small molecules, and acts as a negative regulator of mitogen-activated protein kinase (MAPK)-signaling pathway by forming protein–protein interactions with JNK1 (c-Jun NH2-terminal kinase) [24,62,63]. Interestingly, MAPK is among the signaling pathways activated once SARS-CoV-2 binds ACE2, which potentially elicits the cytokine storm, identified as the major cause of tissue injury. In this line, MAPK inhibitors are recognized as a novel therapeutic approach for modulation of pro-inflammatory cytokine production [64], which might suggest the possible modulatory effect of GSTP1 as well. Additionally, GSTP1 has the potential to form a GSTP1/Nrf2 protein complex, suggesting a possibility that GSTP1 protein might help Nrf2 stabilization and its further actions [65]. So far, GSTP1 variant alleles have been shown to affect the susceptibility and severity of COVID-19. Namely, the GSTP1 Val allele, which is shown to decrease the odds of developing severe COVID-19 [28], decreases the probability of developing long-COVID myalgia in the present study. The frequent occurrence of myalgia, as a clinical manifestation in both acute and long COVID-19, represents a consequence of several interconnected molecular events. Due to the systemic inflammatory response observed in COVID-19 patients, the involvement of upregulated proinflammatory cytokines, especially IL-6 in hyperalgesia, a stated characteristic of myalgia, has been anticipated [66]. Moreover, hypoxic ischemia in the musculoskeletal system associated with hyperlactemia results in diminished ATP production, leading to pain and fatigue. In that context, myalgia could reflect the state of systemic chronic inflammation along with immunological dysfunction observed in long-COVID-19 [67]. In this study, apart from the *GSTP1 Val* allele, the *GSTO1 Asp* allele also exhibits a modifying effect on the probability of the occurrence of myalgia in long-COVID patients, which is even more potentiated in individuals carrying both of these alleles. GSTO1 exhibits diverse sets of functions involved in redox-sensitive signaling pathways and innate immune response [68]. In the first phase of antiviral host defense, a supporting role of GSTO1-1 in the activation of NLRP3 inflammasome has been established [69]. The NLRP3 inflammasome activation is a necessary step for the conversion of pro-IL-1β and pro-IL-18 into respective mature forms, resulting in subsequent IL-6 release [70,71]. Since the *GSTO1 Ala* allele is more efficient in NEK7/NLRP3 inflammasome activation, it is reasonable to speculate that the presence of this GSTO1 allele with higher deglutathionylation activity could represent the underlying mechanism for the associations between this genetic polymorphism and innate immune response [63,72]. Therefore, our results are in accordance with the recent data on highly activated innate immune cells in patients with long-COVID [73].

Long-COVID myalgia can also be associated with genetic variability in glutathione peroxidases. Namely, we have shown that individuals carrying combined *GPX1 Leu* and *GPX3 C* alleles have an increased probability of developing this long-COVID manifestation, while cumulative odds are even higher in a combination of *GSTP1 Ile* and *GSTO1 Ala* alleles. Apart from their significant role in antioxidant defense by preventing detrimental damage caused by H_2O_2 [74], GPX1 has been shown to affect ASK1 activation and interact with TRAF2, further interfering with the formation of the active ASK1 complex [75]. This way, similarly to GSTP1, GPX1 also seems to participate in signaling cascades activated by SARS-CoV-2, with possible effects on long-COVID development [76]. Another glutathione peroxidase analyzed in our study, GPX3, has just recently been investigated in critically ill COVID-19 patients as an important factor in the interplay between inflammation and oxidative stress [77]. Due to the fact that infectious diseases are associated with decreased Se levels and reduced GPX activity, while GPX3 seems to be able to mediate the anti-inflammatory effects in case of GPX1 loss [78], the role of this antioxidant enzyme in COVID-19 has emerged, and Se and Zn supplementation trials in COVID-19 have initiated [79,80].

"Brain fog" in long-COVID is also more probable in the carriers of the *GPX1 Leu* allele. Decreased GPX1 activity is possible due to the fact that the change of proline with leucine (Pro200Leu) in this SNP leads to alterations in the secondary and tertiary structure of GPX1 [81]. The individuals in whom another important antioxidant enzyme, such as GSTM1, is lacking, the odds of "brain fog" are even higher, as observed in this study. What is more, this class of GSTs, precisely the *GSTM1-null* genotype, is associated with an increased risk of inflammatory lung diseases [82]. Therefore, it seems reasonable to assume that apart from contributing to oxidative stress and inflammation in the acute COVID-19, the lack of this enzyme affects long-COVID too. Importantly, Nrf2 genetic variability, precisely the *Nrf2 A* allele, when present in combination with *GPX LeuLeu/GSTM1-null* genotype further increases the probability of long-COVID "brain fog", highlighting its crucial role in redox homeostasis.

It seems that variations in the human genome, apart from age, gender and chronic diseases, represent an important source of variability in determining both COVID-19 severity in the acute phase, but also long-COVID probability. So far, 13 loci in the human genome that affect COVID-19 susceptibility and severity have been identified [83] while the data on the association between genetic variations and long-COVID are insufficient.

This study has several limitations that need to be addressed. Due to a limited number of patients with different neurological manifestations of long-COVID (fatigue, myalgia or "brain fog"), as well as the fact that some of the patients overlap, a multivariable analysis could not be conducted. Unfortunately, due to the same reason, the modifying effect of comorbidities could not be evaluated either. Therefore, our findings should be regarded more as "hypothesis generating" rather than explanatory, demanding a larger study population. Still, although the sample size is rather small, this study may offer some essential information that could be the base for future longitudinal research.

5. Conclusions

Taken together, our results on the association between alterations in antioxidant genetic profile and neurological manifestations in long-COVID are in agreement with the current hypothesis regarding the biological basis and pathophysiological mechanisms in this condition. The fact that certain genetic variants of antioxidant enzymes, independently or in combination, affect the probability of long-COVID manifestations, further emphasizes the involvement of genetic susceptibility both when SARS-CoV-2 infection is initiated in the host cells, and also months after. However, further studies are needed to clarify the exact roles of specific enzymes, as well as, redox-sensitive transcription factors. Our results might even contribute to better identification of potential therapeutic strategies for both prevention and treatment of long-COVID sequelae.

Author Contributions: Conceptualization, Z.B. and M.P.-E.; methodology, M.E., T.D., V.C. and D.J.; software, Z.B.; validation, T.D., V.C. and D.J.; formal analysis, Z.B.; investigation, M.E., M.A., T.D., V.C. and D.J.; resources, G.S., J.R., I.M., N.T., M.A. and M.E.; writing—original draft preparation, M.P.-E., Z.B., M.E. and M.A.; writing—review and editing, M.P.-E., Z.B., M.E., M.A., T.S., A.S.-R., M.M., T.D., V.C., D.J., G.S., J.R., I.M. and N.T.; visualization, T.D., V.C. and D.J.; supervision, Z.B. and M.P.-E.; project administration, T.S. All authors have read and agreed to the published version of the manuscript.

Funding: This work was supported by the Special Research Program on COVID-19, Project No. 7546803, entitled AntioxIdentification, Science Fund of the Republic of Serbia.

Institutional Review Board Statement: The study was conducted according to the guidelines of the Declaration of Helsinki, and approved by the Institutional Review Board (or Ethics Board) of the University Clinical Centre of Serbia, Serbia, approval number 566/01 from 13 July 2020 and 608/01 from 7 August 2020).

Informed Consent Statement: Written informed consent has been obtained from the patients to publish this paper.

Data Availability Statement: The data are partially contained within the article, while complete data supporting reported results are available at the RedCap platform (Research Electronic Data Capture, Vanderbilt University) of the Faculty of Medicine University in Belgrade.

Acknowledgments: We would like to thank Sanja Sekulic for providing valuable practical support in conducting this study.

Conflicts of Interest: The authors declare no conflict of interest.

References

1. Stefanou, M.-I.; Palaiodimou, L.; Bakola, E.; Smyrnis, N.; Papadopoulou, M.; Paraskevas, G.P.; Rizos, E.; Boutati, E.; Grigoriadis, N.; Krogias, C. Neurological manifestations of long-COVID syndrome: A narrative review. *Ther. Adv. Chronic Dis.* **2022**, *13*, 20406223221076890. [CrossRef] [PubMed]

2. Peterson, C.J.; Sarangi, A.; Bangash, F. Neurological sequelae of COVID-19: A review. *Egypt. J. Neurol. Psychiatry Neurosurg.* **2021**, *57*, 1–8. [CrossRef] [PubMed]

3. Nolen, L.T.; Mukerji, S.S.; Mejia, N.I. Post-acute neurological consequences of COVID-19: An unequal burden. *Nat. Med.* **2022**, *28*, 20–23. [CrossRef] [PubMed]

4. Sivan, M.; Taylor, S. NICE guideline on long COVID. *BMJ* **2020**, *371*, m4938. [CrossRef] [PubMed]

5. Callard, F.; Perego, E. How and why patients made Long COVID. *Soc. Sci. Med.* **2021**, *268*, 113426. [CrossRef]

6. Zhang, X.; Wang, F.; Shen, Y.; Zhang, X.; Cen, Y.; Wang, B.; Zhao, S.; Zhou, Y.; Hu, B.; Wang, M. Symptoms and health outcomes among survivors of COVID-19 infection 1 year after discharge from hospitals in Wuhan, China. *JAMA Netw. Open* **2021**, *4*, e2127403. [CrossRef]

7. Fernández-de-Las-Peñas, C.; Palacios-Ceña, D.; Gómez-Mayordomo, V.; Cuadrado, M.L.; Florencio, L.L. Defining post-COVID symptoms (post-acute COVID, long COVID, persistent post-COVID): An integrative classification. *Int. J. Environ. Res. Public Health* **2021**, *18*, 2621. [CrossRef]

8. Long-Term-Effects-of-Coronavirus. Available online: https://www.nhs.uk/conditions/coronavirus-covid-19 (accessed on 24 March 2022).

9. Liguori, C.; Pierantozzi, M.; Spanetta, M.; Sarmati, L.; Cesta, N.; Iannetta, M.; Ora, J.; Mina, G.G.; Puxeddu, E.; Balbi, O. Subjective neurological symptoms frequently occur in patients with SARS-CoV2 infection. *Brain Behav. Immun.* **2020**, *88*, 11–16. [CrossRef]

10. Nuzzo, D.; Picone, P. Potential neurological effects of severe COVID-19 infection. *Neurosci. Res.* **2020**, *158*, 1–5. [CrossRef]

11. Paterson, R.W.; Brown, R.L.; Benjamin, L.; Nortley, R.; Wiethoff, S.; Bharucha, T.; Jayaseelan, D.L.; Kumar, G.; Raftopoulos, R.E.; Zambreanu, L. The emerging spectrum of COVID-19 neurology: Clinical, radiological and laboratory findings. *Brain* **2020**, *143*, 3104–3120. [CrossRef]

12. Chou, S.H.; Beghi, E.; Helbok, R. GCS-NeuroCOVID Consortium and ENERGY Consortium: Global incidence of neurological manifestations among patients hospitalized with COVID-19-a report for the GCS-NeuroCOVID Consortium and the ENERGY Consortium. *JAMA Netw. Open* **2021**, *4*, e2112131. [CrossRef] [PubMed]

13. Russell, A.L.R.; Hardwick, M.; Jeyanantham, A.; White, L.M.; Deb, S.; Burnside, G.; Joy, H.M.; Smith, C.J.; Pollak, T.A.; Nicholson, T.R. Spectrum, risk factors and outcomes of neurological and psychiatric complications of COVID-19: A UK-wide cross-sectional surveillance study. *Brain Commun.* **2021**, *3*, fcab168. [CrossRef] [PubMed]

14. Mao, L.; Jin, H.; Wang, M.; Hu, Y.; Chen, S.; He, Q.; Chang, J.; Hong, C.; Zhou, Y.; Wang, D.; et al. Neurologic Manifestations of Hospitalized Patients with Coronavirus Disease 2019 in Wuhan, China. *JAMA Neurol.* **2020**, *77*, 683–690. [CrossRef] [PubMed]
15. Baig, A.M. Deleterious outcomes in long-hauler COVID-19: The effects of SARS-CoV-2 on the CNS in chronic COVID syndrome. *ACS Chem. Neurosci.* **2020**, *11*, 4017–4020. [CrossRef]
16. Fiani, B.; Covarrubias, C.; Desai, A.; Sekhon, M.; Jarrah, R. A contemporary review of neurological sequelae of COVID-19. *Front. Neurol.* **2020**, *11*, 640. [CrossRef]
17. Tavassoly, O.; Safavi, F.; Tavassoly, I. Seeding brain protein aggregation by SARS-CoV-2 as a possible long-term complication of COVID-19 infection. *ACS Chem. Neurosci.* **2020**, *11*, 3704–3706. [CrossRef]
18. Sun, B.; Tang, N.; Peluso, M.J.; Iyer, N.S.; Torres, L.; Donatelli, J.L.; Munter, S.E.; Nixon, C.C.; Rutishauser, R.L.; Rodriguez-Barraquer, I. Characterization and biomarker analyses of post-COVID-19 complications and neurological manifestations. *Cells* **2021**, *10*, 386. [CrossRef]
19. Fernandes, I.G.; De Brito, C.A.; Dos Reis, V.M.S.; Sato, M.N.; Pereira, N.Z. SARS-CoV-2 and other respiratory viruses: What does oxidative Stress have to do with it? *Oxid. Med. Cell. Longev.* **2020**, *2020*, 8844280. [CrossRef]
20. Chatterjee, S. Oxidative stress, inflammation, and disease. In *Oxidative Stress and Biomaterials*; Elsevier: Amsterdam, The Netherlands, 2016; pp. 35–58.
21. Cheng, Y.; Sun, F.; Wang, L.; Gao, M.; Xie, Y.; Sun, Y.; Liu, H.; Yuan, Y.; Yi, W.; Huang, Z. Virus-induced p38 MAPK activation facilitates viral infection. *Theranostics* **2020**, *10*, 12223. [CrossRef]
22. Kosanovic, T.; Sagic, D.; Djukic, V.; Pljesa-Ercegovac, M.; Savic-Radojevic, A.; Bukumiric, Z.; Lalosevic, M.; Djordjevic, M.; Coric, V.; Simic, T. Time Course of Redox Biomarkers in COVID-19 Pneumonia: Relation with Inflammatory, Multiorgan Impairment Biomarkers and CT Findings. *Antioxidants* **2021**, *10*, 1126. [CrossRef]
23. Wu, B.; Dong, D. Human cytosolic glutathione transferases: Structure, function, and drug discovery. *Trends Pharmacol. Sci.* **2012**, *33*, 656–668. [CrossRef] [PubMed]
24. Pljesa-Ercegovac, M.; Savic-Radojevic, A.; Matic, M.; Coric, V.; Djukic, T.; Radic, T.; Simic, T. Glutathione transferases: Potential targets to overcome chemoresistance in solid tumors. *Int. J. Mol. Sci.* **2018**, *19*, 3785. [CrossRef] [PubMed]
25. Menon, D.; Innes, A.; Oakley, A.J.; Dahlstrom, J.E.; Jensen, L.M.; Brüstle, A.; Tummala, P.; Rooke, M.; Casarotto, M.G.; Baell, J.B. GSTO1-1 plays a pro-inflammatory role in models of inflammation, colitis and obesity. *Sci. Rep.* **2017**, *7*, 17832. [CrossRef] [PubMed]
26. Djukic, T.; Stevanovic, G.; Coric, V.; Bukumiric, Z.; Pljesa-Ercegovac, M.; Matic, M.; Jerotic, D.; Todorovic, N.; Asanin, M.; Ercegovac, M. GSTO1, GSTO2 and ACE2 Polymorphisms Modify Susceptibility to Developing COVID-19. *J. Pers. Med.* **2022**, *12*, 458. [CrossRef] [PubMed]
27. Silvagno, F.; Vernone, A.; Pescarmona, G.P. The role of glutathione in protecting against the severe inflammatory response triggered by COVID-19. *Antioxidants* **2020**, *9*, 624. [CrossRef] [PubMed]
28. Coric, V.; Milosevic, I.; Djukic, T.; Bukumiric, Z.; Savic-Radojevic, A.; Matic, M.; Jerotic, D.; Todorovic, N.; Asanin, M.; Ercegovac, M. GSTP1 and GSTM3 variant alleles affect susceptibility and severity of COVID-19. *Front. Mol. Biosci.* **2021**, *8*, 747493. [CrossRef]
29. Jerotic, D.; Ranin, J.; Bukumiric, Z.; Djukic, T.; Coric, V.; Savic-Radojevic, A.; Todorovic, N.; Asanin, M.; Ercegovac, M.; Milosevic, I. SOD2 rs4880 and GPX1 rs1050450 polymorphisms do not confer risk of COVID-19, but influence inflammation or coagulation parameters in Serbian cohort. *Redox Rep.* **2022**, *27*, 85–91. [CrossRef]
30. Camargo-Martínez, W.; Lozada-Martínez, I.; Escobar-Collazos, A.; Navarro-Coronado, A.; Moscote-Salazar, L.; Pacheco-Hernández, A.; Janjua, T.; Bosque-Varela, P. Post-COVID 19 neurological syndrome: Implications for sequelae's treatment. *J. Clin. Neurosci.* **2021**, *88*, 219–225. [CrossRef]
31. Taquet, M.; Geddes, J.R.; Husain, M.; Luciano, S.; Harrison, P.J. 6-month neurological and psychiatric outcomes in 236,379 survivors of COVID-19: A retrospective cohort study using electronic health records. *Lancet Psychiatry* **2021**, *8*, 416–427. [CrossRef]
32. Qin, Y.; Wu, J.; Chen, T.; Li, J.; Zhang, G.; Wu, D.; Zhou, Y.; Zheng, N.; Cai, A.; Ning, Q. Long-term microstructure and cerebral blood flow changes in patients recovered from COVID-19 without neurological manifestations. *J. Clin. Investig.* **2021**, *131*, e147329. [CrossRef]
33. Shi, M.; Bradner, J.; Bammler, T.K.; Eaton, D.L.; Zhang, J.; Ye, Z.; Wilson, A.M.; Montine, T.J.; Pan, C.; Zhang, J. Identification of glutathione S-transferase pi as a protein involved in Parkinson disease progression. *Am. J. Pathol.* **2009**, *175*, 54–65. [CrossRef]
34. Piacentini, S.; Polimanti, R.; Squitti, R.; Mariani, S.; Migliore, S.; Vernieri, F.; Rossini, P.M.; Manfellotto, D.; Fuciarelli, M. GSTO1*E155del polymorphism associated with increased risk for late-onset Alzheimer's disease: Association hypothesis for an uncommon genetic variant. *Neurosci. Lett.* **2012**, *506*, 203–207. [CrossRef] [PubMed]
35. Ercegovac, M.; Jovic, N.; Sokic, D.; Savic-Radojevic, A.; Coric, V.; Radic, T.; Nikolic, D.; Kecmanovic, M.; Matic, M.; Simic, T. GSTA1, GSTM1, GSTP1 and GSTT1 polymorphisms in progressive myoclonus epilepsy: A Serbian case-control study. *Seizure* **2015**, *32*, 30–36. [CrossRef] [PubMed]
36. Eum, K.-D.; Seals, R.M.; Taylor, K.M.; Grespin, M.; Umbach, D.M.; Hu, H.; Sandler, D.P.; Kamel, F.; Weisskopf, M.G. Modification of the association between lead exposure and amyotrophic lateral sclerosis by iron and oxidative stress related gene polymorphisms. *Amyotroph. Lateral Scler. Front. Degener.* **2015**, *16*, 72–79. [CrossRef] [PubMed]
37. Allocati, N.; Masulli, M.; Di Ilio, C.; Federici, L. Glutathione transferases: Substrates, inihibitors and pro-drugs in cancer and neurodegenerative diseases. *Oncogenesis* **2018**, *7*, 8. [CrossRef]

38. Chartoumpekis, D.V.; Wakabayashi, N.; Kensler, T.W. Keap1/Nrf2 pathway in the frontiers of cancer and non-cancer cell metabolism. *Biochem. Soc. Trans.* **2015**, *43*, 639–644. [CrossRef]

39. Cho, H.-Y.; Marzec, J.; Kleeberger, S.R. Functional polymorphisms in Nrf2: Implications for human disease. *Free Radic. Biol. Med.* **2015**, *88*, 362–372. [CrossRef]

40. Liu, X.; Zhang, X.; Ding, Y.; Zhou, W.; Tao, L.; Lu, P.; Wang, Y.; Hu, R. Nuclear factor E2-related factor-2 negatively regulates NLRP3 inflammasome activity by inhibiting reactive oxygen species-induced NLRP3 priming. *Antioxid. Redox Signal.* **2017**, *26*, 28–43. [CrossRef]

41. Saha, S.; Buttari, B.; Panieri, E.; Profumo, E.; Saso, L. An overview of Nrf2 signaling pathway and its role in inflammation. *Molecules* **2020**, *25*, 5474. [CrossRef]

42. Michielsen, H.J.; De Vries, J.; Van Heck, G.L. Psychometric qualities of a brief self-rated fatigue measure: The Fatigue Assessment Scale. *J. Psychosom. Res.* **2003**, *54*, 345–352. [CrossRef]

43. Ross, A.J.; Medow, M.S.; Rowe, P.C.; Stewart, J.M. What is brain fog? An evaluation of the symptom in postural tachycardia syndrome. *Clin. Auton. Res.* **2013**, *23*, 305–311. [CrossRef]

44. Shimoyama, Y.; Mitsuda, Y.; Tsuruta, Y.; Hamajima, N.; Niwa, T. Polymorphism of Nrf2, an antioxidative gene, is associated with blood pressure and cardiovascular mortality in hemodialysis patients. *Int. J. Med. Sci.* **2014**, *11*, 726. [CrossRef]

45. Jarrott, B.; Head, R.; Pringle, K.G.; Lumbers, E.R.; Martin, J.H. "LONG COVID"—A hypothesis for understanding the biological basis and pharmacological treatment strategy. *Pharmacol. Res. Perspect.* **2022**, *10*, e00911. [CrossRef]

46. Cuadrado, A.; Pajares, M.; Benito, C.; Jiménez-Villegas, J.; Escoll, M.; Fernández-Ginés, R.; Yagüe, A.J.G.; Lastra, D.; Manda, G.; Rojo, A.I. Can activation of NRF2 be a strategy against COVID-19? *Trends Pharmacol. Sci.* **2020**, *41*, 598–610. [CrossRef]

47. Nalbandian, A.; Sehgal, K.; Gupta, A.; Madhavan, M.V.; McGroder, C.; Stevens, J.S.; Cook, J.R.; Nordvig, A.S.; Shalev, D.; Sehrawat, T.S. Post-acute COVID-19 syndrome. *Nat. Med.* **2021**, *27*, 601–615. [CrossRef]

48. Libby, P.; Lüscher, T. COVID-19 is, in the end, an endothelial disease. *Eur. Heart J.* **2020**, *41*, 3038–3044. [CrossRef]

49. Fogarty, H.; Townsend, L.; Morrin, H.; Ahmad, A.; Comerford, C.; Karampini, E.; Englert, H.; Byrne, M.; Bergin, C.; O'Sullivan, J.M. Persistent endotheliopathy in the pathogenesis of long COVID syndrome. *J. Thromb. Haemost.* **2021**, *19*, 2546–2553. [CrossRef]

50. von Meijenfeldt, F.A.; Havervall, S.; Adelmeijer, J.; Lundström, A.; Magnusson, M.; Mackman, N.; Thalin, C.; Lisman, T. Sustained prothrombotic changes in COVID-19 patients 4 months after hospital discharge. *Blood Adv.* **2021**, *5*, 756–759. [CrossRef]

51. Ackermann, M.; Verleden, S.E.; Kuehnel, M.; Haverich, A.; Welte, T.; Laenger, F.; Vanstapel, A.; Werlein, C.; Stark, H.; Tzankov, A.; et al. Pulmonary Vascular Endothelialitis, Thrombosis, and Angiogenesis in COVID-19. *N. Engl. J. Med.* **2020**, *383*, 120–128. [CrossRef]

52. Liu, Q.; Yang, Y.; Fan, X. Microvascular pericytes in brain-associated vascular disease. *Biomed. Pharmacother.* **2020**, *121*, 109633. [CrossRef]

53. Østergaard, L. SARS CoV-2 related microvascular damage and symptoms during and after COVID-19: Consequences of capillary transit-time changes, tissue hypoxia and inflammation. *Physiol. Rep.* **2021**, *9*, e14726. [CrossRef]

54. Marshall, M. COVID and the brain: Researchers zero in on how damage occurs. *Nature* **2021**, *595*, 484–485. [CrossRef] [PubMed]

55. Checconi, P.; De Angelis, M.; Marcocci, M.E.; Fraternale, A.; Magnani, M.; Palamara, A.T.; Nencioni, L. Redox-modulating agents in the treatment of viral infections. *Int. J. Mol. Sci.* **2020**, *21*, 4084. [CrossRef] [PubMed]

56. Ishii, T.; Itoh, K.; Takahashi, S.; Sato, H.; Yanagawa, T.; Katoh, Y.; Bannai, S.; Yamamoto, M. Transcription factor Nrf2 coordinately regulates a group of oxidative stress-inducible genes in macrophages. *J. Biol. Chem.* **2000**, *275*, 16023–16029. [CrossRef] [PubMed]

57. Lippi, G.; Lavie, C.J.; Henry, B.M.; Sanchis-Gomar, F. Do genetic polymorphisms in angiotensin converting enzyme 2 (ACE2) gene play a role in coronavirus disease 2019 (COVID-19)? *Clin. Chem. Lab. Med.* **2020**, *58*, 1415–1422. [CrossRef]

58. Devaux, C.A.; Rolain, J.-M.; Raoult, D. ACE2 receptor polymorphism: Susceptibility to SARS-CoV-2, hypertension, multi-organ failure, and COVID-19 disease outcome. *J. Microbiol. Immunol. Infect.* **2020**, *53*, 425–435. [CrossRef]

59. Paniri, A.; Hosseini, M.M.; Akhavan-Niaki, H. First comprehensive computational analysis of functional consequences of TMPRSS2 SNPs in susceptibility to SARS-CoV-2 among different populations. *J. Biomol. Struct. Dyn.* **2021**, *39*, 3576–3593. [CrossRef]

60. Suzuki, T.; Shibata, T.; Takaya, K.; Shiraishi, K.; Kohno, T.; Kunitoh, H.; Tsuta, K.; Furuta, K.; Goto, K.; Hosoda, F. Regulatory nexus of synthesis and degradation deciphers cellular Nrf2 expression levels. *Mol. Cell. Biol.* **2013**, *33*, 2402–2412. [CrossRef]

61. Bartolini, D.; Galli, F. The functional interactome of GSTP: A regulatory biomolecular network at the interface with the Nrf2 adaption response to oxidative stress. *J. Chromatogr. B Anal. Technol. Biomed. Life Sci.* **2016**, *1019*, 29–44. [CrossRef]

62. Thévenin, A.F.; Zony, C.L.; Bahnson, B.J.; Colman, R.F. GSTpi modulates JNK activity through a direct interaction with JNK substrate, ATF2. *Protein Sci.* **2011**, *20*, 834–848. [CrossRef]

63. Board, P.G.; Menon, D. Glutathione transferases, regulators of cellular metabolism and physiology. *Biochim. Biophys. Acta (BBA)-Gen. Subj.* **2013**, *1830*, 3267–3288. [CrossRef] [PubMed]

64. Rex, D.A.B.; Dagamajalu, S.; Kandasamy, R.K.; Raju, R.; Prasad, T.S. SARS-CoV-2 signaling pathway map: A functional landscape of molecular mechanisms in COVID-19. *J. Cell Commun. Signal.* **2021**, *15*, 601–608. [CrossRef] [PubMed]

65. Bartolini, D.; Commodi, J.; Piroddi, M.; Incipini, L.; Sancineto, L.; Santi, C.; Galli, F. Glutathione S-transferase pi expression regulates the Nrf2-dependent response to hormetic diselenides. *Free Radic. Biol. Med.* **2015**, *88*, 466–480. [CrossRef] [PubMed]

66. Lippi, G.; Wong, J.; Henry, B.M. Myalgia may not be associated with severity of coronavirus disease 2019 (COVID-19). *World J. Emerg. Med.* **2020**, *11*, 193. [CrossRef] [PubMed]

67. Seixas, M.L.G.A.; Mitre, L.P.; Shams, S.; Lanzuolo, G.B.; Bartolomeo, C.S.; Silva, E.A.; Prado, C.M.; Ureshino, R.; Stilhano, R.S. Unraveling Muscle Impairment Associated With COVID-19 and the Role of 3D Culture in Its Investigation. *Front. Nutr.* **2022**, *9*, 115. [CrossRef] [PubMed]

68. Board, P.G.; Coggan, M.; Chelvanayagam, G.; Easteal, S.; Jermiin, L.S.; Schulte, G.K.; Danley, D.E.; Hoth, L.R.; Griffor, M.C.; Kamath, A.V. Identification, characterization, and crystal structure of the Omega class glutathione transferases. *J. Biol. Chem.* **2000**, *275*, 24798–24806. [CrossRef]

69. Hughes, M.M.; Hooftman, A.; Angiari, S.; Tummala, P.; Zaslona, Z.; Runtsch, M.C.; McGettrick, A.F.; Sutton, C.E.; Diskin, C.; Rooke, M. Glutathione transferase omega-1 regulates NLRP3 inflammasome activation through NEK7 deglutathionylation. *Cell Rep.* **2019**, *29*, 151–161. [CrossRef]

70. Zhao, N.; Di, B.; Xu, L. The NLRP3 inflammasome and COVID-19: Activation, pathogenesis and therapeutic strategies. *Cytokine Growth Factor Rev.* **2021**, *61*, 2–15. [CrossRef]

71. McKee, C.M.; Coll, R.C. NLRP3 inflammasome priming: A riddle wrapped in a mystery inside an enigma. *J. Leukoc. Biol.* **2020**, *108*, 937–952. [CrossRef]

72. Tanaka-Kagawa, T.; Jinno, H.; Hasegawa, T.; Makino, Y.; Seko, Y.; Hanioka, N.; Ando, M. Functional characterization of two variant human GSTO 1-1s (Ala140Asp and Thr217Asn). *Biochem. Biophys. Res. Commun.* **2003**, *301*, 516–520. [CrossRef]

73. Phetsouphanh, C.; Darley, D.R.; Wilson, D.B.; Howe, A.; Munier, C.; Patel, S.K.; Juno, J.A.; Burrell, L.M.; Kent, S.J.; Dore, G.J. Immunological dysfunction persists for 8 months following initial mild-to-moderate SARS-CoV-2 infection. *Nat. Immunol.* **2022**, *23*, 210–216. [CrossRef] [PubMed]

74. Ekoue, D.N.; He, C.; Diamond, A.M.; Bonini, M.G. Manganese superoxide dismutase and glutathione peroxidase-1 contribute to the rise and fall of mitochondrial reactive oxygen species which drive oncogenesis. *Biochim. Biophys. Acta (BBA) -Bioenerg.* **2017**, *1858*, 628–632. [CrossRef] [PubMed]

75. Lee, S.; Lee, E.-K.; Kang, D.H.; Lee, J.; Hong, S.H.; Jeong, W.; Kang, S.W. Glutathione peroxidase-1 regulates ASK1-dependent apoptosis via interaction with TRAF2 in RIPK3-negative cancer cells. *Exp. Mol. Med.* **2021**, *53*, 1080–1091. [CrossRef] [PubMed]

76. Hsu, A.C.Y.; Wang, G.; Reid, A.T.; Veerati, P.C.; Pathinayake, P.S.; Daly, K.; Mayall, J.R.; Hansbro, P.M.; Horvat, J.C.; Wang, F. SARS-CoV-2 Spike protein promotes hyper-inflammatory response that can be ameliorated by Spike-antagonistic peptide and FDA-approved ER stress and MAP kinase inhibitors in vitro. *Biorxiv* **2020**. [CrossRef]

77. Notz, Q.; Herrmann, J.; Schlesinger, T.; Helmer, P.; Sudowe, S.; Sun, Q.; Hackler, J.; Roeder, D.; Lotz, C.; Meybohm, P. Clinical significance of micronutrient supplementation in critically ill COVID-19 patients with severe ARDS. *Nutrients* **2021**, *13*, 2113. [CrossRef]

78. Blunt, K.; Jacobse, J.; Allaman, M.; Washington, M.; Goettel, J.; Wilson, K.; Williams, C.; Short, S. GPX3 IS REQUIRED TO MEDIATE PROTECTIVE EFFECTS OF GPX1 LOSS IN COLITIS. *Gastroenterology* **2022**, *162*, S63. [CrossRef]

79. Moghaddam, A.; Heller, R.A.; Sun, Q.; Seelig, J.; Cherkezov, A.; Seibert, L.; Hackler, J.; Seemann, P.; Diegmann, J.; Pilz, M. Selenium deficiency is associated with mortality risk from COVID-19. *Nutrients* **2020**, *12*, 2098. [CrossRef]

80. Heller, R.A.; Sun, Q.; Hackler, J.; Seelig, J.; Seibert, L.; Cherkezov, A.; Minich, W.B.; Seemann, P.; Diegmann, J.; Pilz, M. Prediction of survival odds in COVID-19 by zinc, age and selenoprotein P as composite biomarker. *Redox Biol.* **2021**, *38*, 101764. [CrossRef]

81. Mihailovic, S.; Coric, V.; Radic, T.; Radojevic, A.S.; Matic, M.; Dragicevic, D.; Djokic, M.; Vasic, V.; Dzamic, Z.; Simic, T. The Association of Polymorphisms in Nrf2 and Genes Involved in Redox Homeostasis in the Development and Progression of Clear Cell Renal Cell Carcinoma. *Oxid. Med. Cell. Longev.* **2021**, *2021*, 6617969. [CrossRef]

82. Wu, W.; Peden, D.; Diaz-Sanchez, D. Role of GSTM1 in resistance to lung inflammation. *Free Radic. Biol. Med.* **2012**, *53*, 721–729. [CrossRef]

83. Niemi, M.E.K.; Karjalainen, J.; Liao, R.G.; Neale, B.M.; Daly, M.; Ganna, A.; Pathak, G.A.; Andrews, S.J.; Kanai, M.; Veerapen, K. Mapping the human genetic architecture of COVID-19. *Nature* **2021**, *600*, 472–477.

 *antioxidants*

MDPI

Article

AMPK-Nrf2 Signaling Pathway in Phrenic Motoneurons following Cervical Spinal Cord Injury

Pauline Michel-Flutot, Laurie Efthimiadi, Lynda Djerbal, Therese B. Deramaudt, Marcel Bonay [†] and Stéphane Vinit *,[†]

Université Paris-Saclay, UVSQ, Inserm, END-ICAP, 78000 Versailles, France
* Correspondence: stephane.vinit@uvsq.fr; Tel.: +33-170-429-427
† These authors contributed equally to this work.

Abstract: High spinal cord injuries (SCI) induce the deafferentation of phrenic motoneurons, leading to permanent diaphragm paralysis. This involves secondary injury associated with pathologic and inflammatory processes at the site of injury, and at the level of phrenic motoneurons. In the present study, we evaluated the antioxidant response in phrenic motoneurons involving the AMPK-Nrf2 signaling pathway following C2 spinal cord lateral hemi-section in rats. We showed that there is an abrupt reduction in the expression of phosphorylated AMPK and Nrf2 at one hour post-injury in phrenic motoneurons. A rebound is then observed at one day post-injury, reflecting a return to homeostasis condition. In the total spinal cord around phrenic motoneurons, the increase in phosphorylated AMPK and Nrf2 occurred at three days post-injury, showing the differential antioxidant response between phrenic motoneurons and other cell types. Taken together, our results display the implication of the AMPK-Nrf2 signaling pathway in phrenic motoneurons' response to oxidative stress following high SCI. Harnessing this AMPK-Nrf2 signaling pathway could improve the antioxidant response and help in spinal rewiring to these deafferented phrenic motoneurons to improve diaphragm activity in patients suffering high SCI.

Keywords: Nrf2; spinal cord injury; phrenic motoneuron; AMPK; neuroinflammation; antioxidant; rat

Citation: Michel-Flutot, P.;
Efthimiadi, L.; Djerbal, L.;
Deramaudt, T.B.; Bonay, M.; Vinit, S.
AMPK-Nrf2 Signaling Pathway in
Phrenic Motoneurons following
Cervical Spinal Cord Injury.
Antioxidants **2022**, *11*, 1665.
https://doi.org/10.3390/
antiox11091665

Academic Editor: Stanley Omaye

Received: 28 July 2022
Accepted: 25 August 2022
Published: 26 August 2022

Publisher's Note: MDPI stays neutral
with regard to jurisdictional claims in
published maps and institutional affiliations.

1. Introduction

High spinal cord injuries (SCI) lead to the axotomy of descending neuronal pathways within the spinal cord. This results in a permanent loss of motor function, including respiratory insufficiency [1]. For patients living with such injuries, a mechanical ventilation is necessary for survival, and only a few of them may be weaned of ventilatory assistance over time. This spontaneous plasticity is, however, extremely limited. The use of preclinical models is consequently required to decipher the inflammatory and plasticity processes involved. The rodent cervical C2 lateral hemi-section (C2HS) model is the most widely used to evaluate the impact of high SCI on spinal neuroplasticity and neuroinflammation related to the phrenic system [2–9].

More precisely, descending respiratory pathways originate from the rostral-ventral respiratory group (rVRG) located in the rostro-ventral medulla [10]. They connect monosy-naptically and bilaterally to the phrenic motoneurons located between C3 and C6 spinal segments of the spinal cord in several species, such as human [11], cat [12,13], rat [14], and mouse [15]. The rVRG also indirectly connects phrenic motoneurons via phrenic interneurons [16–18]. Phrenic motoneurons innervate the diaphragm (main inspiratory muscle) bilaterally through the phrenic nerves. In this preclinical model, a unilateral C2HS induces a disruption of the bulbospinal tract from one side of the brainstem to the phrenic motoneurons innervating the diaphragm, leading to hemi-diaphragm paralysis. The con-tralateral side of the C2 hemi-injury remains intact, allowing the survival and recovery of the animal. This also leads to inflammatory processes at the site of injury [19,20], but also in the aera of deafferented motoneurons [9,21].

Glutamate excitotoxicity participates in these inflammatory processes. SCI induce the axotomy of glutamatergic neurons innerving motoneurons, which leads to excessive release of glutamate [22]. This glutamate then binds to its α-amino-3-hydroxy-5-methyl-4-isoxazolepropionic acid (AMPA), Kainate, and N-methyl-D-aspartate (NMDA) receptors. These ionotropic receptors are permeable to monovalent cations (sodium and potassium), as well as to calcium for NMDA and for certain AMPA receptors. An overactivation of these ionotropic receptors leads to a strong influx of calcium into the postsynaptic neuron, resulting, among other things, in oxidative stress [23]. More precisely, this leads to an increase in the production of reactive oxygen species (ROS) by the mitochondria and their accumulation in the cell body [24]. To cope with this cellular stress [25], there is an increase in the production of antioxidant molecules in the cell. This increase occurs through the modulation of AMPK [26]. AMPK is activated in case of increased ATP concentration in the cytoplasm, which happens in case of increased activity of the electron transport chain in the mitochondrion due to ROS formation [24]. AMPK activation is performed through its phosphorylation by LKB1. AMPK can also be activated by the increase in calcium concentration in the cytoplasm [27]. Phosphorylated AMPK (pAMPK) is then involved in transcription factors' phosphorylation, facilitating their translocation to the nucleus where they induce the expression of diverse genes involved in metabolism and antioxidative processes [27], such as nuclear factor erythroid 2-related factor (Nrf2) [28].

Nrf2 is a key regulator of the cellular anti-inflammatory and antioxidant responses [29–31]. Nrf2 is sequestered in the cytoplasm by Kelch-like ECH-associated protein 1 (Keap1) through the Neh2 domain and is degraded through ubiquitin-dependent proteasomal degradation under physiological conditions [32,33]. In case of cellular stress, Nrf2 is released, and then translocates into the nucleus where it binds to antioxidant response elements (ARE), leading to the transcription of related genes coding for anti-inflammatory, antioxidant, and cytoprotective proteins, such as NADPH quinine oxidoreductase 1 (NQO1) and heme oxygenase 1 (HO-1) [34,35]. Following SCI, it has been shown that in mice, a reduction of Nrf2 expression [36] or its absence (Nrf2 knockout mice) [37] leads to an increase in neuronal death associated with locomotor dysfunction [36,37]. Nrf2 indeed has an important role in the reduction of TNF-α and MMP9 production following SCI [38], with these molecules being involved in spinal neuroinflammation [39,40]. Activation of Nrf2 has therefore been considered as a potential treatment following SCI, considering its anti-inflammatory and antioxidant effects [29–31]. The administration of Nrf2 activators has shown beneficial effects in this context, such as a reduction in inflammatory cytokine production, and an increase in Nrf2 expression and its related signaling pathway in neuronal and glial cells [29]. These results display the pivotal role of Nrf2 in neuroinflammation and the resulting functional outcome following SCI.

A better understanding of the role of Nrf2 and its related pathway on inflammation and oxidative stress in this context could help in harnessing its beneficial effects to develop new effective strategies aiming to reduce neuroinflammation and improve the functional outcome. To date, however, no study has evaluated the oxidative stress and Nrf2-mediated antioxidant and anti-inflammatory activities at the level of phrenic motoneurons following cervical SCI. We therefore investigated these inflammatory-related processes in phrenic motoneurons at different time points post-injury (P.I.) in rats following a C2 spinal cord lateral hemi-section.

2. Materials and Methods

2.1. Ethics Statement

All experiments reported in this manuscript conformed to policies laid out by the National Institutes of Health (Bethesda, MD, USA) in the Guide for the Care and Use of Laboratory Animals and the European Communities Council Directive of 22 September 2010 (2010/63/EU, 74) regarding animal experimentation (Apafis #2017111516297308_v3). These experiments were performed on 3–4-month-old male Sprague–Dawley rats (Janvier, France). The animals were dual-housed in individually ventilated cages in a state-of-the-art animal care facility (2CARE animal facility, accreditation A78-322-3, France), with access to food and water ad libitum with a 12 h light/dark cycle. The Ethics Committee of the RBUCE-UP chair of Excellence (University of Paris-Sud, grant agreement No. 246556) and the University of Versailles Saint-Quentin-en-Yvelines approved these experiments.

2.2. Animal Groups and Surgical Preparation

In this study, 63 animals were used, divided into 9 groups: control ($n = 4$), Sham animals (at 1 h ($n = 6$), 1 day ($n = 7$), 3 days ($n = 7$), and 7 days ($n = 7$) post-surgery), and C2-injured animals (at 1 h ($n = 8$), 1 day ($n = 8$), 3 days ($n = 8$), and 7 days ($n = 8$) post-injury). None of the C2-injured animals died during or after the procedure, however one Sham animal died at 1 h due to a technical issue. All animals (except $n = 4$ of each group) received an intrapleural injection with the retrograde tracer Cholera toxin B fragment (CTB, List Biologicals, Campbell, CA, USA) to identify phrenic motor neurons in the cervical spinal cord, similar to that described previously [41–44]. Briefly, each rat was anaesthetized with isoflurane (Iso-vet, Piramal, Voorschoten, Netherlands; ~1.5% in O_2), placed in the supine position on a surgical table while spontaneously breathing into a face mask. The lateral sides of the rib cage were shaved and gently palpated to identify the fifth intercostal space at the anterior axillary line. A sterilized 26G needle connected to a 50 µL Hamilton syringe was used to inject 25 µL of a 0.2% CTB (dissolved in sterile injectable saline) between the 5th and the 6th ribs into the pleural space on each side. The needle was modified so that it could be inserted no further than 6 mm below the surface of the skin. After the injections, the animals were monitored closely for signs of respiratory distress associated with an eventual pneumothorax. Isoflurane was turned off and the rats were monitored for 30 min. None of the rats showed signs of respiratory distress. Three days after CTB injection, rats underwent C2 or sham surgery. As described previously [7], buprenorphine (Buprécare, 50 µg/kg, analgesic, Axience, Pantin, France), carprofen (5 mg/kg, anti-inflammatory, Rymadyl, Zoetis, Malakoff, France), enrofloxacin (4 mg/kg, antibiotics, Baytril, Bayer, Leverkusen, Germany), and medetomidine (0.1 mg/kg, Médétor, Virbac, Carros, France) were administered subcutaneously 10–20 min before inducing isoflurane anesthesia in a closed chamber (in 100% O_2). Rats were orotracheally intubated and ventilated with a rodent ventilator (model 683; Harvard Apparatus, South Natick, MA, USA). Anesthesia was maintained throughout the procedure (1.5–2% isoflurane in 100% O_2). Skin and muscles were retracted, and a C2 laminectomy and durotomy were performed. For the C2-injured animals, the spinal cord was sectioned unilaterally just caudal to the C2 dorsal roots with micro-scissors, followed with a micro-scalpel to ensure the section of all the remaining fibers, as described previously by our team [7]. The wounds and skin were sutured shut. Sham rats underwent the same procedures without hemi-section. An intra-muscular injection of atipamezole (0.5 mg/kg, Revertor, Virbac, Carros, France) was administered to reverse medetomidine. The isoflurane was discontinued, the endotracheal tube was pulled off, and the rats were monitored throughout recovery. All animals received an injection of analgesic (tramadol, 15 mg/kg), antibiotics (enrofloxacin, 4 mg/kg), and anti-inflammatory (carpofen, 5 mg/kg) drugs for 2 days post-surgery.

2.3. Tissue Harvesting and Sample Processing

After the proper time post-surgery (1 h, 1 day, 3 days, or 7 days), the animals were deeply anesthetized (5% isoflurane) and a lethal injection of 1 mL of urethane (0.2 g/mL, i.c., Sigma-Aldrich, Darmstadt, Germany) was performed. Then, the animals were transcardially perfused with cold (4 °C) heparinized saline (5 UI/mL) followed by Antigenfix solution (DIAPATH, Martinengo (BG), Italy). The C1–C6 spinal cords were dissected out and stored in Antigenfix solution overnight at 4 °C, then transferred to an ascending sucrose concentration (20% to 30% sucrose in phosphate-buffered saline (PBS) 1X (BP665-1; Fisher Scientific, Illkirch, France) until they lost buoyancy. The cervical spinal cord was divided into two distinct parts. The C1–C3 spinal cord was sectioned longitudinally (30 µm thickness) with a cryostat (NX70, Thermo Fisher Scientific, Waltham, MA, USA), mounted on slides, and the injury was assessed with cresyl violet staining: 10 min in cresyl violet solution (0.001% cresyl violet acetate (C5042-10G, Sigma-Aldrich) and 0.125% glacial acetic acid (A/0400/PB15, Fisher Scientific) in distilled water), 1 min in 70% ethanol, 1 min in 95% ethanol, 2 × 1 min in 100% ethanol (E/0600DF/17, Fisher Scientific), and 2 min in xylene (X/0100/PB17, Fischer Scientific), and cover-slipped with Eukitt® mounting medium. A brightfield microscopy (Aperio AT2, Leica, Nanterre, France) was used to reconstruct the extent of the injury. Each injury was reported on a stereotaxic C2 metameric transverse plane, as described previously [6,7]. Each reconstructed injury was digitized and analyzed with ImageJ 1.53n software (National Institutes of Health, Bethesda, MD, USA). The percentage of each injured side was evaluated by reference to a complete hemi-section (which is 100%, as previously described [7,45]) (Figure 1).

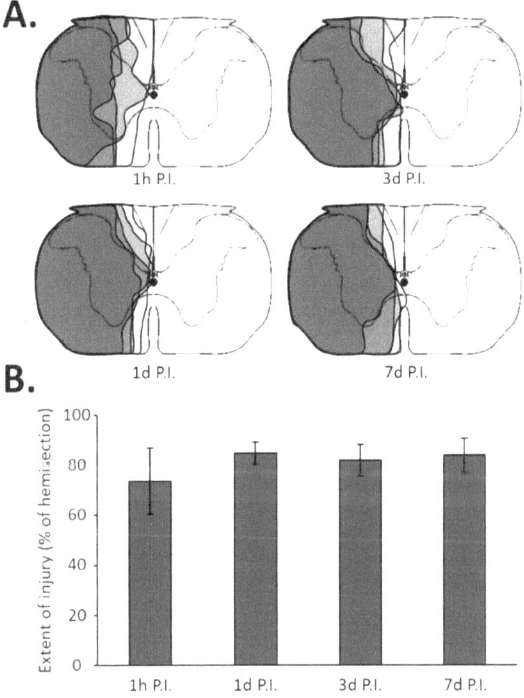

Figure 1. Extent of injury following a C2 spinal cord hemi-section. (**A**) Representation of a transversal view of the C2 spinal cord and representative extent of injury for each group in gray, i.e., at 1 h post-injury (P.I.), 1 day P.I., 3 days P.I., and 7 days P.I. (**B**) Extent of injury quantification in percentage compared with 100% uninjured hemi-spinal cord. The quantification has been made only in the ventral part where the phrenic motoneurons are located. There was no difference between the different groups (one-way ANOVA, $p = 0.277$).

2.4. Immunohistochemistry and In Situ Enzymatic Reaction

The C3–C6 spinal cord was transversally cut (30 μm thickness) with a cryostat, and the sections were stored in a cryoprotectant solution (sucrose 30% (Pharmagrade, 141621, AppliChem, Darmstadt, Germany), ethylene glycol 30% (BP230-4, Fisher Scientific), and polyvinylpyrrolidone 40 (PVP40-100G; Sigma-Aldrich 1% in phosphate-buffered saline PBS 1X)) at −22 °C for further investigations. Sections were prepared for double-labeling with CTB and specific antibodies incubations for: CD11b (1/250, CBL1512, Merck Millipore, Guyancourt, France), CD68 (1/300, MAB1435, Millipore-Merck), iNOS (1/500, AB5382, Merck Millipore), p-AMPK (1/200, 09–290, Merck Millipore), or Nrf2 (1/500, ab31163, Abcam, Cambridge, UK). For dual-labeling with CTB and the molecule of interest, free floating sections were first washed with PBS 0.1 M, incubated in a custom-made blocking solution (5% normal donkey serum, 0.1% triton in PBS 1X for 30 min), and incubated overnight at 4 °C with goat anti-CTB antibody (1/5000, 227040, Calbiochem, Millipore-Merck) and one antibody at the concentration described above. After 3×5 min washes, the sections were incubated with a secondary antibody: Alexa Fluor 594 donkey anti-goat (1/1000, A11058, Invitrogen, Waltham, MA, USA), Alexa Fluor 488 donkey anti-rabbit (A21206), or anti-mouse (A21202), for the other antibodies (1/2000, Molecular probes, Eugene, OR, USA) at room temperature for 2 h and 30 min. The sections were then washed several times in PBS $1\times$ and mounted on slides using an anti-fade solution (Prolong Gold antifade reagent, P36930, Invitrogen). Negative controls in which primary and/or secondary antibodies were excluded from the incubation period were also performed. Images of the sections were taken with a Hamamatsu ORCA-R^2 camera mounted on an Olympus IX83 P2ZF microscope (Tokyo, Japan). Images were analyzed using ImageJ 1.53n software (National Institutes of Health).

For NADPH diaphorase (Nadph-Ox.) histochemistry, free floating sections were incubated in 0.1 M PBS containing 0.2% triton X-100, 1 mg/mL of β-nicotinamide adenine dinucleotide phosphate (β-NADPH, N-1630, Sigma-Aldrich, Darmstadt, Germany), and 1 mg/mL of nitro blue tetrazolium (N-5514, Sigma-Aldrich) during 30 min at 37 °C. Then, CTB immunohistochemistry was performed for identifying the phrenic motoneurons, and the sections were mounted on slides, as previously described. A brightfield microscopy (Aperio AT2, Leica) was used to capture images of the different sections stained with NADPH oxidase.

2.5. Protein Extraction and Western Blotting

After the proper post-lesion time (1 h, 1 day, 3 days, or 7 days, $n = 4$ for each group, including $n = 4$ at each time post-surgery for the uninjured group), animals were euthanized (5% isoflurane in 100% O_2 followed by 1 mL i.c. of urethane, U2500-100 g, Sigma, 0.2 g/mL) and fresh segments of the spinal cord (C3–C6) were collected and frozen immediately in liquid nitrogen. The C1–C3 area of each animal was also collected, and the extent of the injury was verified as described previously. For immunoblot analysis, flash-frozen fresh spinal segments (C3–C6) were lysed with cold RIPA buffer (150 mM of NaCl, 1% Triton X-100, 0.5% sodium deoxycholate, 0.1% SDS, 50 mM of Tris-HCl, pH 7.5, supplemented with a protease inhibitor mixture (Roche Diagnostics, Indianapolis, IN, USA)). Protein concentrations were determined with the DC protein assay kit (5000111EDU, BioRad, Hercules, CA, USA). Twenty µg of total proteins was resolved by SDS-PAGE (4–20% precast gels, BioRad) and transferred to polyvinylidene difluoride membrane (Immobilon-FL, Merck). Membrane blocking was performed in 5% Milk/TBST (10 mM of Tris-HCl, pH 7.4, 150 mM of NaCl, and 0.1% Tween 20) for 1 h prior to incubation with primary antibodies. Primaries antibodies were for Nrf2 (1/1000, sc-13032, Santa-Cruz, Dallas, TX, USA), HO-1 (1/1000, sc-10791, Santa-Cruz), NQO1 (1/1000, AB34176, Abcam), and GAPDH (1/5000, CB1001, Merck Millipore). The corresponding IRDye680- or IRDye800-conjugated secondary antibodies (LI-COR, Bad Homburg vor der Höhe, Germany) were used at dilution $\pm$ 1/4000. Immunoreactivity was visualized using the Odyssey Imaging system and the Image Studio software (LI-COR).

2.6. Data Analyses and Statistics

All the data were presented as mean $\pm$ SD and statistics were considered significant when $p < 0.05$. For immunolabeling analysis, the mean of each motoneuron's values was considered for each animal group. One-way analysis of variance (ANOVA) or Kruskal–Wallis (Fisher LSD method) one-way ANOVA on ranks (when the normality test or variance test failed, Dunn's method) were used to compare the values between the different groups. SigmaPlot 12.5 software (Systat Software, San Jose, CA, USA) was used for all analyses.

3. Results

3.1. Inflammatory Processes in Phrenic Motoneurons

The C2 spinal cord injury leads to the activation of immune cells in the area of phrenic motoneurons, characterized by an increase in the occupied surface by CD11b+ and CD68+ cells until at least 7 days (d) P.I. (Supplementary Figures S1 and S2). This corresponds to previous observations performed following spinal cord injury [21,46].

In phrenic motoneurons specifically, expression of NADPH oxidase, an oxidative stress marker, has been evaluated (Figure 2A). This expression tended to increase in phrenic motoneurons over time until it became significant compared to the uninjured group (36.94 ± 6.32 AU) at 3 d P.I. for the intact side (54.84 ± 3.15 AU, $p < 0.05$). This increase was further amplified at 7 d P.I. for the intact side (83.78 ± 2.98 AU) when compared to the uninjured group and 1 h P.I. (43.18 ± 2.28 AU, $p < 0.05$). For the injured side, the values were significantly increased at 7 d P.I. (79.06 ± 4.63 AU) when compared to the uninjured group ($p < 0.05$), and to 1 h P.I. groups (46.30 ± 2.36 AU, $p < 0.05$) (Figure 2B).

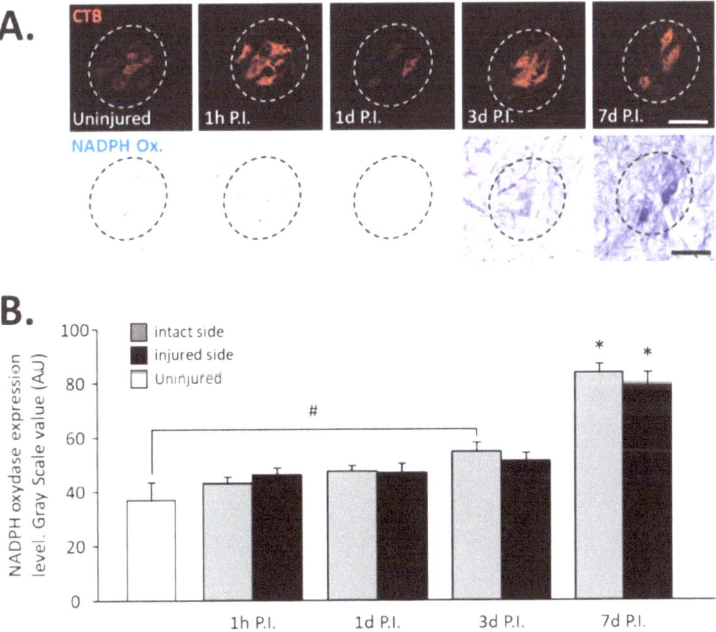

Figure 2. NADPH oxidase expression in phrenic motoneurons following C2 hemi-section. (**A**) Representative images showing expression of NADPH oxidase in phrenic motoneurons labeled with CTB in uninjured animals and in denervated phrenic motoneurons in C2 hemisected rats, 1 hour (h), 1 day (d), 3 d, and 7 d post-injury (P.I.). (**B**) Quantification of NADPH oxidase in phrenic motoneurons for uninjured animals, and intact and injured sides of C2 hemisected animals 1 h, 1 d, 3 d, and 7 d following injury. # 3 d P.I. intact side, compared with uninjured group, $p < 0.05$. * 7 d P.I. compared to uninjured group and 1 h P.I. corresponding side, $p < 0.05$.

In the same way, an increase in the expression of iNOS (Figure 3A), another inflammatory-related enzyme, was observed from 1 d P.I. for the intact side (32.46 ± 7.87 AU) and the injured side (31.90 ± 5.81 AU) compared to uninjured animals (18.83 ± 1.73 AU, $p < 0.05$). This increase remained until at least 7 d P.I. for the intact side (35.46 ± 6.92 AU, $p < 0.05$ vs. uninjured) and the injured side (37.86 ± 6.76 AU, $p < 0.05$ vs. uninjured) (Figure 3B).

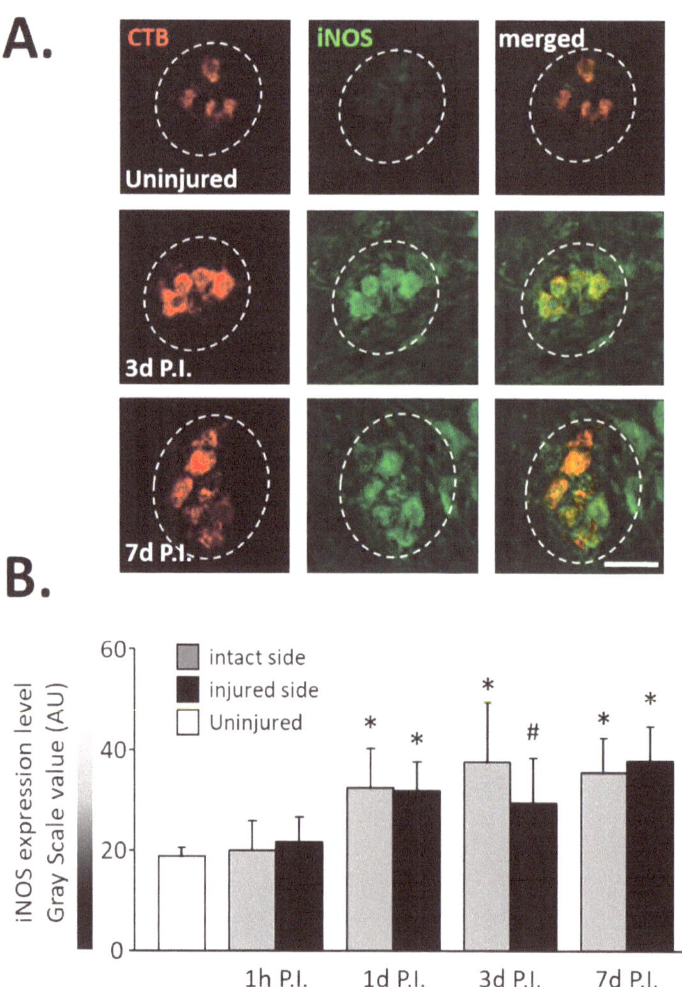

Figure 3. iNOS expression in phrenic motoneurons following C2 hemi-section. (**A**) Representative images showing expression of iNOS in phrenic motoneurons labeled with CTB in uninjured animals and in denervated phrenic motoneurons in C2 hemisected rats, 3 and 7 days post-injury (P.I.). (**B**) Quantification of iNOS in phrenic motoneurons for uninjured animals, and intact and injured sides of C2 hemisected animals 1 h, 1 d, 3 d, and 7 d following injury. # 3 d P.I. injured side, compared with uninjured group, $p < 0.05$. * Compared to uninjured group and 1 h P.I. corresponding side, $p < 0.05$.

3.2. Effect of SCI on AMPK Phosphorylation

The expression of phosphorylated AMPK (p-AMPK) was then observed in phrenic motoneurons as a marker of the response to cellular stress following SCI (Figure 4A).

At 1 h P.I., this expression was significantly reduced for the intact side (40.00 ± 9.32 AU) and the injured side (39.94 ± 9.44 AU) compared to the uninjured group (59.84 ± 13.90 AU, $p < 0.001$). There was, however, a rebound at 1 d P.I. for the injured side (67.02 ± 12.40 AU) compared to 1 h P.I. ($p < 0.001$). p-AMPK expression remained similar to the uninjured group for both sides at later time points (Figure 4B).

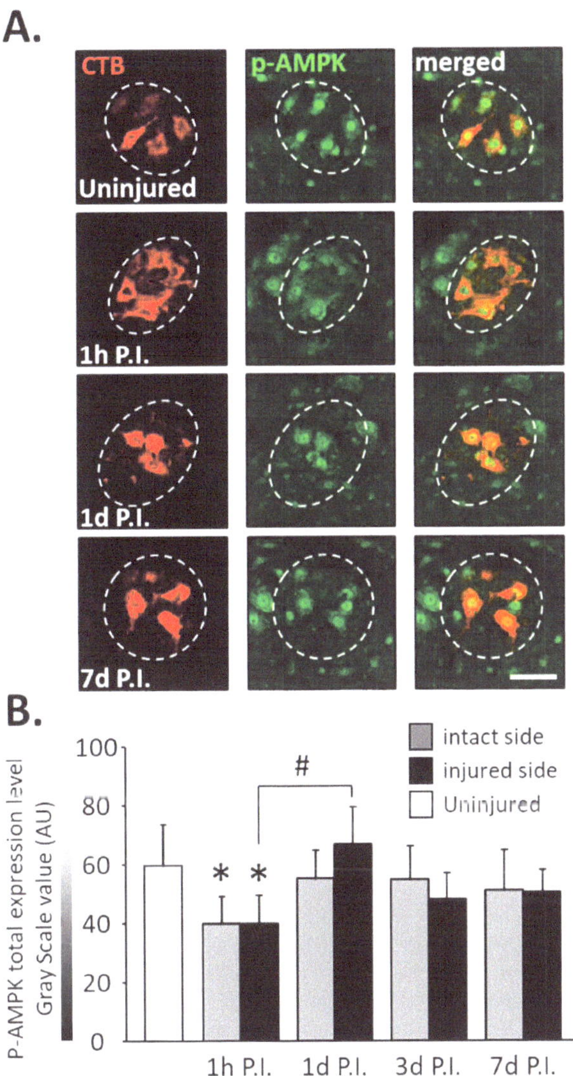

Figure 4. Phosphorylated AMPK (pAMPK) expression in phrenic motoneurons following C2 hemisection. (**A**) Representative images showing the expression of pAMPK in phrenic motoneurons labeled with CTB in uninjured animals and in denervated phrenic motoneurons in C2 hemisected rats, 1 h, 1 d, and 7 d post-injury (P.I.). (**B**) Quantification of pAMPK in phrenic motoneurons for uninjured animals, and intact and injured sides of C2 hemisected animals 1 h, 1 d, 3 d, and 7 d following injury. # 1 d P.I. compared to 1 h P.I. for the injured side, $p < 0.001$. * Compared to the uninjured group, $p < 0.001$.

3.3. Impact on Nrf2 Expression in Phrenic Motoneurons

The expression of total Nrf2 in phrenic motoneurons (Figure 5A), as well the percentage of phrenic motoneurons expressing nuclear Nrf2 (Figure 6A), were then evaluated to determine the antioxidant and anti-inflammatory response in these cells.

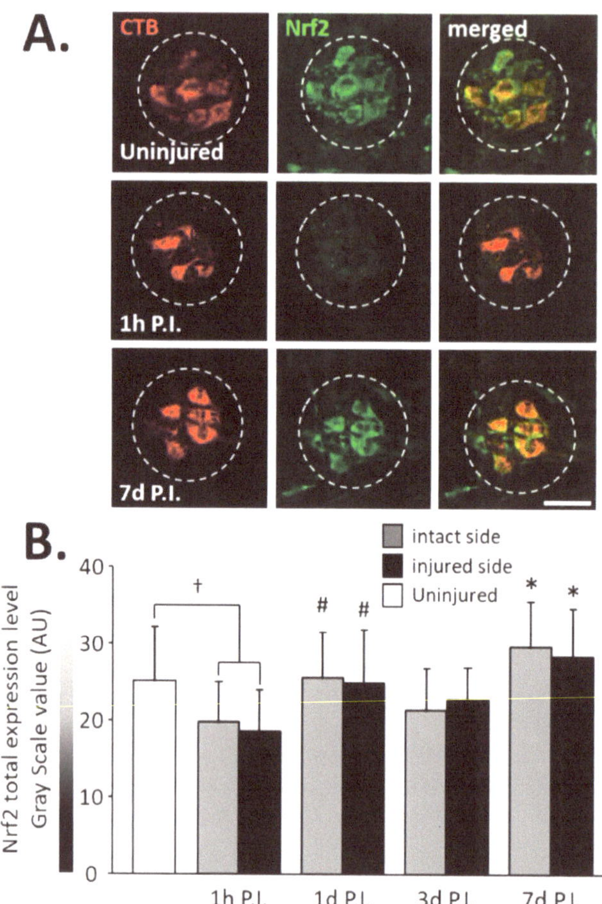

Figure 5. Nrf2 expression in phrenic motoneurons following C2 hemi-section. (**A**) Representative images showing expression of Nrf2 in phrenic motoneurons labeled with CTB in uninjured animals and in denervated phrenic motoneurons in C2 hemisected rats, 1 h and 7 d post-injury (P.I.). (**B**) Quantification of Nrf2 in phrenic motoneurons for uninjured animals, and intact and injured sides of C2 hemisected animals 1 h, 1 d, 3 d, and 7 d following injury. # Compared to 1 h P.I. group, $p < 0.05$. * Compared to uninjured group and 1 h P.I. corresponding side, $p < 0.05$. † Uninjured compared to 1 h intact side and injured side, $p < 0.05$.

A significant decrease in total Nrf2 was observed at 1 h P.I. in the intact side (19.79 ± 5.16 AU) and the injured side (18.64 ± 5.35 AU) compared to that of uninjured rats (25.08 ± 6.99 AU, $p < 0.05$). This expression was then restored at 1 d P.I. for both sides (intact side: 25.55 ± 5.95 AU; injured side: 24.95 ± 6.86 AU) compared to 1 h P.I. ($p < 0.05$). This level of expression was then maintained until at least 7 d P.I. for the intact (29.60 ± 5.90 AU, $p < 0.05$ compared to 1 h P.I.) and the injured (28.44 ± 6.21 AU, $p < 0.05$ compared to 1 h P.I.) sides (Figure 5B).

Concerning the percentage of phrenic motoneurons expressing nuclear Nrf2, a significant increase was observed at the injured side at 1 d P.I. ($0.59 \pm 0.13\%$) compared to 1 h P.I. ($0.05 \pm 0.05\%$, $p < 0.05$). This increased expression persisted until at least 7 d P.I. (0.54 ± 0.10, $p < 0.05$ compared to 1 h P.I.) (Figure 6B). Note, however, that there was a decreasing trend at 1 h P.I. for both sides compared to the uninjured group, and for the intact side, the tendency resembled the injured side (Figure 6B).

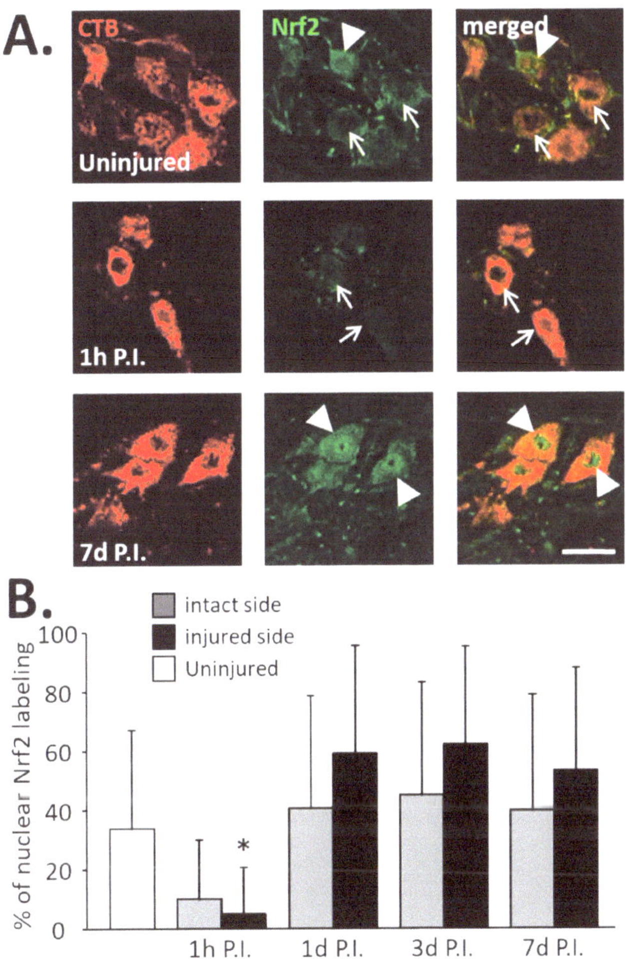

Figure 6. Percentage of phrenic motoneurons expressing nuclear Nrf2 following C2 hemi-section. (**A**) Representative images showing the expression of Nrf2 in phrenic motoneurons labeled with CTB in uninjured animals and in denervated phrenic motoneurons in C2 hemisected rats, 1 h and 7 d post-injury (P.I.). White arrowheads show phrenic motoneurons without nuclear Nrf2 expression. White arrows show phrenic motoneurons with nuclear Nrf2 expression. (**B**) Percentage of phrenic motoneurons expressing nuclear Nrf2 for uninjured animals, and intact and injured sides of C2 hemisected animals 1 h, 1 d, 3 d, and 7 d following injury. * Compared to 1 h P.I., 3 d P.I., and 7 d P.I. corresponding sides, $p < 0.05$.

3.4. Impact on Nrf2 Signaling Pathway in the C3–C6 Spinal Cord

Finally, quantification of Nrf2 and two of its effectors, HO-1 and NQO1, was carried out on the complete C3–C6 spinal cord to evaluate the activation of the Nrf2 pathway globally at the level of phrenic motoneurons (Figure 7). Concerning Nrf2 expression, for the injured side, a reduction was observed at 1 h P.I. (0.79 ± 0.20 AU) compared to the uninjured group (1.34 ± 0.18 AU, $p = 0.045$). A rebound was observed at 3 d P.I. (1.75 ± 0.56 AU) compared to 1 h P.I. ($p < 0.001$). Then, there was a significant decreased at 7 d P.I. (1.07 ± 0.15 AU) compared to 3 d P.I. ($p = 0.016$), and a return to values similar to the uninjured ones ($p > 0.05$) (Figure 7A). For the intact side, an increase was observed between 1 h P.I. (1.17 ± 0.37 AU) and 3 d P.I. (2.77 ± 0.18 AU, $p < 0.001$). Interestingly, this increased expression was significantly higher for the intact side than for the injured side ($p < 0.001$). Values then significantly decreased at 7 d P.I. (1.01 ± 0.14 AU, $p < 0.001$, compared to 3 d P.I.) and returned to those seen in the uninjured group ($p > 0.05$) (Figure 7A).

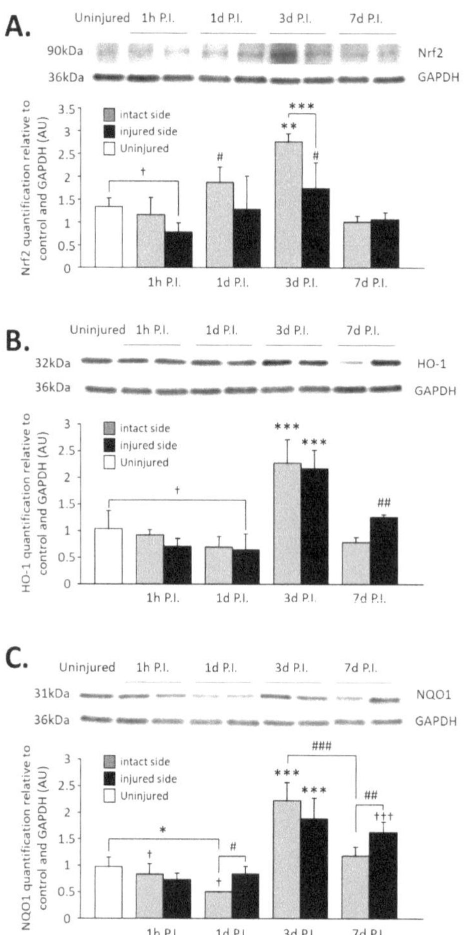

Figure 7. Quantification of Nrf2 and two of its effectors, HO-1 and NQO1, in the C3–C6 spinal cord following C2 hemi-section. (**A**) Representative blot and associated quantification of Nrf2 expression in the C3–C6 spinal cord, 1 h, 1 d, 3 d, and 7 d post-injury (P.I.). ** Compared to uninjured and 1 h, 1 d, and 7 d P.I. corresponding sides, $p < 0.05$. *** 3 d P.I. intact side compared to 3 d P.I. injured side, $p < 0.001$. † Uninjured compared to 1 h P.I. injured side, $p < 0.05$. # Compared to 1 h and 7 d P.I. corresponding side, $p < 0.05$. (**B**) Representative blot and associated quantification of HO-1 expression in the C3–C6 spinal cord, 1 h, 1 d, 3 d, and 7 d post-injury (P.I.). *** Compared to uninjured and 1 h, 1 d, and 7 d P.I. corresponding sides, $p < 0.001$. ## Compared to 1 h and 1 d P.I. corresponding side, $p < 0.01$. † Uninjured compared to 1 d P.I. injured side, $p = 0.039$. (**C**) Representative blot and associated quantification of NQO1 expression in the C3–C6 spinal cord, 1 h, 1 d, 3 d, and 7 d post-injury (P.I.). *** Compared to uninjured and 1 h, 1 d, and 7 d P.I. corresponding sides, $p < 0.001$. ### 3 d P.I. intact side compared to 7 d P.I. intact side, $p < 0.001$. ††† Compared to uninjured and 1 h and 1 d P.I. corresponding sides, $p < 0.001$. ## 7 d P.I. intact side compared to 7 d P.I. injured side, $p < 0.01$. # 1 d P.I. intact side compared to 1 d P.I. injured side, $p < 0.05$. * 1 h P.I. intact side compared to 1 h P.I. injured side, $p < 0.05$. † Compared to 7 d P.I. corresponding side, $p < 0.05$.

Concerning HO-1 expression on the injured side, a significant decrease was observed at 1 d P.I. (0.71 ± 0.14 AU) compared to uninjured animals (1.04 ± 0.34 AU, $p = 0.039$). Similar to Nrf2 expression, a rebound was observed at 3 d P.I. (2.17 ± 0.35 AU, $p < 0.001$ compared to 1 d P.I.), and for the intact side, a drastic increase was observed at 3 d P.I. (2.28 ± 0.44 AU) compared to 1 d P.I. (0.70 ± 0.19 AU, $p < 0.001$). HO-1 expression then decreased at 7 d P.I. for both sides (injured side: 1.26 ± 0.05 AU; intact side: 0.79 ± 0.09 AU) compared to 3 d P.I. ($p < 0.001$) (Figure 7B).

Finally, NQO1 expression was significantly reduced for the intact side at 1 d P.I. (0.84 ± 0.19 AU) compared to uninjured animals (0.97 ± 0.18 AU, $p = 0.006$) and compared to the injured side (0.51 ± 0.001 AU, $p = 0.038$). A drastic increase was then observed at 3 d P.I. for both sides (intact side: 2.23 ± 0.34 AU; injured side: 1.88 ± 0.39 AU) compared to 1 d P.I. ($p < 0.001$). At 7 d P.I, while the expression remained similar to 3 d P.I. for the injured side (1.63 ± 0.19 AU, $p > 0.05$), there was a significant reduction for the intact side (1.18 ± 0.16 AU) compared to 3 d P.I. ($p < 0.001$) and to the injured side ($p = 0.008$) (Figure 7C).

4. Discussion

The present study is the first to evaluate the impact of high SCI on oxidative stress, and AMPK and Nrf2 signaling pathway modulation in phrenic motoneurons in a rat model.

Following SCI, axotomy of descending spinal pathways leads to general neuroinflammation below the lesion site. This is due in great part to the glutamate excitotoxicity phenomenon [23]. We observed an activation and a recruitment of immune cells over time around the identified phrenic motoneurons following C2HS in rats. Our results correlate with other studies showing that the surface occupied by OX42- (microglia/macrophages marker) or GFAP (astrocytes marker)-labeled cells is increased in the C4 spinal cord in the same model [21]. This shows the gradual establishment of neuroinflammatory processes at the level of phrenic motoneurons in the C3–C6 spinal cord. These phrenic motoneurons are also affected by the injury, and successive glutamate excitotoxicity [30]. Here, we showed that expressions of NADPH oxidase and iNOS, two enzymes producing ROS and reactive nitrogen species, respectively, are gradually increased over time in the intact and the injured sides, indicating that phrenic motoneurons' homeostasis is impacted. Others also pointed out molecular modifications in phrenic motoneurons' function. In a model of C2HS in rats, a reduction in NMDA receptor mRNA, but not in AMPA receptor mRNA, was observed at 3 d P.I. in phrenic motoneurons [47]. This could be a response to the overactivation of these receptors due to glutamate excitotoxicity. In a model of C2 hemi-contusion in rats, a decrease in the membrane/cytosolic ratio for KCC2 expression was observed for the intact and the injured sides, which reflects a modification in chloride flux in phrenic motoneurons in rat [48]. All these data show that following cervical SCI, deafferented and non-deafferented phrenic motoneurons' homeostasis is highly modified, however without any impact on their survival [49]. This cellular stress appearing in non-deafferented phrenic motoneurons could be explained by the propagation of inflammatory processes from the injured side, or even by a potential functional adaptation for the loss of activity on the injured side.

Unexpectedly, we observed a significant reduction in pAMPK and total Nrf2 in phrenic motoneurons at 1 h P.I., and a marked tendency for the percentage of phrenic motoneurons expressing nuclear Nrf2. This reduction was, however, only temporary, considering a rebound characterized by a significant increase was observed at 1 d P.I. for these 2 molecules. This phenomenon could be explained by an increased activity of the ubiquitin proteasome system, leading to the degradation of pAMPK and Nrf2. Indeed, a previous study showed that in hippocampal neurons, an upregulation of action potentials leads to an increase in protein degradation via the ubiquitin proteasome system [50]. In our model, glutamate excitotoxicity results in immediate overactivation of phrenic motoneurons, which could lead to a transient increase in ubiquitin proteasome system activity, and subsequent pAMPK and Nrf2 degradation. NADPH oxidase and iNOS were either not expressed or expressed at a very low level physiologically, and the absence of a reduction in their expression at 1 h P.I. was not surprising. However, we are the first to observe such a drastic reduction in metabolism and anti-inflammation/antioxidant-related molecules at 1 h P.I. in phrenic motoneurons.

Phrenic motoneurons have the capacity to respond to the cellular stress caused by glutamate excitotoxicity, involving activation of AMPK and Nrf2 signaling pathways. The reduction in neuron metabolism due to the spinal shock [51] could explain the drastic rebound in expression observed at 3 d following the injury. As AMPK is a key regulator of cellular metabolism [27], its presence is indeed required in case of reduced neuron metabolism. Following SCI, higher AMPK signaling is known to increase autophagy [52] and reduce apoptosis in rats [53,54] and mice [55]. On the contrary, inhibition of AMPK leads to enhanced axonal regeneration following dorsal column crush in mice [56]. This shows the pivotal role of AMPK following SCI and could explain its rebound in expression in phrenic motoneurons at 3 d P.I., considering no study explored AMPK function in phrenic motoneurons in this context. Concerning Nrf2, the cellular stress caused by glutamate excitotoxicity leads to anti-inflammatory and antioxidant responses, which requires Nrf2 signaling [31]. This could explain the rebound also observed for Nrf2 expression using immunolabeling at 1 d P.I., as well as the increased expression of Nrf2, HO-1, and NQO1 observed at 3 d P.I. using immunoblotting. This difference in timing could be explained by the fact that immunoblotting was performed on the total C3–C6 spinal cord, which contains not only phrenic motoneurons but also other neuronal and glial cells. On the contrary, immunolabeling analysis was performed only on phrenic motoneurons in the C3–C6 spinal cord. Considering that other neuronal and glial cells are in higher quantity than phrenic motoneurons, a difference in inflammatory kinetics would erase the increase in Nrf2 expression observed in phrenic motoneurons, specifically when using the immunoblotting method. Nrf2 appeared to be beneficial following injury [30,31]. In mice lacking Nrf2, SCI leads to more severe locomotor dysfunction and neural death compared to control animals [57], whereas activation of the Nrf2 pathway in rats leads to improved locomotor function and neuroprotective effects following SCI [31,37,58]. Nrf2 is also expressed in glial cells, including astrocytes. Hyperactivation of Nrf2 in astrocytes via Keap1 depletion leads to a reduced loss of myelin and oligodendrocytes in mice following spinal cord contusion [59]. These data as well as our results show that Nrf2 pathway activity in spinal neuronal and glial cells participates in the defense response against oxidative stress following spinal cord injury.

Here, we proposed to link glutamate excitotoxicity occurring in phrenic motoneurons following high spinal cord injury [60] with anti-inflammatory and antioxidant responses involving AMPK [52] and Nrf2 signaling [31]. This AMPK-Nrf2 pathway was indeed shown to be associated with neuroprotection [61]. Axotomy in the upper spinal cord (C2) led to glutamate excitotoxicity in phrenic motoneurons located in the lower spinal cord (C3–C6). This induced overactivation of glutamatergic receptors and high calcium influx [60]. It resulted in high production of ROS [62] and activation of AMPK via phosphorylation due to a higher ATP concentration [27]. pAMPK then facilitated Nrf2 nuclear translocation [28]. This allowed transcription of antioxidant and anti-inflammatory molecules (Figure 8).

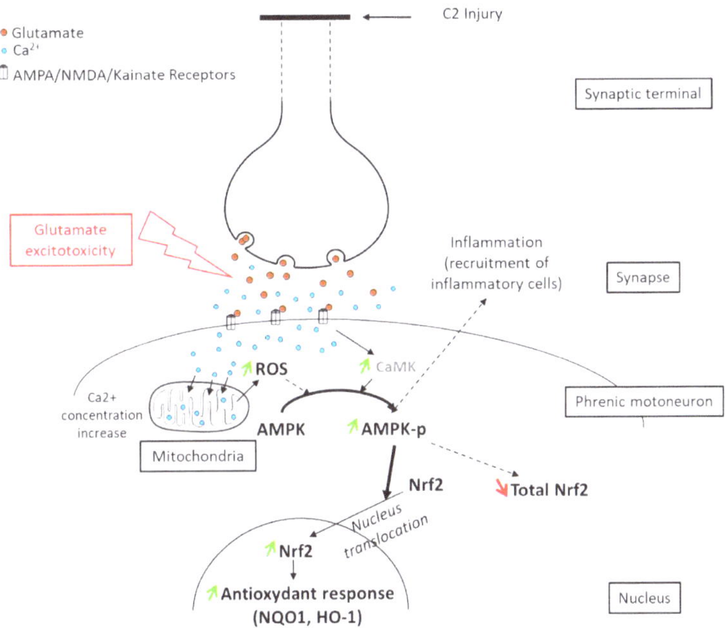

Figure 8. Proposed schematic of Nrf2 signaling pathway activation through phosphorylation of AMPK following spinal cord injury. Following axotomy, synaptic terminals release all their neurotransmitters, including glutamate. This glutamate excessively activates its receptors (AMPA, NMDA, and Kainate), which leads to glutamate excitotoxicity in postsynaptic neurons (here, phrenic motoneurons). Intracellularly, it induces an increase in the calcium concentration, leading to increased reactive oxygen species (ROS) production by mitochondria. These ROS, associated with the increased calcium concentration, provoke AMPK phosphorylation. pAMPK then facilitates Nrf2 nucleus translocation, leading to an increase in the antioxidant response mediated by Nrf2 and its effectors (NQO1, HO-1).

5. Conclusions

Taken together, our results illustrated the implication of the AMPK-Nrf2 pathway in phrenic motoneurons' response to oxidative stress following SCI. This improved our current knowledge about phrenic motoneurons' function and paves the way for future therapies aiming to improve phrenic motoneurons' response to oxidative stress and neuroinflammation. Deciphering the time-dependent changes in AMPK-Nrf2 signaling and its eventual pharmacological modulation could indeed help in spinal rewiring of these deafferented phrenic motoneurons and could lead to improved diaphragm activity in patients suffering high SCI.

Supplementary Materials: The following supporting information can be downloaded at: https://www.mdpi.com/article/10.3390/antiox11091665/s1, Figure S1: CD11b expression around phrenic motoneurons following C2 hemisection; Figure S2: CD68 expression around phrenic motoneurons following C2 hemisection.

Author Contributions: P.M.-F.: investigation, formal analysis, visualization, writing—original draft, writing—review and editing; L.E.: investigation, formal analysis, visualization, writing—review and editing; L.D.: investigation, formal analysis, visualization, writing—review and editing; T.B.D.: investigation, supervision, writing—review and editing; M.B.: conceptualization, supervision, funding acquisition, writing—review and editing; S.V.: conceptualization, experimental design, investigation, supervision, project administration, funding acquisition, writing—review and editing. All authors have read and agreed to the published version of the manuscript.

Funding: This research was funded by the Chancellerie des Universités de Paris (Legs Poix) (S.V., M.B.), the Fondation de France (S.V.), the Fondation Médisite (S.V.), INSERM (M.B., S.V.), and Université de Versailles Saint-Quentin-en-Yvelines (S.V.). The supporters had no role in the study design, data collection and analysis, the decision to publish, or preparation of the manuscript.

Institutional Review Board Statement: The study was conducted according to the guidelines of the Declaration of Helsinki and approved by the Ethics Committee of the University of Versailles Saint-Quentin-en-Yvelines (Comité d'éthique n047) and complied with the French and European laws (EU Directive 2010/63/EU) regarding animal experimentation (Apafis #2017111516297308_v3).

Informed Consent Statement: Not applicable.

Data Availability Statement: The data are contained within the manuscript and Supplementary Files.

Acknowledgments: This work has benefited from the facilities of CYMAGES and histology (UFR SVS, UVSQ, Université Paris-Saclay, 78180 Montigny-le-Bretonneux, France).

Conflicts of Interest: The authors declare no conflict of interest.

References

1. Winslow, C.; Rozovsky, J. Effect of spinal cord injury on the respiratory system. *Am. J. Phys. Med. Rehabil.* **2003**, *82*, 803–814. [CrossRef] [PubMed]
2. Nantwi, K.D.; El-Bohy, A.A.; Schrimsher, G.W.; Reier, P.J.; Goshgarian, H.G. Spontaneous Functional Recovery in a Paralyzed Hemidiaphragm Following Upper Cervical Spinal Cord Injury in Adult Rats. *Neurorehabilit. Neural Repair* **1999**, *13*, 225–234. [CrossRef]
3. Golder, F.J.; Reier, P.J.; Bolser, D.C. Altered respiratory motor drive after spinal cord injury: Supraspinal and bilateral effects of a unilateral lesion. *J. Neurosci.* **2001**, *21*, 8680–8689. [CrossRef] [PubMed]
4. Goshgarian, H.G. Invited Review: The crossed phrenic phenomenon: A model for plasticity in the respiratory pathways following spinal cord injury. *J. Appl. Physiol.* **2003**, *94*, 795–810. [CrossRef] [PubMed]
5. Lane, M.A.; Lee, K.-Z.; Fuller, D.D.; Reier, P.J. Spinal circuitry and respiratory recovery following spinal cord injury. *Respir. Physiol. Neurobiol.* **2009**, *169*, 123–132. [CrossRef] [PubMed]
6. Vinit, S.; Gauthier, P.; Stamegna, J.C.; Kastner, A. High cervical lateral spinal cord injury results in long-term ipsilateral hemidiaphragm paralysis. *J. Neurotrauma* **2006**, *23*, 1137–1146. [CrossRef] [PubMed]
7. Keomani, E.; Deramaudt, T.B.; Petitjean, M.; Bonay, M.; Lofaso, F.; Vinit, S. A murine model of cervical spinal cord injury to study post-lesional respiratory neuroplasticity. *J. Vis. Exp.* **2014**, *87*, e51235. [CrossRef]
8. Porter, W.T. The Path of the Respiratory Impulse from the Bulb to the Phrenic Nuclei. *J. Physiol.* **1895**, *17*, 455–485. [CrossRef]
9. Michel-Flutot, P.; Jesus, I.; Vanhee, V.; Bourcier, C.H.; Emam, L.; Ouguerroudj, A.; Lee, K.-Z.; Zholudeva, L.V.; Lane, M.A.; Mansart, A.; et al. Effects of Chronic High-Frequency rTMS Protocol on Respiratory Neuroplasticity Following C2 Spinal Cord Hemisection in Rats. *Biology* **2022**, *11*, 473. [CrossRef]
10. Feldman, J.L.; Del Negro, C.A.; Gray, P.A. Understanding the rhythm of breathing: So near, yet so far. *Annu. Rev. Physiol.* **2013**, *75*, 423–452. [CrossRef]
11. Gandevia, S.C.; Rothwell, J.C. Activation of the human diaphragm from the motor cortex. *J. Physiol.* **1987**, *384*, 109–118. [CrossRef]
12. Feldman, J.L.; Loewy, A.D.; Speck, D.F. Projections from the ventral respiratory group to phrenic and intercostal motoneurons in cat: An autoradiographic study. *J. Neurosci.* **1985**, *5*, 1993–2000. [CrossRef]
13. Lipski, J.; Bektas, A.; Porter, R. Short latency inputs to phrenic motoneurones from the sensorimotor cortex in the cat. *Exp. Brain Res.* **1986**, *61*, 280–290. [CrossRef]
14. Alexandrov, V.G.; Ivanova, T.G.; Alexandrova, N.P. Prefrontal control of respiration. *J. Physiol. Pharmacol.* **2007**, *58* (Suppl. 5), 17–23.
15. Vandeweerd, J.-M.; Hontoir, F.; De Knoop, A.; De Swert, K.; Nicaise, C. Retrograde Neuroanatomical Tracing of Phrenic Motor Neurons in Mice. *J. Vis. Exp.* **2018**, *132*, e56758. [CrossRef]
16. Lane, M.A.; White, T.E.; Coutts, M.A.; Jones, A.L.; Sandhu, M.S.; Bloom, D.C.; Bolser, D.C.; Yates, B.J.; Fuller, D.D.; Reier, P.J. Cervical prephrenic interneurons in the normal and lesioned spinal cord of the adult rat. *J. Comp. Neurol.* **2008**, *511*, 692–709. [CrossRef]
17. Lane, M.A. Spinal respiratory motoneurons and interneurons. *Respir. Physiol. Neurobiol.* **2011**, *179*, 3–13. [CrossRef]
18. Satkunendrarajah, K.; Karadimas, S.K.; Laliberte, A.M.; Montandon, G.; Fehlings, M.G. Cervical excitatory neurons sustain breathing after spinal cord injury. *Nature* **2018**, *562*, 419–422. [CrossRef]
19. Bradbury, E.J.; Burnside, E.R. Moving beyond the glial scar for spinal cord repair. *Nat. Commun.* **2019**, *10*, 3879. [CrossRef]
20. Hellenbrand, D.J.; Quinn, C.M.; Piper, Z.J.; Morehouse, C.N.; Fixel, J.A.; Hanna, A.S. Inflammation after spinal cord injury: A review of the critical timeline of signaling cues and cellular infiltration. *J. Neuroinflamm.* **2021**, *18*, 284. [CrossRef]
21. Windelborn, J.A.; Mitchell, G.S. Glial activation in the spinal ventral horn caudal to cervical injury. *Respir. Physiol. Neurobiol.* **2012**, *180*, 61–68. [CrossRef] [PubMed]

22. Hulsebosch, C.E. Recent advances in pathophysiology and treatment of spinal cord injury. *Adv. Physiol. Educ.* **2002**, *26*, 238–255. [CrossRef] [PubMed]

23. Jia, M.; Njapo, S.A.; Rastogi, V.; Hedna, V.S. Taming glutamate excitotoxicity: Strategic pathway modulation for neuroprotection. *CNS Drugs* **2015**, *29*, 153–162. [CrossRef] [PubMed]

24. Gorlach, A.; Bertram, K.; Hudecova, S.; Krizanova, O. Calcium and ROS: A mutual interplay. *Redox Biol.* **2015**, *6*, 260–271. [CrossRef] [PubMed]

25. Deramaudt, T.B.; Dill, C.; Bonay, M. Regulation of oxidative stress by Nrf2 in the pathophysiology of infectious diseases. *Med. Mal. Infect.* **2013**, *43*, 100–107. [CrossRef] [PubMed]

26. Shah, S.A.; Amin, F.U.; Khan, M.; Abid, M.N.; Rehman, S.U.; Kim, T.H.; Kim, M.W.; Kim, M.O. Anthocyanins abrogate glutamate-induced AMPK activation, oxidative stress, neuroinflammation, and neurodegeneration in postnatal rat brain. *J. Neuroinflamm.* **2016**, *13*, 286. [CrossRef] [PubMed]

27. Hardie, D.G.; Schaffer, B.E.; Brunet, A. AMPK: An Energy-Sensing Pathway with Multiple Inputs and Outputs. *Trends Cell Biol.* **2016**, *26*, 190–201. [CrossRef] [PubMed]

28. Joo, M.S.; Kim, W.D.; Lee, K.Y.; Kim, J.H.; Koo, J.H.; Kim, S.G. AMPK Facilitates Nuclear Accumulation of Nrf2 by Phosphorylating at Serine 550. *Mol. Cell. Biol.* **2016**, *36*, 1931–1942. [CrossRef]

29. Samarghandian, S.; Pourbagher-Shahri, A.M.; Ashrafizadeh, M.; Khan, H.; Forouzanfar, F.; Aramjoo, H.; Farkhondeh, T. A Pivotal Role of the Nrf2 Signaling Pathway in Spinal Cord Injury: A Prospective Therapeutics Study. *CNS Neurol. Disord. Drug Targets* **2020**, *19*, 207–219. [CrossRef]

30. Guo, X.; Kang, J.; Wang, Z.; Wang, Y.; Liu, M.; Zhu, D.; Yang, F.; Kang, X. Nrf2 signaling in the oxidative stress response after spinal cord injury. *Neuroscience* **2022**, *498*, 311–324. [CrossRef]

31. Jiang, T.; He, Y. Recent Advances in the Role of Nuclear Factor Erythroid-2-Related Factor 2 in Spinal Cord Injury: Regulatory Mechanisms and Therapeutic Options. *Front. Aging Neurosci.* **2022**, *14*, 549. [CrossRef]

32. Hayes, J.D.; Dinkova-Kostova, A.T. The Nrf2 regulatory network provides an interface between redox and intermediary metabolism. *Trends Biochem. Sci.* **2014**, *39*, 199–218. [CrossRef]

33. Itoh, K.; Wakabayashi, N.; Katoh, Y.; Ishii, T.; Igarashi, K.; Engel, J.D.; Yamamoto, M. Keap1 represses nuclear activation of antioxidant responsive elements by Nrf2 through binding to the amino-terminal Neh2 domain. *Genes Dev.* **1999**, *13*, 76–86. [CrossRef]

34. Kansanen, E.; Kuosmanen, S.M.; Leinonen, H.; Levonen, A.-L. The Keap1-Nrf2 pathway: Mechanisms of activation and dysregulation in cancer. *Redox Biol.* **2013**, *1*, 45–49. [CrossRef]

35. Ali, M.; Bonay, M.; Vanhee, V.; Vinit, S.; Deramaudt, T.B. Comparative effectiveness of 4 natural and chemical activators of Nrf2 on inflammation, oxidative stress, macrophage polarization, and bactericidal activity in an in vitro macrophage infection model. *PLoS ONE* **2020**, *15*, e0234484. [CrossRef]

36. Shao, Z.; Lv, G.; Wen, P.; Cao, Y.; Yu, D.; Lu, Y.; Li, G.; Su, Z.; Teng, P.; Gao, K.; et al. Silencing of PHLPP1 promotes neuronal apoptosis and inhibits functional recovery after spinal cord injury in mice. *Life Sci.* **2018**, *209*, 291–299. [CrossRef]

37. Mao, L.; Wang, H.; Wang, X.; Liao, H.; Zhao, X. Transcription factor Nrf2 protects the spinal cord from inflammation produced by spinal cord injury. *J. Surg. Res.* **2011**, *170*, e105–e115. [CrossRef]

38. Guo, Y.; Liu, Y.; Xu, L.; Wu, D.; Wu, H.; Li, C.Y. Reduced Nrf2 and Phase II enzymes expression in immune-mediated spinal cord motor neuron injury. *Neurol. Res.* **2010**, *32*, 460–465. [CrossRef]

39. Tran, A.P.; Warren, P.M.; Silver, J. The Biology of Regeneration Failure and Success after Spinal Cord Injury. *Physiol. Rev.* **2018**, *98*, 881–917. [CrossRef]

40. Zivkovic, S.; Ayazi, M.; Hammel, G.; Ren, Y. For Better or for Worse: A Look into Neutrophils in Traumatic Spinal Cord Injury. *Front. Cell. Neurosci.* **2021**, *15*, 648076. [CrossRef]

41. Mantilla, C.B.; Zhan, W.Z.; Sieck, G.C. Retrograde labeling of phrenic motoneurons by intrapleural injection. *J. Neurosci. Methods* **2009**, *182*, 244–249. [CrossRef]

42. MacFarlane, P.M.; Vinit, S.; Mitchell, G.S. Spinal nNOS regulates phrenic motor facilitation by a 5-HT2B receptor- and NADPH oxidase-dependent mechanism. *Neuroscience* **2014**, *269*, 67–78. [CrossRef]

43. Nichols, N.L.; Vinit, S.; Bauernschmidt, L.; Mitchell, G.S. Respiratory function after selective respiratory motor neuron death from intrapleural CTB-saporin injections. *Exp. Neurol.* **2014**, *267*, 18–29. [CrossRef]

44. MacFarlane, P.M.; Vinit, S.; Mitchell, G.S. Serotonin 2A and 2B receptor-induced phrenic motor facilitation: Differential requirement for spinal NADPH oxidase activity. *Neuroscience* **2011**, *178*, 45–55. [CrossRef]

45. Fuller, D.D.; Sandhu, M.S.; Doperalski, N.J.; Lane, M.A.; White, T.E.; Bishop, M.D.; Reier, P.J. Graded unilateral cervical spinal cord injury and respiratory motor recovery. *Respir. Physiol. Neurobiol.* **2009**, *165*, 245–253. [CrossRef]

46. Bowes, A.L.; Yip, P.K. Modulating inflammatory cell responses to spinal cord injury: All in good time. *J. Neurotrauma* **2014**, *31*, 1753–1766. [CrossRef]

47. Rana, S.; Zhan, W.-Z.; Sieck, G.C.; Mantilla, C.B. Cervical spinal hemisection alters phrenic motor neuron glutamatergic mRNA receptor expression. *Exp. Neurol.* **2022**, *353*, 114030. [CrossRef]

48. Allen, L.L.; Seven, Y.B.; Baker, T.L.; Mitchell, G.S. Cervical spinal contusion alters Na(+)-K(+)-2Cl- and K(+)-Cl- cation-chloride cotransporter expression in phrenic motor neurons. *Respir. Physiol. Neurobiol.* **2019**, *261*, 15–23. [CrossRef]

49. Allen, L.L.; Nichols, N.L.; Asa, Z.A.; Emery, A.T.; Ciesla, M.C.; Santiago, J.V.; Holland, A.E.; Mitchell, G.S.; Gonzalez-Rothi, E.J. Phrenic motor neuron survival below cervical spinal cord hemisection. *Exp. Neurol.* **2021**, *346*, 113832. [CrossRef]

50. Djakovic, S.N.; Schwarz, L.A.; Barylko, B.; DeMartino, G.N.; Patrick, G.N. Regulation of the Proteasome by Neuronal Activity and Calcium/Calmodulin-dependent Protein Kinase II*. *J. Biol. Chem.* **2009**, *284*, 26655–26665. [CrossRef]

51. Ditunno, J.F.; Little, J.W.; Tessler, A.; Burns, A.S. Spinal shock revisited: A four-phase model. *Spinal Cord* **2004**, *42*, 383–395. [CrossRef] [PubMed]

52. Wu, C.; Chen, H.; Zhuang, R.; Zhang, H.; Wang, Y.; Hu, X.; Xu, Y.; Li, J.; Li, Y.; Wang, X.; et al. Betulinic acid inhibits pyroptosis in spinal cord injury by augmenting autophagy via the AMPK-mTOR-TFEB signaling pathway. *Int. J. Biol. Sci.* **2021**, *17*, 1138–1152. [CrossRef] [PubMed]

53. Zhao, H.; Chen, S.; Gao, K.; Zhou, Z.; Wang, C.; Shen, Z.; Guo, Y.; Li, Z.; Wan, Z.; Liu, C.; et al. Resveratrol protects against spinal cord injury by activating autophagy and inhibiting apoptosis mediated by the SIRT1/AMPK signaling pathway. *Neuroscience* **2017**, *348*, 241–251. [CrossRef] [PubMed]

54. Gao, K.; Niu, J.; Dang, X. Neuroprotection of melatonin on spinal cord injury by activating autophagy and inhibiting apoptosis via SIRT1/AMPK signaling pathway. *Biotechnol. Lett.* **2020**, *42*, 2059–2069. [CrossRef]

55. Lin, S.; Tian, H.; Lin, J.; Xu, C.; Yuan, Y.; Gao, S.; Song, C.; Lv, P.; Mei, X. Zinc promotes autophagy and inhibits apoptosis through AMPK/mTOR signaling pathway after spinal cord injury. *Neurosci. Lett.* **2020**, *736*, 135263. [CrossRef]

56. Kong, G.; Zhou, L.; Serger, E.; Palmisano, I.; De Virgiliis, F.; Hutson, T.H.; McLachlan, E.; Freiwald, A.; La Montanara, P.; Shkura, K.; et al. AMPK controls the axonal regenerative ability of dorsal root ganglia sensory neurons after spinal cord injury. *Nat. Metab.* **2020**, *2*, 918–933. [CrossRef]

57. Mao, L.; Wang, H.D.; Wang, X.L.; Tian, L.; Xu, J.Y. Disruption of Nrf2 exacerbated the damage after spinal cord injury in mice. *J. Trauma Acute Care Surg.* **2012**, *72*, 189–198. [CrossRef]

58. Wang, X.; de Rivero Vaccari, J.P.; Wang, H.; Diaz, P.; German, R.; Marcillo, A.E.; Keane, R.W. Activation of the nuclear factor E2-related factor 2/antioxidant response element pathway is neuroprotective after spinal cord injury. *J. Neurotrauma* **2012**, *29*, 936–945. [CrossRef]

59. Zhao, W.; Gasterich, N.; Clarner, T.; Voelz, C.; Behrens, V.; Beyer, C.; Fragoulis, A.; Zendedel, A. Astrocytic Nrf2 expression protects spinal cord from oxidative stress following spinal cord injury in a male mouse model. *J. Neuroinflamm.* **2022**, *19*, 134. [CrossRef]

60. Liu, D.; Xu, G.Y.; Pan, E.; McAdoo, D.J. Neurotoxicity of glutamate at the concentration released upon spinal cord injury. *Neuroscience* **1999**, *93*, 1383–1389. [CrossRef]

61. Wang, Y.; Huang, Y.; Xu, Y.; Ruan, W.; Wang, H.; Zhang, Y.; Saavedra, J.M.; Zhang, L.; Huang, Z.; Pang, T. A Dual AMPK/Nrf2 Activator Reduces Brain Inflammation after Stroke by Enhancing Microglia M2 Polarization. *Antioxid. Redox Signal.* **2018**, *28*, 141–163. [CrossRef]

62. Park, E.; Velumian, A.A.; Fehlings, M.G. The role of excitotoxicity in secondary mechanisms of spinal cord injury: A review with an emphasis on the implications for white matter degeneration. *J. Neurotrauma* **2004**, *21*, 754–774. [CrossRef]

Article

Dipeptide Repeat Pathology in *C9orf72*-ALS Is Associated with Redox, Mitochondrial and NRF2 Pathway Imbalance

José Jiménez-Villegas [1,2,3,4], Janine Kirby [5], Ana Mata [6,7], Susana Cadenas [6,7], Martin R. Turner [8], Andrea Malaspina [9,10], Pamela J. Shaw [5], Antonio Cuadrado [1,2,3,4] and Ana I. Rojo [1,2,3,4,*]

[1]   Department of Biochemistry, Medical College, Autonomous University of Madrid (UAM), 28029 Madrid, Spain
[2]   Instituto de Investigaciones Biomédicas "Alberto Sols" (CSIC/UAM), 28029 Madrid, Spain
[3]   Instituto de Investigación Sanitaria La Paz (IdiPaz), 28029 Madrid, Spain
[4]   Centro de Investigación Biomédica en Red de Enfermedades Neurodegenerativas (CIBERNED), 28029 Madrid, Spain
[5]   Sheffield Institute for Translational Neuroscience, University of Sheffield, Sheffield S10 2HQ, UK
[6]   Centro de Biología Molecular "Severo Ochoa" (CSIC/UAM), 28049 Madrid, Spain
[7]   Instituto de Investigación Sanitaria Princesa (IIS-IP), 28006 Madrid, Spain
[8]   Nuffield Department of Clinical Neurosciences, University of Oxford, Oxford OX3 9DU, UK
[9]   Neuroscience and Trauma Centre, Blizard Institute, Barts and The London School of Medicine & Dentistry, Queen Mary University of London, London E1 2AT, UK
[10] Queen Square Motor Neuron Disease Centre, Neuromuscular Department, Institute of Neurology, University College London, London WC1N 3BG, UK
*   Correspondence: airojo@iib.uam.es

Citation: Jiménez-Villegas, J.; Kirby, J.; Mata, A.; Cadenas, S.; Turner, M.R.; Malaspina, A.; Shaw, P.J., Cuadrado, A.; Rojo, A.I. Dipeptide Repeat Pathology in *C9orf72*-ALS Is Associated with Redox, Mitochondrial and NRF2 Pathway Imbalance. *Antioxidants* **2022**, *11*, 1897. https://doi.org/10.3390/antiox11101897

Academic Editor: Marcel Bonay

Received: 23 August 2022
Accepted: 21 September 2022
Published: 25 September 2022

Publisher's Note: MDPI stays neutral with regard to jurisdictional claims in published maps and institutional affiliations.

Abstract: The hexanucleotide expansion of the *C9orf72* gene is found in 40% of familial amyotrophic lateral sclerosis (ALS) patients. This genetic alteration has been connected with impaired management of reactive oxygen species. In this study, we conducted targeted transcriptional profiling in leukocytes from *C9orf72* patients and control subjects by examining the mRNA levels of 84 redox-related genes. The expression of ten redox genes was altered in samples from *C9orf72* ALS patients compared to healthy controls. Considering that Nuclear factor erythroid 2-Related Factor 2 (NRF2) modulates the expression of a wide range of redox genes, we further investigated its status on an in vitro model of dipeptide repeat (DPR) toxicity. This model mimics the gain of function, toxic mechanisms attributed to *C9orf72* pathology. We found that exposure to DPRs increased superoxide levels and reduced mitochondrial potential as well as cell survival. Importantly, cells overexpressing DPRs exhibited reduced protein levels of NRF2 and its target genes upon inhibition of the proteasome or its canonical repressor, the E3 ligase adapter KEAP1. However, NRF2 activation was sufficient to recover cell viability and redox homeostasis. This study identifies NRF2 as a putative target in precision medicine for the therapy of ALS patients harboring *C9orf72* expansion repeats.

Keywords: NRF2; amyotrophic lateral sclerosis; *C9orf72*; dipeptide repeat proteins

1. Introduction

The hexanucleotide expansion located in the first intron of the *C9orf72* gene affects 40% of familial amyotrophic lateral sclerosis (ALS) patients [1] and causes a devastating disease that involves the progressive degeneration of motor neurons and muscle denervation [2]. Current therapies for ALS, riluzole [3] and edaravone [4], have produced limited clinical benefit [5]. It is imperative to elucidate the primary mechanisms responsible for motor neuron loss in order to design effective therapeutics against the toxicity derived from *C9orf72* expansion.

Haploinsufficiency has been proposed as a toxic mechanism induced by $(G_4C_2)_n$ expansion [6]; however, *C9orf72* null mice do not develop motor neuron dysfunction, suggesting the existence of other mechanisms involved in disease pathophysiology [7].

Two possible gain of function mechanisms have been highlighted, including the production of long repeat RNA and dipeptide repeat proteins (DPRs). DPRs are translated from long repeat RNAs by a non-canonical process, repeat associated-non-AUG (RAN) translation, giving rise to five different species. Poly-GR, poly-GA and poly-GP are produced from the sense strand and poly-PR, poly-GP and poly-PA from the antisense strand [8,9]. Of these, poly-GA, poly-GR, and poly-PR have shown toxicity in cellular models, and these arginine-containing DPRs (Arg-DPRs) are the most toxic [10–12].

Emerging evidence suggests that oxidative stress and mitochondrial dysfunction are important elements of mutant *C9orf72* pathology [13,14]. A central regulator of the antioxidant response in cells is the transcription factor NRF2 (Nuclear factor erythroid 2-Related Factor 2). NRF2 is principally controlled by regulation at the protein level by the ubiquitin–proteasome system due to the presence of at least two degradation domains, exhibiting a half-life of approximately 30 min depending on the cell type [15,16]. The best-characterized regulatory mechanism involves the E3 ubiquitin ligase adapter KEAP1 (Kelch-like-ECH-associated protein 1) which targets NRF2 for ubiquitin/proteasome degradation [17]. Oxidative stress or electrophilic molecules modify several redox-sensitive cysteines in KEAP1 [18,19] in a way that renders it no longer effective in connecting NRF2 with the ubiquitination machinery [20]. Thus, newly synthesized NRF2 escapes degradation, accumulates in the nucleus, and activates its cytoprotective program through the activation of genes containing NRF2-regulated Antioxidant Response Elements (ARE). The repertoire of NRF2-regulated genes includes genes related to redox homeostasis (HMOX1, NQO1, GCLM, GCL, GPX), proteostasis control (PSMB7, ULK1, SQSTM1, LAMP2A) and inflammatory response (IL1β, IL6, MARCO) [21], which are processes thought to be dysregulated in C9orf72 patients. However, the pathogenic role for DPRs in neurotoxicity related to the regulation of the endogenous antioxidant defense remains poorly understood.

In this study, we sought to determine the alterations in redox gene expression in *C9orf72*-ALS patients and the interplay between arginine containing DPRs, control of redox alterations, and NRF2. Our results suggest NRF2 as a candidate therapeutic target for *C9orf72*-ALS patients.

2. Materials and Methods

2.1. Targeted qRT-PCR Analysis in Human Samples

The case-control study was performed on 7 *C9orf72* mutation carriers and 7 healthy controls who were recruited according to the criteria specified by the UK 'AMBRoSIA' ALS biomarker cohort study. *C9orf72* expansion testing was carried out as described in [22]. Venous blood was collected in EDTA-containing tubes for leukocyte RNA isolation. Within 15 min of venipuncture, an EDTA blood bottle was passed through an RNA extraction kit LeukoLOCK™ filter from LeukoLOCK™ Total RNA Isolation System (Invitrogen, Waltham, MA, USA; ID AM1923)). A total of 3 mL of phosphate buffered saline was then passed through the filter, followed by 3 mL of the RNAlater™ solution (Invitrogen, Waltham, MA, USA; ID AM7021). Leukocyte RNA which was bound to the filter was isolated according to the manufacturer instructions and stored at −80 °C until further processing.

RNA integrity was assessed using an RNA 6000 Nano Kit (Agilent, Santa Clara, CA, USA; ID 5067-1511), according to manufacturer's instructions. Briefly, 1 μL of each sample (approximately 200 ng of RNA) was loaded over the Gel-Dye Mix placed previously in the chip wells. Electrophoresis, data analysis and calculation of the RNA integrity number (RIN) were achieved by employing an Agilent 2100 Bioanalyzer (Agilent, Santa Clara, CA, USA). *C9orf72* carriers were matched to healthy controls of a similar age, gender and whose samples had similar RNA integrity values (Table 1).

Table 1. Human samples information.

	Age (Years)	Gender	Site of Onset	RIN [1] Value	FVC [2] (% Predicted)	ALSFRS-R [3]	Progression Rate	Riluzole
	61	Male	-	9.6	-	-	-	-
	56	Male	-	9.5	-	-	-	-
	68	Male	-	9	-	-	-	-
Healthy controls	71	Female	-	8.6	-	-	-	-
	69	Female	-	8.4	-	-	-	-
	57	Female	-	8	-	-	-	-
	63	Female	-	7.3	-	-	-	-
Mean ± SD	63.6 ± 5.9	-	-	8.6 ± 0.8	-	-	-	-
	53	Male	Bulbar	9.4	117	38	3.3	no
	73	Female	Bulbar	9.1	51	32	4.0	no
C9orf72 patients	65	Female	Limb	9	110	41	0.3	no
	67	Female	Limb	8.9	93	42	1.2	no
	58	Female	Limb	8.7	99	n/a	n/a	yes
	61	Male	Limb	8.5	92	41	1.0	no
	70	Female	Limb	8	70	40	0.4	no
Mean ± SD	63.9 ± 7.0	-	-	8.8 ± 0.5	90.3 ± 22.9	39.0 ± 3.7	1.7 ± 1.6	-

[1] RIN, RNA integrity number; [2] FVC, forced vital capacity; [3] ALSFSR-R; revised Amyotrophic Lateral Sclerosis Functional Rating Scale.

Reverse transcription using 400 ng of RNA from each control or *C9orf72* patient sample was performed with an RT2 First Strand Kit (Qiagen, Hilden, Germany; ID 330401)) according to the manufacturer's instructions. The expression of 84 key genes involved in oxidative stress/antioxidant response was analyzed with the RT2 Profiler™ PCR Array Human Oxidative Stress Plus (Qiagen, Hilden, Germany; ID PAHS-065Y)) for each sample. Data from the 84 genes can be found in Supplementary Table S1. The SYBR Green chemistry of the Applied Biosystems 7900 HT FAST Real-Time PCR system (Applied Biosystems, Waltham, MA, USA) was used. The expression level of each gene was normalized with the geometric mean of the housekeeping genes ACTB, B2M, GAPDH, HPRT1, RPLP0 to produce the ΔCt value and this was further normalized with the mean of the control group to eventually calculate $2^{-\Delta\Delta Ct}$ values or the fold of change (FC). Data are represented and statistical analysis was performed with log2(FC). Results are presented as the mean ± standard error of the mean (SEM) from each experimental group. Statistical analysis was performed using a script of the R programming language. Since data were not normally distributed (Shapiro–Wilk test, $p < 0.05$), the differences in gene expression between patients and controls were evaluated using the non-parametric Wilcoxon rank-sum test. The difference in gene expression was considered significant at $p < 0.05$.

2.2. Cell Culture and Reagents

The Mouse Motor Neuron-Like Hybrid Cell Line (NSC34) (Cedarlane Laboratories, Burlington, Canada) was grown in DMEM (Dulbecco's modified Eagle's medium with high glucose; Sigma-Aldrich, San Luis, MO, USA; ID D5648) supplemented with 10% heat inactivated fetal bovine serum (FBS, Biowest, Nuaillé, France), 2 mM glutamine (Gibco, Waltham, MA, USA; ID 25030081) and 80 µg/mL gentamicin (Normon Laboratories, Tres Cantos, Spain). NSC34-Sham and NSC34-$(G_4C_2)_{102}$ are isogenic cells lines, as previously described [23]. These cells contain an Flp-In™ T-Rex™ expression cassette, encoding an FRT recombination site and a tetracycline repressor element, where $(G_4C_2)_{102}$ and Sham sequences are stably integrated from pcDNA5/FRT/TO® and pcDNA/FRT/TO®-$(G_4C_2)_{102}$. NSC34-sham and NSC34-$(G_4C_2)_{102}$ cells were grown in DMEM supplemented with 10% tetracycline-free tested FBS (Takara Bio, Kusatsu, Japan; ID 631106) which was previously heat inactivated, 2 mM glutamine (Gibco, Waltham, MA, USA; ID 25030081), 80 µg/mL gentamicin (Normon Laboratories, Tres Cantos, Spain), 100 µg/mL hygromycin B and 5 µg/mL blasticidin (InvivoGen, San Diego, CA, USA; IDs ant-hg-1 and ant-bl-1). For expression of the hexanucleotide repeat cassette, 1 µg/mL tetracycline (Gibco, Waltham, MA, USA; ID A39246) was added to the medium and refreshed every three days. 3-(4,5-

dimethylthiazol-2-yl)-2,5-diphenyltetrazolium (MTT), dimethyl fumarate, actinomycin D and MG132 were purchased from Sigma Aldrich (San Luis, MO, US; IDs M5655, 242926, A9415, P8833, C2211). Dimethyl fumarate, MG132 and actinomycin D were diluted in dimethyl sulfoxide. Puromycin was diluted in H_2O. HA, HA-GR_{20} and HA-PR_{20} were synthetized by GenicBio Limited (Shangai, China), diluted in H_2O, and kept as 1 mM stocks at $-70\ ^\circ C$.

2.3. Transfection, Lentiviral Production and Transduction

NSC34 cells were transfected with pEGFP-C3, pEGFP-GR_{50} and pEGFP-PR_{50} (kindly provided by Dr. JP Taylor, St. Jude Children's Research Hospital. Memphis, TN, USA) or TK-Renilla (Promega, Madison, CA, USA) and ARE-LUC (Dr. J. Alam, Ochsner Clinic Foundation, New Orleans, LA, USA) using Lipofectamine 2000 (Invitrogen, Waltham, MA, USA; ID 11668019) in OptiMEM (Gibco, Waltham, MA, USA; ID 31985062) at a ratio of 1:1 (µL lipofectamine: µg DNA) according to the manufacturer's instructions. EGFP expression was checked using a fluorescence microscope, ensuring that the transfection efficiency was greater than 70%. Pseudotyped lentiviral vectors were produced in HEK293T cells that were grown in DMEM - high glucose (Dulbecco's modified Eagle's medium with high glucose; Sigma-Aldrich, San Luis, MO, USA; ID D5648) supplemented with 10% heat inactivated fetal bovine serum (FBS; Biowest, Nuaillé, France), 2 mM glutamine (Gibco, Waltham, MA, USA, ID 25030081) and 80 µg/mL gentamicin (Normon Laboratories, Tres Cantos, Spain). HEK293T cells were transiently co-transfected with 10 µg of control or $NRF2^{\Delta ETGE}$-V5 transfer vectors (after cloning into pWPXL lentiviral expression plasmid, (Addgene, ID 12257; deposited by Dr. Didier Trono, École Polytechnique Fédérale de Lausanne, Lausanne, Switzerland), 6 µg of the packaging plasmid pSPAX2 (Addgene, ID 12260) and 6 µg of the VSV-G envelope protein plasmid pMD2G (Addgene, ID 12259; deposited by Dr. Didier Trono) using Lipofectamine reagent and Plus^{TM} Reagent (Invitrogen, Waltham, MA, USA; IDs 18324012 and 11514015) according to the manufacturer's instructions. For lentiviral transduction, NSC34 cells were grown to confluence and then transduced with supernatant from HEK293T-producing lentiviruses plus 10 µg/mL polybrene (Sigma-Aldrich, San Luis, MO, USA; ID TR-1003). The medium was removed 24 h after and expression of the proteins took place during 48 h.

2.4. Immunofluorescence Analysis

NSC34 cells were grown over poly-D-lysine-coated (0.1 mg/mL, Sigma-Aldrich, San Luis, MO, USA; ID P6407) coverslips. Prior to fixation, for SUnSET experiments, cells were treated with 1 µg/mL puromycin during 1 h. The medium was removed from wells, and following a wash with PBS pH 7.4, cells were fixed with 4% paraformaldehyde (Sigma-Aldrich, San Luis, MO, USA; 158127) for 15 min at room temperature. After PBS washes, cells were permeabilized and blocked in PBS containing 0.3% Triton X100 (Sigma-Aldrich, San Luis, MO, USA; ID X100) with 3% bovine serum albumin (NZYTech, Lisboa, Portugal; ID MB046) for 1 h. Coverslips were incubated with the suitable antibodies (Table 2), diluted in blocking solution overnight at 4 °C and washed with PBS. Coverslips were incubated with the appropriate secondary antibodies conjugated with Alexa Fluor dyes (Invitrogen, Waltham, MA, USA) for 1 h at room temperature. Nuclei were counterstained with 1 µM 4'-6-diamino-2-phenylindole (DAPI, Invitrogen, Waltham, MA, USA; ID D1306) diluted in PBS for 10 min and washed with PBS. Coverslips were mounted using ProLong^{TM} Gold Antifade Mountant (Invitrogen, Waltham, MA, USA; ID P36930) and dried for 24 h before visualization. Fluorescence was visualized using a Nikon Eclipse 90i fluorescence microscope (Nikon, Tokyo, Japan) or a Leica TCS SP5 Confocal Microscope (Leica Microsystems, Wetzlar, Germany). Images from confocal microscopy were presented as maximal projections. Image analysis was performed with Fiji software [24].

Table 2. Antibodies used throughout the study.

Epitope	Dilution	Species	Source	Reference
NRF2	1:5000	Rabbit	Homemade	-
HO1	1:2000	Rabbit	Homemade	-
GFP	1:1000	Mouse	Sigma Aldrich	G1546
ACTB	1:4000	Goat	Santa Cruz Biotech.	sc-1616
GAPDH	1:30000	Mouse	Calbiochem	CB1001
HA	1:500	Mouse	Abcam	ab9110
Puromycin	1:1000	Mouse	Sigma Aldrich	MABE343

2.5. Superoxide Generation and Mitochondrial Membrane Potential Measurement

Intracellular and mitochondrial superoxide anion levels were assessed using dihydroethidium (HE, Invitrogen, Waltham, MA, USA; ID D11347) or MitoSOX Red (Invitrogen, Waltham, MA, USA; ID M36008), respectively. The mitochondrial membrane potential was assessed using MitoTracker Red FM (Invitrogen, Waltham, MA, USA; ID M22425). For flow cytometric analysis, cells were incubated with 2 µM HE for 60 min, 1 µM MitoSOX for 30 min or 0.5 µM MitoTracker Red FM for 30 min. Cells were washed once in PBS, detached, and centrifuged for 5 min at $250 \times g$. Pellets were resuspended in an appropriate volume of PBS and dead cells were stained with 1 µM DAPI or 2 µM TO-PRO-3 Iodide (Invitrogen, Waltham, MA, USA; ID T3602). Then, fluorescence was recorded in a BD FACSCanto™ II Cell Analyzer (BD Biosciences, Franklin Lakes, NJ, USA). A data analysis was performed with Kaluza software (Beckman Coulter, Pasadena, CA, USA).

2.6. Cell Viability Assessed by MTT Reduction and Annexin-V/Propidium Iodide

The tetrazolium ring of 3-(4,5-dimethylthiazol-2-yl)-2,5-diphenyltetrazolium bromide (MTT) can be reduced with active dehydrogenases to produce a formazan precipitate. Cells were incubated in serum-free media with MTT (0.2 mg/mL) for 45 min at 37 °C. Thereafter, the media were removed and DMSO added to each well to dissolve the formazan precipitate for 20 min with gentle agitation. A total of 100 µL of the supernatants was analyzed in 96-well multiwell plates at 550 nm using a VERSAmax microplate reader (Molecular Devices, San Jose, CA, USA). For Annexin-V-FITC staining, the cells were washed twice with Annexin-V binding buffer (containing 10 mM HEPES-NaOH, pH 7.4, 150 mM NaCl, 5 mM KCl, 1 mM $MgCl_2$, 1.8 mM $CaCl_2$), resuspended in 100 µl of a 1:100 dilution of TACS Annexin-V-FITC (R&D Systems, Biotechne, Minneapolis, MN, USA; ID 4830-01-K) in Annexin-V binding buffer containing 1 µg/mL Propidium Iodide (PI) and incubated for 15 min at room temperature. Fluorescence was measured with a CytoFLEX flow cytometer (Beckman Coulter, Pasadena, CA, USA), by analyzing at least 10,000 events per condition. The data analysis was performed using Kaluza softwar (Beckman Coulter, Pasadena, CA, USA).

2.7. Cell Cycle Analysis

For EdU analysis, the Click–iT™ EdU Alexa Fluor™ 647 Flow Cytometry Assay Kit (Invitrogen, Waltham, MA, USA; ID C10424) was used following the manufacturer's instructions. Briefly, cells were treated with 10 µM EdU for 2 h, harvested, fixed, and permeabilized. For analysis of DNA content, cells were counterstained with 1µg/mL Hoechst 33342 (Invitrogen, Waltham, MA, USA; ID H3570) for 30 min in darkness. Analysis of EdU incorporation and DNA content was performed with a CytoFLEX flow cytometer (Beckman Coulter, Pasadena, CA, USA), by analyzing at least 10,000 events per condition. The data analysis was performed using Kaluza software (Beckman Coulter, Pasadena, CA, USA).

2.8. Immunoblotting

Cells were washed with cold PBS and homogenized in lysis buffer (50 mM Tris pH 7.6, 400 mM NaCl, 1 mM EDTA, 1 mM EGTA and 1% SDS). Samples were heated at 95 °C for 15 min and sonicated. Protein quantification was performed with the DC^{TM} Protein Assay (Bio-Rad, Hercules, CA, USA; ID 5000112), and protein loading buffer (50 mM Tris-HCl pH 6.8, 2% SDS, 0.1% bromophenol blue, 10 % glycerol, 150 mM β-mercaptoethanol) was added. Samples were boiled at 95 °C and cellular debris was cleared with centrifugation. Proteins were resolved using SDS-PAGE and transferred to 0.45 μm pore size Immobilion-P membranes (Millipore, Burlington, MA, USA; ID IPVH00010). For immunoblotting, membranes were hydrated in methanol, washed in TTBS (20 mM Tris-HCl pH 7.5, 150 mM NaCl and 0.1% Tween 20) buffer and blocked with 5% non-fat dry milk in TTBS. Membranes were incubated with the appropriate dilution of the primary antibodies (Table 2) in 0.4% BSA or 2.5% non-fat-dry milk for 1 h, washed and incubated with 1:10,000 dilution of secondary antibodies coupled to horseradish peroxidase in 0.4% BSA TTBS for 1 h. Proteins were detected by carrying out enhanced chemiluminescence (Amersham^{TM} ECL^{TM} Select Western Blotting Detection Reagent, GE Healthcare, Chicago, IL, USA; ID GERPN2235).

2.9. mRNA Quantification by qRT-PCR

Total RNA was extracted using TRI$^{®}$ Reagent (Invitrogen, Waltham, MA, USA; ID AM9738) according to the manufacturer's instructions. One μg of RNA (quantified using Nanodrop^{TM} 2000 Spectrophotometer) was reverse-transcribed in a 20 μL volume using the High-Capacity RNA to cDNA^{TM} kit (Invitrogen, Waltham, MA, USA; ID 4387406). For qPCR, the reaction was performed in 10 μL using the fluorescent dye Power SYBR^{TM} Green PCR Master Mix (Applied Biosystems, Waltham, MA, USA; ID 4367659) and a mixture of 5 pmol of reverse and forward primers. Primer pairs are listed in Table 3. Quantification was performed using a StepOne detection system (Applied Biosystems, Waltham, MA, USA). PCR cycles proceeded as follows: initial denaturation for 10 min at 95 °C, then 40 cycles of denaturation (15 s, 95 °C), annealing (30 s, 60 °C) and extension (30 s, 60 °C). Data analysis is based on the ΔΔCT method with normalization of the raw data to the geometric mean of housekeeping genes. For intron analysis of the *Nfe2l2* transcript, RNA was pretreated with DNase I (Invitrogen, Waltham, MA, USA; ID 18068015) according to manufacturer's instructions before retro-transcription to avoid genomic DNA contamination. Additionally, intron levels were normalized with total *Nfe2l2* abundance using intron-spanning primers.

Table 3. qPCR primers used throughout the study.

Gene		Forward Primer (5′—3′)	Reverse Primer (5′—3′)
Nfe2l2	Mature	CCCGAAGCACGCTGAAGGCA	CCAGGCGGTGGGTCTCCGTA
	Intron 1	TTTATGACCAAATACCGAGCACA	GGCTCAATGTCTGGTAACATCC
	Intron 2	AGCACAGGGTCACAACGAG	CGCTGTTCGTTAAAGGGGAG
	Intron 3	CCAAGGCATGAAGAACAGACA	TAATATGGAGTCACTGGGAGGA
	Intron 4	TGTGTGTTAAGTGAGTCTGGG	TAGGAGTGTGCTGTTTTCTGC
Actb		CACCCAGCACATTTAGCTAGCTGA	TTCAGAGCAACTGCCCTGAAAGCA
Gapdh		CGACTTCAACAGCAACTCCCACTCTTCC	TGGGTGGTCCAGGGTTTCTTACTCCTT
Tbp		TGCACAGGAGCCAAGAGTGAA	CACATCACAGCTCCCCACCA
Slc1a2		ACAATATGCCCAAGCAGGTAGA	CTTTGGCTCATCGGAGCTGA
Naca		GACAGTGATGAGTCAGTACCAGA	TGCTTGGCTTTACTAACAGGTTC

2.10. Luciferase Assays

Transient transfections of NSC34 cells were performed with the expression vectors for TK-Renilla and ARE-LUC. Cells were seeded on 24-well plates (100,000 cells per well), cultured for 16 h, and transfected Lipofectamine 2000 (Invitrogen, Waltham, MA, USA; ID 11668019) in OptiMEM (Gibco, Waltham, MA, USA; ID 31985062) at a ratio of 1:1 (μL lipofectamine:μg DNA) according to the manufacturer's instructions. After 48 h of recovery from transfection, the cells were lysed and assayed with a dual-luciferase assay system (Promega, Madison, WI, USA; ID E1910) according to the manufacturer's instructions. Relative light units were measured in a GloMax 96 microplate luminometer with dual injectors (Promega, Madison, WI, USA)

2.11. RNA Immunoprecipitation

Transfected NSC34 cells were detached in cold PBS and pelleted by performing centrifugation. Cells were lysed in 1 mL of RNA immunoprecipitation lysis buffer (Tris-HCl pH 7.4 25 mM, KCl 150 mM, EDTA 0.5 mM, NP40 0.5%, RNAseOUT 100 U/mL (Invitrogen, Waltham, MA, USA; ID 10777019), 1 mM phenylmethylsulfonyl fluoride and 1 μg/mL leupeptin (Sigma-Aldrich, San Luis, MO, USA; IDs P7626 and L5793) at 4 °C for 30 min with end-over-end rotation. Lysates were cleared by centrifugation at $13,000 \times g$ for 15 min and supernatants were collected. Lysates were divided in half for immunoprecipitation with 5 μg of anti-GFP antibody or with 5 μg of IgG isotype control overnight at 4 °C with end-over-end rotation. Immunocomplexes were isolated using Protein G Sepharose® 4 Fast Flow (Cytiva, Danaher Corporation, Washington DC, WA, USA; ID GE17-0618-01) during 1 h and beads with immunocomplexes washed three times with 500 μL of RIP lysis buffer. Before elution, a sample was taken to ensure EGFP immunoprecipitation with Western blot. For elution of RNA from immunocomplexes, TRI® Reagent was used according to the manufacturer's instructions. RNA isolates were quantified and treated with appropriate amounts of DNAse I according to the manufacturer's instructions. Retrotranscription and qPCR analysis were performed as described above.

2.12. Statistical Analyses

Unless otherwise indicated, all experiments were performed at least 3 times and all data presented in the graphs are the mean of at least 3 independent samples. Data are presented as mean $\pm$ SEM (standard error of the mean). Statistical differences between groups were assessed using GraphPad Prism 8 by one and two-way analyses of variance, using Dunnett's post-hoc test when comparing transfections/peptide treatments to EGFP or HA controls and Bonferroni's post-hoc test when comparing transductions/stable cell lines to Sham or Control transduction (**** indicates p values <0.0001, ***, $p < 0.001$, ** p textless 0.01 and * $p < 0.05$).

3. Results

3.1. C9orf72 Patient Leukocytes Exhibit Dysfunctional Expression of Several Redox Genes

Blood provides a window for investigating the inter-related systemic and local immune responses, inflammation and redox signaling alterations in chronic pathologies [25,26]. We investigated whether alterations in redox-related genes could be measurable in peripheral blood from *C9orf72*-ALS patients, as a reflection of gene expression changes occurring in the brain and spinal cord. We conducted a case-control study on seven *C9orf72*-ALS patients and seven healthy controls who were highly characterized and matched in terms of age and gender (Table 1). Targeted gene expression screening was performed with qRT-PCR using a pathway-focused array of 84 redox genes on leukocyte RNA (Supplementary Table S2). As shown in the gene expression volcano plot (Figure 1A), we identified nine redox genes that were significantly downregulated and one gene that was upregulated (Wilcoxon rank-sum test, $p < 0.05$) in the leukocytes of patients vs. controls. These genes encode members of the NADPH oxidases family (CYBB and DUOX2) that are involved in ROS production (Figure 1B), proteins that regulate the metabolism of glutathione (GSR and GPX3), or perform other redox functions (DHCR24, FHL2, CCS, FOXM1, TRAPPC6A, NUDT1). These results suggest a global impairment of redox homeostasis in the blood of *C9orf72*-ALS patients.

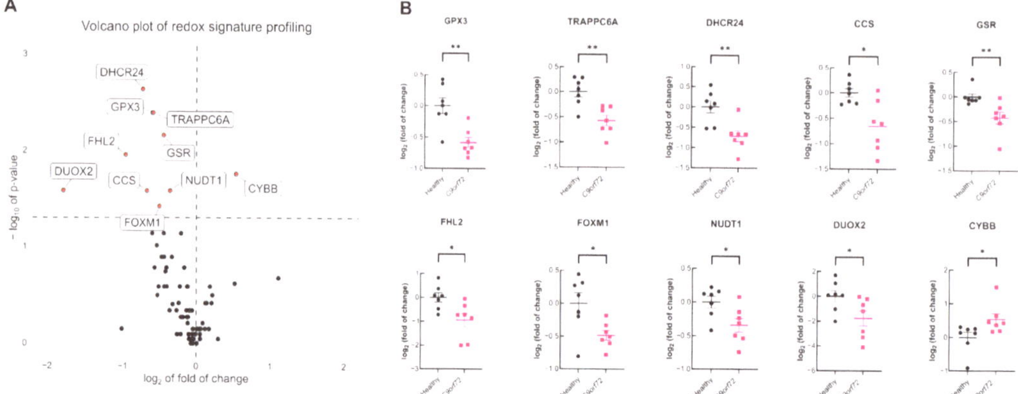

Figure 1. Expression of redox signature is impaired in leukocytes from *C9orf72* patients. mRNA levels for redox genes were determined with qRT-PCR in leukocyte RNA from healthy donors ($n = 7$) or *C9orf72* ALS patients ($n = 7$) and normalized using the geometric mean of the levels of the ACTB, B2M, GAPDH, HPRT1 and RPLP0 as housekeeping genes. (**A**) Volcano plot depicting the mean $\log_2$ of the fold of change of each tested gene and $-\log_{10}$ p-value between *C9orf72* and healthy individuals. Horizontal dashed line indicates $p < 0.05$, and vertical dashed line indicates $\log_2$ fold of change = 0. (**B**) $\log_2$ fold of change of the expression of the genes with statistically significant changes in each individual. Horizontal bars indicate the mean $\log_2$ fold of change value for the mRNA levels in each group and error bars indicate the SEM. Individual points indicate the expression of that gene in each individual. Statistical analysis was performed with the Wilcoxon rank-sum non-parametric test. * $p < 0.05$; **, $p < 0.01$ comparing ALS vs. control group.

3.2. Arg-DPRs Lead to Decreased Mitochondrial Membrane Potential and Cell Death

The hallmark of *C9orf72*-related pathology is the transcription of hexanucleotide expansions that produce toxic dipeptide repeat proteins. In a reverse translational approach, we aimed to correlate the redox alterations in *C9orf72*-ALS patients with this pathological mechanism in a simplified model of DPR expression using cultured the NSC34 motor neuronal cell line. NSC34 cells were transfected with expression vectors for either enhanced green fluorescent protein (EGFP), or EGFP fused to a 50-long dipeptide repeat of poly-GR (EGFP-GR$_{50}$) or poly-PR (EGFP-PR$_{50}$). EGFP immunofluorescence was analyzed 48 h following transfection. EGFP exhibited a diffuse pattern of expression in both cytosol and nuclei, whereas EGFP-GR$_{50}$ and EGFP-PR$_{50}$ were preferentially accumulated in nuclear dots (Figure 2A), mimicking the characteristic accumulation in nucleolar structures of DRPs [27]. Since mitochondria are a central organelle for redox biology, first, we attempted to address the impact of Arg-DPRs on mitochondrial membrane potential ($\Delta\Psi$m). A flow cytometry evaluation of $\Delta\Psi$m with MitoTracker Red FM revealed that the overexpression of EGFP-GR$_{50}$ or EGFP-PR$_{50}$ reduced the mitochondrial potential by 30% compared to EGFP-cells (Figure 2B). As an alternative approach, and by taking advantage of the cell-penetrating properties of arginine-containing peptides [28], we conducted similar experiments in this cell line submitted to 6 µM of either HA control peptide, HA-tagged 20-long dipeptide repeat poly-GR (HA-GR$_{20}$) or poly-PR (HA-PR$_{20}$). Immunodetection with anti-HA antibodies showed puncta accumulation of HA-GR$_{20}$ and HA-PR$_{20}$ in agreement with previous reports [29,30] (Figure 2D). When NSC34 cells were exposed to HA-GR$_{20}$ or HA-PR$_{20}$ peptides during 24 h, a 20% drop in MitoTracker Red FM fluorescence was observed in HA-GR$_{20}$ or HA-PR$_{20}$ compared to HA-treated cells (Figure 2E). Since a long-lasting drop in $\Delta\Psi$m may induce cell death, we evaluated cell viability with an MTT assay. A reduction in the formazan signal, measured with the MTT assay, may derive from an increase in cell death but also an inhibition in proliferation. For this reason, we validated that the Arg-DPRs did not modify NSC34 cell cycle progression by analyzing EdU incorporation and

DNA content. Although cell cycle was not altered in any condition, there was an increase in the percentage of cell death upon Arg-DPRs exposure, which we found by analyzing Annexin-V-FITC and Propidium Iodide (IP) staining (Supplementary Materials; Figure S1). As shown in Figure 2C, the overexpression of EGFP-GR$_{50}$ and EGFP-PR$_{50}$ reduced the survival rate by 40% compared with EGFP-transfected cells. Accordingly, HA-GR$_{20}$ and HA-PR$_{20}$ reduced the survival rate to the same extent compared with HA-treated cells (Figure 2F). Taken together, these findings link cellular death with a decrease/reduction in the mitochondrial membrane potential induced by Arg-DPRs.

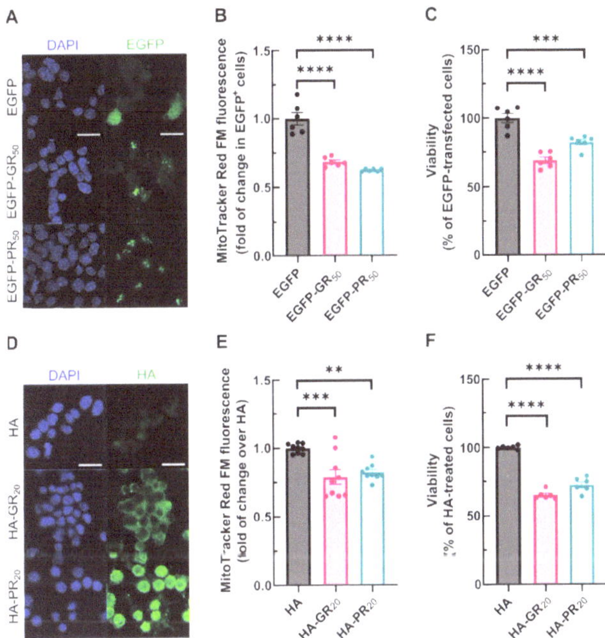

Figure 2. Endogenous and exogenous exposure to poly-GR and poly-PR decrease mitochondrial membrane potential and induce cell death. NSC34 cells were either transiently transfected with EGFP (control), EGFP-GR$_{50}$ or EGFP-PR$_{50}$ during 48 h or treated with 6 μM of HA, HA-GR$_{20}$ or HA-PR$_{20}$ peptides for 6 h (immunofluorescence) or 24 h (flow cytometry and viability assay). (**A**,**D**) EGFP fluorescence or HA immunofluorescence staining was analyzed by microscopy. Nuclei were counterstained with 4′,6-diamidino-2-phenylindole (DAPI). Scale bar: 40 μm. (**B**,**E**) cells were treated with 0.5 μM MitoTracker Red FM during 30 min and fluorescence was measured by flow cytometry analysis. Each point depicts the geometric mean of probe fluorescence in cells from that sample ((**B**), alive EGFP$^+$ cells, (**E**), all alive cells). (**C**,**F**) viability of the cells was measured using an MTT assay. Bar height indicates the mean of the whole group and error bars indicate the SEM. Statistical analysis was performed with one-way ANOVA. Dunnett's post-hoc test. **, $p < 0.01$; ***, $p < 0.001$; ****, $p < 0.0001$.

3.3. Arg-Containing Dipeptide Repeats from the C9orf72 Expansion Disrupt Redox Homeostasis

The dysregulation of mitochondrial membrane potential is closely related to the maintenance of high levels of reactive oxygen species (ROS). Therefore, we next analyzed the levels of superoxide ions inside mitochondria with MitoSOX Red. NSC34 was transfected with either EGFP, EGFP-GR$_{50}$ or EGFP-PR$_{50}$. The flow cytometry analysis revealed that, although the overexpression of EGFP-GR$_{50}$ did not modify superoxide levels, overexpression of EGFP-PR$_{50}$ led to a slight but significant increase in superoxide levels (Figure 3A). Then, we evaluated the effect of exogenous Arg-DPRs peptides on mitochondrial superoxide levels. NSC34 cells were submitted to either HA, HA-GR$_{20}$ or HA-PR$_{20}$ over 2 h and stained with MitoSOX Red. In this experimental setting, mitochondrial superoxide levels were higher in both HA-GR$_{20}$ and HA-PR$_{20}$ compared with HA-treated cells (Figure 3C). Accordingly, the overexpression of EGFP-DPRs (Figure 3B) or treatment with exogenous Arg-DPRs peptides (Figure 3D) led to a similar rise in the fluorescence derived from the oxidation of another redox sensitive probe, hydroethidine (HE). These results provide strong evidence that Arg-DPRs increase ROS levels in NSC34 motor neurons, probably contributing to mitochondrial damage.

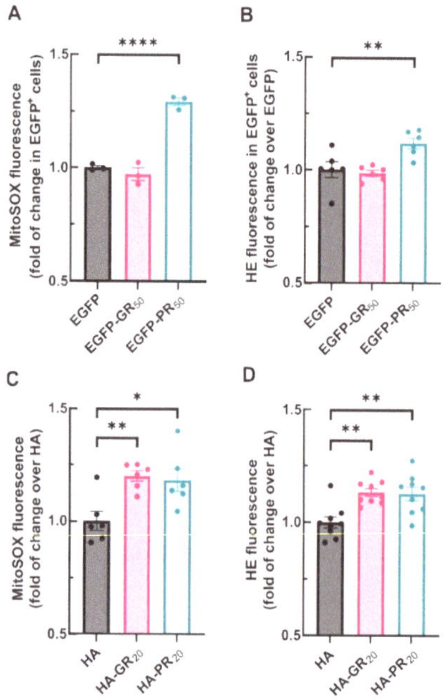

Figure 3. DPRs exposure leads to mitochondrial and intracellular superoxide generation in NSC34 cells. (**A**) NSC34 cells were transiently transfected with EGFP (control), EGFP-GR$_{50}$ or EGFP-PR$_{50}$ during 48 h. or treated with 6 μM of HA, HA-GR$_{20}$ or HA-PR$_{20}$ peptides for 2 h. (**A,C**) Cells were treated with 1 μM MitoSOX during 30 min and fluorescence was measured by flow cytometry analysis. (**B,D**) cells were treated with 2 μM hydroethidine during the last hour of transfection/treatment and probe intensity was measured by flow cytometry analysis. Each point depicts the geometric mean of probe fluorescence in cells from that sample ((**A,B**), alive/EGFP$^+$ cells, (**C,D**), all alive cells). Bar height indicates the mean of the whole group and error bars indicate the SEM. Statistical analysis was performed with one-way ANOVA. Dunnett's post-hoc test. *, $p < 0.05$; **, $p < 0.01$; ****, $p < 0.0001$.

3.4. Activation of NRF2 Is Impeded by DPRs

Next, we investigated the potential impact of Arg-DPRs on NRF2 activation. NSC34 cells were transfected with either EGFP, EGFP-GR$_{50}$ or EGFP-PR$_{50}$; after 48 h, cells were treated with dimethyl fumarate (DMF; 100 μM DMF for 16 h) to inhibit KEAP1. As expected, DMF led to a robust increase in both NRF2 and HO-1 in EGFP-cells. Interestingly, this increase was reduced in EGFP-GR$_{50}$ or EGFP-PR$_{50}$ cells compared to EGFP cells (Figure 4A,B). The increase in the NRF2 protein levels correlated with higher transactivating activity as determined with an ARE-driven luciferase-reporter construct, 3xARE-LUC, created from the NRF2 responsive ARE in the mouse *Hmox1* promoter [31]. NSC34 cells were co-transfected with EGFP, EGFP-GR$_{50}$ or EGFP-PR$_{50}$ together with ARE-firefly luciferase and TK-Renilla reporters. After overnight recovery, they were stimulated for 16 h with DMF. As shown in Figure 4C, DMF increased reporter gene activity by about 5-fold in EGFP control cells. By contrast, the overexpression of either EGFP-GR$_{50}$ or EGFP-PR$_{50}$ yielded a reduced 3-fold increase. Concordantly, the induction of two verified NRF2 targets,

Sqstm1 and *Nqo1*, exhibited reduced transcript levels upon DMF treatment in EGFP-GR$_{50}$ and EGFP-PR$_{50}$ cells compared to EGFP cells (Figure 4D). As an additional confirmation of our results, we analyzed NRF2 activation in a chronic model of DPR exposure based on a tetracycline-inducible cassette that contains 102 repeats of the (G_4C_2) hexanucleotide and produces only the sense strand of the toxic RNA species [32]. To determine the status of NRF2/HO-1, sham and $(G_4C_2)_{102}$ cells were treated with tetracycline for 7 days, and then submitted to 100 µM DMF for 16 h. NRF2 and HO-1 protein levels were measured in response to NRF2 activation. DMF prompted a robust increase in both NRF2 and HO-1 protein levels; however, this induction was significantly decreased in $(G_4C_2)_{102}$ bearing cells (Figure 4E,F). Taken together, these results point to an impairment in NRF2 activation induced by Arg-DPRs.

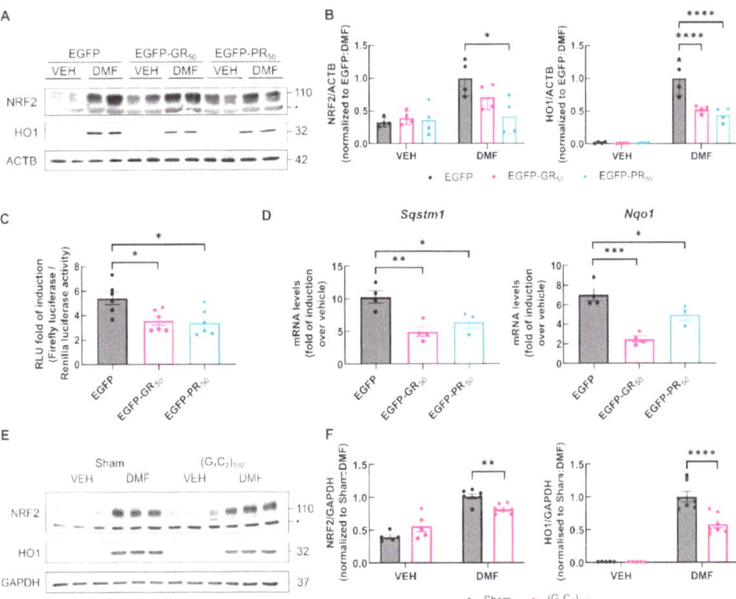

Figure 4. Pharmacological activation of NRF2-dependent transcription by dimethyl fumarate is impaired in EGFP-GR$_{50}$, EGFP-PR$_{50}$ or $(G_4C_2)_{102}$ expressing cells. (**A**) NSC34 cells were transiently transfected with EGFP, EGFP-GR$_{50}$ or EGFP-PR$_{50}$ for 48 h and treated with either VEH (DMSO 0.1%) or DMF 100 µM over the last 16 h. (**A**) The levels of the indicated proteins were measured by performing an immunoblot analysis. ACTB levels were determined as a loading control. (**B**) Densitometric quantification for the immunoblots for NRF2 and HO1 presented in (**A**). (**C**) NSC34 cells were transiently co-transfected with EGFP, EGFP-GR$_{50}$ and EGFP-PR$_{50}$ and plasmids encoding for ARE-luciferase reporter and TK-Renilla luciferase for 48 h and treated as before. Fold of induction over vehicle in relative luciferase units is shown. (**D**) Fold of induction on mRNA levels of the *Sqstm1* and *Nqo1* genes in EGFP-DPRs transfected cells treated with DMF. mRNA levels were normalized using the geometric mean of the housekeeping genes *Gapdh* and *Actb* and to the non-induced control. (**E**) The expression of the sham cassette or the $(G_4C_2)_{102}$ cassette was induced for 7 days with 1 µg/mL tetracycline in NSC34 cells. In the last 16 h, cells were treated with either VEH (DMSO 0.1%) or DMF 100 µM. The levels of the indicated proteins were measured by performing an immunoblot analysis. GAPDH levels were determined as a loading control. (**F**) Densitometric quantification for the immunoblots for NRF2 and HO1 presented in (**C**). Individual points in (**B,D**) represent the protein levels normalized by protein amount in each lane, bar height represents the mean of those protein levels in each group and error bars represent the SEM. Statistical analysis was performed with two-way ANOVA. Dunnett's post-hoc test in (**B–D**). Bonferroni's post-hoc test in (**F**). *, $p < 0.05$; **, $p < 0.01$; ***, $p < 0.001$; ****, $p < 0.0001$.

3.5. Accumulation of NRF2 Transcription Factor Is Impaired by DPRs

Since the transactivation of ARE-containing genes with DMF depends on newly synthetized NRF2 [33], we examined the transcription, splicing and stability of *Nfe2l2* mRNA. NSC34 cells were transiently transfected with EGFP, EGFP-GR$_{50}$ or EGFP-PR$_{50}$. At 48 h following transfection, no change was found in immature or mature mRNA levels of *Nfe2l2* when comparing samples from EGFP-GR$_{50}$ and EGFP-PR$_{50}$ to EGFP cells (Figure 5A,B). The degradation rate of *Nfe2l2* mRNA was also analyzed using an actinomycin D chase assay. Gene transcription was blocked with 5 µg/mL actinomycin D and the levels of *Nfe2l2* were measured with qRT-PCR. As shown in Figure 5C, *Nfe2l2* mRNA exhibited a half-life of 85 min in EGFP [34], which was not altered by the overexpression of EGFP-GR$_{50}$ or EGFP-PR$_{50}$. We also analyzed whether the Arg-DPRs nuclear aggregates might sequester the *Nfe2l2* transcripts. We performed RNA immunoprecipitation by employing anti-GFP or control IgG antibody, and the levels of *Nfe2l2* mRNA bound in the immunocomplexes were determined with qRT-PCR. As shown in Figure 5D, *Slc1a2* and *Naca* transcripts, used as controls [29], were enriched in EGFP-PR$_{50}$ immunoprecipitates compared to the EGFP. In contrast, *Nfe2l2* mRNA was not enriched in the same complexes.

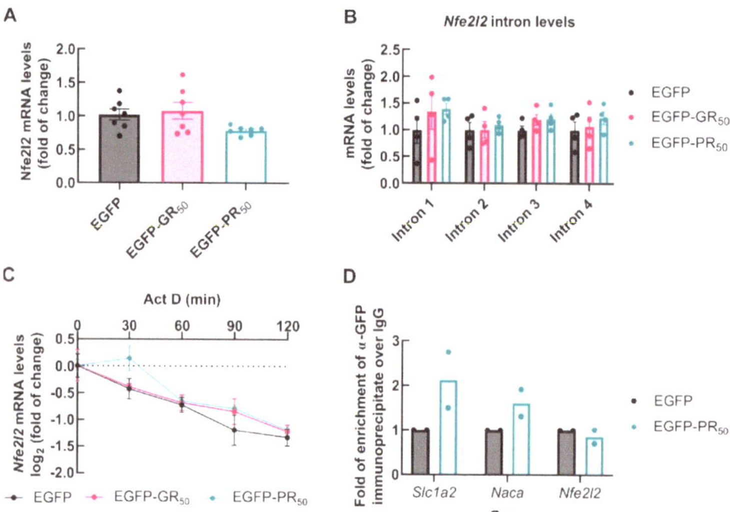

Figure 5. DPRs do not modify *Nfe2l2* mRNA levels, splicing, half-life or sequester them. NSC34 cells were transiently transfected with EGFP (control), EGFP-GR$_{50}$ or EGFP-PR$_{50}$ for 48h. (**A**,**B**) *Nfe2l2* total transcripts levels and intron levels in mRNA were measured by qRT-PCR. (**C**) Transcription was blocked with 5 µg/mL of actinomycin D in transfected NSC34 cells for the indicated times and mRNA levels for *Nfe2l2* were measured by qRT-PCR. (**D**) mRNA levels of the indicated transcripts were measured by qRT-PCR in RNA bound to EGFP-immunocomplexes. In (**A**,**B**) mRNA levels were normalized by the geometric mean of the housekeeping genes *Gapdh*, *Actb* and *Tbp*. In (**C**) mRNA levels were normalized by the geometric mean of the stable transcripts from *Gapdh* and *Actb*. Bar height (**A**,**B**,**D**) and point (**C**) indicates the mean of the whole group and error bars indicate the SEM. Statistical analysis was performed with one-way ANOVA for (**A**,**B**), and two-way ANOVA in (**C**), yielding non-significant changes.

Then, we used a SUnSET assay [35] to evaluate protein translation in EGFP, EGFP-GR$_{50}$ or EGFP-PR$_{50}$ expressing cells. Transfected cells were treated with 1 µg/mL puromycin for 1 h. During this period, puromycin is incorporated into the C-terminus of nascent proteins; then, its levels can be detected by immunocytochemistry. As shown in Figure 6A,B, puromycin fluorescence was diminished in EGFP-GR$_{50}$ and EGFP-PR$_{50}$ expressing cells compared to EGFP. These data are in line with previous reports [29,36], supporting the fact that Arg-DPRs lead to a decrease in general protein synthesis. Then, we blocked NRF2 proteasomal degradation to specifically determine the impact of Arg-DPRs on NRF2 protein synthesis. NSC34 transfected with EGFP, EGFP-GR$_{50}$ or EGFP-PR$_{50}$ were submitted to MG132 at the indicated times. As shown in Figure 6C,D, NRF2 levels increased as expected in EGFP-overexpressing cells; importantly, the expression of EGFP-GR$_{50}$ or EGFP-PR$_{50}$ resulted in significantly smaller NRF2 accumulation. Taken together, these data indicate that the translational impairment induced by Arg-DPRs directly impacts NRF2 synthesis, resulting in reduced transcriptional activity.

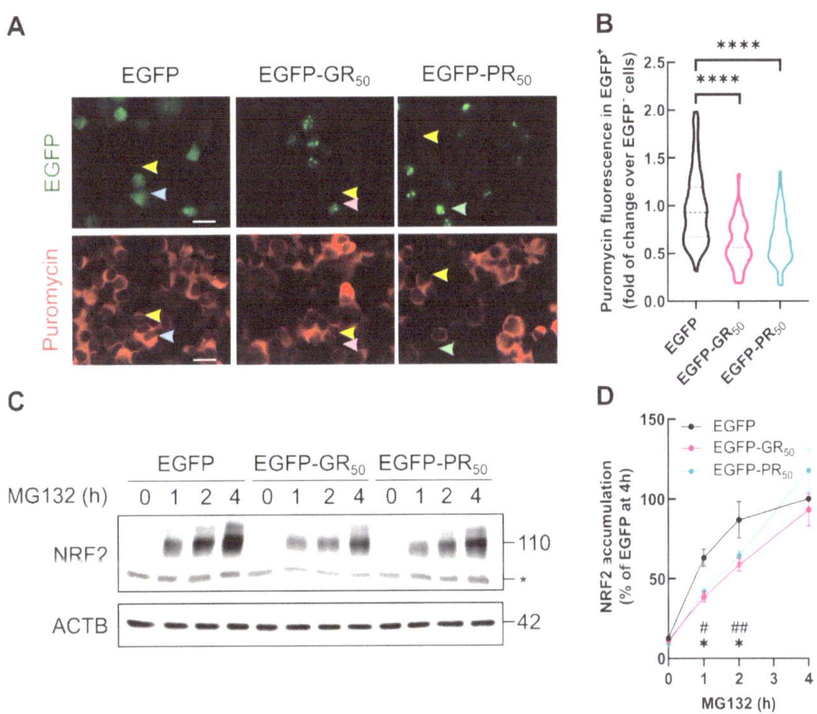

Figure 6. General translational shutdown caused by DPRs impairs NRF2 induction by MG132. NSC34 cells were transiently transfected with EGFP, EGFP-GR$_{50}$ or EGFP-PR$_{50}$. (**A**,**B**) transfected NSC34 cells were treated with 1 µg/mL puromycin for 1 h. (**A**) immunocytochemical detection of puromycin-labelled peptides. Yellow arrows, EGFP$^-$ cells; blue arrows, EGFP$^+$ cells; red arrows, EGFP-GR50$^+$ cells; green arrows, EGFP-PR$_{50}$$^+$ cells. Scale-bar: 40 µm; (B) violin plots depicting the puromycin intensity distribution in at least 100 cells in EGFP, EGFP-GR$_{50}$ and EGFP-PR$_{50}$ conditions. (**C**) transfected cells were treated with MG132 (20 µM) for the indicated times. NRF2 levels were measured by immunoblot. ACTB was measured as a loading control. *, non-specific band. (**D**) densitometric quantification of the immunoblot presented in A. Values are expressed as % NRF2 accumulation at time 4 h. In (B) dotted lines represent the median and the 1st and 3rd quartile. Statistical analysis was performed with one-way ANOVA. Dunnett's post-hoc test. ****, $p < 0.0001$. In (**D**) the point represents the mean and error bars represent SEM. Statistical analysis was performed with two-way ANOVA. Dunnett's post-hoc test. #, EGFP vs. EGFP-GR$_{50}$; *, EGFP vs. EGFP-PR$_{50}$. * or #, $p < 0.05$; ##, $p < 0.01$.

3.6. NRF2 Activation Protects against Arg-DPR Toxicity in NSC34 Cells

Despite the delay in NRF2 activation caused by DPRs, we investigated whether the NRF2 system was responsive enough to counteract ROS induction by Arg-DPRs. To test this hypothesis, NSC34 transfected with EGFP and EGFP-PR$_{50}$ were submitted to 100 µM DMF for 4 h and HE fluorescence was assessed with flow cytometry. As shown in Figure 7A, DMF treatment abrogated the increase in HE fluorescence produced by EGFP-PR$_{50}$ expression. We replicated these results in our stably transfected NSC34 (G$_4$C$_2$)$_{102}$ model induced for 7 days (Figure 7B). NSC34 (G$_4$C$_2$)$_{102}$ showed a slight but significant increase in superoxide levels compared to sham cells, that were rescued upon DMF treatment for 4 h.

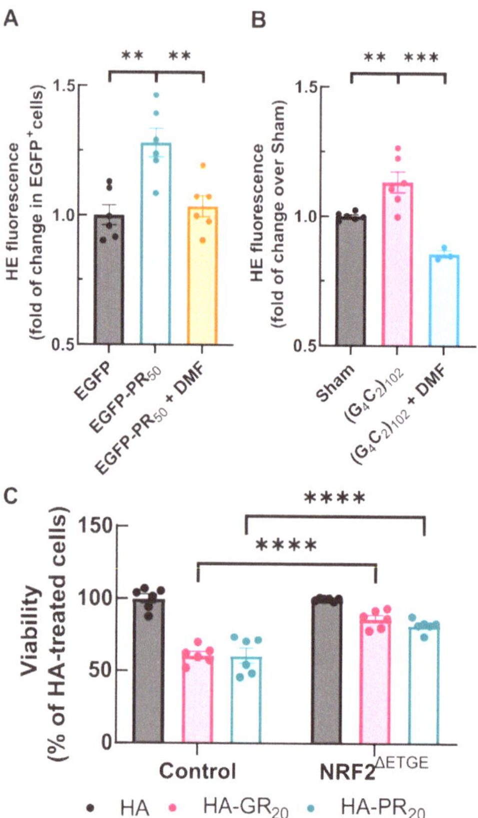

Figure 7. Activation of NRF2 rescues from DPRs toxicity. (**A**,**B**) NSC34 cells were transiently transfected with EGFP, EGFP-GR$_{50}$ or EGFP-PR$_{50}$ for 48 h or the expression of the Sham cassette or the (G$_4$C$_2$)$_{102}$ cassette was induced for 7 days with 1 µg/mL tetracycline in NSC34 cells, respectively. Cells were treated with either VEH (DMSO 0.1%) or DMF 100 µM over the last 4 h and stained with 2 µM hydroethidine during the last hour. Probe intensity was measured by flow cytometry analysis. In (**A**,**B**), each point depicts the geometric mean of probe fluorescence in cells from that sample (**A**), alive EGFP⁺ cells, (**B**), all alive cells). (**C**) NSC34 cells were transduced with either control lentiviral particles or encoding the stable NRF2 version lacking the ETGE KEAP1-binding motif for 48 h and then treated with 6 µM of HA, HA-GR$_{20}$ or HA-PR$_{20}$ peptides for 24 h. Cell viability was measured using a MTT assay Bar height indicates the mean of the whole group and error bars indicate the SEM. Statistical analysis was performed with one-way ANOVA (**A**,**B**), Dunnett's post-hoc test or two-way ANOVA (**C**), Bonferroni's post-hoc test. ** $p < 0.01$; ***, $p < 0.001$; ****, $p < 0.0001$.

Next, we questioned whether supplementation of NRF2 levels with the overexpression of a stable version of NRF2 (NRF2$^{\Delta ETGE}$) might improve the survival rate of NSC34 cells submitted to the Arg-DPR challenge. NSC34 cells were infected with either control lentiviral vectors for NRF2$^{\Delta ETGE}$ for 24 h, and then treated with HA, HA-GR$_{20}$ or HA-PR$_{20}$ peptides for an additional 24 h. As shown in Figure 7C, Arg-DPRs peptides led to a 40% reduction in NSC34 viability, measured with the MTT assay, that was fully recovered by overexpressing NRF2 (Supplementary Figure S2). These results provide a proof of concept indicating that

NRF2 activation could have therapeutic potential in *C9orf72*-related ALS by protecting motor neurons from Arg-DPR toxicity.

4. Discussion

In the present study, we investigated the redox, mitochondrial and NRF2 status in the context of dipeptide repeat protein toxicity resulting from pathological hexanucleotide expansion in the *C9orf72* gene. First, we performed redox signature profiling of leukocyte RNA from *C9orf72* patients, and found a dysregulation in genes involved with defense against oxidative damage compared to healthy volunteers. Secondly, we found that Arg-DPRs present in *C9orf72* patients trigger an increase in ROS at the mitochondrial level, this being coupled with a reduction in mitochondrial membrane potential in the NSC34 cell line. Interestingly, the levels of *FoxM1* and *Trappa6c* mRNA were reduced in C9orf72 patients' leukocytes and in Arg-DPRs mouse NSC34 (data not shown). Thirdly, we described an impairment in the activation of the transcription factor NRF2 in the presence of Arg-DPRs. Finally, we describe how activation of the NRF2 transcriptional response diminishes Arg-DPR toxicity, suggesting that the activation of NRF2 in *C9orf72* may represent a potential therapeutic target.

In this study, we reported the dysregulation of ten key transcripts related to redox biology in leukocytes from *C9orf72* ALS patients compared to healthy controls for first the time. The dysregulated mRNAs could be grouped into those coding for antioxidant enzymes including GSR and peroxidases (GPX3 and DUOX2) and proteins related to super-oxide metabolism (CCS and CYBB). The downregulation of CSS, a chaperone that delivers Cu2+ ions to the SOD1 protein, and up-regulation of CYBB, a membrane-bound subunit of NADPH oxidase, could potentially contribute to superoxide accumulation [37]. Hydrogen peroxide or hydroperoxides are reduced by GPX3 activity, leading to the production of oxidized glutathione (GSSG), which in turn is recycled into GSH by GSR. Adequate expression levels of GPX3 and GSR are crucial to maintain antioxidant defense and the glutathione pool and, therefore, to protect cells from oxidative damage. Moreover, we found a diminished expression of three transcripts encoded by genes whose expression is activated by ROS (DHCR24, FOXM1 and NUDT1) and two transcripts related to ROS-activated pathways (FHL2 and TRAPPC6A). Taken together, these data indicate that the antioxidant response of *C9orf72* patients is impaired compared with control subjects. These data are in agreement with previous reports which described redox dysregulation in ALS. Recently, a meta-analysis which included 41 studies with 4588 ALS patients and 6344 control individuals analyzing 15 oxidative stress markers from the blood suggests that MDA (lipid peroxidation end product), 8-OHdG (a marker for DNA damage), and AOPP (Advanced Oxidation Protein Product) were significantly elevated in the blood of ALS patients when compared with control individuals. On the contrary, this study reveals that the levels of antioxidant glutathione and uric acid were significantly downregulated in patients with ALS [38]. Future research will address whether this impairment is also measurable in leukocytes from patients with sporadic ALS or other types of familial ALS.

Although Arg-DPRs can lead to reduced mitochondrial membrane potential through different routes, the connection of mitochondrial damage, dysregulated ROS and poly-repeat dipeptides is supported by other studies. In fact, zebrafish embryos injected with poly-GR exhibited enhanced levels of mitochondrial superoxide generation [39]. ROS levels were also increased in *C9orf72*-derived neurons in an age-dependent manner. Pharmacological or genetic reduction in oxidative stress partially rescued DNA damage in *C9orf72* neurons and control neurons expressing GR_{80} [40]. Samples from the frontal cortex of mice expressing GR_{80} showed decreased activities of mitochondrial electron chain complexes I and V [41]. Closer to human pathology, higher levels of ROS and susceptibility to oxidative damage have been found in iPSC-derived astrocytes from patients [42] and skeletal myocytes [43], respectively.

The dysfunction in the oxidative stress response prompted us to analyze the response of the transcription factor NRF2, known to regulate several antioxidant enzymes and pathways related to redox metabolism. NRF2 dysfunction has already been found to be present in different ALS settings [44–47]. However, the impact of poly-dipeptide repeat proteins has not been addressed so far. In this regard, the activation of NRF2 and its transcriptional targets by DMF was partially impaired by poly-dipeptide overexpression. This fact directly connects the imbalanced ROS levels with an impaired antioxidant response in the context of Arg-DPRs. In this study, we provide evidence that HO-1 expression is not sufficiently induced in the presence of poly-repeat dipeptides, connecting for the first time HO-1 and *C9orf72* pathology. HO-1 degrades prooxidant heme into equimolar quantities of carbon monoxide, biliverdin, and iron. Biliverdin is immediately converted into bilirubin by biliverdin reductase, while iron is oxidized and sequestered by ferritin. Despite the well-known antioxidant properties of bilirubin, a novel role for HO-1 has recently been proposed of protection against oxidative injury by regulating mitochondrial quality control [48]. The expression of other NRF2-target genes, such as glyoxylase 1, has been shown to be reduced in neurons and astrocytes derived from fibroblasts of *C9orf72*-patients [49]. This correlation could indicate that the activation of NRF2 is hampered by Arg-DPRs in ALS patients, helping to explain the results recently disclosed from a clinical trial aiming to evaluate the safety and efficacy of DMF in ALS [50]. In this study, there was no significant improvement in the primary endpoint (Amyotrophic Lateral Sclerosis Functional Rating Scale-Revised), although there was a mild improvement in the patient's neurophysiological index. In this case, a question arises. Would NRF2 dysfunction in ALS be sufficient to hinder the NRF2 cytoprotective response? We already demonstrated some protection from oxidative stress using DMF in the present study. Given that the impairment of NRF2 by Arg-DPRs is only partial, activation of the NRF2 pathway through more potent inducers could effectively protect against Arg-DPRs toxicity. Importantly, robust NRF2 overexpression protected NSC34 cells from the toxicity induced by Arg-DPRs. Therefore, a window for cytoprotection upon NRF2 stimulation may still be achievable, but further in vivo studies are needed.

Taking into consideration that free newly produced NRF2 is the main functional NRF2 species, synthesis defects likely contribute to its impaired activation, and even more so considering its short half-life. Indeed, brusatol, a drug used as a classical NRF2 inhibitor, exerts its effects through global protein synthesis inhibition [51]. In this study, we provide evidence in a motor neuron cell line that poly-dipeptide repeats downregulate global translation in agreement with previous reports employing HeLa cells or Drosophila neurons [52,53]. In basal conditions, the balance between synthesis and degradation defines the levels of the NRF2 protein. Upon proteasomal inhibition, NRF2 accumulation occurs to a lesser extent when Arg-DPRs are overexpressed, suggesting a significant impact of general translational shutdown upon NRF2 activation.

5. Conclusions

In this study, we characterized redox and mitochondrial disturbances in *C9orf72*-ALS together with impairment of the activation of NRF2. We propose that the general translation shutdown induced by Arg-DPRs reduces NRF2 synthesis, linking general cellular events in *C9orf72*-related ALS with specific molecular consequences. This work serves to underline the importance of NRF2 in *C9orf72*-related ALS and as a proof of concept to emphasize the importance of its activation as a potential therapeutic target, since Arg-DPRs toxicity can be prevented by NRF2 overexpression.

Supplementary Materials: The following supporting information can be downloaded at: https://www.mdpi.com/article/10.3390/antiox11101897/s1; Figure S1: Analysis of Arg-DPRs on cell cycle and survival of NSC34 cell line, Figure S2: Overexpression of NRF2$^{\Delta ETGE}$ through lentiviral delivery; Table S1: RT^2 Profiler™ PCR Array Human Oxidative Stress Plus, PAHS-065Y, Qiagen; Table S2: Complete descriptive statistics and statistical analysis of redox-focused profiling in *C9orf72* ALS-patients vs. healthy controls.

Author Contributions: Conceptualization, J.J.-V. and A.I.R.; Formal analysis, J.J.-V. and A.I.R.; Funding acquisition, S.C., P.J.S., A.C. and A.I.R.; Investigation, J.J.-V., J.K., A.M. (Ana Mata), S.C. and A.I.R.;

Methodology, J.J.-V., J.K., A.M. (Ana Mata), M.R.T., A.M. (Andrea Malaspina) and P.J.S.; Supervision, A.I.R.; Writing—original draft, J.J.-V. and A.I.R.; Writing—review & editing, J.J.-V., J.K., A.M. (Ana Mata), S.C., M.R.T., A.M. (Andrea Malaspina), P.J.S., A.C. and A.I.R. Provision of genetically characterized human biosamples through the AMBRoSIA programme (P.J.S., J.K., M.R.T. and A.M.). All authors have read and agreed to the published version of the manuscript.

Funding: This work was supported by the SAF2019-110061RB-100 grant from the Spanish Ministry of Economy and Competitiveness; P-024-FTPGB 2018 from the Spanish Foundation named "Tatiana de Guzman el Bueno", COEN-2018-Pathfinder phase III financed by the Network of Centers of Excellence in Neurodegeneration including the UK Medical Research Council and PI19/01030 financed by the Spanish "Instituto de Salud Carlos III". J.J.V. hold an FPU PhD fellowship from the Spanish Ministry of Universities (FPU20/03326) and was a holder of a master's degree fellowship ("Ayudas para el fomento de la Investigación en Estudios de Máster-UAM 2020") funded by the Autonomous University of Madrid. This research was also supported by the Motor Neurone Disease Association (AMBRoSIA 972-797 and NECTAR 974-797 awards) and by the NIHR Sheffield Biomedical Research Centre (JS-BRC-1215-20017). This article is based upon work from COST Action CA20121, supported by COST (European Cooperation in Science and Technology).

Institutional Review Board Statement: The study was conducted in accordance with the Declaration of Helsinki, and the protocol was approved by the London-South East Research Ethics Committee (16/LO/2136).

Informed Consent Statement: Informed consent was obtained from all subjects involved in the study.

Data Availability Statement: Data is contained within the article and supplementary material.

Conflicts of Interest: The authors declare no conflict of interest.

References

1. Renton, A.E.; Majounie, E.; Waite, A.; Simón-Sánchez, J.; Rollinson, S.; Gibbs, J.R.; Schymick, J.C.; Laaksovirta, H.; van Swieten, J.C.; Myllykangas, L.; et al. A Hexanucleotide Repeat Expansion in C9ORF72 Is the Cause of Chromosome 9p21-Linked ALS-FTD. *Neuron* **2011**, *72*, 257–268. [CrossRef] [PubMed]
2. Wijesekera, L.C.; Leigh, P.N. Amyotrophic Lateral Sclerosis. *Orphanet J. Rare Dis.* **2009**, *4*, 3. [CrossRef]
3. Riviere, M.; Meininger, V.; Zeisser, P.; Munsat, T. An Analysis of Extended Survival in Patients with Amyotrophic Lateral Sclerosis Treated with Riluzole. *Arch. Neurol.* **1998**, *55*, 526–528. [CrossRef] [PubMed]
4. Yoshino, H.; Kimura, A. Investigation of the Therapeutic Effects of Edaravone, a Free Radical Scavenger, on Amyotrophic Lateral Sclerosis (Phase II Study). *Amyotroph. Lateral Scler. Off. Publ. World Fed. Neurol. Res. Group Mot. Neuron Dis.* **2006**, *7*, 241–245. [CrossRef] [PubMed]
5. Petrov, D.; Mansfield, C.; Moussy, A.; Hermine, O. ALS Clinical Trials Review: 20 Years of Failure. Are We Any Closer to Registering a New Treatment? *Front. Aging Neurosci.* **2017**, *9*, 68. [CrossRef] [PubMed]
6. Xi, Z.; Zinman, L.; Moreno, D.; Schymick, J.; Liang, Y.; Sato, C.; Zheng, Y.; Ghani, M.; Dib, S.; Keith, J.; et al. Hypermethylation of the CpG Island near the G4C2 Repeat in ALS with a C9orf72 Expansion. *Am. J. Hum. Genet.* **2013**, *92*, 981–989. [CrossRef]
7. Sudria-Lopez, E.; Koppers, M.; de Wit, M.; van der Meer, C.; Westeneng, H.-J.; Zundel, C.A.C.; Youssef, S.A.; Harkema, L.; de Bruin, A.; Veldink, J.H.; et al. Full Ablation of C9orf72 in Mice Causes Immune System-Related Pathology and Neoplastic Events but No Motor Neuron Defects. *Acta Neuropathol.* **2016**, *132*, 145–147. [CrossRef]
8. Mori, K.; Weng, S.-M.; Arzberger, T.; May, S.; Rentzsch, K.; Kremmer, E.; Schmid, B.; Kretzschmar, H.A.; Cruts, M.; Van Broeckhoven, C.; et al. The C9orf72 GGGGCC Repeat Is Translated into Aggregating Dipeptide-Repeat Proteins in FTLD/ALS. *Science* **2013**, *339*, 1335–1338. [CrossRef]
9. Mori, K.; Arzberger, T.; Grässer, F.A.; Gijselinck, I.; May, S.; Rentzsch, K.; Weng, S.-M.; Schludi, M.H.; van der Zee, J.; Cruts, M.; et al. Bidirectional Transcripts of the Expanded C9orf72 Hexanucleotide Repeat Are Translated into Aggregating Dipeptide Repeat Proteins. *Acta Neuropathol.* **2013**, *126*, 881–893. [CrossRef]
10. Freibaum, B.D.; Taylor, J.P. The Role of Dipeptide Repeats in C9ORF72-Related ALS-FTD. *Front. Mol. Neurosci.* **2017**, *10*, 35. [CrossRef]
11. Mizielinska, S.; Grönke, S.; Niccoli, T.; Ridler, C.E.; Clayton, E.L.; Devoy, A.; Moens, T.; Norona, F.E.; Woollacott, I.O.C.; Pietrzyk, J.; et al. C9orf72 repeat expansions cause neurodegeneration in Drosophila through arginine-rich proteins. *Science.* **2014**, 3451192–3451194. [CrossRef] [PubMed]
12. Radwan, M.; Ang, C.-S.; Ormsby, A.R.; Cox, D.; Daly, J.C.; Reid, G.E.; Hatters, D.M. Arginine in C9ORF72 Dipolypeptides Mediates Promiscuous Proteome Binding and Multiple Modes of Toxicity. *Mol. Cell. Proteom. MCP* **2020**, *19*, 640–654. [CrossRef] [PubMed]
13. Barber, S.C.; Shaw, P.J. Oxidative Stress in ALS: Key Role in Motor Neuron Injury and Therapeutic Target. *Free Radic. Biol. Med.* **2010**, *48*, 629–641. [CrossRef] [PubMed]

14. Obrador, E.; Salvador-Palmer, R.; López-Blanch, R.; Jihad-Jebbar, A.; Vallés, S.L.; Estrela, J.M. The Link between Oxidative Stress, Redox Status, Bioenergetics and Mitochondria in the Pathophysiology of ALS. *Int. J. Mol. Sci.* **2021**, *22*, 6352. [CrossRef] [PubMed]

15. Rada, P.; Rojo, A.I.; Evrard-Todeschi, N.; Innamorato, N.G.; Cotte, A.; Jaworski, T.; Tobón-Velasco, J.C.; Devijver, H.; García-Mayoral, M.F.; Van Leuven, F.; et al. Structural and Functional Characterization of Nrf2 Degradation by the Glycogen Synthase Kinase 3/β-TrCP Axis. *Mol. Cell. Biol.* **2012**, *32*, 3486–3499. [CrossRef] [PubMed]

16. McMahon, M.; Thomas, N.; Itoh, K.; Yamamoto, M.; Hayes, J.D. Redox-Regulated Turnover of Nrf2 Is Determined by at Least Two Separate Protein Domains, the Redox-Sensitive Neh2 Degron and the Redox-Insensitive Neh6 Degron. *J. Biol. Chem.* **2004**, *279*, 31556–31567. [CrossRef] [PubMed]

17. Itoh, K.; Wakabayashi, N.; Katoh, Y.; Ishii, T.; Igarashi, K.; Engel, J.D.; Yamamoto, M. Keap1 Represses Nuclear Activation of Antioxidant Responsive Elements by Nrf2 through Binding to the Amino-Terminal Neh2 Domain. *Genes Dev.* **1999**, *13*, 76–86. [CrossRef]

18. Dinkova-Kostova, A.T.; Holtzclaw, W.D.; Cole, R.N.; Itoh, K.; Wakabayashi, N.; Katoh, Y.; Yamamoto, M.; Talalay, P. Direct evidence that sulfhydryl groups of Keap1 are the sensors regulating induction of phase 2 enzymes that protect against carcinogens and oxidants. *Proc Natl Acad Sci USA* **2002**, *99*, 11908–11913. [CrossRef]

19. Kobayashi, A.; Kang, M.-I.; Watai, Y.; Tong, K.I.; Shibata, T.; Uchida, K.; Yamamoto, M. Oxidative and Electrophilic Stresses Activate Nrf2 through Inhibition of Ubiquitination Activity of Keap1. *Mol. Cell. Biol.* **2006**, *26*, 221–229. [CrossRef]

20. Eggler, A.L.; Small, E.; Hannink, M.; Mesecar, A.D. Cul3-Mediated Nrf2 Ubiquitination and Antioxidant Response Element (ARE) Activation Are Dependent on the Partial Molar Volume at Position 151 of Keap1. *Biochem. J.* **2009**, *422*, 171–180. [CrossRef]

21. Cuadrado, A.; Manda, G.; Hassan, A.; Alcaraz, M.J.; Barbas, C.; Daiber, A.; Ghezzi, P.; León, R.; López, M.G.; Oliva, B.; et al. Transcription Factor NRF2 as a Therapeutic Target for Chronic Diseases: A Systems Medicine Approach. *Pharmacol. Rev.* **2018**, *70*, 348–383. [CrossRef] [PubMed]

22. Shepheard, S.R.; Parker, M.D.; Cooper-Knock, J.; Verber, N.S.; Tuddenham, L.; Heath, P.; Beauchamp, N.; Place, E.; Sollars, E.S.A.; Turner, M.R.; et al. Value of Systematic Genetic Screening of Patients with Amyotrophic Lateral Sclerosis. *J. Neurol. Neurosurg. Psychiatry* **2021**, *92*, 510–518. [CrossRef] [PubMed]

23. Stopford, M.J.; Higginbottom, A.; Hautbergue, G.M.; Cooper-Knock, J.; Mulcahy, P.J.; De Vos, K.J.; Renton, A.E.; Pliner, H.; Calvo, A.; Chio, A.; et al. C9ORF72 Hexanucleotide Repeat Exerts Toxicity in a Stable, Inducible Motor Neuronal Cell Model, Which Is Rescued by Partial Depletion of Pten. *Hum. Mol. Genet.* **2017**, *26*, 1133–1145. [CrossRef] [PubMed]

24. Schindelin, J.; Arganda-Carreras, I.; Frise, E.; Kaynig, V.; Longair, M.; Pietzsch, T.; Preibisch, S.; Rueden, C.; Saalfeld, S.; Schmid, B.; et al. Fiji: An Open-Source Platform for Biological-Image Analysis. *Nat. Methods* **2012**, *9*, 676–682. [CrossRef] [PubMed]

25. Feng, Y.; Wang, X. Antioxidant Therapies for Alzheimer's Disease. *Oxid. Med. Cell. Longev.* **2012**, *2012*, 472932. [CrossRef]

26. Jha, N.K.; Jha, S.K.; Kar, R.; Nand, P.; Swati, K.; Goswami, V.K. Nuclear Factor-Kappa β as a Therapeutic Target for Alzheimer's Disease. *J. Neurochem.* **2019**, *150*, 113–137. [CrossRef]

27. Kwon, I.; Xiang, S.; Kato, M.; Wu, L.; Theodoropoulos, P.; Wang, T.; Kim, J.; Yun, J.; Xie, Y.; McKnight, S.L. Poly-Dipeptides Encoded by the C9orf72 Repeats Bind Nucleoli, Impede RNA Biogenesis, and Kill Cells. *Science* **2014**, *345*, 1139–1145. [CrossRef]

28. Kanekura, K.; Harada, Y.; Fujimoto, M.; Yagi, T.; Hayamizu, Y.; Nagaoka, K.; Kuroda, M. Characterization of Membrane Penetration and Cytotoxicity of C9orf72-Encoding Arginine-Rich Dipeptides. *Sci. Rep.* **2018**, *8*, 12740. [CrossRef]

29. Kanekura, K.; Yagi, T.; Cammack, A.J.; Mahadevan, J.; Kuroda, M.; Harms, M.B.; Miller, T.M.; Urano, F. Poly-Dipeptides Encoded by the C9ORF72 Repeats Block Global Protein Translation. *Hum. Mol. Genet.* **2016**, *25*, 1803–1813. [CrossRef]

30. Gill, A.L.; Wang, M.Z.; Levine, B.; Premasiri, A.; Vieira, F.G. Primary Neurons and Differentiated NSC-34 Cells Are More Susceptible to Arginine-Rich ALS Dipeptide Repeat Protein-Associated Toxicity than Non-Differentiated NSC-34 and CHO Cells. *Int. J. Mol. Sci.* **2019**, *20*, 6238. [CrossRef]

31. Stewart, D.; Killeen, E.; Naquin, R.; Alam, S.; Alam, J. Degradation of Transcription Factor Nrf2 via the Ubiquitin-Proteasome Pathway and Stabilization by Cadmium. *J. Biol. Chem.* **2003**, *278*, 2396–2402. [CrossRef] [PubMed]

32. Snowden, J.S.; Rollinson, S.; Thompson, J.C.; Harris, J.M.; Stopford, C.L.; Richardson, A.M.T.; Jones, M.; Gerhard, A.; Davidson, Y.S.; Robinson, A.; et al. Distinct Clinical and Pathological Characteristics of Frontotemporal Dementia Associated with C9ORF72 Mutations. *Brain J. Neurol.* **2012**, *135*, 693–708. [CrossRef] [PubMed]

33. Baird, L.; Llères, D.; Swift, S.; Dinkova-Kostova, A.T. Regulatory Flexibility in the Nrf2-Mediated Stress Response Is Conferred by Conformational Cycling of the Keap1-Nrf2 Protein Complex. *Proc. Natl. Acad. Sci. USA* **2013**, *110*, 15259–15264. [CrossRef] [PubMed]

34. Poganik, J.R.; Long, M.J.C.; Disare, M.T.; Liu, X.; Chang, S.-H.; Hla, T.; Aye, Y. Post-Transcriptional Regulation of Nrf2-MRNA by the MRNA-Binding Proteins HuR and AUF1. *FASEB J.* **2019**, *33*, 14636–14652. [CrossRef] [PubMed]

35. Schmidt, E.K.; Clavarino, G.; Ceppi, M.; Pierre, P. SUnSET, a Nonradioactive Method to Monitor Protein Synthesis. *Nat. Methods* **2009**, *6*, 275–277. [CrossRef] [PubMed]

36. Lee, K.-H.; Zhang, P.; Kim, H.J.; Mitrea, D.; Sarkar, M.; Freibaum, B.; Cika, J.; Coughlin, M.; Messing, J.; Molliex, A.; et al. C9orf72 Dipeptide Repeats Impair the Assembly, Dynamics and Function of Membrane-Less Organelles. *Cell* **2016**, *167*, 774–788.e17. [CrossRef] [PubMed]

37. Wong, P.C.; Waggoner, D.; Subramaniam, J.R.; Tessarollo, L.; Bartnikas, T.B.; Culotta, V.C.; Price, D.L.; Rothstein, J.; Gitlin, J.D. Copper Chaperone for Superoxide Dismutase Is Essential to Activate Mammalian Cu/Zn Superoxide Dismutase. *Proc. Natl. Acad. Sci. USA* **2000**, *97*, 2886–2891. [CrossRef] [PubMed]
38. Cenini, G.; Lloret, A.; Cascella, R. Oxidative Stress and Mitochondrial Damage in Neurodegenerative Diseases: From Molecular Mechanisms to Targeted Therapies. *Oxid. Med. Cell. Longev.* **2020**, *2020*, 1270256. [CrossRef]
39. Riemslagh, F.W.; Verhagen, R.F.M.; van der Toorn, E.C.; Smits, D.J.; Quint, W.H.; van der Linde, H.C.; van Ham, T.J.; Willemsen, R. Reduction of Oxidative Stress Suppresses Poly-GR-Mediated Toxicity in Zebrafish Embryos. *Dis. Model. Mech.* **2021**, *14*, dmm049092. [CrossRef]
40. Lopez-Gonzalez, R.; Lu, Y.; Gendron, T.F.; Karydas, A.; Tran, H.; Yang, D.; Petrucelli, L.; Miller, B.L.; Almeida, S.; Gao, F.-B. Poly(GR) in C9ORF72-Related ALS/FTD Compromises Mitochondrial Function and Increases Oxidative Stress and DNA Damage in IPSC-Derived Motor Neurons. *Neuron* **2016**, *92*, 383–391. [CrossRef]
41. Choi, S.Y.; Lopez-Gonzalez, R.; Krishnan, G.; Phillips, H.L.; Li, A.N.; Seeley, W.W.; Yao, W.-D.; Almeida, S.; Gao, F.-B. C9ORF72-ALS/FTD-Associated Poly(GR) Binds Atp5a1 and Compromises Mitochondrial Function in Vivo. *Nat. Neurosci.* **2019**, *22*, 851–862. [CrossRef] [PubMed]
42. Birger, A.; Ben-Dor, I.; Ottolenghi, M.; Turetsky, T.; Gil, Y.; Sweetat, S.; Perez, L.; Belzer, V.; Casden, N.; Steiner, D.; et al. Human IPSC-Derived Astrocytes from ALS Patients with Mutated C9ORF72 Show Increased Oxidative Stress and Neurotoxicity. *eBioMedicine* **2019**, *50*, 274–289. [CrossRef] [PubMed]
43. Lynch, E.; Semrad, T.; Belsito, V.S.; FitzGibbons, C.; Reilly, M.; Hayakawa, K.; Suzuki, M. C9ORF72-Related Cellular Pathology in Skeletal Myocytes Derived from ALS-Patient Induced Pluripotent Stem Cells. *Dis. Model. Mech.* **2019**, *12*, dmm039552. [CrossRef]
44. Sarlette, A.; Krampfl, K.; Grothe, C.; Neuhoff, N.; von Dengler, R.; Petri, S. Nuclear Erythroid 2-Related Factor 2-Antioxidative Response Element Signaling Pathway in Motor Cortex and Spinal Cord in Amyotrophic Lateral Sclerosis. *J. Neuropathol. Exp. Neurol.* **2008**, *67*, 1055–1062. [CrossRef] [PubMed]
45. Deng, Z.; Lim, J.; Wang, Q.; Purtell, K.; Wu, S.; Palomo, G.M.; Tan, H.; Manfredi, G.; Zhao, Y.; Peng, J.; et al. ALS-FTLD-Linked Mutations of SQSTM1/P62 Disrupt Selective Autophagy and NFE2L2/NRF2 Anti-Oxidative Stress Pathway. *Autophagy* **2020**, *16*, 917–931. [CrossRef] [PubMed]
46. Wang, F.; Lu, Y.; Qi, F.; Su, Q.; Wang, L.; You, C.; Che, F.; Yu, J. Effect of the Human SOD1-G93A Gene on the Nrf2/ARE Signaling Pathway in NSC-34 Cells. *Mol. Med. Rep.* **2014**, *9*, 2453–2458. [CrossRef]
47. Moujalled, D.; Grubman, A.; Acevedo, K.; Yang, S.; Ke, Y.D.; Moujalled, D.M.; Duncan, C.; Caragounis, A.; Perera, N.D.; Turner, B.J.; et al. TDP-43 Mutations Causing Amyotrophic Lateral Sclerosis Are Associated with Altered Expression of RNA-Binding Protein HnRNP K and Affect the Nrf2 Antioxidant Pathway. *Hum. Mol. Genet.* **2017**, *26*, 1732–1746. [CrossRef]
48. Hull, T.D.; Boddu, R.; Guo, L.; Tisher, C.C.; Traylor, A.M.; Patel, B.; Joseph, R.; Prabhu, S.D.; Suliman, H.B.; Piantadosi, C.A.; et al. Heme Oxygenase-1 Regulates Mitochondrial Quality Control in the Heart. *JCI Insight* **2016**, *1*, e85817. [CrossRef]
49. Allen, S.P.; Hall, B.; Woof, R.; Francis, L.; Gatto, N.; Shaw, A.C.; Myszczynska, M.; Hemingway, J.; Coldicott, I.; Willcock, A.; et al. C9orf72 Expansion within Astrocytes Reduces Metabolic Flexibility in Amyotrophic Lateral Sclerosis. *Brain J. Neurol.* **2019**, *142*, 3771–3790. [CrossRef]
50. Vucic, S.; Henderson, R.D.; Mathers, S.; Needham, M.; Schultz, D.; Kiernan, M.C. Safety and Efficacy of Dimethyl Fumarate in ALS: Randomised Controlled Study. *Ann. Clin. Transl. Neurol.* **2021**, *8*, 1991–1999. [CrossRef]
51. Harder, B.; Tian, W.; La Clair, J.J.; Tan, A.-C.; Ooi, A.; Chapman, E.; Zhang, D.D. Brusatol Overcomes Chemoresistance through Inhibition of Protein Translation. *Mol. Carcinog.* **2017**, *56*, 1493–1500. [CrossRef] [PubMed]
52. Vanneste, J.; Vercruysse, T.; Boeynaems, S.; Sicart, A.; Van Damme, P.; Daelemans, D.; Van Den Bosch, L. C9orf72-Generated Poly-GR and Poly-PR Do Not Directly Interfere with Nucleocytoplasmic Transport. *Sci. Rep.* **2019**, *9*, 15728. [CrossRef] [PubMed]
53. Moens, T.G.; Niccoli, T.; Wilson, K.M.; Atilano, M.L.; Birsa, N.; Gittings, L.M.; Holbling, B.V.; Dyson, M.C.; Thoeng, A.; Neeves, J.; et al. C9orf72 Arginine-Rich Dipeptide Proteins Interact with Ribosomal Proteins in Vivo to Induce a Toxic Translational Arrest That Is Rescued by EIF1A. *Acta Neuropathol.* **2019**, *137*, 487–500. [CrossRef] [PubMed]

 *antioxidants*

MDPI

Review

The Nrf2 in Obesity: A Friend or Foe?

Yudong Xia, Xiaoying Zhai, Yanning Qiu, Xuemei Lu and Yi Jiao *

State Key Laboratory of Pathogenesis, Prevention and Treatment of High Incidence Diseases in Central Asia, Xinjiang Key Laboratory of Molecular Biology for Endemic Diseases, Department of Biochemistry and Molecular Biology, School of Basic Medical Sciences, Xinjiang Medical University, Urumqi 830011, China
* Correspondence: jymiranda@xjmu.edu.cn

Abstract: Obesity and its complications have become serious global health concerns recently and increasing work has been carried out to explicate the underlying mechanism of the disease development. The recognized correlations suggest oxidative stress and inflammation in expanding adipose tissue with excessive fat accumulation play important roles in the pathogenesis of obesity, as well as its associated metabolic syndromes. In adipose tissue, obesity-mediated insulin resistance strongly correlates with increased oxidative stress and inflammation. Nuclear factor erythroid 2-related factor 2 (Nrf2) has been described as a key modulator of antioxidant signaling, which regulates the transcription of various genes coding antioxidant enzymes and cytoprotective proteins. Furthermore, an increasing number of studies have demonstrated that Nrf2 is a pivotal target of obesity and its related metabolic disorders. However, its effects are controversial and even contradictory. This review aims to clarify the complicated interplay among Nrf2, oxidative stress, lipid metabolism, insulin signaling and chronic inflammation in obesity. Elucidating the implications of Nrf2 modulation on obesity would provide novel insights for potential therapeutic approaches in obesity and its comorbidities.

Keywords: Nrf2; obesity; lipid metabolism; chronic inflammation; oxidative stress

Citation: Xia, Y.; Zhai, X.; Qiu, Y.; Lu, X.; Jiao, Y. The Nrf2 in Obesity: A Friend or Foe? *Antioxidants* **2022**, *11*, 2067. https://doi.org/10.3390/antiox11102067

Academic Editor: Marcel Bonay

Received: 8 October 2022
Accepted: 17 October 2022
Published: 20 October 2022

Publisher's Note: MDPI stays neutral with regard to jurisdictional claims in published maps and institutional affiliations.

1. Introduction

Obesity, characterized by excess adiposity, has become a major public health problem worldwide which adversely affects mental and physical health and imposes a substantial socio-economic burden [1,2]. The prevalence of obesity has doubled in more than 70 countries around the world and has constantly increased since 1980 [3]. Moreover, health issues resulting from overweight and/or obese have affected more than 2 billion people [4]. Obesity increases the risk for an extensive range of diseases, including type 2 diabetes mellitus (T2DM) [5], ischemic heart disease [6,7], hypertension [8] and certain types of cancer [9]. Obesity has been positively correlated with higher all-cause mortality risk [10,11]. What is more, the evidence cooperatively indicates that obesity accelerates ageing [12].

Although a lot of research has focused on obesity and its comorbidities, the underlying mechanisms have not yet been made fully clear. In recent years, the crucial role of obesity in the development of T2DM has become progressively fair due to their association with insulin resistance (IR). The oblique common cause of IR is obesity, while IR is the pathogenic basis of obesity and T2DM [13–15]. Interestingly, obesity-induced IR is strongly associated with chronic low-grade inflammation in white adipose tissue (WAT). Both in vitro and in vivo studies have suggested a relationship between WAT inflammation and the progression of T2DM [16–18]. However, the mechanism linking obesity and WAT inflammation with IR has not been entirely explicated.

Growing studies have shown that oxidative stress, which is defined by excessive endogenous reactive oxygen species (ROS) levels [19,20], plays a critical role in the pathogenesis and development of obesity and T2DM. An extended hyperglycemic condition leads to a periodic and sustained increase in ROS levels [21,22], and elevates the levels of free fatty acids (FFA) released by WAT of obese individuals, promoting ROS production [22,23]. In turn, the hyperglycemia-mediated ROS overproduction and the increased ROS generation within obese WAT leads to the secretion of metabolically adverse adipocytokine patterns, and provokes inflammatory responses, contributing to IR through the interaction between inflammatory molecules (certain adipocytokines) and insulin signaling [16,17,24]. Collectively, they may serve both as the cause and result of each other, thus forming a vicious circle.

Maintaining redox balance depends on a powerful antioxidant system that counteracts oxidative stress. The nuclear factor erythroid 2-related factor 2 (Nrf2), belonging to the family of the cap n' collar transcription factors, is a key regulator of the cellular response to oxidative stress by controlling the expression of antioxidant and detoxification enzymes to eliminate excess ROS [25–27]. Nrf2 is constitutively expressed under homeostatic conditions, localized in the cytoplasm, which binds to the Kelch-like ECH-associated protein 1 (Keap1), facilitating the ubiquitination and proteasomal degradation [28]. Under oxidative stress conditions, certain reactive cysteines in Keap1 are modified by ROS, and then Nrf2 dissociates from the Keap1 and translocates to the nucleus [29–31]. Subsequently, Nrf2 binds to antioxidant response element (ARE; GTGACNNNGC) present in the regulatory region of the Nrf2 target genes that codes antioxidant and detoxification proteins, thereby inducing the gene transcription [32,33] (Figure 1).

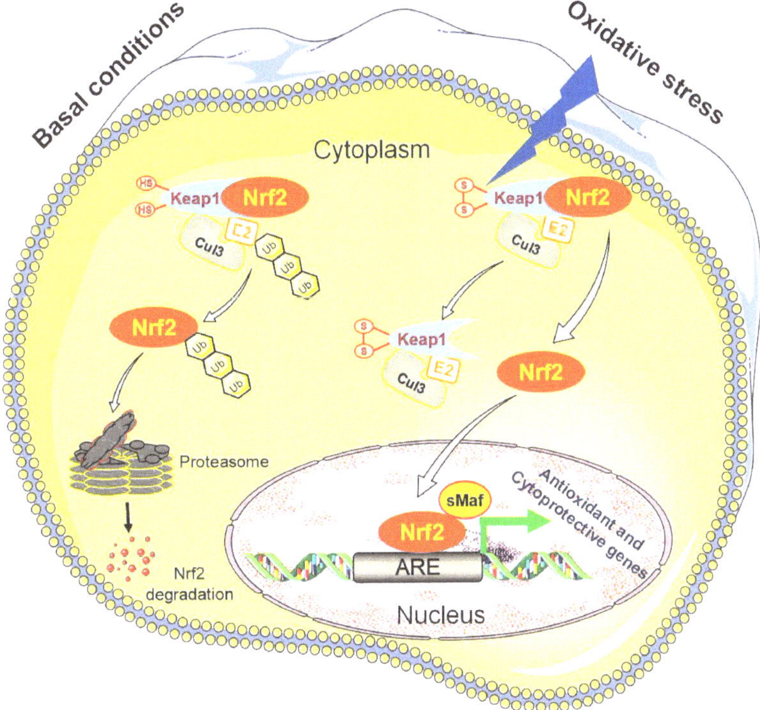

Figure 1. Schematic diagram of Keap1-Nrf2-ARE signaling pathway. Under basal non-stressed conditions, Keap1 binds to cullin 3 (Cul3)-based E3 ubiquitin ligases, forming Keap1-Cul3-Nrf2 complex, leading to Nrf2 ubiquitination and proteasomal degradation. In response to oxidative stress, some cysteine residues of Keap1 are oxidized, resulting in a conformation change in Keap1 and disruption of the Keap1–Cul3 interaction. Nrf2 then dissociates from Keap1 and translocates to the nucleus, where it forms a heterodimer with small Maf proteins (sMaf), binds to ARE and stimulates the transcription of antioxidant and cytoprotective genes.

Nrf2 is like a coin with two sides. Although Nrf2 activation is an important antioxidant modulator in normal cells, this effect is also one of the main factors involved in the enhanced resistance to chemotherapy and radiotherapy in cancer cells since many studies are focused to reduce Nrf2 activation in chemoresistant and radioresistant cancer cells [34–36]. More remarkably, Nrf2 activation by antidiabetic agents has been reported to accelerate tumor metastasis [37].

It has certainly been acknowledged that Nrf2 is indispensable for the regulation of inflammatory responses [18,38,39]. Given that obesity has been closely associated with inflammation and oxidative stress, the potential protective function of Nrf2 is of great interest. However, the effects of Nrf2 are controversial and even contradictory. Table 1 summarizes the recent findings presented on the different roles of Nrf2 signaling in animal models of diet-induced obesity (DIO).

Table 1. Different roles of Nrf2 signaling in animal models of diet-induced obesity (DIO).

Mice	Age	Body Weight (BW)	Diet	Treatment	Impaired Glucose Homeostasis	Obesity	References
C57; WT	5 weeks	17–20 g	45% fat, 28 weeks	HFD plus oltipraz (0.75 g/kg diet)	↓	↓	[40]
C57; WT	6–7 weeks	16–18 g	60% fat, 21 days or 95 days	Oral gavage with CDDO-Im (30 μmol/kg BW) 3 times per week throughout the feeding period	Unreported	↓	[41]
C57; WT	7 weeks	20–25 g	60% fat, 14 weeks	HFD containing 0.3% glucoraphanin	↓	↓	[42]
C57; Nrf2-HeNKO, Nrf2-ANKO	8 weeks	25–30 g	60% fat, 170 days	Nrf2 KO in hepatocytes (HeNKO) or adipocytes (ANKO)	Nrf2-HeNKO↓, Nrf2-ANKO↑	Nrf2-HeNKO↓, Nrf2-ANKO↑	[43]
C57; Nrf2-KO, Keap1-KD	12 weeks	25–30 g	39.7% fat, 12 weeks	Nrf2-KO, Keap1-KD	Nrf2-KO↓, Keap1-KD↑	Unaltered	[44]
C57; Nrf2-KO, Nrf2-mKO	8–12 weeks	22–27 g	60% fat, 10 weeks	Nrf2-KO, myeloid Nrf2-KO (Nrf2-mKO)	Nrf2-KO↓, Nrf2-mKO Unaltered	Nrf2-KO↓, Nrf2-mKO Unaltered	[45]
C57; Nrf2-KO	4 weeks	~20 g	41% fat, 16 weeks	Nrf2-KO	Unreported	↓	[46]
C57; WT	Unreported	12–14 g	60% fat, 12 weeks	Oral gavage with 5 mg Hydroxytyrosol/kg BW/day	Improve the WAT dysfunction		[47]
C57; WT	8 weeks	15–18 g	60% fat, 16 weeks + 15 days	Curcumin was given daily by oral gavage at the dose of 50 mg/kg BW for 15 days	↓	Unaltered	[48]

Table 1. *Cont.*

Mice	Age	Body Weight (BW)	Diet	Treatment	Impaired Glucose Homeostasis	Obesity	References
C57; Nrf2-KO	9–10 weeks	25–28 g	60% fat, 180 days	Nrf2-KO	↓	↓	[49]
C57; WT	7 weeks	22–25 g	60% fat, 12 weeks	Treated with two doses of oral administration of sesamol (100 mg/kg/day or 200 mg/kg/day) for 12 weeks	↓	↓	[50]
C57; Keap1-KD	3 weeks (start HFD)	16–24 g (6 weeks, start measuring BW)	60% fat, 24 weeks	Keap1-KD	↑	↑	[51]
C57; Nrf2-KO	4 weeks	~20 g	41% fat, 24 weeks	Nrf2-KO	Unreported	↓	[52]
C57; Nrf2-KO	12–16 weeks	~20 g	60% fat, 6 weeks	Nrf2-KO	↓	↓	[53]
C57; Nrf2-ANKO	6 weeks	~30 g	60% fat, 14 weeks	Nrf2 KO in adipocytes (ANKO)	↓	↓ in 5–11 weeks and comparable with WT mice after 14 weeks	[54]
C57; Keap1-KD	8 weeks	15–20 g	60% fat, 90 days	Keap1-KD	↓	↓	[55]
ICR; Nrf2-KO	6–8 weeks	22–24 g	10% lard, 2% cholesterol, 0.5% bile salt and 87.5% base forage (Not reported total fat), 8 weeks	Nrf2-KO	↑	↑	[56]
C57; WT	5 weeks	22–24 g	45% fat, 17 weeks	EGCG at 25 or 75 mg/kg were administered via i. p. three times a week over a period of seventeen weeks	↓	↓	[57]
C57; Keap1-KD	9 weeks	22–26 g	60% fat, 36 days	Keap1-KD	Unreported	↓	[58]
C57; WT	6 weeks	20–25 g	60% fat, 14 weeks	PCG at 10 or 100 mg/kg was administered by oral gavage five times a week	↓	Unreported (Inflammation↓)	[59]

Note: Oltipraz, CDDO-Im, glucoraphanin, hydroxytyrosol, curcumin, sesamol, EGCG, and PCG are considered to be the specific activators of Nrf2 or exert their biological functions by activating Nrf2. KO: knockout; KD: knockdown; HFD: high-fat diet. ↓ Indicates decreased/ameliorated; ↑ Indicates increased/deteriorated.

2. Is Nrf2 Up-Regulated or Down-Regulated in Response to Obesity? Is It an Antagonist or a Protector?

Although many studies have established that obesity can increase oxidative stress [22–24,48,60–62], as a master regulator of antioxidant defense, Nrf2 appears to display differential expression patterns in obesity. In a recent study, the WAT of wild-type (WT) mice and ob/ob mice fed high-fat diet (HFD; 41% kcal fat) and normal diet (ND; 11% kcal fat), respectively, for 16 weeks, trended increased protein levels of Nrf2 and nuclear Nrf2 (nNrf2), and the mRNA levels of the Nrf2 downstream genes (such as superoxide dismutase, SOD; and glutathione peroxidase, GPX), compared with control mice [46]. Early research from Xu, et al. [63] reported that the ob/ob mice had higher mRNA expression levels of heme oxygenase 1 (HO1), NAD(P)H: quinone oxidoreductase 1 (NQO1) in liver and HO1 in WAT, compared with WT mice. Moreover, this heightened adaptive antioxidant response was disrupted by Nrf2 ablation. These observations are consistent with research which has reported that feeding WT mice with HFD (60% kcal fat) for 180 days up-regulated the Nrf2 mRNA levels by approximately 50% in the liver, and by 14 times in the WAT [49].

Nevertheless, contradictory results emerged from earlier studies. The mRNA expression levels of Nrf2 and its targets glutathione transferase m6 (GSTm6) and NQO1 were found to be reduced upon 4-week HFD feeding in WT mice [64]. Data from He and colleagues demonstrated that the total and nuclear protein levels of Nrf2 in 18-week HFD (60% kcal fat)-fed mice were almost 50% lower in the skeletal muscle, compared with the ND-fed mice [48]. Moreover, Illesca and his team found the mRNA levels and DNA-binding activities of Nrf2 exhibited significant decreases in WAT with 12-week HFD (60% kcal fat) exposure [47]. These data indicated that functional Nrf2 is actually suppressed in response to HFD, whereas Li, et al. [65] observed that 12-week HFD (60% kcal fat) had no substantial impact on the mRNA levels of Nrf2 and its target genes NQO1 and HO1 in the liver of WT mice or hepatocyte-specific Nrf2 KO mice. What is more, in another study, HFD feeding significantly increased the nuclear but not total protein levels of Nrf2 in the liver. However, from the authors' points of view, Nrf2 deficiency may have induced obesity and steatohepatitis in mice fed with HFD and the activation of Nrf2 could reduce these effects [56].

The inconsistent results concerning the effects of DIO on Nrf2 indicate that the differences in the diet composition (i.e., lard vs. soybean oil), fat content and the length of feeding period may be directly or indirectly correlated with the regulation of Nrf2. Table 2 summarizes the recent findings concerning the influence of HFD composition (fat content) and feeding duration on the activity of Nrf2. Obesity is strongly associated with oxidative stress in humans and rodent models [23]. When the body is exposed to obese danger, Nrf2, as a crucial regulator of antioxidant defense, is up-regulated to counter oxidative stress, suggesting an instinctive compensatory response from the organism. In response to obesity, a certain degree of Nrf2 activation exerts cellular protective effects via anti-oxidative stress, but it may seriously destroy the redox homeostasis and aggravate obesity if its activation exceeds the extent or duration. This may explain why some studies (see Table 1) showed that Nrf2 KO mice exhibited an improved obesity phenotype compared with WT mice. Early studies [66] supported a contention that the benefits of Nrf2 activation by knocking down Keap1 in acute toxicity were conclusive and that continuous Nrf2 activation beyond a certain threshold was rather disadvantageous to long-term survival. The above analysis showed that increased Nrf2 activity within a certain range and duration is protective in obesity. However, it is uncontrollable to fundamentally increase the activity of Nrf2 through genetic modification, so some studies have shown contradictory results (see Table 1). On the other hand, compared with genetic modification, the dosage and duration of treatment with Nrf2 activators can be artificially controlled. Therefore, from the review articles [39,67–69] published in recent years and Table 1 summarized in this paper, we found that Nrf2 activators, always as protectors, alleviate obesity and related metabolic diseases, showing decreased ROS production, inhibited lipid accumulation during adipogenesis, attenuated pro-inflammatory cytokines and improved glucose homeostasis. It may be possible in the future to develop effective and safe Nrf2 activators for the therapy of obesity.

Table 2. Influence of HFD composition (fat content) and feeding duration on Nrf2 activity.

Diet Source and Identifier	Fat Calories	Fat Composition (*w/w*)	Feeding Duration	Effects on Nrf2	References
Bio-Serv, F3282	60%	Lard (36.0%)	10 weeks	Decreased Nrf2 mRNA levels in WAT and liver	[45]
Research Diets, D12492	60%	Lard (31.7%), Soybean Oil (3.2%)	24 weeks	Increased Nrf2 mRNA levels in WAT	[51]
Research Diets, Unreported	60%	Unreported	26 weeks	Increased Nrf2 mRNA levels in WAT and liver	[49]
Research Diets, D12492	60%	Lard (31.7%), Soybean Oil (3.2%)	12 weeks	Decreased Nrf2 DNA binding activity and mRNA levels in eWAT	[47]
Harlan Teklad, TD88137	42%	Anhydrous Milkfat (21.0%)	12 weeks	Increased Nrf2 mRNA levels in liver	[70]
Research Diets, D12042201	41%	Anhydrous Butter (20.0%)	16 weeks	Increased total and nuclear protein levels of Nrf2 in WAT	[46]
Guangdong Animal Center, Unreported	60%	Unreported	18 weeks	Decreased total and nuclear protein levels of Nrf2 in skeletal muscle	[48]
Harlan Teklad, TD10885	45%	Anhydrous Milkfat (21.0%), Soybean Oil (2.0%)	28 weeks	Decreased nuclear but not total protein levels of Nrf2 in WAT	[40]
Unreported, Unreported	36%	Soybean Oil (18.0%)	12 weeks	Increased Nrf2 mRNA levels in liver	[71]
Bio-Serv, F5194	Unreported	Lard (35%), cholesterol (0.15%)	4 weeks	Decreased mRNA levels of Nrf2 and downstream NQO1 and GSTm6 in liver	[64]
Research Diets, D12492	60%	Lard (31.7%), Soybean Oil (3.2%)	12 weeks	No significant changes in mRNA or protein levels of Nrf2	[65]

3. Adipogenesis

The WAT of ob/ob mice exhibited elevated expression levels of fatty acid synthase (FAS) and peroxisome proliferator-activated receptor $\gamma2$ (PPARγ2), which involved in adipogenesis and lipogenesis, in contrast to WT mice, but this augmentation was diminished in global and adipocyte-specific Nrf2 KO ob/ob mice [63]. In an earlier study, the absence of Nrf2 resulted in attenuated expression levels of PPARγ, CCAAT enhancer-binding protein α (C/EBPα) and their downstream genes during adipocyte differentiation in employing 3T3-L1 cells, mouse embryonic fibroblasts (MEFs) or human subcutaneous preadipocytes. Simultaneously, enhanced Nrf2 activity by knocking down Keap1 reversed these changes [52]. HFD exposure also enhanced the mRNA and protein expression levels of PPARγ2 in liver, whereas the up-regulation of PPARγ2 was partially abrogated in Nrf2-deficient hepatocytes [65], suggesting that the regulatory effects of Nrf2 on PPARγ2 in hepatocytes may display a similar pattern as in adipocytes. What is more, the Nrf2 KO hepatocytes treated with palmitate which simulates PPARγ and consequent lipogenesis also showed lower mRNA levels of PPARγ and its downstream lipogenic genes, including the FAS, stearoyl-CoA desaturase 1 (SCD1) and fatty acid binding protein 4 (FABP4), than those in WT hepatocytes [65].

Similar to the aforementioned studies, Sun and colleagues discovered significantly decreased protein expression levels of acetyl-CoA carboxylase (ACC), sterol regulatory element binding protein 1 (SREBP1), diacylglycerol acyltransferase 1 (DGAT1), DGAT2, FAS and perilipin A in the WAT of Nrf2 KO mice, under both ND and HFD conditions [46]. However, it was also shown that the mRNA levels of DGAT2 in epididymal WAT (eWAT) were slightly increased instead of decreased by the loss of Nrf2 in adipocytes [43], and SREBP1 exhibited higher rather than lower mRNA expression levels in HFD-fed hepatocyte-specific Nrf2-deficient mice than the control counterparts [65]. Intriguingly, it was reported that sustaining Nrf2 activation reduced the triglycerides (TG) and free fatty acid (FFA) contents in Keap1-KD ob/ob mice, and weakened the mRNA expression of PPARγ, C/EBPα, SCD1, FABP4, ACC-1 and ACC-2. Moreover, after 36 days of HFD (60% kcal fat) feeding, Keap1-KD mice (non-ob/ob) had a lower body weight, with decreased food intake, compared to control mice [58]. The question emerges whether the improved obesity in Keap1-KD mice is the effect of Nrf2 activation, or an indirect result of decreased food intake caused by Keap1 gene manipulation. Wakabayashi, et al. [72] reported that Keap1-deficient mice (not KD, but KO) died from malnutrition resulting from severe hyperkeratosis in the esophagus and forestomach, indicating that the down-regulation or deletion of Keap1 may have affected the ingestive behavior of animals. In addition to the genetic manipulation of Keap1-Nrf2, recent studies using Nrf2 pharmacological regulators to investigate the role of Nrf2 on adipogenesis have also yielded conflicting results [73–76].

The in vitro experiments conducted by Xu and his team using MEFs suggested that the induction of Nrf2 mRNA expression in the early stages (0–12 h) of the differentiation from MEFs to adipocytes significantly declined in the later stage. What is more, after the early time point when MEFs differentiate into adipocytes, pharmacological Nrf2 activation by L-sulforaphane suppressed further differentiation and lipid accumulation [58]. L-sulforaphane was one of the first studied activators of Nrf2, belonging to the isothiocyanate group, present in broccoli. L-sulforaphane-rich broccoli sprout powder significantly improved serum insulin concentration, glucose-to-insulin ratio and insulin resistance in type 2 diabetic patients [77]. These findings are similar to those from Chartoumpekis, et al. [78]. These data indicated that Nrf2 was essential for adipocyte differentiation, which is a result consistent with most studies demonstrating that Nrf2 deficiency restrains fat accumulation in WAT. However, for differentiated or differentiating adipocytes, Nrf2 may play an antagonistic role in lipid accumulation, which partially explains why various research has shown that Nrf2 activation ameliorates obesity. In another study, treatment with punicalagin, which improves obesity via activating Nrf2, substantially inhibited lipid accumulation in the early stages (0–2 days) of 3T3-L1 cell differentiation. However, these effects were dramatically attenuated after that time point [59]. An intriguing finding was also observed in an in vivo experiment. Adipocyte-specific Nrf2 KO (ANKO) mice exhibited transiently delayed BW growth from week 5 to week 11 of HFD feeding; nonetheless, after 14 weeks, ANKO mice showed comparable BW and body fat content with control mice [50]. In summary, the effects of time specificity of Nrf2 to fat remains unclear and requires further study.

4. Inflammation

Since the discovery [79] that the pro-inflammatory cytokine tumor necrosis factor α (TNFα) induced by obesity was able to promote IR, numerous studies have consistently shown increased inflammation responses in WAT of obese humans and animals [16,24,80–84]. While contributions of oxidative stress to inflammation have been recognized, contradictory reports exist regarding the function of Nrf2 in obesity-induced inflammation. Xue and his team employed global or adipocyte-specific Nrf2-knockout (KO) mice on leptin-deficient (ob/ob) background to investigate the roles of Nrf2 in obesity and its associated disorders. They found augmented inflammatory responses determined by increased expression levels of pro-inflammatory cytokines, including TNFα and interleukin 1β (IL-1β), in WAT of ob/ob mice, compared with non-ob/ob littermates [63]. This is consistent with the view that chronic subclinical inflammation might be involved in the pathogenesis of obesity [16,84–88]. However, either systemic or adipocyte-specific ablation of Nrf2 in ob/ob mice can unexpectedly improve inflammation, as Nrf2 has been shown to be able to suppress inflammation [38,89,90]. In a more recent study, the deficiency of Nrf2 in hepatocytes of mice attenuated inflammation induced by feeding HFD (60% kcal fat) for 12 weeks, as suggested by the suppressed phosphorylation of nuclear factor kappa-B (NF-κB), the mRNA degrees of TNFα and C–C motif chemokine ligand 2 (CCL2) and minimized macrophage infiltration [65]. These findings are also in agreement with a previous study conducted by More and colleagues. They observed that constitutive Nrf2 activation by Keap1 konckdown (KD) exaggerated inflammation through increasing the mRNA expression of TNFα in both WAT and liver of C57BL/6 mice fed with long-term (24 weeks) HFD (60% kcal fat) [51].

Why is it that Nrf2 is widely considered as an anti-inflammatory factor [38,89,91], and yet so many studies have located that the absence of Nrf2 can ameliorate inflammation in metabolic syndrome? Nrf2 has been shown to promote adipogenesis by up-regulating the expression of PPARγ in WAT [52] and liver [65]. Thus, continuous activation of Nrf2 under stress conditions may increase lipid accumulation and lead to lipid peroxidation, which in turn causes tissue injury and inflammation. The pro-inflammation might not be the direct effect of Nrf2, but its secondary effect mediated by promoting adipogenesis. On the other hand, Nrf2 modulates inflammatory responses in a cell type-dependent manner, which may also partially contribute to the contradictory effects of Nrf2 on inflammation. Nrf2 acts as a transcription factor to activate the expression levels of IL-6 in hepatocytes [92], which however block the transcription of IL-6 and IL-1β in macrophages [89] by inhibiting the recruitment of RNA polymerase II. A study from Meher and colleagues found that the mRNA expression levels of monocyte chemoattractant protein 1 (MCP-1) and IL-1β in macrophages were up-regulated when co-culturing with adipocytes, but this up-regulation was significantly suppressed in Nrf2 KO macrophages. However, co-culturing with Nrf2-deficient macrophages induced the mRNA expression of IL-6 and MCP-1 in adipocytes [45]. This study implied that Nrf2 plays an important role in promoting the immunological function of macrophages (participating in inflammation), but the activity of Nrf2 in macrophages did not affect the response of adipocytes to inflammatory signals, and adipose tissue inflammation may have been mainly regulated by Nrf2 in adipocytes rather than macrophages (Figure 2). In addition to the specific roles of Nrf2 in different cell types, the crosstalk between cells may also have a great influence on the effects of Nrf2 on inflammation.

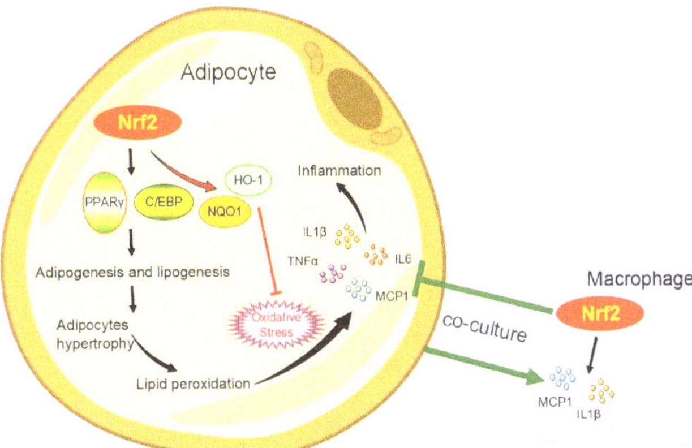

Figure 2. The subtle role of Nrf2 in obesity-related inflammation. Nrf2 promotes adipogenesis and lipogenesis by up-regulating PPARγ and C/EBP in adipocytes. Thus, continuous activation of Nrf2 under obesity-induced stress conditions may uncontrollably increase lipid accumulation and lead to lipid peroxidation, which in turn causes tissue injury and inflammation. At the same time, Nrf2, as a double-edged sword, counteracts oxidative stress by enhancing the expression of NQO1 and HO-1 and subsequently alleviates inflammation. The mRNA levels of MCP-1 and IL-1β in macrophages are up-regulated when co-culturing with adipocytes, but this up-regulation is significantly suppressed in Nrf2 KO macrophages. However, co-culturing with Nrf2-deficient macrophages induces the mRNA expression of IL-6 and MCP-1 in adipocytes.

5. Insulin Resistance

Oxidative stress is a potent inducer of insulin resistance which is a pathologic condition during the development of obesity [15,60,93]. A few studies have confirmed Nrf2 is related to insulin signaling and insulin resistance due to its crucial cytoprotective, ROS scavenging, and anti-inflammatory roles [39,83,93–95]. However, the precise underlying mechanism by which Nrf2 exerts its effects on insulin resistance is still obscure. Uruno, et al. [96] over-expressd Nrf2 globally and specifically in the skeletal muscle (SkM) or liver by Keap1 gene hypomorphic knockdown and SkM- or liver-specific Keap1 knockout. Their results showed that either systemic or SkM-specific KO/KD of Keap1 in db/db mice decreased blood glucose and elevate insulin sensitivity. However, these changes were not seen in liver-specific Keap1-KO db/db mice. Moreover, they employed Keap1-KD mice (non-db/db) fed with HFD (62.2% kcal fat) for 8 weeks, and Nrf2 exhibited the same ameliorated IR effects. Subsequently, it was noted that Nrf2 induction in SkM reduced glycogen content, resulting in improved glucose tolerance [97]. These observations are similar to a report which found that Nrf2 ablation in HFD-treated mice can lead to hepatic IR by activation of the NF-κB signaling pathway [56]. Chartoumpekis, et al. [98] also found that increased Nrf2 activity in Keap1-KD lipodystrophic mice alleviated impaired metabolism and IR via inhibition of lipogenic genes in the liver.

Inconsistent with the above studies, data from More and colleagues suggested that after 24 weeks of HFD (60% kcal fat) feeding, Keap1-KD mice had higher blood glucose with aggravated IR, compared to control mice [51]. The same study showed that Nrf2 activation induced by knocking down Keap1 suppressed insulin signaling and deteriorated IR in ob/ob mice [58]. These findings are similar to those by Meakin, et al. [99] and Chartoumpekis, et al. [49]. They reported that Nrf2 KO mice displayed improved insulin sensitivity when exposed to HFD.

Fibroblast growth factor 21 (FGF21), which is a stress-inducible peptide hormone that regulates energy balance and glucose and lipid homeostasis, has been proven to alleviate hyperglycemia, insulin resistance, dyslipidemia and fatty liver [100–103]. HFD-fed Nrf2-null mice had higher circulating FGF21 levels and increased FGF21 mRNA levels in WAT and liver compared with WT mice. Furthermore, the over-expression of Nrf2 in ST-2 cells, which is a cell line derived from murine bone marrow stroma that expresses functional FGF21 and Nrf2, led to a decline in FGF21 mRNA expression. As a result, the improved obesity and IR phenotype of Nrf2 KO mice may be partly attributed to the up-regulation of FGF21 induced by Nrf2 ablation [49]. However, there have been conflicting results in a more recent study showing that FGF21 is obviously repressed in Nrf2-deficient 3T3-L1 cells, using small interfering RNA (siRNA) [104]. More interestingly, FGF21 has been reported to inhibit inflammation by activating Nrf2 and suppressing the NF-κB signaling pathway [105].

To clarify which tissue contributes to the changes in IR brought by Nrf2, Chartoumpekis and colleagues generated mice with conditional knockout of Nrf2 in hepatocytes (HeNKO) or adipocytes (ANKO). After 170 days of HFD (60% kcal fat) feeding, HeNKO mice and ANKO mice revealed comparable increasing levels in body weight and body fat mass. However, ANKO mice exhibited higher fasting glucose, cholesterol levels and disruptive glucose tolerance compared with control group; deletion of Nrf2 in hepatocytes inversely led to an ameliorated metabolic profile (improved insulin sensitivity and glucose tolerance) without any difference in hepatic steatosis. It is interesting that the FGF21 mRNA levels showed no significant difference in both WAT of ANKO mice and liver of HeNKO mice, and FGF21 plasma levels werealso similar among all genotypes [43].

In summary, the inconsistent results on the role of Nrf2 in IR from the above studies indicate that the difference in genetic manipulation of Nrf2 (Nrf2 Knockout or Keap1 Knockdown), length of HFD feeding period, duration of Nrf2 activation and cell specificity of Nrf2's function, may lead to different effects on IR. The study of Loh, et al. [106] may explain the effects of Nrf2 activation duration on insulin signaling. Although the excess ROS is strongly related to the pathophysiology of many diseases, including diabetes and obesity, actually physiological levels of ROS may be necessary for normal intracellular signaling. Loh and colleagues demonstrated subtle increases in physiological ROS production enhance insulin signaling in vivo. One potential explanation is as follows. Short-term Nrf2 activation improves IR through scavenging the excessive ROS, however, continuous Nrf2 activation impair insulin signaling by excessively removing ROS. More future studies focusing on the regulation of ROS homeostasis by Nrf2 are needed. Any coin has two sides and even the roles of inflammation in IR have inspired different views in recent years [107–110].

6. Conclusions

In this review, we discussed Nrf2's subtle expression patterns in response to obesity and the complex roles of Nrf2 on adipogenesis, inflammation and insulin resistance. In obesity, chronic oxidative stress is considered to be the important factor that leads to the development of pathologies such as lipid disorders, insulin resistance and diabetes. Nrf2 is a key regulator of the cellular response to oxidative stress by controlling the expression of antioxidant and detoxification enzymes to eliminate excess ROS. An increasing number of studies have provided casual evidence that Nrf2 is a pivotal target of obesity and its related metabolic disorders. Of note, previous reports have had many contradictions about the roles of Nrf2 in obesity and its related disorders. As indicated by the above analysis, further study on the spatio-temporal specificity of Nrf2 regulating cell function, the final effects of Nrf2 crosstalk between different cell types on the overall function of the body, the biological effects of different gene manipulation of Nrf2, and the regulation of ROS homeostasis by the activation degree and duration of Nrf2 will undoubtedly help clarify the mentioned inconsistence and lead to a greater understanding of the essential roles of Nrf2 in the modulation of metabolic homeostasis, and would provide novel insights for potential therapeutic approaches in obesity and its comorbidities.

Author Contributions: Conceptualization, Y.X. and Y.J.; writing—original draft preparation, Y.X., X.Z., Y.Q. and X.L.; writing—review and editing, Y.X. and Y.J.; supervision, Y.J. All authors have read and agreed to the published version of the manuscript.

Funding: This research was funded by the National Natural Science Foundation of China, grant number 82260178, the Innovation Team Foundation of the Xinjiang Uyghur Autonomous Region of China, grant number 2022D14009, the Project of State Key Laboratory of Pathogenesis, Prevention, and Treatment of High Incidence Diseases in Central Asia, grant number SKL-HIDCA2021-DX1, and the National College Students Innovation and Entrepreneurship Training Program, grant number 202010760010.

Conflicts of Interest: The authors declare no conflict of interest.

References

1. Wharton, S.; Lau, D.C.W.; Vallis, M.; Sharma, A.M.; Biertho, L.; Campbell-Scherer, D.; Adamo, K.; Alberga, A.; Bell, R.; Boule, N.; et al. Obesity in adults: A clinical practice guideline. *CMAJ* **2020**, *192*, E875–E891. [CrossRef] [PubMed]
2. Heymsfield, S.B.; Wadden, T.A. Mechanisms, Pathophysiology, and Management of Obesity. *N. Engl. J. Med.* **2017**, *376*, 254–266. [CrossRef] [PubMed]
3. Collaborators, G.B.D.O.; Afshin, A.; Forouzanfar, M.H.; Reitsma, M.B.; Sur, P.; Estep, K.; Lee, A.; Marczak, L.; Mokdad, A.H.; Moradi-Lakeh, M.; et al. Health Effects of Overweight and Obesity in 195 Countries over 25 Years. *N. Engl. J. Med.* **2017**, *377*, 13–27. [CrossRef] [PubMed]
4. Friedrich, M.J. Global Obesity Epidemic Worsening. *JAMA* **2017**, *318*, 603. [CrossRef] [PubMed]
5. Schnurr, T.M.; Jakupovic, H.; Carrasquilla, G.D.; Angquist, L.; Grarup, N.; Sorensen, T.I.A.; Tjonneland, A.; Overvad, K.; Pedersen, O.; Hansen, T.; et al. Obesity, unfavourable lifestyle and genetic risk of type 2 diabetes: A case-cohort study. *Diabetologia* **2020**, *63*, 1324–1332. [CrossRef]
6. Aslam, M.I.; Hahn, V.S.; Jani, V.; Hsu, S.; Sharma, K.; Kass, D.A. Reduced Right Ventricular Sarcomere Contractility in HFpEF with Severe Obesity. *Circulation* **2020**, *143*, 965–967. [CrossRef] [PubMed]
7. Mahajan, R.; Stokes, M.; Elliott, A.; Munawar, D.A.; Khokhar, K.B.; Thiyagarajah, A.; Hendriks, J.; Linz, D.; Gallagher, C.; Kaye, D.; et al. Complex interaction of obesity, intentional weight loss and heart failure: A systematic review and meta-analysis. *Heart* **2020**, *106*, 58–68. [CrossRef]

8. Zhao, Y.; Qin, P.; Sun, H.; Liu, Y.; Liu, D.; Zhou, Q.; Guo, C.; Li, Q.; Tian, G.; Wu, X.; et al. Metabolically healthy general and abdominal obesity are associated with increased risk of hypertension. *Br. J. Nutr.* **2020**, *123*, 583–591. [CrossRef] [PubMed]

9. Lauby-Secretan, B.; Scoccianti, C.; Loomis, D.; Grosse, Y.; Bianchini, F.; Straif, K. Body Fatness and Cancer–Viewpoint of the IARC Working Group. *N. Engl. J. Med.* **2016**, *375*, 794–798. [CrossRef] [PubMed]

10. Jayedi, A.; Soltani, S.; Zargar, M.S.; Khan, T.A.; Shab-Bidar, S. Central fatness and risk of all cause mortality: Systematic review and dose-response meta-analysis of 72 prospective cohort studies. *BMJ* **2020**, *370*, m3324. [CrossRef] [PubMed]

11. Kivimäki, M.; Strandberg, T.; Pentti, J.; Nyberg, S.T.; Frank, P.; Jokela, M.; Ervasti, J.; Suominen, S.B.; Vahtera, J.; Sipilä, P.N.; et al. Body-mass index and risk of obesity-related complex multimorbidity: An observational multicohort study. *Lancet Diabetes Endocrinol.* **2022**, *10*, 253–263. [CrossRef]

12. Tam, B.T.; Morais, J.A.; Santosa, S. Obesity and ageing: Two sides of the same coin. *Obes. Rev.* **2020**, *21*, e12991. [CrossRef] [PubMed]

13. Kahn, S.E.; Hull, R.L.; Utzschneider, K.M. Mechanisms linking obesity to insulin resistance and type 2 diabetes. *Nature* **2006**, *444*, 840–846. [CrossRef] [PubMed]

14. Barazzoni, R.; Gortan Cappellari, G.; Ragni, M.; Nisoli, E. Insulin resistance in obesity: An overview of fundamental alterations. *Eat. Weight Disord.* **2018**, *23*, 149–157. [CrossRef] [PubMed]

15. Tong, Y.; Xu, S.; Huang, L.; Chen, C. Obesity and insulin resistance: Pathophysiology and treatment. *Drug Discov. Today* **2022**, *27*, 822–830. [CrossRef]

16. Hotamisligil, G.S. Inflammation and metabolic disorders. *Nature* **2006**, *444*, 860–867. [CrossRef]

17. Prasad, M.; Chen, E.W.; Toh, S.A.; Gascoigne, N.R.J. Autoimmune responses and inflammation in type 2 diabetes. *J. Leukoc. Biol.* **2020**, *107*, 739–748. [CrossRef]

18. Saha, S.; Buttari, B.; Panieri, E.; Profumo, E.; Saso, L. An Overview of Nrf2 Signaling Pathway and Its Role in Inflammation. *Molecules* **2020**, *25*, 5474. [CrossRef] [PubMed]

19. Schieber, M.; Chandel, N.S. ROS Function in Redox Signaling and Oxidative Stress. *Curr. Biol.* **2014**, *24*, R453–R462. [CrossRef] [PubMed]

20. Forman, H.J.; Zhang, H. Targeting oxidative stress in disease: Promise and limitations of antioxidant therapy. *Nat. Rev. Drug Discov.* **2021**, *20*, 689–709. [CrossRef]

21. Yu, T.; Robotham, J.L.; Yoon, Y. Increased production of reactive oxygen species in hyperglycemic conditions requires dynamic change of mitochondrial morphology. *Proc. Natl. Acad. Sci. USA* **2006**, *103*, 2653–2658. [CrossRef]

22. Chattopadhyay, M.; Khemka, V.K.; Chatterjee, G.; Ganguly, A.; Mukhopadhyay, S.; Chakrabarti, S. Enhanced ROS production and oxidative damage in subcutaneous white adipose tissue mitochondria in obese and type 2 diabetes subjects. *Mol. Cell. Biochem.* **2015**, *399*, 95–103. [CrossRef]

23. Furukawa, S.; Fujita, T.; Shimabukuro, M.; Iwaki, M.; Yamada, Y.; Nakajima, Y.; Nakayama, O.; Makishima, M.; Matsuda, M.; Shimomura, I. Increased oxidative stress in obesity and its impact on metabolic syndrome. *J. Clin. Investig.* **2004**, *114*, 1752–1761. [CrossRef]

24. Paltoglou, G.; Schoina, M.; Valsamakis, G.; Salakos, N.; Avloniti, A.; Chatzinikolaou, A.; Margeli, A.; Skevaki, C.; Papagianni, M.; Kanaka-Gantenbein, C.; et al. Interrelations among the adipocytokines leptin and adiponectin, oxidative stress and aseptic inflammation markers in pre- and early-pubertal normal-weight and obese boys. *Endocrine* **2017**, *55*, 925–933. [CrossRef] [PubMed]

25. Kaspar, J.W.; Niture, S.K.; Jaiswal, A.K. Nrf2:INrf2 (Keap1) signaling in oxidative stress. *Free Radic. Biol. Med.* **2009**, *47*, 1304–1309. [CrossRef] [PubMed]

26. Baird, L.; Yamamoto, M. The Molecular Mechanisms Regulating the KEAP1-NRF2 Pathway. *Mol. Cell. Biol.* **2020**, *40*, e00099-20. [CrossRef] [PubMed]

27. He, F.; Ru, X.; Wen, T. NRF2, a Transcription Factor for Stress Response and Beyond. *Int. J. Mol. Sci.* **2020**, *21*, 4777. [CrossRef] [PubMed]

28. Kobayashi, A.; Kang, M.I.; Okawa, H.; Ohtsuji, M.; Zenke, Y.; Chiba, T.; Igarashi, K.; Yamamoto, M. Oxidative stress sensor Keap1 functions as an adaptor for Cul3-based E3 ligase to regulate proteasomal degradation of Nrf2. *Mol. Cell. Biol.* **2004**, *24*, 7130–7139. [CrossRef]

29. Yamamoto, T.; Suzuki, T.; Kobayashi, A.; Wakabayashi, J.; Maher, J.; Motohashi, H.; Yamamoto, M. Physiological significance of reactive cysteine residues of Keap1 in determining Nrf2 activity. *Mol. Cell. Biol.* **2008**, *28*, 2758–2770. [CrossRef] [PubMed]

30. Kopacz, A.; Kloska, D.; Forman, H.J.; Jozkowicz, A.; Grochot-Przeczek, A. Beyond repression of Nrf2: An update on Keap1. *Free Radic. Biol. Med.* **2020**, *157*, 63–74. [CrossRef]

31. Levonen, A.L.; Landar, A.; Ramachandran, A.; Ceaser, E.K.; Dickinson, D.A.; Zanoni, G.; Morrow, J.D.; Darley-Usmar, V.M. Cellular mechanisms of redox cell signalling: Role of cysteine modification in controlling antioxidant defences in response to electrophilic lipid oxidation products. *Biochem. J.* **2004**, *378*, 373–382. [CrossRef] [PubMed]

32. Rushmore, T.H.; Morton, M.R.; Pickett, C.B. The antioxidant responsive element. Activation by oxidative stress and identification of the DNA consensus sequence required for functional activity. *J. Biol. Chem.* **1991**, *266*, 11632–11639. [CrossRef]

33. Moi, P.; Chan, K.; Asunis, I.; Cao, A.; Kan, Y.W. Isolation of NF-E2-related factor 2 (Nrf2), a NF-E2-like basic leucine zipper transcriptional activator that binds to the tandem NF-E2/AP1 repeat of the beta-globin locus control region. *Proc. Natl. Acad. Sci. USA* **1994**, *91*, 9926–9930. [CrossRef] [PubMed]

34. Fabrizio, F.P.; Sparaneo, A.; Muscarella, L.A. NRF2 Regulation by Noncoding RNAs in Cancers: The Present Knowledge and the Way Forward. *Cancers* **2020**, *12*, 3621. [CrossRef] [PubMed]
35. Wang, P.; Long, F.; Lin, H.; Wang, S.; Wang, T. Dietary Phytochemicals Targeting Nrf2 to Enhance the Radiosensitivity of Cancer. *Oxid. Med. Cell. Longev.* **2022**, *2022*, 7848811. [CrossRef]
36. Tossetta, G.; Marzioni, D. Natural and synthetic compounds in Ovarian Cancer: A focus on NRF2/KEAP1 pathway. *Pharmacol. Res.* **2022**, *183*, 106365. [CrossRef]
37. Wang, H.; Liu, X.; Long, M.; Huang, Y.; Zhang, L.; Zhang, R.; Zheng, Y.; Liao, X.; Wang, Y.; Liao, Q.; et al. NRF2 activation by antioxidant antidiabetic agents accelerates tumor metastasis. *Sci. Transl. Med.* **2016**, *8*, 334ra351. [CrossRef] [PubMed]
38. Mohan, S.; Gupta, D. Crosstalk of toll-like receptors signaling and Nrf2 pathway for regulation of inflammation. *Biomed. Pharmacother.* **2018**, *108*, 1866–1878. [CrossRef]
39. Li, S.; Eguchi, N.; Lau, H.; Ichii, H. The Role of the Nrf2 Signaling in Obesity and Insulin Resistance. *Int. J. Mol. Sci.* **2020**, *21*, 6973. [CrossRef]
40. Yu, Z.; Shao, W.; Chiang, Y.; Foltz, W.; Zhang, Z.; Ling, W.; Fantus, I.G.; Jin, T. Oltipraz upregulates the nuclear factor (erythroid-derived 2)-like 2 [corrected](NRF2) antioxidant system and prevents insulin resistance and obesity induced by a high-fat diet in C57BL/6J mice. *Diabetologia* **2011**, *54*, 922–934. [CrossRef] [PubMed]
41. Shin, S.; Wakabayashi, J.; Yates, M.S.; Wakabayashi, N.; Dolan, P.M.; Aja, S.; Liby, K.T.; Sporn, M.B.; Yamamoto, M.; Kensler, T.W. Role of Nrf2 in prevention of high-fat diet-induced obesity by synthetic triterpenoid CDDO-imidazolide. *Eur. J. Pharmacol.* **2009**, *620*, 138–144. [CrossRef] [PubMed]
42. Nagata, N.; Xu, L.; Kohno, S.; Ushida, Y.; Aoki, Y.; Umeda, R.; Fuke, N.; Zhuge, F.; Ni, Y.; Nagashimada, M.; et al. Glucoraphanin Ameliorates Obesity and Insulin Resistance through Adipose Tissue Browning and Reduction of Metabolic Endotoxemia in Mice. *Diabetes* **2017**, *66*, 1222–1236. [CrossRef] [PubMed]
43. Chartoumpekis, D.V.; Palliyaguru, D.L.; Wakabayashi, N.; Fazzari, M.; Khoo, N.K.H.; Schopfer, F.J.; Sipula, I.; Yagishita, Y.; Michalopoulos, G.K.; O'Doherty, R.M.; et al. Nrf2 deletion from adipocytes, but not hepatocytes, potentiates systemic metabolic dysfunction after long-term high-fat diet-induced obesity in mice. *Am. J. Physiol. Endocrinol. Metab.* **2018**, *315*, E180–E195. [CrossRef] [PubMed]
44. Zhang, Y.K.; Wu, K.C.; Liu, J.; Klaassen, C.D. Nrf2 deficiency improves glucose tolerance in mice fed a high-fat diet. *Toxicol. Appl. Pharmacol.* **2012**, *264*, 305–314. [CrossRef] [PubMed]
45. Meher, A.K.; Sharma, P.R.; Lira, V.A.; Yamamoto, M.; Kensler, T.W.; Yan, Z.; Leitinger, N. Nrf2 deficiency in myeloid cells is not sufficient to protect mice from high-fat diet-induced adipose tissue inflammation and insulin resistance. *Free Radic. Biol. Med.* **2012**, *52*, 1708–1715. [CrossRef]
46. Sun, X.; Li, X.; Jia, H.; Wang, H.; Shui, G.; Qin, Y.; Shu, X.; Wang, Y.; Dong, J.; Liu, G.; et al. Nuclear Factor E2-Related Factor 2 Mediates Oxidative Stress-Induced Lipid Accumulation in Adipocytes by Increasing Adipogenesis and Decreasing Lipolysis. *Antioxid Redox Signal.* **2020**, *32*, 173–192. [CrossRef]
47. Illesca, P.; Valenzuela, R.; Espinosa, A.; Echeverria, F.; Soto-Alarcon, S.; Ortiz, M.; Videla, L.A. Hydroxytyrosol supplementation ameliorates the metabolic disturbances in white adipose tissue from mice fed a high-fat diet through recovery of transcription factors Nrf2, SREBP-1c, PPAR-gamma and NF-kappaB. *Biomed. Pharmacother.* **2019**, *109*, 2472–2481. [CrossRef] [PubMed]
48. He, H.J.; Wang, G.Y.; Gao, Y.; Ling, W.H.; Yu, Z.W.; Jin, T.R. Curcumin attenuates Nrf2 signaling defect, oxidative stress in muscle and glucose intolerance in high fat diet-fed mice. *World J. Diabetes* **2012**, *3*, 94–104. [CrossRef]
49. Chartoumpekis, D.V.; Ziros, P.G.; Psyrogiannis, A.I.; Papavassiliou, A.G.; Kyriazopoulou, V.E.; Sykiotis, G.P.; Habeos, I.G. Nrf2 represses FGF21 during long-term high-fat diet-induced obesity in mice. *Diabetes* **2011**, *60*, 2465–2473. [CrossRef] [PubMed]
50. Lee, D.H.; Chang, S.H.; Yang, D.K.; Song, N.J.; Yun, U.J.; Park, K.W. Sesamol Increases Ucp1 Expression in White Adipose Tissues and Stimulates Energy Expenditure in High-Fat Diet-Fed Obese Mice. *Nutrients* **2020**, *12*, 1459. [CrossRef] [PubMed]
51. More, V.R.; Xu, J.; Shimpi, P.C.; Belgrave, C.; Luyendyk, J.P.; Yamamoto, M.; Slitt, A.L. Keap1 knockdown increases markers of metabolic syndrome after long-term high fat diet feeding. *Free Radic. Biol. Med.* **2013**, *61*, 85–94. [CrossRef]
52. Pi, J.; Leung, L.; Xue, P.; Wang, W.; Hou, Y.; Liu, D.; Yehuda-Shnaidman, E.; Lee, C.; Lau, J.; Kurtz, T.W.; et al. Deficiency in the nuclear factor E2-related factor-2 transcription factor results in impaired adipogenesis and protects against diet-induced obesity. *J. Biol. Chem.* **2010**, *285*, 9292–9300. [CrossRef] [PubMed]
53. Schneider, K.; Valdez, J.; Nguyen, J.; Vawter, M.; Galke, B.; Kurtz, T.W.; Chan, J.Y. Increased Energy Expenditure, Ucp1 Expression, and Resistance to Diet-induced Obesity in Mice Lacking Nuclear Factor-Erythroid-2-related Transcription Factor-2 (Nrf2). *J. Biol. Chem.* **2016**, *291*, 7754–7766. [CrossRef] [PubMed]
54. Zhang, L.; Dasuri, K.; Fernandez-Kim, S.O.; Bruce-Keller, A.J.; Keller, J.N. Adipose-specific ablation of Nrf2 transiently delayed high-fat diet-induced obesity by altering glucose, lipid and energy metabolism of male mice. *Am. J. Transl. Res.* **2016**, *8*, 5309–5319.
55. Slocum, S.L.; Skoko, J.J.; Wakabayashi, N.; Aja, S.; Yamamoto, M.; Kensler, T.W.; Chartoumpekis, D.V. Keap1/Nrf2 pathway activation leads to a repressed hepatic gluconeogenic and lipogenic program in mice on a high-fat diet. *Arch. Biochem. Biophys.* **2016**, *591*, 57–65. [CrossRef] [PubMed]
56. Liu, Z.; Dou, W.; Ni, Z.; Wen, Q.; Zhang, R.; Qin, M.; Wang, X.; Tang, H.; Cao, Y.; Wang, J.; et al. Deletion of Nrf2 leads to hepatic insulin resistance via the activation of NF-kappaB in mice fed a high-fat diet. *Mol. Med. Rep.* **2016**, *14*, 1323–1331. [CrossRef] [PubMed]

57. Sampath, C.; Rashid, M.R.; Sang, S.; Ahmedna, M. Green tea epigallocatechin 3-gallate alleviates hyperglycemia and reduces advanced glycation end products via nrf2 pathway in mice with high fat diet-induced obesity. *Biomed. Pharmacother.* **2017**, *87*, 73–81. [CrossRef] [PubMed]

58. Xu, J.; Kulkarni, S.R.; Donepudi, A.C.; More, V.R.; Slitt, A.L. Enhanced Nrf2 activity worsens insulin resistance, impairs lipid accumulation in adipose tissue, and increases hepatic steatosis in leptin-deficient mice. *Diabetes* **2012**, *61*, 3208–3218. [CrossRef]

59. Kang, B.; Kim, C.Y.; Hwang, J.; Jo, K.; Kim, S.; Suh, H.J.; Choi, H.S. Punicalagin, a Pomegranate-Derived Ellagitannin, Suppresses Obesity and Obesity-Induced Inflammatory Responses Via the Nrf2/Keap1 Signaling Pathway. *Mol. Nutr. Food Res.* **2019**, *63*, e1900574. [CrossRef]

60. Hurrle, S.; Hsu, W.H. The etiology of oxidative stress in insulin resistance. *Biomed. J.* **2017**, *40*, 257–262. [CrossRef] [PubMed]

61. Bonnard, C.; Durand, A.; Peyrol, S.; Chanseaume, E.; Chauvin, M.A.; Morio, B.; Vidal, H.; Rieusset, J. Mitochondrial dysfunction results from oxidative stress in the skeletal muscle of diet-induced insulin-resistant mice. *J. Clin. Investig.* **2008**, *118*, 789–800. [CrossRef] [PubMed]

62. Jakubiak, G.K.; Osadnik, K.; Lejawa, M.; Kasperczyk, S.; Osadnik, T.; Pawlas, N. Oxidative Stress in Association with Metabolic Health and Obesity in Young Adults. *Oxid. Med. Cell. Longev.* **2021**, *2021*, 9987352. [CrossRef] [PubMed]

63. Xue, P.; Hou, Y.; Chen, Y.; Yang, B.; Fu, J.; Zheng, H.; Yarborough, K.; Woods, C.G.; Liu, D.; Yamamoto, M.; et al. Adipose deficiency of Nrf2 in ob/ob mice results in severe metabolic syndrome. *Diabetes* **2013**, *62*, 845–854. [CrossRef] [PubMed]

64. Tanaka, Y.; Aleksunes, L.M.; Yeager, R.L.; Gyamfi, M.A.; Esterly, N.; Guo, G.L.; Klaassen, C.D. NF-E2-related factor 2 inhibits lipid accumulation and oxidative stress in mice fed a high-fat diet. *J. Pharmacol. Exp. Ther.* **2008**, *325*, 655–664. [CrossRef]

65. Li, L.; Fu, J.; Liu, D.; Sun, J.; Hou, Y.; Chen, C.; Shao, J.; Wang, L.; Wang, X.; Zhao, R.; et al. Hepatocyte-specific Nrf2 deficiency mitigates high-fat diet-induced hepatic steatosis: Involvement of reduced PPARgamma expression. *Redox Biol.* **2020**, *30*, 101412. [CrossRef] [PubMed]

66. Taguchi, K.; Maher, J.M.; Suzuki, T.; Kawatani, Y.; Motohashi, H.; Yamamoto, M. Genetic analysis of cytoprotective functions supported by graded expression of Keap1. *Mol. Cell. Biol.* **2010**, *30*, 3016–3026. [CrossRef] [PubMed]

67. Schneider, K.S.; Chan, J.Y. Emerging role of Nrf2 in adipocytes and adipose biology. *Adv. Nutr.* **2013**, *4*, 62–66. [CrossRef] [PubMed]

68. Annie-Mathew, A.S.; Prem-Santhosh, S.; Jayasuriya, R.; Ganesh, G.; Ramkumar, K.M.; Sarada, D.V.L. The pivotal role of Nrf2 activators in adipocyte biology. *Pharmacol. Res.* **2021**, *173*, 105853. [CrossRef] [PubMed]

69. Zhang, Z.; Zhou, S.; Jiang, X.; Wang, Y.H.; Li, F.; Wang, Y.G.; Zheng, Y.; Cai, L. The role of the Nrf2/Keap1 pathway in obesity and metabolic syndrome. *Rev. Endocr. Metab. Disord.* **2015**, *16*, 35–45. [CrossRef]

70. Huang, J.; Tabbi-Anneni, I.; Gunda, V.; Wang, L. Transcription factor Nrf2 regulates SHP and lipogenic gene expression in hepatic lipid metabolism. *Am. J. Physiol. Gastrointest. Liver Physiol.* **2010**, *299*, G1211–G1221. [CrossRef]

71. Kim, S.; Sohn, I.; Ahn, J.I.; Lee, K.H.; Lee, Y.S.; Lee, Y.S. Hepatic gene expression profiles in a long-term high-fat diet-induced obesity mouse model. *Gene* **2004**, *340*, 99–109. [CrossRef]

72. Wakabayashi, N.; Itoh, K.; Wakabayashi, J.; Motohashi, H.; Noda, S.; Takahashi, S.; Imakado, S.; Kotsuji, T.; Otsuka, F.; Roop, D.R.; et al. Keap1-null mutation leads to postnatal lethality due to constitutive Nrf2 activation. *Nat. Genet.* **2003**, *35*, 238–245. [CrossRef]

73. Gao, T.; Lai, M.; Zhu, X.; Ren, S.; Yin, Y.; Wang, Z.; Liu, Z.; Zuo, Z.; Hou, Y.; Pi, J.; et al. Rifampicin impairs adipogenesis by suppressing NRF2-ARE activity in mice fed a high-fat diet. *Toxicol. Appl. Pharmacol.* **2021**, *413*, 115393. [CrossRef]

74. Oh, Y.; Ahn, C.B.; Je, J.Y. Low molecular weight blue mussel hydrolysates inhibit adipogenesis in mouse mesenchymal stem cells through upregulating HO-1/Nrf2 pathway. *Food Res. Int.* **2020**, *136*, 109603. [CrossRef]

75. Xu, J.; Shimpi, P.; Armstrong, L.; Salter, D.; Slitt, A.L. PFOS induces adipogenesis and glucose uptake in association with activation of Nrf2 signaling pathway. *Toxicol. Appl. Pharmacol.* **2016**, *290*, 21–30. [CrossRef]

76. Qiu, M.; Xiao, F.; Wang, T.; Piao, S.; Zhao, W.; Shao, S.; Yan, M.; Zhao, D. Protective effect of Hedansanqi Tiaozhi Tang against non-alcoholic fatty liver disease in vitro and in vivo through activating Nrf2/HO-1 antioxidant signaling pathway. *Phytomedicine* **2020**, *67*, 153140. [CrossRef] [PubMed]

77. Szczesny-Malysiak, E.; Stojak, M.; Campagna, R.; Grosicki, M.; Jamrozik, M.; Kaczara, P.; Chlopicki, S. Bardoxolone Methyl Displays Detrimental Effects on Endothelial Bioenergetics, Suppresses Endothelial ET-1 Release, and Increases Endothelial Permeability in Human Microvascular Endothelium. *Oxid. Med. Cell. Longev.* **2020**, *2020*, 4678252. [CrossRef]

78. Chartoumpekis, D.V.; Ziros, P.G.; Sykiotis, G.P.; Zaravinos, A.; Psyrogiannis, A.I.; Kyriazopoulou, V.E.; Spandidos, D.A.; Habeos, I.G. Nrf2 activation diminishes during adipocyte differentiation of ST2 cells. *Int. J. Mol. Med.* **2011**, *28*, 823–828. [CrossRef] [PubMed]

79. Hotamisligil, G.S.; Shargill, N.S.; Spiegelman, B.M. Adipose expression of tumor necrosis factor-alpha: Direct role in obesity-linked insulin resistance. *Science* **1993**, *259*, 87–91. [CrossRef] [PubMed]

80. Weisberg, S.P.; Leibel, R.; Tortoriello, D.V. Dietary curcumin significantly improves obesity-associated inflammation and diabetes in mouse models of diabesity. *Endocrinology* **2008**, *149*, 3549–3558. [CrossRef]

81. Shoelson, S.E.; Lee, J.; Goldfine, A.B. Inflammation and insulin resistance. *J. Clin. Investig.* **2006**, *116*, 1793–1801. [CrossRef]

82. Wu, H.; Ballantyne, C.M. Metabolic Inflammation and Insulin Resistance in Obesity. *Circ. Res.* **2020**, *126*, 1549–1564. [CrossRef] [PubMed]

83. Jia, Q.; Morgan-Bathke, M.E.; Jensen, M.D. Adipose tissue macrophage burden, systemic inflammation, and insulin resistance. *Am. J. Physiol. Endocrinol. Metab.* **2020**, *319*, E254–E264. [CrossRef]
84. Villarroya, F.; Cereijo, R.; Gavaldà-Navarro, A.; Villarroya, J.; Giralt, M. Inflammation of brown/beige adipose tissues in obesity and metabolic disease. *J. Intern. Med.* **2018**, *284*, 492–504. [CrossRef] [PubMed]
85. Rakotoarivelo, V.; Variya, B.; Ilangumaran, S.; Langlois, M.F.; Ramanathan, S. Inflammation in human adipose tissues-Shades of gray, rather than white and brown. *Cytokine Growth Factor Rev.* **2018**, *44*, 28–37. [CrossRef] [PubMed]
86. Kawai, T.; Autieri, M.V.; Scalia, R. Adipose tissue inflammation and metabolic dysfunction in obesity. *Am. J. Physiol. Cell Physiol.* **2021**, *320*, C375–C391. [CrossRef] [PubMed]
87. Nagashimada, M.; Sawamoto, K.; Ni, Y.; Kitade, H.; Nagata, N.; Xu, L.; Kobori, M.; Mukaida, N.; Yamashita, T.; Kaneko, S.; et al. CX3CL1-CX3CR1 Signaling Deficiency Exacerbates Obesity-induced Inflammation and Insulin Resistance in Male Mice. *Endocrinology* **2021**, *162*, bqab064. [CrossRef] [PubMed]
88. Nishimura, S.; Manabe, I.; Nagasaki, M.; Eto, K.; Yamashita, H.; Ohsugi, M.; Otsu, M.; Hara, K.; Ueki, K.; Sugiura, S.; et al. CD8+ effector T cells contribute to macrophage recruitment and adipose tissue inflammation in obesity. *Nat. Med.* **2009**, *15*, 914–920. [CrossRef]
89. Kobayashi, E.H.; Suzuki, T.; Funayama, R.; Nagashima, T.; Hayashi, M.; Sekine, H.; Tanaka, N.; Moriguchi, T.; Motohashi, H.; Nakayama, K.; et al. Nrf2 suppresses macrophage inflammatory response by blocking proinflammatory cytokine transcription. *Nat. Commun.* **2016**, *7*, 11624. [CrossRef]
90. Ahmed, S.M.; Luo, L.; Namani, A.; Wang, X.J.; Tang, X. Nrf2 signaling pathway: Pivotal roles in inflammation. *Biochim. Biophys. Acta Mol. Basis Dis.* **2017**, *1863*, 585–597. [CrossRef]
91. Guo, C.; Bi, J.; Li, X.; Lyu, J.; Liu, X.; Wu, X.; Liu, J. Immunomodulation effects of polyphenols from thinned peach treated by different drying methods on RAW264.7 cells through the NF-kappaB and Nrf2 pathways. *Food Chem.* **2021**, *340*, 127931. [CrossRef] [PubMed]
92. Wruck, C.J.; Streetz, K.; Pavic, G.; Gotz, M.E.; Tohidnezhad, M.; Brandenburg, L.O.; Varoga, D.; Eickelberg, O.; Herdegen, T.; Trautwein, C.; et al. Nrf2 induces interleukin-6 (IL-6) expression via an antioxidant response element within the IL-6 promoter. *J. Biol. Chem.* **2011**, *286*, 4493–4499. [CrossRef] [PubMed]
93. Bloch-Damti, A.; Bashan, N. Proposed mechanisms for the induction of insulin resistance by oxidative stress. *Antioxid Redox Signal.* **2005**, *7*, 1553–1567. [CrossRef] [PubMed]
94. Kasai, S.; Shimizu, S.; Tatara, Y.; Mimura, J.; Itoh, K. Regulation of Nrf2 by Mitochondrial Reactive Oxygen Species in Physiology and Pathology. *Biomolecules* **2020**, *10*, 320. [CrossRef]
95. Wang, Z.; Zuo, Z.; Li, L.; Ren, S.; Gao, T.; Fu, J.; Hou, Y.; Chen, Y.; Pi, J. Nrf2 in adipocytes. *Arch. Pharm. Res.* **2020**, *43*, 350–360. [CrossRef] [PubMed]
96. Uruno, A.; Furusawa, Y.; Yagishita, Y.; Fukutomi, T.; Muramatsu, H.; Negishi, T.; Sugawara, A.; Kensler, T.W.; Yamamoto, M. The Keap1-Nrf2 system prevents onset of diabetes mellitus. *Mol. Cell. Biol.* **2013**, *33*, 2996–3010. [CrossRef]
97. Uruno, A.; Yagishita, Y.; Katsuoka, F.; Kitajima, Y.; Nunomiya, A.; Nagatomi, R.; Pi, J.; Biswal, S.S.; Yamamoto, M. Nrf2-Mediated Regulation of Skeletal Muscle Glycogen Metabolism. *Mol. Cell. Biol.* **2016**, *36*, 1655–1672. [CrossRef] [PubMed]
98. Chartoumpekis, D.V.; Yagishita, Y.; Fazzari, M.; Palliyaguru, D.L.; Rao, U.N.; Zaravinos, A.; Khoo, N.K.; Schopfer, F.J.; Weiss, K.R.; Michalopoulos, G.K.; et al. Nrf2 prevents Notch-induced insulin resistance and tumorigenesis in mice. *JCI Insight* **2018**, *3*, e97735. [CrossRef] [PubMed]
99. Meakin, P.J.; Chowdhry, S.; Sharma, R.S.; Ashford, F.B.; Walsh, S.V.; McCrimmon, R.J.; Dinkova-Kostova, A.T.; Dillon, J.F.; Hayes, J.D.; Ashford, M.L. Susceptibility of Nrf2-null mice to steatohepatitis and cirrhosis upon consumption of a high-fat diet is associated with oxidative stress, perturbation of the unfolded protein response, and disturbance in the expression of metabolic enzymes but not with insulin resistance. *Mol. Cell. Biol.* **2014**, *34*, 3305–3320. [CrossRef]
100. Geng, L.; Lam, K.S.L.; Xu, A. The therapeutic potential of FGF21 in metabolic diseases: From bench to clinic. *Nat. Rev. Endocrinol.* **2020**, *16*, 654–667. [CrossRef] [PubMed]
101. Kim, K.H.; Jeong, Y.T.; Oh, H.; Kim, S.H.; Cho, J.M.; Kim, Y.N.; Kim, S.S.; Kim, D.H.; Hur, K.Y.; Kim, H.K.; et al. Autophagy deficiency leads to protection from obesity and insulin resistance by inducing Fgf21 as a mitokine. *Nat. Med.* **2013**, *19*, 83–92. [CrossRef] [PubMed]
102. Lin, Z.; Tian, H.; Lam, K.S.; Lin, S.; Hoo, R.C.; Konishi, M.; Itoh, N.; Wang, Y.; Bornstein, S.R.; Xu, A.; et al. Adiponectin mediates the metabolic effects of FGF21 on glucose homeostasis and insulin sensitivity in mice. *Cell. Metab.* **2013**, *17*, 779–789. [CrossRef] [PubMed]
103. Jimenez, V.; Jambrina, C.; Casana, E.; Sacristan, V.; Muñoz, S.; Darriba, S.; Rodó, J.; Mallol, C.; Garcia, M.; León, X.; et al. FGF21 gene therapy as treatment for obesity and insulin resistance. *EMBO Mol. Med.* **2018**, *10*, e8791. [CrossRef]
104. Kim, B.R.; Lee, G.Y.; Yu, H.; Maeng, H.J.; Oh, T.J.; Kim, K.M.; Moon, J.H.; Lim, S.; Jang, H.C.; Choi, S.H. Suppression of Nrf2 attenuates adipogenesis and decreases FGF21 expression through PPAR gamma in 3T3-L1 cells. *Biochem. Biophys. Res. Commun.* **2018**, *497*, 1149–1153. [CrossRef]
105. Yu, Y.; He, J.; Li, S.; Song, L.; Guo, X.; Yao, W.; Zou, D.; Gao, X.; Liu, Y.; Bai, F.; et al. Fibroblast growth factor 21 (FGF21) inhibits macrophage-mediated inflammation by activating Nrf2 and suppressing the NF-kappaB signaling pathway. *Int. Immunopharmacol.* **2016**, *38*, 144–152. [CrossRef] [PubMed]

106. Loh, K.; Deng, H.; Fukushima, A.; Cai, X.; Boivin, B.; Galic, S.; Bruce, C.; Shields, B.J.; Skiba, B.; Ooms, L.M.; et al. Reactive oxygen species enhance insulin sensitivity. *Cell. Metab.* **2009**, *10*, 260–272. [CrossRef] [PubMed]
107. Zhu, Q.; An, Y.A.; Kim, M.; Zhang, Z.; Zhao, S.; Zhu, Y.; Asterholm, I.W.; Kusminski, C.M.; Scherer, P.E. Suppressing adipocyte inflammation promotes insulin resistance in mice. *Mol. Metab.* **2020**, *39*, 101010. [CrossRef]
108. Wernstedt Asterholm, I.; Tao, C.; Morley, T.S.; Wang, Q.A.; Delgado-Lopez, F.; Wang, Z.V.; Scherer, P.E. Adipocyte inflammation is essential for healthy adipose tissue expansion and remodeling. *Cell. Metab.* **2014**, *20*, 103–118. [CrossRef]
109. Baumel-Alterzon, S.; Katz, L.S.; Brill, G.; Garcia-Ocana, A.; Scott, D.K. Nrf2: The Master and Captain of Beta Cell Fate. *Trends Endocrinol. Metab.* **2021**, *32*, 7–19. [CrossRef]
110. Azzimato, V.; Jager, J.; Chen, P.; Morgantini, C.; Levi, L.; Barreby, E.; Sulen, A.; Oses, C.; Willerbrords, J.; Xu, C.; et al. Liver macrophages inhibit the endogenous antioxidant response in obesity-associated insulin resistance. *Sci. Transl. Med.* **2020**, *12*, eaaw9709. [CrossRef]

Article

Broccoli-Derived Glucoraphanin Activates AMPK/PGC1α/NRF2 Pathway and Ameliorates Dextran-Sulphate-Sodium-Induced Colitis in Mice

Qiyu Tian [1,2], Zhixin Xu [1], Qi Sun [1], Alejandro Bravo Iniguez [1], Min Du [2] and Mei-Jun Zhu [1,*]

1 School of Food Science, Washington State University, Pullman, WA 99164, USA
2 Department of Animal Sciences, Washington State University, Pullman, WA 99164, USA
* Correspondence: meijun.zhu@wsu.edu

Abstract: As the prevalence of inflammatory bowel diseases (IBD) rises, the etiology of IBD draws increasing attention. Glucoraphanin (GRP), enriched in cruciferous vegetables, is a precursor of sulforaphane, known to have anti-inflammatory and antioxidative effects. We hypothesized that dietary GRP supplementation can prevent mitochondrial dysfunction and oxidative stress in an acute colitis mouse model induced by dextran sulfate sodium (DSS). Eight-week-old mice were fed a regular rodent diet either supplemented with or without GRP. After 4 weeks of dietary treatments, half of the mice within each dietary group were subjected to 2.5% DSS treatment to induce colitis. Dietary GRP decreased DSS-induced body weight loss, disease activity index, and colon shortening. Glucoraphanin supplementation protected the colonic histological structure, suppressed inflammatory cytokines, interleukin (IL)-1β, IL-18, and tumor necrosis factor-α (TNF-α), and reduced macrophage infiltration in colonic tissues. Consistently, dietary GRP activated AMP-activated protein kinase (AMPK), peroxisome proliferator-activated receptor-gamma coactivator (PGC)-1α, and nuclear factor erythroid 2-related factor 2 (NRF2) pathways in the colonic tissues of DSS-treated mice, which was associated with increased mitochondrial DNA and decreased content of the oxidative product 8-hydroxydeoxyguanosine (8-OHDG), a nucleotide oxidative product of DNA. In conclusion, dietary GRP attenuated mitochondrial dysfunction, inflammatory response, and oxidative stress induced by DSS, suggesting that dietary GRP provides a dietary strategy to alleviate IBD symptoms.

Keywords: glucoraphanin; inflammatory bowel diseases; DSS; NRF2; oxidative stress; mitochondrial homeostasis

Citation: Tian, Q.; Xu, Z.; Sun, Q.; Iniguez, A.B.; Du, M.; Zhu, M.-J. Broccoli Derived Glucoraphanin Activates AMPK/PGC1α/NRF2 Pathway and Ameliorates Dextran-Sulphate-Sodium-Induced Colitis in Mice. *Antioxidants* **2022**, *11*, 2404. https://doi.org/10.3390/antiox11122404

Academic Editor: Marcel Bonay

Received: 4 November 2022
Accepted: 1 December 2022
Published: 4 December 2022

Publisher's Note: MDPI stays neutral with regard to jurisdictional claims in published maps and institutional affiliations.

1. Introduction

Inflammatory bowel diseases (IBDs), such as Crohn's disease and ulcerative colitis, are chronic inflammatory disorders that affect the gastrointestinal tract, and their incidence and clinical severity have increased worldwide [1]. IBDs have a very complicated etiology, which is closely associated with immune system dysregulation (caused by genetic or environmental factors), impaired intestinal homeostasis, mitochondrial dysfunction, and dysbiosis of gut microbiota. Since the gastrointestinal tract is exposed to numerous commensal bacteria and opportunistic pathogens, defects in the intestinal mucosal barrier allow immunogenic factors to infiltrate into the mucosa, causing chronic inflammation [2,3].

A wide variety of dietary bioactive compounds, such as polyphenols, are known to have anti-inflammatory effects [4,5]. Grape seed extract, containing high amounts of flavonoids and phenolic compounds, reduced the risk of IBD through suppressing inflammation [6]. Similarly, Goji berries suppressed inflammation and ameliorated dextran sulfate sodium (DSS)-induced colitis in mice [7].

As the precursor of phytochemical sulforaphane (SFN), glucoraphanin (GRP) is enriched in cruciferous vegetables such as bok choy, broccoli, and cabbage. The inactive

---

precursor GRP can be hydrolyzed by myrosinase to produce SFN. Heating broccoli inactivates myrosinase, and microbiota-produced myrosinase plays an important role in hydrolyzing GRP [8]. SFN has been tested in various disease models for its preventive effects on inflammation, liver injury and steatosis [9,10]. Previous work has shown that sulforaphane alleviated colitis severity in mice induced by DSS, supported by suppressed inflammatory cytokine expression, enhanced M2 macrophage polarization, and increased probiotic *Butyricicoccus* abundance [11–13].

SFN robustly activates the Kelch-like ECH-associated protein 1- Nuclear factor erythroid 2- related factor 2 (KEAP1-NRF2) pathway [14,15]. NRF2 is a transcription factor that plays an important role in redox homeostasis. In normal conditions, KEAP1 is constantly expressed and located in the cytoplasm, and it tethers the transcription factor NRF2 to repress its function [16]. When oxidation occurs, NRF2 is released and translocated into the nucleus, where it binds to antioxidant response elements (ARE) in the promoters of genes involved in cellular stress responses to initiate their expression [17]. Phase 2 and antioxidative enzymes play an important role in detoxifying harmful chemicals and intermediates, which serve as a key cellular defense mechanism against carcinogens [18]. Several cytoprotective phase 2 enzymes are upregulated by NRF2, including heme oxygenase-1 (HO-1), NAD(P)H: quinone oxidoreductase (NQO-1), and glutathione S-transferases (GSTs) [19,20]. These enzymes have anti-inflammatory and antioxidative activities [21]. The NRF2 pathway also decreases the production of pro-inflammatory cytokines and suppresses inflammation [22], while *Nrf2*-deficient mice have decreased phase 2 enzyme activities, such as GST and NQO-1 [23]. As a result, *Nrf2* knockout mice were more sensitive to DSS-induced colitis, demonstrating severe colitis symptoms, with increased pro-inflammatory cytokine levels and decreased antioxidative enzyme expression [24]. These data show that NRF2 is a potential therapeutic target for IBD prevention and treatment. In addition, the risk of colorectal cancer is dramatically increased in patients diagnosed with colitis [25]. SFN exhibits anticancer properties in cell culture and animal models through different pathways and mechanisms [26,27]. SFN can block HT29 human colon cancer cells in G_2-M Phase and induce cell apoptosis [28].

The NRF2 pathway also protects against mitochondrial dysfunction in neurodegenerative diseases [29–31]. Mitochondrial dysfunction is also associated with IBD progression [32]. However, the effects of SFN/GRP on mitochondrial homeostasis in DSS-induced colitis have not been examined. Here, we hypothesized that broccoli-derived GRP attenuates DSS-induced colitis through activation of the NRF2 pathway, suppressing oxidative stress and inflammation while improving mitochondrial homeostasis.

2. Materials and Methods

2.1. Animals and Experimental Design

C57BL/6 mice were purchased from Jackson Laboratory (Bar Harbor, ME, USA). All animal procedures were approved by the Washington State University Institutional Animal Use and Care Committee (IAUCC). All mice were housed in a temperature-controlled room with a 12-h light/dark cycle. Mice had free access to food and water. Feed was changed daily to minimize the oxidation of functional compounds. Eight-week-old mice were randomly assigned to 2 groups, receiving either a control diet (AIN-93G purified diet, Harlan Laboratory) or a diet supplemented with glucoraphanin (600 ppm, Thorne Research) for 4 weeks. Then, mice in each dietary group were further randomly divided into two subgroups receiving no DSS, or 2.5% (*w/v*) DSS (Millipore, Billerica, MA, USA) for 9 days to induce colitis, followed by a 9-day recovery period with regular drinking water. Body weight was monitored daily during DSS treatment.

2.2. Assessment of Colitis Symptoms and Disease Activity Index

The disease activity index (DAI) score was assessed according to the previously described method [33]. Body weight loss (scored as 0–4), stool consistency (scored as 0–4), and fecal bleeding (scored as 0–4) were recorded daily during the DSS induction and recovery period. DAI score was recorded as the summation of body weight loss, stool consistency, and fecal bleeding.

2.3. Colon Sample Collection and Processing

At necropsy, mice were anesthetized with CO_2 inhalation, followed by cervical dislocation. The colon section was dissected, and a 5 mm segment of the distal colon was fixed in 4% (w/v) paraformaldehyde (pH 7.0), processed, and embedded in paraffin for histological analysis. The remaining colon tissue was rinsed with PBS, frozen in liquid nitrogen, and stored at $-80\ ^\circ$C for molecular and biochemical analysis.

2.4. Histological Evaluation

Histological evaluation was performed as previously described [7]. Paraffin-embedded distal colonic tissues were sectioned at 5 μm thickness, deparaffinized, and subjected to hematoxylin and eosin (H&E) staining. Histological examination and imaging were performed using a Lecia DM2000 LED light microscope (Leica Microsystems, Chicago, IL, USA). Each colonic section was scored blindly using previously published score criteria. The scores of crypt damage (0–4 scale), the severity of inflammation (0–3 scale), and depth of injury (0–3 scale) were recorded for each mouse. The pathobiological score is the summation of these scores on a scale from 0 to 10.

2.5. Immunoblotting Analyses

Powdered colonic samples were homogenized in a Precellys homogenizer (Rockville, MD, USA) with lysis buffer. The colonic protein extracts were separated by 10% SDS-polyacrylamide gel electrophoresis and transferred to nitrocellulose membranes. After blocking with 5% w/v bovine serum albumin, membranes were incubated overnight with the selected primary antibodies at 4 $^\circ$C. The membranes were rinsed three times using Tris-buffered saline (TBS) with 0.5% Tween 20, followed by incubation with either IRDye 680 goat anti-mouse or IRDye 800CW goat anti-rabbit secondary antibodies (Li-Cor Biosciences, Lincoln, NE, USA). Finally, the bands were visualized and quantified using the Odyssey Infrared Imaging System and Image Studio Lite software (Li-Cor Biosciences, Lincoln, NE, USA). Band density was normalized to the β-tubulin content. Antibodies against interleukin (IL)-1β (#12242), TNF-α (#3707), phos-AMP-activated protein kinase (AMPK) (#2535), AMPK (#5831), NRF2 (#12721), OPA1 Mitochondrial Dynamin Like GTPase (OPA1) (#80471), dynamin-related protein 1 (DRP1) (#8570) and HO-1 (#5853) were purchased from Cell Signaling Technology (Beverly, MA, USA), and antibodies against superoxide dismutase (SOD) 1 (sc-17767) and xanthine oxidase (sc-398548) were purchased from Santa Cruz Biotechnology Inc. (Dallas, TX, USA). Antibody against peroxisome proliferator-activated receptor-gamma coactivator (PGC-1α) (#66369-I) was purchased from Protintech (Rosemont, IL, USA). The antibodies against β-tubulin (#E7) and IL-18 (CPTC-IL18) were obtained from the Developmental Studies Hybridoma Bank, University of Iowa (Iowa City, IA, USA). All antibodies were used at 1:1000, except NRF2, SOD1, and xanthine oxidase which were used at 1:500.

2.6. Immunohistochemical Staining

Deparaffinized and rehydrated colonic tissue sections were permeabilized in Tris-buffered saline (TBS) containing 0.3% Triton X-100 at room temperature for 30 min, boiled in sodium citrate buffer (pH 6.0) for 20 min, blocked with 10% goat serum (Vector Laboratories, Burlingame, CA, USA) in TBS (pH 7.4) for 1 h to reduce non-specific binding, then incubated overnight with primary antibodies against F4/80 (Bio-Rad Laboratories, Hercules, CA, USA, MCA497R, 1:200) or 8-hydroxydeoxyguanosine (8-OHDG) (Santa Cruz Biotechnology, sc-39387, 1:100) at 4 $^\circ$C. Tissue sections were then incubated with biotinylated secondary antibodies (1:200, Vector Laboratories) at room temperature for 30 min. The positive cells were visualized using a Vectastain ABC and DAB peroxidase (HRP) substrate kit (Vector

Laboratories), followed by hematoxylin counterstaining. Images were obtained using a Lecia DM2000 LED light microscope (Leica Microsystems).

2.7. Mitochondrial DNA Copy Number

Total DNA was isolated from colon tissue. Mitochondrial DNA (mtDNA) copy number was measured by quantitative PCR (qPCR) and normalized using nuclear 18S rRNA gene copies. Relative mtDNA copy number was calculated by using the $2^{-\Delta\Delta CT}$ method.

2.8. Statistical Analysis

Data were analyzed as a complete randomized design using the General Linear Model and Statistical Analysis System (SAS Institute Inc., Cary, NC, USA). Unpaired Student's *t*-test and two-way ANOVA were used for identifying the significant difference. Data were presented as mean ± standard error of the means (SEM). A significant difference was considered as $p \leq 0.05$.

3. Results

3.1. Glucoraphanin Ameliorates the Symptoms of DSS-Induced Colitis

Dietary GRP supplementation reduced the body weight loss induced by DSS (Figure 1A), significantly improved fecal consistency and reduced blood in the stool (Figure 1B,C). Overall, dietary GRP improved the DAI score from day 9 to day 18 in mice challenged with DSS (Figure 1D). Glucoraphanin supplementation prevented colon length shortening in DSS-treated mice (Figure 1E,F). These data showed that GRP supplementation has protective effects against DSS-induced colitis.

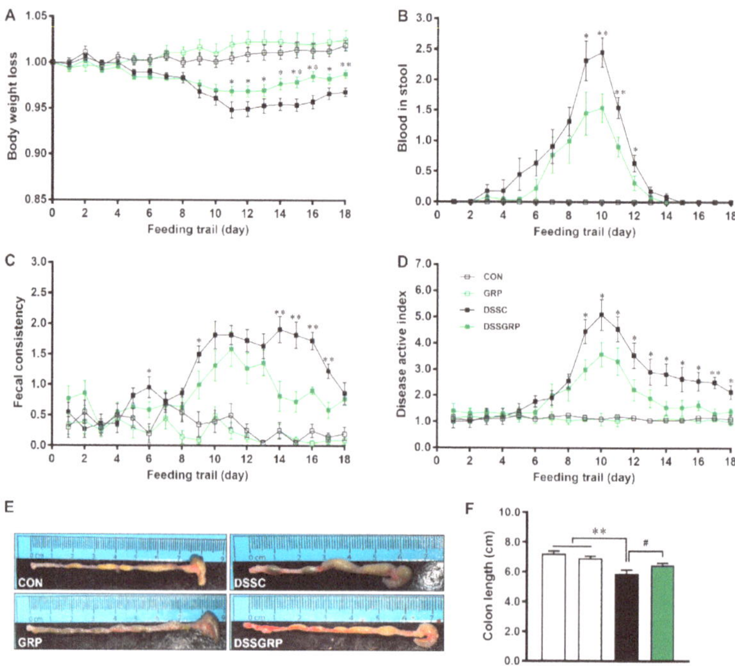

Figure 1. GRP supplementation suppresses disease symptoms in DSS-induced colitis mice. (**A**) Body weight loss. (**B**) Blood in stool. (**C**) Fecal consistency. (**D**) Disease activity index. (**E**) The representative images of the dissected colon. (**F**) Colon length. CON: Control, GRP: Sulforaphane glucosinolate, DSSC: DSS-treated control mice, DSSGRP: DSS-treated mice with GRP supplementation. Mean ± SEM, *n* = 10. $^{\#} p < 0.10$, $^{*} p \leq 0.05$, $^{**} p \leq 0.01$.

3.2. Glucoraphanin Alleviates Histological Damage and Suppresses Inflammation in DSS-Induced Colitis

Histologically, DSS elicited notable distortion of crypts in the intestine, which was rescued by GRP supplementation (Figure 2A,B). F4/80 is a well-established marker for macrophages [34]. The F4/80-positive cells detected by IHC staining were decreased by dietary GRP in mice with DSS-induced colitis (Figure 3A,B), suggesting that GRP supplementation reduced macrophage infiltration in DSS-treated mice. Furthermore, GRP supplementation reduced the expression of inflammatory cytokines IL-1β, IL-18, and TNF-α (Figure 3C). Data suggested that GRP protects the colonic structure and prevents inflammatory response in DSS-induced colitis.

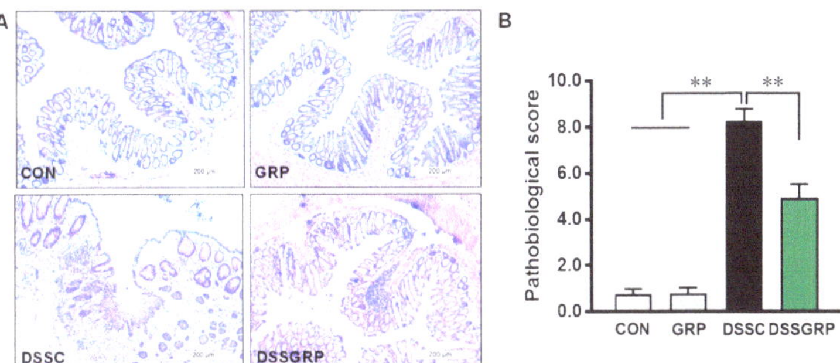

Figure 2. GRP supplementation decreases pathological scores in the colonic tissues of mice. (**A**) Representative hematoxylin and eosin staining. (**B**) The pathological score of distal colonic tissues. CON: Control, GRP: Sulforaphane glucosinolate, DSSC: DSS-treated control mice, DSSGRP: DSS-treated mice with GRP supplementation. n = 8–9 mice per group. ** $p \leq 0.01$.

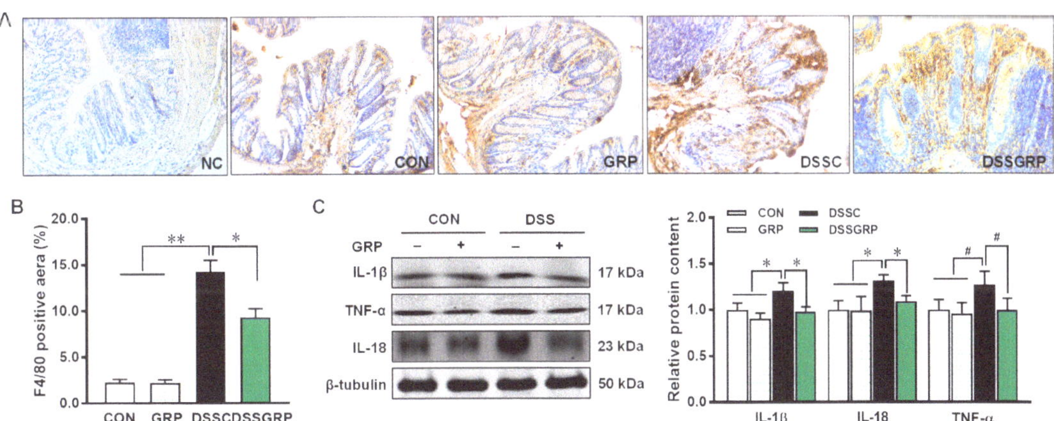

Figure 3. GRP supplementation inhibits macrophage infiltration and suppresses inflammation in the colonic tissues of mice with DSS-induced colitis. (**A**) Representative images of F4/80 staining. (**B**) The statistics of F4/80 staining. (**C**) Relative protein contents of IL-1β, TNF-α, and IL18. NC: Negative control, CON: Control, GRP: Sulforaphane glucosinolate, DSSC: DSS-treated control mice, DSSGRP: DSS-treated mice with GRP supplementation. n = 8–9 mice per group. # $p < 0.10$, * $p \leq 0.05$, ** $p \leq 0.01$.

3.3. Glucoraphanin Activates AMPK and Mitochondrial Biogenesis in DSS-Induced Colitis

Hyperactivation of the inflammatory response is associated with DSS-induced colitis. AMPK is known for its anti-inflammatory effects [35]. GRP supplementation elevated AMPK activity as shown by increased AMPK phosphorylation (Figure 4A). There is a strong correlation between AMPK and PGC-1α [36,37]. As the major regulator of mitochondria biogenesis, PGC-1α expression was also increased by GRP intake in DSS-treated mice (Figure 4B). Furthermore, GRP supplementation increased mtDNA copy number in DSS-treated mice (Figure 4C). No difference was found in the protein expression of OPA1 which mediates mitochondrial fusion [38]. DRP1, which regulates mitochondrial fission [39], was significantly increased in DSS-treated mice, but suppressed by dietary GRP. These data suggested that GRP supplementation activates AMPK and upregulates mitochondrial biogenesis in DSS-challenged mice.

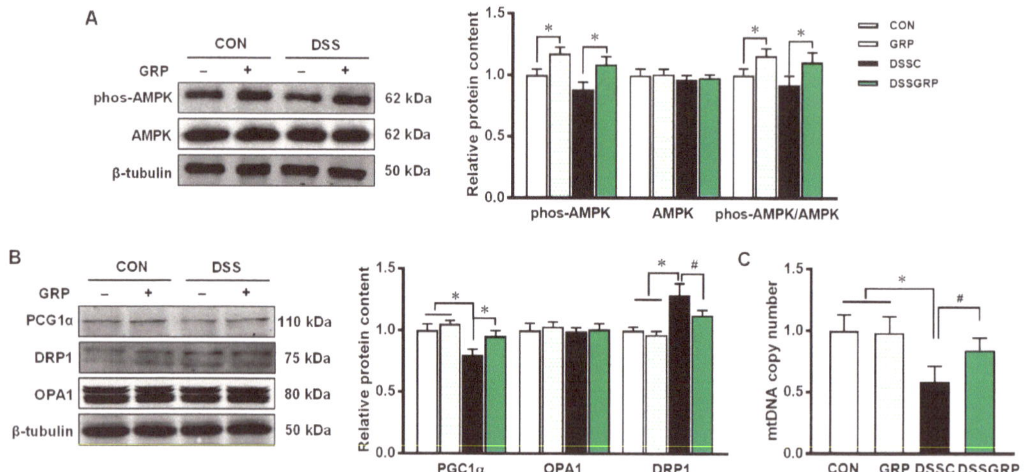

Figure 4. GRP supplementation activates AMPK and increases mitochondrial biogenesis. (**A**) Relative protein contents of phospho-AMPK and AMPK. (**B**) Relative protein contents of PGC-1α, DRP1, and OPA1. (**C**) Mitochondrial DNA copy number. CON: Control, GRP: Sulforaphane glucosinolate, DSSC: DSS-treated control mice, DSSGRP: DSS-treated mice with GRP supplementation. $n = 8$–9 mice per group. [#] $p < 0.10$, [*] $p \leq 0.05$.

3.4. Glucoraphanin Supplementation Activates NRF2 and Alleviates Oxidative Stress

Oxidative stress plays a significant role in the etiology of colitis and other inflammatory diseases. NRF2 is the master regulator of antioxidative responses that is activated by SFN [40]. Consistently, NRF2 was elevated by GRP supplementation in mice regardless of DSS induction (Figure 5A). Its downstream target HO-1 was lower in DSS-treated mice compared with control mice without DSS induction but restored by GRP supplementation. The content of xanthine oxidase was higher in DSS-treated mice compared to CON mice supplemented with or without GRP (Figure 5A). However, supplementation of GRP did decrease the level of xanthine oxidase compared to non DSS-treated mice. No difference was observed in SOD1 protein content between CON and DSS-treated mice (Figure 5A). The levels of 8-OHDG, a marker for DNA damage, were markedly increased in DSS-treated mice (Figure 5BC) but decreased by GRP supplementation. The results suggested the protective effects of GRP against DSS-induced DNA damage.

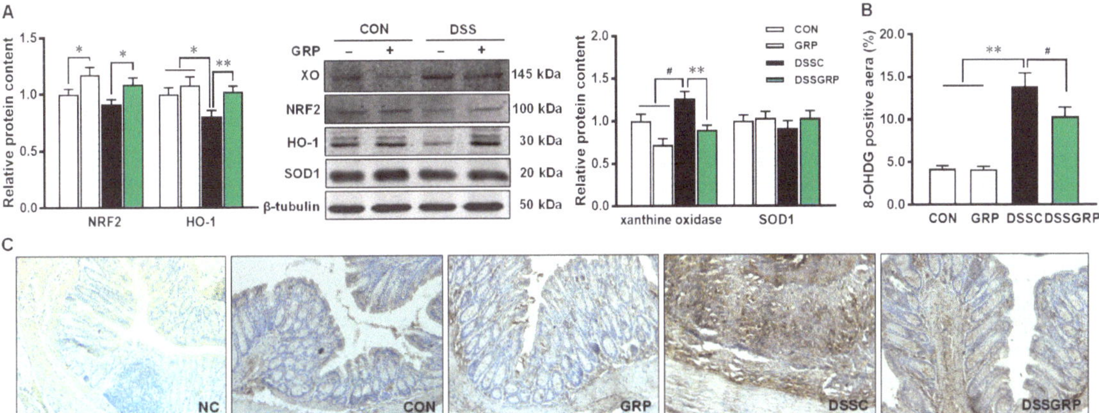

Figure 5. GRP supplementation activates NRF2 signaling and attenuates oxidative stress in the colonic tissues of mice with DSS-induced colitis. (**A**) Relative protein contents of NRF2, HO-1, xanthine oxidase (XO), and SOD1. (**B**) 8-OHdG staining statistics. (**C**) Representative images of 8-OHdG staining. NC: Negative control, CON: Control, GRP: Sulforaphane glucosinolate, DSSC: DSS-treated control mice, DSSGRP: DSS-treated mice with GRP supplementation. n = 8–9 mice per group. [#] $p < 0.10$, [*] $p \leq 0.05$, [**] $p \leq 0.01$.

4. Discussion

The incidence and prevalence of IBD are increasing across the world. North America and Europe have the highest incidence and prevalence of IBDs, which might be associated with the westernized diet [41]. However, the incidence of IBDs has also risen in developing countries in recent decades [1,42]. To prevent the relapse or improve the quality of life for IBD patients, various therapeutic approaches to IBD have been investigated, such as using anti-tumor necrosis factor (TNF) antibodies, Janus kinase inhibitors, and immunosuppressant drugs [43,44]. When patients received these drugs, they frequently experienced side effects such as asthma, diarrhea, and skin rash [45]. Natural bioactive compounds, enriched in foods and medicinal plants, provide an alternative for the treatment of chronic inflammatory diseases. Red raspberries reduced inflammation and protected mice from DSS-induced colitis through decreased immune cell infiltration, increased antioxidant enzyme expression, and enhanced intestinal barrier function [46]. Panaxynol extracted from American ginseng also showed beneficial effects on DSS-induced colitis, which exerted anti-inflammatory properties and prevented DNA damage in macrophages [47]. In a previous study, SFN administration prevented *H. pylori*-induced gastritis and suppressed non-steroidal anti-inflammatory drugs (NSAIDs)-induced oxidative stress [48]. Moreover, SFN inhibited histone deacetylase (HDAC) activity, which enhanced histone acetylation on the *bax* and *p21* promoters and their expression levels in prostate epithelial cells [49]. In DSS-induced colitis mice, dietary SFN supplementation decreased body weight loss [12], alleviated the severity of colitis induced by DSS, and prevented associated dysbiosis [11]. Consistently, we found that dietary GRP rescued the DSS-induced body weight loss, colon shortening, and decreased DAI score.

Inflammatory mediators and cytokines play a crucial role in IBD development and progression. IL-18 impairs the development and maturation of colonic goblet cells. The deletion of IL-18 conferred resistance against colitis. On the other hand, activation of IL-18 signaling exaggerated the progression of DSS-induced colitis associated with the depletion of goblet cells [50]. TNF-α and members of the IL family, such as IL-1, IL-6, IL-12, and IL-18, are produced by macrophages, dendritic cells, and epithelial cells, which promote inflammation [51,52]. Macrophages, neutrophils, and eosinophils were found to accumulate in DSS-induced acute colitis [53]. In this study, we found that GRP supplementation decreased macrophage infiltration in colonic tissue and downregulated inflammatory cytokines. Thus, dietary GRP protects against DSS-induced colitis through the suppression of the inflammatory response and immune cell infiltration.

AMPK is a key regulator of energy homeostasis and enhances intestinal epithelial differentiation and barrier function [54]. Metformin, an indirect AMPK activator, improves ileal epithelial barrier function in IL-10 knockout mice [55] and rescues intestinal barrier dysfunction induced by lipopolysaccharide [56]. AMPK can be activated by bioactive compounds such as resveratrol and polyphenol-rich foods such as purple potato and raspberry [57–59]. Our data showed that dietary GRP increased AMPK phosphorylation in DSS-induced mice. The PGC-1α level, a well-known AMPK downstream target [36], was also elevated in GRP-supplemented mice, indicating that GRP protected against colitis partially through stimulating the AMPK/PGC-1α pathway.

Mitochondrial function is indispensable for proper intestinal stem cell differentiation and function [60,61]; PGC-1α is a master regulator of mitochondrial biogenesis and respiration [62]. In ulcerative colitis patients, the intestine shows reduced mitochondrial oxidative phosphorylation (OXPHOS) [63,64] and decreased expression of genes involved in mitochondrial function [65]. In this study, dietary GRP increased PGC-1α, suggesting improved mitochondria biogenesis in DSS-induced colitis. In agreement, mtDNA copy number was significantly decreased in mice treated with DSS, which was restored by GRP supplementation. Mitochondrial fission is mediated by the dynamin-related GTPase, DRP1, which cooperates with other mitochondrial outer membrane proteins, such as Fis1 and Mff in mitochondrial fission [66]. Sulforaphane enhances the hyperfusion of mitochondria and impairs fission machinery in an NRF2-independent manner in human retinal pigment epithelial cells [67]. To remove dysfunctional mitochondria, fusion and fission are tightly regulated, in which fission is responsible for mitochondrial autophagy [68]. A recent study reported that DRP1 levels were increased in DSS-treated mice [69]. However, the underlying mechanisms of mitochondrial fusion and fission events in IBD are not fully understood. Here, we found that DRP1 was upregulated in DSS-treated mice, which was mitigated by GRP supplementation, without the involvement of OPA1. Our data suggested that GRP maintained mitochondrial homeostasis in DSS-induced colitis through suppression of mitochondrial fission.

Besides inflammation, oxidative stress is another major pathogenic factor in inflamed tissue; reactive oxygen species (ROS) associated with oxidative stress produces oxidation products and causes DNA damage [70]. The NRF2 pathway plays an important role in maintaining redox homeostasis [71]. When NRF2 is inactivated, it is trapped and degraded in the cytoplasm. The activated NRF2 translocates to the nuclei and initiates the expression of its target genes, including antioxidative and anti-inflammatory mediators and enzymes [72]. The endogenous antioxidant defense mechanism consists of several phase II detoxification enzymes, including GST and HO-1 [73,74]. The expression of these antioxidant enzymes is regulated by promoters containing ARE, which is bound by the NRF2 transcription factor [75]. In agreement with previous studies, we found that NRF2 and xanthine oxidase were highly elevated in GRP-supplemented DSS-treated mice. HO-1, as a rate-limiting enzyme in heme catabolism, possesses a wide spectrum of protective properties including antioxidative, anti-inflammatory, and antiapoptotic effects [76]. However, either an increase or decrease of HO-1 can be found in mice treated with DSS [77,78]. Our data showed increased HO-1 by GRP supplementation, consistent with increased NRF2. GRP supplementation did not change SOD1 levels, suggesting the preventative effect of GRP against oxidative stress might be through other mechanisms. DNA damage is associated with the initiation and promotion of carcinogenesis. 8-OHdG is one of the major products generated by free-radical-induced oxidation that causes DNA damage [79]. Consistent with the enhanced NRF2 and antioxidant enzymes, GRP supplementation decreased 8-OHdG levels in the colonic tissue of DSS-treated mice.

Inflammatory bowel diseases are constantly associated with dysbiosis of the gut microbiota. Dietary GRP was previously found to modulate gut microbiota in a high-fat diet mouse model. Body weight and fat mass were significantly decreased by GRP supplementation, associated with an increased abundance of *Akkermansia* and *Alloprevotella* [80]. In a DSS-induced colitis mouse model, SFN was reported to increase the abundance of a butyrate-producing bacteria, *Butyricicoccus,* and improve the symptoms of colitis [11]. As the precursor of SFN, GRP can be partially hydrolyzed by gut microbiota. Additional studies are required to evaluate the effects of GRP on IBD symptoms and alterations of the gut microbiota.

5. Conclusions

The results demonstrated that dietary GRP ameliorated the pathological score and colon shortening induced by DSS. The supplementation of GRP reduced the inflammatory response, activated NRF2, and alleviated oxidative stress in the colon of DSS-treated mice. Mitochondrial homeostasis was maintained by dietary GRP in a mouse model of DSS-induced colitis. Therefore, GRP provides an alternative strategy for alleviating IBD symptoms.

Author Contributions: Q.T.: conceptualization, methodology, data curation, investigation, writing—original draft. Z.X.: data curation. Q.S.: data curation. A.B.I.: data curation. M.D.: writing—review and editing. M.-J.Z.: conceptualization, supervision, writing—review and editing. All authors have read and agreed to the published version of the manuscript.

Funding: This work was financially supported by the National Institutes of USDA-National Institute of Food and Agriculture (USDA-NIFA) (2018-67017-27517), and Washington State University Agricultural Research Center Emerging Research Issues Competitive Grant.

Institutional Review Board Statement: All animal procedures were approved by the Washington State University Institutional Animal Use and Care Committee (IAUCC) (ethical code: ASAF #6484).

Informed Consent Statement: Not applicable.

Data Availability Statement: The data presented in this study are available in the article.

Conflicts of Interest: The authors declare no conflict of interest.

References

1. Molodecky, N.A.; Soon, S.; Rabi, D.M.; Ghali, W.A.; Ferris, M.; Chernoff, G.; Benchimol, E.I.; Panaccione, R.; Ghosh, S.; Barkema, H.W. Increasing incidence and prevalence of the inflammatory bowel diseases with time, based on systematic review. *Gastroenterology* **2012**, *142*, 46–54.e42. [PubMed]
2. Kaplan, M.; Mutlu, E.A.; Benson, M.; Fields, J.Z.; Banan, A.; Keshavarzian, A. Use of herbal preparations in the treatment of oxidant-mediated inflammatory disorders. *Complementary Ther. Med.* **2007**, *15*, 207–216. [CrossRef]
3. Antoni, L.; Nuding, S.; Wehkamp, J.; Stange, E.F. Intestinal barrier in inflammatory bowel disease. *World J. Gastroenterol.* **2014**, *20*, 1165. [CrossRef]
4. Scarano, A.; Butelli, E.; De Santis, S.; Cavalcanti, E.; Hill, L.; De Angelis, M.; Giovinazzo, G.; Chieppa, M.; Martin, C.; Santino, A. Combined dietary anthocyanins, flavonols, and stilbenoids alleviate inflammatory bowel disease symptoms in mice. *Front. Nutr.* **2018**, *4*, 75. [CrossRef]
5. Martin, D.A.; Bolling, B.W. A review of the efficacy of dietary polyphenols in experimental models of inflammatory bowel diseases. *Food Funct.* **2015**, *6*, 1773–1786. [CrossRef] [PubMed]
6. Wang, H.; Xue, Y.; Zhang, H.; Huang, Y.; Yang, G.; Du, M.; Zhu, M.J. Dietary grape seed extract ameliorates symptoms of inflammatory bowel disease in IL 10-deficient mice. *Mol. Nutr. Food Res.* **2013**, *57*, 2253–2257. [CrossRef] [PubMed]
7. Kang, Y.; Xue, Y.; Du, M.; Zhu, M.J. Preventive effects of Goji berry on dextran-sulfate-sodium-induced colitis in mice. *J. Nutr. Biochem.* **2017**, *40*, 70–76. [CrossRef]
8. Bheemreddy, R.M.; Jeffery, E.H. The metabolic fate of purified glucoraphanin in F344 rats. *J. Agric. Food Chem.* **2007**, *55*, 2861–2866. [CrossRef]
9. Heiss, E.; Herhaus, C.; Klimo, K.; Bartsch, H.; Gerhäuser, C. Nuclear factor κB is a molecular target for sulforaphane-mediated anti-inflammatory mechanisms. *J. Biol. Chem.* **2001**, *276*, 32008–32015. [CrossRef]
10. Zhou, R.; Lin, J.; Wu, D. Sulforaphane induces Nrf2 and protects against CYP2E1-dependent binge alcohol-induced liver steatosis. *Biochim. Biophys. Acta (BBA)-Gen. Subj.* **2014**, *1840*, 209–218. [CrossRef]
11. Zhang, Y.; Tan, L.; Li, C.; Wu, H.; Ran, D.; Zhang, Z. Sulforaphane alter the microbiota and mitigate colitis severity on mice ulcerative colitis induced by DSS. *AMB Express* **2020**, *10*, 1–9. [CrossRef] [PubMed]
12. Wagner, A.E.; Will, O.; Sturm, C.; Lipinski, S.; Rosenstiel, P.; Rimbach, G. DSS-induced acute colitis in C57BL/6 mice is mitigated by sulforaphane pre-treatment. *J. Nutr. Biochem.* **2013**, *24*, 2085–2091. [CrossRef] [PubMed]

13. Sun, Y.; Tang, J.; Li, C.; Liu, J.; Liu, H. Sulforaphane attenuates dextran sodium sulphate induced intestinal inflammation via IL-10/STAT3 signaling mediated macrophage phenotype switching. *Food Sci. Hum. Wellness* **2022**, *11*, 129–142. [CrossRef]

14. Bai, Y.; Cui, W.; Xin, Y.; Miao, X.; Barati, M.T.; Zhang, C.; Chen, Q.; Tan, Y.; Cui, T.; Zheng, Y. Prevention by sulforaphane of diabetic cardiomyopathy is associated with up-regulation of Nrf2 expression and transcription activation. *J. Mol. Cell. Cardiol.* **2013**, *57*, 82–95. [CrossRef] [PubMed]

15. Kensler, T.W.; Egner, P.A.; Agyeman, A.S.; Visvanathan, K.; Groopman, J.D.; Chen, J.G.; Chen, T.Y.; Fahey, J.W.; Talalay, P. Keap1–nrf2 signaling: A target for cancer prevention by sulforaphane. In *Natural Products in Cancer Prevention and Therapy*; Springer: Berlin/Heidelberg, Germany, 2012; pp. 163–177.

16. Baird, L.; Dinkova-Kostova, A.T. The cytoprotective role of the Keap1–Nrf2 pathway. *Arch. Toxicol.* **2011**, *85*, 241–272.

17. Motohashi, H.; Yamamoto, M. Nrf2–Keap1 defines a physiologically important stress response mechanism. *Trends Mol. Med.* **2004**, *10*, 549–557. [CrossRef]

18. Yu, X.; Kensler, T. Nrf2 as a target for cancer chemoprevention. *Mutat. Res. Fundam. Mol. Mech. Mutagen.* **2005**, *591*, 93–102. [CrossRef]

19. Kim, H.J.; Zheng, M.; Kim, S.K.; Cho, J.J.; Shin, C.H.; Joe, Y.; Chung, H.T. CO/HO-1 induces NQO-1 expression via Nrf2 activation. *Immune Netw.* **2011**, *11*, 376–382. [CrossRef]

20. Piao, M.S.; Park, J.J.; Choi, J.Y.; Lee, D.H.; Yun, S.J.; Lee, J.B.; Lee, S.C. Nrf2-dependent and Nrf2-independent induction of phase 2 detoxifying and antioxidant enzymes during keratinocyte differentiation. *Arch. Dermatol. Res.* **2012**, *304*, 387–395. [CrossRef]

21. Morse, D.; Choi, A.M. Heme oxygenase-1: The "emerging molecule" has arrived. *Am. J. Respir. Cell Mol. Biol.* **2002**, *27*, 8–16. [CrossRef]

22. Kim, J.; Cha, Y.N.; Surh, Y.J. A protective role of nuclear factor-erythroid 2-related factor-2 (Nrf2) in inflammatory disorders. *Mutat. Res. Fundam. Mol. Mech. Mutagen.* **2010**, *690*, 12–23. [CrossRef]

23. Ramos-Gomez, M.; Kwak, M.K.; Dolan, P.M.; Itoh, K.; Yamamoto, M.; Talalay, P.; Kensler, T.W. Sensitivity to carcinogenesis is increased and chemoprotective efficacy of enzyme inducers is lost in nrf2 transcription factor-deficient mice. *Proc. Natl. Acad. Sci. USA* **2001**, *98*, 3410–3415. [CrossRef]

24. Khor, T.O.; Huang, M.T.; Kwon, K.H.; Chan, J.Y.; Reddy, B.S.; Kong, A.N. Nrf2-deficient mice have an increased susceptibility to dextran sulfate sodium–induced colitis. *Cancer Res.* **2006**, *66*, 11580–11584. [CrossRef]

25. Rogler, G. Chronic ulcerative colitis and colorectal cancer. *Cancer Lett.* **2014**, *345*, 235–241. [CrossRef]

26. Saavedra-Leos, M.Z.; Jordan-Alejandre, E.; Puente-Rivera, J.; Silva-Cázares, M.B. Molecular Pathways Related to Sulforaphane as Adjuvant Treatment: A Nanomedicine Perspective in Breast Cancer. *Medicina* **2022**, *58*, 1377. [CrossRef]

27. Clarke, J.D.; Dashwood, R.H.; Ho, E. Multi-targeted prevention of cancer by sulforaphane. *Cancer Lett.* **2008**, *269*, 291–304. [CrossRef] [PubMed]

28. Gamet-Payrastre, L.; Li, P.; Lumeau, S.; Cassar, G.; Dupont, M.A.; Chevolleau, S.; Gasc, N.; Tulliez, J.; Tercé, F. Sulforaphane, a naturally occurring isothiocyanate, induces cell cycle arrest and apoptosis in HT29 human colon cancer cells. *Cancer Res.* **2000**, *60*, 1426–1433. [PubMed]

29. Tufekci, K.U.; Civi Bayin, E.; Genc, S.; Genc, K. The Nrf2/ARE pathway: A promising target to counteract mitochondrial dysfunction in Parkinson's disease. *Parkinson's Dis.* **2011**, *2011*, 314082. [CrossRef] [PubMed]

30. Abdalkader, M.; Lampinen, R.; Kanninen, K.M.; Malm, T.M.; Liddell, J.R. Targeting Nrf2 to suppress ferroptosis and mitochondrial dysfunction in neurodegeneration. *Front. Neurosci.* **2018**, *12*, 466. [CrossRef] [PubMed]

31. Bento-Pereira, C.; Dinkova-Kostova, A.T. Activation of transcription factor Nrf2 to counteract mitochondrial dysfunction in Parkinson's disease. *Med. Res. Rev.* **2021**, *41*, 785–802. [CrossRef]

32. Novak, E.A.; Mollen, K.P. Mitochondrial dysfunction in inflammatory bowel disease. *Front. Cell. Dev. Biol.* **2015**, *3*, 62. [PubMed]

33. Bibi, S.; Kang, Y.; Du, M.; Zhu, M.J. Maternal high-fat diet consumption enhances offspring susceptibility to DSS-induced colitis in mice. *Obesity* **2017**, *25*, 901–908. [CrossRef]

34. dos Anjos Cassado, A. F4/80 as a major macrophage marker: The case of the peritoneum and spleen. *Macrophages* **2017**, *62*, 161–179. [CrossRef]

35. Guragain, D.; Gurung, P.; Chang, J.H.; Katila, N.; Chang, H.W.; Jeong, B.S.; Choi, D.Y.; Kim, J.A. AMPK is essential for IL-10 expression and for maintaining balance between inflammatory and cytoprotective signaling. *Biochim. Biophys. Acta (BBA)-Gen. Subj.* **2020**, *1864*, 129631. [CrossRef]

36. Chaube, B.; Malvi, P.; Singh, S.V.; Mohammad, N.; Viollet, B.; Bhat, M.K. AMPK maintains energy homeostasis and survival in cancer cells via regulating p38/PGC-1α-mediated mitochondrial biogenesis. *Cell Death Discov.* **2015**, *1*, 15063. [CrossRef] [PubMed]

37. Jäger, S.; Handschin, C.; St-Pierre, J.; Spiegelman, B.M. AMP-activated protein kinase (AMPK) action in skeletal muscle via direct phosphorylation of PGC-1α. *Proc. Natl. Acad. Sci. USA* **2007**, *104*, 12017–12022. [CrossRef]

38. Song, Z.; Ghochani, M.; McCaffery, J.M.; Frey, T.G.; Chan, D.C. Mitofusins and OPA1 mediate sequential steps in mitochondrial membrane fusion. *Mol. Biol. Cell* **2009**, *20*, 3525–3532. [CrossRef]

39. Losón, O.C.; Song, Z.; Chen, H.; Chan, D.C. Fis1, Mff, MiD49, and MiD51 mediate Drp1 recruitment in mitochondrial fission. *Mol. Biol. Cell* **2013**, *24*, 659–667. [CrossRef] [PubMed]

40. Zhang, C.; Su, Z.Y.; Khor, T.O.; Shu, L.; Kong, A.N.T. Sulforaphane enhances Nrf2 expression in prostate cancer TRAMP C1 cells through epigenetic regulation. *Biochem. Pharmacol.* **2013**, *85*, 1398–1404. [CrossRef] [PubMed]

41. Shoda, R.; Matsueda, K.; Yamato, S.; Umeda, N. Epidemiologic analysis of Crohn disease in Japan: Increased dietary intake of n-6 polyunsaturated fatty acids and animal protein relates to the increased incidence of Crohn disease in Japan. *Am. J. Clin. Nutr.* **1996**, *63*, 741–745. [CrossRef] [PubMed]

42. Kaplan, G.G.; Windsor, J.W. The four epidemiological stages in the global evolution of inflammatory bowel disease. *Nat. Rev. Gastroenterol. Hepatol.* **2021**, *18*, 56–66. [CrossRef]

43. Zenlea, T.; Peppercorn, M.A. Immunosuppressive therapies for inflammatory bowel disease. *World J. Gastroenterol.* **2014**, *20*, 3146. [CrossRef]

44. Hirten, R.P.; Sands, B.E. New Therapeutics for Ulcerative Colitis. *Annu. Rev. Med.* **2021**, *72*, 199–213. [CrossRef] [PubMed]

45. Seibold, F.; Fournier, N.; Beglinger, C.; Mottet, C.; Pittet, V.; Rogler, G.; Group, S.I.C.S. Topical therapy is underused in patients with ulcerative colitis. *J. Crohn's Colitis* **2014**, *8*, 56–63. [CrossRef] [PubMed]

46. Bibi, S.; Kang, Y.; Du, M.; Zhu, M.J. Dietary red raspberries attenuate dextran sulfate sodium-induced acute colitis. *J. Nutr. Biochem.* **2018**, *51*, 40–46. [CrossRef]

47. Chaparala, A.; Poudyal, D.; Tashkandi, H.; Witalison, E.E.; Chumanevich, A.A.; Hofseth, J.L.; Nguyen, I.; Hardy, O.; Pittman, D.L.; Wyatt, M.D. Panaxynol, a bioactive component of American ginseng, targets macrophages and suppresses colitis in mice. *Oncotarget* **2020**, *11*, 2026. [CrossRef] [PubMed]

48. Yanaka, A. Role of sulforaphane in protection of gastrointestinal tract against H. pylori and NSAID-induced oxidative stress. *Curr. Pharm. Des.* **2017**, *23*, 4066–4075. [CrossRef]

49. Myzak, M.C.; Hardin, K.; Wang, R.; Dashwood, R.H.; Ho, E. Sulforaphane inhibits histone deacetylase activity in BPH-1, LnCaP and PC-3 prostate epithelial cells. *Carcinogenesis* **2006**, *27*, 811–819. [CrossRef] [PubMed]

50. Nowarski, R.; Jackson, R.; Gagliani, N.; de Zoete, M.R.; Palm, N.W.; Bailis, W.; Low, J.S.; Harman, C.C.; Graham, M.; Elinav, E. Epithelial IL-18 equilibrium controls barrier function in colitis. *Cell* **2015**, *163*, 1444–1456. [CrossRef]

51. Obregon, C.; Rothen-Rutishauser, B.; Gerber, P.; Gehr, P.; Nicod, L.P. Active uptake of dendritic cell-derived exovesicles by epithelial cells induces the release of inflammatory mediators through a TNF-α-mediated pathway. *Am. J. Pathol.* **2009**, *175*, 696–705. [CrossRef]

52. Kelsall, B. Recent progress in understanding the phenotype and function of intestinal dendritic cells and macrophages. *Mucosal Immunol.* **2008**, *1*, 460–469. [CrossRef] [PubMed]

53. Stevceva, L.; Pavli, P.; Husband, A.J.; Doe, W.F. The inflammatory infiltrate in the acute stage of the dextran sulphate sodium induced colitis: B cell response differs depending on the percentage of DSS used to induce it. *BMC Clin. Pathol.* **2001**, *1*, 3. [CrossRef]

54. Sun, X.; Yang, Q.; Rogers, C.J.; Du, M.; Zhu, M.J. AMPK improves gut epithelial differentiation and barrier function via regulating Cdx2 expression. *Cell Death Differ.* **2017**, *24*, 819–831. [CrossRef]

55. Xue, Y.; Zhang, H.; Sun, X.; Zhu, M.J. Metformin improves ileal epithelial barrier function in interleukin-10 deficient mice. *PLoS ONE* **2016**, *11*, e0168670. [CrossRef] [PubMed]

56. Wu, W.; Wang, S.; Liu, Q.; Shan, T.; Wang, Y. Metformin protects against LPS-induced intestinal barrier dysfunction by activating AMPK pathway. *Mol. Pharm.* **2018**, *15*, 3272–3284. [CrossRef] [PubMed]

57. Zou, T.; Wang, B.; Yang, Q.; de Avila, J.M.; Zhu, M.J.; You, J.; Chen, D.; Du, M. Raspberry promotes brown and beige adipocyte development in mice fed high-fat diet through activation of AMP-activated protein kinase (AMPK) α1. *J. Nutr. Biochem.* **2018**, *55*, 157–164. [CrossRef] [PubMed]

58. Dasgupta, B.; Milbrandt, J. Resveratrol stimulates AMP kinase activity in neurons. *Proc. Natl. Acad. Sci. USA* **2007**, *104*, 7217–7222. [CrossRef]

59. Sun, X.; Du, M.; Navarre, D.A.; Zhu, M.J. Purple potato extract promotes intestinal epithelial differentiation and barrier function by activating AMP-activated protein kinase. *Mol. Nutr. Food Res.* **2018**, *62*, 1700536. [CrossRef]

60. Schell, J.C.; Wisidagama, D.R.; Bensard, C.; Zhao, H.; Wei, P.; Tanner, J.; Flores, A.; Mohlman, J.; Sorensen, L.K.; Earl, C.S. Control of intestinal stem cell function and proliferation by mitochondrial pyruvate metabolism. *Nat. Cell Biol.* **2017**, *19*, 1027–1036. [CrossRef]

61. Ludikhuize, M.C.; Meerlo, M.; Gallego, M.P.; Xanthakis, D.; Julià, M.B.; Nguyen, N.T.; Brombacher, E.C.; Liv, N.; Maurice, M.M.; Paik, J.-H. Mitochondria define intestinal stem cell differentiation downstream of a FOXO/Notch axis. *Cell Metab.* **2020**, *32*, 889–900.e887. [CrossRef] [PubMed]

62. Ventura-Clapier, R.; Garnier, A.; Veksler, V. Transcriptional control of mitochondrial biogenesis: The central role of PGC-1α. *Cardiovasc. Res.* **2008**, *79*, 208–217. [CrossRef] [PubMed]

63. Bär, F.; Bochmann, W.; Widok, A.; Von Medem, K.; Pagel, R.; Hirose, M.; Yu, X.; Kalies, K.; König, P.; Böhm, R. Mitochondrial gene polymorphisms that protect mice from colitis. *Gastroenterology* **2013**, *145*, 1055–1063.e1053. [CrossRef] [PubMed]

64. Sifroni, K.G.; Damiani, C.R.; Stoffel, C.; Cardoso, M.R.; Ferreira, G.K.; Jeremias, I.C.; Rezin, G.T.; Scaini, G.; Schuck, P.F.; Dal-Pizzol, F. Mitochondrial respiratory chain in the colonic mucosal of patients with ulcerative colitis. *Mol. Cell. Biochem.* **2010**, *342*, 111–115. [CrossRef] [PubMed]

65. Fukushima, K.; Fiocchi, C. Paradoxical decrease of mitochondrial DNA deletions in epithelial cells of active ulcerative colitis patients. *Am. J. Physiol.-Gastrointest. Liver Physiol.* **2004**, *286*, G804–G813. [CrossRef] [PubMed]

66. Chang, C.R.; Blackstone, C. Dynamic regulation of mitochondrial fission through modification of the dynamin-related protein Drp1. *Ann. N. Y. Acad. Sci.* **2010**, *1201*, 34. [CrossRef] [PubMed]

67. O'Mealey, G.B.; Berry, W.L.; Plafker, S.M. Sulforaphane is a Nrf2-independent inhibitor of mitochondrial fission. *Redox Biol.* **2017**, *11*, 103–110. [CrossRef] [PubMed]

68. Twig, G.; Elorza, A.; Molina, A.J.; Mohamed, H.; Wikstrom, J.D.; Walzer, G.; Stiles, L.; Haigh, S.E.; Katz, S.; Las, G. Fission and selective fusion govern mitochondrial segregation and elimination by autophagy. *EMBO J.* **2008**, *27*, 433–446. [CrossRef]

69. Mancini, N.L.; Goudie, L.; Xu, W.; Sabouny, R.; Rajeev, S.; Wang, A.; Esquerre, N.; Al Rajabi, A.; Jayme, T.S.; van Tilburg Bernandes, E. Perturbed mitochondrial dynamics is a novel feature of colitis that can be targeted to lessen disease. *Cell. Mol. Gastroenterol. Hepatol.* **2020**, *10*, 287–307. [CrossRef] [PubMed]

70. Wiseman, H.; Halliwell, B. Damage to DNA by reactive oxygen and nitrogen species: Role in inflammatory disease and progression to cancer. *Biochem. J.* **1996**, *313*, 17. [CrossRef]

71. Hayes, J.D.; Dinkova-Kostova, A.T. The Nrf2 regulatory network provides an interface between redox and intermediary metabolism. *Trends Biochem. Sci.* **2014**, *39*, 199–218. [CrossRef] [PubMed]

72. Kobayashi, M.; Yamamoto, M. Molecular mechanisms activating the Nrf2-Keap1 pathway of antioxidant gene regulation. *Antioxid. Redox Signal.* **2005**, *7*, 385–394. [CrossRef] [PubMed]

73. Reuland, D.J.; Khademi, S.; Castle, C.J.; Irwin, D.C.; McCord, J.M.; Miller, B.F.; Hamilton, K.L. Upregulation of phase II enzymes through phytochemical activation of Nrf2 protects cardiomyocytes against oxidant stress. *Free. Radic. Biol. Med.* **2013**, *56*, 102–111. [CrossRef]

74. Zhu, L.; Liu, Z.; Feng, Z.; Hao, J.; Shen, W.; Li, X.; Sun, L.; Sharman, E.; Wang, Y.; Wertz, K. Hydroxytyrosol protects against oxidative damage by simultaneous activation of mitochondrial biogenesis and phase II detoxifying enzyme systems in retinal pigment epithelial cells. *J. Nutr. Biochem.* **2010**, *21*, 1089–1098. [CrossRef] [PubMed]

75. Sohel, M.M.H.; Amin, A.; Prastowo, S.; Linares-Otoya, L.; Hoelker, M.; Schellander, K.; Tesfaye, D. Sulforaphane protects granulosa cells against oxidative stress via activation of NRF2-ARE pathway. *Cell Tissue Res.* **2018**, *374*, 629–641. [CrossRef]

76. Araujo, J.A.; Zhang, M.; Yin, F. Heme oxygenase-1, oxidation, inflammation, and atherosclerosis. *Front. Pharmacol.* **2012**, *3*, 119. [CrossRef] [PubMed]

77. Fan, H.; Chen, W.; Zhu, J.; Zhang, J.; Peng, S. Toosendanin alleviates dextran sulfate sodium-induced colitis by inhibiting M1 macrophage polarization and regulating NLRP3 inflammasome and Nrf2/HO-1 signaling. *Int. Immunopharmacol.* **2019**, *76*, 105909. [CrossRef]

78. Zhong, W.; Xia, Z.; Hinrichs, D.; Rosenbaum, J.T.; Wegmann, K.W.; Meyrowitz, J.; Zhang, Z. Hemin exerts multiple protective mechanisms and attenuates dextran sulfate sodium–induced colitis. *J. Pediatr. Gastroenterol. Nutr.* **2010**, *50*, 132–139. [CrossRef] [PubMed]

79. Valavanidis, A.; Vlachogianni, T.; Fiotakis, C. 8-hydroxy-2′-deoxyguanosine (8-OHdG): A critical biomarker of oxidative stress and carcinogenesis. *J. Environ. Sci. Health C* **2009**, *27*, 120–139. [CrossRef]

80. Xu, X.; Dai, M.; Lao, F.; Chen, F.; Hu, X.; Liu, Y.; Wu, J. Effect of glucoraphanin from broccoli seeds on lipid levels and gut microbiota in high-fat diet-fed mice. *J. Funct. Foods* **2020**, *68*, 103858. [CrossRef]

 *antioxidants*

Article

Quantitative Proteomics Identifies Novel Nrf2-Mediated Adaptive Signaling Pathways in Skeletal Muscle Following Exercise Training

Anjali Bhat [1], Rafay Abu [2,3], Sankarasubramanian Jagadesan [4], Neetha Nanoth Vellichirammal [4], Ved Vasishtha Pendyala [1], Li Yu [5], Tara L. Rudebush [5], Chittibabu Guda [4], Irving H. Zucker [5], Vikas Kumar [2,4,*] and Lie Gao [1,*]

[1] Department of Anesthesiology, University of Nebraska Medical Center, Omaha, NE 68198, USA
[2] Mass Spectrometry and Proteomics Core Facility, University of Nebraska Medical Center, Omaha, NE 68198, USA
[3] Department of Biochemistry, Glocal University, Saharanpur 247121, Uttar Pradesh, India
[4] Department of Genetics, Cell Biology & Anatomy, University of Nebraska Medical Center, Omaha, NE 68198, USA
[5] Department of Cellular & Integrative Physiology, University of Nebraska Medical Center, Omaha, NE 68198, USA
[*] Correspondence: vikas.kumar@unmc.edu (V.K.); lgao@unmc.edu (L.G.)

Citation: Bhat, A.; Abu, R.; Jagadesan, S.; Vellichirammal, N.N.; Pendyala, V.V.; Yu, L.; Rudebush, T.L.; Guda, C.; Zucker, I.H.; Kumar, V.; et al. Quantitative Proteomics Identifies Novel Nrf2-Mediated Adaptive Signaling Pathways in Skeletal Muscle Following Exercise Training. *Antioxidants* **2023**, *12*, 151. https://doi.org/10.3390/antiox12010151

Academic Editor: Marcel Bonay

Received: 5 December 2022
Revised: 31 December 2022
Accepted: 6 January 2023
Published: 9 January 2023

Abstract: Exercise training (ExT) improves skeletal muscle health via multiple adaptive pathways. Nrf2 is a principal antioxidant transcription factor responsible for maintaining intracellular redox homeostasis. In this study, we hypothesized that Nrf2 is essential for adaptive responses to ExT and thus beneficial for muscle. Experiments were carried out on male wild type (WT) and iMS-$Nrf2^{flox/flox}$ inducible muscle-specific Nrf2 (KO) mice, which were randomly assigned to serve as sedentary controls (Sed) or underwent 3 weeks of treadmill ExT thus generating four groups: WT-Sed, WT-ExT, KO-Sed, and KO-ExT groups. Mice were examined for exercise performance and in situ tibialis anterior (TA) contractility, followed by mass spectrometry-based proteomics and bioinformatics to identify differentially expressed proteins and signaling pathways. We found that maximal running distance was significantly longer in the WT-ExT group compared to the WT-Sed group, whereas this capacity was impaired in KO-ExT mice. Force generation and fatigue tolerance of the TA were enhanced in WT-ExT, but reduced in KO-ExT, compared to Sed controls. Proteomic analysis further revealed that ExT upregulated 576 proteins in WT but downregulated 207 proteins in KO mice. These proteins represent pathways in redox homeostasis, mitochondrial respiration, and proteomic adaptation of muscle to ExT. In summary, our data suggest a critical role of Nrf2 in the beneficial effects of SkM and adaptation to ExT.

Keywords: Nrf2; skeletal muscle; exercise; proteomics

1. Introduction

Physical activity is a fundamental component of human health [1]. Exercise training (ExT) effectively prevents, treats, or reverses many pathological conditions in vital organs from the brain to the kidney [2,3]. Skeletal muscle (SkM) provides contractile force for body movement and exhibits functional enhancement following physical activity. Repetitive contraction–relaxation significantly impacts muscular metabolism, function, and structure by inducing mitochondrial biogenesis, hyperplasia, hypertrophy, fiber type transformation, and angiogenesis [4]. These phenotypic changes are the result of an array of gene expressions and intracellular signaling transduction pathways in response to contraction-evoked biophysical and biochemical stimuli among which excess reactive oxygen species (ROS) play a major role [5]. Excess ROS oxidize cellular components, altering their native structure which can affect their homeostatic function. On the other hand, low levels of ROS serve as physiological messengers essential for activating intracellular signaling pathways under normal conditions (oxidative eustress) [6]. The level of ROS depends on the balance between production of ROS and the ability of the cell to scavenge ROS by antioxidant proteins. Consequently, oxidative distress occurs when excess ROS accumulates. However, when cellular ROS is diminished to below physiological levels and their signaling function is perturbed, additional pathological conditions associated with redox biology, known as 'reductive stress', are prevalent [7]. It is imperative for healthy cells to maintain a delicate balance between ROS generation and elimination.

Nuclear factor-erythroid factor 2-related factor 2 (Nrf2) is a crucial master transcription factor, affecting nearly 500 genes that encode proteins associated with redox equilibrium, energy metabolism, drug detoxification, and stress responses [8–10]. The function of Nrf2 is strictly regulated by its negative regulator Keap1, a protein classified as BTB-Kelch family, which binds to Cullin 3 and Rbx1 to form a multisubunit Cullins-RING ligase for protein ubiquitination. Two C-terminal Kelch-domains of the Keap1 dimer separately bind to the ETGE and DLG motifs of Nrf2 Neh2-domain, whose lysine residues are therefore ubiquitinylated by Cullin ligase, labeling Nrf2 for proteasomal degradation [11]. This negative regulatory process is dependent on ROS-induced modification of three sensor cysteines (Cys151/Cys273/Cys288) in Keap1. These perturbations lead to Nrf2-mediated antioxidant defense activation triggered by DNA binding to antioxidant response elements (AREs) of appropriate genes [12]. Together, ROS and the Nrf2/Keap1 system form a delicate feedback regulatory mechanism critical for the maintenance of intracellular redox homeostasis.

To investigate the functional significance of Nrf2 in SkM, we created two transgenic mouse lines—iMS-*Nrf2*$^{flox/flox}$ and iMS-*Keap1*$^{flox/flox}$—which allowed us to selectively delete (i.e., Nrf2 gene knockout) or upregulate (i.e., Keap1 gene knockout) SkM Nrf2 in a Tamoxifen-inducible manner [13]. Utilizing label-free mass spectrometry-based proteomic profiling of these mice, we identified over 200 cytoprotective proteins that were upregulated and four intracellular signaling pathways activated in SkM by Nrf2 [13]. We further explored the redox changes by employing iodoacetyl tandem mass tag (iodoTMT^{TM})-labeled cysteine quantitation. We found 34 redox sensitive proteins harboring reversible cysteines, which are associated with mitochondrial oxidative phosphorylation, energy metabolism, and extracellular matrix structure [14]. The data strongly suggest that Nrf2 plays a critical role in SkM function by regulating gene expression and inducing protein post-translational modification. In the present study, we hypothesized that ExT enhances SkM function via activation of Nrf2 signaling and that the beneficial adaptations of SkM to ExT are attenuated after Nrf2 gene deletion.

2. Materials and Methods

All animal procedures were conducted in accordance with the guidelines of the National Institutes of Health Guide for the Care and Use of Laboratory Animals and conformed to ARRIVE Guidelines (https://www.nc3rs.org.uk/arrive-guidelines, accessed on 9 September 2021), as approved by the Animal Care and Use Committee of the University of Nebraska Medical Centre (UNMC-IACUC Protocol #18-174-02).

2.1. Animal Preparation

Forty-eight male iMS-*Nrf2*$^{flox/flox}$ mice between the ages of 12 and 20 weeks were used in the present study. The iMS-*Nrf2*$^{flox/flox}$ model is a Tamoxifen inducible- skeletal muscle specific- Nrf2 knockout model created in our laboratory by crossing the HSA-MCM line [15] (Jackson Laboratory Stock No. 025750) with a Nrf2$^{flox/flox}$ line [16] (provided by Dr. Shyam Biswal of Johns Hopkins University, Baltimore, USA). This model has been previously tested and validated by genotyping, RT-PCR, and Western blotting and characterized by proteomic and bioinformatic analysis [13]. Mice were assigned to four groups (12/group): wildtype sedentary (WT-Sed), wildtype exercise training (WT-ExT), Nrf2 knockout sedentary (KO-Sed), and Nrf2 knockout exercise training (KO-ExT). In the KO groups, the skeletal muscle Nrf2 gene was inactivated by intraperitoneal injections of Tamoxifen at the age of 12 weeks for five consecutive days (2 mg/0.2 mL day^{-1}, Sigma-Aldrich, St. Louis, MO, USA; Cat. No. T5648). Control mice (Nrf2-intact control) received vehicle administration (15% ethanol in sunflower seed oil, 0.2 mL day^{-1} for 5 days). Four weeks post Tamoxifen, muscle Nrf2 gene KO mice and controls underwent the following experiments.

2.2. In Vivo Mouse Experiments

Exercise training—Mice in ExT groups received three consecutive weeks of training on a treadmill without incline as previously described [17], with a slight modification. The training protocol consisted of one warm-up period at the beginning and one cool-down period at the end of the training session at a speed of 6 m/min for 5 min, between which was a 50 min training period at 60% maximal workload. The maximal workload was calculated using the protocol described in the following section. Due to significant differences in initial exercise capacity among groups, the training intensity was set at 14 m/min for the WT-ExT group and 9 m/min for the KO-ExT group.

Exercise capacity evaluation—All mice received two tests (before and after ExT/Sed interventions) for maximal running distance on a treadmill (Model no. Exer-3/6 Treadmill; Columbus Instruments, Columbus, OH, USA), as previously described from our laboratory [18]. Over the first three days, the mice were placed on the treadmill for 20 min/day without running to acclimatize them to the test environment. On day 4, the treadmill was run at an inclination of 15° with the rear shocking grid on, starting at a speed of 6 m/min for 6 min, followed by an increase of 3 m/min every 3 min until exhaustion. Exhaustion was defined when the mice remained on the shocking grid (0.2 V, 1 Hz) for 5 s without attempting to reengage the treadmill.

In Situ Muscle Experiments—In situ tibialis anterior (TA) contractility was examined for both muscle force generation and fatigue tolerance, according to previous methods used on soleus and extensor digitorum longus muscles in our laboratory [18]. Under 2% isoflurane anesthesia, mice were placed on a metal heating pad in the supine position. A small incision on the right anterior ankle skin was made where the TA tendon was identified, severed, and tied with a no. 6 silk suture attached to a force transducer (MLT1030/A, ADInstruments; Colorado Springs, CO, USA). A silver bipolar electrode was then inserted into the right TA belly and an intermittent tetanic stimulation with trains of square wave pulses (2.5 V, 0.3 s at 50 Hz per 3 s for a total of 20 min) was delivered by a pulse generator (A310, WPI; Sarasota, FL, USA). Force generation was determined by averaging the force of the initial three tetanic contractions. Fatigue tolerance was then calculated as the percent force generated by the last three tetanic contractions to the initial three tetanic contractions. During the experiment, the mice were kept warm using an isothermal pad and heat lamp, while the TA belly and tendon were moisturized by periodic administration of warm saline. Once the functional assessment was completed, the mice were euthanized by inhalation of CO_2. The left TA belly was collected, snap-frozen in liquid nitrogen, and then saved in a −80 °C freezer for the following ex vivo analyses. We chose the left, rather than right, TA muscle for molecular biological assays to avoid potential contraction-induced influences on protein expression.

2.3. Ex Vivo Experiments

Mass spectrometry-based proteomics—Proteomic analysis of TA muscle was carried out as previously described [13]. Briefly, 30 mg muscle mass/sample was homogenized in Pierce RIPA buffer (Thermo Scientific, Rockford, IL 61101, USA) with 1% protease inhibitor cocktail (Abcam, ab65621) and then centrifuged at 20,000× *g* 20 min at 4 °C for protein extraction. Each sample produced 100 μg protein, which was equally distributed for proteomic analysis and Western blot assay. 50 μg protein was first reduced and alkylated with DTT and iodoacetamide respectively, following removal of detergent using chloroform/methanol extraction. The protein pellet obtained was re-suspended in 50 mM ammonium bicarbonate and digested with MS-grade trypsin to obtain peptides, which were re-suspended in 2% acetonitrile and 0.1% formic acid. 500 ng of peptide was loaded onto a trap column (Acclaim PepMap 100 75 μm × 2 cm C18 LC, Thermo Scientific, Rockford, IL 61101, USA) at a flow rate of 4 μL min^{-1} then separated with a Thermo RSLC Ultimate 3000 (Thermo Scientific) on a separate column (Thermo Easy-Spray PepMap RSLC C18 75 μm × 50 cm C-18 2 μm, Thermo Scientific, Rockford, IL 61101, USA) with a step gradient of 4–25% solvent B (0.1% FAin 80%ACN) for 10–130 min and 25–45% solvent B for 130–145 min at 300 nl min^{-1} and 50 °C. Eluted peptides were analyzed by a Thermo Orbitrap Fusion Lumos Tribrid (Thermo Scientific) mass spectrometer in a data-dependent acquisition mode. A survey full scan MS (m/z 350–1800) was acquired in the Orbitrap with a resolution of 120,000. The automatic gain control (AGC) target for precursor ion scan (MS1) was set as 4 × 10^5 and ion filling time set at 100 ms. The most intense ions with charge state 2–6 were isolated in 3 s cycles and fragmented using higher energy collisional dissociation fragmentation with 35% normalized collision energy and detected at a mass resolution of 30,000 at 200 m/z. The AGC target for MS/MS was set at 5 × 10^4 and ion filling time at 60 ms; dynamic exclusion was set for 30 s with a 10 ppm mass window. Label free quantitative analysis was performed using progenesis QI proteomics 4.2 (Nonlinear Dynamics, Milford, MA, USA). Individual raw files for biological replicates (n = 6) for each condition were imported and automatically aligned to a reference standard run by the software. A threshold alignment score of 75% and above was chosen as a cut off value. After peak picking, all the detected features were normalized to a reference run followed by ion abundance quantification. Quantified peaks were exported and searched for peptide identification against the Swiss-prot *Mus musculus* protein database downloaded on 13 February 2019 using the in-house mascot 2.6.2 (Matrix Science, Boston, MA, USA) search engine. The search was set up for full tryptic peptides with a maximum of two missed cleavage sites. Acetylation of protein N-terminus and oxidized methionine were included to serve as variable modifications; carbamidomethylation of cysteine was set as fixed modification. The precursor mass tolerance threshold was set to 10 ppm and the maximum fragment mass error was 0.02 Da. The significance threshold of the ion score was calculated based on a false discovery rate of 1%. Qualitative analysis was performed using protein expression fold changes between WT-ExT vs. WT-Sed and KO-ExT vs. KO-Sed were represented as Log$_2$FC$^{\text{WT-ExT}}$ and Log$_2$FC$^{\text{KO-ExT}}$, respectively.

Western blot analysis—Western blotting was then employed to confirm hits identified from the proteomics screen. Briefly, 50 μg protein/sample was boiled for 5 min and then loaded on a 7.5% SDS-PAGE gel (30 μg protein/10 μL per well) following electrophoresis using a Bio-Rad mini gel apparatus at 40 mA/gel for 45 min. The fractionated protein on the gel was stained by Ponceau S and then electrically transferred onto a polyvinyl difluoride membrane (Millipore). The membrane was first probed with following antibodies: NQO1 (ab80588, Abcam), SOD2 (sc-30080, Santa Cruz Biotechnology, Dallas, Texas 75220, USA), GSTM1 (12412-1-AP, Proteintech, Rosemont, IL60018, USA), and GSTM3 (15214-1-AP, Proteintech). The secondary antibodies were HRP Goat Anti-Rabbit IgG and HRP Goat anti-Mouse IgG (Thermo-Fisher Scientific, Rockford, IL 61101, USA). After three washes with TBST, the membrane was treated with an enhanced chemiluminescence substrate (Pierce; Rockford, IL, USA) for 5 min. The blots on the membrane were visualized and analyzed using a UVP BioImaging System (EpiChemi II Darkroom). The final reported data were normalized through Ponceau S staining.

Differential proteomic and pathway enrichment analyses—Proteins identified by mass spectrometry were quantified to identify differentially expressed proteins between each experimental and control condition. ANOVA *p*-value and absolute fold changes were used to identify differentially expressed proteins between sedentary and training mice. A protein was considered to be differentially expressed if the *p*-value was ≤0.05 and the absolute fold change ≥1.5. Gene enrichment analyses of differentially regulated proteins to identify known functions, pathways, and networks affected were performed using Ingenuity Pathway Analysis (IPA) (Ingenuity Systems; Mountain View, CA, USA).

Dihydroethidium (DHE) Staining—DHE staining was employed to evaluate muscle ROS. The unfixed frozen mouse TA muscle was cut into 30-μm sections using a cryostat

(Leica CM 1850) and placed on glass slides, which were immersed in 2×10^{-6} mol/L DHE diluted with DMSO and acetone. After incubation in a light-protected, humidified chamber at 37 °C for 30 min, extra staining solution was removed, followed by three rinses with PBS. The fluorescence generated by oxidized DHE was detected using a laser confocal microscope (Leica TSC STED) using 488-nm excitation wavelength and a 585-nm filter. The relative fluorescent intensity of image was quantified using Image-J software (National Institutes of Health).

2.4. Statistical Analyses

Physiological and western blot data—Data are expressed as mean ± SD. A *t*-test was used for analyzing the differences between gene knockout mice with WT using SigmaPlot software. A p value of <0.05 was statistically significant.

Proteomic and bioinformatic analyses—For all comparisons, the ANOVA p-value was used. The cut-off used for the differential expression analysis summary was $p \leq 0.05$, and for the absolute fold change, the cut off was ≥ 1.5. IPA pathway analysis was also performed on genes with the same cut-off. For Volcano plots, the cut-off used to add gene names to differentially expressed proteins were absolute Log2Fold change >1 and ANOVA p-value ≤ 0.05. For each replicate, we identified proteins that exhibit statistically significant changes in expression based on Benjamini–Hochberg (BH) method adjusted p-values at FDR of 0.05 to account for multiple comparisons, and volcano plots were generated using Partek Genomics Suite 7.0 (Partek Inc. Chesterfield, MO 63005, USA).

3. Results

3.1. Exercise Capacity and In Situ TA Contractility

Exercise capacity was determined by measuring maximal running distance on a treadmill at two time points, pre- and post-ExT/Sed period, in all groups (Figure 1). There was no difference at the first test (pre-ExT/Sed) within WT groups (ExT vs. Sed) or KO groups (ExT vs. Sed). In the second test, ExT-WT ran significantly further than Sed-WT (763.3 ± 74.9 vs. 552.8 ± 49.9 m, $p < 0.001$; n = 6/group), suggesting improved exercise performance following ExT when SkM-Nrf2 was intact. In contrast, when SkM-Nrf2 was deleted, there was significant impairment in exercise performance (ExT-KO 132.1 ± 39.5 vs. Sed-KO 213.7 ± 28.8 m, $p < 0.01$; n = 6/group), suggesting a critical role for Nrf2 in muscle adaptation to exercise. Surprisingly, in KO-Sed mice, we found a slightly—but significantly—reduced running distance after 3 weeks of sedentary activity. These data suggest that Nrf2 is, in fact, essential for the normal function of SkM.

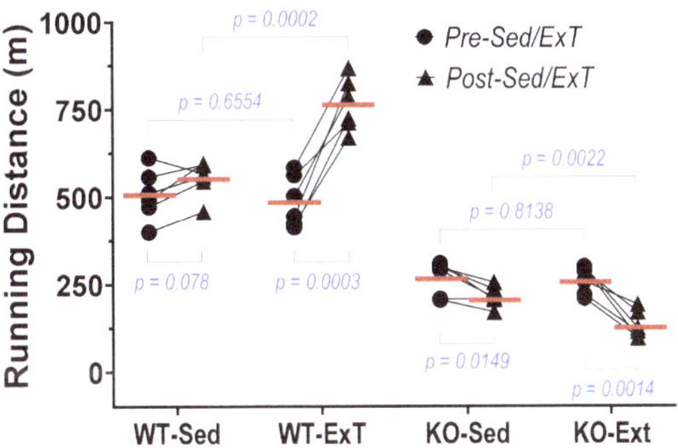

Figure 1. Effects of ExT on exercise performance of mice with SkM-Nrf2 intact (WT) and deleted (KO). n = 6/group.

Right TA contractility was induced by electrically stimulating muscle in situ for 20 min to determine force generation (g) and fatigue tolerance (% of ending force to initial force) (Figure 2A–C). The WT-ExT group displayed a significantly greater initial force generation

(ExT-WT 16.18 ± 0.77 vs. Sed-WT 12.32 ± 1.06 g, $p < 0.001$; n = 6/group) and greater fatigue tolerance (ExT-WT 69.37 ± 5.42 vs. Sed-WT 56.78 ± 6.59 %, $p < 0.01$; n = 6/group) compared with WT-Sed mice. These data support the view of an enhancement by ExT on muscle contractility. However, these two physiological parameters were reduced in KO mice undergoing ExT (Force Generation: ExT-KO 6.37 ± 0.74 vs. Sed-KO 7.82 ± 0.92 g, $p < 0.05$; Fatigue Tolerance: ExT-KO 23.97 ± 7.48 vs. Sed-KO 46.24 ± 7.95 %, $p < 0.001$; n = 6/group), suggesting a critical role of Nrf2 in muscle adaptation to exercise. Correspondingly, ROS level was reduced in WT-ExT group (7.13 ± 0.47 vs. 9.39 ± 0.25, $p < 0.005$) but increased in KO-ExT group (13.49 ± 0.65 vs. 11.18 ± 0.49, $p < 0.05$) as compared to their Sed controls (Figure 2D).

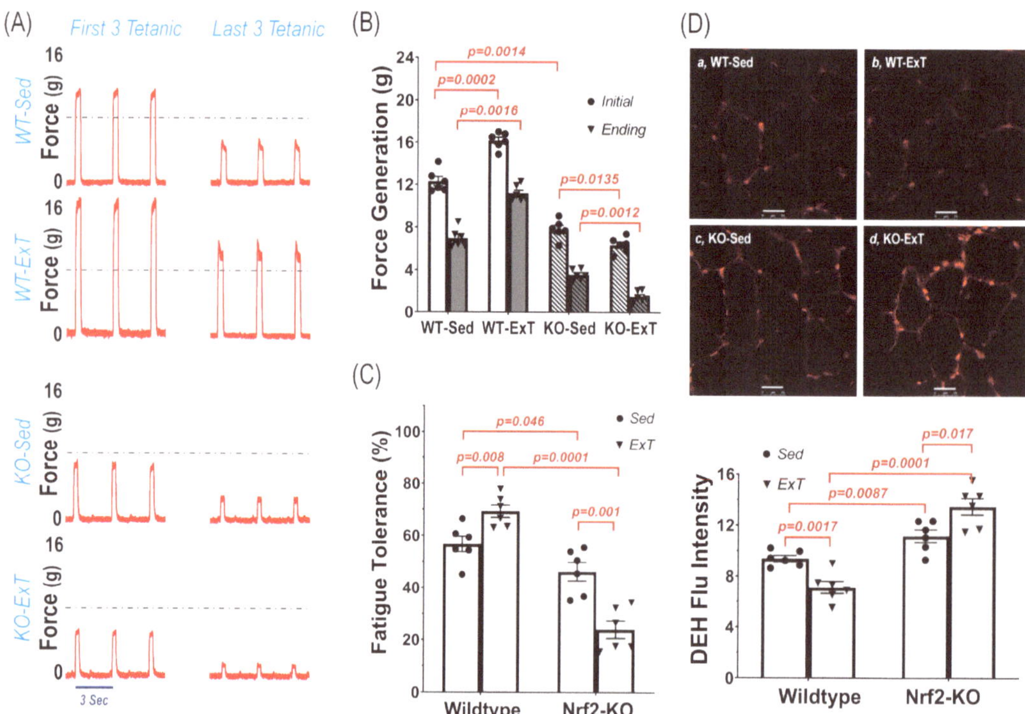

Figure 2. Effects of ExT on tibialis anterior contractility and ROS production in WT and SkM-Nrf2 KO mice. Contraction was induced by intermittent tetanic stimulation (2.5 V, 0.3 s at 50 Hz per 3 s for 20 min). (**A**) Original recording of the first and last three tetanic contractions. (**B**) Group data of force generation and (**C**) fatigue tolerance. (**D**) ROS level measured by DHE staining; Scale bar = 25 μm. n = 6/group.

3.2. Proteomics Analyses

The protein profile of the TA was determined by LCMS-based label-free quantitative mass spectrometry. From the samples of WT mice, we identified and quantified 2008 proteins, of which 592 (29.48%) were significantly altered in their expression by ExT (576 upregulation and 16 downregulation; Supplementary Table S1). In the samples from KO mice, 2055 proteins were identified and quantified with 279 proteins (13.58%) expression significantly changed by ExT (72 upregulation and 207 downregulation; Supplementary Table S2). This dataset demonstrates that the predominant effect of ExT on SkM at the molecular level is to upregulate protein expression, most of which was abolished when Nrf2 was deleted. This suggests that Nrf2 plays an important role in mediating the molecular adaptation of SkM to exercise. Principal component analysis (PCA) demonstrated that separation occurred between ExT and sedentary samples with 60.59% in WT and 45.89% in Nrf2-KO of variation explained by PC1 (Figure 3a,d). Volcano plots (Figure 3b,e) indicate significantly upregulated (red dots) or downregulated (green dots) proteins following ExT with fold changes and the expression profiles of unique and overlapping proteins between ExT and Sed groups (WT-ExT vs. WT-Sed and KO-ExT vs. KO-Sed). Heatmap analysis of the top 50 protein profiles revealed a relatively good consistency of protein profiles between Sed and ExT in WT and Nrf2-KO mice (Figure 3c,f).

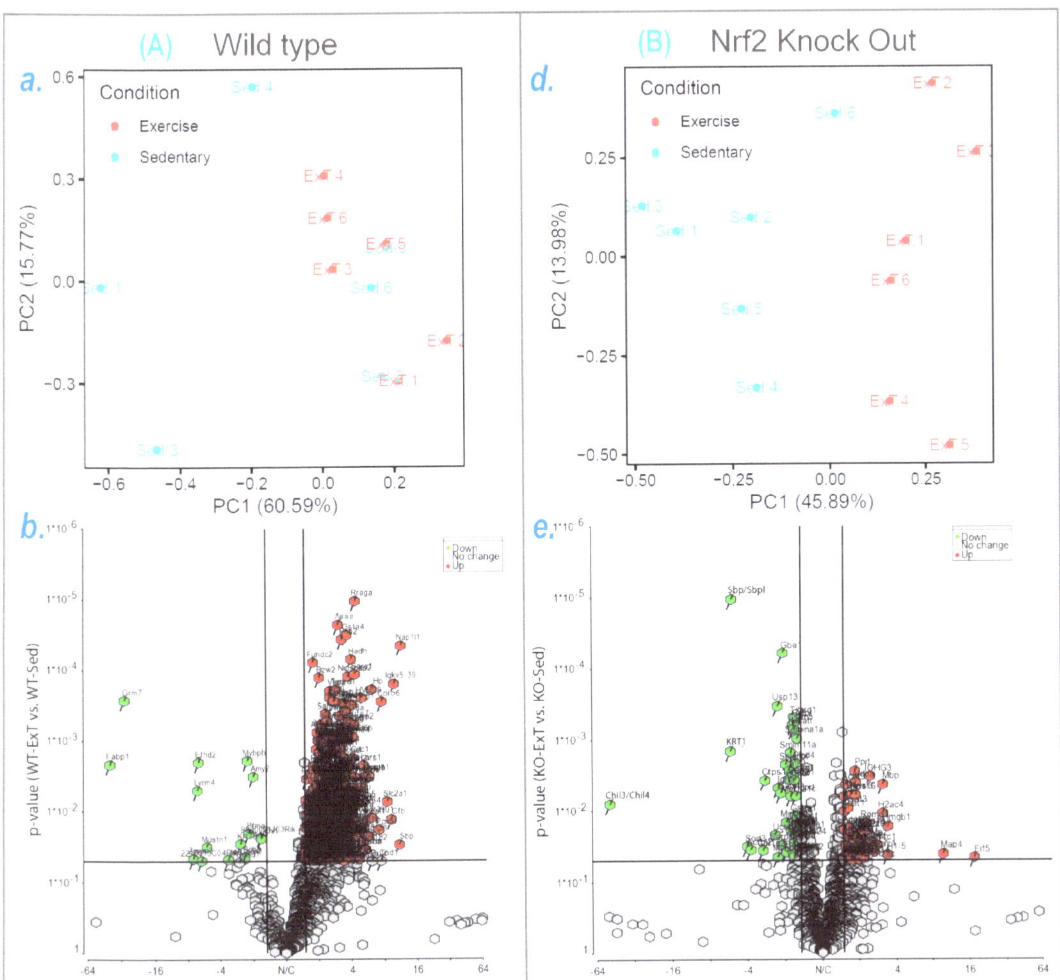

Figure 3. *Cont.*

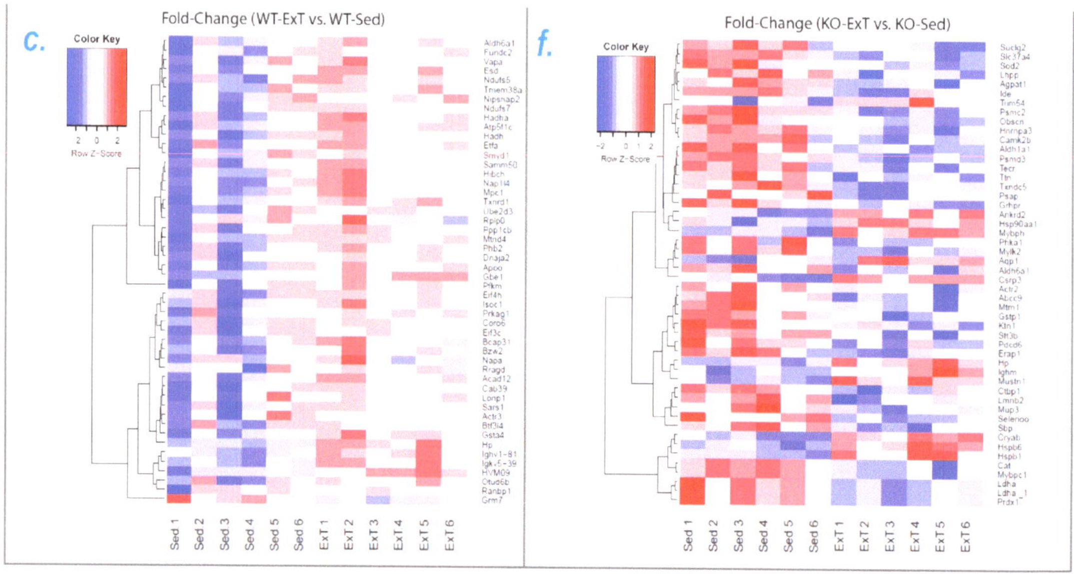

Figure 3. Differentially expressed proteins in TA muscles detected by mass spectrometry-based proteomics analysis with comparing WT-ExT to WT-Sed (**A**) and KO-ExT to KO-Sed (**B**) mice. (**a,d**) principal component analysis (PCA). The percent variance explained by each principal component is indicated on the PCA plot axes. (**b,e**) Volcano plots showing the log2 fold change (ExT/Sed) plotted against the −log10 p value, highlighting significantly changed proteins (red and green; $p < 0.05$ and an absolute fold change of 1.5, n = 6/group, moderated t-test). The vertical lines correspond to the absolute fold change of 1.5, and the horizontal line represents a p value of 0.05. Log$_2$ fold changes in WT-ExT and KO-ExT are represented as Log2FC$^{\text{WT-ExT}}$ and Log2FC$^{\text{KO-ExT}}$, respectively. (**c,f**), heatmaps show the log2 of the top 50 differentially expressed proteins.

3.3. Canonical Pathway Analyses

By performing Ingenuity Pathway Analysis (IPA) of the above proteomic profiles, we found that ExT changed 56 intracellular pathway statuses (54 activations and 2 inhibitions) in WT mice, such as the sirtuin signaling pathway, oxidative phosphorylation, Nrf2-mediated oxidative stress responses, NRE-mediated mRNA degradation pathway, HIF1a signaling, and others (Figure 4A). On the contrary, in Nrf2-KO mice, we found that ExT altered the status of 34 pathways (30 inhibitions and 4 activations), such as Nrf2-mediated oxidative stress responses, xenobiotic metabolism PXR signaling pathway, xenobiotic metabolism AHR signaling pathway, necroptosis signaling pathway, dilated cardiomyopathy signaling pathway, glutathione-mediated detoxification, and others. Figure 4 shows the top 20 signaling pathways modified by ExT in WT mice (Panel A) and Nrf2-KO mice (Panel B). This shows the total gene number/identified gene percentage for each pathway indicated in Figure 4A(a) and the full name/significance p-value for each pathway indicated in Figure 4A(b). The entire canonical pathway analysis data are shown in Supplementary Tables S3–S6. A comparative account of different signaling pathways involved in energy production, regulation of redox status, immune signaling, and pathways involved in general physiology in WT-ExT and in the KO-ExT mice are shown in Figure 5. Expression of proteins involved in important signaling mechanisms showed a significant reduction in Nrf2 KO mice after ExT suggesting the importance of an intact Nrf2 system in exercise mediated benefits.

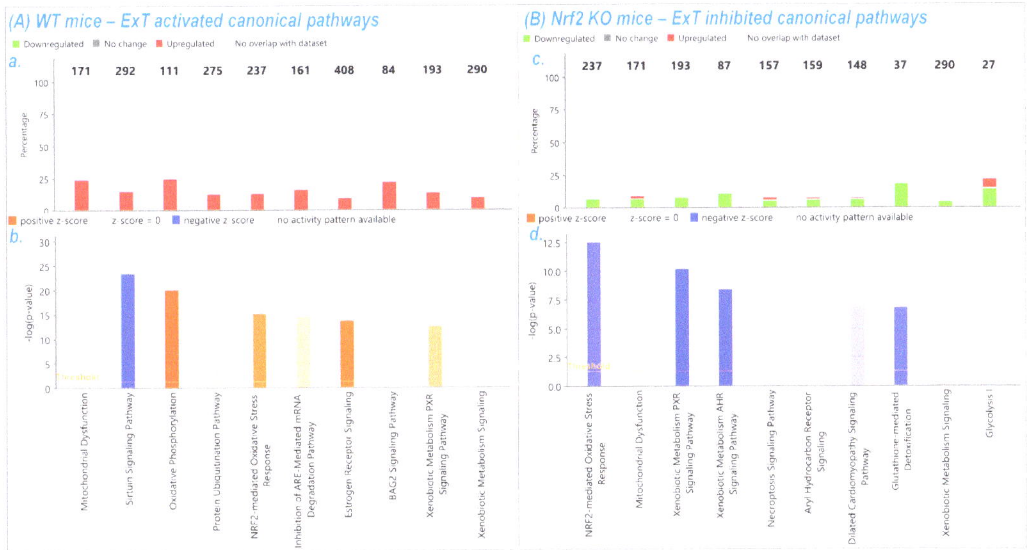

Figure 4. Stacked bar charts displaying the top 10 canonical pathways found to be differentially represented in comparing protein expression between ExT and Sed groups of WT mice (**A**) and Nrf2-KO mice (**B**). The bars are sorted so that the most significant pathways are at the left of the chart. (**a,c**) The total number of genes in each pathway is displayed above the bar. Percentage was calculated by dividing the number of identified genes by the total number of genes that map to the same pathway. (**b,d**) The pathway name is indicated under bar. The significance *p*-values are calculated by the right-tailed Fisher's Exact Test, such that taller bars equate to increased significance.

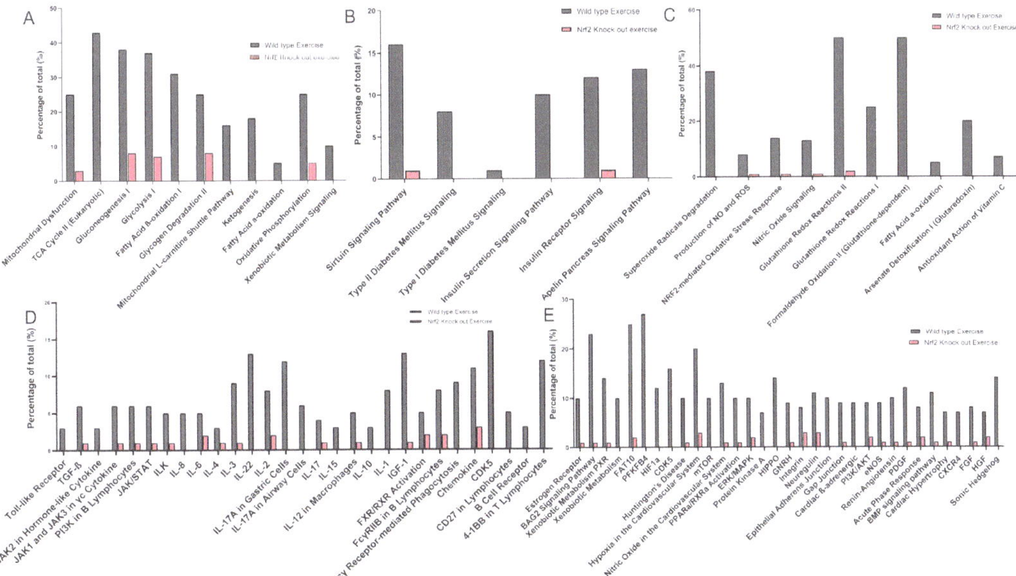

Figure 5. Percentage protein expression in pathways related to (**A**) energy production, (**B**) insulin signaling, (**C**) redox status, (**D**) immune signaling, and (**E**) other physiological pathways in WT-ExT mice in comparison to Nrf2-KO mice after ExT. Gray bars indicate proteins upregulated in WT mice and red bars indicate their downregulated expression levels in Nrf2-KO mice.

3.4. Nrf2 Specific Pathway Analyses

While IPA revealed approximately 90 intracellular pathways altered by ExT, our interest was focused primarily on Nrf2-mediated signaling. In WT mice, ExT upregulated 33 of 237 known Nrf2-associated proteins (14%; Figure 4A(a); Supplementary Table S3), whereas in Nrf2-KO mice, ExT downregulated 18 (8%) and upregulated only 3 (1%) of 237 Nrf2-proteins (Figure 4B(c); Supplementary Table S5). These results are shown in Figure 6 where it can be seen which Nrf2 target proteins were upregulated by ExT in WT mice (Purple symbols in Figure 4A) and downregulated by ExT in Nrf2-KO mice (Green symbols in Figure 4B). While the majority of these Nrf2-target proteins overlap between the two groups (see Venn Diagram), several proteins were identified only in WT or Nrf2-KO mice. These include USP14, HIP2, CCT7, FTL, FTH1, and SQSTM1, most of which are involved in ubiquitination, proteasomal degradation, protein repair, and protein removal (inserted Table in Figure 6).

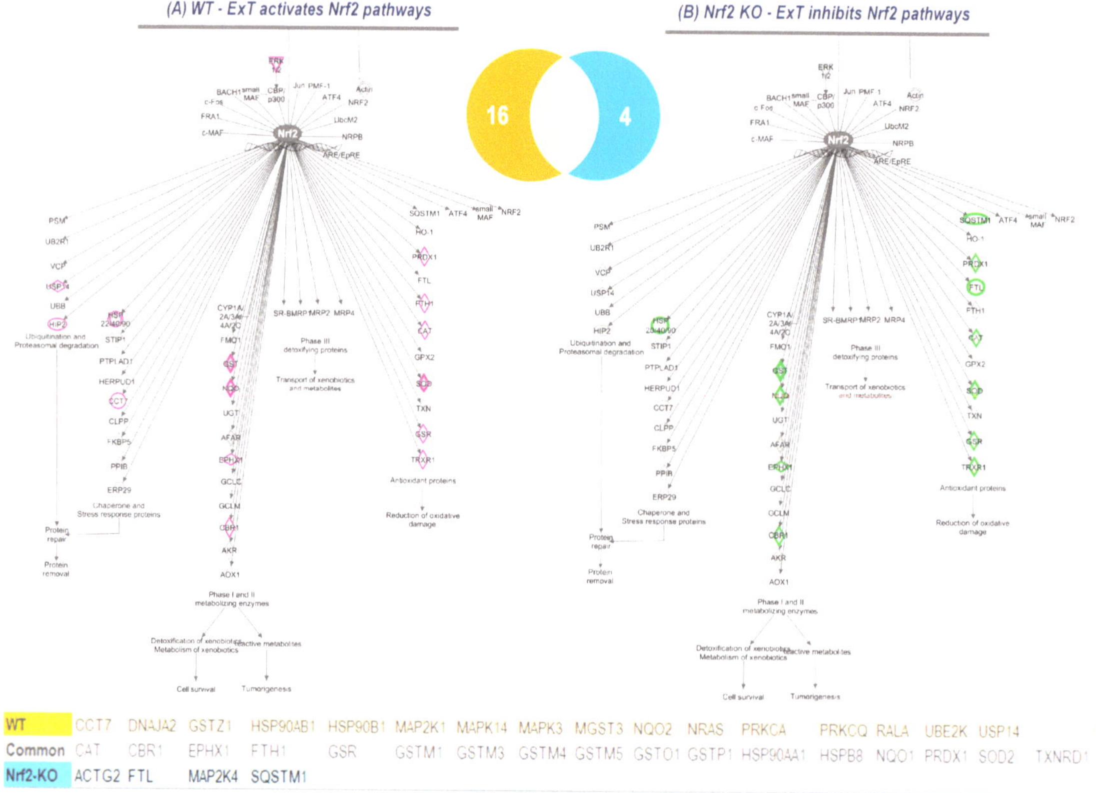

Figure 6. Nrf2 downstream proteins upregulated in WT mice (Purple symbols in (**A**)) and downregulated in Nrf2-KO mice (Green symbols in (**B**)) after ExT. The Venn diagram shows overlap of Nrf2-associated proteins identified in WT and Nrf2-KO mice. Their gene IDs are listed in the inserted table with yellow as WT unique, grey as the common, and blue as Nrf2-KO unique.

3.5. Mitochondrial Function and Oxidative Phosphorylation Analyses

One of the most important adaptations to endurance training is an increase in process of ATP production in skeletal muscle. We therefore performed the IPA analysis of proteomics data to determine ExT-induced mitochondrial function and oxidative phosphorylation. As can be seen in Figure 7, all of the five complex functions were enhanced in WT mice after ExT (Figure 7A). However, in Nrf2-KO mice, ExT reduced complex I function, enhanced complex IV and V function, but did not change complex II and III function (Figure 7B).

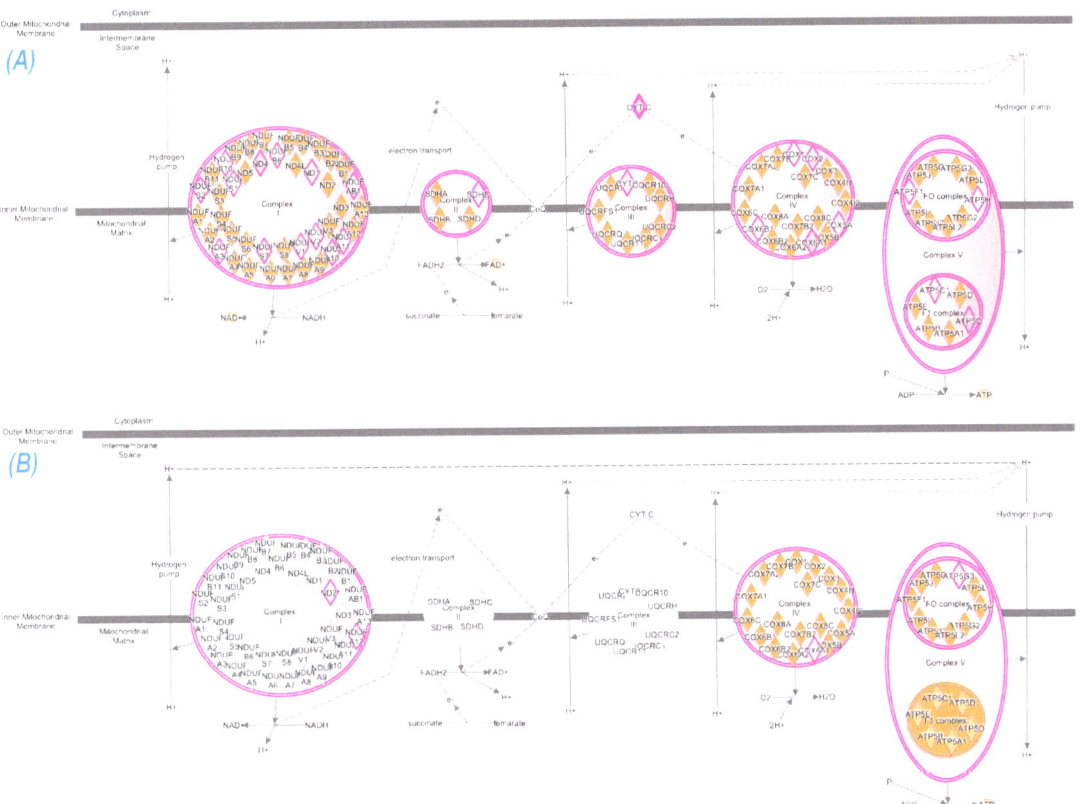

Figure 7. Mitochondrial function and oxidative phosphorylation in WT (**A**) and Nrf2 KO (**B**) mice following ExT. The purple outlined mitochondrial complexes with orange or green tinting are predicated to be functionally enhanced or inhibited, respectively.

3.6. Western Blotting Analysis

Western blotting was utilized to validate the sensitivity and accuracy of mass spectrometry by evaluating the expression levels of NQO1, SOD2, GSTM1, and GSTM3—four principal antioxidant enzymes selected from Figure 6 which were upregulated in WT mice and downregulated in Nrf2-KO mice following ExT. As can be seen from Figure 8, in WT mice all proteins were dramatically upregulated following ExT (NQO1: 8.28 ± 0.43 vs. 4.24 ± 0.46, $p = 0.0002$; SOD2: 5.55 ± 0.31 vs. 3.86 ± 0.36, $p = 0.0075$; GSTM1: 3.09 ± 0.08 vs. 1.83 ± 0.13, $p = 0.0034$; GSTM3: 2.03 ± 0.08 vs. 1.32 ± 0.05, $p = 0.0075$. n = 5/group). In KO-Sed mice however, expression of these proteins was lower as compared with WT-Sed mice, which was further downregulated following ExT (NQO1: 1.62 ± 0.16 vs. 2.36 ± 0.22, $p = 0.026$; SOD2: 2.19 ± 0.32 vs. 2.96 ± 0.07, $p = 0.049$; GSTM1: 0.41 ± 0.04 vs. 1.06 ± 0.07, $p = 0.0015$; GSTM3: 0.42 ± 0.02 vs. 0.78 ± 0.01, $p = 0.0065$. n = 5/group). These changes were consistent with the mass spectrometry data shown in Figure 6.

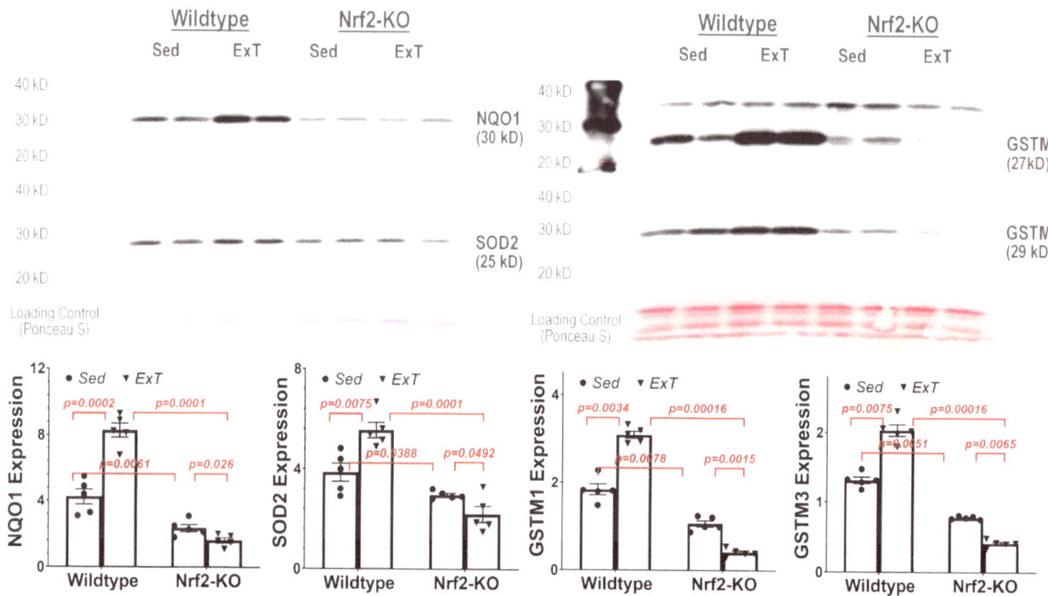

Figure 8. Western blotting shows that NQO1, SOD2, GSTM1, and GSTM3 are upregulated in WT-ExT and downregulated in KO-ExT as compared to Sed controls. n = 5/group.

4. Discussion

As a master antioxidant transcription factor, Nrf2 is known to regulate muscle metabolism, function, and structure under both physiological and pathophysiological conditions [19,20]. Using proteomic and bioinformatic analyses in muscle specific Nrf2 knockout and overexpression mouse lines, iMS-*Nrf2$^{flox/flox}$* and iMS-*Keap1$^{flox/flox}$*, we previously identified over 200 Nrf2-targeted proteins in muscle, which were actively involved in oxido-reduction coenzyme metabolism, purine ribonucleoside triphosphate metabolism, ATP metabolism, propanoate metabolism, cellular detoxification, NADP metabolism glutathione metabolism, electron transport chain, and other functions [13]. Using a redox proteomics technique, we further quantified 34 cysteine redox peptide-bearing proteins, which were associated with mitochondrial oxidative phosphorylation, energy metabolism, and extracellular matrix [14]. On the other hand, in a permanent coronary artery ligation-induced heart failure mouse model with reduced ejection fraction (HFrEF), we showed a markedly weakened Nrf2/antioxidant defense, impaired muscle function/structure, and reduced exercise capacity, which were ameliorated by the Nrf2 activator curcumin [18]. Because of these crucial roles of Nrf2 in SkM, we hypothesized that during exercise, the internal environment of myocytes is disturbed by contraction-induced multiple stressors including accumulated ROS, in which Nrf2 is activated and functions as a hub to maintain cellular homeostasis by upregulating antioxidant enzymes and other cytoprotective proteins. This Nrf2-mediated response underpins SkM adaptation to exercise and exercise-associated benefits.

We found a significant increase in maximal running distance in WT mice after 3-weeks of training, while this exercise capacity did not change in 3-week sedentary mice as a time course control. On the other hand, evaluation of TA contraction further indicated that both force generation and fatigue tolerance were significantly increased in ExT mice compared with non-ExT controls. The data from the in vivo and in situ experiments suggest that ExT markedly promotes treadmill running performance that is, at least in part, due to enhanced muscle contractility and other functions. In contrast with WT mice, SkM-Nrf2 deficient mice showed a significantly shorter running distance after ExT, which accompanied a markedly reduced muscle force generation and impaired fatigue tolerance. This evidence strongly suggests that Nrf2 plays a crucial role in exercise-induced beneficial adaptation in SkM. Without Nrf2-supported antioxidant defense and cytoprotective mechanisms in response to increased oxidative stress during exercise, the ROS surge that accompanies muscle contraction tends to damage skeletal myocytes. Furthermore, we found that exercise performance of mice with Nrf2 deficiency was reduced, even in the 3-week sedentary period (Post-Sed vs. Pre-Sed in KO-Sed group, Figure 1), suggesting that a compensatory mechanism to Nrf2-deletion was not apparent at week 3 after Nrf2 was deleted by Tamoxifen administration. However, due to technical limitations, we were not able to determine if the in-situ TA contractility in the same Nrf2-KO mouse was further impaired after 3-weeks in sedentary conditions.

To explore the molecular mechanisms underlying functional alterations after ExT and the role of Nrf2 deletion, we utilized mass spectrometry to determine the protein profile of the TA, followed by downstream bioinformatic analysis to reveal signaling pathways and integrative networks involved under these conditions. In WT mice, we identified 576 proteins in TA that were upregulated by ExT, most of which are newly reported to be associated with exercise (Supplementary Table S1). The protein with the greatest increase was adenylyltransferase (gene name: SELENOO), which was expressed at a very low level in Sed muscle, but was 8211-fold upregulated after exercise (3651.64 ± 760.45 vs. 231.98 ± 87.77, $p < 0.0062$; n = 6/group). This enzyme catalyzes transfer of adenosine $5'$-monophosphate (AMP) to Ser, Thr, and Tyr residues of target proteins and results in protein adenylylation (AMPylation), playing a role in protein post-translational modification [21]. There are two protein AMP transferases. We identified the highly conserved pseudokinase selenoprotein-O (SELO) located inside mitochondria [22]. SELO plays a critical role in the response to oxidative stress and regulates global S-glutathionylation levels via AMPylation in conjunction with glutaredoxin [23]. However, its functional significance in SkM and exercise remains to be determined. Another interesting protein we identified was haptoglobin, whose expression was high in muscle tissues even under basal conditions and was subsequently upregulated 6-fold after ExT ($1,014,955.2 \pm 200,279.7$ vs. $158,233.3 \pm 9191.8$, $p < 0.0062$; n = 6/group). This protein has been demonstrated to play a key function in SkM to prevent oxidative stress/inflammation and maintain muscle mass/function [24]. Absence of haptoglobin causes muscle atrophy and weakness due to the activation of an atrophic program, which could be further exacerbated by acute exercise or high fat diet [24]. A recent study showed that intravenous infusion of exogenous haptoglobin in rats attenuates hemoglobin-impaired muscle contraction, and it is therefore recommended for improved muscle metabolic control and exercise tolerance in patients suffering from acute or chronic hemolysis [25]. Other studies indicate that exercise induces global histone modifications which mediate chromatin remodeling and transcriptional activation in SkM [26–28].

Exercise trained wild type mice also showed upregulated levels of SLC2A1, (also referred as Solute Carrier Family 2 Member 1). Abnormal expression of this protein demonstrates various neurological dysfunctions such as Dystonia 9, Stomatin-Deficient Cryohydrocytosis, and Glucose transporter type 1 deficiency syndrome (patients suffer from severe body movement disorders and developmental and varying degrees of cognitive impairments [29–31]).

In SkM-Nrf2 KO mice, we found that 207 proteins were downregulated following ExT (Supplementary Table S2). The highest downregulated protein was chitinase-3-like protein 1, a glycoprotein consisting of 383 amino acids with a molecular mass of 40 kDa, which counteracts TNFα-mediated inflammation and insulin resistance in SkM cells [25]. Additionally, one of the Nrf2 downstream target proteins, glutathione S-transferase Mu 5 (Gstm5) was listed as the one of the top 6 downregulated proteins in SkM with Nrf2 deletion after exercise. This protein belongs to one of eight GST supergene families: the GSTAs, GSTMs, GSTTs, GSTP, GSTZ, GSTS, GSTK, and GSTO, located on seven chromosomes [32]. It has been reported that Gstm5 expression was significantly downregulated in ovarian cancer [33] and lung adenocarcinoma [34]. In colorectal cancer, the expression of Gstm5 was reinforced by miR-20b-3p and actively regulates miR-20b-3p/GSTM5/AKT-mTOR signaling axis [35].

Using IPA analysis of the protein profiles, we found that the functional status of 56 intracellular signaling pathways in WT and 34 pathways in KO mice were significantly modified by ExT. The most noticeable of these was the Nrf2-mediated oxidative stress response, which was accentuated in WT but suppressed in KO mice following exercise, clearly suggesting a critical role of Nrf2 in ExT-evoked muscle adaptation. We have previously reported a reduction in Nrf2 signaling in SkM when Nrf2 was knocked out [13,14]. In the current study, we demonstrated that this impaired pathway was further suppressed after exercise. This negative response of Nrf2-associated signaling to exercise, is believed to be partly responsible for the ExT-induced impairment of exercise capacity and muscle contractility in mice with SkM-Nrf2 deficiency. Among the 237 proteins associated with Nrf2-mediated signaling pathways, ExT upregulated 33 proteins in WT mice and changed 21 proteins (18 downregulated and 3 upregulated) in KO mice. A total of 17 proteins were found overlapping between groups. These proteins, we propose, represent the loyal targets of Nrf2 in muscle, which are exclusively regulated by this master antioxidant transcription factor. Without the positive regulation from Nrf2, these proteins dramatically decline in SkM following ExT. In Nrf2 KO mice, ExT-induced elevation in fatty acid binding protein (FABP) expression may also indicate adrenergic overdrive, acute myocardial infarction, and ventricular tachyarrhythmia [36]. Nrf2 KO mice also showed elevated levels of proliferation and apoptosis adaptor protein 15 (PEA-15), a ubiquitously expressed small protein in different types of cells such as fibroblasts, skeletal muscle, and adipose tissues. Its overexpression has been associated with reduced insulin sensitivity, often leading to type 2 diabetes [37,38]. The signaling of PEA-15 is also actively involved in cancer development and progression [39]. Nrf2 deletion also upregulated levels of CHI3L1, which is a chitinase 3-like protein belonging to the glycosyl hydrolase family. Nrf2 has been shown to negatively regulate CHI3L1, in a post-traumatic osteoarthritis disease model [40].

In addition to Nrf2-signaling, several other antioxidant mechanisms were also activated by exercise, such as HIF1α signaling, nitric oxide signaling, glutathione-mediated detoxification, production of nitric oxide and reactive oxygen species, and glutathione redox reactions. IPA analysis also identified intracellular pathways other than those associated with antioxidant defense. These pathways were, in general, activated in WT, but inactivated in Nrf2-KO. Following ExT, these are predominantly associated with the regulation of overall physiological response—such as insulin production, immune activation, and redox status. As expected, our data demonstrated that many metabolic pathways associated with ATP generation were activated in WT mice following ExT, including oxidative phosphorylation, tricarboxylic cycle, gluconeogenesis, glycolysis, and fatty acid β-oxidation. These upregulated metabolic pathways are essential for establishing and maintaining muscle contraction during exercise, due in part to the relatively small intramuscular stores of ATP [41].

In summary, the present study demonstrated that ExT enhanced exercise performance and muscle contractility in mice. These effects were markedly abrogated when SkM-Nrf2 was deleted. We further observed that the majority of the proteins we identified and the intracellular signaling pathways were upregulated and activated in WT mice but downregulated and suppressed in KO mice following exercise. These data suggest a critical role of Nrf2 in the exercise-induced beneficial effects on SkM. In addition, these identified proteins and signaling pathways provide insight into novel mechanisms underlying SkM adaptation to exercise and a potential therapeutic strategy to treat diseases associated with myopathy.

Supplementary Materials: The supporting information can be downloaded at: https://www.mdpi.com/article/10.3390/antiox12010151/s1.

Author Contributions: L.G. conceived the project with input from I.H.Z. and V.K.; L.G. and A.B. co-wrote the article; L.G. carried out animal experiments and A.B. conducted data analyses; R.A. performed proteomic analysis under the guidance of V.K.; S.J. and N.N.V. performed bioinformatics analysis with consults from C.G.; T.L.R. and L.Y. created the transgenic mouse model. Animal breeding was managed by T.L.R.; Genotyping assay and Western blot analyses were performed by L.Y. and V.V.P. All authors have read and agreed to the published version of the manuscript.

Funding: This study was supported by National Institutes of Health (NIH) grant R01HL160820 awarded to L.G. I.H.Z. is supported, in part, by the Theodore F. Hubbard Foundation. The Bioinformatics and Systems Biology Core is partially supported by NIH awards to C.G. (5P20GM103427, 5P30CA036727, 5P30MH062261, 5U54GM115458).

Institutional Review Board Statement: The animal study protocol was approved by the Institutional Animal Care and Use Committee (IACUC) of University of Nebraska Medical Center (protocol code 18-174-02-FC; approved on 22 February 2019).

Informed Consent Statement: Not applicable.

Data Availability Statement: The data are contained within the article and supplementary materials.

Conflicts of Interest: The authors declare no conflict of interest.

References

1. Warburton, D.E.R.; Bredin, S.S.D. Health benefits of physical activity: A systematic review of current systematic reviews. *Curr. Opin. Cardiol.* **2017**, *32*, 541–556. [CrossRef] [PubMed]
2. Ruegsegger, G.N.; Booth, F.W. Health Benefits of Exercise. *Cold Spring Harb. Perspect. Med.* **2018**, *8*, a029694. [CrossRef] [PubMed]
3. Pedersen, B.K.; Saltin, B. Exercise as medicine—Evidence for prescribing exercise as therapy in 26 different chronic diseases. *Scand. J. Med. Sci. Sports.* **2015**, *25* (Suppl. 3), 1–72. [PubMed]
4. Yan, Z.; Okutsu, M.; Akhtar, Y.N.; Lira, V.A. Regulation of exercise-induced fiber type transformation, mitochondrial biogenesis, and angiogenesis in skeletal muscle. *J. Appl. Physiol.* **2011**, *110*, 264–274. [CrossRef]
5. Egan, B.; Zierath, J.R. Exercise metabolism and the molecular regulation of skeletal muscle adaptation. *Cell Metab.* **2013**, *17*, 162–184. [CrossRef]
6. Sies, H.; Jones, D.P. Reactive oxygen species (ROS) as pleiotropic physiological signalling agents. *Nat. Rev. Mol. Cell. Biol.* **2020**, *21*, 363–383.
7. Xiao, W.; Loscalzo, J. Metabolic Responses to Reductive Stress. *Antioxid. Redox Signal.* **2020**, *32*, 1330–1347. [CrossRef]
8. Yang, L.; Palliyaguru, D.L.; Kensler, T.W. Frugal chemoprevention: Targeting Nrf2 with foods rich in sulforaphane. *Semin. Oncol.* **2016**, *43*, 146–153. [CrossRef]
9. Hahn, M.E.; Timme-Laragy, A.R.; Karchner, S.I.; Stegeman, J.J. Nrf2 and Nrf2-related proteins in development and developmental toxicity: Insights from studies in zebrafish (*Danio rerio*). *Free Radic. Biol. Med.* **2015**, *88*, 275–289.
10. Fuse, Y.; Kobayashi, M. Conservation of the Keap1-Nrf2 System: An Evolutionary Journey through Stressful Space and Time. *Molecules* **2017**, *22*, 436. [CrossRef]
11. Canning, P.; Sorrell, F.J.; Bullock, A.N. Structural basis of Keap1 interactions with Nrf2. *Free Radic. Biol. Med.* **2015**, *88*, 101–107. [CrossRef] [PubMed]
12. Suzuki, T.; Muramatsu, A.; Saito, R.; Iso, T.; Shibata, T.; Kuwata, K.; Kawaguchi, S.I.; Iwawaki, T.; Adachi, S.; Suda, H.; et al. Molecular Mechanism of Cellular Oxidative Stress Sensing by Keap1. *Cell Rep.* **2019**, *28*, 746–758.e4. [CrossRef] [PubMed]
13. Gao, L.; Kumar, V.; Vellichirammal, N.N.; Park, S.Y.; Rudebush, T.L.; Yu, L.; Son, W.M.; Pekas, E.J.; Wafi, A.M.; Hong, J.; et al. Functional, proteomic and bioinformatic analyses of Nrf2- and Keap1- null skeletal muscle. *J. Physiol.* **2020**, *598*, 5427–5451. [CrossRef] [PubMed]
14. Abu, R.; Yu, L.; Kumar, A.; Gao, L.; Kumar, V. A Quantitative Proteomics Approach to Gain Insight into NRF2-KEAP1 Skeletal Muscle System and Its Cysteine Redox Regulation. *Genes* **2021**, *12*, 1655. [CrossRef] [PubMed]
15. McCarthy, J.J.; Srikuea, R.; Kirby, T.J.; Peterson, C.A.; Esser, K.A. Inducible Cre transgenic mouse strain for skeletal muscle-specific gene targeting. *Skelet. Muscle* **2012**, *2*, 8. [CrossRef]
16. Kong, X.; Thimmulappa, R.; Craciun, F.; Harvey, C.; Singh, A.; Kombairaju, P.; Reddy, S.P.; Remick, D.; Biswal, S. Enhancing Nrf2 pathway by disruption of Keap1 in myeloid leukocytes protects against sepsis. *Am. J. Respir. Crit Care Med.* **2011**, *184*, 928–938. [CrossRef]
17. Wafi, A.M.; Yu, L.; Gao, L.; Zucker, I.H. Exercise training upregulates Nrf2 protein in the rostral ventrolateral medulla of mice with heart failure. *J. Appl. Physiol.* **2019**, *127*, 1349–1359. [CrossRef]
18. Wafi, A.M.; Hong, J.; Rudebush, T.L.; Yu, L.; Hackfort, B.; Wang, H.; Schultz, H.D.; Zucker, I.H.; Gao, L. Curcumin improves exercise performance of mice with coronary artery ligation-induced HFrEF: Nrf2 and antioxidant mechanisms in skeletal muscle. *J. Appl. Physiol.* **2019**, *126*, 477–486. [CrossRef]
19. Done, A.J.; Traustadottir, T. Nrf2 mediates redox adaptations to exercise. *Redox Biol.* **2016**, *10*, 191–199. [CrossRef]

20. Kitaoka, Y. The Role of Nrf2 in Skeletal Muscle on Exercise Capacity. *Antioxidants* **2021**, *10*, 1712. [CrossRef]
21. Sieber, S.A.; Cappello, S.; Kielkowski, P. From Young to Old: AMPylation Hits the Brain. *Cell Chem. Biol.* **2020**, *27*, 773–779. [CrossRef] [PubMed]
22. Sreelatha, A.; Yee, S.S.; Lopez, V.A.; Park, B.C.; Kinch, L.N.; Pilch, S.; Servage, K.A.; Zhang, J.; Jiou, J.; Karasiewicz-Urbanska, M.; et al. Protein AMPylation by an Evolutionarily Conserved Pseudokinase. *Cell* **2018**, *175*, 809–821.e19. [CrossRef] [PubMed]
23. Zhang, Y.; Roh, Y.J.; Han, S.J.; Park, I.; Lee, H.M.; Ok, Y.S.; Lee, B.C.; Lee, S.R. Role of Selenoproteins in Redox Regulation of Signaling and the Antioxidant System: A Review. *Antioxidants* **2020**, *9*, 383. [CrossRef] [PubMed]
24. Bertaggia, E.; Scabia, G.; Dalise, S.; Lo Verso, F.; Santini, F.; Vitti, P.; Chisari, C.; Sandri, M.; Maffei, M. Haptoglobin is required to prevent oxidative stress and muscle atrophy. *PLoS ONE* **2014**, *9*, e100745. [CrossRef]
25. Ferguson, S.K.; Harral, J.W.; Pak, D.I.; Redinius, K.M.; Stenmark, K.R.; Schaer, D.J.; Buehler, P.W.; Irwin, D.C. Impact of cell-free hemoglobin on contracting skeletal muscle microvascular oxygen pressure dynamics. *Nitric Oxide* **2018**, *76*, 29–36. [CrossRef]
26. Shimizu, J.; Kawano, F. Exercise-induced histone H3 trimethylation at lysine 27 facilitates the adaptation of skeletal muscle to exercise in mice. *J. Physiol.* **2022**, *600*, 3331–3353. [CrossRef]
27. McGee, S.L.; Hargreaves, M. Histone modifications and exercise adaptations. *J. Appl. Physiol.* **2011**, *110*, 258–263. [CrossRef]
28. McGee, S.L.; Fairlie, E.; Garnham, A.P.; Hargreaves, M. Exercise-induced histone modifications in human skeletal muscle. *J. Physiol.* **2009**, *587*, 5951–5958. [CrossRef]
29. Daniele, L.L.; Han, J.Y.S.; Samuels, I.S.; Komirisetty, R.; Mehta, N.; McCord, J.L.; Yu, M.; Wang, Y.; Boesze-Battaglia, K.; Bell, B.A.; et al. Glucose uptake by GLUT1 in photoreceptors is essential for outer segment renewal and rod photoreceptor survival. *FASEB J.* **2022**, *36*, e22428. [CrossRef]
30. Wang, D.; Pascual, J.M.; De Vivo, D. Glucose Transporter Type 1 Deficiency Syndrome. In *GeneReviews((R))*; Adam, M.P., Everman, D., Mirzaa, G., Pagon, R., Wallace, S., Bean, L., Gripp, K., Amemiya, A., Eds.; University of Washington: Seattle, WA, USA, 1993.
31. Suls, A.; Dedeken, P.; Goffin, K.; Van Esch, H.; Dupont, P.; Cassiman, D.; Kempfle, J.; Wuttke, T.V.; Weber, Y.; Lerche, H.; et al. Paroxysmal exercise-induced dyskinesia and epilepsy is due to mutations in SLC2A1, encoding the glucose transporter GLUT1. *Brain* **2008**, *131*, 1831–1844. [CrossRef]
32. Strange, R.C.; Spiteri, M.A.; Ramachandran, S.; Fryer, A.A. Glutathione-S-transferase family of enzymes. *Mutat Res.* **2001**, *482*, 21–26. [CrossRef] [PubMed]
33. Zhang, J.; Li, Y.; Zou, J.; Lai, C.T.; Zeng, T.; Peng, J.; Zou, W.D.; Cao, B.; Liu, D.; Zhu, L.Y.; et al. Comprehensive analysis of the glutathione S-transferase Mu (GSTM) gene family in ovarian cancer identifies prognostic and expression significance. *Front. Oncol.* **2022**, *12*, 968547. [CrossRef] [PubMed]
34. Hao, X.; Zhang, J.; Chen, G.; Cao, W.; Chen, H.; Chen, S. Aberrant expression of GSTM5 in lung adenocarcinoma is associated with DNA hypermethylation and poor prognosis. *BMC Cancer* **2022**, *22*, 685. [CrossRef] [PubMed]
35. Liu, F.; Xiao, X.L.; Liu, Y.J.; Xu, R.H.; Zhou, W.J.; Xu, H.C.; Zhao, A.G.; Xu, Y.X.; Dang, Y.Q.; Ji, G. CircRNA_0084927 promotes colorectal cancer progression by regulating miRNA-20b-3p/glutathione S-transferase mu 5 axis. *World J. Gastroenterol.* **2021**, *27*, 6064–6078. [CrossRef] [PubMed]
36. Iso, T.; Sunaga, H.; Matsui, H.; Kasama, S.; Oshima, N.; Haruyama, H.; Furukawa, N.; Nakajima, K.; Machida, T.; Murakami, M.; et al. Serum levels of fatty acid binding protein 4 and fat metabolic markers in relation to catecholamines following exercise. *Clin. Biochem.* **2017**, *50*, 896–902. [CrossRef] [PubMed]
37. Valentino, R.; Lupoli, G.A.; Raciti, G.A.; Oriente, F.; Farinaro, E.; Della Valle, E.; Salomone, M.; Riccardi, G.; Vaccaro, O.; Donnarumma, G.; et al. The PEA15 gene is overexpressed and related to insulin resistance in healthy first-degree relatives of patients with type 2 diabetes. *Diabetologia* **2006**, *49*, 3058–3066. [CrossRef] [PubMed]
38. Condorelli, G.; Vigliotta, G.; Iavarone, C.; Caruso, M.; Tocchetti, C.G.; Andreozzi, F.; Cafieri, A.; Tecce, M.F.; Formisano, P.; Beguinot, L.; et al. PED/PEA-15 gene controls glucose transport and is overexpressed in type 2 diabetes mellitus. *EMBO J.* **1998**, *17*, 3858–3866. [CrossRef] [PubMed]
39. Greig, F.H.; Nixon, G.F. Phosphoprotein enriched in astrocytes (PEA)-15: A potential therapeutic target in multiple disease states. *Pharmacol Ther.* **2014**, *143*, 265–274. [CrossRef]
40. Song, Y.; Hao, D.; Jiang, H.; Huang, M.; Du, Q.; Lin, Y.; Liu, F.; Chen, B. Nrf2 Regulates CHI3L1 to Suppress Inflammation and Improve Post-Traumatic Osteoarthritis. *J. Inflamm. Res.* **2021**, *14*, 4079–4088. [CrossRef]
41. Hargreaves, M.; Spriet, L.L. Skeletal muscle energy metabolism during exercise. *Nat. Metab.* **2020**, *2*, 817–828. [CrossRef]

Article

Transcriptomic Profiling and Pathway Analysis of Mesenchymal Stem Cells Following Low Dose-Rate Radiation Exposure

John E. Slaven [1], Matthew Wilkerson [2,3], Anthony R. Soltis [2,3], W. Bradley Rittase [1], Dmitry T. Bradfield [1], Michelle Bylicky [1,†], Lynnette Cary [4], Alena Tsioplaya [4], Roxane Bouten [1], Clifton Dalgard [5,6] and Regina M. Day [1,*]

[1] Department of Pharmacology and Molecular Therapeutics, Uniformed Services University of the Health Sciences, 4301 Jones Bridge Rd., Bethesda, MD 20814, USA

[2] Collaborative Health Initiative Research Program, Uniformed Services University of the Health Sciences, Bethesda, MD 20814, USA

[3] Henry M. Jackson Foundation for the Advancement of Military Medicine, Bethesda, MD 20817, USA

[4] Armed Forces Radiobiology Research Institute, Uniformed Services University of the Health Sciences, Bethesda, MD 20814, USA

[5] The American Genome Center, Uniformed Services University of the Health Sciences, Bethesda, MD 20814, USA

[6] Department of Anatomy, Physiology and Genetics, Uniformed Services University of the Health Sciences, Bethesda, MD 20814, USA

[*] Correspondence: regina.day@usuhs.edu; Tel.: +1-301-295-3236; Fax: +1-301-295-3220

[†] Current address: Radiation Oncology Branch, National Cancer Institute, National Institutes of Health, Bethesda, MD 20892, USA.

Citation: Slaven, J.E.; Wilkerson, M.; Soltis, A.R.; Rittase, W.B.; Bradfield, D.T.; Bylicky, M.; Cary, L.; Tsioplaya, A.; Bouten, R.; Dalgard, C.; et al. Transcriptomic Profiling and Pathway Analysis of Mesenchymal Stem Cells Following Low Dose-Rate Radiation Exposure. *Antioxidants* **2023**, *12*, 241. https://doi.org/10.3390/antiox12020241

Academic Editor: Marcel Bonay

Received: 21 December 2022
Revised: 17 January 2023
Accepted: 19 January 2023
Published: 21 January 2023

Abstract: Low dose-rate radiation exposure can occur in medical imaging, as background from environmental or industrial radiation, and is a hazard of space travel. In contrast with high dose-rate radiation exposure that can induce acute life-threatening syndromes, chronic low-dose radiation is associated with Chronic Radiation Syndrome (CRS), which can alter environmental sensitivity. Secondary effects of chronic low dose-rate radiation exposure include circulatory, digestive, cardiovascular, and neurological diseases, as well as cancer. Here, we investigated 1–2 Gy, 0.66 cGy/h, ^{60}Co radiation effects on primary human mesenchymal stem cells (hMSC). There was no significant induction of apoptosis or DNA damage, and cells continued to proliferate. Gene ontology (GO) analysis of transcriptome changes revealed alterations in pathways related to cellular metabolism (cholesterol, fatty acid, and glucose metabolism), extracellular matrix modification and cell adhesion/migration, and regulation of vasoconstriction and inflammation. Interestingly, there was increased hypoxia signaling and increased activation of pathways regulated by iron deficiency, but Nrf2 and related genes were reduced. The data were validated in hMSC and human lung microvascular endothelial cells using targeted qPCR and Western blotting. Notably absent in the GO analysis were alteration pathways for DNA damage response, cell cycle inhibition, senescence, and pro-inflammatory response that we previously observed for high dose-rate radiation exposure. Our findings suggest that cellular gene transcription response to low dose-rate ionizing radiation is fundamentally different compared to high-dose-rate exposure. We hypothesize that cellular response to hypoxia and iron deficiency are driving processes, upstream of the other pathway regulation.

Keywords: radiation; low dose-rate; mesenchymal stem cells; human microvascular endothelial cells; RNAseq; gene regulation

1. Introduction

Redox homeostasis has been defined as the balance between cellular generation of nonradical reactive oxygen species (ROS) and cellular antioxidant defenses [1–4]. Redox homeostasis has been demonstrated to contribute to normal cellular processes, including physiological redox signaling, oxidation and reduction of metabolic substrates, and oxidative phosphorylation [1–3]. However, oxidative damage to biological molecules can occur as the result of a redox imbalance, which can result in necrosis, apoptosis, and/or accelerated senescence [3,5]. Redox toxicity can occur as the result of exposure to oxidizing agents (ultraviolet and ionizing radiation, heavy metals, chemical toxins, etc.), from free radical production by normal physiological processes (mitochondrial oxidative phosphorylation, arachidonic acid metabolism, etc.), or from a deficit of antioxidant capacity of the cell [3,5–7]. Redox imbalances have been implicated in a wide variety of chronic diseases, such as cancer, cardiovascular disease, kidney disease, fibrotic remodeling diseases, and a variety of neurological diseases [2,6,8–10]. The cellular biological response to redox stress has been shown to depend upon the quantity, quality, duration, and total area of reactive species exposure [3,4,11–14].

Radiation damages biological macromolecules directly through energy deposition and indirectly through the generation of free radicals. Although free radicals have half-lives of milliseconds [14], they can induce chain reactions of oxidation or nitrosylation events, culminating in macromolecule modification and destruction [15]. Macromolecules oxidized following radiation exposure include DNA, proteins, lipids, and carbohydrates, each with specific downstream biological effects on cellular function [5,16–19]. DNA and proteins have been considered critical targets of radiation-induced oxidative damage [15,20,21], although all macromolecular damage has biological consequences. The biological response of a cell to radiation can be either repair and survival, several types of programmed cell death, necrosis, or accelerated senescence, depending on the qualities of radiation and specific antioxidant and repair defenses active in the cell [5,14,22,23].

The biological effects of radiation have been demonstrated to be dependent upon total dose, dose rate, and linear energy transfer values. High-dose-rate radiation is considered to be ≥ 0.1 Gy/min [24,25]. High-dose-rate ionizing radiation exposure can occur during cancer radiotherapy or due to accidental radiation exposure. A collection of potentially life-threatening syndromes—Acute Radiation Syndrome (ARS) and Delayed Effects of Acute Radiation Exposure (DEARE)—can occur from high-dose radiation exposure, depending upon the total dose of radiation, the dose rate, and the area of radiation exposure [26,27]. In contrast, low dose-rate irradiation has been defined as between 0.006–0.1 Gy/h [28]. Low dose-rate radiation exposure can occur in medical imaging, as a result of low levels of ionizing radiation in the environment, or as a hazard of space travel [29]. In contrast with high-dose-rate radiation exposure, the effects of chronic low-dose radiation are associated with Chronic Radiation Syndrome (CRS). The characteristics of CRS are not immediately life-threatening, but result in alterations in environmental sensitivity. Early events in CRS include increased olfactory and taste thresholds, sensitivity to vibration, and changes in systemic immunity [30]. Secondary effects of chronic radiation exposure have been observed in the circulatory and digestive systems, although these have been shown to spontaneously revert when radiation exposure ends [30]. Cardiovascular and neurological diseases are of concern for long-distance space travel [31,32]. Cancer is believed to be a serious risk of low-dose radiation, but the response of individuals to low-dose radiation for the prediction of cancer risk is still poorly defined [29].

In general, acute high-dose radiation exposure induces greater macromolecular damage and results in greater suppression of cell proliferation compared with low dose-rate radiation [33]. Cellular responses to high dose-rate irradiation include the immediate activation of DNA damage response and DNA repair pathways, upregulation of cell cycle inhibitors, ER stress response, and the unfolded protein response pathway [21,23]. These responses have been demonstrated to follow the robust activation of specific enzymes and transcription factors responsive to oxidative stress and DNA damage [14,34]. In comparison, the effects of low dose-rate radiation are less well understood [29,30,35].

Radiation effects have been shown to have cell-type and tissue-type specificity [36]. Mesenchymal stem cells (MSCs) are multipotent adult stem cells, originating in the bone marrow, with the capacity to enter through the circulation, migrate to injured tissue, engraft, and differentiate into different phenotypes depending upon their environment [37,38]. MSCs were previously demonstrated to be relatively resistant to radiation-induced damage, requiring 2 Gy to reduce the surviving fraction of cells to 37% [39]. 10 Gy was demonstrated in MSCs to reduce the surviving population to <1% [39]. Because of the importance of MSC in repair following tissue injury, we investigated the effects of chronic low dose-rate (0.66 cGy/h) radiation on MSCs. We confirmed our findings in primary human lung microvascular endothelial cells. Our findings indicate that the pathways regulated by chronic low dose-rate radiation differ greatly compared with those regulated by acute high dose-rate radiation exposure.

2. Methods

2.1. Reagents

Chemicals and reagents were purchased from MilliporeSigma (St. Louis, MO, USA) except where indicated.

2.2. Cell Culture and Irradiation

Human bone marrow mesenchymal stem cells (MSCs) were purchased from Lonza (Morristown, NJ, USA) and cultured in mesenchymal stem cell growth medium (Lonza). Human lung microvascular endothelial cells (HLMVECs) were purchased from Cell Applications (San Diego, CA, USA) and cultured on plates treated with endothelial cell attachment factor in Microvascular Endothelial Cell Growth Medium (Cell Applications). Cells were grown and irradiated in a humidified environment of 5% CO_2/95% air at 37 °C, according to the manufacturer's instructions. Cells were used within seven passages for all experiments. For irradiations, cells were plated in 25 ml flasks or in Lab-Tek Flaskette Chamber slides (ThermoFisher Scientific, Waltham, MA, USA), and grown to 60–70% confluence. Cells were irradiated in the Armed Forces Radiobiology Research Institute (AFRRI) Low Level Cobalt Facility at 0.66 cGy/h to reach a total irradiation of 0.3, 0.7, 1, or 2 Gy. Dosimetry for Low Level Cobalt Facility was performed by preliminary mapping of irradiation field inside the cell culture incubator. The irradiation field from the bare source was measured using an A12 ion chamber calibrated at ADCL, University of Wisconsin. The dose rate was measured at the position of each flask (21 points of measurements, 7 flasks on 3 shelves) under the same conditions as actual cell irradiation conditions. To achieve desired dose rate, the 4-times attenuator was mounted and field uniformity was measured at the same positions with ion chamber PTW TN 75 cc. The dose rate on a date of mapping of the field was measured as 0.6646 cGy/h with field uniformity 0.28%. Control dishes were cultured in parallel in a separate incubator without radiation exposure. The 10 Gy irradiation was performed on cells at 70–90% confluence using an RS2000 Biological Irradiator (Rad Source Technologies, Alpharetta, GA, USA) at a dose rate of 1.15 Gy/min (160 kV, 25 mA) for a total dose of 10 Gy, as previously described with previously described dosimetry [40].

2.3. Cellular and Nuclear Morphology

Cells were plated on Nunc Lab-Tek flask on a slide (ThermoFisher Scientific). Cell irradiations were initiated once cells reached 60–70% confluent, and analyses were conducted in triplicate after 7 and 14 days of chronic low dose-rate radiation exposure, or at 3 and 7 days after acute high-dose-rate radiation exposure. Dishes were washed twice with phosphate-buffered saline (PBS), and fixed with 3.7% formaldehyde in PBS for 5 min at room temperature, washed twice more with PBS, and then mounted on slides in Prolong Gold antifade reagent with DAPI (Invitrogen, Eugene, OR, USA). Cells were imaged as for DAPI staining. All nuclei were analyzed in 5 random images, at least 100 nuclei per slide and three slides per time point, by a researcher blinded to treatment groups [40].

2.4. Gamma-H2AX Immunohistochemistry

Cells were fixed with 4% paraformaldehyde for 20 min at room temperature, then washed in PBS for 5 min, 3 times. Next, cells were permeabilized using 70% ethanol for

5 min at room temperature. Cells were blocked for 1 h in 5% normal donkey serum in PBS and then incubated in anti-γ-H2AX (#9718S, Cell Signaling, Danvers, MA, USA), 1:400 dilution in blocking solution for 1 h. Blocking and primary incubation were performed in a humidified chamber at 37 °C. Cells were then washed 3× in PBS before incubating for 1 h at room temperature in secondary antibody (Thermo Fisher, A21206), 1:2000 in PBS. Cells were again washed 3× in PBS and coverslipped with Pro-Long Gold antifade with DAPI. A Nikon Eclipse Ti microscope with a Nikon DSRi2 camera was used for digital images (Nikon Instruments, Inc., Melville, NY, USA). All images were obtained using the same exposure conditions to avoid false positives and false negatives in the scoring and images. Scoring was performed by a researcher blinded to the conditions. γ-H2AX foci were counted in DAPI-positive cells by scoring all cells within a field, and counting at least 100 cells per slide.

2.5. Quantitative PCR Analysis

Immediately following irradiation, total RNA was isolated from MSC or HLMVEC using the RNeasy Mini Kit with on-column DNase digestion (Qiagen, Valencia, CA, USA) according to manufacturer's protocol. RNA was quantified spectroscopically (ND-1000 Spectrophotometer, Nano-Drop, Wilmington, DE, USA) and 1.0 µg was reverse transcribed using iScript cDNA synthesis kit (Bio-Rad, Hercules, CA, USA), according to the manufacturer's protocol. RT-qPCRs were performed in technical duplicates using iTaq™ Universal SYBR Green Supermix (Bio-Rad), on a CFX96 Touch Real-Time PCR Detection System (Bio-Rad) as described [40]. Primers for qRT-PCR were designed using NCBI/Primer-BLAST and purchased from Integrated DNA Technologies (Coralville, IA, USA). Forward and reverse primer sequences are shown in Table 1. Relative gene expression to the reference genes was calculated using the ΔΔCq method using CFX Maestro software, 2.0 (Bio-Rad) [41,42].

Table 1. Primers for qPCR.

HGNC Gene Symbol	Forward Primer	Reverse Primer
GAPDH	5′-AGCCACATCGCTCAGACAC-3′	5′-GCCCAATACGACCAAATCC-3′
HMGCS1	5′-TTGTGCCCGAAGGAGGAAAC-3′	5′-CTGGCCCAAGCCAATGGTAT-3′
PTGS2	5′-CTGATGATTGCCCGACTCCC-3′	5′-CGCAGTTTACGCTGTCTAGC-3′
GTPBP4	5′-GAAAATTACGGTGGTGCCGTC-3′	5′-GCCCCAGAGCCAACTTGTA-3′
EGR1	5′-CCCCGACTACCTGTTTCCAC-3′	5′-TGGGTTTGATGAGCTGGGAC-3′
DUSP1	5′- CAGAGCCCCATTACGACCTC-3′	5′-TTGGTCCCGAATGTGCTGAG-3′
KRT34	5′-TCAGAAGCAAGTACCAGACGGA-3′	5′-CTGACTCCTGGTCTCGTTCAG-3′
NRG1	5′-CTGGTGATCGCTGCCAAAAC-3′	5′-GTAGGCCACCACACACATGA-3′
CLCA2	5′-ACTGTGGGCAACGACACTATG-3′	5′-TTCAGGGTGTAAGTCCAGTGC-3′
END1	5′-CTGCCTTTTCTCCCCGTTAAA-3′	5′-GGACTGGGAGTGGGTTTCTC-3′
OXTR	5′-TCCTGTACCCATCCAGCGA-3′	5′-TCCGCAGGCGAACCTAAAG-3′
NPR3	5′-CTGAGTACTCGCACCTCACG-3′	5′-TCACTGCTCGCACACATGAT-3′
LIPG	5′-AGTTGTGGTTGACTGGCTCC-3′	5′-TGTGATTGCTGTGATTCGGC-3′
ACSF2	5′-TTCAGTTCCCAGTAGCTTCACT-3′	5′-CTCCTTGAGTTGGGCAAAGGT-3′
LOXL4	5′-TCTGCGGATCACATGGACTG-3′	5′-AAAGTTGGCACATGCGTAGC-3′
HAS2	5′-TCCCGGTGAGACAGATGAGT-3′	5′-GGCTGGGTCAAGCATAGTGT-3′
COL1A2	5′-TGTGGATACGCGGACTTTGT-3′	5′-CAGCAAAGTTCCCACCGAGA-3′
CYBRD1	5′-AGGGCATCGCTTCTTTCAGGTTT-3′	5′-ACGAAAACACCTTCTGGCGG-3′
ADA	5′-GGAACCAGGCTGAACTGGTC-3′	5′-GCCGCTCTGTCTTGAGTATGT-3′

Table 1. *Cont.*

HGNC Gene Symbol	Forward Primer	Reverse Primer
SOD2	5′-CTGTTGGTGTCCAAGGCTCA-3′	5′-GTAGTAAGCGTGCTCCCACA-3′
PTGER4	5′-CGCTCGTGGTGCGAGTATT-3′	5′-GGGAGATGAAGGAGCGAGAGT-3′
GPX4	5′-GCCTTTGCCGCCTACTGA-3′	5′-CTTGGCGGAAAACTCGTGC-3′
HMOX1	5′-TGCGTTCCTGCTCAACATCC-3′	5′-AGTGTAAGGACCCATCGGAGA-3′
TGFB1	5′-TGGACATCAACGGGTTCACT-3′	5′-GAAGTTGGCATGGTAGCCCT-3′
IL6	5′-TCCTTCTCCACAAACATGTAACAA-3′	5′-TCACCAGGCAAGTCTCCTCA -3′
APOE	5′-GGGGCCTCTAGAAAGAGCTGG-3′	5′-TAATCCCAAAAGCGACCCAGT-3′
MSMO1	5′-GGTTCCGAGGTTGGAACACCT-3′	5′-TTCAAATCTCTGCAGACAGCCT-3′
KISS1	5′-CCACTTTGGGGAGCCATTAGA-3′	5′-CAGTTGTAGTTCGGCAGGTC-3′
NFE2L2	5′-TTCGGCTACGTTTCAGTCAC-3′	5′-TGTCCTGTTGCATACCGTCT-3′

Gene ID numbers: GAPDH #2597; HMGCS1 #3157; PTGS2 #5743; GTPBP4 #23560; EGR1 #1958; DUSP1 #1843; KRT34 #3885; NRG1 #3084; CLCA2 #9635; END1 #55823; OXTR #5021; NPR3 #4883; LIPG #9388; ACSF2 #80221; LOXL4 #84171; HAS2 #3037; COL1A2 #1278; CYBRD1 #79901; ADA #100; SOD2 #6648; PTGER4 #5734; GPX4 #2879; HMOX2 #3162; TGFB1 #7040; IL6 #3569; APOE #348; MSMO1 #6307; KISS1 #3814; NFE2L2 #4780.

2.6. Western Blotting

Immediately following irradiation, cells were lysed for 20 min at 4 °C in RIPA buffer (ThermoFisher Scientific) with protease and phosphatase inhibitors (#A32953 and #A32957, ThermoFisher Scientific). Lysates were centrifuged at 7000 RCF for 7 min at 4 °C. Proteins were separated by polyacrylamide gel electrophoresis and transferred to nitrocellulose membranes (MilliporeSigma) as previously described [40]. Nitrocellulose membranes were blocked in Tris-buffered saline with 5% BSA for 1.5 h. Primary antibodies were diluted in 5% BSA in Tris-buffered saline: β-actin (MilliporeSigma #A1982, 1:5000); DUSP1 (Cell Signaling #35217; 1:1000); total and phosphorylated p44/42 MAPK (Cell Signaling, Danvers, MA, USA, #4696 and #4370; 1:2000); total and Ser473 phosphorylated Akt (Cell Signaling #2920 and #4060; 1:2000); EGR1 (Cell Signaling #4154; 1:1000); full-length and cleaved caspase-3 (Cell Signaling #14220 and #9661; both 1:1000). Conjugated secondary antibodies (LI-COR, Lincoln, NE, USA; 1:10,000) were used for detection using the Odyssey system (LI-COR). β-actin was used as a loading control and for normalization of sample concentrations.

2.7. Transcriptome Profiling by RNA Sequencing

Immediately following irradiation, total RNA was isolated from MSC using the RNeasy Mini Kit with on-column DNase digestion (Qiagen) according to manufacturer's protocol. RNA was quantified spectroscopically (ND-1000 Spectrophotometer, Nano-Drop, Wilmington, DE, USA). The total RNA integrity was assessed using automated capillary electrophoresis with a Fragment Analyzer (Roche, Pleasanton, CA, USA). For all samples with an RNA quality indicator (RQI) > 8.0, a total of >75 ng RNA was used as the input for library preparation using the TruSeq Stranded mRNA Library Preparation Kit (Illumina, San Diego, CA, USA). The sequencing libraries were quantified by Real-Time PCR on a Roche LightCycler 480 Instrument II using a KAPA Library Quantification Kit for NGS (Kapa, Wilmington, MA, USA). The size distribution was assessed by automated capillary-based gel electrophoresis with a Fragment Analyzer to confirm the absence of free adapters or adapter dimers. The sequencing libraries were pooled and sequenced on a NovaSeq 6000 Sequencer (Illumina) using a NovaSeq 6000 SP Reagent Kit (300 cycles) within one flowcell lane using an XP workflow with 101 + 8 + 8 + 101 cycle parameters with paired-end reads of 75 bp in length. Raw sequencing reads were demuxed using bcl2fastq2 (v2.20) and aligned to the human reference genome (hg38) with MapSplice (v2.2.2) [43]. Gene-level quantification was performed with HTSeq (v0.9.1) [44] against GENCODE (v28) basic gene annotations. Read alignment statistics and sample quality features were calculated with Samtools and RseQC [45–47]. Sequencing quality was verified by manual inspection of sample-wise characteristics: total reads, mapping percentages, pairing percentages, transcript integrity number (TIN), $5'$ to $3'$ gene body read coverage slopes, and ribosomal RNA content [48]. The transcript abundance quantitation data were deposited in the NCBI Gene Expression Omnibus (GSE222541). Differential expression analysis was performed with DESeq2 (v1.16.1) [49] on raw gene counts. We defined significant differentially expressed genes (DEGs) between irradiated and control samples as those with a False Discovery Rate (FDR) q-value < 0.05, an absolute fold change > 1.5 (i.e., $|\log_2$ (fold-change)$| > 0.585$), and mean transcripts per million (TPM) ≥ 1 across samples.

2.8. Gene Ontology, Pathway Enrichment Analysis, and Heat Map Construction

Gene Ontology (GO) and Kyoto Encyclopedia of Genes and Genomes (KEGG) pathway analyses were performed using the Database for Annotation, Visualization, Integrated Discovery (DAVID), version 6.8, with the medium classification stringency, an enrichment threshold of 0.05, and the Bonferroni method of adjustment for multiple testing (Laboratory of Human Retrovirology and Immunoinformatics, Frederick, MD, USA) [50,51]. GO was also performed using Gene Ontology enRIchment anaLysis and visuaLizAtion tool (GOrilla), version 4.1 (http://cbl-gorilla.cs.technion. ac.il/accessed on 15 October 2022) [52,53]. Venn diagrams were constructed using Venny, version 2.1 (Juan Carlos Oliveros, BioInfoGP Service, Centro Nacional de Biotecnologia, Madrid, Spain; https://bioinfogp.cnb.csic.es/tools/venny/ accessed on 9 July 2022). Pathway interconnection were determined using Metascape (https://metascape.org accessed on 15 November 2022 [54]).

2.9. Statistics

Statistical analyses of assays were performed using Graphpad Prism 7 (San Diego, CA, USA) or Excel. For RNA-seq and qPCR analysis, one-way ANOVA with a post-test analysis was used for comparing multiple data sets. For Western blot analysis, two-way ANOVA with either Tukey's or Sidak's post hoc tests for multiple comparisons were used.

3. Results

3.1. Low Dose-Rate Radiation Effects on Cellular Morphology, Apoptosis, and Double-Stranded DNA Breaks in Mesenchymal Stem Cells (MSC)

Our previous studies showed that high-dose/high-dose-rate (10 Gy/0.989–1.15 Gy/min) X-ray irradiation primarily induces accelerated senescence in primary pulmonary artery endothelial cells (PAEC), primary human lung microvascular endothelial cells (HLMVEC), and primary mesenchymal stem cells (MSCs) [22,23,40]. We compared irradiation of MSC at low- and high-dose-rates. The morphology of MSCs following 0.66 Gy/h ^{60}Co radiation showed that the cells maintained a consistent morphology after 7 days (~1 Gy) and 14 days (~2 Gy) (Figure 1A). Additionally, cells appeared to increase in density over the time course of the experiment. In contrast, MSCs exposed to 10 Gy X-ray irradiation (1.15 Gy/min) displayed flattened "fried egg" morphology with increased cellular area and reduced cell numbers, consistent with cellular senescence, at 2 weeks post-irradiation (Figure 1A).

We previously found that low levels of apoptosis were present in 10 Gy (0.989–1.15 Gy/min) X-ray irradiated PAECs, HLMVECs, and MSCs, although in most cases the increase in apoptosis did not reach significance compared with control levels [22,23,40]. We investigated the induction of apoptosis in low- and high-dose-rate irradiation in MSCs using nuclear morphological analysis (Figure 1B) [40]. Nuclear blebbing is consistent with late apoptotic events. We did not observe any significant increase in nuclear morphological changes in any of the irradiated cells at the time points examined.

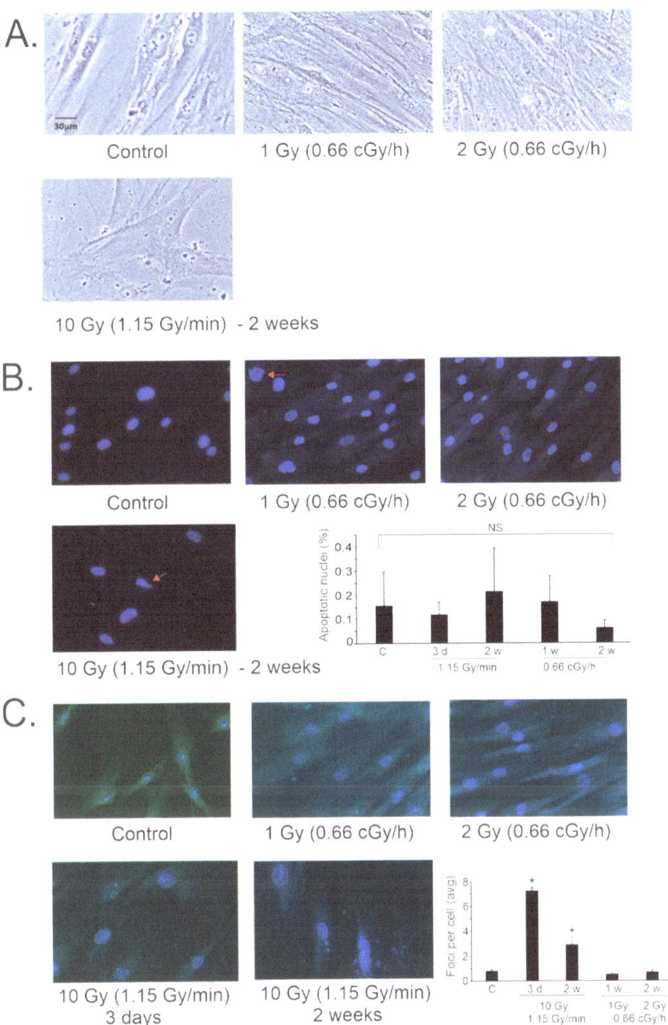

Figure 1. Effects of high and low dose-rate X-ray irradiation on cell morphology, apoptosis, and γ-H2AX foci in MSCs. MSCs were grown to 60% confluence and exposed to ^{60}Co irradiation at 0.66 cGy/h for 1 Gy (1 week) or 2 Gy (2 weeks). Alternatively, cells were exposed to 10 Gy X-ray irradiation (1.15 Gy/min) and assayed at 3 days and 1 week post-irradiation. Cells were fixed at the indicated times and stained with DAPI, and immunohistochemistry was performed for γ-H2AX. (**A**). Light microscopy was used to examine cell morphology. Representative images are shown, 20× magnification. (**B**). DAPI was used to examine the nucluear morphology of the fixed cells at the indicated times. Arrows (10 Gy irradiation) indicate nuclear blebbing, a late apoptotic event. Representative images are shown from each condition, 20× magnification. Nuclei were scored from all cells in random fields to determine percentage of apoptotic nuclei at 3 days (3 d), 1 week (1 w) or 2 weeks (2 w) post-irradiation. Graph shows average of percent apoptosis ± SEM; NS = not significant compared with control (C). (**C**). γ-H2AX immunohistochemistry was used to detect foci surrounding double-stranded DNA breaks in the fixed cells at the indicated times. Representative images are shown from each condition, 20× magnification. Foci were scored in all cells from random fields to determine numbers of foci per cell. Graph shows average of nuclear foci ± SEM; * indicates $p < 0.05$ compared with sham-irradiated control cells.

We examined DNA damage in the cells after 1 and 2 Gy (0.66 cGy/h, ^{60}Co) or 10 Gy (1.15 Gy/min, X-ray) exposures. DNA damage initiates signaling pathways that result in the phosphorylation of serine 139 on histone H2AX to form γ-H2AX, which is present in complexes surrounding double-stranded DNA breaks [55]. Immunohistochemistry for γ-H2AX complexes in the nuclei of MSCs following low- and high-dose-rate radiation exposure showed a significant increase in foci at 3 days and 2 weeks following 10 Gy/1.15 Gy/min X-ray irradiation (Figure 1C). Interestingly, exposure to low dose-rate irradiation did not show a significant increase in γ-H2AX foci compared with basal levels.

3.2. Genome-Wide Transcriptional Responses to Low Dose-Rate Radiation

To expand the understanding of overall gene expression changes in primary MSCs in response to low dose-rate radiation, we used comprehensive transcriptome profiling by RNA-seq. Gene expression profiles from sham-irradiated (control) MSCs were compared with MSCs irradiated at 0.66 cGy/h for 1 Gy total (~1 week) and 2 Gy total (~2 weeks). Comparative differential expression analysis identified 862 genes differentially expressed between 1 Gy irradiation samples and matched controls (q-value < 0.05, absolute fold change > 1.5) (Table S1). For 2 Gy irradiation, comparative differential expression analysis identified 725 differential genes (q-value < 0.05, absolute fold change > 1.5) (Table S2). A heatmap of the differentially expressed genes (DEGs) of all samples is shown in Figure 2A. At 1 Gy, 472 genes were downregulated compared with the matched control, and 390 genes were upregulated. At 2 Gy, 427 genes were downregulated compared with the matched control, and 298 genes were upregulated. A comparison of the gene sets from 1- and 2-Gy-regulated genes showed an overlap of ~39% of genes that were regulated at both doses of radiation (Figure 2B).

GO analyses were focused on terms relevant to cellular biological processes (BPs) and not disease states. The BP graphs show clusters of pathways with enrichment scores ≥ 1.6 (Figure 3, Supplemental Tables S3–S6). DAVID analysis showed that following 1 Gy low dose-rate irradiation (1 week, 0.66 cGy/h), the largest changes (both up- and down-regulation) in BP terms were vasoconstriction and blood pressure, metabolic processes (including glucose, cholesterol, fatty acid/lipid metabolism, and cellular response to starvation), proliferation, cellular response to hypoxia/reactive oxygen species and iron ion responses, apoptosis, and adhesion, and extracellular matrix modification (Figure 3A,B). In contrast, following 2 Gy low dose-rate irradiation (2 weeks, 0.66 cGy/h), DAVID analysis showed positive regulation of proliferation, continued metabolic processes (cholesterol and fatty acid biosynthesis/metabolism, response to starvation), continued cellular response to hypoxia, extracellular matrix modification (especially collagen), and apoptotic signaling (Figure 3C,D). Metascape (https://metascape.org accessed on 10 August 2022) [54] was used to create an image of clustered GO terms present in 1300 genes with the lowest q-value (up- and down-regulated, from both 1 Gy and 2 Gy) (Figure 3E). Relationships were identified between the pathway functions, notably the signaling genes for MAPK, protein phosphorylation and enzyme-linked receptors with genes that regulate cellular adhesion and locomotion, vascular development, tissue morphogenesis, and skeletal system development. Additional links were identified between regulation of genes for extracellular matrix organization, supramolecular fiber organization, and overall changes in genes encoding the core proteins making up the extracellular matrix (ECM; NABA core matrisome) and the ECM-associated proteins (NABA-matrisome-associated). GOrilla analysis of the ranked DEGs with q value ≤ 10^{-5} also identified enrichment in cellular processes, including cholesterol, lipid, sterol, and alcohol metabolism, protein catabolism, polysaccharide metabolism, responses to hormones and oxidative stress, extracellular matrix reorganization and cell motility, and developmental processes, including cell differentiation, and regulation of tissue remodeling, including vascular smooth muscle and bone remodeling (data not shown). GOrilla analysis also showed major changes in pathways regulating proliferation, inflammation, and programmed cell death (data not shown).

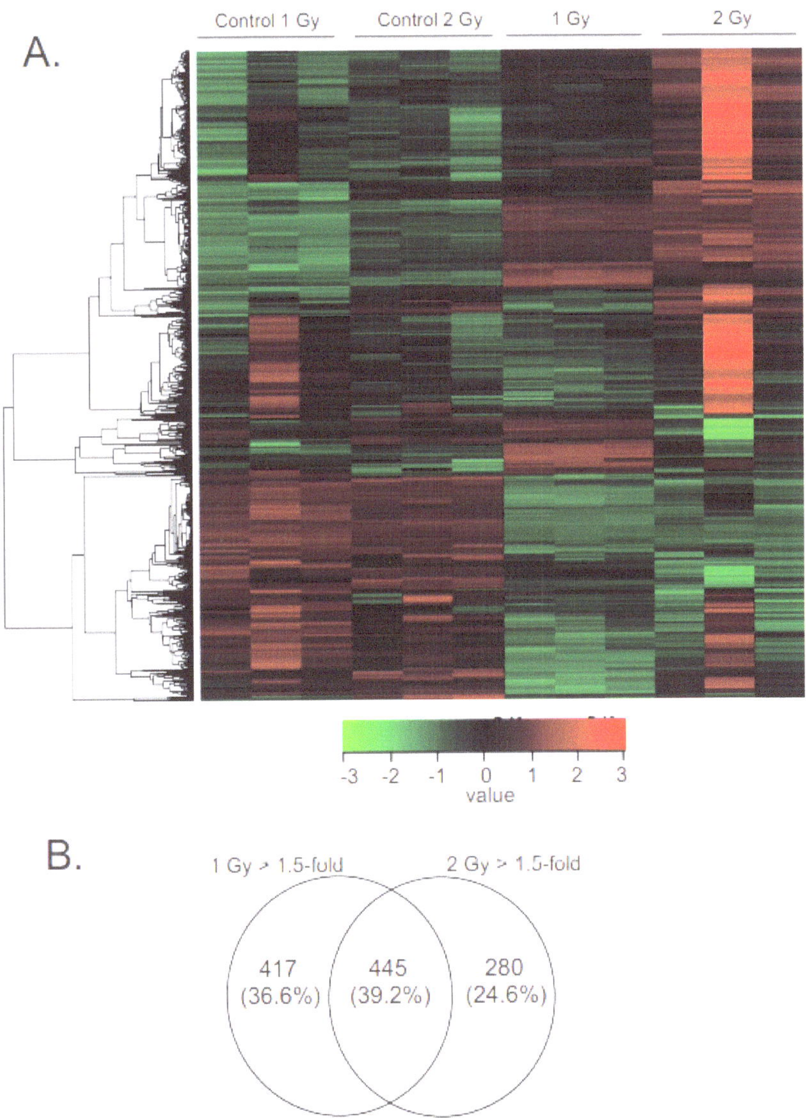

Figure 2. Gene expression changes in irradiated MSCs. MSCs were grown to 60% confluence and exposed to ^{60}Co irradiation at 0.66 cGy/h for 1 Gy (1 week) or 2 Gy (2 weeks). Control cells were cultured under identical conditions for each time point. RNAseq was performed using N = 3 samples for each condition. (**A**). Heat map indicates gene expression patterns following radiation exposure. (**B**). Venn diagram illustrating the number of genes with altered expression at each time point, q < 0.05, absolute fold change >1.5.

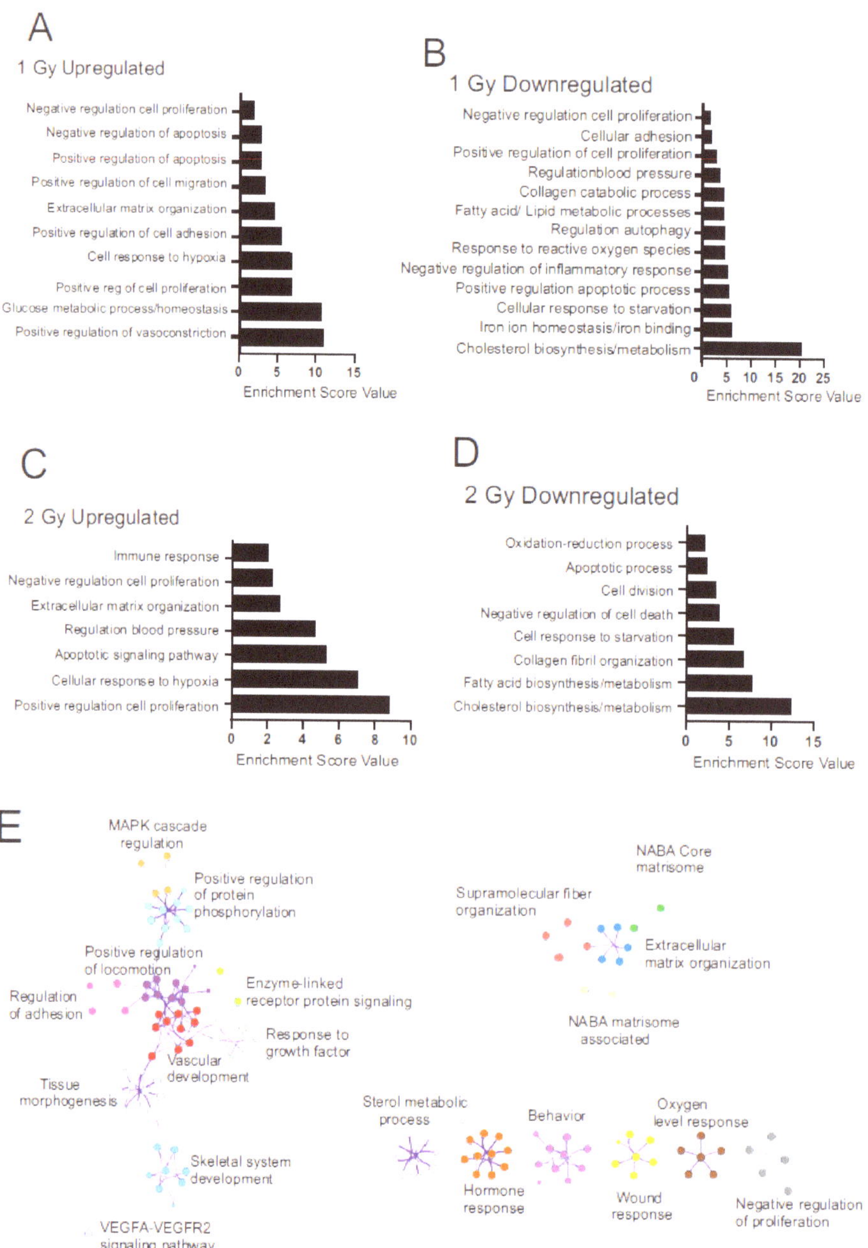

Figure 3. GO term cluster, KEGG pathway enrichment, and Metascape analyses of differentially expressed genes in MSCs following chronic low dose-rate irradiation. MSCs were grown to 60% confluence and exposed to ^{60}Co irradiation at 0.66 cGy/h for 1 Gy (1 week) or 2 Gy (2 weeks). Control cells were cultured under identical conditions for 1 or 2 weeks. Irradiated and control cells were lysed at the same time, and RNA was prepared for RNAseq. Pathway regulation was compared for all conditions. A,B. 1 Gy irradiation, upregulated pathways (**A**) and downregulated pathways (**B**). C,D. 2 Gy irradiation, upregulated pathways (**C**) and downregulated pathways (**D**). (**E**). Clustered GO terms using Metascape using genes with q < 0.05 to visualize pathway relationships.

3.3. Focused Heatmap Analysis of Chronic Low-Dose Radiation Gene Regulation

According to the GO analysis by DAVID and GOrilla, we evaluated gene regulation in the pathways and processes found to be most affected by chronic low-dose radiation. The focused analyses included genes selected by the DAVID and GOrilla, and included additional genes that we curated through literature searches for genes involved in each process that were also regulated in our study. qPCR and/or Western blotting was used to validate the pathways identified by RNAseq.

3.3.1. Alteration of Cellular Metabolism

Chronic low-dose radiation affected a number of metabolic pathways in the MSCs, including downregulation of cholesterol synthesis, upregulation of glycolysis over oxidative phosphorylation, and a reduction in fatty acid biosynthesis and modification (Figure 4A). qPCR was performed to validate at least one gene in each pathway (Figure 4B). Both 1 Gy and 2 Gy low dose-rate radiation showed downregulation of almost all genes encoding cholesterol synthesis enzymes, including for the synthesis of squalene from acetyl-CoA and for the synthesis of cholesterol from squalene (Figure 4A, left panel). These genes include the enzyme for the initiation of cholesterol synthesis (acetyl-CoA acetyltransferase 2, ACAT2), the rate-limiting enzyme (3-hydroxy-3-methylglutaryl-CoA synthase 1, HMGCS1), through to the final enzyme in the pathway (24-dehydrocholesterol reductase, DHCR24). In addition to the downregulation of cholesterol synthesis enzymes, we also observed the suppression of two major regulators of cholesterol biosynthesis: insulin-induced gene (INSIG1) and sterol regulatory element binding transcription factor-1 and -2 (SREBF1 and SREBF2). Finally, we observed downregulation of Niemann-Pick Type C disease 1 (NPC1), which regulates intracellular cholesterol transport and esterification.

Low-dose chronic radiation also resulted in the regulation of metabolic pathways that favored glucose metabolism over mitochondrial oxidative phosphorylation (Figure 4A, middle panel). Genes associated with increased glycolysis were upregulated: 3-phosphate dehydrogenase (GAPDH), pyruvate dehydrogenase kinases (PDK1, 3, and 4), aldo-keto reductase family 1 member C3 (AKR1C3), leptin (LEP), phosphofructo-2-kinase/fructose-2,6-bisphosphatase 3 (PFKFB3), proprotein convertase subtilisin (PCSK9), FOXO1, and NUAK2. PCSK9, which can negatively regulate glucose metabolism, was decreased five-fold. PDK1, -3, and -4, were each upregulated ~two-fold, inhibiting pyruvate dehydrogenase, and reducing the production of acetyl-coenzyme A from pyruvate. The upregulation of the PKDs can result in decreased activity of the tricarboxylic acid (TCA) cycle, decreased oxidative phosphorylation, and increasing the production of lactate as a final downstream function of glycolysis. We also observed an increase in lactate dehydrogenase, which catalyzes the conversion of pyruvate to lactate, again suggesting that pyruvate is being diverted away from the TCA cycle. FOXO1, a transcription factor responsible for increased gluconeogenesis, was reduced. Nu [novel] AMPK-related protein kinase-2 (NUAK2), which is responsive to increased AMP/decreased ATP, low glucose, and oxidative or endoplasmic reticulum stress, signals to suppress cell death by glucose starvation, was increased.

There was a general decrease in genes encoding enzymes for fatty acid (FA) metabolism and processing. Low-density lipoprotein receptor (LDLR), which can take up lipids from the environment, was decreased. There were decreases in folliculin (FLCN) and folliculin-interacting proteins (FNIP1 and FNIP2) regulators of AMP-dependent protein kinase (AMPK), a master regulator of FA metabolism, antioxidant responses, and mitochondrial and lysosome biogenesis. Decreases were observed in a number of types of FA acid synthesis enzymes: fatty acid synthase (FASN), a central regulator of lipid metabolism; acetyl-CoA carboxylase, which catalyzes the rate-limiting step in long-chain FA biosynthesis; acyl-CoA synthetase family member 2 (ACSF2), which enables medium-chain FA ligase activity; desaturase enzymes (FADS1 and 2), which produce highly unsaturated FA (HUFA); and stearoyl-coenzyme A desaturase (SCD), which synthesizes monounsaturated FA. Enzymes for the processing of FA were decreased, including for FA desaturation (FA desaturase-1 and 2, FADS1 and 2) and transport (FA binding protein 3, FABP3). Additionally, enzymes for FA degradation were also reduced: HADH, ASAH1, PLA1A, and GBA.

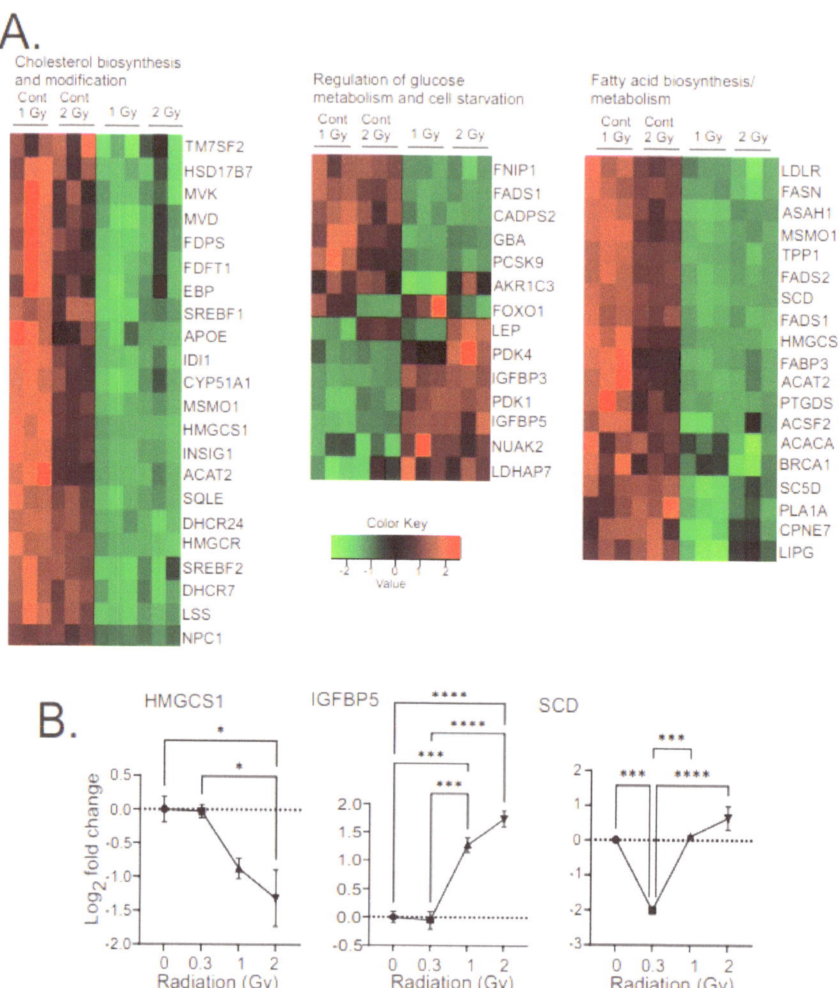

Figure 4. Heatmaps of gene expression changes in pathways for cholesterol biosynthesis and modification, glucose metabolism and cell starvation, and fatty acid biosynthesis and metabolism. (**A**). RNAseq was used to identify gene expression changes in primary human MSCs following 1 or 2 Gy (0.66 cGy/h) ^{60}Co irradiation. Control cells were cultured under identical conditions for 1 or 2 weeks. GO analysis was performed using DAVID and heatmaps were generated using genes with q < 0.05. Rows are centered and unit variance is applied to rows. (**B**). qPCR gene regulation using log base 2 scale fold change of genes supporting heatmaps of pathways. Data show averages ± SEM N = 3 biological replicates with two technical repeats; * indicates $p < 0.05$, *** indicates $p < 0.001$, **** indicates $p < 0.0001$, respectively, compared with sham irradiated control cells.

3.3.2. Regulation of Proliferation and Cell Division

We observed mixed regulation of genes related to cell proliferation. Focused heatmaps of proliferation and cell cycle genes are shown in Figure 5A; three of the regulated genes were validated by qPCR (Figure 5B). However, the cell numbers and morphology determined by light microscopy suggested that the MSCs continued to proliferate over the course of the 1 and 2 Gy exposures, without significant apoptosis or accelerated senescence (see Figure 1). We observed upregulation of proliferation-inducing genes. Upregulated growth factors included fibroblast growth factor 1 (FGF1), transform in growth factor A (TGFA), and vascular endothelial growth factor A (VEGFA). Upregulated transcription factors included Odd-skipped related transcription factor (OSR1) and c-Jun (JUN). We also observed downregulation of other growth factors, such as colony stimulating factor (CSF1) and pleiotrophin (PTN), as well as some transcription factors, such as transcription factor AP4 (TFAP4). There was also mixed regulation of factors that regulate apoptosis, including the upregulation of Baculovirus inhibitor of apoptosis repeat containing 5 (BIRC5), which suppresses apoptosis and promotes proliferation, but also the downregulated proteins that inhibit apoptosis, such as SFRP4 and IFIT3.

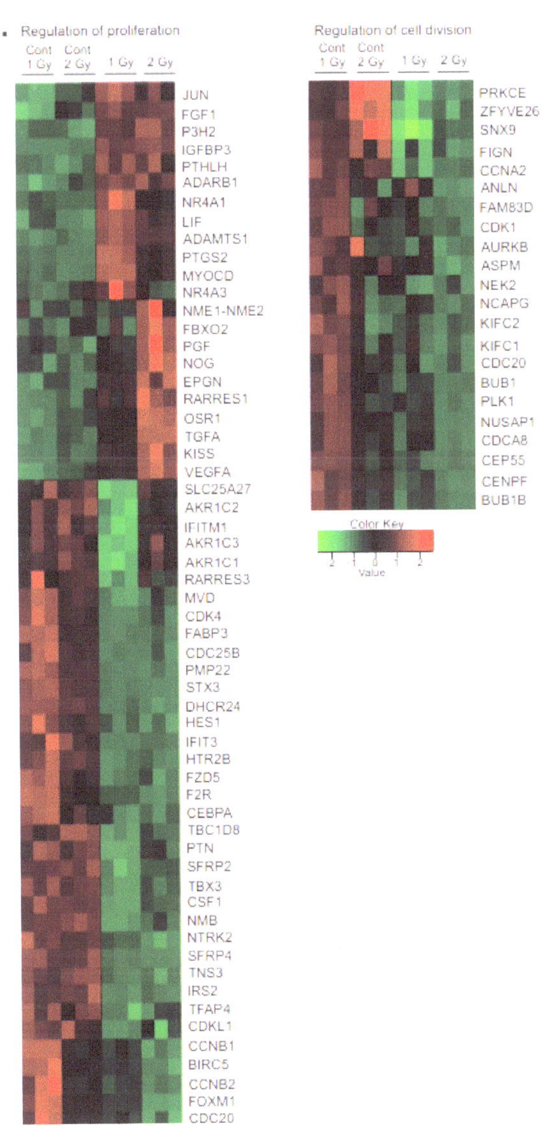

Figure 5. *Cont.*

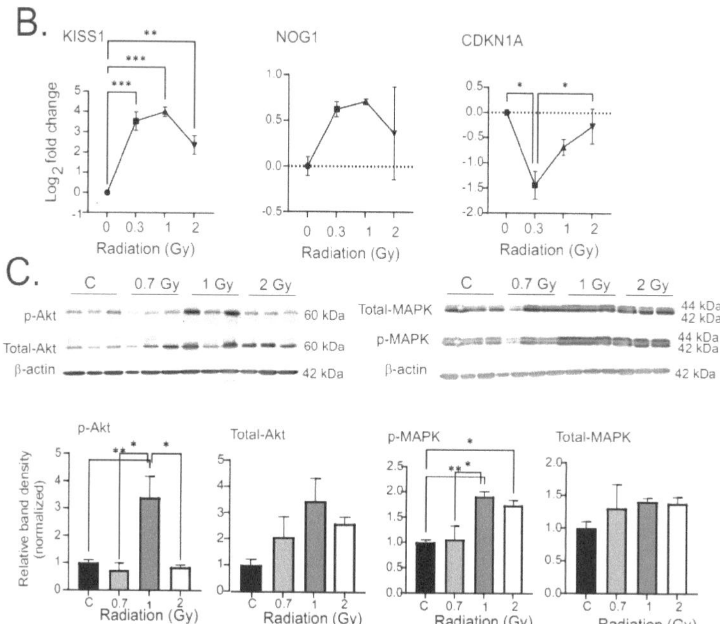

Figure 5. Heatmaps of gene expression changes for pathways in regulation of proliferation and cell division. (A). RNAseq was used to identify gene expression changes in primary human MSCs following 1 or 2 Gy (0.66 cGy/h) ^{60}Co irradiation. Control cells were cultured under identical conditions for 1 or 2 weeks. GO analysis was performed using DAVID and heatmaps were generated using genes with q < 0.05. Rows are centered and unit variance is applied to rows. **(B).** qPCR gene regulation using log base 2 scale fold change of genes supporting heatmaps of pathways. Data show averages ± SEM N = 3 biological replicates with two technical repeats; * indicates $p < 0.05$, ** indicates $p < 0.01$, *** indicates $p < 0.001$, respectively, compared with sham-irradiated control cells. **(C).** Western blot data showing regulation of AKT and MAPK (phosphorylated and total). Western blots were performed on N = 3 biological repeats. Bar graphs show average band densities normalized to β-actin. Graphs show means ± SEM; * indicates $p < 0.05$ and ** indicates $p < 0.01$, respectively, compared with sham-irradiated control cells.

There was a general downregulation of cell-cycle regulatory proteins, including cyclin-dependent kinase 1 (CDK1), cyclin-dependent kinase 4 (CDK4), cyclin A2 (CCNA2), cyclin-dependent kinase-like 1 (CDKL1), cyclins B1 and 2 (CCNB1 and 2), and cell-division-cycle-associated A8, 20 and 25B (CDCA8,20, 25B). Proteins interacting with the chromosome, centromere, and mitotic spindle were also downregulated, including kinesin family members C1 and C2 (KIFC1, 2), nucleolar-spindle-associated protein (NUSAP1), centrosomal protein 55 (CEP55), centromere protein F (CENPF), and condensin subunit CAP g (NCAPG). These proteins are known to have roles in centrosome stabilization, chromosome condensation, and mitotic spindle formation, all required for cell division.

Interestingly, the downregulation of these proteins occurred without the induction of apoptosis or accelerated senescence (See Figure 1). The phosphorylation of Akt and p42/p44 MAPK are associated with proliferation and cell survival, and we observed significant increases in both phosphorylated Akt and p42/p44 MAPK in the 1 and 2 Gy exposures (Figure 5C). This suggests that the mixed regulation of proliferation and cell cycle genes favors cell survival and proliferation, or potentially, cell cycle arrest without senescence.

3.3.3. Regulation of Apoptosis, Cell Death, and Autophagy

A variety of forms of cell death and autophagy are induced in cells in response to high-dose acute ionizing radiation, and we previously observed the upregulation of pro-apoptotic and pro-senescence pathways [5,40]. In contrast, we did not observe significant levels of apoptosis at any time points following chronic low-dose radiation (see Figure 1), and GO analysis showed apoptotic pathway regulation predominantly favoring the inhibition of apoptosis by a number of mechanisms (Figure 6). The regulation of this group of genes was confirmed by qPCR of EGR1 and DUSP1 and Western blotting of Egr1 and DUSP1 (Figure 6B,C).

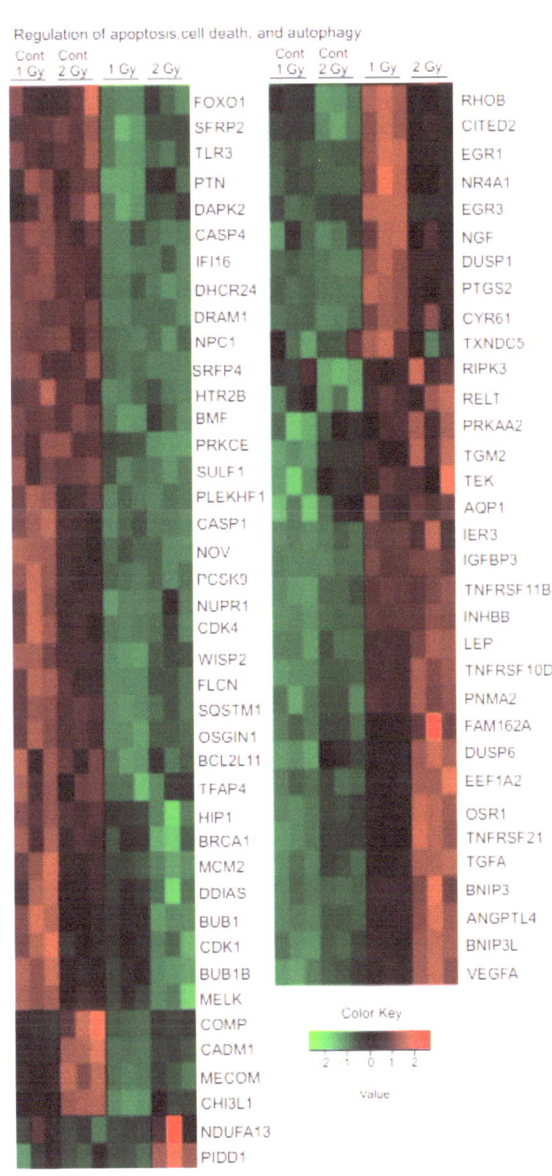

Figure 6. *Cont.*

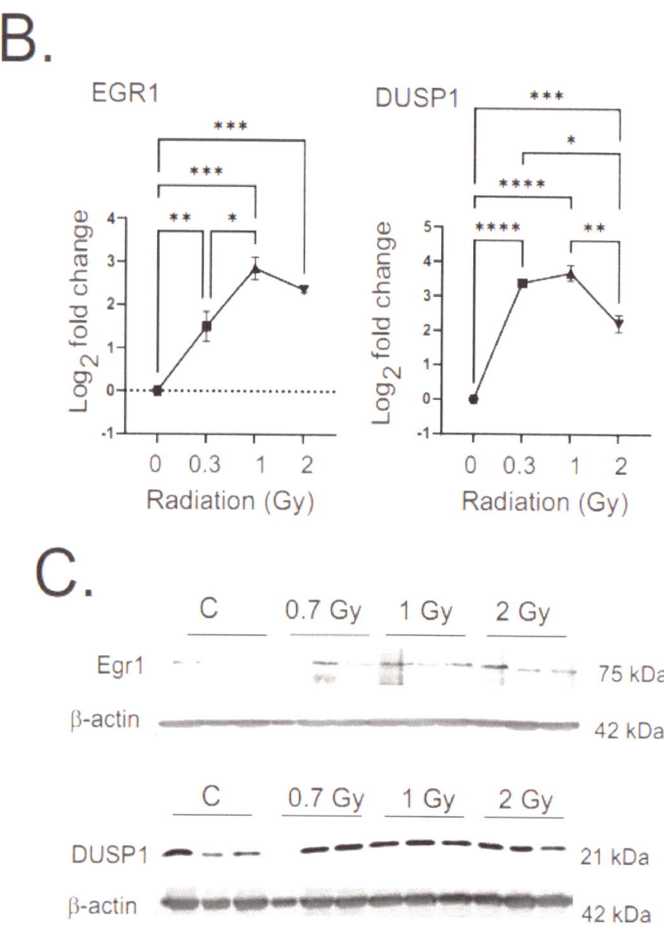

Figure 6. Heatmaps of gene expression changes for pathways in regulation of apoptosis, cell death, and autophagy. (A). RNAseq was used to identify gene expression changes in primary human MSCs following 1 or 2 Gy (0.66 cGy/h) ^{60}Co irradiation. Control cells were cultured under identical conditions for 1 or 2 weeks. GO analysis was performed using DAVID and heatmaps were generated using genes with q < 0.05. Rows are centered and unit variance is applied to rows. **(B)**. qPCR gene regulation using log base 2 scale fold change of genes supporting heatmaps of pathways. Data show averages ± SEM from N = 3 biological replicates with two technical repeats. * indicates $p < 0.05$, ** indicates $p < 0.01$, *** indicates $p < 0.001$, **** indicates $p < 0.0001$, respectively, compared with sham-irradiated control cells. **(C)**. Western blot data showing regulation of Egr1 and DUSP1. Western blots were performed on N = 3 biological repeats. Bar graphs show average band densities normalized to β-actin.

Gene regulation was observed in the pathways for intrinsic and extrinsic apoptosis, p53-pathway-regulated apoptosis, and Wnt signaling-induced apoptosis. Two pro-apoptotic master regulators were downregulated: forkhead box protein O1 (FOXO1) and nephroblastoma-overexpressed protein (NOV). Three anti-apoptotic regulators were up-regulated: early growth response 1 (EGR1), leptin (LEP), and eukaryotic elongation factor 1A2 (EEF1A2). There was most notably a reduction of a number of genes in the p53-mediated pathway of apoptosis, including DNA-damage-regulated autophagy modulator 1 (DRAM1), interferon gamma-inducible protein 16 (IFI16), breast cancer suppressor protein (BRCA1), NADH:ubiquinone oxidoreductase subunit A13 (NDUFA13), maternal embryonic leucine zipper kinase (MELK), p53-induced death domain protein 1 (PIDD1), and oxidative-stress-induced growth inhibitor 1 (OSGIN1). Several pro-apoptotic genes in the extrinsic/death-receptor-initiated apoptosis pathway were downregulated: Toll-like receptor 3 (TLR3), death-associated protein kinase 2 (DAPK2), Huntingtin-interacting protein 1 (HIP1), and sequestosome-1 (SQSTM1). At the same time, two anti-apoptotics for the extrinsic apoptotic pathway were upregulated: tumor necrosis factor receptor superfamily member 11B (TNFSF11B) and family member 10D (TNFRSF10D).

Interestingly, we did observe regulation of pro-apoptotic pathways associated with DNA-damage-, oxidative-stress-, or hypoxia-induced apoptosis. Examples of these included downregulation of anti-apoptotic genes for secreted frizzled related proteins 2 and 4 (SFRP2 and 4), 24-decydrocholesterol reductase (DHCR24), DNA-damage-induced apoptosis (DDIAS), and chitinase-3-like protein 1 (CH13L1), as well as increased expression of pro-apoptotic dual-specificity protein phosphatase 1 and 6 (DUSP1 and 6), immediate early response 3 (IER3), family with sequence similarity 162 member A (FAM162A), osmotic-stress-resistance protein (OSR), and BCL2/adenovirus E1B 19 kDa protein-interacting protein 3 (BNIP3). As stated previously, we did not observe significant apoptosis in the cells, suggesting that the overall effect of these mixed regulations favored cell survival.

3.3.4. Regulation of Pathways for Extracellular Matrix, and Cell Attachment and Migration

GO analysis of the gene changes showed significant radiation-induced alterations in pathways related to synthesis, breakdown, and organization of the extracellular matrix (ECM), as well as for cell attachment to the extracellular matrix (Figure 7A). Two genes from these pathways, keratin 34 (KRT34) and neuregulin 1 (NRG1), were validated by qPCR (Figure 7B).

Changes in gene expression related to proteins of the ECM were characterized by upregulation of genes associated with wound repair and/or fibrosis. The upregulated genes included markers of fibrotic remodeling (collagen 4A1 and 5A1 [COL4A1, COL5A1], prolyl 4-hydroxylase subunit alpha 1 [P4H1], procollagen-lysine, 2-oxoglutarate 5-dioxygenase 2 [PLOD2], lysyl oxidase-like 4 [LOXL4], prolyl 3-hydroxylase 2 [P3H2], a disintegrin and metalloproteinase with thrombospondin motifs 2 [ADAMTS2]) as well as known drivers of fibrosis (hyaluronic acid synthase 1 and 2 [HAS1, HAS2], prolyl 4-hydroxylase subunit alpha 2 [P4H2], lysyl oxidase [LOX], ADAMTS6, and procollagen C-endopeptidase enhancer 1 [PCOLCEL]). The downregulated genes also included some other markers of fibrosis (vitronectin [VTN], fibrillin 1 [FBN1], cartilage oligomeric matrix protein [COMP], integrin subunit alpha 4 [ITGA4], secreted frizzled related protein 2 [SFRP2], proline and arginine rich end leucine rich repeat protein [PRELP], stearoyl-CoA desaturase 1 [SCD1], and thrombospondin 3 [THBS3]) as well as genes with antifibrotic function (disintegrin and metalloproteinase domain-containing protein 12 [ADAMS12], and cathepsins D and K [CTSD, CTSK, two proteases]). These changes were accompanied by decreased expression of some proteins for ECM breakdown and decreased proteins that functioned to maintain or increase barrier function: filaggrin (FLG), interleukin 32 (IL32), intercellular adhesion 1 (ICAM1), wingless/integration 1 (WNT1) inducible-signaling pathway protein 2 (WISP2), and trophinin-associated protein (TROAP). The combination of these alterations of proteins of the ECM could function to increase the matrix stiffness.

The regulation of pathways for cell adhesion included a number of genes predicted to increase cellular motility and/or invasiveness: neuregulin 1 (NRG1), fibroblast growth factor 1 (FGF1), WNT family member 5B (WNT5B), heparin-binding epidermal growth factor-like growth factor (HBEGF), nestin (NES), semaphoring 7A (SEMA7A), endothelin 1 (EDN1), cysteine-rich angiogenic inducer 61 (CYR61), erythrocyte membrane protein band 4.1-like B4 (EPB41L4B), cluster of differentiation 274 (CD274, also known as programmed cell death 1 ligand 1), desmin (DES), sushi repeat-containing protein X-linked 1 (SRPX2), transient receptor potential cation channel subfamily M member 8 channel-associated factor 2 (TCAF2), and glutamine gamma-glutamyltransferase 2 (TGM2). Additional changes in cellular intermediate fillaments included increased expression of a number of keratin

(KRT) proteins, including KRT7, 8, 14, 16, 18, 34, 80, and 81, which also can increase cellular stiffness and promote motility. We did not observe upregulation of any proteins associated with inhibition of motility, but we did observe the downregulation of several proteins that would increase cellular adhesion: secreted phosphoprotein 1 (SPP1), desmoplakin (DSP), and metastasis suppressor protein 1 (MTSS1).

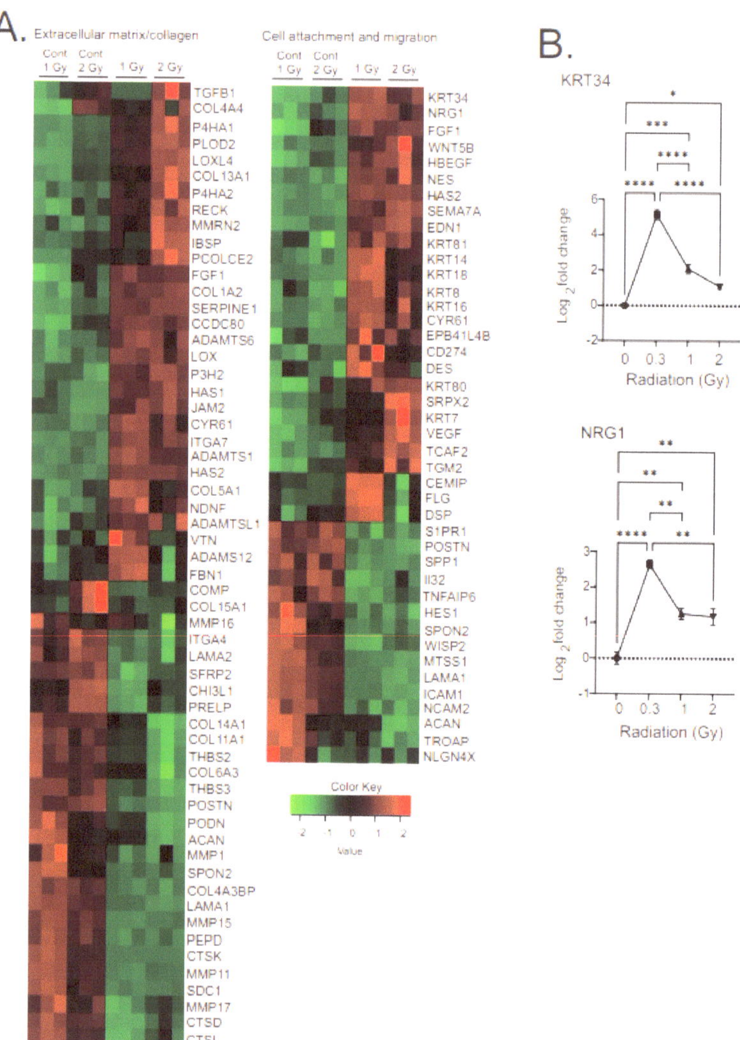

Figure 7. Heatmaps of gene expression changes for pathways involved in the regulation of extracellular matrix and collagen synthesis and cell attachment and migration. (**A**). RNAseq was used to identify gene expression changes in primary human MSCs following 1 or 2 Gy (0.66 cGy/h) ^{60}Co irradiation. Control cells were cultured under identical conditions for 1 or 2 weeks. GO analysis was performed using DAVID and heatmaps were generated using genes with q < 0.05. Rows are centered and unit variance is applied to rows. (**B**). qPCR gene regulation using log base 2 scale fold change of genes supporting heatmaps of pathways. Data show averages ± SEM from N = 3 biological replicates with two technical repeats. * indicates $p < 0.05$, ** indicates $p < 0.01$, *** indicates $p < 0.001$, **** indicates $p < 0.0001$, respectively, compared with sham-irradiated control cells.

3.3.5. Regulation of Pathways for Vascular Constriction and Inflammation

GO analysis of the transcriptome revealed alterations in pathways for vascular constriction and inflammation (Figure 8A). Three of these genes were validated using qPCR: OXTR, HMOX1, and NPR3, which showed similar alterations as the transcriptomic data (Figure 8B). The regulation of vascular constriction pathways showed an increase in the number of genes associated with increased vasoconstriction: 5-hydroxytrypamine receptor 2A (HTR2A), adrenoreceptor alpha 1B and 1D (ADRA1B and 1D), oxytocin receptor (OXTR), prostaglandin–endoperoxide synthase 2 (PTGS2, also called cyclooxygenase-2), gap junction alpha 5 (GJA5), leptin (LEP), apelin (APLN), and endothelin 1 (EDN1), which has a specific role in pulmonary hypertension. At the same time, there was a decrease in expression of several genes associated with vasodilation: superoxide dismutase 2 (SOD2) and heme oxygenase 1 (HMOX1).

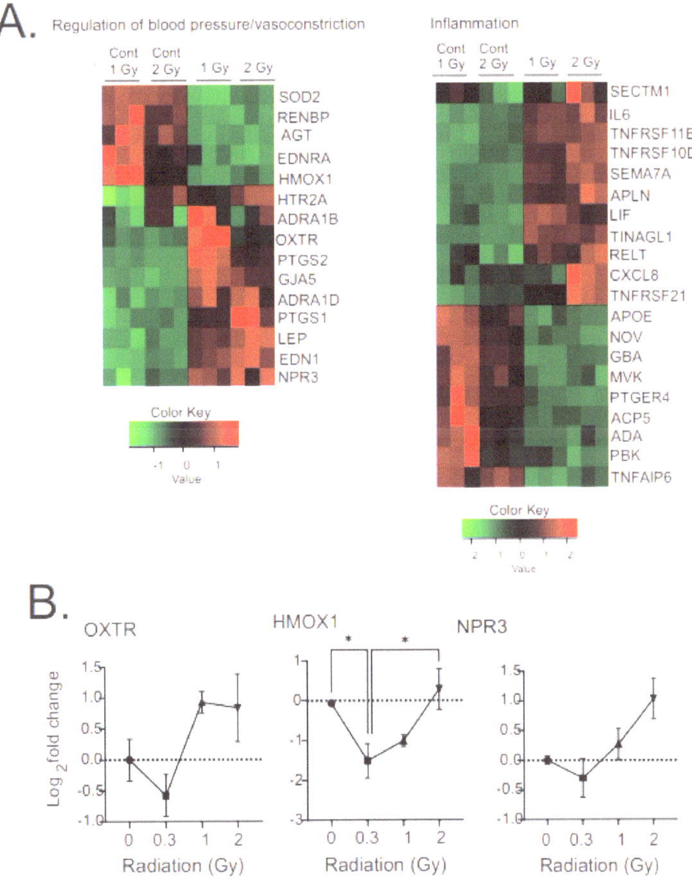

Figure 8. Heatmaps of gene expression changes for pathways involved in the regulation of blood pressure or vasoconstriction and inflammation. (**A**). RNAseq was used to identify gene expression changes in primary human MSCs following 1 or 2 Gy (0.66 cGy/h) ^{60}Co irradiation. Control cells were cultured under identical conditions for 1 or 2 weeks. GO analysis was performed using DAVID and heatmaps were generated using genes with q < 0.05. Rows are centered and unit variance is applied to rows. (**B**). qPCR gene regulation using log base 2 scale fold change of genes supporting heatmaps of pathways. Data show averages ± SEM from N = 3 biological replicates with two technical repeats. * indicates $p < 0.05$ compared with sham-irradiated control cells.

Radiation-induced changes in expression of inflammatory factors and activators showed mixed up- and down-regulation of pro- and anti-inflammatory factors. Upregulated anti-inflammatory factors included: secreted and transmembrane 1 (SECTM1), osteoprotegerin (TNFRSF11B), tumor necrosis factor receptor superfamily member 10D (TNFRSF10D), leukemia inhibitory factor (LIF), semaphoring 7A (SEMA7A). At the same time, we observed downregulation of some pro-inflammatory factors: TNFRSF21, apolipoprotein E (APOE), nephroblastoma-overexpressed protein (NOV), and PDZ-binding protein (PBK).

3.3.6. Regulation of PATHWAYs for Cell Response to Hypoxia and Iron Homeostasis

GO analysis revealed the regulation of pathways related to cellular response to hypoxia as well as to iron homeostasis and iron-binding proteins (Figure 9A). Four of the genes in these groups were validated by qPCR (Figure 9B). Interestingly, the regulation of genes to hypoxia included a number of genes that are also regulated by redox stress [56], and a number of genes related to iron homeostasis and iron-binding can also be regulated by redox stress [57].

Surprisingly, we observed downregulation of a number of enzymes that would mitigate redox stress, including superoxide dismutase 1 and 3 (SOD1, 3) and glutathione peroxidase 1 and 4 (GPX1, 4). We also observed the downregulation of the oxidative stress-activated transcription factor nuclear factor erythoid-derived 2-like 1 and 2 (NFE2L1 and 2), as well as its regulator, kelch-like ECH-associate protein 1 (KEAP1). In contrast, we observed an upregulation of almost 20 genes previously shown to be regulated by hypoxia-inducible factor-1α (HIF-1α): pyruvate dehydrogenase kinase 4 (PDK4), (FAM162A), (MGARP), stanniocalcin-1 (STC1), procollagen-lysine, 2-oxoglutarate 5-dioxygenase 2 (PLOD2), lactate dehydrogenase-A (LDHA), endothelin 1 (EDN1), BCL2/adenovirus E1B 19 kDa protein-interacting protein 3 (BNIP3), angiopoietin-like 4 (ANGPTL4), vascular endothelial growth factor A (VEGFA), solute carrier family 2, facilitated glucose transporter member 8 (SLC2A8), aldehyde oxidase (AOX1), aquaporin-1 (AQP1), apolipoprotein L domain-containing protein 1 (APOLD1), potassium two-pore-domain channel subfamily K member 3 (KCNK3), sodium calcium exchanger 1(NCX1 or SLC8A1), prostaglandin–endoperoxide synthase 2 (cyclooxygenase, PTGS2), and neuron-derived neurotrophic factor (NDNF). These genes have a variety of functions including the regulation of glycolysis, cholesterol and lipid biosynthesis, increased proliferation, inhibition of apoptosis, extracellular matrix modification, and cell motility. The regulation of hypoxia-responsive membrane channels, together with a large group of genes regulated by HIF-1α, strongly suggest the induction of cellular adaptation to a hypoxic environment.

Iron-regulated pathways were also identified by GO analysis. Two major proteins for iron storage were downregulated: ferritin light chain (FTL) and ferritin heavy chain 1 (FTH1). At the same time, there was a downregulation of a number of iron-dependent enzymes, including a number of enzymes involved in fatty acid and cholesterol metabolism: fatty acid desaturase 1 and 2 (FADS1, FADS2), lanosterol 14-alpha demethylase (CYP51A1), hydroxysteroid 17-beta dehydrogenase-14 and 11-beta dehydrogenase 1 (HSD17B14 and HSD11B1), 7-dehydrocholesterol reductase (DHCR7), cytochrome B5 reductase-like (CYB5RL), stearoyl –CoA-desaturase 1 and 5 (SCD, SC5D), and flavin containing dimethylaniline monooxygenase 3 and 4 (FMO3, FMO4). Besides these, four genes were regulated with iron valence modification activity: cytochrome B reductase 1 (CYBRD1), six-transmembrane epithelial antigen of prostate 1(STEAP1), cytoglobin (CYGB), and ferric-chelate reductase 1 (FRRS1). Interestingly, three of the four upregulated genes were all related to cellular response iron deficiency: family with sequence similarity 162 member A (FAM162A), endothelin 1 (EDN1), and stanniocalcin 1 (STC1). Together, these gene regulation patterns suggests cellular responses to iron deficiency and downstream signaling.

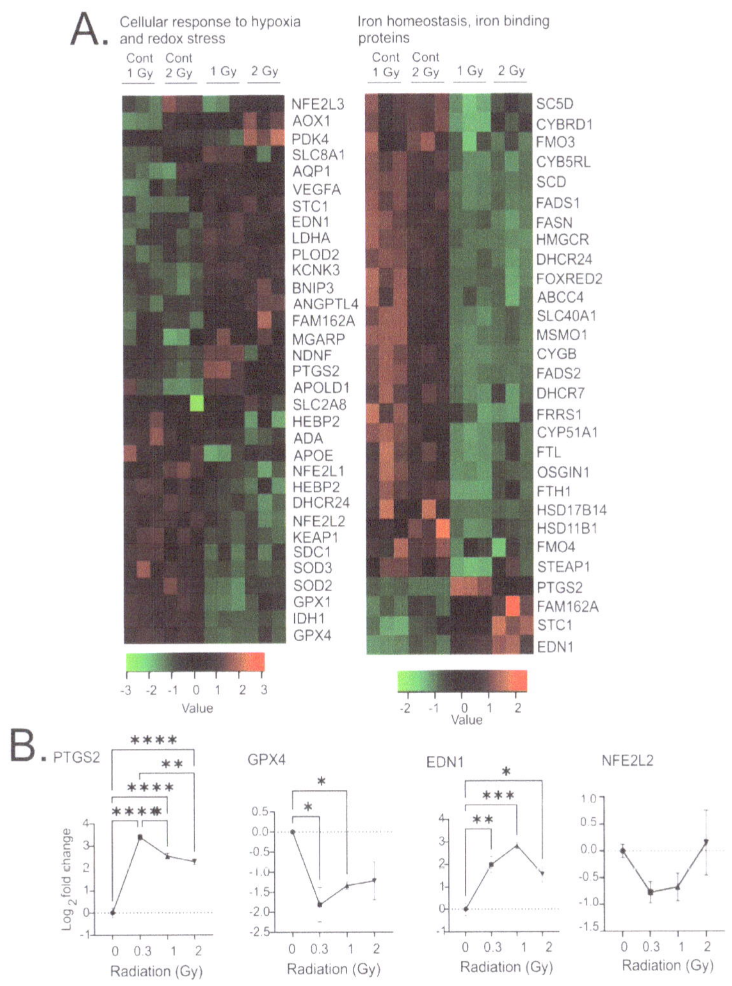

Figure 9. Heatmaps of gene expression changes for pathways involved in cellular response to hypoxia or redox stress and iron homeostasis and iron-binding proteins. (**A**). RNAseq was used to identify gene expression changes in primary human MSCs following 1 or 2 Gy (0.66 cGy/h) ^{60}Co irradiation. Control cells were cultured under identical conditions for 1 or 2 weeks. GO analysis was performed using DAVID and heatmaps were generated using genes with q < 0.05. Rows are centered and unit variance is applied to rows. (**B**). qPCR gene regulation using log base 2 scale fold change of genes supporting heatmaps of pathways. Data show averages ± SEM from N = 3 biological replicates with two technical repeats. * indicates $p < 0.05$, ** indicates $p < 0.01$, *** indicates $p < 0.001$, **** indicates $p < 0.0001$, respectively, compared with sham-irradiated control cells.

3.4. Effect of Chronic Low Dose-Rate Irradiation on Primary Human Lung Microvascular Endothelial Cells

The vascular endothelium is hypothesized to be a primary mediator of radiation injuries [58]. We therefore investigated the effect of low dose-rate radiation on primary human lung microvascular endothelial cells in culture using the same targets identified in MSCs (Figure 10). We found similar regulation for genes in cholesterol and fatty acid biosynthesis (HMGCS1 and MSMO1), for cell survival and suppression of accelerated senescence (EGR1 and CDKN1A and MAPK and Akt activation), for reduced antioxidant signaling (SOD2), and we observed a trend for extracellular matrix modification (COL1A2).

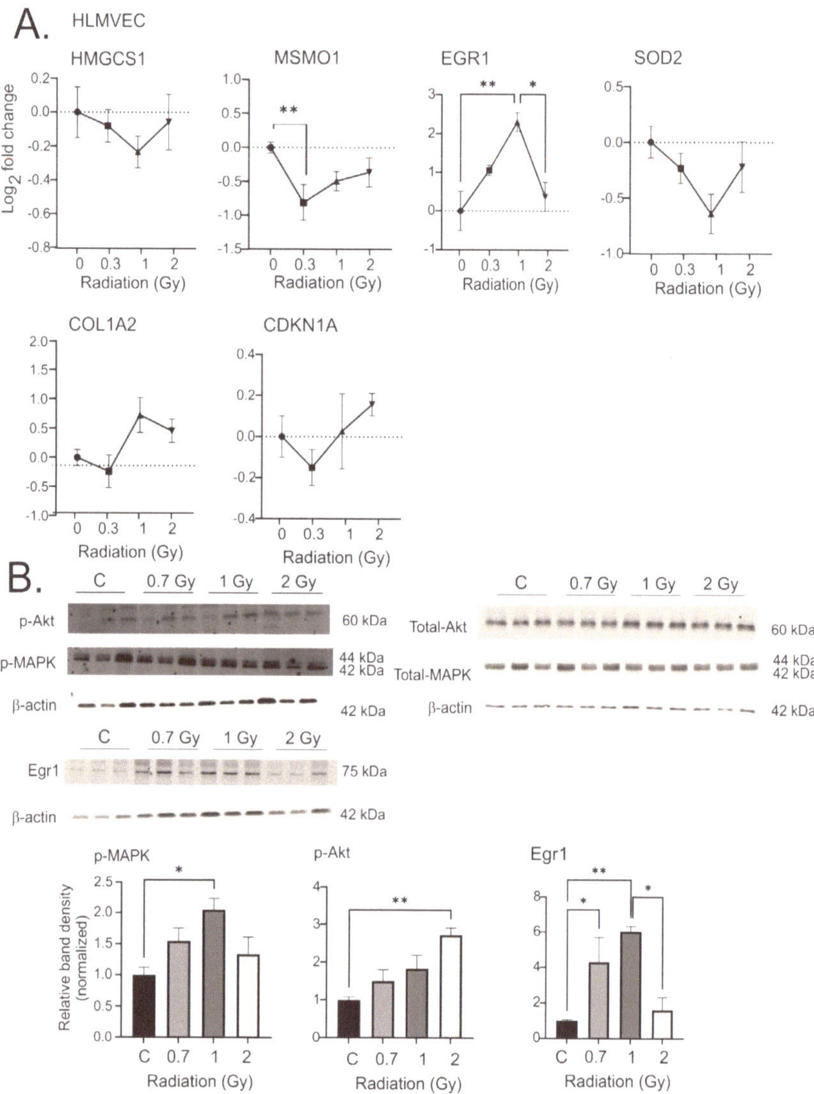

Figure 10. Response of human lung microvascular endothelial cells (HLMVECs) to chronic low dose-rate irradiation. HLMVECs were grown to 50% confluence and exposed to ^{60}Co irradiation at 0.64 cGy/h for 1 Gy (1 week) or 2 Gy (2 weeks). Control cells were cultured under identical conditions. Irradiated and control cells were lysed at the same time, and either frozen for protein analysis or placed in RNAlater. (**A**). qPCR gene regulations are represented on a log base 2 fold change scale, using N = 3 biological replicates with two technical repeats. Graphs show means $\pm$ SEM; * indicates $p < 0.05$ and ** indicates $p < 0.01$, respectively, compared with sham-irradiated control cells. (**B**). Western blots of phosphorylated and total MAPK and Akt, and Egr1. Western blots were performed on N = 3 biological repeats. Bar graphs show average band densities normalized to β-actin. Graphs show means $\pm$ SEM; * indicates $p < 0.05$ and ** indicates $p < 0.01$, respectively, compared with sham-irradiated control cells.

4. Discussion

The potential for accidental radiation exposure has increased with the increasing use of medical and industrial radiation, and with the potential increase in nuclear energy generation and the potential for military use of radioactive weapons. Current proposals for space travel will also result in chronic radiation exposure. Therefore, an increased understanding of the cellular effects of chronic low dose-rate radiation is needed. Here, we demonstrate in primary human MSCs that exposure to 0.3–2 Gy gamma radiation (0.66 cGy/h) did not result in growth arrest, the induction of significant apoptosis, or accelerated senescence. GO analysis of the transcriptomic changes showed that low dose-rate radiation exposure resulted in changes in gene expression related to cellular metabolic activity, pro-survival and proliferation signaling, alterations in the extracellular matrix and cellular motility, increased expression of factors related to vascular constriction, and increased signaling related to hypoxia and iron deficiency. We confirmed our findings using targeted qPCR and Western blotting in primary HLMVEC. Noticeably absent from the identified DAVID and Metascape pathway analyses were the upregulation of DNA damage responses, strong pro-apoptotic and senescence pathway activation, oxidative stress-related signaling, and marked pro-inflammatory activation pathways, that our laboratory previously identified as major pathway responses to acute exposure to 10 Gy irradiation (1.15 Gy/min) [40]. Together, these data suggest that acute high-dose/high-dose-rate radiation exposure and chronic low dose-rate radiation exposure activate distinct signaling events, resulting in fundamentally different cellular outcomes.

Several previous studies investigated the sensitivity and responses of MSCs to radiation. One study utilized 40–2000 mGy (dose-rate not disclosed), and showed increased apoptosis, senescence, and autophagy, with evidence of DNA damage response at 1–48 h post-irradiation [7]. A second study exposed MSCs to 1–9 Gy radiation at 300 cGy/min, and examined the cells from 12 h to several weeks after irradiation [39]. This study also showed DNA damage response and loss of viability. A third study performed comprehensive analysis of transcriptome changes in MSCs exposed to 0.01 to 1 Gy (0.79 Gy/min), using microarrays to identify pathway regulation over a short time course from 1–48 h post-irradiation [59]. In contrast with our study, which exposed cells to 0.66 cGy/h, the irradiation of the MSCs in all of these studies was performed as acute exposures. Our study did not identify significant DNA damage response, apoptosis or senescence, although it is possible that these processes occurred at a low level earlier in the time course of the exposure, and that a percentage of the remaining cells recovered and proliferated over the time course. Our analysis at the 1- and 2-week time points suggests that the cells continued to proliferate (due to cellular density) and maintained normal (non-senescent) morphology. A previous study using very-low-dose X-ray irradiation of fibroblasts also identified an initial cell cycle pause, followed by resumed proliferative response [60]. The authors concluded that the early pause in cell cycle could have been associated with DNA repair [60]. Future work to investigate very early responses of MSCs to chronic low-dose radiation is needed to determine whether an early response may include a transient pause in the cell cycle.

In our experiments, MSCs responded to chronic low dose-rate radiation by downregulating cholesterol and lipid biosynthesis, as well as downregulating some enzymes in the pathway for oxidative phosphorylation while upregulating enzymes for glycolysis. A shift to aerobic glycolytic metabolism was previously demonstrated in vivo following fractionated high-dose radiation (5 Gy/day × 3 days, 702 cGy/min) [61]. Whether the changes observed in the chronic low-dose exposure is also aerobic or anaerobic glycolysis requires further investigation. With regard to changes in lipid and cholesterol metabolism, a number of reports showed *increased* cholesterol biosynthesis following acute, high-dose radiation exposure, which has been hypothesized to be a potential mechanism for radiation-induced cardiovascular disease and carcinogenesis [62]. In contrast, relatively low-dose radiation (25–50 mGy, 1.0 mGy/min) was shown to reduce atherosclerosis lesions in a murine model predisposed to atherosclerosis, although the mechanism of this is unknown [63].

At both 1 and 2 Gy exposures, we observed changes in the expression of ECM proteins, as well as proteins involved in cellular attachment and motility. The specific alterations in the expression of the ECM proteins and cellular intermediate filaments could have the combined effect of increasing cellular stiffness and increasing motility [64]. In some cases, the changes in proteins that increase cell motility, including some keratins, have also been linked to increased cellular proliferation [65]. Further studies of MSCs following chronic radiation are required to determine whether cellular motility is altered and also to measure specific changes in ECM stiffness.

Although GO analysis of altered gene expression identified a number of altered pathways, analysis of potential pathway hierarchies led us to hypothesize that hypoxia and iron deficiency pathways lie upstream of the other pathways that are regulated. Hypoxia can be associated with increased cellular proliferation and inhibition of apoptosis, in some cases through AQP1 regulation [66]. Additionally, cellular responses to hypoxia, and activation of hypoxia-inducible factors (HIFs), can lead to downstream regulation of cholesterol and lipid metabolism, changes in ECM protein synthesis and cell migration, increased cell survival and proliferation gene expression, and changes in glycolysis [61,67]. In vivo, pulmonary hypoxia leads to decreased cholesterol synthesis, and favors certain types of pulmonary vascular constriction [68]. In agreement with the potent regulation of hypoxia (directly or indirectly), GO analysis identified at least 15 genes known to be activated by hypoxia-inducible factors. Interestingly, we also found *downregulation* of genes associated with oxidative protective mechanisms, including SOD and Nrf2, suggesting that redox stress is not strongly regulated by chronic low dose-rate radiation at the time points that were studied.

Radiation-induced oxygen depletion in aqueous environments was identified over 40 years ago [69,70]. High-dose radiation (15 Gy, 67 cGy/min) induces hypoxia-related gene expression at 24 h post-irradiation [71], and recent advances in FLASH radiotherapy have rekindled interest in this effect [72]. The biological effects of radiation that would lead to cellular iron deficiency or iron-deficiency-like signaling are not known. Although our laboratory has described the effects of radiation on iron in vivo, these effects are initiated by red blood cell and reticulocyte hemolysis [73,74].

We also observed an increase in pathway regulation related to iron deficiency. High levels of iron can inhibit activation of HIFs, whereas low iron can increase HIF activation [75]. Iron deficiency can independently modulate glycolysis, regulate genes involved in cholesterol and lipid metabolism, and increase Egr-1 signaling [76,77]. Studies have shown that hypoxia can affect iron homeostasis and absorption under some conditions [78]. The link between radiation and cellular iron is not known. In cultured breast cancer cells, radiation-induced autophagic cell death was associated with iron accumulation, increased levels of transferrin receptor, and increased ferritin following acute exposure to 1–8 Gy X-ray irradiation (1.0 Gy/min) [79]. One possibility is that autophagy-induced release of cellular iron occurs in chronic low-dose radiation at an early time point, perhaps during an adaptation phase. Such events could result in iron deficiency, but future studies with early time points are needed to determine whether some cellular adaptation occurred early during the irradiation.

Studies of the biological response to low levels of radiation have revealed that there is a large uncertainty in determining actual health risks [80,81]. Several studies have suggested that there may be a non-linear biological response to radiation at low doses, with some data showing complex cellular responses that are not always detrimental [80]. According to the hormesis hypothesis, the dose–response relationship can be non-linear, in which low-dose stress can result in an optimal outcome [82]. Some animal model data from low-dose radiation studies do support a non-linear response [80]. However, the factors that determine in vivo responses to radiation, in animal models or in humans, are not sufficiently understood to predict individual outcomes of low level radiation exposure [35]. Further understanding of pathway regulation by low-dose radiation may allow the identification of markers to predict biological outcomes and also to identify countermeasures for adverse effects of low-dose radiation.

Supplementary Materials: The following are available online at https://www.mdpi.com/article/10.3390/antiox12020241/s1, Table S1: Differentially expressed MSC genes at 1 Gy, 0.66 cGy/h, Table S2: Differentially expressed MSC genes at 2 Gy, 0.66 cGy/h, Table S3: Biological Processes Gene Ontology for Upregulated Genes in 1 Gy Irradiated Human Mesenchymal Stem Cells, Table S4: Biological Processes Gene Ontology for Downregulated Genes in 1 Gy Irradiated Human Mesenchymal Stem Cells, Table S5: Biological Processes Gene Ontology for Upregulated Genes in 2 Gy Irradiated Human Mesenchymal Stem Cells, Table S6: Biological Processes Gene Ontology for Downregulated Genes in 2 Gy Irradiated Human Mesenchymal Stem Cells.

Author Contributions: Conceptualization, R.M.D.; methodology, W.B.R., D.T.B., M.B., L.C., A.T., R.B. and J.E.S.; validation, W.B.R., J.E.S., D.T.B., A.T. and R.M.D.; formal analysis, R.M.D., W.B.R., J.E.S., D.T.B., M.W., A.R.S. and C.D.; investigation, R.M.D., W.B.R., J.E.S., D.T.B., M.B., R.B., A.T., M.W., A.R.S. and L.C.; data curation, W.B.R., J.E.S., D.T.B., A.R.S. and M.W.; writing—original draft preparation, R.M.D.; writing—review and editing, R.M.D., W.B.R., J.E.S., D.T.B., M.W., A.R.S., A.T. C.D., R.B., M.B. and L.C.; supervision, R.M.D. and C.D.; project administration, R.M.D. and C.D.; funding acquisition, R.M.D. and C.D. All authors have read and agreed to the published version of the manuscript.

Funding: This work was supported by a pilot grant from the Opportunity Funds Management Core of the Centers for Medical Countermeasures against Radiation, National Institutes of Health (NIH), National Institutes of Allergy and Infectious Diseases, grant number U19AI067773 (Award Number 5U19AI067773-14, P.I. Regina M. Day), and by Award Number DM178018 (P.I. Regina M. Day) from the Defense Medical Research and Materiel Command, Radiation Health Effects Research Program, Joint Program Committee 7. Pilot studies for this project were supported by a NASA fellowship to Michelle Bylicky.

Institutional Review Board Statement: Not applicable.

Informed Consent Statement: Not applicable.

Data Availability Statement: All transcriptomics data will be made available upon request.

Acknowledgments: For this work we are grateful to Michael Woolbert for the support of the RS2000 at USUHS. Some of the authors are employees of the U.S. Government, and this work was prepared as part of their official duties. Title 17 U.S.C. §105 provides that 'Copyright protection under this title is not available for any work of the United States Government.' Title 17 U.S.C §101 defines a U.S. Government work as a work prepared by a military service member or employees of the U.S. Government as part of that person's official duties. The views in this article are those of the authors and do not necessarily reflect the views, official policy, or position of the National Institute of Allergy and Infectious Diseases or the National Institutes of Health, the Uniformed Services University of the Health Sciences, the Armed Forces Radiobiology Research Institute, Department of the Navy, Department of Defense or the U.S. Federal Government.

Conflicts of Interest: The authors declare no conflict of interest.

References

1. Zuo, J.; Zhang, Z.; Luo, L.; Nice, E.; Zhang, W.; Wang, C.; Huang, C. Redox signaling at the crossroads of human health and disease. *MedComm* **2022**, *3*, e127. [CrossRef]
2. Tretter, V.; Hochreiter, B.; Zach, M.; Krenn, K.; Klein, K. Understanding cellular redox homeostasis: A challenge for precision medicine. *Int. J. Mol. Sci.* **2021**, *23*, 106. [CrossRef]
3. Sies, H.; Jones, D.P. Reactive oxygen species (ROS) as pleiotropic physiological signalling agents. *Nat. Rev. Mol. Cell Biol.* **2020**, *21*, 363–383. [CrossRef]
4. Lennicke, C.; Cocheme, H. Redox metabolism: ROS as specific molecular regulators of cell signaling and function. *Mol. Cell.* **2021**, *81*, 3691–3707. [CrossRef] [PubMed]
5. Panganiban, R.A.; Snow, A.; Day, R. Mechanisms of radiation toxicity in transformed and non-transformed cells. *Int. J. Mol. Sci.* **2013**, *14*, 15931–15958. [CrossRef] [PubMed]
6. Pizzino, G.; Irrera, N.; Cucinotta, M.; Pallio, G.; Mannino, F.; Arcoraci, V.; Squadrito, F.; Altavilla, D.; Bitto, A. Oxidative Stress: Harms and Benefits for Human Health. *Oxid. Med. Cell Longev.* **2017**, *2017*, 8416763. [CrossRef] [PubMed]

7. Alessio, N.; Del Gaudio, S.; Capasso, S.; Di Bernardo, G.; Cappabianca, S.; Cipollaro, M.; Peluso, G.; Galderisi, U. Low dose radiation induced senescence of human mesenchymal stromal cells and impaired the autophagy process. *Oncotarget* **2015**, *6*, 8155–8166. [CrossRef]

8. Otoupalova, E.; Smith, S.; Cheng, G.; Thannickal, V. Oxidative Stress in Pulmonary Fibrosis. *Compr. Physiol.* **2020**, *10*, 509–547.

9. Masarone, M.; Rosato, V.; Dallio, M.; Gravina, A.; Aglitti, A.; Loguercio, C.; Federico, A.; Persico, M. Role of Oxidative Stress in Pathophysiology of Nonalcoholic Fatty Liver Disease. *Oxid. Med. Cell Longev.* **2018**, *2018*, 9547613. [CrossRef]

10. Zullo, A.; Guida, R.; Sciarrillo, R.; Mancini, F. Redox Homeostasis in Cardiovascular Disease: The Role of Mitochondrial Sirtuins. *Front. Endocrinol. (Lausanne)* **2022**, *13*, 858330. [CrossRef]

11. Citrin, D.; Cotrim, A.; Hyodo, F.; Baum, B.; Krishna, M.; Mitchell, J. Radioprotectors and mitigators of radiation-induced normal tissue injury. *Oncologist* **2010**, *15*, 360–371. [CrossRef] [PubMed]

12. Day, R.M.; Suzuki, Y. Cell proliferation, reactive oxygen and cellular glutathione. *Dose Response* **2006**, *3*, 425–442. [CrossRef]

13. Suzuki, Y.J.; Jain, V.; Park, A.; Day, R. Oxidative stress and oxidant signaling in obstructive sleep apnea and associated cardiovascular diseases. *Free Radic. Biol. Med.* **2006**, *40*, 1683–1692. [CrossRef] [PubMed]

14. Liu, R.; Bian, Y.; Liu, L.; Liu, L.; Liu, X.; Ma, S. Molecular pathways associated with oxidative stress and their potential applications in radiotherapy (review). *Int. J. Mol. Med.* **2022**, *49*, 65. [CrossRef] [PubMed]

15. Ward, J.F. The complexity of DNA damage: Relevance to biological consequences. *Int. J. Radiat. Biol.* **1994**, *66*, 427–432. [CrossRef]

16. Barshishat-Kupper, M.; McCart, E.; Freedy, J.; Tipton, A.; Nagy, V.; Kim, S.; Landauer, M.; Mueller, G.; Day, R. Protein oxidation in the lungs of C57BL/6J Mice following X-irradiation. *Proteomes* **2015**, *3*, 249–265. [CrossRef]

17. Barshishat-Kupper, M.; Tipton, A.; McCart, E.; McCue, J.; Mueller, G.; Day, R. Effect of ionizing radiation on liver protein oxidation and metabolic function in C57BL/6J mice. *Int. J. Radiat. Biol.* **2014**, *90*, 1169–1178. [CrossRef]

18. Jakubczyk, K.; Dec, K.; Kaldunska, J.; Kawczuga, D.; Kochman, J.; Janda, K. Reactive oxygen species—Sources, functions, oxidative damage. *Pol. Merkur. Lekarski.* **2020**, *48*, 124–127.

19. Kagan, V.E.; Tyurina, Y.; Sun, W.; Vlasova, I.I.; Dar, H.; Tyurin, V.; Amoscato, A.; Mallampalli, R.; van der Wel, P.; He, R.; et al. Redox phospholipidomics of enzymatically generated oxygenated phospholipids as specific signals of programmed cell death. *Free Radic. Biol. Med.* **2020**, *147*, 231–241. [CrossRef]

20. Ward, J.F. DNA damage as the cause of ionizing radiation-induced gene activation. *Radiat. Res.* **1994**, *138*, S85–S88. [CrossRef]

21. Daly, M.J. Death by protein damage in irradiated cells. *DNA Repair. (Amst.)* **2012**, *11*, 12–21. [CrossRef] [PubMed]

22. Bylicky, M.A.; Mueller, G.; Day, R. Radiation resistance of normal human astrocytes: The role of non-homologous end joining DNA repair activity. *J. Radiat. Res.* **2019**, *60*, 37–50. [CrossRef]

23. Panganiban, R.A.; Mungunsukh, O.; Day, R. X-irradiation induces ER stress, apoptosis, and senescence in pulmonary artery endothelial cells. *Int. J. Radiat. Biol.* **2013**, *89*, 656–667. [CrossRef] [PubMed]

24. United State Nuclear Regulatory Commission. Backgrounder on Biological Effects of Radiation. 2020. Available online: https://www.nrc.gov/reading-rm/doc-collections/fact-sheets/bio-effects-radiation.html (accessed on 20 July 2022).

25. Wakeford, R.; Tawn, E. The meaning of low dose and low dose-rate. *J. Radiol. Prot.* **2010**, *30*, 1–3. [CrossRef] [PubMed]

26. Xiao, M.; Whitnall, M. Pharmacological countermeasures for the acute radiation syndrome. *Curr. Mol. Pharmacol.* **2009**, *2*, 122–133. [CrossRef]

27. Macia, I.G.M.; Calduch, A.L.; Lopez, E. Radiobiology of the acute radiation syndrome. *Rep. Pract. Oncol. Radiother.* **2011**, *16*, 123–130. [CrossRef] [PubMed]

28. Tang, D.; Chen, X.; Kang, R.; Kroemer, G. Ferroptosis: Molecular mechanisms and health implications. *Cell Res.* **2021**, *31*, 107–125. [CrossRef]

29. Sasaki, M.S.; Tachibana, A.; Takeda, S. Cancer risk at low doses of ionizing radiation: Artificial neural networks inference from atomic bomb survivors. *J. Radiat. Res.* **2014**, *55*, 391–406. [CrossRef]

30. Akleyev, A.V. Early signs of chronic radiation syndrome in residents of the Techa riverside settlements. *Radiat. Environ. Biophys.* **2021**, *60*, 203–212. [CrossRef]

31. Meerman, M.; Gartner, T.B.; Buikema, J.; Wu, S.; Siddiqi, S.; Bouten, C.; Grande-Allen, K.; Suyker, W.; Hjortnaes, J. Myocardial Disease and Long-Distance Space Travel: Solving the Radiation Problem. *Front. Cardiovasc. Med.* **2021**, *8*, 631985. [CrossRef]

32. Straume, T.; Amundson, S.; Blakely, W.; Burns, F.; Chen, A.; Dainiak, N.; Franklin, S.; Leary, J.; Loftus, D.; Morgan, W.; et al. NASA Radiation Biomarker Workshop, September 27–28, 2007. *Radiat. Res.* **2008**, *170*, 393–405. [CrossRef] [PubMed]

33. Magae, J.; Furukawa, C.; Ogata, H. Dose-rate effect on proliferation suppression in human cell lines continuously exposed to gamma rays. *Radiat. Res.* **2011**, *176*, 447–458. [CrossRef]

34. Tamulevicius, P.; Wang, M.; Iliakis, G. Homology-directed repair is required for the development of radioresistance during S phase: Interplay between double-strand break repair and checkpoint response. *Radiat. Res.* **2007**, *167*, 1–11. [CrossRef]

35. Applegate, K.E.; Ruhm, W.; Wojcik, A.; Bourguignon, M.; Brenner, A.; Hamasaki, K.; Imai, T.; Imaizumi, M.; Imaoka, T.; Kakinuma, S.; et al. Individual response of humans to ionising radiation: Governing factors and importance for radiological protection. *Radiat. Environ. Biophys.* **2020**, *59*, 185–209. [CrossRef] [PubMed]

36. Prasanna, P.G.; Stone, H.; Wong, R.; Capala, J.; Bernhard, E.; Vikram, B.; Coleman, C. Normal tissue protection for improving radiotherapy: Where are the gaps? *Transl. Cancer Res.* **2012**, *1*, 35–48. [PubMed]

37. Horwitz, E.M.; Gordon, P.; Koo, W.; Marx, J.; Neel, M.; McNall, R.; Muul, L.; Hofmann, T. Isolated allogeneic bone marrow-derived mesenchymal cells engraft and stimulate growth in children with osteogenesis imperfecta: Implications for cell therapy of bone. *Proc. Natl. Acad. Sci. USA* **2002**, *99*, 8932–8937. [CrossRef]
38. Bara, J.J.; Richards, R.; Alini, M.; Stoddart, M. Concise review: Bone marrow-derived mesenchymal stem cells change phenotype following in vitro culture: Implications for basic research and the clinic. *Stem Cells* **2014**, *32*, 1713–1723. [CrossRef]
39. Chen, M.F.; Lin, C.; Chen, W.; Yang, C.; Chen, C.; Liao, S.; Liu, J.; Lu, C.; Lee, K. The sensitivity of human mesenchymal stem cells to ionizing radiation. *Int. J. Radiat. Oncol. Biol. Phys.* **2006**, *66*, 244–253. [CrossRef]
40. Bouten, R.M.; Dalgard, C.; Soltis, A.; Slaven, J.; Day, R. Transcriptomic profiling and pathway analysis of cultured human lung microvascular endothelial cells following ionizing radiation exposure. *Sci. Rep.* **2021**, *11*, 24214. [CrossRef]
41. Schmittgen, T.D.; Livak, K.J. Analyzing real-time PCR data by the comparative C(T) method. *Nat. Protoc.* **2008**, *3*, 1101–1108. [CrossRef]
42. Pfaffl, M.W. A new mathematical model for relative quantification in real-time RT-PCR. *Nucleic Acids Res.* **2001**, *29*, e45. [CrossRef]
43. Wang, K.; Singh, D.; Zeng, Z.; Coleman, S.; Huang, Y.; Savich, G.; He, X.; Mieczkowski, P.; Grimm, S.; Perou, C.; et al. MapSplice: Accurate mapping of RNA-seq reads for splice junction discovery. *Nucleic Acids Res.* **2010**, *38*, e178. [CrossRef] [PubMed]
44. Anders, S.; Pyl, P.; Huber, W. HTSeq—A Python framework to work with high-throughput sequencing data. *Bioinformatics* **2015**, *31*, 166–169. [CrossRef] [PubMed]
45. Wang, L.; Wang, S.; Li, W. RSeQC: Quality control of RNA-seq experiments. *Bioinformatics* **2012**, *28*, 2184–2185. [CrossRef] [PubMed]
46. Li, H.; Handsaker, B.; Wysoker, A.; Fennell, T.; Ruan, J.; Homer, N.; Marth, G.; Abecasis, G.; Durbin, R. 1000 Genome Project Data Processing Subgroup. The sequence alignment/map format and SAMtools. *Bioinformatics* **2009**, *25*, 2078–2079.
47. Danecek, P.; Bonfield, J.; Liddle, J.; Marshall, J.; Ohan, V.; Pollard, M.; Whitwham, A.; Keane, T.; McCarthy, S.; Davies, R.; et al. Twelve years of SAMtools and BCFtools. *Gigascience* **2021**, *10*, giab008. [CrossRef]
48. Kumar, P.; Chakraborty, J.; Sukumar, G.; Dalgard, C.; Chatterjee, R.; Biswas, R. Comparative RNA-seq analysis reveals dysregulation of major canonical pathways in ERG-inducible LNCaP cell progression model of prostate cancer. *Oncotarget* **2019**, *10*, 4290–4306. [CrossRef]
49. Love, M.I.; Huber, W.; Anders, S. Moderated estimation of fold change and dispersion for RNA-seq data with DESeq2. *Genome Biol.* **2014**, *15*, 550. [CrossRef]
50. Huang, D.W.; Sherman, B.; Lempicki, R. Systematic and integrative analysis of large gene lists using DAVID bioinformatics resources. *Nat. Protoc.* **2009**, *4*, 44–57. [CrossRef] [PubMed]
51. Huang da, W.; Sherman, B.; Lempicki, R. Bioinformatics enrichment tools: Paths toward the comprehensive functional analysis of large gene lists. *Nucleic Acids Res.* **2009**, *37*, 1–13. [CrossRef]
52. Eden, E.; Navon, R.; Steinfeld, I.; Lipson, D.; Yakhini, Z. GOrilla: A tool for discovery and visualization of enriched GO terms in ranked gene lists. *BMC Bioinform.* **2009**, *10*, 48. [CrossRef]
53. Eden, E.; Lipson, D.; Yogev, S.; Yakhini, Z. Discovering motifs in ranked lists of DNA sequences. *PLoS Comput. Biol.* **2007**, *3*, e39. [CrossRef] [PubMed]
54. Zhou, Y.; Zhou, B.; Pache, L.; Chang, M.; Khodabakhshi, A.; Tanaseichuk, O.; Benner, C.; Chanda, S. Metascape provides a biologist-oriented resource for the analysis of systems-level datasets. *Nat. Commun.* **2019**, *10*, 1523. [CrossRef]
55. Kuo, L.J.; Yang, L. Gamma-H2AX—A novel biomarker for DNA double-strand breaks. *In Vivo* **2008**, *22*, 305–309. [PubMed]
56. Halvarsson, C.; Rorby, E.; Eliasson, P.; Lang, S.; Soneji, S.; Jonsson, J. Putative role of nuclear factor-kappa b but not hypoxia-inducible factor-1alpha in hypoxia-dependent regulation of oxidative stress in hematopoietic stem and progenitor cells. *Antioxid. Redox Signal.* **2019**, *31*, 211–226. [CrossRef]
57. Pantopoulos, K.; Hentze, M. Rapid responses to oxidative stress mediated by iron regulatory protein. *EMBO J.* **1995**, *14*, 2917–2924. [CrossRef]
58. Bouten, R.M.; Young, E.; Selwyn, R.; Iacono, D.; Rittase, W.; Day, R. Effects of radiation on endothelial barrier and vascular integrity. In *Tissue Barriers in Disease, Injury and Regeneration*; Gorbunov, N.V., Ed.; Elsevier: Amsterdam, The Netherlands, 2021; pp. 43–94.
59. Jin, Y.W.; Na, Y.; Lee, Y.; An, S.; Lee, J.; Jung, M.; Kim, H.; Nam, S.; Kim, C.; Yang, K.; et al. Comprehensive analysis of time- and dose-dependent patterns of gene expression in a human mesenchymal stem cell line exposed to low-dose ionizing radiation. *Oncol. Rep.* **2008**, *19*, 135–144. [CrossRef]
60. Truong, K.; Bradley, S.; Baginski, B.; Wilson, J.; Medlin, D.; Zheng, L.; Wilson, R.; Rusin, M.; Takacs, E.; Dean, D. The effect of well-characterized, very low-dose x-ray radiation on fibroblasts. *PLoS ONE* **2018**, *13*, e0190330. [CrossRef] [PubMed]
61. Zhong, J.; Rajaram, N.; Brizel, D.; Frees, A.; Ramanujam, N.; Batinic-Haberle, I.; Dewhirst, M. Radiation induces aerobic glycolysis through reactive oxygen species. *Radiother. Oncol.* **2013**, *106*, 390–396. [CrossRef]
62. Werner, E.; Alter, A.; Deng, Q.; Dammer, E.; Wang, Y.; Yu, D.; Duong, D.; Seyfried, N.; Doetsch, P. Ionizing Radiation induction of cholesterol biosynthesis in Lung tissue. *Sci. Rep.* **2019**, *9*, 12546. [CrossRef]
63. Mitchel, R.E.; Hasu, M.; Bugden, M.; Wyatt, H.; Little, M.; Gola, A.; Hildebrandt, G.; Priest, N.; Whitman, S. Low-dose radiation exposure and atherosclerosis in ApoE(-)/(-) mice. *Radiat. Res.* **2011**, *175*, 665–676. [CrossRef]
64. Chung, B.M.; Rotty, J.; Coulombe, P. Networking galore: Intermediate filaments and cell migration. *Curr. Opin. Cell Biol.* **2013**, *25*, 600–612. [CrossRef] [PubMed]

65. Karantza, V. Keratins in health and cancer: More than mere epithelial cell markers. *Oncogene* **2011**, *30*, 127–138. [CrossRef]
66. Liu, M.; Liu, Q.; Pei, Y.; Gong, M.; Cui, X.; Pan, J.; Zhang, Y.; Liu, Y.; Liu, Y.; Yuan, X.; et al. Aqp-1 Gene Knockout Attenuates Hypoxic Pulmonary Hypertension of Mice. *Arterioscler. Thromb. Vasc. Biol.* **2019**, *39*, 48–62. [CrossRef] [PubMed]
67. Dekker, Y.; Le Devedec, S.; Danen, E.; Liu, Q. Crosstalk between Hypoxia and Extracellular Matrix in the Tumor Microenvironment in Breast Cancer. *Genes* **2022**, *13*, 1585. [CrossRef] [PubMed]
68. Norton, C.E.; Weise-Cross, L.; Ahmadian, R.; Yan, S.; Jernigan, N.; Paffett, M.; Naik, J.; Walker, B.; Resta, T. Altered Lipid Domains Facilitate Enhanced Pulmonary Vasoconstriction after Chronic Hypoxia. *Am. J. Respir. Cell Mol. Biol.* **2020**, *62*, 709–718. [CrossRef]
69. Ling, C.C. Time scale of radiation-induced oxygen depletion and decay kinetics of oxygen-dependent damage in cells irradiated at ultrahigh dose rates. *Radiat. Res.* **1975**, *63*, 455–467. [CrossRef]
70. Michaels, H.B. Oxygen depletion in irradiated aqueous solutions containing electron affinic hypoxic cell radiosensitizers. *Int. J. Radiat. Oncol. Biol. Phys.* **1986**, *12*, 1055–1058. [CrossRef]
71. Jackson, I.L.; Zhang, X.; Hadley, C.; Rabbani, Z.; Zhang, Y.; Marks, S.; Vujaskovic, Z. Temporal expression of hypoxia-regulated genes is associated with early changes in redox status in irradiated lung. *Free Radic. Biol. Med.* **2012**, *53*, 337–346. [CrossRef]
72. Moon, E.J.; Petersson, K.; Olcina, M. The importance of hypoxia in radiotherapy for the immune response, metastatic potential and FLASH-RT. *Int. J. Radiat. Biol.* **2022**, *98*, 439–451. [CrossRef]
73. Rittase, W.B.; Muir, J.; Slaven, J.; Bouten, R.; Bylicky, M.; Wilkins, W.; Day, R. Deposition of iron in the bone narrow of a nurine nodel of hematopoietic acute radiation syndrome. *Exp. Hematol.* **2020**, *84*, 54–66. [CrossRef] [PubMed]
74. Rittase, W.B.; Slaven, J.; Suzuki, Y.; Muir, J.; Lee, S.; Rusnak, M.; Brehm, G.; Bradfield, D.; Symes, A.; Day, R. Iron Deposition and Ferroptosis in the Spleen in a Murine Model of Acute Radiation Syndrome. *Int. J. Mol. Sci.* **2022**, *23*, 11029. [CrossRef]
75. Shah, Y.M.; Xie, L. Hypoxia-inducible factors link iron homeostasis and erythropoiesis. *Gastroenterology* **2014**, *146*, 630–642. [CrossRef] [PubMed]
76. Lee, S.M.; Lee, S.; Prywes, R.; Vulpe, C. Iron deficiency upregulates Egr1 expression. *Genes Nutr.* **2015**, *10*, 468. [CrossRef] [PubMed]
77. Rockfield, S.; Chhabra, R.; Robertson, M.; Rehman, N.; Bisht, R.; Nanjundan, M. Links Between Iron and Lipids: Implications in Some Major Human Diseases. *Pharmaceuticals* **2018**, *11*, 18. [CrossRef]
78. Romney, S.J.; Newman, B.; Thacker, C.; Leibold, E. HIF-1 regulates iron homeostasis in *Caenorhabditis elegans* by activation and inhibition of genes involved in iron uptake and storage. *PLoS Genet.* **2011**, *7*, e1002394. [CrossRef]
79. Ma, S.; Fu, X.; Liu, L.; Liu, Y.; Feng, H.; Jiang, H.; Liu, X.; Liu, R.; Liang, Z.; Li, M.; et al. Iron-Dependent Autophagic Cell Death Induced by Radiation in MDA-MB-231 Breast Cancer Cells. *Front. Cell Dev. Biol.* **2021**, *9*, 723801. [CrossRef]
80. Belli, M.; Indovina, L. The Response of Living Organisms to Low Radiation Environment and Its Implications in Radiation Protection. *Front. Public Health* **2020**, *8*, 601711. [CrossRef]
81. Lowe, D.; Roy, L.; Tabocchini, M.; Ruhm, W.; Wakeford, R.; Woloschak, G.; Laurier, D. Radiation dose rate effects: What is new and what is needed? *Radiat. Environ. Biophys.* **2022**, *61*, 507–543. [CrossRef]
82. Schirrmacher, V. Less Can Be More: The hormesis theory of stress adaptation in the global biosphere and its implications. *Biomedicines* **2021**, *9*, 293. [CrossRef]

Review

Stress Activated MAP Kinases and Cyclin-Dependent Kinase 5 Mediate Nuclear Translocation of Nrf2 via Hsp90α-Pin1-Dynein Motor Transport Machinery

Tetsuro Ishii [1,*], **Eiji Warabi** [1] **and Giovanni E. Mann** [2]

[1] School of Medicine, University of Tsukuba, Tsukuba 305-8577, Japan
[2] King's British Heart Foundation Centre of Research Excellence, School of Cardiovascular and Metabolic Medicine, Faculty of Life Sciences & Medicine, King's College London, 150 Stamford Street, London SE1 9NH, UK
[*] Correspondence: ishiitetsuro305@gmail.com

Abstract: Non-lethal low levels of oxidative stress leads to rapid activation of the transcription factor nuclear factor-E2-related factor 2 (Nrf2), which upregulates the expression of genes important for detoxification, glutathione synthesis, and defense against oxidative damage. Stress-activated MAP kinases p38, ERK, and JNK cooperate in the efficient nuclear accumulation of Nrf2 in a cell-type-dependent manner. Activation of p38 induces membrane trafficking of a glutathione sensor neutral sphingomyelinase 2, which generates ceramide upon depletion of cellular glutathione. We previously proposed that caveolin-1 in lipid rafts provides a signaling hub for the phosphorylation of Nrf2 by ceramide-activated PKCζ and casein kinase 2 to stabilize Nrf2 and mask a nuclear export signal. We further propose a mechanism of facilitated Nrf2 nuclear translocation by ERK and JNK. ERK and JNK phosphorylation of Nrf2 induces the association of prolyl cis/trans isomerase Pin1, which specifically recognizes phosphorylated serine or threonine immediately preceding a proline residue. Pin1-induced structural changes allow importin-α5 to associate with Nrf2. Pin1 is a co-chaperone of Hsp90α and mediates the association of the Nrf2-Pin1-Hsp90α complex with the dynein motor complex, which is involved in transporting the signaling complex to the nucleus along microtubules. In addition to ERK and JNK, cyclin-dependent kinase 5 could phosphorylate Nrf2 and mediate the transport of Nrf2 to the nucleus via the Pin1-Hsp90α system. Some other ERK target proteins, such as pyruvate kinase M2 and hypoxia-inducible transcription factor-1, are also transported to the nucleus via the Pin1-Hsp90α system to modulate gene expression and energy metabolism. Notably, as malignant tumors often express enhanced Pin1-Hsp90α signaling pathways, this provides a potential therapeutic target for tumors.

Keywords: Nrf2; Hsp90; ERK; JNK; Cdk5; Pin1; HO-1

Citation: Ishii, T.; Warabi, E.; Mann, G.E. Stress Activated MAP Kinases and Cyclin-Dependent Kinase 5 Mediate Nuclear Translocation of Nrf2 via Hsp90α-Pin1-Dynein Motor Transport Machinery. *Antioxidants* **2023**, *12*, 274. https://doi.org/10.3390/antiox12020274

Academic Editor: Marcel Bonay

Received: 25 December 2022
Revised: 19 January 2023
Accepted: 22 January 2023
Published: 26 January 2023

1. Introduction

Cells respond to mild oxidative stress, such as non-lethal levels of hydrogen peroxide (H_2O_2), and rapidly upregulate defense systems to afford protection against oxidative damage (reviewed in [1]). Among the redox-sensitive transcription factors, nuclear factor-E2-related factor 2 (Nrf2) plays an important role in the expression of genes involved in detoxification, glutathione (GSH) synthesis, and defenses against oxidative damage [2–4]. Upregulation and activation of Nrf2 are controlled through a complex transcriptional, translational, and post-translational network that ensures an increase in its activity during redox perturbation, inflammation, growth factor stimulation, and nutrient/energy fluxes, thereby enabling Nrf2 to orchestrate adaptive responses to diverse forms of stress (reviewed in [5–9]).

The rapid activation of Nrf2 by stress agents depends on both inhibition of degradation and/or enhanced Nrf2 translation in the cytoplasm and facilitated nuclear translocation of Nrf2 to the nucleus (Figure 1A). Nrf2 is unstable under unstressed conditions due to constant degradation via the 26S proteasome, which is mediated by its cytoplasmic binding partner Kelch-like ECH-associated protein 1 (Keap1). Keap1 has cysteine residues highly reactive with various types of electrophiles, which form Michael adducts with Keap1 to inhibit interaction with Nrf2 resulting in Nrf2 stabilization and accumulation (reviewed in [10–12]). Degradation of Nrf2 is also controlled by phosphorylation at the Neh6 domain by glycogen synthase kinase-3 (GSK-3) (reviewed in [6–8]). It enhances the degradation of Nrf2 mediated by β-transducin repeat-containing protein, present in the ubiquitin ligase complex. Therefore, inhibition of GSK-3 by PI3-K (phosphatidylinositol 3-kinase)/Akt (PKB) signaling is a prerequisite to induce Keap1-mediated Nrf2 stabilization by electrophiles (reviewed in [6–8]).

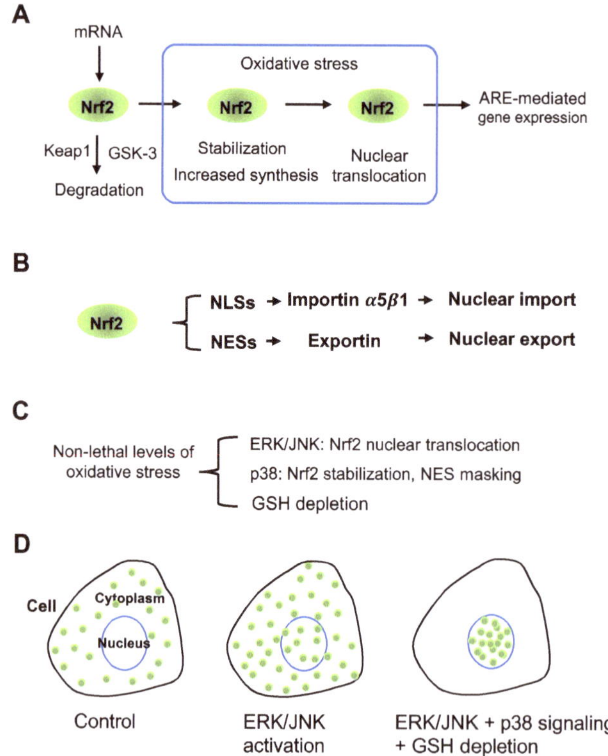

Figure 1. MAP kinases p38, ERK, JNK, and GSH depletion control activation of the transcription factor Nrf2. (**A**) Rapid activation of Nrf2 under oxidative stress depends on two separate steps, stabilization and upregulation of Nrf2 protein and nuclear translocation. (**B**) Nuclear import of Nrf2 is controlled by the association of importin α5β1 to its nuclear localization signals (NLSs), while nuclear export of Nrf2 is controlled by exportin binding to its nuclear export signals (NESs). (**C**) Non-lethal levels of oxidative stress, such as H_2O_2, induce GSH depletion and activation of MAP kinases (reviewed in [1]). ERK and JNK control Nrf2 nuclear translocation, and p38 signaling induces stabilization of Nrf2 and masking of an NES. (**D**) In unstressed conditions, Nrf2 (green dots) levels are low and mainly localized in the cytoplasm but not in the nucleus due to functional nuclear export signals. ERK/JNK activation could cause nuclear transport of Nrf2 across the nuclear membrane but does not enable Nrf2 to accumulate in the nucleus due to the functional nuclear export signals. As p38 signaling masks a nuclear export signal, simultaneous activation of ERK and p38 signaling pathways under GSH-depleting conditions induces effective nuclear accumulation/activation of Nrf2.

In addition to Nrf2 stabilization, oxidative stress induces rapid Nrf2 protein synthesis [13]. Treatment of rat cardiomyocytes, human HeLa, and other cells with H_2O_2 induces rapid upregulation of translational Nrf2 protein synthesis independent of Nrf2 protein stabilization in these cells [13,14]. Notably, Nrf2 mRNA has an internal ribosomal entry site within the 5′ untranslated region (5′UTR) of human Nrf2 mRNA [14], and H_2O_2 treatments have been shown to cause rapid translocation of La autoantigen (Sjögren Syndrome Antigen B) from the nucleus to perinuclear space to associate with ribosomes. Binding La autoantigen to 5′UTR of Nrf2 mRNA stabilizes it and induces an Internal Ribosome Entry Site mediated protein translation in HeLa cells [15]. However, increased Nrf2 levels in the cytoplasm do not automatically lead to nuclear accumulation of Nrf2, as the transport of Nrf2 into the nucleus is tightly controlled, requiring additional regulators such as importins [16].

Nuclear import of proteins is tightly controlled by the interaction with a cytosolic protein termed importin composed of α and β subunits [17–19]. Importin α recognizes and binds nuclear localization signals (NLSs) of karyophilic proteins, and importin β helps bind them to the nuclear envelope, with energy-dependent, small GTPase Ran-mediated translocation through the pore resulting in the accumulation of import substrate and importin-α in the nucleus [18–20]. Nrf2 has three functional Lys/Arg-rich NLS motifs in Neh1, Neh2, and Neh3 domains, respectively, and mutations of the three NLS motifs significantly impair nuclear translocation of Nrf2 in HepG2 cells, and ARE-reporter gene expression in human leukemia K562 cells treated with *tertiary*-butylhydroquinone (tBHQ) [16]. These authors further showed that anti-importin $\alpha5$ and anti-importin $\beta1$, respectively, co-immunoprecipitated with Nrf2 shortly after treating K562 cells with tBHQ, and that the amount of Nrf2 co-immunoprecipitated by these antibodies in nuclear fractions increased over 30–60 min [16]. This study shows that the transcriptional activation of Nrf2 depends largely on importin-mediated nucleocytoplasmic shuttling. It is also noted that exportin mediates nuclear export through interaction with nuclear export signals (NESs) [21,22]. Nrf2 has at least two functional NESs in the leucine zipper domain [23] and the transactivation domain [24]. The former Nrf2 NES can be masked by the formation of heterodimers with functional partner small Maf proteins, and the latter is a redox-sensitive NES. Thus, maximal Nrf2 activation or nuclear accumulation is achieved by enhanced association with importin $\alpha5\beta1$ in the cytoplasm to facilitate nuclear translocation and inhibition of exportin binding in the nucleus (Figure 1B).

Although the precise mechanism of importin-dependent Nrf2 nuclear translocation remains unclear, previous studies have established the importance of stress-activated mitogen-activated protein kinases (MAPKs), p38, extracellular signal-regulated kinase (ERK), and c-Jun N-terminal kinase (JNK) in the nuclear translocation and activation of Nrf2 (reviewed in [25,26]). MAPKs are proline-directed serine/threonine protein kinases activated by dual phosphorylation on threonine and tyrosine residues in response to a wide array of extracellular stimuli. They are essential components of signaling pathways that convert various extracellular signals into intracellular responses through serial phosphorylation cascades (reviewed in [27–29]). ERK, known alternatively as microtubule-associated protein-2/myelin basic protein kinase, is activated by numerous hormones, growth factors, and other extracellular stimuli (reviewed in [30,31]), whereas JNK and p38 are activated by distinct and overlapping sets of stress-related stimuli, including heat shock, inflammatory cytokines, ultraviolet, gamma irradiation, and hyperosmolarity (reviewed in [32]). Interestingly, tBHQ activates ERK2 and JNK1 in HepG2 and HeLa cells [33], and phenethyl isothiocyanate activates ERK2 and JNK1, which phosphorylate Nrf2, resulting in nuclear translocation in human prostate cancer PC-3 cells [34] (Figure 1C). Table 1 summarizes the role of MAPKs in Nrf2 activation and ARE-mediated reporter gene expression in different cell types stimulated with various stress agents reported between 2000 to 2012 [34–54]. Evidently, the dependence of Nrf2 activation on ERK, JNK, and p38 MAPKs differs significantly among cells and stress agents (see Table 1). The dependence of Nrf2 activation on MAP kinases is widely confirmed by later numerous studies using different types of cultured cells stimulated with other various natural and synthetic chemical agents that activate Nrf2. However, molecular mechanisms of MAP kinase-dependent Nrf2 activation remain unsolved.

Table 1. MAP kinase-dependent Nrf2 activation.

Cells	Activators	MAPK Dependence	References
Human hepatoma HepG2	Sodium arsenite and mercury chloride	JNK-dependent ARE reporter gene and HO-1 expression	[35]
	Pyrrolidine dithiocarbamate	ERK and p38 inhibitors PD98059 and SB202190 reduced about 50% in γ-glutamylcystein synthetase expression	[36]
	Diallyl sulfide	ERK- and p38- dependent Nrf2 nuclear translocation and HO-1 expression	[37]
	Gallic acid	p38 inhibitor reduced ARE-dependent P-form of phenol sulfotransferase expression	[38]
Human hepatocyte	Quercetin	p38- and ERK-dependent Nrf2 activation and HO-1 expression	[39]
Human mammary epithelia MCF-7	Cadmium chloride	p38-dependent but ERK-independent HO-1 expression	[40]
Human HeLa	Phenethyl isothiocyanate	JNK-dependent ARE-reporter gene expression	[41]
Human monocytic THP-1	α-Lipoic acid	p38 inhibitor significantly reduced Nrf2 dependent HO-1 expression	[42]
Human aortic smooth muscle	Oxidized low-density lipoprotein	ERK, p38, and JNK inhibitors respectively reduced HO-1 expression and Nrf2 nuclear translocation	[43]
Human prostate carcinoma PC-3	Phenethyl isothiocyanate	ERK and JNK phosphorylate Nrf2 and induce nuclear translocation of Nrf2	[34]
Human monocyte	Curcumin	p38 inhibitor but not ERK inhibitor reduced ARE-dependent GCLM and HO-1 mRNA expression	[44]
Mouse alveolar epithelial C10	Hyperoxia	Hyperoxia activates NADPH oxidase, which results in ERK-dependent Nrf2 activation	[45]
Mouse macrophage RAW 264.7	Lipopolysaccharide	p38 inhibitor significantly reduced Nrf2 dependent HO-1 expression	[46]
Mouse cochlear	Piperine	JNK inhibitor significantly reduced ARE-reporter gene expression and HO-1 expression	[47]
Mouse keratinocyte	3H-1,2-dithiole-3-thione	ERK inhibitor but not p38 inhibitor suppressed Nrf2 nuclear translocation and ARE-reporter gene expression	[48]
Rat epithelial L2	4-hydroxynonenal	ERK- and p38-dependent EPRE-mediated γ-glutamyl transpeptidase expression	[49]
Rat vascular smooth muscle	15d-PGJ$_2$	p38 inhibitor abolished Nrf2 dependent HO-1 expression	[50]
Rat primary hepatocytes	Methionine restriction	ERK-dependent Nrf2 nuclear translocation and GSH-*S*-transferase π expression	[51]
Rat kidney epithelial NRK-52E	Curcumin	p38 inhibitor reduced about 50% of HO-1 expression, but ERK and JNK inhibitors did not suppress HO-1 expression	[52]
Bovine aortic endothelial	Spermine NONOate (NO donor)	p38 and ERK inhibitors SB203580 and PD98059 respectively reduce HO-1 expression	[53]
Guinea pig gastric mucosal	Indomethacin	p38 inhibitor significantly reduced Nrf2 nuclear accumulation and HO-1 expression	[54]

Abbreviations: ARE, antioxidant response element; HO-1, heme oxygenase-1; GCLM, glutamate-cysteine ligase modifier subunit; EPRE, electrophile response element.

In this review, we critically evaluate the differential role of p38, JNK, and ERK on the activation of Nrf2 by stress agents. We hypothesize that ERK/JNK signaling contributes to the nuclear translocation of Nrf2 and that simultaneous activation of p38 signaling activated under glutathione (GSH) deletion leads to maximal accumulation of Nrf2 in the nucleus (Figure 1D). We previously proposed that p38 signaling leads to the phosphorylation of Nrf2 by PKCζ and casein kinase 2 (CK2) to stabilize it and masks a nuclear exporting signal (NES) [1] as described in Section 2. Notably, previous studies show that stress-activated ERK/JNK signaling facilitates Nrf2 nuclear translocation [34,48,50]. We discuss the possible functional partners of ERK that regulate facilitated Nrf2 nuclear translocation, and propose a novel mechanism of Nrf2 nuclear translocation mediated by ERK, involving peptidylprolyl cis/trans isomerase (PPIase) NIMA-interacting 1 (Pin1) and the molecular chaperon heat shock protein 90α (Hsp90α) as discussed in Sections 3–6. We further discuss cyclin-dependent kinase 5 (Cdk5)-mediated Nrf2 nuclear translocation in Section 7.

2. p38 Controls Glutathione Sensor Neutral Sphingomyelinase 2

Maintenance of the small antioxidant GSH at high levels is essential for protecting cells against oxidative damage. Cells respond to the downregulation of GSH under oxidative stress by activating Nrf2, which upregulates the expression of genes required for GSH synthesis to restore cellular GSH levels. Notably, neutral sphingomyelinase 2 (nSMase2) senses GSH levels, and its activity is inhibited by high levels of cellular GSH (>3mM), but depletion of cellular GSH induces nSMase2 activation to generate the lipid signaling molecule ceramide [55–59]. We previously proposed that depletion of GSH induces ceramide/PKCζ/CK2 signaling leading to phosphorylation of Nrf2 by these kinases [60–62]. Interestingly, oxidative stress induces activation of p38 MAP kinase, which causes trafficking of nSMase2 from perinuclear regions to the plasma membrane [63,64], thereby enhancing ceramide generation by nSMase2 under oxidative stress. We further speculated that ceramide/PKCζ/CK2 signaling phosphorylates Nrf2 tethered to caveolin 1 (Cav1) in membrane lipid rafts/caveolae and that phosphorylation by these kinases stabilizes Nrf2 and masks an NES which could favor Nrf2 nuclear localization [1] (Figure 2A).

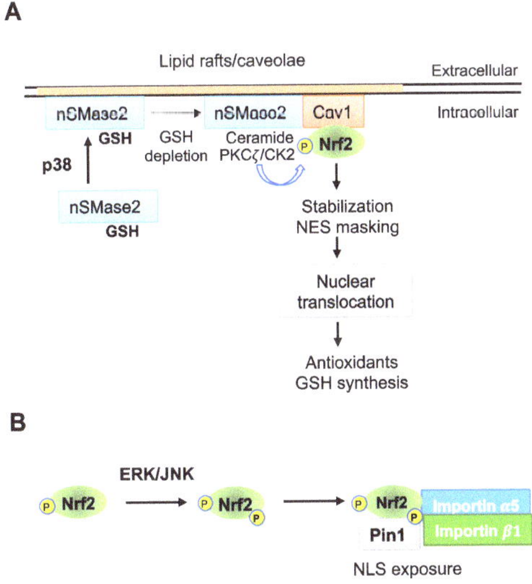

Figure 2. **Stress-activated p38 controls nSMase2 activation, and ERK and JNK mediate Pin1-dependent Nrf2 nuclear translocation**. (**A**) MAP kinase p38 induces the transfer of neutral sphingomyelinase 2 (nSMase2) from the perinuclear region to the plasma membrane. Glutathione (GSH) depletion causes ceramide/PKCζ/CK2 signaling, which induces Nrf2 phosphorylation and stabilization and masks a nuclear export signal [1]. (**B**) Phosphorylation of Nrf2 by ERK/JNK leads to an association with Pin1, which causes a cis/trans structural change at the preceding proline residue (pSer/Thr-Pro), allowing an association with importin α5 at the exposed nuclear localization signal. Then, importin β1 associates with importin α5, allowing Nrf2 to translocate to the nucleus.

However, this hypothesis requires an additional mechanism to facilitate the translocation of stabilized Nrf2 from cell membrane compartments to the nucleus. Facilitated nuclear translocation of Nrf2 requires importin binding to the NLSs and directional movement through the cytoplasm to nuclear membrane pores. Upon arrival at the nuclear pores, the complexes are transferred to the nuclear interior by importin-dependent facilitated diffusion [20]. Mechanisms and functional partners for the facilitated movement of Nrf2 from the cell periphery to nuclear pores remain unclear. We discuss the possible functional partners and mechanism of facilitated Nrf2 nuclear translocation in the following sections.

3. ERK/JNK and PPIase Pin1 Control Nrf2 Nuclear Translocation

Xu et al. [34] previously showed that ERK and JNK modulate the nuclear translocation of Nrf2 in human prostate cancer PC-3 cells. These authors pre-treated the cells in 0.5% serum-containing medium overnight, which induced sequestration of Nrf2 in the cytoplasm. Upon transfection of JNK1 and its activating kinase MKK into PC-3 cells, Nrf2 was localized in both the cytoplasm and nucleus. Similar results were obtained when PC-3 cells were transfected with ERK2 and its activating kinase MEK1. These authors suggested that JNK1 and ERK2 can phosphorylate Nrf2 and induce nuclear translocation.

Recent studies show that the PPIase Pin1 upregulates nuclear accumulation of Nrf2 [65–67]. PPIase catalyzes the conversion between cis and trans conformations of proline imidic peptide bonds, playing a role in protein folding, signal transduction, trafficking, assembly, and cell cycle regulation [68]. The three classes of PPIase are cyclophilins, FK506-binding proteins (FKBPs), and parvulins [68], and Pin1 belongs to the parvulin class of PPIase. Cyclophilins and FKBPs are called immunophilins as immunosuppressive drugs such as cyclosporin A, FK506, and rapamycin directly bind and inhibit these PPIases [68]. Previous studies show that immunophilins associate with steroid hormone receptors to modulate their functions (reviewed in [69]).

Pin1 is a unique PPIase that specifically recognizes phosphorylated serine or threonine immediately preceding a proline residue (pSer/Thr-Pro), isomerizes the peptide bond, and is known to play an important role in cell cycle progression (reviewed in [68,70]). Liang et al. [65] showed that Pin1 contributes to Nrf2 activation in pancreatic ductal adenocarcinoma cells with high K-ras activity [65]. Saeidi et al. [66] found that enhanced H-ras signaling in human breast cancer cells induces the association of Pin1 with Nrf2 to protect Nrf2 from Keap1-mediated degradation. Saeidi et al. [66] further showed that phosphorylation of human Nrf2 at Ser-215, -408, and -577 is essential for its interaction with Pin1 [66]. These authors showed that among MAP kinases, ERK and JNK, but not p38, phosphorylate these serine residues, suggesting Ras-ERK signaling promotes Pin1 association with ERK-phosphorylated Nrf2 to facilitate translocation to the nucleus. We speculated, based on these studies, that ERK2 and JNK initially phosphorylate Nrf2, resulting in the association of Pin1 to the p-Ser/Thr-Pro site(s). It seems plausible that a Pin1-mediated structural change in Nrf2 could expose NLS(s) for association with importin $\alpha 5$, which then recruits importin $\beta 1$ to facilitate nuclear translocation (Figure 2B). We next discuss the possible additional partners for the ERK/JNK-Pin1-mediated Nrf2 nuclear translocation.

4. PPIase and Hsp90 Cooperate in the Nuclear Transport of Signaling Molecules

Another important function of some PPIases is their interaction with heat shock protein 90 (Hsp90) and the dynein/dynactin complex [71,72]. Hsp90 is the major molecular chaperone protecting many client proteins from denaturation and aggregation (reviewed in [73–75]). Notably, in addition to chaperone activity, Hsp90 controls nucleocytoplasmic trafficking of signaling molecules (reviewed in [76,77]). PPIases are regarded as co-chaperones of Hsp90 and are crucial for translocating hormone receptors, transcription factors, and signaling molecules [77–81].

Hsp90 has three functional domains, an N-terminal domain with ATPase, a middle domain for binding co-chaperones and clients, and a C-terminal domain for dimerization [82] (Figure 3A). The Hsp90 dimer forms two alternative structures, ATP-free open and ATP-bound closed structures [83,84] (Figure 3A). Notably, ATP hydrolysis is coupled with the chaperone activity and accompanies the release of associated clients. There are many co-chaperone proteins (cofactors) that regulate Hsp90 functions, including its ATPase activity [85]. Importantly, to keep a client associated with the benefits of long-distance translocation, a co-chaperone p23 inhibits ATPase activity [86] and is recruited to the Hsp90-FKBP52 complex [81,87].

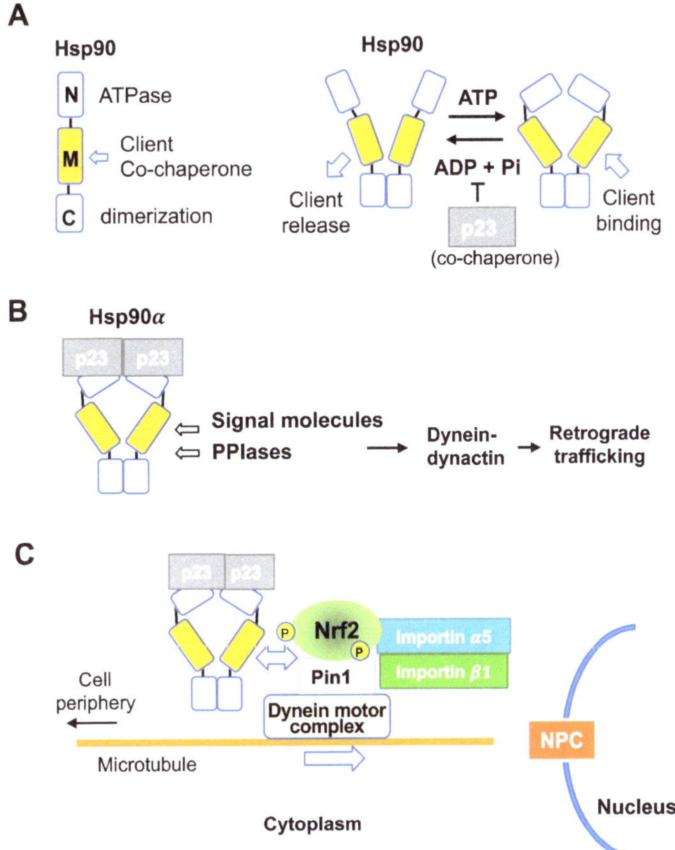

Figure 3. Proposed mechanism of Hsp90-mediated nuclear translocation of Nrf2. (**A**) Hsp90 is composed of three domains, an N-terminal domain with ATPase activity, a middle domain for binding co-chaperones and clients, and a C-terminal domain for dimerization. Hsp90 dimer takes two forms, e.g., open and ATP-bound closed ring forms. Co-chaperone p23 inhibits Hsp90 ATPase activity. (**B**) Hsp90α dimer-p23 complex is the basal structure for transporting PPIase-associated signal molecules such as steroid receptors to the nucleus. (**C**) Hsp90α-p23-Pin1-Nrf2-importin-α5β1 containing complex associates with a dynein motor complex, which carries the cargo toward the nuclear pore complex (NPC) along microtubules.

Two isoforms of Hsp90, α and β, are abundantly expressed in the cytoplasm with similar chaperone activities. Proliferating cells express higher levels of Hsp90α than Hsp90β, as Hsp90α gene expression is controlled by c-Myc downstream of growth factor signaling [88]. As Hsp90α has higher dimer-forming potential than Hsp90β [89,90], and p23 co-chaperone preferentially associates with Hsp90α compared to Hsp90β [84], we propose that the Hsp90α dimer and p23 co-chaperone compose a backbone for the nuclear transport machinery for co-transport of PPIase and client proteins (Figure 3B).

5. Functional Interaction of Nrf2 with Hsp90

Ngo et al. [91] recently showed that Hsp90 (isoforms not identified) directly interacts with Nrf2 in a yeast model expression system and HeLa cells, and with Hsp90 preventing overexpressed Nrf2 from forming protein aggregates. This study shows that Nrf2 is a client of Hsp90 and suggests a possibility that under the normal/low levels of Nrf2, the Pin1-Nrf2 associates with Hsp90α dimer-p23, forming a multiprotein complex that can be transported efficiently to the nuclear pore complex via the dynein motor complex along microtubules (Figure 3C). To verify this hypothesis, further studies are required to detect stable interaction of the Nrf2-Pin1 complex with Hsp90α and to demonstrate the Pin1-mediated linkage of the signaling complex with the dynein motor system.

Interestingly, previous studies suggest functional interactions of Nrf2 with Hsp90. Jia et al. [92] observed that an Hsp90 ATPase inhibitor, 17-dimethylaminoethylamino-17-demethoxygeldanamycin (17-DMAG), upregulated nuclear Nrf2 and the expression of HO-1 in neuronal HT22 cells subjected to hypoxia/reoxygenation. Lazaro et al. [93] also observed that 17-DMAG upregulated Nrf2 activation in macrophages and vascular smooth muscle cells in atherosclerotic plaques in diabetic apolipoprotein E-deficient mice. In contrast to these reports, Hsp90 ATPase inhibitors, including 17-allylamino-demethoxygeldanamycin (17-AAG), caused the gradual death of human cancer cells expressing high levels of Nrf2 [94]. The Hsp90 inhibitor exhibits no toxicity to other cancer cells with normal levels of Nrf2 but induces toxicity in cells after treatment with diethyl maleate (100 μM) to upregulate Nrf2 levels [94]. This toxic effect of 17-AAG may reflect the importance of Hsp90 chaperone activity to prevent Nrf2 protein aggregation in Nrf2-overexpressing cells [91].

6. Pin1 Controls the Nuclear Translocation of Other ERK Substrates

In addition to Nrf2, the ERK-Pin1 system naturally controls the nuclear translocation of several other ERK substrates, which are important for energy metabolism and proliferation. The nuclear translocation of pyruvate kinase M2 (PKM2) is controlled by the ERK-Pin1 system [95]. PKM2 promotes glucose metabolism by aerobic glycolysis and contributes to anabolic metabolism [96]. PKM2 expression is upregulated in multiple cancer types and contributes to the Warburg effect (reviewed in [97–101]). Yang et al. [102] showed that activated ERK2 binds directly to PKM2 through the ERK2 docking groove and phosphorylates PKM2 at Ser-37, inducing a structural change of PKM2 from tetramer to dimer or monomer. PKM2 dimer recruits Pin1 for cis-trans isomerization of PKM2 and binding to importin α5 and translocation to the Hsp90 containing complex to the nucleus [102], supporting our hypothesis of ERK-Pin1-Hsp90 dependent Nrf2 nuclear translocation (Figure 3C).

The ERK-Pin1-Hsp90α machinery also controls the nuclear transport of hypoxia-inducible transcription factor-1 (HIF-1α). HIF-1α mediates the activation of networks of target genes involved in angiogenesis, erythropoiesis, and glycolysis (reviewed in [103–105]). Besides hypoxic conditions, phosphorylation of HIF-1α at Ser-451 by ERK is another central post-translational modification, which regulates its stability under both hypoxia and physiological normoxia (reviewed in [106]), and plays a crucial role in promoting tumor growth [107–109]. Jalouli et al. [108] showed that Pin1 associates with the p-Ser-Pro motif and regulates HIF-1α transcriptional activity. HIF-1α is a client of Hsp90α [110] and contains NLSs for importin α binding [111–113]. Mylonis et al. [114] further showed that phosphorylation of Ser-641/643 by ERK inhibits the association of exportin to the NES, resulting in the accumulation of HIF-1α in the nucleus. These studies indicate that the nuclear translocation of HIF-1α is largely dependent on the ERK-mediated Pin1-Hsp90α system.

7. Cdk5 Controls Nrf2 Nuclear Translocation through Pin1

In addition to ERK and JNK, another proline-directed serine/threonine kinase, cyclin-dependent kinase 5 (Cdk5), also controls Nrf2 activation. Jimenez-Blasco et al. [115] showed that treating astrocytes from fetal rat brains with 20 µM N-methyl-d-aspartate (NMDA) for 8 h induced Nrf2-dependent activation of antioxidant genes. These authors further showed that NMDA induces the phospholipase C-mediated endoplasmic reticulum release of Ca^{2+} and activation of PKCδ, which phosphorylates and stabilizes Cdk5 cofactor p35. Furthermore, the active p35/Cdk5 complex phosphorylated Nrf2 leading to Nrf2 nuclear translocation and ARE-mediated gene expression [115]. In another study, Lee et al. [116] showed that tBHQ induces Nrf2 activation and expression of NQO1 in IMR-32 human neuroblastoma cells independent of ERK. Interestingly, oxidative stress induces the upregulation of Cdk5 catalytic subunit p35 in IMR-32 cells [117], suggesting Cdk5/p35 instead of ERK plays a role in Pin1-mediated Nrf2 nuclear translocation in neuroblastoma cells. It has been reported that Cdk5 phosphorylates ubiquitin ligase TRIM59 leading to Pin1 and importin α5 association resulting in nuclear translocation [118]. These results suggest that Cdk5/Nrf2/Pin1 axis contributes to Nrf2 nuclear translocation, like ERK/Nrf2/Pin1 axis, irrespective of whether Cdk5 phosphorylates different Ser/Thr-Pro sites of Nrf2. Cdk5 regulates neuronal functions but is also associated with cancer development and has been considered a potential target for cancer treatment (reviewed in [119–123]).

8. Summary and Conclusions

MAP kinases contribute to the activation of the transcription factor Nrf2 (reviewed by Kong et al., 2001, 2002), noting that activation is dependent on the cell type and stress agent (see Table 1). We proposed that p38-mediated signaling could stabilize Nrf2 via ceramide-activated PKCζ phosphorylation and mask an NES by CK2 phosphorylation under oxidative stress accompanying GSH depletion (Figure 2A) [1]. In contrast, ERK2 and JNK1 induce Nrf2 translocation into the nucleus [34]. Concerning the mechanism underlying ERK/JNK-dependent Nrf2 nuclear translocation, we propose a novel concept that direct phosphorylation of Nrf2 by ERK/JNK induces assembly of the Hsp90α-Pin1-Nrf2 complex. Pin1 causes prolyl-isomerization of Nrf2, which allows importin α5 to associate the Hsp90α-Pin1-Nrf2 complex. Then, the Hsp90α-Pin1-Nrf2-importin containing signaling complex is carried by the dynein motor system toward the nuclear pore complex along microtubules (Figure 3C). In addition to ERK and JNK, Cdk5 also phosphorylates Nrf2 and helps Nrf2 nuclear translocation via Hsp90α-Pin1-dynein machinery. Importantly, Hsp90 contributes to Nrf2 activation in two ways, inhibition of denaturation or aggregation when Nrf2 is over-expressed [91] and facilitation of Nrf2 nuclear import. The Hsp90 chaperone activity depends on ATP hydrolysis, but it accompanies client release. However, ATP hydrolysis is suppressed during nuclear transport of Hsp90α-Pin1-Nrf2 containing multiprotein complex via association of co-chaperon p23 (Figure 3C). Thus, under oxidative stress, p38 and ERK/JNK and/or Cdk5 could work together to stabilize Nrf2 to facilitate the maximal accumulation of Nrf2 in the nucleus (Figure 1C,D).

Hsp90α is required for the proliferation, migration, and invasion of cancer cells in culture [124]. Hsp90 is considered a druggable target for cancer treatment [125,126]. Accumulating evidence indicates that Pin1 plays a key role in various cancers. Pin1-mediated β-catenin accumulation occurs in about 70% of hepatocellular carcinoma [127]. Intriguingly, cell proliferation, migration, and invasion are significantly inhibited in Pin1-silenced Hep-2 cells [128]. Silencing of Pin1 causes down-regulation of β-catenin and cyclin D1 expression [128] and significantly increases the sensitivity to cisplatin in HeLa cells [129]. Thus, ERK- and Cdk5-signaling coupled with Hsp90α-Pin1-dynein machinery may be a prime target of chemotherapy for tumors. Developing chemical agents that inhibit Pin1 rather than Hap90 ATPase activity could be a promising approach for cancer chemotherapy (reviewed in [130–132]). For instance, a Pin1 inhibitor all-trans retinoic acid has been used to treat acute promyelocytic leukemia in animal models and human patients [133], and all-trans retinoic acid reduced the growth of transplanted tamoxifen-resistant human breast cancer cells in mice [134]. Dubiella et al. [135] showed that a Pin1 inhibitor Sulfopin reduced tumor progression and conferred survival benefits in animal models, while Liu et al. [136] developed a delivery system of a Pin1 inhibitor AG177724 targeting cancer-associated fibroblasts and observed the inhibition of tumor growth in mice.

Kim et al. [137] raised a question concerning the role of Pin1 in the expression of HO-1 induced by nitric oxide (NO) in mouse vascular smooth muscle cells and embryonic fibroblasts. These authors observed that treatment of the cells with NO donor nitroprusside (3 mM) for 8 h upregulated HO-1 levels in control cells to a higher extent than in Pin1 deficient cells and argued that Nrf2/ARE-mediated transcriptional activity is negatively controlled by Pin1 in fibroblasts [137]. However, the expression of HO-1 is controlled by both Nrf2 and AP-1 via partially overlapping AREs and TPA-responsive elements in the gene promotor [138], and Mouawad et al. [139] showed that NO-dependent expression of HO-1 is controlled by transcription factors C/EBPβ and AP-1 in macrophages [139]. Therefore, we suggest the possibility that NO-activated AP-1 could induce HO-1 gene expression more efficiently in the absence of Pin1-mediated Nrf2 nuclear translocation.

Low molecular kinase inhibitors are widely used to examine the function of kinases, but some inhibitors exhibit unexpected side effects. For instance, the p38 inhibitor SB203580 is a potent agonist of aryl hydrocarbon receptor (AhR) [140,141], which is a ligand-activated transcription factor and key regulator of xenobiotic metabolism, and AhR activation induces Cyp1a1 gene expression via xenobiotic responsive element (XRE) in Hepa 1c1c7 and HepG2 cells [142]. A complex effect of SB203580 is observed in HepG2 cells, which express high AhR levels [143]. Yu et al. [143] observed that tBHQ (100 μM) activated p38 and upregulated NQO1 expression in HepG2 cells, suggesting the role of the p38/Nrf2/ARE axis for the NQO1 expression. However, these authors observed that the addition of the p38 inhibitor SB203580 (5 μM) with tBHQ upregulated expression of NQO1 levels higher than tBHQ alone in 24 h, and argued that the p38 kinase pathway functions as a negative regulator in the ARE-mediated induction of phase II detoxifying enzymes [143]. We believe this conclusion may be misleading due to neglecting the influence of AhR activation by SB203580. Notably, NQO1 gene expression can be controlled by both Nrf2/ARE and AhR/XRE [144], and Nrf2 gene expression can be upregulated by AhR/XRE signaling [145]. Treatment of hepatoma 1c1c7 cells with AhR agonist 2,3,7,8-tetrachlorodibenzo-*p*-dioxin (10 nM) leads to upregulation of Nrf2 mRNA in 2 h and Nrf2 protein levels in 6 h [145]. Notably, tBHQ is a ligand for AhR. It is known to induce expression of Cyp1a1 in hepatoma 1c1c7 and HepG2 cells [146], suggesting the possibility that SB203580 and tBHQ synergistically activate AhR/XRE signaling, leading to rapid NQO1 gene expression with delayed Nrf2/ARE-mediated NQO1 expression. Thus, Phase I and II xenobiotic metabolism is coordinately regulated by the cross-talk between AhR and Nrf2 via ARE and XRE elements, respectively (reviewed in [147]).

In summary, we propose that ERK, JNK, and Cdk5 control Nrf2 phosphorylation inducing a formation of a multiprotein complex containing Hsp90α-p23-Pin1-Nrf2-importins to associate with dynein motors to move from cell membrane signaling compartments toward nuclear pores along microtubules (Figure 3C). Therefore, the integrity of microtubules is important to ensure the nuclear translocation of Nrf2 and other signaling molecules. It is important to note that tau family proteins control the assembly and maintenance of the structural stability of microtubules and that proline-directed kinases, MAP kinases, GSK-3β, and Cdk5, can phosphorylate tau proteins and affect their functions (reviewed in [148–150]). GSK-3β phosphorylated tau reduces the affinity to microtubules, causing the dysfunction of microtubules [151,152]. Interestingly, Pin1 interacts with phosphorylated tau proteins, restores the ability of tau to bind microtubules and promote assembly in vitro [153], and downregulates the GSK-3β-mediated phosphorylation of tau [154,155]. These studies suggest that Pin1/tau axis is also important for microtubule-mediated facilitated translocation of Nrf2 to the nucleus. As tau controls axonal transport and neurite outgrowth in neurons, defects in tau function could lead to neurodegeneration [156]. Thus,

Pin1 plays an important but opposite role in the pathogenesis of Alzheimer's disease and many human cancers (reviewed in [157,158]).

Funding: The Great Britain SASAKAWA Foundation for a Butterfield Award (B131) supporting our UK-Japan collaboration between the University of Tsukuba (T.I., E.W.) and King's College London (G.E.M.). The Japanese Society for Promotion of Science, KAKENHI (E.W. JP20K07421), British Heart Foundation (G.E.M.), and COST Action 'BenBedPhar' CA20121 (G.E.M.).

Conflicts of Interest: The authors declare they have no potential conflict of interest.

Abbreviations

AhR, aryl hydrocarbon receptor; ERK, extracellular signal-regulated kinase; FKBP, FK506-binding protein; GSH, glutathione; GSK-3, glycogen synthase kinase-3; Keap1, Kelch-like ECH-associated protein 1; HIF-1α, hypoxia-inducible transcription factor-1α; HO-1, heme oxygenase 1; Hsp90, heat shock protein 90; MAPK, mitogen-activated protein kinase; NES, nuclear export signal; NLS, nuclear localization signal; NMDA, N-methyl-d-aspartate; Nrf2, nuclear factor-E2-related factor 2; Pin1, peptidyl-prolyl cis/trans isomerase NIMA-interacting 1; PI3-K, phosphatidylinositol 3-kinase; PKM2, pyruvate kinase M2; PPIase, peptidyl-prolyl cis/trans isomerase; tBHQ, *tert*-butylhydroquinone; 5'UTR, 5' untranslated region.

References

1. Ishii, T.; Warabi, E.; Mann, G.E. Mechanisms underlying Nrf2 nuclear translocation by non-lethal levels of hydrogen peroxide: p38 MAPK-dependent neutral sphingomyelinase2 membrane trafficking and ceramide/PKCζ/CK2 signaling. *Free Radic. Biol. Med.* **2022**, *191*, 191–202. [CrossRef]
2. Itoh, K.; Chiba, T.; Takahashi, S.; Ishii, T.; Igarashi, K.; Katoh, Y.; Oyake, T.; Hayashi, N.; Satoh, K.; Hatayama, I.; et al. An Nrf2/small Maf heterodimer mediates the induction of phase II detoxifying enzyme genes through antioxidant response elements. *Biochem. Biophys. Res. Commun.* **1997**, *236*, 313–322. [CrossRef]
3. Ishii, T.; Itoh, K.; Takahashi, S.; Sato, H.; Yanagawa, T.; Katoh, Y.; Bannai, S.; Yamamoto, M. Transcription factor Nrf2 coordinately regulates a group of oxidative stress-inducible genes in macrophages. *J. Biol. Chem.* **2000**, *275*, 16023–16029. [CrossRef] [PubMed]
4. Chan, J.Y.; Kwong, M. Impaired expression of glutathione synthetic enzyme genes in mice with targeted deletion of the Nrf2 basic-leucine zipper protein. *Biochim. Biophys. Acta* **2000**, *1517*, 19–26. [CrossRef] [PubMed]
5. Hayes, J.D.; Dinkova-Kostova, A.T. The Nrf2 regulatory network provides an interface between redox and intermediary metabolism. *Trends Biochem. Sci.* **2014**, *39*, 199–218. [CrossRef] [PubMed]
6. Hayes, J.D.; Chowdhry, S.; Dinkova-Kostova, A.T.; Sutherland, C. Dual regulation of transcription factor Nrf2 by Keap1 and by the combined actions of β-TrCP and GSK-3. *Biochem. Soc. Trans.* **2015**, *43*, 611–620. [CrossRef]
7. Cuadrado, A. Structural and functional characterization of Nrf2 degradation by glycogen synthase kinase 3/β-TrCP. *Free Radic. Biol. Med.* **2015**, *88*, 147–157. [CrossRef]
8. Tebay, L.E.; Robertson, H.; Durant, S.T.; Vitale, S.R.; Penning, T.M.; Dinkova-Kostova, A.T.; Hayes, J.D. Mechanisms of activation of the transcription factor Nrf2 by redox stressors, nutrient cues, and energy status and the pathways through which it attenuates degenerative disease. *Free Radic. Biol. Med.* **2015**, *88 Pt B*, 108–146. [CrossRef]
9. Yamamoto, M.; Kensler, T.W.; Motohashi, H. The KEAP1-NRF2 System: A Thiol-Based Sensor-Effector Apparatus for Maintaining Redox Homeostasis. *Physiol. Rev.* **2018**, *98*, 1169–1203. [CrossRef]
10. Itoh, K.; Tong, K.I.; Yamamoto, M. Molecular mechanism activating Nrf2-Keap1 pathway in regulation of adaptive response to electrophiles. *Free Radic. Biol. Med.* **2004**, *36*, 1208–1213. [CrossRef] [PubMed]
11. Kwak, M.K.; Wakabayashi, N.; Kensler, T.W. Chemoprevention through the Keap1-Nrf2 signaling pathway by phase 2 enzyme inducers. *Mutat. Res.* **2004**, *555*, 133–148. [CrossRef]
12. Itoh, K.; Mimura, J.; Yamamoto, M. Discovery of the negative regulator of Nrf2, Keap1: A historical overview. *Antioxid. Redox Signal.* **2010**, *13*, 1665–1678. [CrossRef] [PubMed]
13. Purdom-Dickinson, S.E.; Sheveleva, E.V.; Sun, H.; Chen, Q.M. Translational control of nrf2 protein in activation of antioxidant response by oxidants. *Mol. Pharmacol.* **2007**, *72*, 1074–1081. [CrossRef] [PubMed]
14. Li, W.; Thakor, N.; Xu, E.Y.; Huang, Y.; Chen, C.; Yu, R.; Holcik, M.; Kong, A.N. An internal ribosomal entry site mediates redox-sensitive translation of Nrf2. *Nucleic Acids Res.* **2010**, *38*, 778–788. [CrossRef]
15. Zhang, J.; Dinh, T.N.; Kappeler, K.; Tsaprailis, G.; Chen, Q.M. La autoantigen mediates oxidant induced de novo Nrf2 protein translation. *Mol. Cell Proteom.* **2012**, *11*, M111.015032. [CrossRef] [PubMed]
16. Theodore, M.; Kawai, Y.; Yang, J.; Kleshchenko, Y.; Reddy, S.P.; Villalta, F.; Arinze, I.J. Multiple nuclear localization signals function in the nuclear import of the transcription factor Nrf2. *J. Biol. Chem.* **2008**, *283*, 8984–8994, Erratum in *J. Biol. Chem.* **2008**, *283*, 14176. [CrossRef] [PubMed]

17. Görlich, D.; Prehn, S.; Laskey, R.A.; Hartmann, E. Isolation of a protein that is essential for the first step of nuclear protein import. *Cell* **1994**, *79*, 767–778. [CrossRef] [PubMed]
18. Görlich, D.; Vogel, F.; Mills, A.D.; Hartmann, E.; Laskey, R.A. Distinct functions for the two importin subunits in nuclear protein import. *Nature* **1995**, *377*, 246–248. [CrossRef] [PubMed]
19. Görlich, D.; Mattaj, I.W. Nucleocytoplasmic transport. *Science* **1996**, *271*, 1513–1518. [CrossRef] [PubMed]
20. Rexach, M.; Blobel, G. Protein import into nuclei: Association and dissociation reactions involving transport substrate, transport factors, and nucleoporins. *Cell* **1995**, *83*, 683–692. [CrossRef]
21. Ullman, K.S.; Powers, M.A.; Forbes, D.J. Nuclear export receptors: From importin to exportin. *Cell* **1997**, *90*, 967–970. [CrossRef] [PubMed]
22. Weis, K. Importins and exportins: How to get in and out of the nucleus. *Trends Biochem. Sci.* **1998**, *23*, 185–189, Erratum in *Trends Biochem. Sci.* **1998**, *23*, 235. [CrossRef]
23. Li, W.; Jain, M.R.; Chen, C.; Yue, X.; Hebbar, V.; Zhou, R.; Kong, A.N. Nrf2 Possesses a redox-insensitive nuclear export signal overlapping with the leucine zipper motif. *J. Biol. Chem.* **2005**, *280*, 28430–28438. [CrossRef]
24. Li, W.; Yu, S.W.; Kong, A.N. Nrf2 possesses a redox-sensitive nuclear exporting signal in the Neh5 transactivation domain. *J. Biol. Chem.* **2006**, *281*, 27251–27263. [CrossRef]
25. Kong, A.N.; Yu, R.; Chen, C.; Mandlekar, S.; Primiano, T. Signal transduction events elicited by natural products: Role of MAPK and caspase pathways in homeostatic response and induction of apoptosis. *Arch. Pharm. Res.* **2000**, *23*, 1–16. [CrossRef] [PubMed]
26. Kong, A.N.; Owuor, E.; Yu, R.; Hebbar, V.; Chen, C.; Hu, R.; Mandlekar, S. Induction of xenobiotic enzymes by the MAP kinase pathway and the antioxidant or electrophile response element (ARE/EpRE). *Drug Metab. Rev.* **2001**, *33*, 255–271. [CrossRef]
27. Wagner, E.F.; Nebreda, A.R. Signal integration by JNK and p38 MAPK pathways in cancer development. *Nat. Rev. Cancer* **2009**, *9*, 537–549. [CrossRef] [PubMed]
28. Cuadrado, A.; Nebreda, A.R. Mechanisms and functions of p38 MAPK signalling. *Biochem. J.* **2010**, *429*, 403–417. [CrossRef]
29. Guo, Y.J.; Pan, W.W.; Liu, S.B.; Shen, Z.F.; Xu, Y.; Hu, L.L. ERK/MAPK signalling pathway and tumorigenesis. *Exp. Ther. Med.* **2020**, *19*, 1997–2007. [CrossRef] [PubMed]
30. Cobb, M.H.; Robbins, D.J.; Boulton, T.G. ERKs, extracellular signal-regulated MAP-2 kinases. *Curr. Opin. Cell Biol.* **1991**, *3*, 1025–1032. [CrossRef]
31. Ahn, N.G.; Seger, R.; Bratlien, R.L.; Krebs, E.G. Growth factor-stimulated phosphorylation cascades: Activation of growth factor-stimulated MAP kinase. *Ciba Found. Symp.* **1992**, *164*, 113–126; discussion 126–131. [CrossRef] [PubMed]
32. Paul, A.; Wilson, S.; Belham, C.M.; Robinson, C.J.; Scott, P.H.; Gould, G.W.; Plevin, R. Stress-activated protein kinases: Activation, regulation and function. *Cell Signal.* **1997**, *9*, 403–410. [CrossRef]
33. Yu, R.; Tan, T.H.; Kong, A.N. Butylated hydroxyanisole and its metabolite tert-butylhydroquinone differentially regulate mitogen-activated protein kinases. The role of oxidative stress in the activation of mitogen-activated protein kinases by phenolic antioxidants. *J. Biol. Chem.* **1997**, *272*, 28962–28970. [CrossRef] [PubMed]
34. Xu, C.; Yuan, X.; Pan, Z.; Shen, G.; Kim, J.H.; Yu, S.; Khor, T.O.; Li, W.; Ma, J.; Kong, A.N. Mechanism of action of isothiocyanates: The induction of ARE-regulated genes is associated with activation of ERK and JNK and the phosphorylation and nuclear translocation of Nrf2. *Mol. Cancer Ther.* **2006**, *5*, 1918–1926. [CrossRef] [PubMed]
35. Yu, R.; Chen, C.; Mo, Y.Y.; Hebbar, V.; Owuor, E.D.; Tan, T.H.; Kong, A.N. Activation of mitogen-activated protein kinase pathways induces antioxidant response element-mediated gene expression via a Nrf2-dependent mechanism. *J. Biol. Chem.* **2000**, *275*, 39907–39913. [CrossRef]
36. Zipper, L.M.; Mulcahy, R.T. Inhibition of ERK and p38 MAP kinases inhibits binding of Nrf2 and induction of GCS genes. *Biochem. Biophys. Res. Commun.* **2000**, *278*, 484–492. [CrossRef]
37. Gong, P.; Hu, B.; Cederbaum, A.I. Diallyl sulfide induces heme oxygenase-1 through MAPK pathway. *Arch. Biochem. Biophys.* **2004**, *432*, 252–260. [CrossRef]
38. Yeh, C.T.; Yen, G.C. Involvement of p38 MAPK and Nrf2 in phenolic acid-induced P-form phenol sulfotransferase expression in human hepatoma HepG2 cells. *Carcinogenesis* **2006**, *27*, 1008–1017. [CrossRef]
39. Yao, P.; Nussler, A.; Liu, L.; Hao, L.; Song, F.; Schirmeier, A.; Nussler, N. Quercetin protects human hepatocytes from ethanol-derived oxidative stress by inducing heme oxygenase-1 via the MAPK/Nrf2 pathways. *J. Hepatol.* **2007**, *47*, 253–261. [CrossRef]
40. Alam, J.; Wicks, C.; Stewart, D.; Gong, P.; Touchard, C.; Otterbein, S.; Choi, A.M.; Burow, M.E.; Tou, J. Mechanism of heme oxygenase-1 gene activation by cadmium in MCF-7 mammary epithelial cells. Role of p38 kinase and Nrf2 transcription factor. *J. Biol. Chem.* **2000**, *275*, 27694–27702. [CrossRef]
41. Keum, Y.S.; Owuor, E.D.; Kim, B.R.; Hu, R.; Kong, A.N. Involvement of Nrf2 and JNK1 in the activation of antioxidant responsive element (ARE) by chemopreventive agent phenethyl isothiocyanate (PEITC). *Pharm. Res.* **2003**, *20*, 1351–1356. [CrossRef]
42. Ogborne, R.M.; Rushworth, S.A.; O'Connell, M.A. Alpha-lipoic acid-induced heme oxygenase-1 expression is mediated by nuclear factor erythroid 2-related factor 2 and p38 mitogen-activated protein kinase in human monocytic cells. *Arterioscler. Thromb. Vasc Biol.* **2005**, *25*, 2100–2105. [CrossRef]
43. Anwar, A.A.; Li, F.Y.; Leake, D.S.; Ishii, T.; Mann, G.E.; Siow, R.C. Induction of heme oxygenase 1 by moderately oxidized low-density lipoproteins in human vascular smooth muscle cells: Role of mitogen-activated protein kinases and Nrf2. *Free Radic. Biol. Med.* **2005**, *39*, 227–236. [CrossRef] [PubMed]

44. Rushworth, S.A.; Ogborne, R.M.; Charalambos, C.A.; O'Connell, M.A. Role of protein kinase C delta in curcumin-induced antioxidant response element-mediated gene expression in human monocytes. *Biochem. Biophys. Res. Commun.* **2006**, *341*, 1007–1016. [CrossRef] [PubMed]
45. Papaiahgari, S.; Kleeberger, S.R.; Cho, H.Y.; Kalvakolanu, D.V.; Reddy, S.P. NADPH oxidase and ERK signaling regulates hyperoxia-induced Nrf2-ARE transcriptional response in pulmonary epithelial cells. *J. Biol. Chem.* **2004**, *279*, 42302–42312. [CrossRef] [PubMed]
46. Pischke, S.E.; Zhou, Z.; Song, R.; Ning, W.; Alam, J.; Ryter, S.W.; Choi, A.M. Phosphatidylinositol 3-kinase/Akt pathway mediates heme oxygenase-1 regulation by lipopolysaccharide. *Cell Mol. Biol. (Noisy-le-Grand)* **2005**, *51*, 461–470.
47. Choi, B.M.; Kim, S.M.; Park, T.K.; Li, G.; Hong, S.J.; Park, R.; Chung, H.T.; Kim, B.R. Piperine protects cisplatin-induced apoptosis via heme oxygenase-1 induction in auditory cells. *J. Nutr. Biochem.* **2007**, *18*, 615–622. [CrossRef]
48. Manandhar, S.; Cho, J.M.; Kim, J.A.; Kensler, T.W.; Kwak, M.K. Induction of Nrf2-regulated genes by 3H-1, 2-dithiole-3-thione through the ERK signaling pathway in murine keratinocytes. *Eur. J. Pharmacol.* **2007**, *577*, 17–27. [CrossRef]
49. Zhang, H.; Liu, H.; Iles, K.E.; Liu, R.M.; Postlethwait, E.M.; Laperche, Y.; Forman, H.J. 4-Hydroxynonenal induces rat gamma-glutamyl transpeptidase through mitogen-activated protein kinase-mediated electrophile response element/nuclear factor erythroid 2-related factor 2 signaling. *Am. J. Respir. Cell Mol. Biol.* **2006**, *34*, 174–181. [CrossRef]
50. Lim, H.J.; Lee, K.S.; Lee, S.; Park, J.H.; Choi, H.E.; Go, S.H.; Kwak, H.J.; Park, H.Y. 15d-PGJ2 stimulates HO-1 expression through p38 MAP kinase and Nrf-2 pathway in rat vascular smooth muscle cells. *Toxicol. Appl. Pharmacol.* **2007**, *223*, 20–27. [CrossRef]
51. Lin, A.H.; Chen, H.W.; Liu, C.T.; Tsai, C.W.; Lii, C.K. Activation of Nrf2 is required for up-regulation of the π class of glutathione S-transferase in rat primary hepatocytes with L-methionine starvation. *J. Agric. Food Chem.* **2012**, *60*, 6537–6545. [CrossRef]
52. Balogun, E.; Hoque, M.; Gong, P.; Killeen, E.; Green, C.J.; Foresti, R.; Alam, J.; Motterlini, R. Curcumin activates the haem oxygenase-1 gene via regulation of Nrf2 and the antioxidant-responsive element. *Biochem. J.* **2003**, *371*, 887–895. [CrossRef] [PubMed]
53. Buckley, B.J.; Marshall, Z.M.; Whorton, A.R. Nitric oxide stimulates Nrf2 nuclear translocation in vascular endothelium. *Biochem. Biophys. Res. Commun.* **2003**, *307*, 973–979. [CrossRef] [PubMed]
54. Aburaya, M.; Tanaka, K.; Hoshino, T.; Tsutsumi, S.; Suzuki, K.; Makise, M.; Akagi, R.; Mizushima, T. Heme oxygenase-1 protects gastric mucosal cells against non-steroidal anti-inflammatory drugs. *J. Biol. Chem.* **2006**, *281*, 33422–33432. [CrossRef]
55. Liu, B.; Hannun, Y.A. Inhibition of the neutral magnesium-dependent sphingomyelinase by glutathione. *J. Biol. Chem.* **1997**, *272*, 16281–16287. [CrossRef] [PubMed]
56. Yoshimura, S.; Banno, Y.; Nakashima, S.; Hayashi, K.; Yamakawa, H.; Sawada, M.; Sakai, N.; Nozawa, Y. Inhibition of neutral sphingomyelinase activation and ceramide formation by glutathione in hypoxic PC12 cell death. *J. Neurochem.* **1999**, *73*, 675–683. [CrossRef]
57. Chatterjee, S.; Han, H.; Rollins, S.; Cleveland, T. Molecular cloning, characterization, and expression of a novel human neutral sphingomyelinase. *J. Biol. Chem.* **1999**, *274*, 37407–37412. [CrossRef]
58. Bernardo, K.; Krut, O.; Wiegmann, K.; Kreder, D.; Micheli, M.; Schäfer, R.; Sickman, A.; Schmidt, W.E.; Schröder, J.M.; Meyer, H.E.; et al. Purification and characterization of a magnesium-dependent neutral sphingomyelinase from bovine brain. *J. Biol. Chem.* **2000**, *275*, 7641–7647. [CrossRef]
59. Lavrentiadou, S.N.; Chan, C.; Kawcak, T.; Ravid, T.; Tsaba, A.; van der Vliet, A.; Rasooly, R.; Goldkorn, T. Ceramide-mediated apoptosis in lung epithelial cells is regulated by glutathione. *Am. J. Respir. Cell Mol. Biol.* **2001**, *25*, 676–684. [CrossRef]
60. Ishii, T.; Warabi, E.; Mann, G.E. Circadian control of BDNF-mediated Nrf2 activation in astrocytes protects dopaminergic neurons from ferroptosis. *Free Radic. Biol. Med.* **2019**, *133*, 169–178. [CrossRef]
61. Ishii, T.; Warabi, E. Mechanism of Rapid Nuclear Factor-E2-Related Factor 2 (Nrf2) Activation via Membrane-Associated Estrogen Receptors: Roles of NADPH Oxidase 1, Neutral Sphingomyelinase 2 and Epidermal Growth Factor Receptor (EGFR). *Antioxidants* **2019**, *8*, 69. [CrossRef]
62. Ishii, T.; Warabi, E.; Mann, G.E. Mechanisms underlying unidirectional laminar shear stress-mediated Nrf2 activation in endothelial cells: Amplification of low shear stress signaling by primary cilia. *Redox Biol.* **2021**, *46*, 102103. [CrossRef] [PubMed]
63. Levy, M.; Castillo, S.S.; Goldkorn, T. nSMase2 activation and trafficking are modulated by oxidative stress to induce apoptosis. *Biochem. Biophys. Res. Commun.* **2006**, *344*, 900–905. [CrossRef] [PubMed]
64. Clarke, C.J.; Truong, T.G.; Hannun, Y.A. Role for neutral sphingomyelinase-2 in tumor necrosis factor alpha-stimulated expression of vascular cell adhesion molecule-1 (VCAM) and intercellular adhesion molecule-1 (ICAM) in lung epithelial cells: p38 MAPK is an upstream regulator of nSMase2. *J. Biol. Chem.* **2007**, *282*, 1384–1396. [CrossRef]
65. Liang, C.; Shi, S.; Liu, M.; Qin, Y.; Meng, Q.; Hua, J.; Ji, S.; Zhang, Y.; Yang, J.; Xu, J.; et al. PIN1 Maintains Redox Balance via the c-Myc/NRF2 Axis to Counteract Kras-Induced Mitochondrial Respiratory Injury in Pancreatic Cancer Cells. *Cancer Res.* **2019**, *79*, 133–145. [CrossRef]
66. Saeidi, S.; Kim, S.J.; Han, H.J.; Kim, S.H.; Zheng, J.; Lee, H.B.; Han, W.; Noh, D.Y.; Na, H.K.; Surh, Y.J. H-Ras induces Nrf2-Pin1 interaction: Implications for breast cancer progression. *Toxicol. Appl. Pharmacol.* **2020**, *402*, 115121. [CrossRef] [PubMed]
67. Saeidi, S.; Kim, S.J.; Guillen-Quispe, Y.N.; Jagadeesh, A.S.V.; Han, H.J.; Kim, S.H.; Zhong, X.; Piao, J.Y.; Kim, S.J.; Jeong, J.; et al. Peptidylprolyl cis-trans isomerase NIMA-interacting 1 directly binds and stabilizes Nrf2 in breast cancer. *FASEB J.* **2022**, *36*, e22068. [CrossRef]

68. Göthel, S.F.; Marahiel, M.A. Peptidylprolyl cis-trans isomerases, a superfamily of ubiquitous folding catalysts. *Cell Mol. Life Sci.* **1999**, *55*, 423–436. [CrossRef]
69. Ratajczak, T.; Cluning, C.; Ward, B.K. Steroid Receptor-Associated Immunophilins: A Gateway to Steroid Signalling. *Clin. Biochem Rev.* **2015**, *36*, 31–52.
70. Rostam, M.A.; Piva, T.J.; Rezaei, H.B.; Kamato, D.; Little, P.J.; Zheng, W.; Osman, N. Peptidylprolyl isomerases: Functionality and potential therapeutic targets in cardiovascular disease. *Clin. Exp. Pharmacol. Physiol.* **2015**, *42*, 117–124. [CrossRef]
71. Galigniana, M.D.; Radanyi, C.; Renoir, J.M.; Housley, P.R.; Pratt, W.B. Evidence that the peptidylprolyl isomerase domain of the hsp90-binding immunophilin FKBP52 is involved in both dynein interaction and glucocorticoid receptor movement to the nucleus. *J. Biol. Chem.* **2001**, *276*, 14884–14889. [CrossRef] [PubMed]
72. Galigniana, M.D.; Harrell, J.M.; Murphy, P.J.; Chinkers, M.; Radanyi, C.; Renoir, J.M.; Zhang, M.; Pratt, W.B. Binding of hsp90-associated immunophilins to cytoplasmic dynein: Direct binding and in vivo evidence that the peptidylprolyl isomerase domain is a dynein interaction domain. *Biochemistry* **2002**, *41*, 13602–13610. [CrossRef] [PubMed]
73. Pearl, L.H.; Prodromou, C. Structure and mechanism of the Hsp90 molecular chaperone machinery. *Annu Rev. Biochem.* **2006**, *75*, 271–294. [CrossRef]
74. Li, J.; Soroka, J.; Buchner, J. The Hsp90 chaperone machinery: Conformational dynamics and regulation by co-chaperones. *Biochim. Biophys. Acta* **2012**, *1823*, 624–635. [CrossRef] [PubMed]
75. Schopf, F.H.; Biebl, M.M.; Buchner, J. The HSP90 chaperone machinery. *Nat. Rev. Mol. Cell Biol.* **2017**, *18*, 345–360. [CrossRef] [PubMed]
76. Pratt, W.B.; Toft, D.O. Regulation of signaling protein function and trafficking by the hsp90/hsp70-based chaperone machinery. *Exp. Biol Med. (Maywood)* **2003**, *228*, 111–133. [CrossRef]
77. Pratt, W.B.; Galigniana, M.D.; Harrell, J.M.; DeFranco, D.B. Role of hsp90 and the hsp90-binding immunophilins in signalling protein movement. *Cell Signal.* **2004**, *16*, 857–872. [CrossRef]
78. Galigniana, M.D.; Echeverría, P.C.; Erlejman, A.G.; Piwien-Pilipuk, G. Role of molecular chaperones and TPR-domain proteins in the cytoplasmic transport of steroid receptors and their passage through the nuclear pore. *Nucleus* **2010**, *1*, 299–308. [CrossRef]
79. Chan, S.C.; Li, Y.; Dehm, S.M. Androgen receptor splice variants activate androgen receptor target genes and support aberrant prostate cancer cell growth independent of canonical androgen receptor nuclear localization signal. *J. Biol. Chem.* **2012**, *287*, 19736–19749. [CrossRef]
80. Lee, J.; An, Y.S.; Kim, M.R.; Kim, Y.A.; Lee, J.K.; Hwang, C.S.; Chung, E.; Park, I.C.; Yi, J.Y. Heat Shock Protein 90 Regulates Subcellular Localization of Smads in Mv1Lu Cells. *J. Cell Biochem.* **2016**, *117*, 230–238. [CrossRef] [PubMed]
81. Mazaira, G.I.; Piwien Pilipuk, G.; Galigniana, M.D. Corticosteroid receptors as a model for the Hsp90•immunophilin-based transport machinery. *Trends Endocrinol Metab.* **2021**, *32*, 827–838. [CrossRef] [PubMed]
82. Li, J.; Buchner, J. Structure, function and regulation of the hsp90 machinery. *Biomed. J.* **2013**, *36*, 106–117. [CrossRef] [PubMed]
83. Pearl, L.H.; Prodromou, C.; Workman, P. The Hsp90 molecular chaperone: An open and shut case for treatment. *Biochem. J.* **2008**, *410*, 439–453. [CrossRef]
84. Synoradzki, K.; Miszta, P.; Kazlauskas, E.; Mickevičiūtė, A.; Michailovienė, V.; Matulis, D.; Filipek, S.; Bieganowski, P. Interaction of the middle domains stabilizes Hsp90α dimer in a closed conformation with high affinity for p23. *Biol. Chem.* **2018**, *399*, 337–345. [CrossRef]
85. Taipale, M.; Tucker, G.; Peng, J.; Krykbaeva, I.; Lin, Z.Y.; Larsen, B.; Choi, H.; Berger, B.; Gingras, A.C.; Lindquist, S. A quantitative chaperone interaction network reveals the architecture of cellular protein homeostasis pathways. *Cell* **2014**, *158*, 434–448. [CrossRef]
86. Rehn, A.B.; Buchner, J. p23 and Aha1. *Subcell Biochem.* **2015**, *78*, 113–131. [CrossRef] [PubMed]
87. Hildenbrand, Z.L.; Molugu, S.K.; Herrera, N.; Ramirez, C.; Xiao, C.; Bernal, R.A. Hsp90 can accommodate the simultaneous binding of the FKBP52 and HOP proteins. *Oncotarget* **2011**, *2*, 43–58. [CrossRef] [PubMed]
88. Li, W.; Miao, X.; Qi, Z.; Zeng, W.; Liang, J.; Liang, Z. Hepatitis B virus X protein upregulates HSP90alpha expression via activation of c-Myc in human hepatocarcinoma cell line, HepG2. *Virol. J.* **2010**, *7*, 45. [CrossRef]
89. Nemoto, T.; Sato, N. Oligomeric forms of the 90-kDa heat shock protein. *Biochem. J.* **1998**, *330 Pt 2*, 989–995. [CrossRef]
90. Kobayakawa, T.; Yamada, S.; Mizuno, A.; Nemoto, T.K. Substitution of only two residues of human Hsp90alpha causes impeded dimerization of Hsp90beta. *Cell Stress Chaperones.* **2008**, *13*, 97–104. [CrossRef]
91. Ngo, V.; Brickenden, A.; Liu, H.; Yeung, C.; Choy, W.Y.; Duennwald, M.L. A novel yeast model detects Nrf2 and Keap1 interactions with Hsp90. *Dis. Model Mech.* **2022**, *15*, dmm.049258. [CrossRef]
92. Jia, Z.; Dong, A.; Che, H.; Zhang, Y. 17-DMAG Protects Against Hypoxia-/Reoxygenation-Induced Cell Injury in HT22 Cells Through Akt/Nrf2/HO-1 Pathway. *DNA Cell Biol.* **2017**, *36*, 95–102. [CrossRef]
93. Lazaro, I.; Oguiza, A.; Recio, C.; Lopez-Sanz, L.; Bernal, S.; Egido, J.; Gomez-Guerrero, C. Interplay between HSP90 and Nrf2 pathways in diabetes-associated atherosclerosis. *Clin. Investig. Arterioscler.* **2017**, *29*, 51–59. [CrossRef]
94. Baird, L.; Suzuki, T.; Takahashi, Y.; Hishinuma, E.; Saigusa, D.; Yamamoto, M. Geldanamycin-Derived HSP90 Inhibitors Are Synthetic Lethal with NRF2. *Mol. Cell Biol.* **2020**, *40*, e00377-20. [CrossRef]
95. Yang, W.; Lu, Z. Nuclear PKM2 regulates the Warburg effect. *Cell Cycle* **2013**, *12*, 3154–3158. [CrossRef]

96. Vander Heiden, M.G.; Locasale, J.W.; Swanson, K.D.; Sharfi, H.; Heffron, G.J.; Amador-Noguez, D.; Christofk, H.R.; Wagner, G.; Rabinowitz, J.D.; Asara, J.M.; et al. Evidence for an alternative glycolytic pathway in rapidly proliferating cells. *Science* **2010**, *329*, 1492–1499. [CrossRef] [PubMed]

97. Chen, M.; Zhang, J.; Manley, J.L. Turning on a fuel switch of cancer: hnRNP proteins regulate alternative splicing of pyruvate kinase mRNA. *Cancer Res.* **2010**, *70*, 8977–8980. [CrossRef] [PubMed]

98. Bayley, J.P.; Devilee, P. The Warburg effect in 2012. *Curr. Opin. Oncol.* **2012**, *24*, 62–67. [CrossRef]

99. Tamada, M.; Suematsu, M.; Saya, H. Pyruvate kinase M2: Multiple faces for conferring benefits on cancer cells. *Clin. Cancer Res.* **2012**, *18*, 5554–5561. [CrossRef] [PubMed]

100. Filipp, F.V. Cancer metabolism meets systems biology: Pyruvate kinase isoform PKM2 is a metabolic master regulator. *J. Carcinog.* **2013**, *12*, 14. [CrossRef] [PubMed]

101. Wong, N.; Ojo, D.; Yan, J.; Tang, D. PKM2 contributes to cancer metabolism. *Cancer Lett.* **2015**, *356*, 184–191. [CrossRef] [PubMed]

102. Yang, W.; Zheng, Y.; Xia, Y.; Ji, H.; Chen, X.; Guo, F.; Lyssiotis, C.A.; Aldape, K.; Cantley, L.C.; Lu, Z. ERK1/2-dependent phosphorylation and nuclear translocation of PKM2 promotes the Warburg effect. *Nat. Cell Biol.* **2012**, *14*, 1295–1304, Erratum in *Nat. Cell Biol.* **2013**, *15*, 124. [CrossRef] [PubMed]

103. Schofield, C.J.; Ratcliffe, P.J. Oxygen sensing by HIF hydroxylases. *Nat. Rev. Mol. Cell Biol.* **2004**, *5*, 343–354. [CrossRef]

104. Kaelin, W.G. Proline hydroxylation and gene expression. *Annu. Rev. Biochem.* **2005**, *74*, 115–128. [CrossRef] [PubMed]

105. Semenza, G.L. Hypoxia-inducible factor 1 (HIF-1) pathway. *Sci STKE.* **2007**, *2007*, cm8. [CrossRef]

106. Keeley, T.P.; Mann, G.E. Defining Physiological Normoxia for Improved Translation of Cell Physiology to Animal Models and Humans. *Physiol Rev.* **2019**, *99*, 161–234. [CrossRef]

107. Kallio, P.J.; Okamoto, K.; O'Brien, S.; Carrero, P.; Makino, Y.; Tanaka, H.; Poellinger, L. Signal transduction in hypoxic cells: Inducible nuclear translocation and recruitment of the CBP/p300 coactivator by the hypoxia-inducible factor-1alpha. *EMBO J.* **1998**, *17*, 6573–6586. [CrossRef]

108. Jalouli, M.; Déry, M.A.; Lafleur, V.N.; Lamalice, L.; Zhou, X.Z.; Lu, K.P.; Richard, D.E. The prolyl isomerase Pin1 regulates hypoxia-inducible transcription factor (HIF) activity. *Cell Signal.* **2014**, *26*, 1649–1656. [CrossRef]

109. Han, H.J.; Saeidi, S.; Kim, S.J.; Piao, J.Y.; Lim, S.; Guillen-Quispe, Y.N.; Choi, B.Y.; Surh, Y.J. Alternative regulation of HIF-1α stability through Phosphorylation on Ser451. *Biochem. Biophys. Res. Commun.* **2021**, *545*, 150–156. [CrossRef]

110. Tang, X.; Chang, C.; Hao, M.; Chen, M.; Woodley, D.T.; Schönthal, A.H.; Li, W. Heat shock protein-90alpha (Hsp90α) stabilizes hypoxia-inducible factor-1α (HIF-1α) in support of spermatogenesis and tumorigenesis. *Cancer Gene Ther.* **2021**, *28*, 1058–1070, Erratum in *Cancer Gene Ther.* **2021**, *28*, 1071–1072. [CrossRef]

111. Depping, R.; Steinhoff, A.; Schindler, S.G.; Friedrich, B.; Fagerlund, R.; Metzen, E.; Hartmann, E.; Köhler, M. Nuclear translocation of hypoxia-inducible factors (HIFs): Involvement of the classical importin alpha/beta pathway. *Biochim. Biophys. Acta* **2008**, *1783*, 394–404. [CrossRef] [PubMed]

112. Chachami, G.; Paraskeva, E.; Mingot, J.M.; Braliou, G.G.; Görlich, D.; Simos, G. Transport of hypoxia-inducible factor HIF-1alpha into the nucleus involves importins 4 and 7. *Biochem. Biophys. Res. Commun.* **2009**, *390*, 235–240. [CrossRef] [PubMed]

113. Depping, R.; Jelkmann, W.; Kosyna, F.K. Nuclear-cytoplasmatic shuttling of proteins in control of cellular oxygen sensing. *J. Mol. Med.* **2015**, *93*, 599–608. [CrossRef]

114. Mylonis, I.; Chachami, G.; Paraskeva, E.; Simos, G. Atypical CRM1-dependent nuclear export signal mediates regulation of hypoxia-inducible factor-1alpha by MAPK. *J. Biol. Chem.* **2008**, *283*, 27620–27627. [CrossRef]

115. Jimenez-Blasco, D.; Santofimia-Castaño, P.; Gonzalez, A.; Almeida, A.; Bolaños, J.P. Astrocyte NMDA receptors' activity sustains neuronal survival through a Cdk5-Nrf2 pathway. *Cell Death Differ.* **2015**, *22*, 1877–1889. [CrossRef] [PubMed]

116. Lee, J.M.; Hanson, J.M.; Chu, W.A.; Johnson, J.A. Phosphatidylinositol 3-kinase, not extracellular signal-regulated kinase, regulates activation of the antioxidant-responsive element in IMR-32 human neuroblastoma cells. *J. Biol. Chem.* **2001**, *276*, 20011–20016. [CrossRef]

117. Strocchi, P.; Pession, A.; Dozza, B. Up-regulation of cDK5/p35 by oxidative stress in human neuroblastoma IMR-32 cells. *J. Cell Biochem.* **2003**, *88*, 758–765. [CrossRef]

118. Sang, Y.; Li, Y.; Zhang, Y.; Alvarez, A.A.; Yu, B.; Zhang, W.; Hu, B.; Cheng, S.Y.; Feng, H. CDK5-dependent phosphorylation and nuclear translocation of TRIM59 promotes macroH2A1 ubiquitination and tumorigenicity. *Nat. Commun.* **2019**, *10*, 4013. [CrossRef]

119. Pozo, K.; Bibb, J.A. The Emerging Role of Cdk5 in Cancer. *Trends Cancer.* **2016**, *2*, 606–618. [CrossRef] [PubMed]

120. Liu, W.; Li, J.; Song, Y.S.; Li, Y.; Jia, Y.H.; Zhao, H.D. Cdk5 links with DNA damage response and cancer. *Mol. Cancer.* **2017**, *16*, 60. [CrossRef]

121. Oner, M.; Lin, E.; Chen, M.C.; Hsu, F.N.; Shazzad Hossain Prince, G.M.; Chiu, K.Y.; Teng, C.J.; Yang, T.Y.; Wang, H.Y.; Yue, C.H.; et al. Future Aspects of CDK5 in Prostate Cancer: From Pathogenesis to Therapeutic Implications. *Int. J. Mol. Sci.* **2019**, *20*, 3881. [CrossRef]

122. Do, P.A.; Lee, C.H. The Role of CDK5 in Tumours and Tumour Microenvironments. *Cancers* **2020**, *13*, 101. [CrossRef] [PubMed]

123. Sharma, S.; Sicinski, P. A kinase of many talents: Non-neuronal functions of CDK5 in development and disease. *Open Biol.* **2020**, *10*, 190287. [CrossRef] [PubMed]

124. Vartholomaiou, E.; Madon-Simon, M.; Hagmann, S.; Mühlebach, G.; Wurst, W.; Floss, T.; Picard, D. Cytosolic Hsp90α and its mitochondrial isoform Trap1 are differentially required in a breast cancer model. *Oncotarget* **2017**, *8*, 17428–17442. [CrossRef] [PubMed]

125. Barginear, M.F.; Van Poznak, C.; Rosen, N.; Modi, S.; Hudis, C.A.; Budman, D.R. The heat shock protein 90 chaperone complex: An evolving therapeutic target. *Curr. Cancer Drug Targets* **2008**, *8*, 522–532. [CrossRef]

126. Barrott, J.J.; Haystead, T.A. Hsp90, an unlikely ally in the war on cancer. *FEBS J.* **2013**, *280*, 1381–1396. [CrossRef] [PubMed]

127. Pang, R.; Yuen, J.; Yuen, M.F.; Lai, C.L.; Lee, T.K.; Man, K.; Poon, R.T.; Fan, S.T.; Wong, C.M.; Ng, I.O.; et al. PIN1 overexpression and beta-catenin gene mutations are distinct oncogenic events in human hepatocellular carcinoma. *Oncogene* **2004**, *23*, 4182–4186. [CrossRef]

128. Fan, G.; Wang, L.; Xu, J.; Jiang, P.; Wang, W.; Huang, Y.; Lv, M.; Liu, S. Knockdown of the prolyl isomerase Pin1 inhibits Hep-2 cell growth, migration, and invasion by targeting the β-catenin signaling pathway. *Biochem. Cell Biol.* **2018**, *96*, 734–741. [CrossRef]

129. Wang, T.; Liu, Z.; Shi, F.; Wang, J. Pin1 modulates chemo-resistance by up-regulating FoxM1 and the involvements of Wnt/β-catenin signaling pathway in cervical cancer. *Mol. Cell Biochem.* **2016**, *413*, 179–187. [CrossRef]

130. Xu, G.G.; Etzkorn, F.A. Pin1 as an anticancer drug target. *Drug News Perspect.* **2009**, *22*, 399–407. [CrossRef]

131. Zhou, X.Z.; Lu, K.P. The isomerase PIN1 controls numerous cancer-driving pathways and is a unique drug target. *Nat. Rev. Cancer* **2016**, *16*, 463–478. [CrossRef] [PubMed]

132. Wu, W.; Xue, X.; Chen, Y.; Zheng, N.; Wang, J. Targeting prolyl isomerase Pin1 as a promising strategy to overcome resistance to cancer therapies. *Pharmacol. Res.* **2022**, *184*, 106456. [CrossRef] [PubMed]

133. Wei, S.; Kozono, S.; Kats, L.; Nechama, M.; Li, W.; Guarnerio, J.; Luo, M.; You, M.H.; Yao, Y.; Kondo, A.; et al. Active Pin1 is a key target of all-trans retinoic acid in acute promyelocytic leukemia and breast cancer. *Nat. Med.* **2015**, *21*, 457–466. [CrossRef] [PubMed]

134. Huang, S.; Chen, Y.; Liang, Z.M.; Li, N.N.; Liu, Y.; Zhu, Y.; Liao, D.; Zhou, X.Z.; Lu, K.P.; Yao, Y.; et al. Targeting Pin1 by All-Trans Retinoic Acid (ATRA) Overcomes Tamoxifen Resistance in Breast Cancer via Multifactorial Mechanisms. *Front. Cell Dev. Biol.* **2019**, *7*, 322. [CrossRef] [PubMed]

135. Dubiella, C.; Pinch, B.J.; Koikawa, K.; Zaidman, D.; Poon, E.; Manz, T.D.; Nabet, B.; He, S.; Resnick, E.; Rogel, A.; et al. Sulfopin is a covalent inhibitor of Pin1 that blocks Myc-driven tumors in vivo. *Nat. Chem Biol.* **2021**, *17*, 954–963. [CrossRef]

136. Liu, J.; Wang, Y.; Mu, C.; Li, M.; Li, K.; Li, S.; Wu, W.; Du, L.; Zhang, X.; Li, C.; et al. Pancreatic tumor eradication via selective Pin1 inhibition in cancer-associated fibroblasts and T lymphocytes engagement. *Nat. Commun.* **2022**, *13*, 4308. [CrossRef]

137. Kim, S.E.; Lee, M.Y.; Lim, S.C.; Hien, T.T.; Kim, J.W.; Ahn, S.G.; Yoon, J.H.; Kim, S.K.; Choi, H.S.; Kang, K.W. Role of Pin1 in neointima formation: Down-regulation of Nrf2-dependent heme oxygenase-1 expression by Pin1. *Free Radic. Biol. Med.* **2010**, *48*, 1644–1653. [CrossRef]

138. Alam, J.; Cook, J.L. How many transcription factors does it take to turn on the heme oxygenase-1 gene? *Am. J. Respir. Cell Mol. Biol.* **2007**, *36*, 166–174. [CrossRef]

139. Mouawad, C.A.; Mrad, M.F.; Al-Hariri, M.; Soussi, H.; Hamade, E.; Alam, J.; Habib, A. Role of nitric oxide and CCAAT/enhancer-binding protein transcription factor in statin-dependent induction of heme oxygenase-1 in mouse macrophages. *PLoS ONE.* **2013**, *8*, e64092. [CrossRef]

140. Murray, I.A.; Patterson, A.D.; Perdew, G.H. Aryl hydrocarbon receptor ligands in cancer: Friend and foe. *Nat. Rev. Cancer* **2014**, *14*, 801–814. [CrossRef]

141. Pollet, M.; Krutmann, J.; Haarmann-Stemmann, T. Commentary: Usage of Mitogen-Activated Protein Kinase Small Molecule Inhibitors: More Than Just Inhibition! *Front. Pharmacol.* **2018**, *9*, 935. [CrossRef]

142. Korashy, H.M.; Anwar-Mohamed, A.; Soshilov, A.A.; Denison, M.S.; El-Kadi, A.O. The p38 MAPK inhibitor SB203580 induces cytochrome P450 1A1 gene expression in murine and human hepatoma cell lines through ligand-dependent aryl hydrocarbon receptor activation. *Chem Res. Toxicol.* **2011**, *24*, 1540–1548. [CrossRef]

143. Yu, R.; Mandlekar, S.; Lei, W.; Fahl, W.E.; Tan, T.H.; Kong, A.N. p38 mitogen-activated protein kinase negatively regulates the induction of phase II drug-metabolizing enzymes that detoxify carcinogens. *J. Biol. Chem.* **2000**, *275*, 2322–2327. [CrossRef]

144. Nioi, P.; Hayes, J.D. Contribution of NAD(P)H:quinone oxidoreductase 1 to protection against carcinogenesis, and regulation of its gene by the Nrf2 basic-region leucine zipper and the arylhydrocarbon receptor basic helix-loop-helix transcription factors. *Mutat Res.* **2004**, *555*, 149–171. [CrossRef]

145. Miao, W.; Hu, L.; Scrivens, P.J.; Batist, G. Transcriptional regulation of NF-E2 p45-related factor (NRF2) expression by the aryl hydrocarbon receptor-xenobiotic response element signaling pathway: Direct cross-talk between phase I and II drug-metabolizing enzymes. *J. Biol. Chem.* **2005**, *280*, 20340–20348. [CrossRef]

146. Gharavi, N.; El-Kadi, A.O. tert-Butylhydroquinone is a novel aryl hydrocarbon receptor ligand. *Drug Metab. Dispos.* **2005**, *33*, 365–372. [CrossRef]

147. Köhle, C.; Bock, K.W. Coordinate regulation of Phase I and II xenobiotic metabolisms by the Ah receptor and Nrf2. *Biochem Pharmacol.* **2007**, *73*, 1853–1862. [CrossRef] [PubMed]

148. Mandelkow, E.M.; Biernat, J.; Drewes, G.; Gustke, N.; Trinczek, B.; Mandelkow, E. Tau domains, phosphorylation, and interactions with microtubules. *Neurobiol. Aging.* **1995**, *16*, 355–362; discussion 362–363. [CrossRef]

149. Johnson, G.V.; Stoothoff, W.H. Tau phosphorylation in neuronal cell function and dysfunction. *J. Cell Sci.* **2004**, *117*, 5721–5729. [CrossRef] [PubMed]

150. Dehmelt, L.; Halpain, S. The MAP2/Tau family of microtubule-associated proteins. *Genome Biol.* **2005**, *6*, 204. [CrossRef] [PubMed]
151. Wagner, U.; Utton, M.; Gallo, J.M.; Miller, C.C. Cellular phosphorylation of tau by GSK-3 beta influences tau binding to microtubules and microtubule organisation. *J. Cell Sci.* **1996**, *109 Pt 6*, 1537–1543. [CrossRef] [PubMed]
152. Lovestone, S.; Hartley, C.L.; Pearce, J.; Anderton, B.H. Phosphorylation of tau by glycogen synthase kinase-3 beta in intact mammalian cells: The effects on the organization and stability of microtubules. *Neuroscience* **1996**, *73*, 1145–1157. [CrossRef] [PubMed]
153. Lu, P.J.; Wulf, G.; Zhou, X.Z.; Davies, P.; Lu, K.P. The prolyl isomerase Pin1 restores the function of Alzheimer-associated phosphorylated tau protein. *Nature* **1999**, *399*, 784–788. [CrossRef]
154. Min, S.H.; Cho, J.S.; Oh, J.H.; Shim, S.B.; Hwang, D.Y.; Lee, S.H.; Jee, S.W.; Lim, H.J.; Kim, M.Y.; Sheen, Y.Y.; et al. Tau and GSK3beta dephosphorylations are required for regulating Pin1 phosphorylation. *Neurochem. Res.* **2005**, *30*, 955–961. [CrossRef]
155. Ma, S.L.; Pastorino, L.; Zhou, X.Z.; Lu, K.P. Prolyl isomerase Pin1 promotes amyloid precursor protein (APP) turnover by inhibiting glycogen synthase kinase-3β (GSK3β) activity: Novel mechanism for Pin1 to protect against Alzheimer disease. *J. Biol. Chem.* **2012**, *287*, 6969–6973. [CrossRef]
156. Santacruz, K.; Lewis, J.; Spires, T.; Paulson, J.; Kotilinek, L.; Ingelsson, M.; Guimaraes, A.; DeTure, M.; Ramsden, M.; McGowan, E.; et al. Tau suppression in a neurodegenerative mouse model improves memory function. *Science* **2005**, *309*, 476–481. [CrossRef]
157. Driver, J.A.; Zhou, X.Z.; Lu, K.P. Pin1 dysregulation helps to explain the inverse association between cancer and Alzheimer's disease. *Biochim. Biophys. Acta* **2015**, *1850*, 2069–2076. [CrossRef] [PubMed]
158. Lanni, C.; Masi, M.; Racchi, M.; Govoni, S. Cancer and Alzheimer's disease inverse relationship: An age-associated diverging derailment of shared pathways. *Mol. Psychiatry* **2021**, *26*, 280–295. [CrossRef]

antioxidants

MDPI

Review

Phenotypic Modulation of Cancer-Associated Antioxidant NQO1 Activity by Post-Translational Modifications and the Natural Diversity of the Human Genome

Angel L. Pey

Departamento de Química Física, Unidad de Excelencia de Química Aplicada a Biomedicina y Medioambiente e Instituto de Biotecnología, Facultad de Ciencias, Universidad de Granada, Av. Fuentenueva s/n, 18071 Granada, Spain; angelpey@ugr.es

Abstract: Human NAD(P)H:quinone oxidoreductase 1 (hNQO1) is a multifunctional and antioxidant stress protein whose expression is controlled by the Nrf2 signaling pathway. hNQO1 dysregulation is associated with cancer and neurological disorders. Recent works have shown that its activity is also modulated by different post-translational modifications (PTMs), such as phosphorylation, acetylation and ubiquitination, and these may synergize with naturally-occurring and inactivating polymorphisms and mutations. Herein, I describe recent advances in the study of the effect of PTMs and genetic variations on the structure and function of hNQO1 and their relationship with disease development in different genetic backgrounds, as well as the physiological roles of these modifications. I pay particular attention to the long-range allosteric effects exerted by PTMs and natural variation on the multiple functions of hNQO1.

Keywords: phosphorylation; acetylation; ubiquitination; intracellular degradation; ligand-dependent stability; cancer; neurological disorders

Citation: Pey, A.L. Phenotypic Modulation of Cancer-Associated Antioxidant NQO1 Activity by Post-Translational Modifications and the Natural Diversity of the Human Genome. *Antioxidants* **2023**, *12*, 379. https://doi.org/10.3390/antiox12020379

Academic Editor: Marcel Bonay

Received: 15 December 2022
Revised: 31 January 2023
Accepted: 2 February 2023
Published: 4 February 2023

1. Human NQO1: A Stress-Protein Associated with Disease

Human NAD(P)H:quinone oxidoreductase 1 (UniProt ID: P15559) is a soluble, typically cytosolic, dimeric protein at the hub of the antioxidant defense and stabilization of up to 50 different proteins, including p53 and HIF-1α [1,2]. Although hNQO1 has historically been labeled as a cytosolic enzyme, it is likely found in multiple subcellular locations [2,3]. As an enzyme, it catalyzes the two-electron reduction of a wide range of quinones to hydroquinones using NAD(P)H as a coenzyme (see Table A1 in [2]), displaying negative cooperativity regarding catalysis and FAD binding [4–7] and also detoxifying superoxide radicals [8]. Its many functions have been recently reviewed, and I refer to these excellent reviews for further information [1,8].

Naturally-occurring variants, as well as mimetics of post-translational modifications (PTMs) in NQO1, can lead to protein loss-of-function through different mechanisms [9–13]. In this review, I summarize current advancements in the study of NQO1 functionality depending upon post-translational modifications on its protein sequence, as well as by mutations and polymorphisms found in cancer cell lines and human populations with uncertain associations with disease (in contrast to the very common, largely inactivating and cancer-associated polymorphism P187S) [1,13–15]), which may highly determine NQO1 activity in different individuals.

Recent advances in DNA sequencing technologies are revealing significant genetic variability among human populations [16]. About half of these mutations are of uncertain significance [17], found in only a single individual [18], and are missense mutations (vs. the human consensus genome) [18]. This genetic variability also causes variable responses to pharmacological treatments [19–21]. It is partially accessible from databases such as gnomAD (with information of over 140,000 human genomes/exomes; https://gnomad.

broadinstitute.org/ accessed on 15 December 2022) and more disease-oriented ones such as ClinVar (https://www.ncbi.nlm.nih.gov/clinvar/ accessed on 15 December 2022) or dbGaP (https://www.ncbi.nlm.nih.gov/gap/ accessed on 15 December 2022).

NQO1 is highly inducible upon stress through the Nrf2 or Ah pathways [1,8]. The Nrf2 regulatory pathway mediates the delicate balance between oxidative signaling and antioxidant defense, and it is likely that the NQO1 antioxidant properties and its modulation of reduced/oxidized forms of NAD^+ play important roles [3,8,22–25]. The Nrf-2 pathway is associated with multiple human pathologies, including alcohol-induced liver disease, cigarette smoking, cancer and neurodegeneration [25]. The molecular details and physio-pathological implications of Nrf2 signaling have been extensively reviewed in the past [25]. As is essential for this manuscript, it must be noted that alterations in NQO1 activity are also associated with certain diseases linked, to different extents, with oxidative stress, such as cancer, Alzheimer's disease, Parkinson's disease and atherosclerosis [1,8]. Remarkably, certain genetic variations in NQO1 have been associated with cancer development, possibly due to a loss of activity and stability [1,26–30]. It is plausible that somatic or germline mutations may cause a predisposition to develop diseases such as cancer [30]. However, the roles of natural genetic diversity and post-translational modifications in NQO1, and their relationship with disease, are unclear. The main aim of this manuscript is to provide an update on these issues.

2. NQO1: A Simple Dimer with Complex Behavior

Evolution has likely selected many mammalian proteins as oligomers to allow them to display complex regulatory behaviors. NQO1 seems to be an excellent example. Simply, as a dimer, NQO1 contains two active sites formed upon interaction of the large N-terminal domains (NTD), but requiring the small C-terminal domain (CTD) of the other subunit for efficient catalysis [28,31]. Both active sites, as well as the two domains, communicate in different functional ligation states along its catalytic cycle [4,9]. Detailed titration calorimetry experiments have revealed genuine negative cooperativity for FAD equilibrium binding to the apo-NQO1 protein [5]. In addition, binding of the competitive inhibitor dicoumarol (Dic) in calorimetric and steady-state activity assays also displays, in some instances, a degree of negative binding cooperativity [6,7]. Extensive functional and hydrogen–deuterium exchange (HDX) studies, as well as theoretical calculations, have supported the existence of long-range communication of local perturbations due to single amino acid changes and ligand binding to sites as distant as 30 Å [9,10,12,13,31,32] (Figure 1). This long-range conformational communication likely underlies the cooperative effects described upon ligand binding.

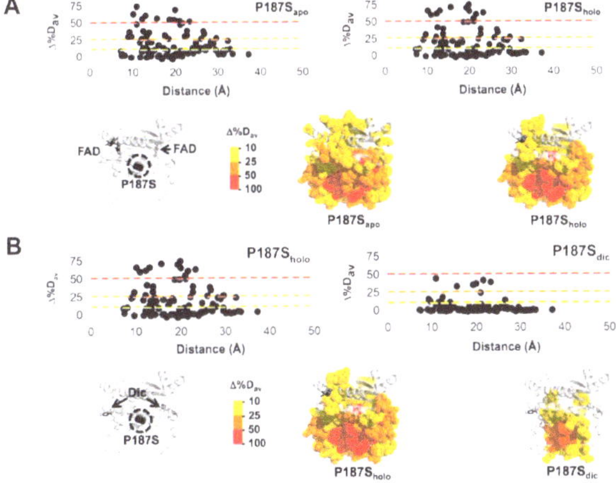

Figure 1. Long-range structural effects due to the cancer-associated P187S polymorphism in different

ligation states (apo, no ligand bound; holo, with FAD bound; dic, with FAD and dicoumarol bound). Data are reproduced from [13]. The figure shows the effect of P187S (vs. the WT protein) in different ligation states (**A**, apo vs. holo; **B**, holo vs. Dic) considering over 100 protein segments by HDX ($\Delta\%D_{av}$, a semiquantitative parameter calculated from the maximal differences in HDX; a positive value indicates a destabilizing effect) and regarding the distance to the mutated P187 residue. All figures were generated using the PDB code 2F1O [33].

3. Post-Translational Modifications (PTMs) in hNQO1

By 11 December 2022, the Phosphosite plus® site (https://www.phosphosite.org/proteinAction.action?id=14721&showAllSites=true accessed on 11 December 2022) contained 12 phosphorylation sites (S13, Y20, S40, Y43, Y68, Y76, S82, Y127, T128, Y129, Y133 and S255), 9 acetylation sites (K31, K59, K61, K77, K90, K209, K210, K251 and K262) and 18 ubiquitination sites (K23, K31, K33, K54, K59, K61, K77, K90, K91, K135, K209, K210, K241, K248, K251, K262 and K271) for hNQO1. As can be seen, these sites are well-spread across the entire protein structure, and in the case of ubiquitination and acetylation, they often overlap (Figure 2). I must note that the functional consequences of only a few of these sites have been characterized in detail. These studies are described and discussed in this section.

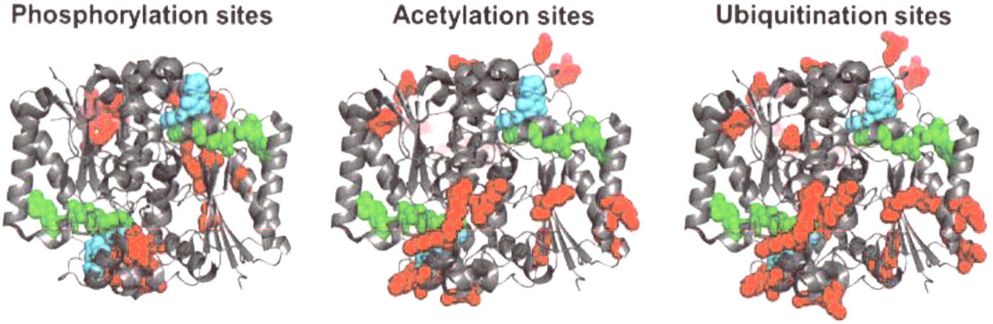

Figure 2. Structural location of phosphorylation, acetylation and ubiquitination sites of hNQO1 based on the data compiled in Phosphosite Plus® [34]. Residues in red show those which are modified based on high-throughput analyses. I show the Dic molecules in cyan and the FAD in green. This figure was made based on the structure with PDB code 2F1O [33].

3.1. Phosphorylation

There are 12 reported phosphorylation sites in hNQO1 (Figure 2). It is interesting to note that most of these sites are not highly solvent-exposed, but are located in regions with moderate to low structural stability, which is highly dependent on ligand binding, based on a recent HDX study (Figure 3) [9]. These analyses suggest that phosphorylation might depend strongly on ligand binding and consequent changes in protein dynamics, or might occur cotranslationally.

The functional consequences of phosphorylation at sites S40, S82 and T128 have been addressed recently by the use of phosphomimetic mutations [10,13,35]. The outcomes of these studies have been very revealing, because the functional effects are widely different depending on the site modified. Phosphorylation at S82 causes the strongest effects, with a remarkable decrease in FAD binding affinity, and its local destabilization extends across the hNQO1 structure, affecting the catalytic cycle and intracellular stability [10,13,35]. The effect of phosphorylation at S82 synergizes with that of the polymorphism P187S, leading to an almost 1000-fold decrease in FAD binding affinity [35]. A network of recently diverged electrostatic interactions in the vicinity of S82 has been shown to explain the different response of human and rat NQO1 to phosphorylation at S82. This is due to a single mutation (R80H) that occurred about 20 million years ago during primate speciation [11,35,36]. Phosphorylation at sites S40 and T128 has milder functional effects, affecting enzyme kinetics and structure much more weakly [10]. Importantly, phosphorylation of T128 by AKT is also associated with enhanced ubiquitination by Parkin and subsequent degradation of hNQO1, supporting the notion that phosphorylation, ubiquitination and intracellular stability of hNQO1 might be intertwined [37]. We are currently characterizing phospho-mimetic mutations at Y127 and Y129, located in the hNQO1 active site, and our preliminary results support that phosphorylation at Y127 may perturb binding of FAD to the active site of hNQO1, whereas Y129 could be implicated in the conformational heterogeneity likely associated with functional negative cooperativity in the holo-protein (Pacheco-García JL, Martín-García JM, Medina M and Pey AL, unpublished observations).

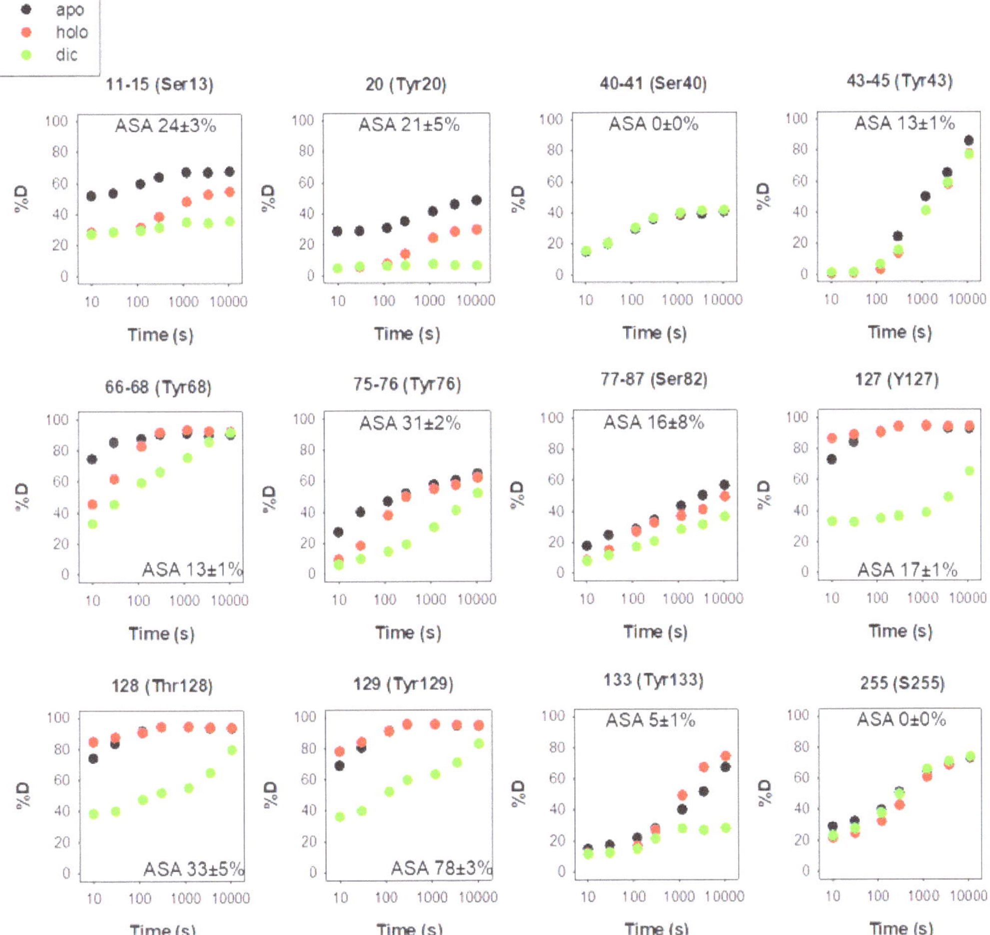

Figure 3. Structural stability of the phosphorylation sites in hNQO1 in different ligation states (apo, no ligand bound; holo, saturated with FAD; dic, saturated with FAD and dicoumarol). Data show the time-dependence (*x*-axis) of the peptide backbone hydrogen–deuterium exchange (%D, *y*-axis) as determined by mass spectrometry (reproduced from [9]). The solvent accessible surface (ASA, %) is indicated in parentheses and calculated using GetArea and the PDB 2F1O [33] as the average ± s.d. from eight monomers.

3.2. Ubiquitination

The intracellular stability of WT, and particularly of the polymorphic P187S hNQO1, is tightly associated to their C-terminal dynamics through ubiquitin-dependent proteasomal degradation of its CTD as an initiation site [15,31,32,38]. While this accelerated degradation of P187S might be due to enhanced dynamics of the CTD of the polymorphic variant P187S [13,31,39], it seems that the apo-state (ligand-free) of hNQO1 is particularly suitable for proteasomal degradation upon ubiquitin tagging even in the WT variant [38]. Thus, the sites K241, K248, K251, K262 and K271 are likely responsible for most ubiquitin-dependent degradation of hNQO1 in cells [38]. A vast majority of the ubiquitination sites are solvent-exposed and found in protein segments with moderate to low structural stability (Figure 4), and thus, are readily accessible for ubiquitin tagging upon interaction with a suitable ubiquitin-ligase (such as CHIP, [24]).

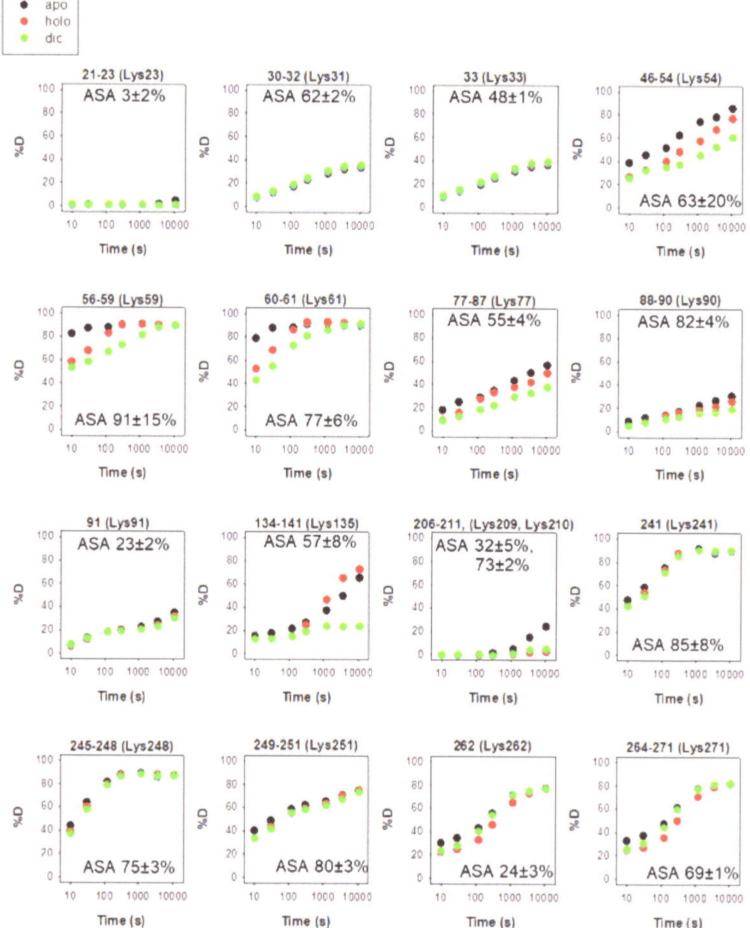

Figure 4. Structural stability of the ubiquitination sites in hNQO1 in different ligation states (apo, no ligand bound; holo, saturated with FAD; dic, saturated with FAD and dicoumarol). Additional details on data representation (reproduced from [9]) can be found in the legend of Figure 2. The solvent accessible surface (ASA, %) is indicated in parentheses and calculated using GetArea and the PDB 2F1O [33] as the average ± s.d. from eight monomers.

3.3. Acetylation

Until recently, acetylation of hNQO1 had been characterized only by high-throughput means, identifying nine sites with generally very high solvent exposure (logically, since Lys residues are often found on the protein surface) and typically in regions with moderate-to-low stability (Figure 5). However, Siegel and coworkers have recently described that acetylation of K33, K59, K61, K77, K90, K209, K251, K262 and K271 readily occurs upon in vitro exposure of recombinant hNQO1 to acetic anhydride or S-acetylglutathione [40]. It is interesting that these authors found helix 7 (residues 65–78) as one of the main targets for hNQO1 acetylation, since this region is next to the phosphosite S82 and the evolutionarily divergent site H80. Since all three phenomena (acetylation, phosphorylation and the variations R80 and H80) involve alterations in the electrostatic network important for FAD binding [11,35,36], I speculate that there could be some crosstalk between them in the modulation of NQO1 functionality in different mammalian species.

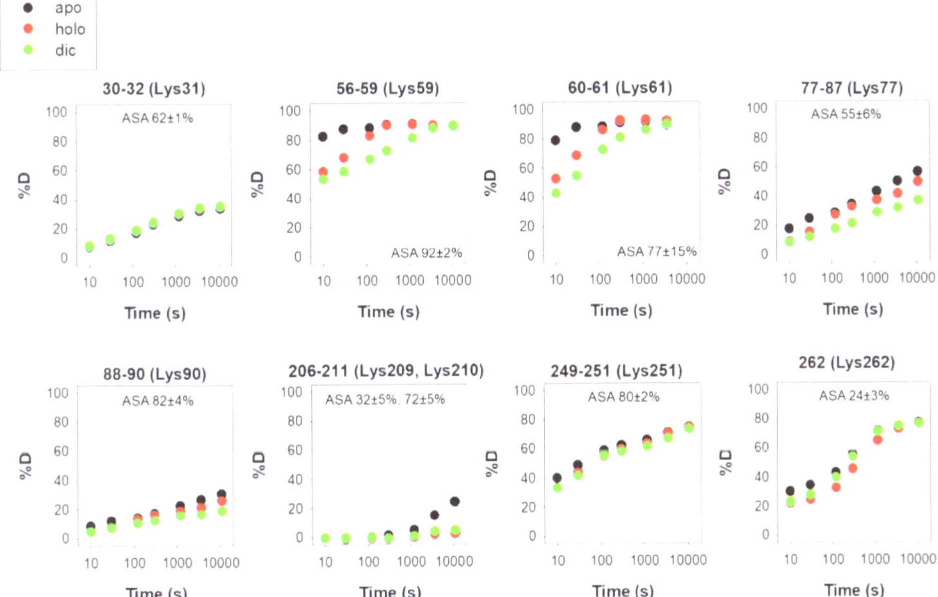

Figure 5. Structural stability of the acetylation sites in hNQO1 in different ligation states (apo, no ligand bound; holo, saturated with FAD; dic, saturated with FAD and dicoumarol). Additional details on data representation (reproduced from [9]) can be found in the legend of Figure 2 [9]. The solvent accessible surface area (ASA, %) is indicated in parenthesis and calculated using GetArea and the PDB 2F1O [33] as the average ± s.d. from eight monomers.

In vitro acetylation of hNQO1 led to a 37% decrease in steady-state catalytic function [40]. Importantly, acetylation was highly susceptible to functional ligand binding (NADH mediated reduction of FAD largely prevented acetylation), and deacetylation at K262 and K271 was quickly catalyzed by different sirtuins in vitro [40]. This phenomenon is likely associated with the NADH-dependent localization of hNQO1 around microtubules [3], and also highlights the plasticity of hNQO1 functionality in different ligation states and different subcellular locations [2,3,13].

Studies from our laboratory using mimetic mutations (at K31 and K209) have shown that the effect of acetylation on hNQO1 activity may be mainly ascribed to the K31 site (see the mutant K31Q in Table 1). This mutant also showed a mild decrease in thermal stability, both as holo- and apo-protein (Table 2). The role of electrostatic interactions in the effects of acetylating K31 was also supported by the greater effects found for the charge-reversal K31E mutant (Tables 1 and 2).

Table 1. Steady-state enzyme kinetic parameters for the reduction in DCPIP by hNQO1 WT and mutants at K31 and K209. Proteins were expressed and purified according to [13]. Activity was measured according to [31] using 20 μM DCPIP and 0–2 mM NADH at 25 °C. Data were fitted to the Michaelis–Menten equation.

NQO1 Variant	k_{cat} (s^{-1})	$K_{M\ (NADH)}$ (mM)	k_{cat}/K_M (s^{-1}·mM^{-1})
WT	50 ± 4	0.54 ± 0.10	91 ± 18
K31Q	36 ± 4	0.44 ± 0.10	82 ± 20
K31E	32 ± 5	0.38 ± 0.08	86 ± 19
K209Q	43 ± 6	0.29 ± 0.10	147 ± 50
K209E	42 ± 5	0.30 ± 0.08	142 ± 40

Table 2. Thermal stability of hNQO1 variants at K31 and K209. Thermal denaturation was carried out by intrinsic fluorescence emission spectroscopy, as described in [36].

NQO1 Variant	Ligation State [1]	T_m (°C) [2]
WT	Holo	55.8 ± 0.5
	Apo	52.0 ± 0.6
K31Q	Holo	54.5 ± 0.1
	Apo	50.5 ± 0.7
K31E	Holo	53.3 ± 0.1
	Apo	49.6 ± 0.6
K209Q	Holo	55.4 ± 0.1
	Apo	51.9 ± 0.3
K209E	Holo	53.5 ± 0.1
	Apo	50.8 ± 0.6

[1] Apo indicates samples in which FAD has been stripped. Holo indicates proteins that were purified in the presence of 20 μM FAD. Protein concentration was 1 μM. [2] Average ± s.d. from four replicates.

Interestingly, these studies also highlight the versatility of certain K sites (such as K31, K59, K61, K209, K210, K251, K262 and K271) that can be modified differently (ubiquitination vs. acetylation) depending on the cellular conditions [38,40].

4. Naturally-Occurring and Artificial Mutations in NQO1: Allosteric Communication of Mutational Effects in a Multifunctional Stress Protein

Historically, two common polymorphisms in hNQO1 (namely rs1800566/c.C609T/p.P187S and rs1131341/c.C465T/p.R139W) have attracted most of the attention due to their frequency and association with cancer susceptibility [6,13,15,28,31,32,36,39,41–44]. These detailed studies have shown the long-range communication of mutational effects in different ligation states, a behavior which has been systematically corroborated by the recent characterization of naturally-occurring and artificial mutations [12,30,45]. We describe the insights provided by these mutational studies and their relationship with allosteric effects on hNQO1 in this section.

4.1. Polymorphic Variants

The c.C609T/p.P187S polymorphism occurs in the human population with a frequency of ~0.25 based on the gnomAD database, and almost reaches a frequency of 0.5 in the East Asian population, with about 5% of human population being homozygotes (https://gnomad.broadinstitute.org/variant/16-69745145-G-A?dataset=gnomad_r2_1 accessed on 15 December 2022). Different studies have supported its association, particularly in homozygosis, with cancer development [8,14,46]. Early studies showed nearly null activity-protein levels in homozygosis and substantially reduced activity in heterozygosis in cells and cancer samples [42,47]. Pro187 is buried in the protein structure, and its change to Ser causes a strong local structural destabilization that propagates through the entire NQO1 structure differently depending on the ligation state (Figure 6). Several functional features of hNQO1 are affected by P187S. First, the affinity for FAD is reduced by 10–40-fold by this polymorphism, depending on the experimental conditions and binding model employed. This is due to a long-range propagation of the variant effect to the FAD binding site in both the apo- and holo-states [5,28,35,36,44,48] (Figure 5). In addition, the effects of P187S also propagate to the CTD, affecting both catalysis and intracellular stability. Binding of the NADH cofactor or the inhibitor Dic are severely reduced, and thermodynamic analyses support that this is due to partial unfolding of the CTD in this variant [13,32,33]. Destabilization of the CTD also causes accelerated degradation of the NQO1 protein through enhanced ubiquitination of the CTD [15,31,32,36,38], and, consequently, CTD thermodynamic stabilization upon Dic binding leads to its intracellular stabilization in cell cultures [13,49]. Gradual structural perturbation at the P187 lead to different effects at different functional sites, supporting that the propagation of the local structural–energetic effects of P187S are anisotropic and long-range [13,32,48]. The large structural destabilization caused by P187S in the apo- and holo-states might also explained its enhanced coaggregation with other proteins in cell cultures under normal and low riboflavin supplies [50].

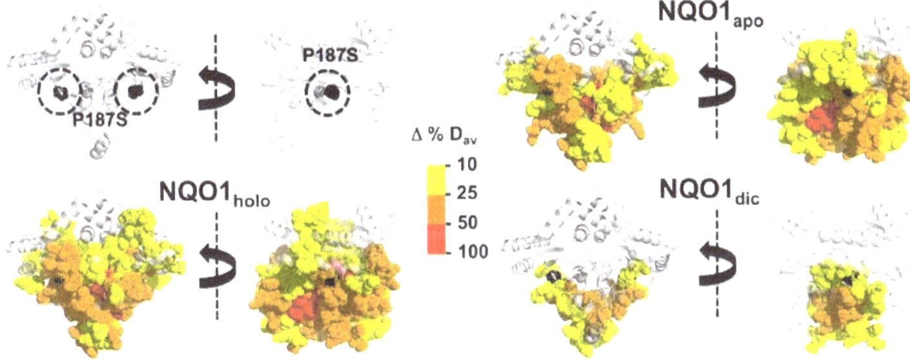

Figure 6. Changes in structural stability of hNQO1 due to the polymorphism P187S in different functional states (NQO1$_{apo}$, no ligand bound; NQO1$_{holo}$, saturated with FAD; NQO1$_{dic}$, saturated with FAD and dicoumarol). Destabilization (as $\Delta\%D_{av}$, compared to the WT in the same ligation state) is represented as indicated by the color scale. Data are taken from [13].

The c.C465T/p.R139W polymorphism is much less frequent, with an overall frequency of ~0.03 that increases up to ~0.06 in South Asian population, and is rarely found in homozygosis (https://gnomad.broadinstitute.org/variant/16-69748869-G-A?dataset=gnomad_r2_1 accessed on 15 December 2022). This polymorphism is associated with decreased intracellular activity and increased cancer risk [41,51,52]. This polymorphism causes dual effects, leading to a missense variation in the hNQO1 sequence (R139W), and in parallel, it massively causes skipping of exon 4 (residues 102–139), which destroys the binding sites of FAD and the substrate, yielding an unstable protein form [41]. Since the effects of the missense variation R139W are quite mild at the protein level [28,44,49], the main loss-of-function mechanism seems to arise from aberrant splicing.

4.2. COSMIC Variants

By 11 December 2022, there were 107 different mutations in hNQO1 compiled in the Catalogue **O**f **S**o**Mat**I**c** Cancer Cell lines catalogue (**COSMIC** database; https://cancer.sanger.ac.uk/cosmic/gene/analysis?ln=NQO1_ENST00000561500 accessed on 11 December 2022), of which 73 were missense mutations. The effects of 10 of these mutations (G3D,

L7P, A29T, D41Y, M45L, M45I, W106C, M155I, H162N and K240Q) have been recently characterized in some detail in vitro [30,31,45,48,53] (Figure 7 and Table 3). This set of mutations mostly target the NTD (Figure 7), and in some cases, affect active site residues (W106C, M155I and H162N).

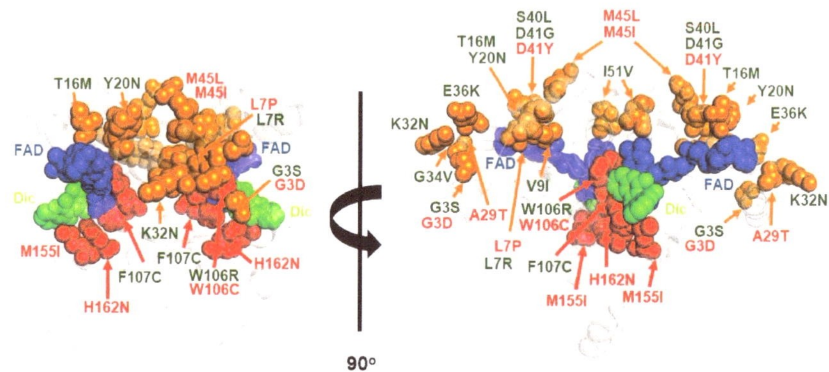

Figure 7. Structural location of COSMIC and gnomAD mutations, characterized experimentally. Variants labeled in red are those from COSMIC and in green are those found in gnomAD. Active site mutants are shown as red spheres. Reproduced from [30].

Table 3. Experimental characterization of the effects of natural variants on NQO1. Mutations were retrieved from COSMIC or gnomAD databases. Three different parameters are reported and normalized using the WT protein: Soluble protein levels upon expression in *E. coli* at 37 °C, change in denaturation temperature (ΔT_m) and change in the dissociation constant for FAD (as the ratio of mutant/WT constants). N.Det. indicates not determined. Original and un-normalized data can be found in [30,45,48].

NQO1 Variant	Soluble Protein Levels (vs. WT)	ΔT_m (vs. WT) (°C)	$K_{d\,(FAD)}$ (WT-Fold)
WT	1.0 ± 0.1	0.0 ± 0.6	1.0 ± 0.2
G3S	2.0 ± 0.3	-0.4 ± 0.8	1.4 ± 0.3
G3D	3.8 ± 1.1	-1.5 ± 0.7	0.9 ± 0.4
L7P	0.5 ± 0.1	N.Det.	N.Det.
L7R	$>>0.1$	N.Det.	N.Det.
V9I	0.8 ± 0.6	-1.7 ± 0.6	1.4 ± 0.3
T16M	0.4 ± 0.2	-4.3 ± 0.7	10.8 ± 0.2
Y20N	0.6 ± 0.1	-5.1 ± 0.6	2.9 ± 0.5
A29T	1.3 ± 0.3	0.1 ± 0.7	5.0 ± 0.4
K32N	0.8 ± 0.2	-0.2 ± 0.6	0.8 ± 0.4
G34V	$>>0.1$	N.Det.	N.Det.
E36K	0.9 ± 0.2	0.0 ± 0.8	0.8 ± 0.5
S40L	$>>0.1$	N.Det.	N.Det.
D41G	$>>0.1$	-7.9 ± 0.5	N.Det.
D41Y	$>>0.1$	-9.7 ± 0.6	N.Det.
M45L	0.4 ± 0.5	-3.7 ± 0.5	0.5 ± 1.0
M45I	0.2 ± 0.4	-3.4 ± 0.6	0.6 ± 0.8
I51V	0.4 ± 0.3	-5.0 ± 0.5	8.9 ± 0.3
W106R	>0.1	-6.4 ± 0.7	~ 500
W106C	0.2 ± 0.5	-2.5 ± 0.7	0.9 ± 0.4
F107C	0.1 ± 0.7	0.6 ± 0.7	~ 0.2
M155I	>0.1	-0.9 ± 0.6	45 ± 1
H162N	0.3 ± 0.9	-0.7 ± 1.0	27 ± 1
K240Q	1.3 ± 0.3	0.5 ± 0.6	2.8 ± 0.6

We must note that the presence of a mutation in a cancer cell line (i.e., COSMIC) does not unambiguously imply that it is a cancer driver mutation (as discussed in [30]). Nevertheless, we found alterations in several NQO1 functional features due to these COSMIC mutations. All mutations, except G3D and K240Q, showed moderate to severe alterations in their foldability (formation of dimers or enhanced aggregation) or in the thermal stability of the folded dimer [30,32,48,53] (Table 3). The mutants M155I and H162N also caused severe defects in FAD binding and catalytic efficiency, whereas the effects of the W106C and K240Q on these functional features were much milder [30,32,48,53] (Figure 7 and Table 3). High resolution stability analyses have shown that impaired FAD binding affinity and catalysis in the M155I and H162N mutants likely stem from extensive and specific destabilization of the structure of the holo-hNQO1 [53].

4.3. gnomAD Variants

By 11 December 2022, there were over 150 variations described in the NQO1 gene in non-disease associated, human population, large-scale sequencing initiatives (gnomAD database; https://gnomad.broadinstitute.org/gene/ENSG00000181019?dataset=gnomad_r2_1 accessed on 11 December 2022), of which 106 are missense variants. 13 of these have been characterized experimentally (G3S, L7R, V9I, T16M, Y20N, K32N, G34V, E36K, S40L, D41G, I51V, W106R, F107C; Figure 7) [30,45]. The study of variants present in the human population is relevant because the presence of germline mutations may cause predisposition to additional somatic mutations and facilitate cancer development [30] (Table 3). Importantly, several of these rare mutations moderately to largely destabilized the protein or reduced NQO1 foldability (L7R, T16M, Y20N, G34V, S40L, D41G, I51V, W106R and F107C). In addition, several mutations affected FAD binding with moderate (~3-fold lower affinity, Y20N; ~10-fold, T16M and I51V) to large effects (~500-fold lower affinity, W106R) [30,53]. The kinetic characterization of the W106R and F107C mutants, two highly non-conservative mutations at the active site of NQO1, provided counterintuitive results; the former had catastrophic consequences on enzyme activity, while the latter had very mild effects [53]. Interestingly, the strongest effect on local stability due to W106R occurred in the NQO1$_{dic}$ state, possibly reflecting heavy effects on catalytic intermediate states [53]. One of the most revealing outcomes of these studies was the finding that variations found in disease-focused databases (i.e., COSMIC) can be as deleterious for NQO1 function as those found in the normal population. (Figure 8).

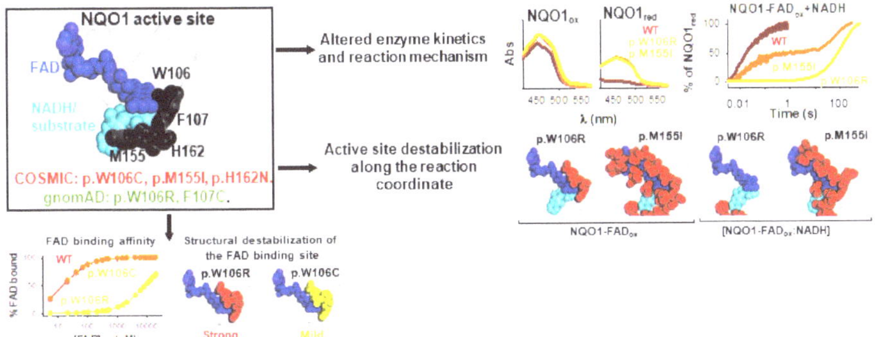

Figure 8. Molecular characterization of active site mutants from COSMIC and gnomAD databases in hNQO1. Experimental analysis include enzyme kinetics and analysis of the reaction mechanism by stopped-flow absorption spectroscopy and effects on the structural stability by HDX- and FAD-binding affinities. Adapted from [53].

4.4. Artificial Variants Aimed at Evaluating the Propagation of Stability Effects and the Structural Basis of Allosterism

Two sets of artificial mutants have recently been generated and characterized in vitro [12,32,48]. The first one was intended to determine whether propagation of mutational effects at site P187 and K240 depended on the magnitude of the local stability effects and single vs. multiple nucleotide changes [32,48]. The latter was intended to perturb the hydrophobic core of hNQO1 at different locations and to different extents [12]. These studies have highlighted the significant plasticity of hNQO1 towards mutational effects in different ligation states.

In the first work, we observed that unnatural variants at the P187 and K240 sites may lead to even more deleterious effects on several hNQO1 functional features than natural variants (i.e., P187S and K240Q), in some cases through a single nucleotide change. Using functional assays, this study provided the first large-scale evidence that propagation of local mutational effects in hNQO1 may occur to distant sites (up to 20 Å) in a general manner, not restricted to the P187S polymorphism [48].

The second work was intended to rationalize these allosteric effects at different sites at the NTD by truncating fully buried hydrophobic residues (L7, L10 and L30) to smaller residues (to V or A) and even residues truncating+increasing conformational entropy (to G) [12] (Figure 9). All three mutations to Gly caused a remarkable decrease in protein foldability, which was extended to any mutation at L30, highlighting the site-dependence of the propagation of mutational effects. Perturbation of the L10 site (mutant L10A) also led to a remarkable ~20-fold decrease in FAD binding affinity, likely caused by the propagation of this structural perturbation by ~10 Å to the FAD binding site [12]. Generally, all of these cavity-making mutations had insignificant catalytic effects, indicating that most of the effects affected protein foldability and/or FAD binding [12]. Most stability effects were found only in the hNQO1-apo-state. Some mutants (L7V and L7A, L10V and L10A), did not prevent formation of dimers, but notably destabilized them [12] (Figure 9).

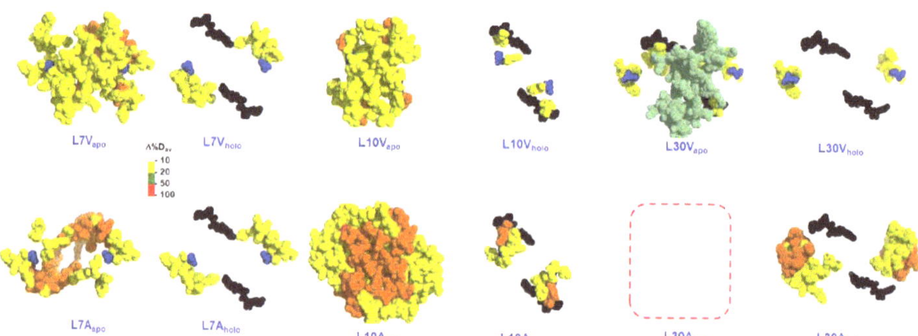

Figure 9. Cavity-making artificial mutations mostly target the structural stability of apo-hNQO1. HDX data are shown as $\Delta\%D_{av}$, as described in [12]. Residues in blue correspond to those that are mutated. Data were displayed using the structure with PDB code 2F1O [33]. Data for L30A$_{apo}$ were not acquired due to the instability of this sample.

These studies elegantly showed that energetic perturbations due to missense variations in hNQO1 may target several functions in different manners through specific effects on different functional (ligation) states.

5. Outlook and Future Perspectives

Two critical aspects for personalized medicine and pharmacogenomics are the roles of PTMs and the emerging genetic variability among individuals, which can determine their susceptibility to disease and response to therapeutics. Although recent advances have been achieved, in silico prediction methods are still underperforming in experimental evaluation, particularly for multifunctional proteins such as hNQO1 [30,53]. Artificial intelligence based approaches are revolutionizing the research in protein chemistry, biochemistry and structural biology, but these are still far from providing good predictors of genotype–phenotype relationships, particularly in multifunctional proteins [54,55]. A multidisciplinary and holistic experimental approach may help in training AI tools to improve the predictions of these relationships on a large scale [56]. In this review, I have briefly updated the information on these relationships using experimental approaches, clearly showing how genetic diversity and interindividual differences in PTMs might be critical for developing oxidative stress-related diseases associated with changes in hNQO1 functionality. Although, in a recent study, we were able to predict 2/3 of the experimental phenotypes, this is still far from being sufficient for appropriate in silico genotype–phenotype correlations [30]. We aim to continue working on the genotype–phenotype correlations in protein multifunctionality, as well as genetic diversity and its modulation by PTMs, using hNQO1 as an excellent model of an antioxidant and multifunctional disease-associated protein.

Funding: This work was funded by the ERDF/Spanish Ministry of Science, Innovation and Universities—State Research Agency [Grant number RTI2018-096246-B-I00], Consejería de Economía, Conocimiento, Empresas y Universidad, Junta de Andalucía [Grant number P18-RT-2413], ERDF/Counseling of Economic transformation, Industry, Knowledge and Universities (Grant B-BIO-84-UGR20) and Comunidad Valenciana (Grant number CIAICO/2021/135).

Institutional Review Board Statement: Not applicable.

Informed Consent Statement: Not applicable.

Data Availability Statement: All data not published will be shared with the readership upon adequate request.

Acknowledgments: A.L.P. thanks all his coworkers over the last decade on research for hNQO1, particularly the labs headed by Petr Man, Milagros Medina, Athi N. Naganathan, Kresten Lindorff-Larsen, Eduardo Salido and J.L.Neira for creating an excellent, multidisciplinary and productive network on hNQO1 research. A.L.P. will be always in debt to David J. Timson, who introduced him to hNQO1 in 2012, and who sadly passed away in August 2022. All the information presented in this work is under CC-BY license and properly cited.

Conflicts of Interest: The author declares no conflict of interest. The funders had no role in the design of the study; in the collection, analyses, or interpretation of data; in the writing of the manuscript; or in the decision to publish the results.

References

1. Beaver, S.K.; Mesa-Torres, N.; Pey, A.L.; Timson, D.J. NQO1: A target for the treatment of cancer and neurological diseases, and a model to understand loss of function disease mechanisms. *Biochim. Biophys. Acta (BBA)-Proteins Proteom.* **2019**, *1867*, 663–676. [CrossRef] [PubMed]
2. Salido, E.; Timson, D.J.; Betancor-Fernández, I.; Palomino-Morales, R.; Anoz-Carbonell, E.; Pacheco-García, J.L.; Medina, M.; Pey, A.L. Targeting HIF-1α Function in Cancer through the Chaperone Action of NQO1: Implications of Genetic Diversity of NQO1. *J. Pers. Med.* **2022**, *12*, 747. [CrossRef]
3. Siegel, D.; Bersie, S.; Harris, P.; Di Francesco, A.; Armstrong, M.; Reisdorph, N.; Bernier, M.; de Cabo, R.; Fritz, K.; Ross, D. A redox-mediated conformational change in NQO1 controls binding to microtubules and α-tubulin acetylation. *Redox Biol.* **2021**, *39*, 101840. [CrossRef]
4. Carbonell, E.A.; Timson, D.J.; Pey, A.L.; Medina, M. The Catalytic Cycle of the Antioxidant and Cancer-Associated Human NQO1 Enzyme: Hydride Transfer, Conformational Dynamics and Functional Cooperativity. *Antioxidants* **2020**, *9*, 772. [CrossRef] [PubMed]
5. Clavería-Gimeno, R.; Velazquez-Campoy, A.; Pey, A.L. Thermodynamics of cooperative binding of FAD to human NQO1: Implications to understanding cofactor-dependent function and stability of the flavoproteome. *Arch. Biochem. Biophys.* **2017**, *636*, 17–27. [CrossRef]
6. Megarity, C.F.; Timson, D.J. Cancer-associated variants of human NQO1: Impacts on inhibitor binding and cooperativity. *Biosci. Rep.* **2019**, *39*, BSR20191874. [CrossRef] [PubMed]
7. Megarity, C.F.; Bettley, H.A.; Caraher, M.C.; Scott, K.A.; Whitehead, R.C.; Jowitt, T.A.; Gutierrez, A.; Bryce, R.A.; Nolan, K.A.; Stratford, I.J.; et al. Negative Cooperativity in NAD(P)H Quinone Oxidoreductase 1 (NQO1). *Chembiochem* **2019**, *20*, 2841–2849. [CrossRef]

8. Ross, D.; Siegel, D. The diverse functionality of NQO1 and its roles in redox control. *Redox Biol.* **2021**, *41*, 101950. [CrossRef]
9. Vankova, P.; Salido, E.; Timson, D.J.; Man, P.; Pey, A.L. A Dynamic Core in Human NQO1 Controls the Functional and Stability Effects of Ligand Binding and Their Communication across the Enzyme Dimer. *Biomolecules* **2019**, *9*, 728. [CrossRef]
10. Pacheco-Garcia, J.L.; Anoz-Carbonell, E.; Loginov, D.S.; Vankova, P.; Salido, E.; Man, P.; Medina, M.; Palomino-Morales, R.; Pey, A.L. Different phenotypic outcome due to site-specific phosphorylation in the cancer-associated NQO1 enzyme studied by phosphomimetic mutations. *Arch. Biochem. Biophys.* **2022**, *729*, 109392. [CrossRef]
11. Pacheco-Garcia, J.L.; Loginov, D.; Rizzuti, B.; Vankova, P.; Neira, J.L.; Kavan, D.; Mesa-Torres, N.; Guzzi, R.; Man, P.; Pey, A.L. A single evolutionarily divergent mutation determines the different FAD-binding affinities of human and rat NQO1 due to site-specific phosphorylation. *FEBS Lett.* **2022**, *596*, 29–41. [CrossRef]
12. Pacheco-Garcia, J.L.; Loginov, D.S.; Anoz-Carbonell, E.; Vankova, P.; Palomino-Morales, R.; Salido, E.; Man, P.; Medina, M.; Naganathan, A.N.; Pey, A.L. Allosteric Communication in the Multifunctional and Redox NQO1 Protein Studied by Cavity-Making Mutations. *Antioxidants* **2022**, *11*, 1110. [CrossRef]
13. Pacheco-Garcia, J.L.; Anoz-Carbonell, E.; Vankova, P.; Kannan, A.; Palomino-Morales, R.; Mesa-Torres, N.; Salido, E.; Man, P.; Medina, M.; Naganathan, A.N.; et al. Structural basis of the pleiotropic and specific phenotypic consequences of missense mutations in the multifunctional NAD(P)H:quinone oxidoreductase 1 and their pharmacological rescue. *Redox Biol.* **2021**, *46*, 102112. [CrossRef]
14. Lajin, B.; Alachkar, A. The NQO1 polymorphism C609T (Pro187Ser) and cancer susceptibility: A comprehensive meta-analysis. *Br. J. Cancer* **2013**, *109*, 1325–1337. [CrossRef] [PubMed]
15. Siegel, D.; Anwar, A.; Winski, S.L.; Kepa, J.K.; Zolman, K.L.; Ross, D. Rapid Polyubiquitination and Proteasomal Degradation of a Mutant Form of NAD(P)H:Quinone Oxidoreductase 1. *Mol. Pharmacol.* **2001**, *59*, 263–268. [CrossRef] [PubMed]
16. Shendure, J.; Akey, J.M. The Origins, Determinants, and Consequences of Human Mutations. *Science* **2015**, *349*, 1478–1483. [CrossRef]
17. Manolio, T.A.; Fowler, D.M.; Starita, L.M.; Haendel, M.A.; MacArthur, D.G.; Biesecker, L.G.; Worthey, E.; Chisholm, R.L.; Green, E.D.; Jacob, H.J.; et al. Bedside Back to Bench: Building Bridges between Basic and Clinical Genomic Research. *Cell* **2017**, *169*, 6–12. [CrossRef] [PubMed]
18. Lek, M.; Karczewski, K.J.; Minikel, E.V.; Samocha, K.E.; Banks, E.; Fennell, T.; O'Donnell-Luria, A.H.; Ware, J.S.; Hill, A.J.; Cummings, B.B.; et al. Analysis of protein-coding genetic variation in 60,706 humans. *Nature* **2016**, *536*, 285–291. [CrossRef]
19. McInnes, G.; Sharo, A.G.; Koleske, M.L.; Brown, J.E.; Norstad, M.; Adhikari, A.N.; Wang, S.; Brenner, S.E.; Halpern, J.; Koenig, B.A.; et al. Opportunities and challenges for the computational interpretation of rare variation in clinically important genes. *Am. J. Hum. Genet.* **2021**, *108*, 535–548. [CrossRef] [PubMed]
20. Bagdasaryan, A.A.; Chubarev, V.N.; Smolyarchuk, E.A.; Drozdov, V.N.; Krasnyuk, I.I.; Liu, J.; Fan, R.; Tse, E.; Shikh, E.V.; Sukocheva, O.A. Pharmacogenetics of Drug Metabolism: The Role of Gene Polymorphism in the Regulation of Doxorubicin Safety and Efficacy. *Cancers* **2022**, *14*, 5436. [CrossRef] [PubMed]
21. van der Lee, M.; Allard, W.G.; Vossen, R.H.A.M.; Baak-Pablo, R.F.; Menafra, R.; Deiman, B.A.L.M.; Deenen, M.J.; Neven, P.; Johansson, I.; Gastaldello, S.; et al. Toward predicting CYP2D6-mediated variable drug response from *CYP2D6* gene sequencing data. *Sci. Transl. Med.* **2021**, *13*, eabf3637. [CrossRef]
22. Venugopal, R.; Jaiswal, A.K. Nrf1 and Nrf2 positively and c-Fos and Fra1 negatively regulate the human antioxidant response element-mediated expression of NAD(P)H:quinone oxidoreductase$_1$ gene. *Proc. Natl. Acad. Sci. USA* **1996**, *93*, 14960–14965. [CrossRef]
23. Jaiswal, A.K. Regulation of genes encoding NAD(P)H:quinone oxidoreductases. *Free. Radic. Biol. Med.* **2000**, *29*, 254–262. [CrossRef]
24. Ross, D.; Siegel, D. Functions of NQO1 in Cellular Protection and CoQ10 Metabolism and its Potential Role as a Redox Sensitive Molecular Switch. *Front. Physiol.* **2017**, *8*, 595. [CrossRef] [PubMed]
25. Ma, Q. Role of Nrf2 in Oxidative Stress and Toxicity. *Annu. Rev. Pharmacol. Toxicol.* **2013**, *53*, 401–426. [CrossRef] [PubMed]
26. Fowke, J.H.; Shu, X.-O.; Dai, Q.; Jin, F.; Cai, Q.; Gao, Y.-T.; Zheng, W. Oral Contraceptive Use and Breast Cancer Risk: Modification by NAD(P)H:Quinone Oxoreductase (*NQO1*) Genetic Polymorphisms. *Cancer Epidemiol. Biomark. Prev.* **2004**, *13*, 1308–1315. [CrossRef]
27. Hamachi, T.; Tajima, O.; Uezono, K.; Tabata, S.; Abe, H.; Ohnaka, K.; Kono, S. *CYP1A1*, *GSTM1*, *GSTT1* and *NQO1* polymorphisms and colorectal adenomas in Japanese men. *World J. Gastroenterol.* **2013**, *19*, 4023–4030. [CrossRef]
28. Lienhart, W.; Gudipati, V.; Uhl, M.K.; Binter, A.; Pulido, S.A.; Saf, R.; Zangger, K.; Gruber, K.; Macheroux, P. Collapse of the native structure caused by a single amino acid exchange in human NAD(P)H:quinone oxidoreductase^1. *FEBS J.* **2014**, *281*, 4691–4704. [CrossRef]
29. Lienhart, W.; Strandback, E.; Gudipati, V.; Koch, K.; Binter, A.; Uhl, M.K.; Rantasa, D.M.; Bourgeois, B.; Madl, T.; Zangger, K.; et al. Catalytic competence, structure and stability of the cancer-associated R139W variant of the human NAD (P)H:quinone oxidoreductase 1 (NQO 1). *FEBS J.* **2017**, *284*, 1233–1245. [CrossRef]
30. Pacheco-Garcia, J.L.; Cagiada, M.; Tienne-Matos, K.; Salido, E.; Lindorff-Larsen, K.; Pey, A.L. Effect of naturally-occurring mutations on the stability and function of cancer-associated NQO1: Comparison of experiments and computation. *Front. Mol. Biosci.* **2022**, *9*, 1063620. [CrossRef] [PubMed]

31. Medina-Carmona, E.; Neira, J.L.; Salido, E.; Fuchs, J.E.; Palomino-Morales, R.; Timson, D.J.; Pey, A.L. Site-to-site interdomain communication may mediate different loss-of-function mechanisms in a cancer-associated NQO1 polymorphism. *Sci. Rep.* **2017**, *7*, srep44532. [CrossRef]

32. Carmona, E.M.; Betancor-Fernández, I.; Santos, J.; Mesa-Torres, N.; Grottelli, S.; Batlle, C.; Naganathan, A.N.; Oppici, E.; Cellini, B.; Ventura, S.; et al. Insight into the specificity and severity of pathogenic mechanisms associated with missense mutations through experimental and structural perturbation analyses. *Hum. Mol. Genet.* **2019**, *28*, 1–15. [CrossRef] [PubMed]

33. Asher, G.; Dym, O.; Tsvetkov, P.; Adler, J.; Shaul, Y. The Crystal Structure of NAD(P)H Quinone Oxidoreductase 1 in Complex with Its Potent Inhibitor Dicoumarol. *Biochemistry* **2006**, *45*, 6372–6378. [CrossRef] [PubMed]

34. Hornbeck, P.V.; Zhang, B.; Murray, B.; Kornhauser, J.M.; Latham, V.; Skrzypek, E. PhosphoSitePlus, 2014: Mutations, PTMs and recalibrations. *Nucleic Acids Res.* **2015**, *43*, D512–D520. [CrossRef] [PubMed]

35. Medina-Carmona, E.; Rizzuti, B.; Martín-Escolano, R.; Pacheco-García, J.L.; Mesa-Torres, N.; Neira, J.L.; Guzzi, R.; Pey, A.L. Phosphorylation compromises FAD binding and intracellular stability of wild-type and cancer-associated NQO1: Insights into flavo-proteome stability. *Int. J. Biol. Macromol.* **2019**, *125*, 1275–1288. [CrossRef] [PubMed]

36. Carmona, E.M.; Fuchs, J.E.; Gavira, J.A.; Mesa-Torres, N.; Neira, J.L.; Salido, E.; Palomino-Morales, R.; Burgos, M.; Timson, D.; Pey, A.L. Enhanced vulnerability of human proteins towards disease-associated inactivation through divergent evolution. *Hum. Mol. Genet.* **2017**, *26*, 3531–3544. [CrossRef]

37. Luo, S.; Kang, S.S.; Wang, Z.-H.; Liu, X.; Day, J.X.; Wu, Z.; Peng, J.; Xiang, D.; Springer, W.; Ye, K. Akt Phosphorylates NQO1 and Triggers its Degradation, Abolishing Its Antioxidative Activities in Parkinson's Disease. *J. Neurosci.* **2019**, *39*, 7291–7305. [CrossRef]

38. Martínez-Limón, A.; Alriquet, M.; Lang, W.-H.; Calloni, G.; Wittig, I.; Vabulas, R.M. Recognition of enzymes lacking bound cofactor by protein quality control. *Proc. Natl. Acad. Sci. USA* **2016**, *113*, 12156–12161. [CrossRef]

39. Muñoz, I.G.; Morel, B.; Medina-Carmona, E.; Pey, A.L. A mechanism for cancer-associated inactivation of NQO1 due to P187S and its reactivation by the consensus mutation H80R. *FEBS Lett.* **2017**, *591*, 2826–2835. [CrossRef]

40. Siegel, D.; Harris, P.S.; Michel, C.R.; de Cabo, R.; Fritz, K.S.; Ross, D. Redox state and the sirtuin deacetylases are major factors that regulate the acetylation status of the stress protein NQO1. *Front. Pharmacol.* **2022**, *13*, 1015642. [CrossRef]

41. Pan, S.S.; Forrest, G.L.; Akman, S.A.; Hu, L.T. NAD(P)H:quinone oxidoreductase expression and mitomycin C resistance developed by human colon cancer HCT 116 cells. *Cancer Res.* **1995**, *55*, 330–335.

42. Traver, R.D.; Siegel, D.; Beall, H.D.; Phillips, R.M.; Gibson, N.W.; Franklin, W.A.; Ross, D.T. Characterization of a polymorphism in NAD(P)H: Quinone oxidoreductase (DT-diaphorase). *Br. J. Cancer* **1997**, *75*, 69–75. [CrossRef]

43. Traver, R.D.; Horikoshi, T.; Danenberg, K.D.; Stadlbauer, T.H.; Danenberg, P.V.; Ross, D.; Gibson, N.W. NAD(P)H:quinone oxidoreductase gene expression in human colon carcinoma cells: Characterization of a mutation which modulates DT-diaphorase activity and mitomycin sensitivity. *Cancer Res* **1992**, *52*, 797–802.

44. Pey, A.L.; Megarity, C.F.; Timson, D.J. FAD binding overcomes defects in activity and stability displayed by cancer-associated variants of human NQO1. *Biochim. Biophys. Acta (BBA)-Mol. Basis Dis.* **2014**, *1842*, 2163–2173. [CrossRef] [PubMed]

45. Pacheco-García, J.; Cano-Muñoz, M.; Sánchez-Ramos, I.; Salido, E.; Pey, A. Naturally-Occurring Rare Mutations Cause Mild to Catastrophic Effects in the Multifunctional and Cancer-Associated NQO1 Protein. *J. Pers. Med.* **2020**, *10*, 207. [CrossRef]

46. Zhou, H.; Wan, H.; Zhu, L.; Mi, Y. Research on the effects of rs1800566 C/T polymorphism of NAD(P)H quinone oxidoreductase 1 gene on cancer risk involves analysis of 43,736 cancer cases and 56,173 controls. *Front. Oncol.* **2022**, *12*, 980897. [CrossRef] [PubMed]

47. Siegel, D.; McGuinness, S.M.; Winski, S.L.; Ross, D. Genotype-phenotype relationships in studies of a polymorphism in NAD(P)H. *Pharmacogenetics* **1999**, *9*, 113–122. [CrossRef]

48. Pey, A.L. Biophysical and functional perturbation analyses at cancer-associated P187 and K240 sites of the multifunctional NADP(H):quinone oxidoreductase 1. *Int. J. Biol. Macromol.* **2018**, *118*, 1912–1923. [CrossRef]

49. Medina-Carmona, E.; Palomino-Morales, R.J.; Fuchs, J.E.; Padín-Gonzalez, E.; Mesa-Torres, N.; Salido, E.; Timson, D.J.; Pey, A.L. Erratum: Conformational Dynamics Is Key to Understanding Loss-of-Function of NQO1 Cancer-Associated Polymorphisms and Its Correction by Pharmacological Ligands. *Sci. Rep.* **2016**, *6*, 21939. [CrossRef] [PubMed]

50. Martínez-Limón, A.; Calloni, G.; Ernst, R.; Vabulas, R.M. Flavin dependency undermines proteome stability, lipid metabolism and cellular proliferation during vitamin B2 deficiency. *Cell Death Dis.* **2020**, *11*, 725. [CrossRef]

51. Eguchi-Ishimae, M.; Eguchi, M.; Ishii, E.; Knight, D.; Sadakane, Y.; Isoyama, K.; Yabe, H.; Mizutani, S.; Greaves, M. The association of a distinctive allele of NAD(P)H:quinone oxidoreductase with pediatric acute lymphoblastic leukemias with MLL fusion genes in Japan. *Haematologica* **2005**, *90*, 1511–1515.

52. Pan, S.-S.; Han, Y.; Farabaugh, P.; Xia, H. Implication of alternative splicing for expression of a variant NAD(P)H:quinone oxidoreductase-1 with a single nucleotide polymorphism at 465C>T. *Pharmacogenetics* **2002**, *12*, 479–488. [CrossRef]

53. Pacheco-García, J.L.; Anoz-Carbonell, E.; Loginov, D.S.; Kavan, D.; Salido, E.; Man, P.; Medina, M.; Pey, A.L. Counterintuitive structural and functional effects due to naturally occurring mutations targeting the active site of the disease-associated NQO1 enzyme. *FEBS J.* **2022**. [CrossRef] [PubMed]

54. Callaway, E. What's next for AlphaFold and the AI protein-folding revolution. *Nature* **2022**, *604*, 234–238. [CrossRef] [PubMed]

55. Akdel, M.; Pires, D.E.V.; Pardo, E.P.; Jänes, J.; Zalevsky, A.O.; Mészáros, B.; Bryant, P.; Good, L.L.; Laskowski, R.A.; Pozzati, G.; et al. A structural biology community assessment of AlphaFold2 applications. *Nat. Struct. Mol. Biol.* **2022**, *29*, 1056–1067. [CrossRef] [PubMed]
56. Høie, M.H.; Cagiada, M.; Frederiksen, A.H.B.; Stein, A.; Lindorff-Larsen, K. Predicting and interpreting large-scale mutagenesis data using analyses of protein stability and conservation. *Cell Rep.* **2022**, *38*, 110207. [CrossRef]

antioxidants

Article

Profiling the miRNA from Exosomes of Non-Pigmented Ciliary Epithelium-Derived Identifies Key Gene Targets Relevant to Primary Open-Angle Glaucoma

Padmanabhan Paranji Pattabiraman [1],*, Valeria Feinstein [2] and Elie Beit-Yannai [2],*

[1] Glick Eye Institute, Department of Ophthalmology, Indiana University School of Medicine, 1160 West Michigan Street, Indianapolis, IN 46202-5209, USA
[2] Clinical Biochemistry and Pharmacology Department, Ben-Gurion University of the Negev, Beer-Sheva 84105, Israel
* Correspondence: ppattabi@iu.edu (P.P.P.); bye@bgu.ac.il (E.B.-Y.)

Abstract: Oxidative stress (OS) on tissues is a major pathological insult leading to elevated intraocular pressure (IOP) and primary open-angle glaucoma (POAG). Aqueous humor (AH) produced by the non-pigmentary ciliary epithelium (NPCE) drains out via the trabecular meshwork (TM) outflow pathway in the anterior chamber. The exosomes are major constituents of AH, and exosomes can modulate the signaling events, as well as the responses of their target TM tissue. Despite the presence of molecular mechanisms to negate OS, oxidative damage directly, as well as indirectly, influences TM health, AH drainage, and IOP. We proposed that the expression of microRNA (miRNAs) carried by exosomes in the AH can be affected by OS, and this can modulate the pathways in target cells. To assess this, we subjected NPCE to acute and chronic OS (A-OS and C-OS), enriched miRNAs, performed miRNA microarray chip analyses, and miRNA-based gene targeting pathway prediction analysis. We found that various miRNA families, including miR27, miR199, miR23, miR130b, and miR200, changed significantly. Based on pathway prediction analysis, we found that these miRNAs can regulate the genes including *Nrf2*, *Keap1*, *GSK3B*, and serine/threonine-protein phosphatase2A (*PP2A*). We propose that OS on the NPCE exosomal miRNA cargo can modulate the functionality of the TM tissue.

Keywords: oxidative stress; exosomes; miRNA; non-pigmented ciliary epithelium; trabecular meshwork; primary open-angle glaucoma

Citation: Pattabiraman, P.P.; Feinstein, V.; Beit-Yannai, E. Profiling the miRNA from Exosomes of Non-Pigmented Ciliary Epithelium-Derived Identifies Key Gene Targets Relevant to Primary Open-Angle Glaucoma. *Antioxidants* **2023**, *12*, 405. https://doi.org/10.3390/antiox12020405

Academic Editor: Marcel Bonay

Received: 31 December 2022
Revised: 2 February 2023
Accepted: 4 February 2023
Published: 7 February 2023

1. Introduction

Oxidative stress (OS) is defined as an imbalance between the production and accumulation of reactive species in cells. Cells and tissues possess the ability to detoxify these reactive products [1]. In fact, OS has been proven to be a key factor in biological regulation under normal and pathological conditions [2]. The initial cause of OS is diverse and results in a series of common cellular events, which include modifications to protein, lipid, and nucleotide that can be tracked and measured [3]. Upon oxidative insult, the cellular defense mechanisms are activated, including increased expression of antioxidant enzymes, such as superoxide dismutase (SOD), catalase, and glutathione peroxidase (GPX), as well as augmenting their activity. In parallel, there is an increased consumption of low molecular weight antioxidants that includes ascorbic acid, glutathione, and uric acid [4]. A basal level of reactive oxygen species is essential to maintain various biological processes, including cell proliferation and differentiation [5]. When the degree of OS crosses a certain level, cellular and tissue damage responses occur. Several cellular signaling events occur in response to OS, including MAPK, PI3K, p53, Notch1, and nuclear factor erythroid 2–related factor 2/Kelch-like ECH-associated protein 1 (Nrf2-Keap1) signaling pathways [6]. These pathways have a rate, strength, and duration that vary depending on the source of the

OS, and the number of cells and tissue exposed to it. The Nrf2/Keap1 pathway stands out among the pathways in its centrality, having the ability to regulate defense mechanisms against OS. The general regulators of Nrf2/Keap1 are detailed in many publications describing its role in different tissues and diseases [7–9]

Extracellular Vesicles (EVs) comprise a range of lipid-bound sacs with different properties and can be divided according to their cellular origin (cytosol or membrane), size (30 nm to 1 μm), and macromolecular contents [10]. The three major EV groups are exosomes, microvesicles, and apoptotic bodies [11]. Exosomes are a nano-scale (30–180 nm) cup-shaped subgroup of EVs, having a lipid bilayer membrane originating from cytoplasmic and specifically from multivesicular bodies (MVBs), mostly from all cell types. The ESCRT (endosomal sorting complexes required for transport) is the dominant machinery responsible for controlling sorted protein into the MVB. Exosomes are released into the extracellular space when the MVBs fuse with the plasma membrane [12]. Exosomes cargo includes miRNA, mRNA, ncRNA, ssDNA, and cytoplasmic and membrane proteins, thus playing a role in cell-to-cell communication. Exosomes, besides having a local effect on the same or neighboring tissues, can spread from their originating tissue via the bloodstream, lymph system, and extracellular fluids, reaching their distal targets [13]. The ability of exosomes to cross biological barriers due to their native lipophilic membrane allows them to diffuse passively through the BBB, placenta, and fat tissue [14]. Specific recognition of exosome membrane surface proteins and uptake mechanisms suggest the targeted delivery of messages by exosomes bring about physiological homeostasis and potentially function under pathological conditions [15].

Reactive oxygen species play a key role in the pathogenesis of primary open-angle glaucoma (POAG), a chronic optic neuropathy. It predominantly starts within the trabecular meshwork (TM) tissue impairment to sense the increase in the intraocular pressure (IOP). This IOP elevation induces apoptosis of the retinal ganglion cells (RGC), leading to blindness [16]. The AH produced by NPCE has a unique short one-way flow. Much of the AH is drained by the TM outflow pathway, including the TM, juxtacanalicular tissue, and, finally, via the Schlemm's canal. The IOP is regulated primarily by fluid resistance to AH outflow. The extracellular matrix (ECM) of the TM is thought to be important in the regulation of IOP [17]. The TM tissue in the ocular drainage system is unique by being continuously exposed to OS [18]. However, we speculate that this exposure can vary the effects, due to either acute OS or chronic OS developed during elevated IOP. Both chronic and acute OS may result in cellular adaptations elicited by the exosome-mediated signals. In the present study, we will examine the changes taking place in NPCE-derived exosomes exposed to acute or chronic OS, further establishing its relevance in OS response genes. The AH carries functional EVs originating from all tissues making up the outflow pathway [19]. Our laboratory recently showed that EVs derived from oxidized NPCE cells protected the TM cells from the direct OS by significantly inducing Nrf2 [20]. In that study, we showed that treatment with EVs released by oxidatively stressed NPCE cells resulted in an increase in the Nrf2 staining in the TM cytoplasm and nucleus and induced Nrf2 protein levels significantly. Contrarily, EVs extracted from the NPCE cells unexposed to OS did not induce Nrf2 changes in the treated TM cells. The TM cells directly exposed to OS showed a significant change in Nrf2. Further, the downstream response included an increase in antioxidant genes and protein expression, including SOD1, SOD2, GPX1, heme oxygenase 1 (HMOX1), Nrf2, and increased catalase and SOD activities [20]. These suggest a strong link in the modulation of OS signaling in TM by EVs from NPCE. Moreover, recently mesenchymal stem cell-derived exosomes were shown to protect TM from OS [21]. The canonical Wnt pathway plays a major role in regulating AH drainage homeostasis. Key proteins involved in this pathway are pGSK, TGFβ, the PP2A phosphatase, and MMPs. Interestingly, these proteins are involved in controlling the activity of Nrf2/Keap1. Taken together, we predict that the major OS pathway, the Nrf2/Keap1, plays an important role in exosome-mediated OS cell responses in TM. This can potentially contribute to changes in the TM affecting IOP and participating in POAG pathology. Thus, the biology of OS and exosomes have merged our understanding of the paracrine function of exosomes, as well as the cellular responses by NPCE and TM to OS challenges [22]. As mentioned earlier, miRNAs are carried by the EVs, and such miRNAs can regulate the Nrf2 antioxidant pathway [23]. The miRNAs are involved in the regulation of various cellular processes in all biological tissues including the ocular tissues. Although there has been no single miRNA or a family that has been solely indicated in the pathogenesis of elevated IOP and POAG, there are multiple studies indicating the role of miRNAs in TM regulated by OS and supplementation of the miRNAs reversing OS, thus favoring a significant role of miRNA in modifying the oxidative damage to TM [24]. Additionally, the involvement of miRNAs in POAG has been investigated using models of mechanical stress, hypoxia, inflammation, apoptosis, and a combination of these models [25,26]. A set of miRNAs were indicated to participate in different signaling pathways known to affect TM cells, including, but not limited to the involvement of mTOR, MEK/ERK, TGF-β, PI3K/AKT, Wnt/β-Catenin, MMP-9 and Nrf2 [27]. However, very little information exists on the miRNAs regulated by OS in the NPCE and their resultant signaling changes on their target tissue, such as the TM. The present study aims to identify miRNA carried as exosomal cargo with the potential to affect the Nrf2/Keap1 pathway and relevant genes important in POAG pathogenesis when the NPCE tissue is exposed to OS.

2. Material and Methods

2.1. Cell Line

The human NPCE cell line was kindly supplied by Prof. Miguel Coca-Prados, Yale University [28]. The NPCE cell lines authentication test was performed at the Genomics Center of Biomedical Core Facility, Technion, Israel, using the Promega GenePrint 24 System. The Authentication report can be found in the supplementary material (Supplementary S1).

2.2. Oxidative Stress

To induce OS in NPCE cells, we utilized 2,2′-Azobis (2-amidinopropane) dihydrochloride (AAPH), which is a free radical-generating azo compound. AAPH is capable of initiating oxidation reactions by continuous production of peroxyl radical followed by alkoxyl radical via nucleophilic and free radical mechanisms [29]. NPCE cells were exposed to an acute OS 15 mM AAPH for 90 min for acute OS (A-OS) or 1.5 mM AAPH for 24 h as chronic OS (C-OS) stimulation.

2.3. Exosomes Isolation

Exosome-depleted media was prepared by first mixing Dulbecco's modified Eagle's medium with regular fetal bovine serum (FBS). This media was subjected to ultracentrifugation for 16 h at 110,000× g overnight at 4 °C using an SW28 rotor. Further filtration through a 0.22 μm filter (Millipore Express PLUS (PES) membrane) was carried out to completely deplete the exosomes. This media was used for downstream experiments. Upon conclusion of the experiments, exosomes secreted into the condition media (CM) from the cultured cells were prepared. Exosomes were isolated from CM of cultured NPCE cell line using a series of ultracentrifugation steps. At first, centrifugation at 300× g for 10 min to get rid of the cells was performed. Then, 2000× g for 10 min to get rid of the dead cells, followed by centrifugation at 10,000× g for 30 min to get rid of the cell debris. The final steps were at 100,000× g for 70 min, repeated twice, as detailed previously [30].

2.4. Exosomes Size and Concentration Analysis

Isolated exosomes were analyzed using the NanoSight NS500 instrument (Malvern Panalytical, Malvern UK) equipped with a blue laser (405 nm). Nanoparticles were illuminated by the laser and their movement under Brownian motion was captured for 60 s. Videos were then subjected to nanoparticle tracking analysis (NTA) using the NanoSight particle tracking software (NTA 2.0). At least three videos were captured for each sample to provide a representative concentration measurement, and all analysis settings were kept constant within each experiment. The settings were: camera level of 16–17; automatic functions for all post-acquisition settings; and camera focus were adjusted to make the particles appear as sharp dots. Using the script control function, three 30 s videos for each sample were recorded. Size distribution profiles obtained from NTA were averaged within each sample across the video replicates and then averaged across samples to provide representative size distribution profiles. These distribution profiles were then normalized to total nanoparticle concentrations or final cell counts. The data obtained by the NanoSight particle tracking software (NTA 2.0) is presented as mean exosome size and the most common exosome size (Mode). All experiments were performed at 1:1000 dilutions, yielding particle concentrations of $\approx 6 \times 10^7$/mL.

2.5. Exosomes miRNA Extraction and Characterization

Total RNA extraction from exosomes was executed with a Total exosomal RNA purification kit (cat: 17200, Norgen Biotek Corp. Thorold, Ontario, Canada), as recommended by the manufacturer. In brief, 200 µL of exosomes in PBS were lysed for 1 min in lysis buffer by vortexing. Spin columns were used for miRNA separation at $3500 \times g$ for 1 min. Columns were washed four times at $14,000 \times g$ for 1 min, following elution in 50 µL, at $600 \times g$ for 2 min. miRNA quality and integrity were assessed with Bioanalyzer 2100 (Agilent Technologies, Santa Clara, CA, USA). Samples with mRNA concentrations higher than 130 ng/µL were further analyzed. Exosomal miRNA were characterized with the suitable microarray chip analysis GeneChip® miRNA 4.0 Array and Flashtag™ Bundle–ThermoFisher (cat 902445).

2.6. miRNA-Based Gene Targeting Pathway Prediction Analysis

We collected information about miRNAs that were reported or predicted to target nuclear factor erythroid-derived 2-like 2 (NFE2L2) or Nrf2 from the following resources: miRDB [31], high-throughput datasets generated with cross-linking ligation and sequencing of hybrids (CLASH) method [32], TarBase [33], and miRTarBase [34]. miRNA expression array files (CEL) were uploaded into Transcriptome Analysis Console (TAC) software v4.0.3 to identify miRNAs that are differentially expressed in OS versus control samples (Fold-change (OS vs. control) < −2, p-adj < 0.1).

Lists of *NFE2L2*, *Keap1*, *GSK3B*, and *PPP2CA* targeting miRNAs were intersected with differentially expressed miRNAs to identify miRNAs whose decrease in exosomes could be directly linked to the activation of *NFE2L2*, *Keap1*, *GSK3B*, and *PPP2CA*.

2.7. Statistical Analysis

Raw data were extracted automatically in Affymetrix data extraction protocol using the software provided by Affymetrix GeneChip® Command Console® 4.3.2 (AGCC) Software. The CEL files were imported and changes in miRNA levels were performed using Affymetrix® Expression Console™ Software. Array data were filtered by probes annotated species. Comparative analysis was carried out between test and control samples using fold-change and independent T-test, in which the null hypothesis was that no difference exists between the two groups. The false discovery rate (FDR) was controlled by adjusting the *p*-value using the Benjamini–Hochberg algorithm.

3. Results

3.1. NPCE Isolated Exosomes Characterization

Extracted NPCE-derived exosomes following acute and chronic OS were compared to the NPCE exosomes derived from the control treatment for differences in size and concentration using NTA. The A-OS or C-OS exposure on the NPCE cells did not affect the exosome size or concentration compared to the control or between treatments. The mean size was found to be 149.1 to 150 nm. There was no significant change in the mode, suggesting that the exosomes released were not different in size (Table 1 and Figure 1). The data represents the average of n = 3 (Table 1).

Table 1. Characterization comparison of NPCE-derived exosomes following acute and chronic oxidative stress.

	Mean (nm)	Mode (nm)	Concentration (Particles/mL)
Control	149.1 ± 6.3	97.5 ± 5.3	$2.34 \times 10^7 \pm 6.80 \times 10^8$
Acute Oxidative stress	149.6 ± 1.8	103.8 ± 6.2	$3.04 \times 10^7 \pm 7.30 \times 10^8$
Chronic Oxidative stress	150.0 ± 2.4	115.6 ± 6.8	$4.99 \times 10^7 \pm 6.04 \times 10^8$

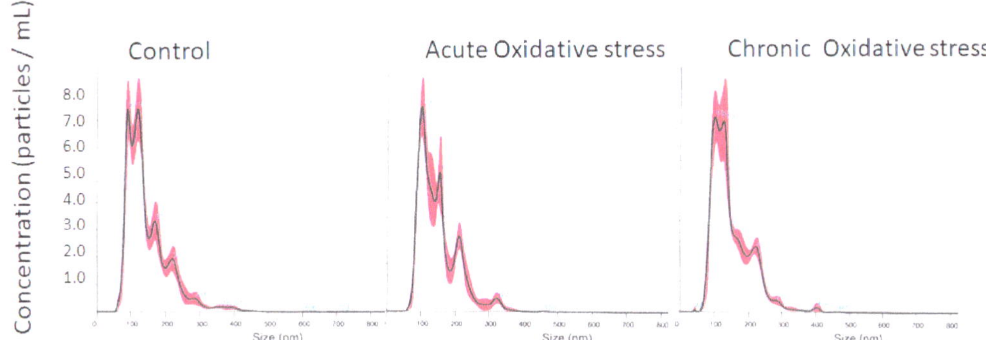

Figure 1. Nanoparticle tracking analysis (NTA) of NPCE exosomes isolated by ultracentrifugation. Data represent the average size distribution profile of n = 3 for each treatment derived from three different videos and analyses.

3.2. miRNA Predicted Nrf2 Gene Target Analysis

Four databases were used for *Nrf2* gene target prediction. TarBase analysis suggested 96 potential miRNAs; miRDB analysis suggested 64 potential miRNAs and miRTarBase analysis revealed 9 miRNAs; whereas the CLASH data could not predict any miRNA predicted for the *Nrf2* gene.

Among all the potential miRNAs predicted to affect the *Nrf2* gene, three miRNAs—hsa-miR27b-3p, hsa-miR199a-5p, and hsa-miR199a-3p—were found in the three data-bases. Moreover, these three miRNAs were found following A-OS and C-OS. A significant increase in signal from the NPCE-derived exosomes of hsa-miR27b-3p was found in A-OS exposed NPCE cells from control and C-OS (Figure 2).

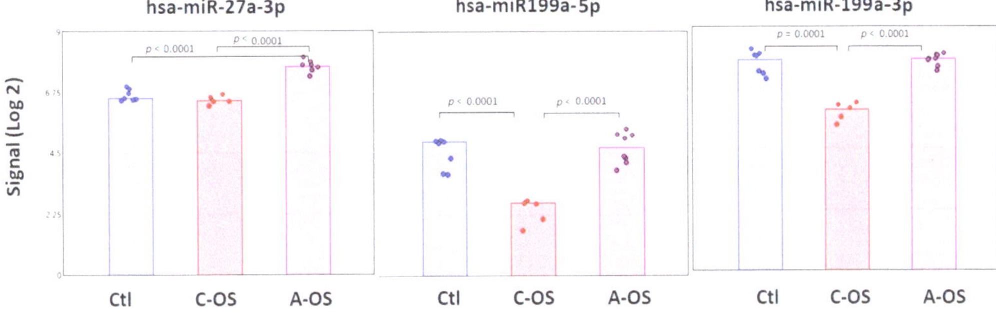

Figure 2. Individual changes in expression of exosomal miRNAs predicted to target the *Nrf2*. Bar plots depict the Affymetrix signal of hsa-miR27a-3p, hsa-miR199a-5p, and hsa-miR199a-3p in NPCE cells-derived exosome following C-OS or A-OS. Each group contains a minimum of five independent replicates. One-way ANOVA was performed to identify the *p* < 0.001 among these three groups.

The hsa-miR199a-5p and hsa-miR199a-3p levels were not different from the control following A-OS. However, following C-OS there was a significant decrease in the miRNA levels compared to control, as well as A-OS (Figure 2).

3.3. miRNA Predicted GSK3B Gene Target Analysis

Among the proteins that control the phosphorylation of Nrf2 is GSK3B [35]. Therefore, we looked for miRNAs present in the NPCE-derived exosomes, which can modify *GSK3B* expression that is differentially expressed under OS versus control samples.

The hsa-miR199a-3p, which was predicted to modify the *Nrf2* gene, can also modify the *GSK* gene expression. There is a significant signal increase following A-OS compared to control and C-OS. A similar pattern was found for hsa-miR27a-3p, whose signal is significantly higher after A-OS compared to control and C-OS. However, the degree of hsa-miR27a-3p signal in control and under OS was 50% further decreased than hsa-miR-199a-3p. The hsa-miR24-3p signal was significantly higher in A-OS versus C-OS, with a relatively high signals in control and OS group relative to the other miRNA-predicted *GSK3B* gene targets analyzed. The hsa-miR23b-3p and hsa-miR29b-3p were predicted *GSK3B* gene targets without significant differences between control, C-OS, and A-OS (Figure 3).

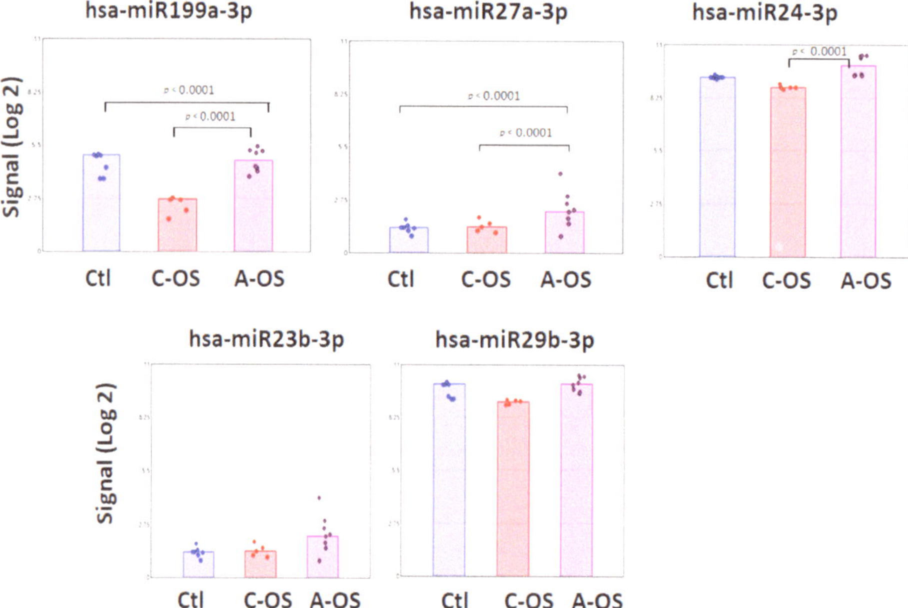

Figure 3. Individual changes in expression of exosomal miRNAs predicted to target *GSK*. Bar plots depict the Affymetrix signal of hsa-miR199a-3p, hsa-miR27a-3p, hsa-miR24-3p, hsa-miR23b-3p, and hsa-miR29b-3p in NPCE cells-derived exosome following C-OS or A-OS. Each group contains a minimum of five independent replicates. One-way ANOVA was performed to identify the $p < 0.001$ among these three groups.

3.4. miRNA Predicted PPP2CA Gene Target Analysis

The interrelation and regulation between PP2A and Nrf2 proteins are associated with OS [36]. We examined miRNAs presented in the NPCE-derived exosomes that are differentially expressed under OS versus control samples, which we believe can modify *PPP2CA* expression. All three detected miRNAs—hsa-miR197-3p, hsa-miR130b-3p, and hsa-miR125-5p—showed an increased signal after A-OS compared to C-OS. This response was particularly noticeable for hsa-miR197-3p, with a significant signal increase compared to the control (Figure 4).

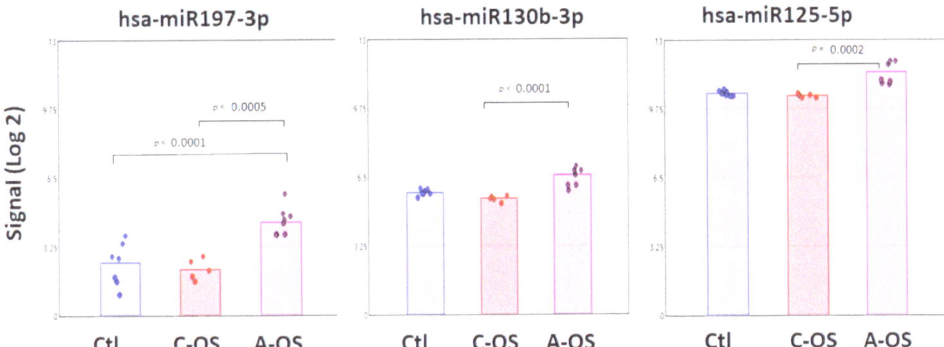

Figure 4. Individual changes in expression of exosomal miRNAs predicted to target *PPP2CA*. Bar plots depict the Affymetrix signal of hsa-miR197a-3p, hsa-miR130b-3p, and hsa-miR125b-5p in NPCE cells derived-exosome following C-OS or A-OS. Each group contains a minimum of five independent replicates. One-way ANOVA was performed to identify the $p < 0.001$ among to these three groups.

3.5. miRNA Predicted Keap1 Gene Target Analysis

Nrf2 in the cytoplasm binds to Keap1, which, in turn, facilitates the ubiquitination and subsequent proteolysis of Nrf2 [37]. Upon oxidative stress, the Nrf2-Keap1 complex depredates and Nrf2 can translocate to the nucleus and regulates the cellular resistance to oxidants. We looked for miRNAs presented in the NPCE-derived exosomes that are differentially expressed under OS versus control samples, as this can modify *Keap1* expression. Only one miRNA, hsa-miR200a-3p, came up in three databases: miRDB, data TarBase, and miRTarBase. Under C-OS, a trend of reduction in the hsa-miR200a-3p signal was found, with a further trend of reduction under A-OS (Figure 5).

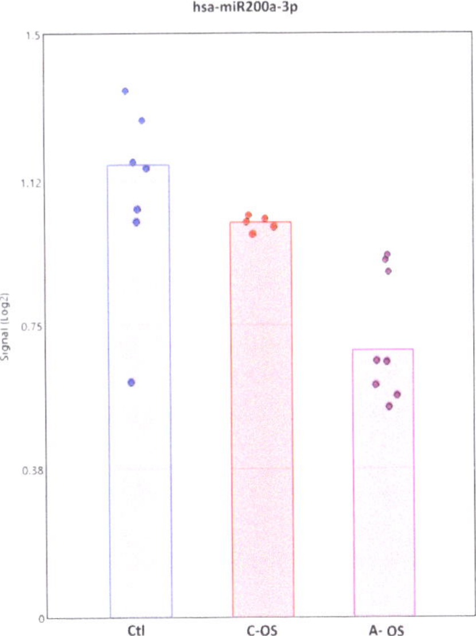

Figure 5. Individual changes in expression of exosomal miRNAs predicted to target *Keap1*. Bar plots depict the Affymetrix signal of hsa-miR200a-3p in NPCE cells-derived exosome following C-OS or A-OS. Each group contains a minimum of five independent replicates. One-way ANOVA was performed to identify the $p < 0.001$ among these three groups.

4. Discussion

Here, we have taken a systems biology approach to successfully examine the contributions of OS on the changes in NPCE exosomal miRNAs. Further, we have predicted how these changes in miRNAs can affect the key genes related to the OS response pathway: *Nrf2*, *Keap1*, *PPP2CA*, and *GSK3B*.

Our experiments comparing NPCE exosome size following A-OS or C-OS followed the general acceptance of lack of size changes in EVs under stress. In general, we were curious to see if OS affected exosome size and/or concentration. Earlier, this phenomenon was addressed by F.J Romero's lab, which reported that stressed retinal pigment epithelium (ARPE-19) cells released a higher number of exosomes compared to the controls without significantly changing the exosome size [38]. In another study, thermal stress and OS increased the release of exosomes by Jurkat and Raji cells without affecting the exosome size [39]. Additionally, the stressors, including hypoxia, TNF-α-induced activation, and high glucose, did not affect exosome size [40]. In contrast, osmotic stress was shown to increase the size and enhance the release of exosomes derived from fibroblasts. Today, it is accepted that some OS might trigger an increase in exosome release, with evidence for OS influencing the exosome cargo loading [22]. Interestingly, there is no evidence of exosomal size changes due to OS.

Our lab has been studying naïve exosomes derived from NPCE cells and their effects on modulating the canonical Wnt signaling pathway and ECM remodeling [41–44]. Earlier, we were able to show the involvement of key proteins in the canonical Wnt pathway and the PDK1/Akt/mTOR signaling pathways [20]. Additionally, we found that the phosphatase PP2A, known to catalyze the dephosphorylation of Akt, was activated in TM cells upon treatment with NPCE-derived exosomes [41]. PP2A aids in the partial deactivation of Akt and activation of GSK3β by dephosphorylating [20]. These proteins were found to be related to control Nrf2 and to be involved in the Wnt pathway, which are important in the pathogenesis of POAG [45,46]. In the current study, enhancement of the miR27a signal was found following A-OS. Increased miR27a expression was found following *in vitro* human TM cells exposed to OS, resulting in the activation of PI3K/AKT and Wnt/β-catenin pathways [47]. The NPCE-derived exosomal miR199a-3p and miR199a-5p signals significantly decreased exclusively in C-OS compared to the control. This supports our understanding that miR199a-3p and miR199a-5p has a potential role to inhibit *Nrf2* expression under C-OS. Recently, Tian et al. showed a similar role of miR199b-3p in regulating the *Nrf2* pathway in a kidney injury model [48]. Interestingly, miR199-5p was suggested to be downregulated by OS-mediated TGFβ2 expression [49]. Elevated levels of TGFβ2 in the POAG patient's AH, is a key growth factor known to be involved in the induction of both TM and optic nerve changes in POAG [50]. In POAG patients, the TM demonstrates altered actin cytoskeleton, ECM composition and this is accompanied by dysregulation of multiple signaling pathways, including but not limited to TGFβ [51] and Wnt/β-catenin, which is reduced under OS [52].

GSK3B is a key protein in the control and regulation of the Wnt and Nrf2 pathways [52]. The miRNAs regulating GSK3B expression and potentially its activity has a wide impact on cellular response against OS. This can extend into modulation of the extracellular matrix (ECM) remodeling in TM [53]. The ECM remodeling in TM plays an important role in the regulation of IOP [54]. For the first time, we were able to report the presence of members of two of the miR199 family predicted to affect GSK3B. The miR199a-5p signal changed significantly under OS compared to the control. Interestingly, miR199 was not found in the POAG human AH exosomes in the earlier report [19,55], but was detected in AH from keratoconus eyes [56]. The significance of the NPCE-derived exosomes on TM among other exosome sources within anterior segment that gets into the AH needs further investigation.

miR29b is involved in the regulation of ECM by controlling the synthesis and deposition of ECM proteins in TM cells [57]. Chronic OS can significantly decrease miR29b and an induction in the expression of miR29b has been shown to upregulate ECM genes, including collagens, laminin, and fibrillin. In our study, for the first time, we show that hsa-miR27a-3p is present in NPCE-derived exosomes and that its signal significantly increased under A-OS. This helps supporting the hypothesis that miR27a-3p can induce modulate Wnt pathway. This is based on recently published data which suggests that miR27a-3p modulates the Wnt pathway via the sFRP-1 antagonist [58].

The Serine/threonine-protein phosphatase 2A is encoded by the *PPP2CA* gene and is reported to be significantly higher in AH samples from the POAG patients relative to cataract controls [59]. PP2A has a role in the modulation of GSK and negatively regulates Wnt signaling networks [20]. In addition, PP2A activity is reduced following OS via AKT signaling. This makes PP2A an important hub between the OS and the Wnt pathway. In the present study hsa-miR197-3, hsa-miR130b and hsa-miR125-5p, which are predicted to modulate the *PPP2CA* gene, were found in NPCE-derived exosomes and their signals significantly increased following A-OS. Limited data exist regarding the role of these miRNAs in the ocular drainage system. which we plan to investigate in the future.

The hsa-miR200a-3p was the only miRNA predicted by three databases to affect the *Keap1* gene. Through the ability to regulate *Nrf2* activation by targeting Keap1, the miR-200a-3p may play a critical role in modulating cellular antioxidant signals. Evidence for its relevance in the ocular drainage system can be found in miR200a upregulation, which has been shown to positively affect retinal ganglion cell pathologies [60].

The major limitation of our study is the lack of experimental target validation. Further experimental evidence will help us decipher, in greater detail, the effects of these oxidatively stressed exosomes on changes in biological pathways in the target cells/tissue in question, which is the TM.

To summarize, the present research contributes another layer to the understanding of how NPCE-derived exosomes can potentially affect the TM, while focusing on miRNAs that can regulate the activity of genes central to the pathophysiology of glaucoma.

Supplementary Materials: The following supporting information can be downloaded at: https://www.mdpi.com/article/10.3390/antiox12020405/s1, Supplementary S1.

Author Contributions: Conceptualization, E.B.-Y.; methodology, E.B.-Y.; software, V.F.; validation, E.B.-Y., V.F. and P.P.P.; formal analysis, V.F. and E.B.-Y.; investigation, E.B.-Y.; resources, E.B.-Y.; data curation, E.B.-Y.; writing—original draft preparation, E.B.-Y. and P.P.P.; writing—review and editing, E.B.-Y. and P.P.P.; visualization, E.B.-Y, P.P.P. and V.F.; supervision, E.B.-Y.; project administration, E.B.-Y. and P.P.P.; funding acquisition, E.B.-Y. and P.P.P. All authors have read and agreed to the published version of the manuscript.

Funding: This research was supported by the ISRAEL SCIENCE FOUNDATION (grant No. 1545/20) (EBY) and by the National Institutes of Health/National Eye Institute (R01EY029320) (PPP) and a grant from Research to Prevent Blindness to IU. The sponsor or funding organization had no role in the research.

Institutional Review Board Statement: Not applicable.

Informed Consent Statement: Not applicable.

Data Availability Statement: The dataset published in this manuscript is part of a larger dataset that has not been shared onto data repository. Requests for material can be made to the corresponding authors.

Acknowledgments: The authors would like to thank Isana Veksler-Lublinsky from the Department of Software and Information Systems Engineering at the Ben-Gurion University of the Negev for her excellent assistance in miRNAs-predicted gene target analysis.

Conflicts of Interest: The authors declare no conflict of interest.

References

1. Pizzino, G.; Irrera, N.; Cucinotta, M.; Pallio, G.; Mannino, F.; Arcoraci, V.; Squadrito, F.; Altavilla, D.; Bitto, A. Oxidative stress: Harms and benefits for human health. *Oxidative Med. Cell. Longev.* **2017**, *2017*, 8416763. [CrossRef]
2. Schieber, M.; Chandel, N.S. ROS function in redox signaling and oxidative stress. *Curr. Biol.* **2014**, *24*, R453–R462. [CrossRef]
3. Frijhoff, J.; Winyard, P.G.; Zarkovic, N.; Davies, S.S.; Stocker, R.; Cheng, D.; Knight, A.R.; Taylor, E.L.; Oettrich, J.; Ruskovska, T. Clinical relevance of biomarkers of oxidative stress. *Antioxid. Redox Signal.* **2015**, *23*, 1144–1170. [CrossRef]
4. Sies, H.; Berndt, C.; Jones, D.P. Oxidative stress. *Annu. Rev. Biochem.* **2017**, *86*, 715–748. [CrossRef]
5. Sena, L.A.; Chandel, N.S. Physiological roles of mitochondrial reactive oxygen species. *Mol. Cell* **2012**, *48*, 158–167. [CrossRef]
6. Ray, P.D.; Huang, B.-W.; Tsuji, Y. Reactive oxygen species (ROS) homeostasis and redox regulation in cellular signaling. *Cell. Signal.* **2012**, *24*, 981–990. [CrossRef]
7. Cuadrado, A.; Rojo, A.I.; Wells, G.; Hayes, J.D.; Cousin, S.P.; Rumsey, W.L.; Attucks, O.C.; Franklin, S.; Levonen, A.-L.; Kensler, T.W. Therapeutic targeting of the NRF2 and KEAP1 partnership in chronic diseases. *Nat. Rev. Drug Discov.* **2019**, *18*, 295–317. [CrossRef]
8. Dodson, M.; De La Vega, M.R.; Cholanians, A.B.; Schmidlin, C.J.; Chapman, E.; Zhang, D.D. Modulating NRF2 in disease: Timing is everything. *Annu. Rev. Pharmacol. Toxicol.* **2019**, *59*, 555. [CrossRef]
9. He, F.; Ru, X.; Wen, T. NRF2, a transcription factor for stress response and beyond. *Int. J. Mol. Sci.* **2020**, *21*, 4777. [CrossRef]
10. Gardiner, C.; Vizio, D.D.; Sahoo, S.; Théry, C.; Witwer, K.W.; Wauben, M.; Hill, A.F. Techniques used for the isolation and characterization of extracellular vesicles: Results of a worldwide survey. *J. Extracell. Vesicles* **2016**, *5*, 32945. [CrossRef]
11. Raposo, G.; Stoorvogel, W. Extracellular vesicles: Exosomes, microvesicles, and friends. *J. Cell Biol.* **2013**, *200*, 373–383. [CrossRef]
12. Colombo, M.; Raposo, G.; Théry, C. Biogenesis, secretion, and intercellular interactions of exosomes and other extracellular vesicles. *Annu. Rev. Cell Dev. Biol.* **2014**, *30*, 255–289. [CrossRef]
13. Zhang, H.-G.; Grizzle, W.E. Exosomes: A novel pathway of local and distant intercellular communication that facilitates the growth and metastasis of neoplastic lesions. *Am. J. Pathol.* **2014**, *184*, 28–41. [CrossRef] [PubMed]
14. Elliott, R.O.; He, M. Unlocking the power of exosomes for crossing biological barriers in drug delivery. *Pharmaceutics* **2021**, *13*, 122. [CrossRef]
15. Gurung, S.; Perocheau, D.; Touramanidou, L.; Baruteau, J. The exosome journey: From biogenesis to uptake and intracellular signalling. *Cell Commun. Signal.* **2021**, *19*, 47. [CrossRef]
16. Zanon-Moreno, V.; Marco-Ventura, P.; Lleo-Perez, A.; Pons-Vazquez, S.; Garcia-Medina, J.J.; Vinuesa-Silva, I.; Moreno-Nadal, M.A.; Pinazo-Duran, M.D. Oxidative stress in primary open-angle glaucoma. *J. Glaucoma* **2008**, *17*, 263–268. [CrossRef] [PubMed]
17. Acott, T.S.; Kelley, M.J. Extracellular matrix in the trabecular meshwork. *Exp. Eye Res.* **2008**, *86*, 543–561. [CrossRef]
18. Kumar, D.M.; Agarwal, N. Oxidative stress in glaucoma: A burden of evidence. *J. Glaucoma* **2007**, *16*, 334–343. [CrossRef]
19. Dismuke, W.M.; Challa, P.; Navarro, I.; Stamer, W.D.; Liu, Y. Human aqueous humor exosomes. *Exp. Eye Res.* **2015**, *132*, 73–77. [CrossRef] [PubMed]
20. Lerner, N.; Chen, I.; Schreiber-Avissar, S.; Beit-Yannai, E. Extracellular Vesicles Mediate Anti-Oxidative Response—In Vitro Study in the Ocular Drainage System. *Int. J. Mol. Sci.* **2020**, *21*, 6105. [CrossRef]
21. Li, Y.-C.; Zheng, J.; Wang, X.-Z.; Wang, X.; Liu, W.-J.; Gao, J.-L. Mesenchymal stem cell-derived exosomes protect trabecular meshwork from oxidative stress. *Sci. Rep.* **2021**, *11*, 14863. [CrossRef]
22. Zhang, W.; Liu, R.; Chen, Y.; Wang, M.; Du, J. Crosstalk between Oxidative Stress and Exosomes. *Oxidative Med. Cell Longev.* **2022**, *2022*, 3553617. [CrossRef] [PubMed]
23. Cheng, X.; Ku, C.-H.; Siow, R.C. Regulation of the Nrf2 antioxidant pathway by microRNAs: New players in micromanaging redox homeostasis. *Free Radic. Biol. Med.* **2013**, *64*, 4–11. [CrossRef]
24. Rong, R.; Wang, M.; You, M.; Li, H.; Xia, X.; Ji, D. Pathogenesis and prospects for therapeutic clinical application of noncoding RNAs in glaucoma: Systematic perspectives. *J. Cell Physiol.* **2021**, *236*, 7097–7116. [CrossRef]
25. Molasy, M.; Walczak, A.; Szaflik, J.; Szaflik, J.P.; Majsterek, I. MicroRNAs in glaucoma and neurodegenerative diseases. *J. Hum. Genet.* **2017**, *62*, 105–112. [CrossRef] [PubMed]
26. Greene, K.M.; Stamer, W.D.; Liu, Y. The role of microRNAs in glaucoma. *Exp. Eye Res.* **2021**, *215*, 108909. [CrossRef] [PubMed]
27. Tabak, S.; Schreiber-Avissar, S.; Beit-Yannai, E. Crosstalk between MicroRNA and oxidative stress in primary open-angle glaucoma. *Int. J. Mol. Sci.* **2021**, *22*, 2421. [CrossRef]
28. Coca-Prados, M.; Wax, M.B. Transformation of human ciliary epithelial cells by simian virus 40: Induction of cell proliferation and retention of beta 2-adrenergic receptors. *Proc. Natl. Acad. Sci. USA* **1986**, *83*, 8754–8758. [CrossRef]
29. Werber, J.; Wang, Y.J.; Milligan, M.; Li, X.; Ji, J.A. Analysis of 2, 2′-azobis (2-amidinopropane) dihydrochloride degradation and hydrolysis in aqueous solutions. *J. Pharm. Sci.* **2011**, *100*, 3307–3315. [CrossRef]
30. Théry, C.; Amigorena, S.; Raposo, G.; Clayton, A. Isolation and characterization of exosomes from cell culture supernatants and biological fluids. *Curr. Protoc. Cell Biol.* **2006**, *30*, 3.22. 21–23.22. 29. [CrossRef]
31. Wong, N.; Wang, X. miRDB: An online resource for microRNA target prediction and functional annotations. *Nucleic Acids Res.* **2015**, *43*, D146–D152. [CrossRef] [PubMed]
32. Ben Or, G.; Veksler-Lublinsky, I. Comprehensive machine-learning-based analysis of microRNA–target interactions reveals variable transferability of interaction rules across species. *BMC Bioinform.* **2021**, *22*, 264. [CrossRef]

33. Karagkouni, D.; Paraskevopoulou, M.D.; Chatzopoulos, S.; Vlachos, I.S.; Tastsoglou, S.; Kanellos, I.; Papadimitriou, D.; Kavakiotis, I.; Maniou, S.; Skoufos, G. DIANA-TarBase v8: A decade-long collection of experimentally supported miRNA–gene interactions. *Nucleic Acids Res.* **2018**, *46*, D239–D245. [CrossRef] [PubMed]

34. Huang, H.-Y.; Lin, Y.-C.-D.; Li, J.; Huang, K.-Y.; Shrestha, S.; Hong, H.-C.; Tang, Y.; Chen, Y.-G.; Jin, C.-N.; Yu, Y. miRTarBase 2020: Updates to the experimentally validated microRNA–target interaction database. *Nucleic Acids Res.* **2020**, *48*, D148–D154. [CrossRef]

35. Wang, M.-x.; Zhao, J.; Zhang, H.; Li, K.; Niu, L.-z.; Wang, Y.-p.; Zheng, Y.-j. Potential protective and therapeutic roles of the Nrf2 pathway in ocular diseases: An update. *Oxidative Med. Cell. Longev.* **2020**, *2020*, 9410952. [CrossRef]

36. Lerner, N. *Non-Pigmented Ciliary Epithelium Derived Exosomes and Their Role within the Drainage System as a Pharmacological Intervention Target for Glaucoma*; Ben-Gurion University of the Negev: Beersheba, Israel, 2019.

37. Kobayashi, M.; Yamamoto, M. Molecular mechanisms activating the Nrf2-Keap1 pathway of antioxidant gene regulation. *Antioxid. Redox Signal.* **2005**, *7*, 385–394. [CrossRef]

38. Huber, W.; Von Heydebreck, A.; Sültmann, H.; Poustka, A.; Vingron, M. Variance stabilization applied to microarray data calibration and to the quantification of differential expression. *Bioinformatics* **2002**, *18*, S96–S104. [CrossRef]

39. Hedlund, M.; Nagaeva, O.; Kargl, D.; Baranov, V.; Mincheva-Nilsson, L. Thermal-and oxidative stress causes enhanced release of NKG2D ligand-bearing immunosuppressive exosomes in leukemia/lymphoma T and B cells. *PLoS ONE* **2011**, *6*, e16899. [CrossRef]

40. de Jong, O.G.; Verhaar, M.C.; Chen, Y.; Vader, P.; Gremmels, H.; Posthuma, G.; Schiffelers, R.M.; Gucek, M.; Van Balkom, B.W. Cellular stress conditions are reflected in the protein and RNA content of endothelial cell-derived exosomes. *J. Extracell. Vesicles* **2012**, *1*, 18396. [CrossRef]

41. Lerner, N.; Schreiber-Avissar, S.; Beit-Yannai, E. Extracellular vesicle-mediated crosstalk between NPCE cells and TM cells result in modulation of Wnt signalling pathway and ECM remodelling. *J. Cell. Mol. Med.* **2020**, *24*, 4646–4658. [CrossRef]

42. Tabak, S.; Feinshtein, V.; Schreiber-Avissar, S.; Beit-Yannai, E. Non-Pigmented Ciliary Epithelium-Derived Extracellular Vesicles Loaded with SMAD7 siRNA Attenuate Wnt Signaling in Trabecular Meshwork Cells In Vitro. *Pharmaceuticals* **2021**, *14*, 858. [CrossRef] [PubMed]

43. Tabak, S.; Hadad, U.; Schreiber-Avissar, S.; Beit-Yannai, E. Non-pigmented ciliary epithelium derived extracellular vesicles uptake mechanism by the trabecular meshwork. *FASEB J.* **2021**, *35*, e21188. [CrossRef] [PubMed]

44. Tabak, S.; Schreiber-Avissar, S.; Beit-Yannai, E. Trabecular meshwork's collagen network formation is inhibited by non-pigmented ciliary epithelium-derived extracellular vesicles. *J. Cell. Mol. Med.* **2021**, *25*, 3339–3347. [CrossRef]

45. Morgan, J.T.; Raghunathan, V.K.; Chang, Y.R.; Murphy, C.J.; Russell, P. Wnt inhibition induces persistent increases in intrinsic stiffness of human trabecular meshwork cells. *Exp. Eye Res.* **2015**, *132*, 174–178. [CrossRef] [PubMed]

46. Webber, H.C.; Bermudez, J.Y.; Sethi, A.; Clark, A.F.; Mao, W. Crosstalk between TGFbeta and Wnt signaling pathways in the human trabecular meshwork. *Exp. Eye Res.* **2016**, *148*, 97–102. [CrossRef]

47. Zhao, J.; Du, X.; Wang, M.; Yang, P.; Zhang, J. Salidroside mitigates hydrogen peroxide-induced injury by enhancement of microRNA-27a in human trabecular meshwork cells. *Artif. Cells Nanomed. Biotechnol.* **2019**, *47*, 1758–1765. [CrossRef]

48. Tian, X.; Liu, Y.; Wang, H.; Zhang, J.; Xie, L.; Huo, Y.; Ma, W.; Li, H.; Chen, X.; Shi, P. The role of miR-199b-3p in regulating Nrf2 pathway by dihydromyricetin to alleviate septic acute kidney injury. *Free Radic. Res.* **2021**, *55*, 842–852. [CrossRef]

49. Wang, M.; Zheng, Y. Oxidative stress and antioxidants in the trabecular meshwork. *PeerJ* **2019**, *7*, e8121. [CrossRef]

50. Agarwal, P.; Daher, A.M.; Agarwal, R. Aqueous humor TGF-beta2 levels in patients with open-angle glaucoma: A meta-analysis. *Mol. Vis.* **2015**, *21*, 612–620.

51. Clark, A.F.; Miggans, S.T.; Wilson, K.; Browder, S.; McCartney, M.D. Cytoskeletal changes in cultured human glaucoma trabecular meshwork cells. *J. Glaucoma* **1995**, *4*, 183–188. [CrossRef]

52. Brigelius-Flohe, R.; Kipp, A.P. Selenium in the redox regulation of the Nrf2 and the Wnt pathway. *Methods Enzym.* **2013**, *527*, 65–86. [CrossRef]

53. Villarreal, G., Jr.; Chatterjee, A.; Oh, S.S.; Oh, D.J.; Kang, M.H.; Rhee, D.J. Canonical wnt signaling regulates extracellular matrix expression in the trabecular meshwork. *Investig. Ophthalmol. Vis. Sci.* **2014**, *55*, 7433–7440. [CrossRef] [PubMed]

54. Vranka, J.A.; Kelley, M.J.; Acott, T.S.; Keller, K.E. Extracellular matrix in the trabecular meshwork: Intraocular pressure regulation and dysregulation in glaucoma. *Exp. Eye Res.* **2015**, *133*, 112–125. [CrossRef] [PubMed]

55. Tanaka, Y.; Tsuda, S.; Kunikata, H.; Sato, J.; Kokubun, T.; Yasuda, M.; Nishiguchi, K.M.; Inada, T.; Nakazawa, T. Profiles of extracellular miRNAs in the aqueous humor of glaucoma patients assessed with a microarray system. *Sci. Rep.* **2014**, *4*, 5089. [CrossRef]

56. Zhang, Y.; Che, D.; Cao, Y.; Yue, Y.; He, T.; Zhu, Y.; Zhou, J. MicroRNA Profiling in the Aqueous Humor of Keratoconus Eyes. *Transl. Vis. Sci. Technol.* **2022**, *11*, 5. [CrossRef]

57. Dunmire, J.J.; Lagouros, E.; Bouhenni, R.A.; Jones, M.; Edward, D.P. MicroRNA in aqueous humor from patients with cataract. *Exp. Eye Res.* **2013**, *108*, 68–71. [CrossRef]

58. Henderson, J.; Wilkinson, S.; Przyborski, S.; Stratton, R.; O'Reilly, S. microRNA27a-3p mediates reduction of the Wnt antagonist sFRP-1 in systemic sclerosis. *Epigenetics* **2021**, *16*, 808–817. [CrossRef]

59. Latarya, G.; Mansour, A.; Epstein, I.; Cotlear, D.; Pikkel, J.; Levartovsky, S.; Yulish, M.; Beit-Yannai, E. Human aqueous humor phosphatase activity in cataract and glaucoma. *Investig. Ophthalmol. Vis. Sci.* **2012**, *53*, 1679–1684. [CrossRef]
60. Yu, S.; Tam, A.L.; Campbell, R.; Renwick, N. Emerging Evidence of Noncoding RNAs in Bleb Scarring after Glaucoma Filtration Surgery. *Cells* **2022**, *11*, 1301. [CrossRef]

antioxidants

MDPI

Review

Activators of Nrf2 to Counteract Neurodegenerative Diseases

Rosa Amoroso [1], Cristina Maccallini [1,*] and Ilaria Bellezza [2]

[1] Department of Pharmacy, University "G.d'Annunzio" of Chieti-Pescara, Via dei Vestini, 31, 66100 Chieti, Italy; rosa.amoroso@unich.it

[2] Department of Medicine and Surgery, University of Perugia, Polo Unico Sant'Andrea delle Fratte, P.e Lucio Severi 1, 06132 Perugia, Italy; ilaria.bellezza@unipg.it

* Correspondence: cristina.maccallini@unich.it

Abstract: Neurodegenerative diseases are incurable and debilitating conditions that result in progressive degeneration and loss of nerve cells. Oxidative stress has been proposed as one factor that plays a potential role in the pathogenesis of neurodegenerative disorders since neuron cells are particularly vulnerable to oxidative damage. Nuclear factor (erythroid-derived 2)-like 2 (Nrf2) is strictly related to anti-inflammatory and antioxidative cell response; therefore, its activation and the consequent enhancement of the related cellular pathways have been proposed as a potential therapeutic approach. Several Nrf2 activators with different mechanisms and diverse structures have been reported, but those applied for neurodisorders are still limited. However, in the very last few years, interesting progress has been made, particularly in enhancing the blood–brain barrier penetration, to make Nrf2 activators effective drugs, and in designing Nrf2-based multitarget-directed ligands to affect multiple pathways involved in the pathology of neurodegenerative diseases. The present review gives an overview of the most representative findings in this research area.

Keywords: Alzheimer's disease; antioxidant response; electrophilic activators; multitargeting compounds; Nrf2 inducers; protein–protein interaction inhibitors; oxidative stress; Parkinson's disease

Citation: Amoroso, R.; Maccallini, C.; Bellezza, I. Activators of Nrf2 to Counteract Neurodegenerative Diseases. *Antioxidants* **2023**, *12*, 778. https://doi.org/10.3390/antiox12030778

Academic Editor: Marcel Bonay

Received: 10 February 2023
Revised: 15 March 2023
Accepted: 18 March 2023
Published: 22 March 2023

1. Introduction

Neurodegenerative diseases affect millions of people worldwide and are a leading cause of disability and a major cause of mortality [1]. The treatment of these pathological conditions is only palliative; therefore, there is an urgent need for effective therapeutic agents, as well as a deeper understanding of the molecular changes affecting neuronal cells during the disease progression. Neuroinflammation is a hallmark in the development of neurodegenerative diseases, as well as nitroxidative stress, which is due to the unbalanced production of both reactive oxygen species (ROS) and reactive nitrogen species (RNS) [2,3]. The key role played by ROS in the onset of age-related neurodegenerative diseases indicates that erythroid-derived 2-like 2 (Nrf2), the master regulator of redox homeostasis [4], may be a promising target for therapeutic interventions. Several Nrf2 activators with different mechanisms and diverse structures have been reported in the literature, but those applied to neurodegenerative diseases are still limited [5]. However, in the very last few years, interesting progress has been made in this field, particularly in enhancing the blood–brain barrier (BBB) permeability of Nrf2 activators, to make them effective drugs, and in designing Nrf2-based multitarget-directed ligands to affect multiple pathways involved in the pathology of neurodegenerative diseases. In the present review, the implications of oxidative stress and Nrf2 activation in the therapy of neurodegeneration are revised, with a particular focus on Alzheimer's disease (AD) and Parkinson's disease (PD), which are the two most prevalent age-related neurodegenerative diseases [1]. Moreover, progress in the development of new Nrf2 activators able to counteract neuroinflammation is discussed.

2. Implication of Oxidative Stress and Nrf2 Activation in Neurodegenerative Diseases

ROS are a byproduct of several cellular metabolic pathways and enzymatic reactions, and they are classified as radicals, including superoxide anion ($O_2^{\bullet-}$) and hydroxyl ($HO^{\bullet-}$), as well as nonradical species such as hydrogen peroxide (H_2O_2) [6]. ROS are reactive molecules with a very short half-life. H_2O_2, the most stable ROS, has a cellular half-life of 10^{-3} s, 1000 times higher than other ROS [7,8]. Although ROS production might be due to environmental factors, it mainly derives from metabolic activities. During mitochondrial respiration, in the electron transport chain (ETC), approximately 1–2% of O_2 is not reduced to water, leading to the generation of $O_2^{\bullet-}$ and H_2O_2. Moreover, cytosolic oxidoreductases such as NADPH oxidases (NOX), cytochrome P450 (CYP) oxidases, cyclooxygenases (COX), and monoamine oxidases (MAO) may contribute to ROS production [8]. Physiological ROS levels are maintained by a plethora of exogenous and endogenous antioxidant defenses. Among the exogenous antioxidants, ascorbic acid (vitamin C), α-tocopherol (vitamin E), and carotenoids play a pivotal role. Endogenous antioxidants include enzymatic antioxidants, e.g., superoxide dismutase (SOD), glutathione peroxidase (GPX), and catalase (CAT), as well as nonenzymatic scavengers, e.g., glutathione (GSH) (Figure 1) [9].

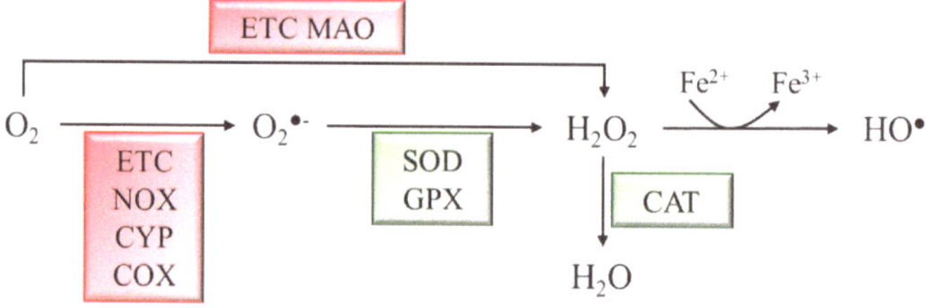

Figure 1. Schematic representation of ROS generation and scavenging. Molecular oxygen (O_2) can be partially reduced to superoxide anion ($O_2^{\bullet-}$) by the ETC, NADPH oxidases (NOX), cytochrome P450 (CYP) oxidases, and cyclooxygenases (COX), or to hydrogen peroxide (H_2O_2) by the ETC and monoamine oxidases (MAO). H_2O_2 can be converted into hydroxyl radical ($HO^{\bullet-}$) through the Fenton rection. Enzymatic antioxidants scavenge ROS to produce less reactive molecules. Superoxide dismutase (SOD) and glutathione peroxidase (GPX) convert $O_2^{\bullet-}$ to H_2O_2, which is converted into H_2O by catalase (CAT).

Redox homeostasis should be strictly controlled since ROS play fundamental biological roles in guaranteeing redox signaling, a transduction system in which reversible electron transfer reactions involving ROS to effector target proteins culminate in the regulation of numerous physiological functions, including neuronal development and function, cellular proliferation and differentiation, and aging prevention [10]. For example, due to its ability to cross the phospholipidic bilayer of the cellular membrane, H_2O_2 can act in both an autocrine and a paracrine manner [7]. When the production of ROS exceeds cellular detoxification capacity, redox balance gets compromised, and oxidative stress insurges. In these conditions, high ROS levels can oxidize nucleic acids, proteins, and lipids, thus leading to cell dysfunction and eventually cell death [9,11].

The master regulator of redox homeostasis is erythroid-derived 2-like 2 (Nrf2) [4]. Nrf2 is a cap 'n' collar (CNC) basic leucine zipper (bZIP) transcription factor responsible for the expression of genes containing the antioxidant responsive element (ARE) sequence in their promoter region, including genes linked to the synthesis or use of GSH [12]. Under basal conditions, Nrf2 has a rapid turnover, with a half-life of approximately 20 min. In fact, Nrf2 activity is controlled by a cytoplasmic repressor protein Keap1, which sequesters Nrf2 in the cytosol and, by recruiting CUL3-dependent E3-ubiquitin ligase, leads to its ubiquitination and proteasomal degradation [13,14]. Under oxidative stress conditions, two of the 27 Cys residues in Keap1 become oxidized, causing a conformational change that, according to the "hinge and latch" model [15,16], impedes the correct orientation of Nrf2 and inhibits its ubiquitination and degradation [13]. The newly synthesized Nrf2 can in turn translocate to the nucleus to exert its functions [12]. Outside of Keap-1, Nrf2 activity is also controlled by other regulators. For example, glycogen synthase kinase-3β (GSK-3β) phosphorylates Nrf2, aiding its ubiquitination by β-transducin repeat-containing protein (β -TrCP)/Cullin-1 E3 ubiquitin ligase in a Keap-1-independent manner [17]. GSK-3β also phosphorylates the protein kinase Fyn, which translocates into the nucleus where it phosphorylates Nrf2, leading to its nuclear export and ubiquitin-dependent proteasomal degradation [18]. To add a further layer of complexity, the BTB and CNC homology transcription factors (BACH1 and BACH2) repress Nrf2 activity by competing for ARE binding [19]. On the other hand, p62/sequestosome 1 (p62/SQSTM1), a ubiquitin-binding protein, competing with Nrf2 for Keap-1 binding, leads to Nrf2 stabilization and, hence, activation (Figure 2) [20].

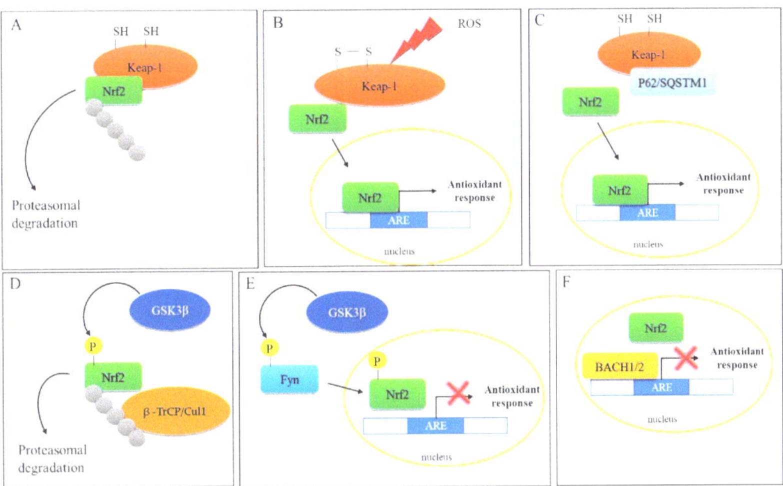

Figure 2. Schematic representation of Nrf2 activation. (**A**) Under basal conditions, Nrf2 activity is controlled by the cytoplasmic repressor protein Keap1, which sequesters Nrf2 in the cytosol, leading to its ubiquitination and proteasomal degradation. (**B**) Under oxidative stress conditions, two of the Cys residues in Keap1 become oxidized, impeding the correct orientation of Nrf2 and inhibiting its ubiquitination and degradation. The newly synthesized Nrf2 translocates to the nucleus to induce the expression of antioxidant response genes. (**C**) P62/sequestosome 1 (p62/SQSTM1) competes with Nrf2 for Keap-1 binding and leads to Nrf2 activation and nuclear translocation. (**D**) Glycogen synthase kinase-3β (GSK-3β) phosphorylates Nrf2, which is in turn ubiquitinated by β-transducin repeat-containing protein/Cullin-1 E3 ubiquitin ligase (β–TrCP/Cul1). (**E**) GSK-3β phosphorylates the protein kinase Fyn, which phosphorylates Nrf2, leading to its nuclear export and culminating in inhibition of gene expression (red cross). (**F**) The BTB and CNC homology transcription factors (BACH1 and BACH2) repress Nrf2 activity (red cross) by competing for ARE binding.

Due to the high oxygen consumption, the presence of polyunsaturated fatty acids, and auto-oxidation of neurotransmitters (e.g., dopamine), the brain is highly vulnerable to ROS-mediated damage [21]. Thus, any imbalance of redox homeostasis could affect brain cells. For example, aging and age-related diseases such as neurodegenerative disorders have been linked to the progressive dysfunction of redox control mechanisms [22]. The disruption of redox balance during healthy aging correlates with the finding that Nrf2 activity in brain decreases with age [23]. Brain expression of Nrf2 is higher in glial cells (astrocytes and microglia) than in neurons, suggesting that glial cells can be causally involved in neurodegenerative diseases despite not being the cells primarily affected by the disease. This concept, known as a non-cell-autonomous mechanism, is recognized to essentially contribute to neurodegeneration. It has been proposed that microglia and astrocytes causally participate in the pathogenesis and progression of several neurodegenerative diseases [24,25].

Microglial cells behave as the innate immune cells of the CNS [26]. Resting microglial cells have a ramified morphology; however, when activated by a danger signal, they undergo a morphological change, becoming amoeboid. While resting microglial cells continuously monitor brain activities to maintain homeostasis, activated microglial cells release cytokines and proinflammatory mediators to eliminate the threat. This acute response of microglial cells is protective, but overactivation of microglia, by inducing severe oxidative stress and neuroinflammation, leads to many neurodegenerative diseases [26]. In aged microglial cells, migration ability and phagocytosis are compromised, and the release of proinflammatory cytokines lasts for prolonged periods, strongly contributing to neurodegeneration [27]. Disturbance of Nrf2 activity and, hence, increased oxidative stress have been linked to several neurodegenerative diseases such as Parkinson's disease and Alzheimer's disease.

2.1. Nrf2 and Parkinson's Disease

PD is a progressive neurological movement disorder characterized by tremor, bradykinesia, rigid muscles, and loss of postural balance [28]. The clinical feature of PD is the depletion of dopaminergic (DArgic) neurons in the substantia nigra pars compacta (SNpc), which results in dopamine (DA) deficiency [29]. PD is prevalently an idiopathic disease, with only 10–15% of cases having a family history. However, Lewy bodies found in both sporadic and familial cases of PD contain aggregated forms of α-synuclein (α-syn). The α-syn oligomers are cytotoxic to DArgic neurons [30]. The toxic effects of α-syn oligomers can be ascribed to α-syn enzymatic ferrireductase activity. Aggregation indeed inhibits α-syn enzymatic activity, leading to an accumulation of oxidized Fe, which can participate in the Fenton reaction, culminating in oxidative stress induction [31]. Moreover, mitochondrial α-syn accumulation can suppress mitochondrial respiratory complex I, leading to ROS production and, thus, to oxidative stress [32]. The crucial role of mitochondria-generated ROS in PD pathogenesis was determined from the finding that the 1-methyl-4-phenyl 1,2,3,6-tetrahydropyridine (MPTP) metabolite 1-methyl-4-phenylpyridinium (MPP$^+$) hampers mitochondrial complex I, causing electron leakage and ROS production [33]. PD-like symptoms indeed arose in individuals using illegal drugs contaminated by MPTP, and brain postmortem examinations revealed damage to the DArgic neurons in the SNpc.

Other PD genetic risk factors comprise genes involved in the control of mitochondrial redox balance including Parkin, PTEN-induced kinase 1 (PINK1), parkinsonism-associated deglycase (PARK7), and leucine-rich repeat serine/threonine protein kinase 2 (LRRK2) [34]. Parkin, an E3 ubiquitin ligase, and PINK1, a serine/threonine kinase, form a signal transduction pathway exerting a pivotal role in the removal of damaged mitochondria by the mitophagic process [35]; PARK7 encodes for DJ-1, a protein that, by binding the mitochondrial complex I, improves its activity [36]. Mutations in PINK1, DJ-1, and LRRK2 result in mitochondrial dysfunction with impaired bioenergetics and lead to uncontrolled ROS generation [36,37]. It should be emphasized that DA metabolism itself is a source of oxidative stress. In fact, excess DA can be catabolized by monoamine oxidase (MAO), producing H_2O_2 or undergoing auto-oxidation to produce highly reactive dopamine quinones (DAQs) and semiquinones [38].

The crucial role played by ROS in PD indicates Nrf2 implication in this disease. In vivo studies showed that both treatment with MPTP and overexpression of α-syn increased DArgic cell death in Nrf2 KO mice [39,40]. Moreover, the Nrf2-induced cytoprotective gene NQO1 (NADPH-quinone oxidoreductase) is partially sequestered in Lewy bodies, together with p62/SQSTM1, a known activator of Nrf2 [41]. Moreover, Bach1 knockout protects mice against MPTP-induced dopaminergic cell death [42]. Moreover, the brain of LRRK2-transgenic mice and the LRRK2-overexpressing neuronal cell line show reduced expression of Nrf2 and its target genes via GSK3β activation [43]. Studies in Drosophila showed that Nrf2 activation induces mitophagy and counteracts neuronal degeneration in Parkin/Pink1 knockdown flies [44]. In addition to direct neuronal toxicity, α-syn oligomers can activate microglial cells promoting neuroinflammation, which, by further increasing ROS production via NADPH oxydase 2 (NOX2) activation, augments oxidative stress [34,41]. In this context, DJ-1, by interacting with the p47phox subunit of NOX2, inhibits its action, reducing ROS production [45]. From all the above, PD genetic risk factors are connected to aberrant Nrf2 signaling which might contribute to oxidative stress (Figure 3).

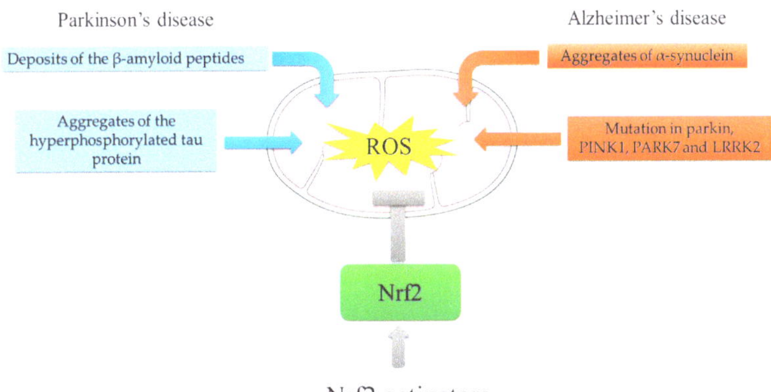

Figure 3. Mitochondrial ROS generation by PD and AD risk factors, and protective effects of Nrf2 activating compounds. ROS generation in AD: deposition of Aβ plaques induces mitochondrial stress; aggregates of the hyperphosphorylated tau protein impair mitochondria distribution and function. ROS generation in PD: aggregated α-syn accumulation suppresses mitochondrial respiratory complex I; mutations in Parkin, PTEN-induced kinase 1 (PINK1), parkinsonism-associated deglycase (PARK7), and leucine-rich repeat serine/threonine protein kinase 2 (LRRK2) result in mitochondrial dysfunction with impaired bioenergetics. The Nrf2-mediated increase in antioxidant response genes can reduce ROS burst. In this light, Nrf2 activators might counteract the ROS-induced damage that leads to neurodegeneration.

2.2. Nrf2 and Alzheimer's Disease

Alzheimer's disease (AD) is the prevalent type of dementia in the elderly population. AD is characterized by memory loss and deterioration of other cognitive functions including comprehension, judgment, and orientation due to synaptic loss and selective neuronal death [46]. AD brains are characterized by the accumulation of extracellular senile plaques composed of deposits of β-amyloid peptides Aβ (1–40) and Aβ (1–42), as well as intracellular neurofibrillary tangles (NFTs) composed of aggregates of the hyperphosphorylated tau protein [40]. Most AD cases are classified as sporadic late-onset AD (onset after 65 years of age), whereas only 5% are autosomal-dominant early-onset AD [47,48]. Early-onset AD is caused by a mutation in three proteins involved in the amyloidogenic pathway: APP (amyloid precursor protein), γ-secretase presenilin 1 (PSEN1), and presenilin 2 (PSEN2). The key role played by ROS in aging and the fact that idiopathic AD is an age-related disease strongly suggest that ROS are involved in AD pathogenesis. This hypothesis is corroborated by the finding that Aβ binds to mitochondrial membranes, thus affecting mitochondrial function and accelerating ROS production [49]. The pivotal role of ROS in AD is based on the findings that Aβ (1–42) induces mitochondrial stress accompanied by ROS generation in AD patients [50,51], and that Aβ plaque deposition leads to oxidative stress, which, propagating spatially over time, leads to neuronal cell death [52,53]. Not only Aβ, but also tau protein is connected to mitochondrial ROS generation in AD. For example, compared to wildtype mice, tau knockout presented reduced oxidative damage, improved memory skills, and improved mitochondrial bioenergetics [54]. Furthermore, ROS induce tau polymerization [55], and tau overexpression induces oxidative stress in vitro and in mice [56,57]. Moreover, Nrf2 activation prevents tau protein aggregation in transgenic mice expressing mutant tau protein [58]. Nrf2-KO transgenic mice carrying both mutated APP and tau died prematurely (approximately 12 months of age) and were characterized by motor deficits, neuroinflammation [59], increased oxidative stress, and high levels of Aβ and tau aggregates [60].

A decrease in expression or inactivation of the p62/SQTRSM1 gene in mice causes neurodegeneration with AD-related symptoms and induces the formation of neurofibrillary tangles [61]. GSK3β is also involved in AD pathology since its hyperactivation has been demonstrated in AD brains [62], and GSK3β-mediated phosphorylation of PSN1 reduces neuronal viability and synaptic plasticity [63]. All these data indicate that AD-critical features, i.e., Aβ and tau protein aggregation, and oxidative stress are deeply related and cooperate in the development of pathology (Figure 3).

3. Small Molecules Inducing Nrf2

So far, the preclinical and clinical studies involving Nrf2 activators have mostly focused on cancer and inflammatory diseases therapies [5]. To date, the use of Nrf2 activators in neurological diseases is quite limited due to the unfavorable pharmacokinetic properties and toxicity issues of these compounds [5]. In the very last few years, advancements have been made in the research of new small-molecular Nrf2 activators endowed with CNS activity, which can be broadly classified into two distinct groups: (1) electrophilic activators that can covalently modify Keap1; (2) activators interfering with the Keap1–Nrf2 protein–protein interactions (PPIs) (Table 1).

3.1. Electrophilic Activators

As previously described, Keap1 ensures that Nrf2 concentration is kept very low thanks to oxidative stress sensors; however, if the oxidative stress increases, the ubiquitination and subsequent degradation of Nrf2 are blocked as a result of the reaction of electrophilic species with the cysteine residue in Keap1 protein [64]. According to this mechanism of induction of Nrf2, the use of electrophiles as drugs able to activate Nrf2 has gained increasing scientific interest, and, in the last few years, many electrophiles able to trigger the Nrf2 pathway have been reported. For example, molecules containing α,β-unsaturated ketones and esters have been shown to react with one or more cysteine residues in Keap1 [65,66], activating Nrf2. Indeed, these compounds are Michael acceptors, and they can undergo the nucleophilic addition of an alkyl thiol to their α,β-unsaturated system.

Chalcone contains an electrophilic α,β-unsaturated ketone that can react with the sulfhydryl groups of thioredoxin and cysteine residues in proteins [67]. Therefore, the α,β-unsaturated ketone electrophilic activity is responsible for the various biological activities of chalcones. One of the very first structure–activity relationship (SAR) studies on chalcone-based structures able to activate Nrf2 was reported in 2011 by Kumar et al. [68]. A series of 59 chalcone derivatives was synthesized, and compound **1** emerged as the most interesting molecule (Figure 4).

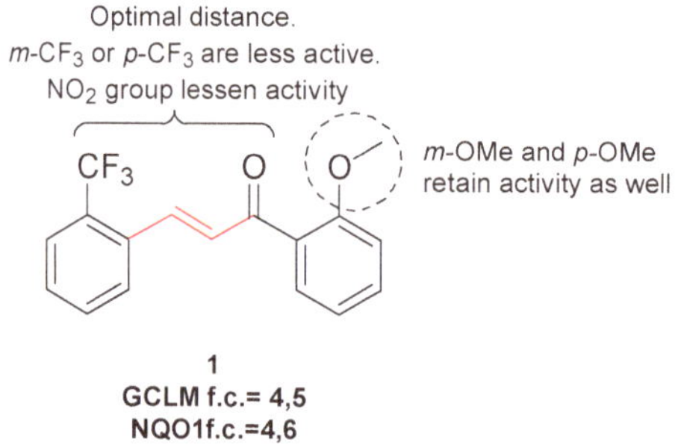

Figure 4. Compound **1**'s chemical structure and activity. The electrophilic α,β-unsaturated moiety of **1** is in red. This compound enhanced the expression of the antioxidant genes GCLM and NADPH-NQO1. The increase is reported as the relative fold change (f.c.) [68].

On the basis of this promising finding, the chalcone leading scaffold, i.e., the 1,3-diphenylprop-2-en-1-one core structure, was recently further investigated including heterocyclic amines such as morpholine, pyrrolidine, and N-methylpiperazine (Figure 5) [69]. In particular, the introduction of a propylmorphline group into the 2-, 3-, and 4-OH of ring A gave interesting results, especially for the 4-position substituted derivatives [69]. An additional enhancement in the Nrf2 activation was observed by replacing the 4-hydroxyl-propylmorpholine moiety with a 4-hydroxyl-propylpyrrolidine one. In addition, it was found that 2′-position substituted derivatives exhibited better activities than the corresponding 3′- and 4′-position substituted ones. The introduction of a N-methylpiperazine group into the 4-OH of ring A led to compound **2** (Figure 5), which showed the best Nrf2-activating potency in this series of compounds, resulting in an amelioration with respect to sulforaphane (SFN), a well-known strong Nrf2 activator (2 EC_{50} = 0.63 μM vs. SFN EC_{50} = 0.87 μM) [69]. Compound **2** was evaluated in BV-2 microglial cells in both normal and proinflammatory conditions, where it stimulated the expression of Nrf2-dependent antioxidant enzymes and the reduction of ROS, inhibiting inflammatory responses induced by the LPS. Moreover, compound **2** was in vivo studied in the scopolamine-induced amnesia model, demonstrating the ability to restore SOD activity and memory performance in mice.

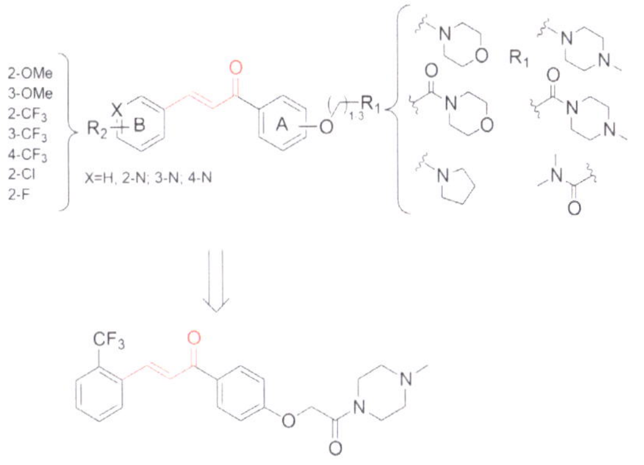

Figure 5. Novel chalcone derivatives as Nrf2 activators. The electrophilic α,β-unsaturated ketone moiety of the chalcone leading scaffold is in red. The introduction of a N-methylpiperazine group into the 4-OH of ring A led to compound **2**, a potent Nrf2 activator (EC_{50} = 0.63 μM) [69].

In another study, the structure of compound **1** was modified by introducing a vinyl-sulfoxide or vinyl-sulfone group to the α,β- unsaturated carbonyl entity of chalcone (Figure 6) [70].

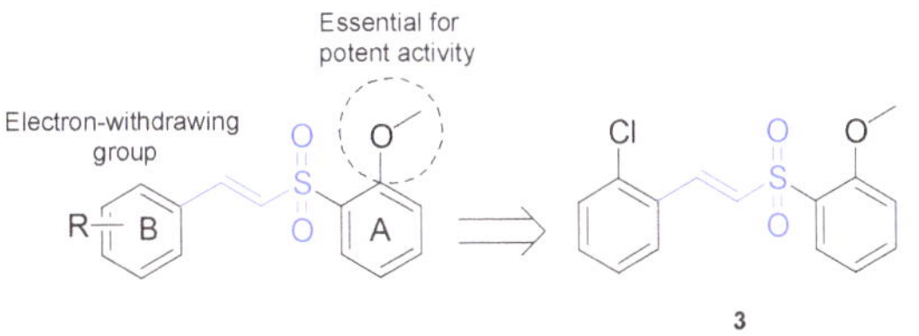

Figure 6. First generation of vinyl-sulfone-based Nrf2 activators. The vinyl-sulfone moiety is in blue. From the structure–activity relationships (SAR) study, it emerged that the presence of an electron-withdrawing group on the B ring favored the Nrf2 activation, while the presence of a 2-methoxyl group on the A ring was essential for the activity. Compound **3** emerged as the most active compound [70].

The introduction of the vinyl sulfone group led to the most potent activities in terms of HO-1 activation compared to the corresponding carbonyl and sulfoxide derivatives. On the basis of its potency of action, compound **3** was selected for further evaluations in DAergic neuronal cells, where it activated Nrf2 as expected, as well as stimulated the expression of the Nrf2-dependent antioxidant enzymes NQO1, GCLC, GLCM, and HO-1. Moreover, compound **3** was evaluated both in vitro and in MPTP-induced in vivo models of PD, demonstrating the attenuation of PD-associated behavioral deficits in the mouse model [70]. However, this molecule was not further investigated due to its poor pharmacokinetic properties; indeed, it showed low solubility, metabolic instability, and toxicity issues, such as cytochrome P (CYP) inhibition and human ether-a-go-go-related gene (hERG) activation. Recently, the molecular structural determinants responsible for vinyl-sulfone derivatives as Nrf2 activators endowed with neuroprotective activity on PC12 cells were investigated via the synthesis of a library of 47 small compounds (Figure 7) [71].

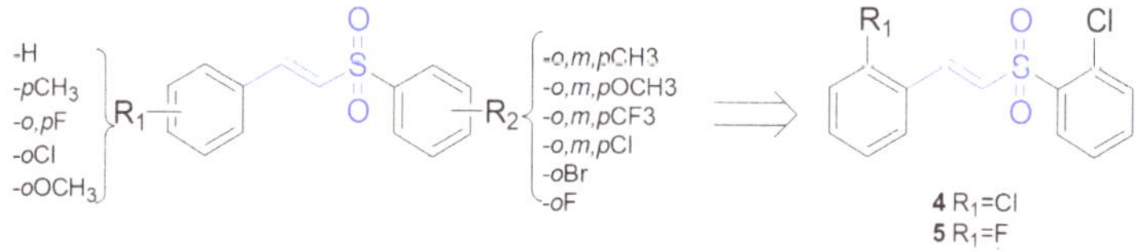

Figure 7. Optimization of vinyl-sulfone-based Nrf2 activators. The vinyl-sulfone moiety is in blue. Tuning the electrophilicity and steric hindrance of the aromatic substituents led to new Nrf2 activators. Compounds **4** and **5** were the most promising molecules of this series [71].

This study found that best results in terms of neuroprotection were obtained by molecules bearing aromatic *o*-substituents, and that, by tuning their electrophilicity and steric hindrance, potent Nrf2 activators can be obtained. In particular, the most interesting compounds of the considered series were **4** and **5** (Figure 7), which protected PC12 cells from H_2O_2-mediated insults, promoting the translocation of Nrf2 and the consequent activation of ARE, with the final transcriptional activation of antioxidant genes and the consequent upregulation of a diverse range of antioxidant species (TrxR, NQO1, GSH, and HO-1). Some years later, in order to improve the druglike properties of compound **3**, the B phenyl-ring was replaced by a pyridyl-moiety, while a morpholinyl group was linked to ring A to improve microsomal stability and increase solubility through the HCl salt form (**6**, Figure 8) [72].

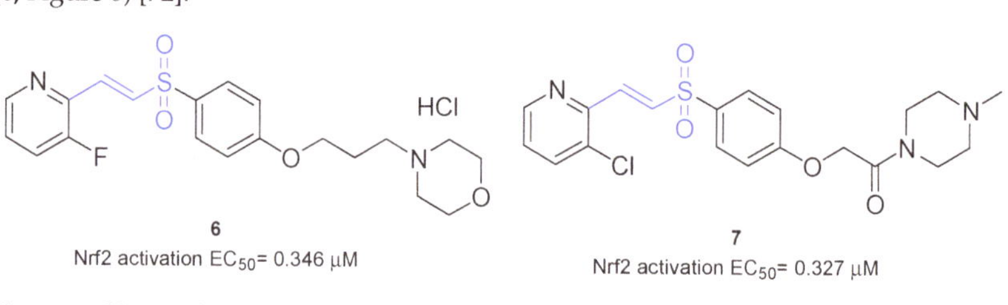

Figure 8. Chemical structure of vinyl-sulfone-based Nrf2 activators with improved activity and drug-like properties. The vinyl-sulfone moiety is in blue. The introduction of both a pyridyl group in the B ring and a morpholinyl or piperazinyl group in the A ring improved the microsomal stability and solubility of these compounds [72].

Moreover, considering the potency of morpholine compounds with 3′-F or 3′-Cl on *o*-pyridine, a 4-piperazine group was introduced instead of the morpholino one (**7**, Figure 8). Compound **6** demonstrated excellent Nrf2 activation potency, safety, and stability, as well as elicited neuroprotective effects against DAergic neuronal cell death in PD, suppressing microglial activation related to neuroinflammation [72]. In a subsequent study, a series of pyridyl-vinyl-sulfones bearing a halogenated A-ring was synthesized in an attempt to maximize the potency of action of this class of Nrf2 activators [73]. Compound **8**

(Figure 9) emerged as the most promising derivative, being able to activate Nrf2 at nanomolar concentrations, with an ameliorated potency of action with respect to the previous developed compounds. Moreover, compound **8** retained positive drug-like properties and an interesting biological profile in both in vitro and in vivo models of PD [73].

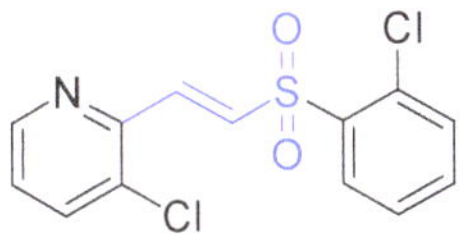

Figure 9. Chemical structure of a potent halogenated vinyl-sulfone-based Nrf2 activator. The vinyl-sulfone moiety is in blue. Compound **8** showed potent Nrf2 activation (EC$_{50}$ = 26 nM) and drug-like properties [73].

3.2. PPI Interfering Compounds: Azole-Based Compounds

Interference with the protein–protein interaction between the KEAP1 Kelch domain and Nrf2 by means of non-electrophilic noncovalent inhibitors is increasingly being explored as an innovative approach to generate compounds activating Nrf2. These molecules are predicted to be safer with respect to electrophilic compounds, whose action is often associated with side-effects due to off-target actions, derived from their simultaneous reaction with different types of nucleophiles commonly present in many biological molecules. Several promising PPI inhibitors have been disclosed to date that are able to enhance Nrf2 signaling with different potential therapeutic applications [74], and diaryl-azole-based compounds in particular have been demonstrated as useful agents in neurological diseases. Starting from oxadiazole derivatives **9–12** (Figure 10) which were able to increase Nrf2 nuclear translocation up to 3.46-fold at 100 nM [75,76], similar triazole derivatives were designed, and compound **13** (Figure 10) was the most interesting one, showing promising results as neuroprotective agents [77,78]. In this compound, the chemical reactivity of the S-group with cysteine nucleophiles is essential for Nrf2 activation, as well as the presence of an electron-withdrawing group and an electron-deficient aromatic system. Very recently, diaryl-triazole derivatives were further investigated as Nrf2 activators able to counteract neuroinflammation. In particular, based on the leading scaffold of compound **11**, i.e., the diaryl-oxadiazole core structure (Figure 10), a library of 26 triazole-based compounds was synthesized (Figure 11), and the in vitro evaluation revealed that these molecules exerted better neuroprotective effects than **11** [79]. From the SAR analysis it emerged that alkyl groups were important to augment the protective effect; in particular, compound **14** (Figure 11) gave the best results. From the in-depth biological evaluation of this molecule, it emerged as a potential neuroprotectant for the treatment of ischemic stroke [79].

Further modifications of compound's **11** azole core led to a series of 1,3,4-oxadiazole or 1,3,4-thiadiazole derivatives (Figure 12), displaying a potency of action as Nrf2 activators similar to the 1,2,4-oxadiazole derivatives (Figure 10) [80]. On the contrary, physicochemical properties were improved, since the presence of the 1,3,4-oxadiazole moiety led to lower lipophilicity, and it improved the aqueous solubility of these molecules. Compound **15** (Figure 12), displaying 1.8-fold induction of Nrf2 in a luciferase reporter assay at 100 nM, was selected for further evaluation. It was able to increase Nrf2 levels and its downstream genes in H$_2$O$_2$-treated PC-12 cells, protecting them from oxidative damage.

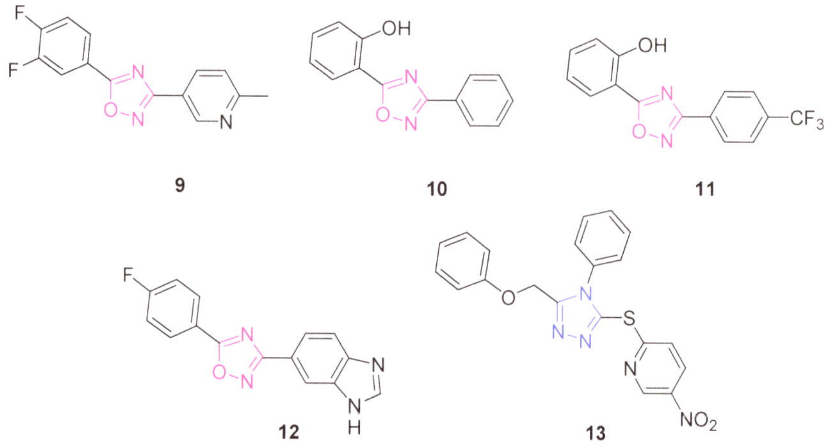

Figure 10. Oxadiazole- and triazole-based Nrf2 activators. The oxadiazole moiety is reported in pink, while the triazole moiety is in blue. These compounds were able to activate Nrf2 nuclear translocation [75–78].

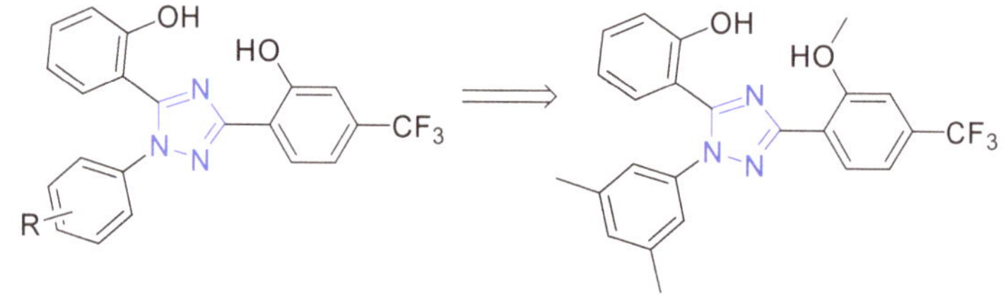

Figure 11. Development of 1,2,4-triazole derivatives as Nrf2 activators. The triazole moiety is in blue. The presence of alkyl groups on the diaryl-triazole scaffold was important for the activity, and **14** emerged as a potential neuroprotectant for the treatment of cerebral ischemic injury [79].

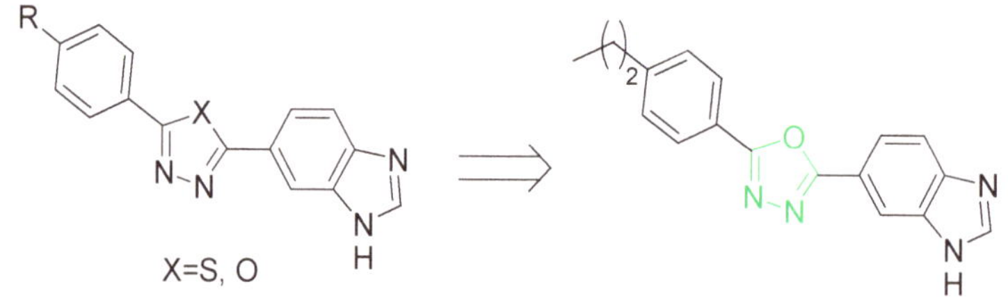

Figure 12. Nrf2 activators with a 1,3,4-oxa/thiadiazole core. The 1,3,4-oxadiazole moiety of compound **15** is in green. This molecule increased the Nrf2 levels and protected PC-12 cells from oxidative damage [80].

Very recently, pyrazole hit compounds as Keap1/Nrf2 complex disruptors were reported by means of a fragment-guided discovery approach [81]. Three "hotspots" for binding were identified in the Nrf2 binding site, i.e., for acidic groups, aromatic systems with a hydrogen bond acceptor ("planar acceptors"), and sulfonamides. From the screening of a collection of carboxylic acid derivatives, compound **16** (Figure 13) was selected for further optimization. The triazole-cyclopropyl derivative **17** then emerged as an interesting derivative, showing an IC$_{50}$ of 41 nM in the displacement of the Nrf2 peptide from the KEAP1 Kelch protein fluorescence polarization assay. However, due to its high polarity, it failed the subsequent cell-based assay. Therefore, its structure was further optimized by introducing a larger amide moiety able to interact with lipophilic areas of the protein, and compound **18** was finally disclosed as the most promising agent ($IC_{50} < 15$ nM). Its high-affinity binding to the Kelch domain of KEAP1 was also confirmed by surface plasmon resonance (SPR) evaluation, with Kd = 2.5 nM [81]. The docking analysis of compound **18** in complex with KEAP1 showed that all the three hotspots identified from the initial fragment screen were occupied by the inhibitor.

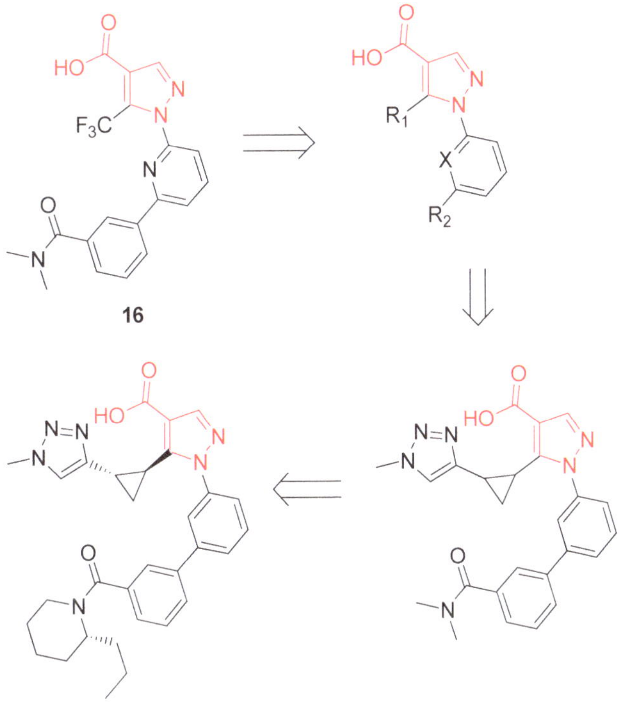

Figure 13. Development of pyrazole carboxylic acid inhibitors of the Keap1/Nrf2 complex protein–protein interactions. The pyrazole-carboxylic acid moiety is in red. Starting from the leading scaffold of **16**, i.e., the pyridyl-pyrazole carboxylic acid core structure, the triazole-cyclopropyl-derivative **17** was developed, showing a potent Nrf2 activation. This compound was further optimized by introducing a larger amide group, and the Nrf2 activator **18** was obtained, which was able to bind the Keap1 Kelch domain with increased affinity [81].

Table 1. Main activity of the Nrf2 inducers discussed in Section 3. EC_{50} is the half maximal effective concentration.

Compound	Activity	Cell Line	Ref.
2	Activation of Nrf2 nuclear translocation ($EC_{50} = 0.63$ μM)	BV-2 microglial cells	[68]
3	Activation of Nrf2 nuclear translocation ($EC_{50} = 0.53$ μM)	CATH.a, mouse DAergic neuronal cell line	[70]
4, 5	Activation of Nrf2 nuclear translocation; transcriptional activation of NRF2-responsive ARE genes	PC12 cell line	[71]
6	Activation of Nrf2 nuclear translocation ($EC_{50} = 0.346$ μM)	BV-2 microglial cells	[72]
7	Activation of Nrf2 nuclear translocation ($EC_{50} = 0.327$ μM)	BV-2 microglial cells	[72]

Table 1. *Cont.*

Compound	Activity	Cell Line	Ref.
8	Activation of Nrf2 nuclear translocation (EC$_{50}$ = 0.327 μM)	U2OS cells and BV-2 microglial cells	[73]
13	Induction of NQO1 and GCLM proteins; transcriptional activation of NRF2-responsive ARE genes	Primary mouse neurons; primary corticostriatal neuronal cocultures	[77,78]
14	Activation of Nrf2 nuclear translocation; induction of HO-1, NQO1, and GCLM proteins; Nrf2 displacement from the Keap1 Kealch domain (K$_D$ = 22.70 μM)	PC12 cell line	[79]
15	ARE inducing activity fold increase (f.i. = 1.8 @100 nM)	PC12 cell line	[80]
18	Stimulation of NQO1 activity (EC$_{50}$ = 43 nM); Nrf2 displacement from the Keap1 Kealch domain (IC$_{50}$ < 15 nM); transcriptional activation of NRF2-responsive ARE genes	BEAS-2B cells; normal human bronchial epithelial cells	[81]

4. Natural Nrf2 Activators

Thymoquinone (TQ, 2-isopropyl-5-methyl-1, 4-benzoquinone, Figure 14) is a monoterpenoid hydrocarbon, and it is the major bioactive compound of the volatile black oil from Nigella sativa. This molecule is endowed with antioxidant and anti-inflammatory properties [82], in addition to counteracting apoptosis in primary dopaminergic cells after their treatment with 1-methyl-4-phenylpyridinium (MPP+) and rotenone [83,84]. Recently, it was found that TQ prevents dopaminergic neurodegeneration in a PD mouse model, activating the nuclear translocation of Nrf2 for binding to ARE. As a consequence, the induction of HO-1, NQO1, and GST expression, as well as of anti-oxidative enzymes, including SOD and GSH-Px, was observed [85]. From a mechanistic viewpoint, TQ is an electrophilic activator that covalently binds the thiol group of the KEAP1 protein, which is the main target of TQ when its biological effects are mediated by Nrf2 [86].

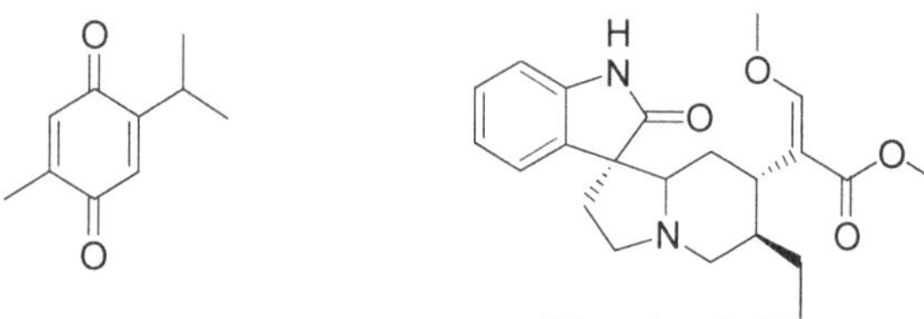

Figure 14. Chemical structure of natural Nrf2 activators with neuroprotectant activity. Thymoquinone prevents dopaminergic neurodegeneration in a PD mouse model, activating the nuclear translocation of Nrf2 for binding to ARE [85]. Rhynchophylline has shown neuroprotective effects in animal models of AD, as well as antioxidant properties [87–89].

Uncaria rhynchophylla (UR) is a component of the traditional Japanese kampo medicines chotosan and yokukansan, and rhynchophylline (Rhy, Figure 14) is a primary oxindole alkaloid obtained from UR. This natural compound has shown neuroprotective effects in animal models of AD, as well as antioxidant properties [87–89]. In an in vivo model of Aβ1–42-induced AD, Rhy was able to attenuate the disease symptoms and the in vitro investigations in SH-SY5Y cells revealed that this compound activates Nrf2 nuclear translocation, triggering the expression of its downstream antioxidant enzymes. However, how Rhy activates the Nrf2–ARE pathway to exert neuroprotection remains to be clarified [90].

5. Multitargeting Nrf2 Activators

Because of the multifactorial etiology of neurodegenerative diseases, in the last few years, the design of multifunctional ligands has emerged as an attractive therapeutic strategy. This multifactorial approach, involving the use of multitarget-directed ligands (MTDL), certainly reflects the enormous and crucial advances in understanding the mechanisms and implications in neurodegenerative diseases [91]. In this context, multitarget ligands have been developed very recently, exhibiting significant Nrf2 inducibility and an additional activity on other targets related to neurodegeneration.

Considering the significance of both BACH1 and Nrf2 in the modulation of heme oxygenase 1 (HMOX1) expression, which protects cells from neurotoxicity [92], the combination of Nrf2 activation with BACH1 inhibition could potentially result in an antioxidant response and a neuroprotective outcome. Recently, the O-methyl-p-cannabidiolquinone (**19**, Figure 15), a derivative of cannabidiol, was found at the same time as a potent inhibitor of BACH1 and an Nrf2 activator.

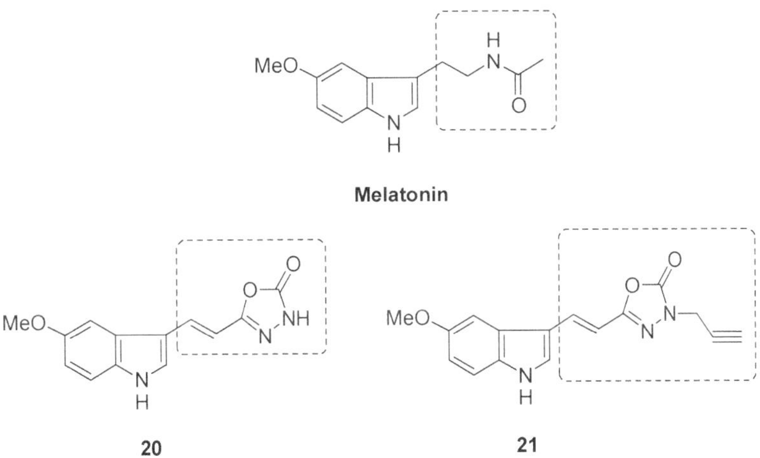

Figure 15. BACH1 inhibitor and Nrf2 inducer compounds. Chemical structure of the cannabidiol derivative **19** [92].

These activities were validated in HaCaT and Hepa1c1c7 cell lines through luciferase assays, Western blot, and RT-qPCR [93]. Compound **19** was also tested in macrophage-like THP1 and neuroblastoma SH-SY-5Y cell lines, confirming a decrease in BACH1, an induction of HMOX1, and a stabilization of Nrf2, as well as a decrease in ROS levels and neuroprotection in a model for Huntington disease. In terms of molecular structure, it is likely that compound **19** contributed to the regulation of redox homeostasis of cells through its quinone moieties with antioxidant properties [94].

Melatonin is a hormone that binds the G protein-coupled receptors MT1 and MT2, as well as the cytosolic receptor MT3/quinone reductase 2 (QR2). QR2 catalyzes the reduction of electrophilic quinones to hydroquinones, which can trigger an overproduction of free radicals. For this reason, this enzyme can be considered a molecular target involved in neurotoxic cascades [95]. Within the framework of a study of oxadiazolone-based bioisosteres of melatonin, two compounds (**20** and **21**, Figure 16) were developed, showing an induction of signaling mediated by Nrf2 (induction capability of 15.1 and 1.76, respectively) and an inhibition of QR2 (Ki 6.6 nM and 3.2 nM, respectively) in a neuronal phenotype [96].

Figure 16. QR2 inhibitor and Nrf2 inducer dual agents. Chemical structures of the melatonin derivatives **20** and **21** [96].

Interestingly, the absorption, distribution, metabolism, excretion, and toxicity (ADMET) and antioxidant properties of these compounds, also evaluated with the oxygen radical absorbance capacity (ORAC) assay, were satisfactory [96]. Quantum mechanics studies revealed that the ring systems contained in the indole-NH-oxadiazolone **20** are coplanar. This molecular arrangement is involved in the stacking interaction with the flavin adenine dinucleotide (FAD) in QR2, allowing the indole nitrogen to establish an interaction with the carbonyl oxygen of Gly174, reminiscent of that observed for polyphenol hydroxyls. Moreover, molecular docking studies and molecular dynamics simulations showed that compounds **20** and **21** bind to the KEAP1 Kelch domain, and their indole core and double bond establish important interactions with essential residues involved in the binding with Nrf2.

Beta-site amyloid precursor protein (APP)-cleaving enzyme 1 (BACE1) is an aspartyl protease widely expressed in the brain, essential for the synthesis of monomeric amyloid-β (Aβ), which spontaneously self-aggregates to form the insoluble fibers known as senile plaques, initiating toxicity in AD [97]. BACE1 is considered a well-validated therapeutic target for AD, and this approach was used for obtaining potent BACE1 inhibitors, such as verubecestat (MK8931, Figure 17) [98]. Within a series of selenium-containing compounds, including the key pharmacophores of verubecestat and ebselen (a potent antioxidant activating the Nrf2 pathway, Figure 17), the derivative **22** (Figure 17) was selected as a promising multifunctional candidate for pharmacological therapy of AD [99]. Biological evaluation showed that compound **22** exhibited good BACE-1 inhibition and Keap1–Nrf2–ARE pathway activation. Furthermore, it alleviates oxidative cell damage induced by H_2O_2 or 6-OHDA, reduced Aβ1-40 in HEK APPswe 293T cells, and was able to cross the BBB.

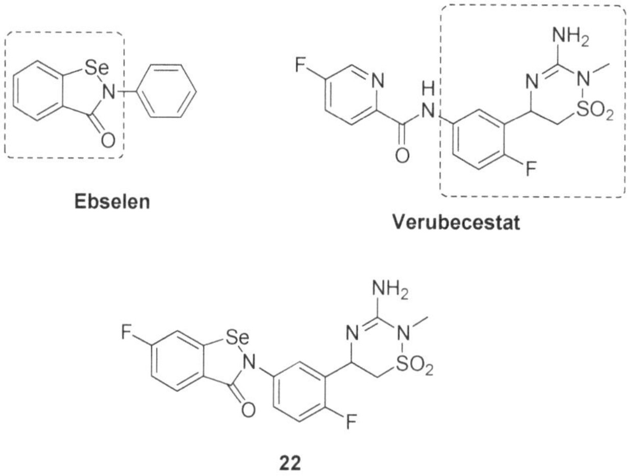

Figure 17. BACE-1 inhibitor and Keap1–Nrf2–ARE pathway activator. Hybridization of ebselen and verubecestat led to **22**, a candidate for the treatment of AD [99].

In the context of AD, another multitarget approach was achieved through the preparation of merged ligands with dual pharmacological activity and subsequent structural optimization [100].

The cholinergic hypothesis of AD suggests that the cognitive decline observed in the disease can be attributed to a selective loss of cholinergic neurons of the basal forebrain. Acetylcholinesterase (AChE) hydrolyzes acetylcholine (ACh), terminating the cholinergic transmission. AChE inhibitors, considered in the therapeutic armamentarium of AD, prevent the degradation of the neurotransmitter, allowing to prolong the action of the deficient neurotransmitter in the brain [101]. Three scaffolds were combined to form a series of molecules containing a flexible carbon chain: (i) dimethyl fumarate, containing an α,β-unsaturated ketone, a representative structure for Nrf2 activation, (ii) tranilast, an old antiallergic drug with anti-inflammatory properties patented for treating AD [102], and (iii) dithiocarbamate, able to interact with the catalytic active site of AChE. Among the tested derivatives, the most potent inhibitor of human AChE was compound **23** (Figure 18, IC$_{50}$ 53 nM) which also showed activation of the Keap1–Nrf2–ARE pathway, antioxidant and anti-inflammatory effects, and permeation of the BBB [103]. From in vivo studies, it emerged that it ameliorated cognitive deficits in the scopolamine-induced mouse model of AD [103].

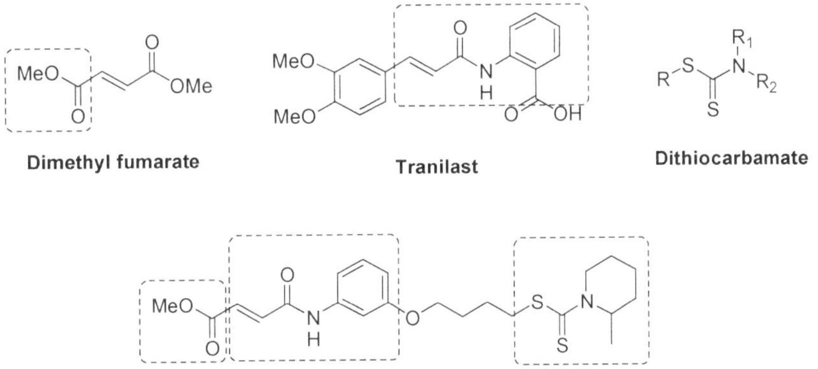

Dimethyl fumarate

Tranilast

Dithiocarbamate

23

Figure 18. AChE inhibitor and Keap1–Nrf2–ARE pathway activator. Chemical structure of the candidate for the treatment of AD **23**, which resulted from the combination of different pharmacophoric elements [103].

A further combination of Nrf2 activator and AChE inhibitor was recently proposed, characterized by the presence of a 1,2,4-oxadiazole core and benzylpiperidine [103]. Among the various derivatives synthesized, compound **24** was selected as the representative molecule (Figure 19).

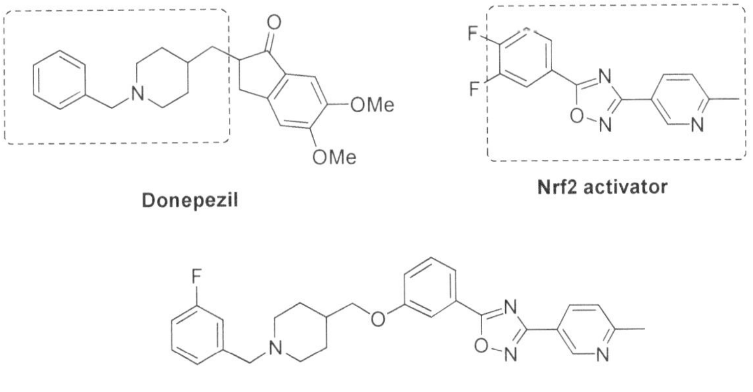

Donepezil

Nrf2 activator

24

Figure 19. AChE inhibitor and Nrf2 activator dual agent. Chemical structure of the hybrid donepezil/1,2,4-oxadiazole compound **24**, a candidate for the treatment of AD [75].

It exhibited excellent human AChE inhibition (hAChE IC$_{50}$ = 0.38 μM) in an ARE luciferase reporter test and it was not cytotoxic against PC12 cells. The inhibitory AchE was ascribed to the benzylpiperidine motif from the AChE inhibitor donepezil, which fits into in the catalytic anionic site of enzyme. Compound **24** also presented Nrf2-inducing activity, determined with an ARE-luciferase reporter assay, as well as an upregulation effect on downstream proteins HO-1, NQO1, and GCLM, with significant antioxidative and anti-inflammatory potency. The 1,2,4-oxadiazole core was identified to have inductivity of Nrf2 [75].

The administration of this bifunctional agent led to a more marked improvement in cognition and inflammation than the combination of an AChE inhibitor and an Nrf2 activator.

6. Conclusions

The potential clinical application of Nrf2 activators has been widely reported for the treatment of different diseases, basically having in common oxidative stress and chronic inflammation as hallmarks. However, in very recent years, the interest in the potential therapeutic role of Nrf2 activators for the treatment of neurodegenerative diseases has grown, and several small molecules have been disclosed, also aiming toward a polypharmacologic approach, based on the multifactorial etiology of these pathological conditions. In this review, we reported the latest and most important findings in the research on compounds able to upregulate the Keap1–Nrf2–ARE pathway, considering activators, Nrf2/Keap1 protein–protein degraders, and multitargeting compounds, all showing a usefulness in counteracting neurodegeneration. The drug design of the reported molecules and their activity have been discussed, with a focus on the amelioration of their pharmacokinetic parameters. In fact, despite the many promising results shown by Nrf2 activators, their clinical use has been limited by issues regarding the interaction with off-targets and their lack of selectivity, as well as by poor BBB penetration [5]. All in all, although the future of Nrf2 activators seems promising, more research is necessary in this field to translate these compounds into the clinical use.

Author Contributions: Conceptualization, C.M. and I.B.; methodology, R.A.; formal analysis, R.A., C.M. and I.B.; investigation, R.A., C.M. and I.B.; resources, R.A., C.M. and I.B.; data curation, C.M.; writing—original draft preparation, R.A., C.M. and I.B.; writing—review and editing, R.A., C.M. and I.B.; visualization, I.B.; supervision, C.M. All authors have read and agreed to the published version of the manuscript.

Funding: This research received no external funding.

Institutional Review Board Statement: Not applicable.

Informed Consent Statement: Not applicable.

Data Availability Statement: Data sharing not applicable.

Conflicts of Interest: The authors declare no conflict of interest.

References

1. GBD 2015 Neurological Disorders Collaborator Group. Global, regional, and national burden of neurological disorders during 1990–2015: A systematic analysis for the Global Burden of Disease Study 2015. *Lancet Neurol.* **2017**, *16*, 877–897. [CrossRef] [PubMed]
2. Barnham, K.; Masters, C.; Bush, A. Neurodegenerative diseases and oxidative stress. *Nat. Rev. Drug Discov.* **2004**, *3*, 205–214. [CrossRef] [PubMed]
3. Maccallini, C.; Amoroso, R. Targeting neuronal nitric oxide synthase as a valuable strategy for the therapy of neurological disorders. *Neural Regen. Res.* **2016**, *11*, 1731–1734. [CrossRef]
4. Boas, S.M.; Joyce, K.L.; Cowell, R.M. The Nrf2-Dependent Transcriptional Regulation of Antioxidant Defense Pathways: Relevance for Cell Type-Specific Vulnerability to Neurodegeneration and Therapeutic Intervention. *Antioxidants* **2021**, *11*, 8. [CrossRef] [PubMed]
5. Cuadrado, A.; Rojo, A.I.; Wells, G.; Hayes, J.D.; Cousin, S.P.; Rumsey, W.L.; Attucks, O.C.; Franklin, S.; Levonen, A.L.; Kensler, T.W.; et al. Therapeutic targeting of the NRF2 and KEAP1 partnership in chronic diseases. *Nat. Rev. Drug Discov.* **2019**, *18*, 295–317. [CrossRef]
6. Trachootham, D.; Lu, W.; Ogasawara, M.A.; Nilsa, R.D.; Huang, P. Redox regulation of cell survival. *Antioxid. Redox Signal* **2008**, *10*, 1343–1374. [CrossRef] [PubMed]
7. Bienert, G.P.; Schjoerring, J.K.; Jahn, T.P. Membrane transport of hydrogen peroxide. *Biochim. Biophys. Acta* **2006**, *1758*, 994–1003. [CrossRef]
8. Heurtaux, T.; Bouvier, D.S.; Benani, A.; Helgueta Romero, S.; Frauenknecht, K.B.M.; Mittelbronn, M.; Sinkkonen, L. Normal and Pathological Nrf2 Signalling in the Central Nervous System. *Antioxidants* **2022**, *11*, 1426. [CrossRef]

9.   Snezhkina, A.V.; Kudryavtseva, A.V.; Kardymon, O.L.; Savvateeva, M.V.; Melnikova, N.V.; Krasnov, G.S.; Dmitriev, A.A. ROS Generation and Antioxidant Defense Systems in Normal and Malignant Cells. *Oxid. Med. Cell. Longev.* **2019**, *2019*, 6175804. [CrossRef]

10. Sinenko, S.A.; Starkova, T.Y.; Kuzmin, A.A.; Tomilin, A.N. Physiological Signaling Functions of Reactive Oxygen Species in Stem Cells: From Flies to Man. *Front. Cell Dev. Biol.* **2021**, *A9*, 714370. [CrossRef]

11. Stadtman, E.R.; Levine, R.L. Free radical-mediated oxidation of free amino acids and amino acid residues in proteins. *Amino Acids* **2003**, *250*, 207–218. [CrossRef] [PubMed]

12. Suzuki, T.; Yamamoto, M. Stress-sensing mechanisms and the physiological roles of the Keap1-Nrf2 system during cellular stress. *J. Biol. Chem.* **2017**, *292*, 16817–16824. [CrossRef] [PubMed]

13. Itoh, K.; Ishii, T.; Wakabayashi, N.; Yamamoto, M. Regulatory mechanisms of cellular response to oxidative stress. *Free Radic. Res.* **1999**, *31*, 319–324. [CrossRef]

14. Cullinan, S.B.; Gordan, J.D.; Jin, J.; Harper, J.W.; Diehl, J.A. The Keap1-BTB protein is an adaptor that bridges Nrf2 to a Cul3-based E3 ligase: Oxidative stress sensing by a Cul3-Keap1 ligase. *Mol. Cell. Biol.* **2004**, *24*, 8477–8486. [CrossRef] [PubMed]

15. Tong, K.I.; Kobayashi, A.; Katsuoka, F.; Yamamoto, M. Two-site substrate recognition model for the Keap1-Nrf2 system: A hinge and latch mechanism. *Biol. Chem.* **2006**, *387*, 1311–1320. [CrossRef]

16. Horie, Y.; Suzuki, T.; Inoue, J.; Iso, T.; Wells, G.; Moore, T.W.; Mizushima, T.; Dinkova-Kostova, A.T.; Kasai, T.; Kamei, T.; et al. Molecular basis for the disruption of Keap1-Nrf2 interaction via Hinge & Latch mechanism. *Commun. Biol.* **2021**, *4*, 576.

17. Rada, P.; Rojo, A.I.; Chowdhry, S.; McMahon, M.; Hayes, J.D.; Cuadrado, A. SCF/β-TrCP promotes glycogen synthase kinase 3-dependent degradation of the Nrf2 transcription factor in a Keap1-independent manner. *Mol. Cell. Biol.* **2011**, *31*, 1121–1133. [CrossRef]

18. Kanninen, K.; White, A.R.; Koistinaho, J.; Malm, T. Targeting Glycogen Synthase Kinase-3β for Therapeutic Benefit against Oxidative Stress in Alzheimer's Disease: Involvement of the Nrf2-ARE Pathway. *Int. J. Alzheimer's Dis.* **2011**, *2011*, 985085.

19. Tonelli, C.; Chio, I.I.C.; Tuveson, D.A. Transcriptional Regulation by Nrf2. *Antioxid. Redox Signal.* **2018**, *29*, 1727–1745. [CrossRef]

20. Komatsu, M.; Kurokawa, H.; Waguri, S.; Taguchi, K.; Kobayashi, A.; Ichimura, Y.; Sou, Y.S.; Ueno, I.; Sakamoto, A.; Tong, K.I.; et al. The selective autophagy substrate p62 activates the stress responsive transcription factor Nrf2 through inactivation of Keap1. *Nat. Cell Biol.* **2010**, *12*, 213–223. [CrossRef]

21. Cobley, J.N.; Fiorello, M.L.; Bailey, D.M. 13 reasons why the brain is susceptible to oxidative stress. *Redox Biol.* **2018**, *15*, 490–503. [CrossRef]

22. Tan, B.L.; Norhaizan, M.E.; Liew, W.P.; Sulaiman Rahman, H. Antioxidant and Oxidative Stress: A Mutual Interplay in Age-Related Diseases. *Front. Pharmacol.* **2018**, *9*, 1162. [CrossRef]

23. Matsumaru, D.; Motohashi, H. The KEAP1-Nrf2 System in Healthy Aging and Longevity. *Antioxidants* **2021**, *10*, 1929. [CrossRef]

24. Hickman, S.; Izzy, S.; Sen, P.; Morsett, L.; El Khoury, J. Microglia in neurodegeneration. *Nat. Neurosci.* **2018**, *21*, 1359–1369. [CrossRef] [PubMed]

25. Acioglu, C.; Li, L.; Elkabes, S. Contribution of astrocytes to neuropathology of neurodegenerative diseases. *Brain Res.* **2021**, *1758*, 147291. [CrossRef]

26. Lazdon, E.; Stolero, N.; Frenkel, D. Microglia and Parkinson's disease: Footprints to pathology. *J. Neural Transm.* **2020**, *127*, 149–158. [CrossRef] [PubMed]

27. Angelova, D.M.; Brown, D.R. Microglia and the aging brain: Are senescent microglia the key to neurodegeneration? *J. Neurochem.* **2019**, *151*, 676–688. [CrossRef]

28. Todorovic, M.; Wood, S.A.; Mellick, G.D. Nrf2: A modulator of Parkinson's disease? *J. Neural Transm.* **2016**, *123*, 611–619. [CrossRef] [PubMed]

29. Poewe, W.; Seppi, K.; Tanner, C.M.; Halliday, G.M.; Brundin, P.; Volkmann, J.; Schrag, A.E.; Lang, A.E. Parkinson disease. *Nat. Rev. Dis. Prim.* **2017**, *3*, 17013. [CrossRef]

30. Mehra, S.; Sahay, S.; Maji, S.K. α-Synuclein misfolding and aggregation: Implications in Parkinson's disease pathogenesis. *Biochim. Biophys. Acta Proteins Proteom.* **2019**, *1867*, 890–908. [CrossRef]

31. Brown, D.R. α-Synuclein as a ferrireductase. *Biochem. Soc. Trans.* **2013**, *41*, 1513–1517. [CrossRef] [PubMed]

32. Chinta, S.J.; Mallajosyula, J.K.; Rane, A.; Andersen, J.K. Mitochondrial α-synuclein accumulation impairs complex I function in dopaminergic neurons and results in increased mitophagy in vivo. *Neurosci. Lett.* **2010**, *486*, 235–239. [CrossRef] [PubMed]

33. Richardson, J.R.; Caudle, W.M.; Guillot, T.S.; Watson, J.L.; Nakamaru-Ogiso, E.; Seo, B.B.; Sherer, T.B.; Greenamyre, J.T.; Yagi, T.; Matsuno-Yagi, A.; et al. Obligatory role for complex I inhibition in the dopaminergic neurotoxicity of 1-methyl-4-phenyl-1,2,3,6-tetrahydropyridine (MPTP). *Toxicol. Sci.* **2007**, *95*, 196–204. [CrossRef] [PubMed]

34. Chakkittukandiyil, A.; Sajini, D.V.; Karuppaiah, A.; Selvaraj, D. The principal molecular mechanisms behind the activation of Keap1/Nrf2/ARE pathway leading to neuroprotective action in Parkinson's disease. *Neurochem. Int.* **2022**, *156*, 105325. [CrossRef] [PubMed]

35. Harper, J.W.; Ordureau, A.; Heo, J.M. Building and decoding ubiquitin chains for mitophagy. *Nat. Rev. Mol. Cell Biol.* **2018**, *19*, 93–108. [CrossRef] [PubMed]

36. Sai, Y.; Zou, Z.; Peng, K.; Dong, Z. The Parkinson's disease-related genes act in mitochondrial homeostasis. *Neurosci. Biobehav. Rev.* **2012**, *36*, 2034–2043. [CrossRef] [PubMed]

37. Dorszewska, J.; Kowalska, M.; Prendecki, M.; Piekut, T.; Kozłowska, J.; Kozubski, W. Oxidative stress factors in Parkinson's disease. *Neural Regen. Res.* **2021**, *16*, 1383–1391. [CrossRef]

38. Hermida-Ameijeiras, A.; Méndez-Alvarez, E.; Sánchez-Iglesias, S.; Sanmartín-Suárez, C.; Soto-Otero, R. Autoxidation and MAO-mediated metabolism of dopamine as a potential cause of oxidative stress: Role of ferrous and ferric ions. *Neurochem. Int.* **2004**, *45*, 103–116. [CrossRef]

39. Innamorato, N.G.; Jazwa, A.; Rojo, A.I.; García, C.; Fernández-Ruiz, J.; Grochot-Przeczek, A.; Stachurska, A.; Jozkowicz, A.; Dulak, J.; Cuadrado, A. Different susceptibility to the Parkinson's toxin MPTP in mice lacking the redox master regulator Nrf2 or its target gene heme oxygenase-1. *PLoS ONE* **2010**, *5*, e11838. [CrossRef]

40. Lastres-Becker, I.; Ulusoy, A.; Innamorato, N.G.; Sahin, G.; Rábano, A.; Kirik, D.; Cuadrado, A. α-Synuclein expression and Nrf2 deficiency cooperate to aggravate protein aggregation, neuronal death and inflammation in early-stage Parkinson's disease. *Hum. Mol. Genet.* **2012**, *J21*, 3173–3192. [CrossRef]

41. Lastres-Becker, I.; García-Yagüe, A.J.; Scannevin, R.H.; Casarejos, M.J.; Kügler, S.; Rábano, A.; Cuadrado, A. Repurposing the Nrf2 Activator Dimethyl Fumarate as Therapy Against Synucleinopathy in Parkinson's Disease. *Antioxid. Redox Signal* **2016**, *25*, 61–77. [CrossRef]

42. Ahuja, M.; Ammal Kaidery, N.; Attucks, O.C.; McDade, E.; Hushpulian, D.M.; Gaisin, A.; Gaisina, I.; Ahn, Y.H.; Nikulin, S.; Poloznikov, A.; et al. Bach1 derepression is neuroprotective in a mouse model of Parkinson's disease. *Proc. Natl. Acad. Sci. USA* **2021**, *118*, e2111643118. [CrossRef] [PubMed]

43. Kawakami, F.; Imai, M.; Tamaki, S.; Ohta, E.; Kawashima, R.; Maekawa, T.; Kurosaki, Y.; Ohba, K.; Ichikawa, T. Nrf2 Expression Is Decreased in LRRK2 Transgenic Mouse Brain and LRRK2 Overexpressing SH-SY5Y Cells. *Biol. Pharm. Bull.* **2023**, *46*, 123–127. [CrossRef] [PubMed]

44. Gumeni, S.; Papanagnou, E.D.; Manola, M.S.; Trougakos, I.P. Nrf2 activation induces mitophagy and reverses Parkin/Pink1 knock down-mediated neuronal and muscle degeneration phenotypes. *Cell Death Dis.* **2021**, *12*, 671. [CrossRef] [PubMed]

45. Belarbi, K.; Cuvelier, E.; Destée, A.; Gressier, B.; Chartier-Harlin, M.C. NADPH oxidases in Parkinson's disease: A systematic review. *Mol. Neurodegener.* **2017**, *12*, 84. [CrossRef]

46. Fu, W.Y.; Ip, N.Y. The role of genetic risk factors of Alzheimer's disease in synaptic dysfunction. *Semin. Cell Dev. Biol.* **2023**, *139*, 3–12. [CrossRef]

47. Qiu, C.; Kivipelto, M.; von Strauss, E. Epidemiology of Alzheimer's disease: Occurrence, determinants, and strategies toward intervention. *Dialogues Clin. Neurosci.* **2009**, *11*, 111–128. [CrossRef]

48. Blennow, K.; de Leon, M.J.; Zetterberg, H. Alzheimer's disease. *Lancet* **2006**, *368*, 387–403. [CrossRef]

49. Wang, X.; Su, B.; Zheng, L.; Perry, G.; Smith, M.A.; Zhu, X. The role of abnormal mitochondrial dynamics in the pathogenesis of Alzheimer's disease. *J. Neurochem.* **2009**, *109* (Suppl. 1), 153–159. [CrossRef]

50. Hensley, K.; Hall, N.; Subramaniam, R.; Cole, P.; Harris, M.; Aksenov, M.; Aksenova, M.; Gabbita, S.P.; Wu, J.F.; Carney, J.M.; et al. Brain regional correspondence between Alzheimer's disease histopathology and biomarkers of protein oxidation. *J. Neurochem.* **1995**, *65*, 2146–2156. [CrossRef]

51. Chen, Z.; Zhong, C. Oxidative stress in Alzheimer's disease. *Neurosci. Bull.* **2014**, *30*, 271–281. [CrossRef] [PubMed]

52. Xie, H.; Hou, S.; Jiang, J.; Sekutowicz, M.; Kelly, J.; Bacskai, B.J. Rapid cell death is preceded by amyloid plaque-mediated oxidative stress. *Proc. Natl. Acad. Sci. USA* **2013**, *110*, 7904–7909. [CrossRef]

53. Cheignon, C.; Tomas, M.; Bonnefont-Rousselot, D.; Faller, P.; Hureau, C.; Collin, F. Oxidative stress and the amyloid beta peptide in Alzheimer's disease. *Redox Biol.* **2018**, *14*, 450–464. [CrossRef] [PubMed]

54. Jara, C.; Aránguiz, A.; Cerpa, W.; Tapia-Rojas, C.; Quintanilla, R.A. Genetic ablation of tau improves mitochondrial function and cognitive abilities in the hippocampus. *Redox Biol.* **2018**, *18*, 279–294. [CrossRef] [PubMed]

55. Gamblin, T.-C.; King, M.-E.; Kuret, J.; Berry, R.W.; Binder, L.I. Oxidative regulation of fatty acid-induced tau polymerization. *Biochemistry* **2000**, *39*, 14203–14210. [CrossRef]

56. Stamer, K.; Vogel, R.; Thies, E.; Mandelkow, E.; Mandelkow, E.M. Tau blocks traffic of organelles, neurofilaments, and APP vesicles in neurons and enhances oxidative stress. *J. Cell Biol.* **2002**, *156*, 1051–1063. [CrossRef] [PubMed]

57. Cente, M.; Filipcik, P.; Pevalova, M.; Novak, M. Expression of a truncated tau protein induces oxidative stress in a rodent model of tauopathy. *Eur. J. Neurosci.* **2006**, *24*, 1085–1090. [CrossRef] [PubMed]

58. Stack, C.; Jainuddin, S.; Elipenahli, C.; Gerges, M.; Starkova, N.; Starkov, A.A.; Jové, M.; Portero-Otin, M.; Launay, N.; Pujol, A.; et al. Methylene blue upregulates Nrf2/ARE genes and prevents tau-related neurotoxicity. *Hum. Mol. Genet.* **2014**, *J23*, 3716–3732. [CrossRef] [PubMed]

59. Rojo, A.I.; Pajares, M.; García-Yagüe, A.J.; Buendia, I.; Van Leuven, F.; Yamamoto, M.; López, M.G.; Cuadrado, A. Deficiency in the transcription factor Nrf2 worsens inflammatory parameters in a mouse model with combined tauopathy and amyloidopathy. *Redox Biol.* **2018**, *18*, 173–180. [CrossRef] [PubMed]

60. Rojo, A.I.; Pajares, M.; Rada, P.; Nuñez, A.; Nevado-Holgado, A.J.; Killik, R.; Van Leuven, F.; Ribe, E.; Lovestone, S.; Yamamoto, M.; et al. Nrf2 deficiency replicates transcriptomic changes in Alzheimer's patients and worsens APP and TAU pathology. *Redox Biol.* **2017**, *13*, 444–451. [CrossRef]

61. Ramesh Babu, J.; Lamar Seibenhener, M.; Peng, J.; Strom, A.-L.; Kemppainen, R.; Cox, N.; Zhu, H.; Wooten, M.-C.; Diaz-Meco, M.-T.; Moscat, J.; et al. Genetic inactivation of p62 leads to accumulation of hyperphosphorylated tau and neurodegeneration. *J. Neurochem.* **2008**, *106*, 107–120. [CrossRef] [PubMed]

62. Leroy, K.; Yilmaz, Z.; Brion, J.P. Increased level of active GSK-3beta in Alzheimer's disease and accumulation in argyrophilic grains and in neurones at different stages of neurofibrillary degeneration. *Neuropathol. Appl. Neurobiol.* **2007**, *33*, 43–55. [CrossRef]
63. Uemura, K.; Kuzuya, A.; Shimozono, Y.; Aoyagi, N.; Ando, K.; Shimohama, S.; Kinoshita, A. GSK3beta activity modifies the localization and function of presenilin 1. *J. Biol. Chem.* **2007**, *282*, 15823–15832. [CrossRef]
64. Wilson, A.J.; Kerns, J.K.; Callahan, J.F.; Moody, C.J. Keap calm, and carry on covalently. *J. Med. Chem.* **2013**, *56*, 7463–7476. [CrossRef] [PubMed]
65. Ma, Q.; He, X. Molecular basis of electrophilic and oxidative defense: Promises and perils of Nrf2. *Pharmacol. Rev.* **2012**, *64*, 1055–1081. [CrossRef] [PubMed]
66. Lee, S.; Hu, L. Nrf2 activation through the inhibition of Keap1–Nrf2 protein–protein interaction. *Med. Chem. Res.* **2020**, *29*, 846–867. [CrossRef] [PubMed]
67. Liu, Y.C.; Hsieh, C.W.; Wu, C.C.; Wung, B.S. Chalcone inhibits the activation of NF-kappa B and STAT3 in endothelial cells via endogenous electrophile. *Life Sci.* **2007**, *80*, 1420–1430. [CrossRef]
68. Kumar, V.; Kumar, S.; Hassan, M.; Wu, H.; Thimmulappa, R.K.; Kumar, A.; Sharma, S.K.; Parmar, V.S.; Biswal, S.; Malhotra, S.V. Novel chalcone derivatives as potent Nrf2 activators in mice and human lung epithelial cells. *J. Med. Chem.* **2011**, *54*, 4147–4159. [CrossRef]
69. Kim, H.J.; Jang, B.K.; Park, J.H.; Choi, J.W.; Park, S.J.; Byeon, S.R.; Pae, A.N.; Lee, Y.S.; Cheong, E.; Park, K.D. A novel chalcone derivative as Nrf2 activator attenuates learning and memory impairment in a scopolamine-induced mouse model. *Eur. J. Med. Chem.* **2022**, *185*, 111777. [CrossRef]
70. Woo, S.Y.; Kim, J.H.; Moon, M.K.; Han, S.H.; Yeon, S.K.; Choi, J.W.; Jang, B.K.; Song, H.J.; Kang, Y.G.; Kim, J.W.; et al. Discovery of Vinyl Sulfones as a Novel Class of Neuroprotective Agents toward Parkinson's Disease Therapy. *J. Med. Chem.* **2014**, *57*, 1473–1487. [CrossRef]
71. Song, Z.L.; Hou, Y.; Bai, F.; Fang, J. Generation of potent Nrf2 activators via tuning the electrophilicity and steric hindrance of vinyl sulfones for neuroprotection. *Bioorg. Chem.* **2021**, *107*, 104520. [CrossRef]
72. Choi, J.W.; Kim, S.; Park, J.H.; Kim, H.J.; Shin, S.J.; Kim, J.W.; Woo, S.Y.; Lee, C.; Han, S.M.; Lee, J.; et al. Optimization of Vinyl Sulfone Derivatives as Potent Nuclear Factor Erythroid 2-Related Factor 2 (Nrf2) Activators for Parkinson's Disease Therapy. *J. Med. Chem.* **2019**, *62*, 811–830. [CrossRef]
73. Choi, J.W.; Kim, S.; Yoo, J.S.; Kim, H.J.; Kim, H.J.; Kim, B.E.; Lee, E.H.; Lee, Y.S.; Park, J.H.; Park, K.D. Development and optimization of halogenated vinyl sulfones as Nrf2 activators for the treatment of Parkinson's disease. *Eur. J. Med. Chem.* **2021**, *212*, 113103. [CrossRef]
74. Crisman, E.; Duarte, P.; Dauden, E.; Cuadrado, A.; Rodríguez-Franco, M.I.; López, M.G.; León, R. KEAP1-Nrf2 protein–protein interaction inhibitors: Design, pharmacological properties and therapeutic potential. *Med. Res. Rev.* **2023**, *43*, 237–287. [CrossRef]
75. Xu, L.L.; Zhu, J.F.; Xu, X.L.; Zhu, J.; Li, L.; Xi, M.Y.; Jiang, Z.Y.; Zhang, M.Y.; Liu, F.; Lu, M.C.; et al. Discovery and modification of in vivo active Nrf2 activators with 1,2,4-oxadiazole core: Hits identification and structure-activity relationship study. *J. Med. Chem.* **2015**, *58*, 5419–5436. [CrossRef] [PubMed]
76. Ayoup, M.S.; Abu-Serie, M.M.; Abdel-Hamid, H.; Teleb, M. Beyond direct Nrf2 activation; reinvestigating 1,2,4-oxadiazole scaffold as a master key unlocking the antioxidant cellular machinery for cancer therapy. *Eur. J. Med. Chem.* **2021**, *220*, 113475. [CrossRef] [PubMed]
77. Quinti, L.; Casale, M.; Moniot, S.; Pais, T.F.; Van Kanegan, M.J.; Kaltenbach, L.S.; Pallos, J.; Lim, R.G.; Naidu, S.D.; Runne, H.; et al. SIRT2- and NRF2-Targeting Thiazole-Containing Compound with Therapeutic Activity in Huntington's Disease Models. *Cell Chem. Biol.* **2016**, *23*, 849–861. [CrossRef] [PubMed]
78. Quinti, L.; Dayalan, N.S.; Träger, U.; Chen, X.; Kegel-Gleason, K.; Llères, D.; Connolly, C.; Chopra, V.; Low, C.; Moniot, S.; et al. KEAP1-modifying small molecule reveals muted Nrf2 signaling responses in neural stem cells from Huntington's disease patients. *Proc. Natl. Acad. Sci. USA* **2017**, *114*, E4676–E4685. [CrossRef] [PubMed]
79. Lao, Y.; Wang, Y.; Chen, J.; Huang, P.; Su, R.; Shi, J.; Jiang, C.; Zhang, J. Synthesis and biological evaluation of 1,2,4-triazole derivatives as potential Nrf2 activators for the treatment of cerebral ischemic injury. *Eur. J. Med. Chem.* **2022**, *236*, 114315. [CrossRef]
80. Lin, H.; Qiao, Y.; Yang, H.; Li, Q.; Chen, Y.; Qu, W.; Liu, W.; Feng, F.; Sun, H. Design and evaluation of Nrf2 activators with 1,3,4-oxa/thiadiazole core as neuro-protective agents against oxidative stress in PC-12 cells. *Bioorg. Med. Chem. Lett.* **2020**, *30*, 126853. [CrossRef]
81. Norton, D.; Bonnette, W.G.; Callahan, J.F.; Carr, M.G.; Griffiths-Jones, C.M.; Heightman, T.D.; Kerns, J.K.; Nie, H.; Rich, S.J.; Richardson, C.; et al. Fragment-Guided Discovery of Pyrazole Carboxylic Acid Inhibitors of the Kelch-like ECH-Associated Protein 1: Nuclear Factor Erythroid 2 Related Factor 2 (KEAP1:NRF2) Protein-Protein Interaction. *J. Med. Chem.* **2021**, *64*, 15949–15972. [CrossRef]
82. Darakhshan, S.; Bidmeshki, P.A.; Hosseinzadeh, C.A.; Sisakhtnezhad, S. Thymoquinone and its therapeutic potentials. *Pharmacol. Res.* **2015**, *95–96*, 138–158. [CrossRef] [PubMed]
83. Radad, K.; Moldzio, R.; Taha, M.; Rausch, W.D. Thymoquinone protects dopaminergic neurons against MPP+ and rotenone. *Phytother. Res.* **2009**, *23*, 696–700. [CrossRef] [PubMed]
84. Radad, K.S.; Al-Shraim, M.M.; Moustafa, M.F.; Rausch, W.D. Neuroprotective role of thymoquinone against 1-methyl-4-phenylpyridiniuminduced dopaminergic cell death in primary mesencephalic cell culture. *Neurosciences* **2015**, *20*, 10–16. [PubMed]

85. Ardah, M.T.; Merghani, M.M.; Haque, M.E. Thymoquinone prevents neurodegeneration against MPTP in vivo and modulates α-synuclein aggregation in vitro. *Neurochem. Int.* **2019**, *128*, 115–126. [CrossRef] [PubMed]

86. Talebi, M.; Talebi, M.; Farkhondeh, T.; Samarghandian, S. Biological and therapeutic activities of thymoquinone: Focus on the Nrf2 signaling pathway. *Phytother. Res.* **2021**, *35*, 1739–1753. [CrossRef]

87. Fujiwara, H.; Iwasaki, K.; Furukawa, K.; Seki, T.; He, M.; Maruyama, M.; Tomita, N.; Kudo, Y.; Higuchi, M.; Saido, T.C.; et al. Uncaria rhynchophylla, a Chinese medicinal herb, has potent antiaggregation effects on Alzheimer's β-amyloid proteins. *J. Neurosci. Res.* **2006**, *84*, 427–433. [CrossRef] [PubMed]

88. Xu, R.; Wang, J.; Xu, J.; Song, X.; Huang, H.; Feng, Y.; Fu, C. Rhynchophylline loaded-mPEG-PLGA nanoparticles coated with Tween-80 for preliminary study in Alzheimer's disease. *Int. J. Nanomed.* **2020**, *15*, 1149–1160. [CrossRef]

89. Qin, Q.J.; Cui, L.Q.; Li, P.; Wang, Y.B.; Zhang, X.Z.; Guo, M.L. Rhynchophylline ameliorates myocardial ischemia/reperfusion injury through the modulation of mitochondrial mechanisms to mediate myocardial apoptosis. *Mol. Med. Rep.* **2019**, *19*, 2581–2590. [CrossRef]

90. Jiang, P.; Chen, L.; Xu, J.; Liu, W.; Feng, F.; Qu, W. Neuroprotective Effects of Rhynchophylline Against Aβ1–42-Induced Oxidative Stress, Neurodegeneration, and Memory Impairment Via Nrf2–ARE Activation. *Neurochem. Res.* **2021**, *46*, 2439–2450. [CrossRef]

91. Cavalli, A.; Bolognesi, M.L.; Minarini, A.; Rosini, M.; Tumiatti, V.; Recanatini, M.; Carlo Melchiorre, C. Multi-target-directed ligands to combat neurodegenerative diseases. *J. Med. Chem.* **2008**, *51*, 347–372. [CrossRef] [PubMed]

92. Zhang, X.; Guo, J.; Wei, X.; Niu, C.; Jia, M.; Qinhan Li, Q.; Meng, D. Bach1: Function, regulation, and involvement in disease. *Oxid. Med. Cell. Longev.* **2018**, *2018*, 1347969. [CrossRef]

93. Casares, L.; Unciti-Broceta, J.D.; Prados, M.E.; Caprioglio, D.; Mattoteia, D.; Higgins, M.; Apendino, G.; Dinkova-Kostova, A.T.; Munoz, E.; de la Vega, L. Isomeric O-methyl cannabidiolquinones with dual BACH1/Nrf2 activity. *Redox Biol.* **2020**, *37*, 101689. [CrossRef] [PubMed]

94. Satoh, T.; McKercher, S.R.; Lipton, S.A. Nrf2/ARE-mediated antioxidant actions of pro-electrophilic drugs. *Free Radic. Biol. Med.* **2013**, *65*, 645–657. [CrossRef]

95. Nosjean, O.; Ferro, M.; Cogé, F.; Beauverger, P.; Henlin, J.-M.; Lefoulon, F.; Fauchère, J.-L.; Delagrange, P.; Canet, E.; Boutin, J.A. Identification of the melatonin-binding site MT 3 as the quinone reductase 2. *J. Biol. Chem.* **2000**, *275*, 31311–31317. [CrossRef] [PubMed]

96. Herrera-Arozamena, C.; Estrada-Valencia, M.; Pérez, C.; Lagartera, L.; Morales-García, J.A.; Pérez-Castillo, A.; Franco-Gonzalez, J.F.; Michalska, P.; Duarte, P.; Leòn, R.; et al. Tuning melatonin receptor subtype selectivity in oxadiazolone-based analogues: Discovery of QR2 ligands and Nrf2 activators with neurogenic properties. *Eur. J. Med. Chem.* **2020**, *190*, 112090. [CrossRef] [PubMed]

97. Harald Hampel, H.; Robert Vassar, R.; Bart De Strooper, B.; John Hardy, J.; Michael Willem, M.; Neeraj Singh, N.; Zhou, J.; Yan, R.; Vanmechelen, E.; De Vos, A.; et al. The β-secretase BACE1 in Alzheimer's Disease. *Biol. Psychiatry* **2021**, *89*, 745–756. [CrossRef]

98. Kennedy, M.E.; Stamford, A.W.; Chen, X.; Cox, K.; Cumming, J.N.; Dockendorf, M.F.; Egan, M.; Ereshefsky, L.; Hodgson, R.A.; Hyde, L.A.; et al. The BACE1 inhibitor verubecestat (MK-8931) reduces CNS β-amyloid in animal models and in Alzheimer's disease patients. *Sci. Transl. Med.* **2016**, *8*, 363ra150. [CrossRef]

99. Qu, L.; Ji, L.; Wang, C.; Luo, H.; Li, S.; Peng, W.; Yin, F.; Lu, D.; Liu, X.; Kong, L.; et al. Synthesis and evaluation of multi-target-directed ligands with BACE-1 inhibitory and Nrf2 agonist activities as potential agents against Alzheimer's disease. *Eur. J. Med. Chem.* **2021**, *219*, 113441. [CrossRef]

100. Guo, J.; Cheng, M.; Liu, P.; Cao, D.; Luo, J.; Wan, Y.; Fang, Y.; Jin, Y.; Xie, S.-S.; Liu, J. A multi-target directed ligands strategy for the treatment of Alzheimer's disease: Dimethyl fumarate plus Tranilast modified Dithiocarbate as AchE inhibitor and Nrf2 activator. *Eur. J. Med. Chem.* **2022**, *242*, 114630. [CrossRef]

101. Stanciu, G.D.; Luca, A.; Rusu, R.N.; Bild, V.; Beschea Chiriac, S.I.; Solcan, C.; Bild, W.; Ababei, D.C. Alzheimer's Disease Pharmacotherapy in relation to cholinergic system involvement. *Biomolecules* **2020**, *10*, 40. [CrossRef] [PubMed]

102. Thapak, P.; Bishnoi, M.; Sharma, S.S. Tranilast, a Transient Receptor Potential Vanilloid 2 Channel (TRPV2) inhibitor attenuates amyloid β-induced cognitive impairment: Possible mechanisms. *Neuromol. Med.* **2022**, *24*, 183–194. [CrossRef] [PubMed]

103. Wang, Y.; Xiong, B.; Lin, H.; Li, Q.; Yang, H.; Qiao, Y.; Li, Q.; Xu, Z.; Lyu, W.; Qu, W.; et al. Design, synthesis and evaluation of fused hybrids with acetylcholinesterase inhibiting and Nrf2 activating functions for Alzheimer's disease. *Eur. J. Med. Chem.* **2022**, *244*, 114806. [CrossRef] [PubMed]

Systematic Review

The Nrf2 Pathway in Depressive Disorders: A Systematic Review of Animal and Human Studies

Gabriele Sani [1,2,*], Stella Margoni [1], Andrea Brugnami [1], Ottavia Marianna Ferrara [1], Evelina Bernardi [1], Alessio Simonetti [1,2,3,4], Laura Monti [5], Marianna Mazza [1,2], Delfina Janiri [1,2], Lorenzo Moccia [1,2], Georgios D. Kotzalidis [1,6], Daniela Pia Rosaria Chieffo [5] and Luigi Janiri [1,2]

[1] Institute of Psychiatry, Department of Neuroscience, Catholic University of the Sacred Hearth, Rome, Largo Francesco Vito 1, 00168 Rome, Italy

[2] Department of Psychiatry, Department of Neuroscience, Head, Neck and Thorax, Fondazione Policlinico Universitario Agostino Gemelli IRCCS, Largo Agostino Gemelli 1, 00168 Rome, Italy

[3] Menninger Department of Psychiatry and Behavioral Sciences, Baylor College of Medicine, 1 Baylor Plaza, Houston, TX 77030, USA

[4] Centro Lucio Bini, Via Crescenzio 42, 00193 Rome, Italy

[5] UOS Clinical Psychology, Clinical Government, Fondazione Policlinico Universitario Agostino Gemelli IRCCS, Largo Agostino Gemelli 1, 00168 Rome, Italy

[6] NESMOS Department, Faculty of Medicine and Psychology, Sant'Andrea University Hospital, University of Rome La Sapienza, Via di Grottarossa, 1035-1039, 00189 Rome, Italy

* Correspondence: gabriele.sani@unicatt.it

Citation: Sani, G.; Margoni, S.; Brugnami, A.; Ferrara, O.M.; Bernardi, E.; Simonetti, A.; Monti, L.; Mazza, M.; Janiri, D.; Moccia, L.; et al. The Nrf2 Pathway in Depressive Disorders: A Systematic Review of Animal and Human Studies. *Antioxidants* **2023**, *12*, 817. https://doi.org/10.3390/antiox12040817

Academic Editor: Marcel Bonay

Received: 13 February 2023
Revised: 12 March 2023
Accepted: 14 March 2023
Published: 27 March 2023

Abstract: There is increasing interest in the involvement of antioxidative systems in protecting from depression. Among these, Nrf2 occupies a central place. We aimed to review the role of Nrf2 in depression. For this reason, we conducted a PubMed search using as search strategy (psychiatr*[ti] OR schizo*[ti] OR psychot*[ti] OR psychos*[ti] OR depress*[ti] OR MDD[ti] OR BD[ti] OR bipolar[ti] OR Anxiety[ti] OR antidepress*[ti] OR panic[ti] OR obsess*[ti] OR compulsio*[ti] OR "mood disord*"[ti] OR phobi*[ti] OR agoraphob*[ti] OR anorex*[ti] OR anorect*[ti] OR bulimi*[ti] OR "eating disorder*"[ti] OR neurodevelopm*[ti] OR retardation[ti] OR autism[ti] OR autistic[ti] OR ASM[ti] OR adhd[ti] OR "attention-deficit"[ti]) AND nrf2, which on the 9th of March produced 208 results of which 89 were eligible for our purposes. Eligible articles were studies reporting data of Nrf2 manipulations or content by any treatment in human patients or animals with any animal model of depression. Most studies were on mice only (N = 58), 20 on rats only, and three on both rats and mice. There were two studies on cell lines (*in vitro*) and one each on nematodes and fish. Only four studies were conducted in humans, one of which was *post mortem*. Most studies were conducted on male animals; however, human studies were carried out on both men and women. The results indicate that Nrf2 is lower in depression and that antidepressant methods (drugs or other methods) increase it. Antioxidant systems and plasticity-promoting molecules, such as those in the Nrf2–HO-1, BDNF–TrkB, and cyclic AMP–CREB pathways, could protect from depression, while glycogen synthase kinase-3β and nuclear factor κB oppose these actions, thus increasing depressive-like behaviours. Since Nrf2 is also endowed with tumorigenic and atherogenic potential, the balance between benefits and harms must be taken into account in designing novel drugs aiming at increasing the intracellular content of Nrf2.

Keywords: depression; Nuclear factor erythroid-2 (Nrf2); pathophysiology; antioxidant pathways; Haemoxygenase (HO-1); Nuclear factor kappa B (NF-κB)

1. Introduction

Recently, much research has been devoted to the study of inflammation and its role in the context of a wide variety of pathological conditions. Inflammation and oxidative stress were found to enhance each other, thus establishing a pathological state [1], which may be found in various psychiatric disorders. The variety of external stimuli to which organisms are subjected triggers adaptive responses, which are designed to restore homoeostasis through a fine balance between oxidation and antioxidant activity [2].

An imbalance between the generation of reactive oxygen species (ROS) and antioxidant defences results in increased oxidative stress [3], with a consequent increase in neuroinflammation, mitochondrial dysfunction, and cell degeneration processes such as apoptosis and ferroptosis, which proved to be crucial in many psychiatric disorders. This is not surprising if we consider that the brain consumes massive doses of oxygen and contains high concentrations of oxidative lipids, thus being extremely vulnerable to oxidative stress-induced damage [2].

In this framework, it is important to draw attention to nuclear factor erythroid-2 (Nrf2), the main endogenous negative regulator of oxidation [4]. Its activation determines the expression of numerous antioxidants and cytoprotective genes capable of modulating oxidative stress. Nrf2 also presides over the regulation of genes involved in the oxidative stress-related pathological processes mentioned above. Consequently, dysregulation in Nrf2 expression with reduced cortical levels may contribute to the aetiopathogenesis of numerous pathological conditions, including psychiatric disorders and neurodegenerative diseases.

Nrf2 is a protein transcription factor composed of 605 amino acids, encoded by the *NFE2L2* gene and belonging to the Cap'n'collar (CNC) family of transcription factors [5]. It contains seven highly conserved functional domains called Nrf2-ECH homology 1 (Neh1-Neh7). Neh1 and Neh3 interact with specific DNA sequences called antioxidant response elements (ARE) [6], thus promoting the transcription of enzymes with antioxidant activity. Neh2 interacts with Kelch-like ECH-associated protein 1 (Keap1), the main negative regulator of Nrf2. Neh4 and Neh5 interact with the cyclic adenosine monophosphate (cAMP)-response element-binding protein (CREB), which also promotes transcriptional activation [7]. In contrast, Neh6 can link to β-transducin repeat-containing protein (β-TrCP) and is involved in Keap1-independent degradation of Nrf2. Finally, Neh7 inhibits the Nrf2-ARE signalling pathway by binding to retinoic X receptor alpha (RXRα) [8]. Therefore, the activity of Nrf2 is subjected to both Keap1-dependent and Keap1-independent regulation.

Keap 1 represents the main Nrf2 suppressor; it forms a homodimer capable of binding ETGE and DLG motifs (stronger and weaker binding sites of Keap-1, respectively; the former is located in the loop region of the antiparallel β-sheet, while the latter is N-terminal to the α-helix [9]) included in the Neh2 domain of Nrf2. Under physiological conditions, the Keap1-Nrf2 complex combines with the E3-ubiquitin ligase Cullin 3 (Cul3) complex, leading to ubiquitination and proteasomal degradation of Nrf2 [10]. Negative regulation is also mediated by the phosphorylation of Nrf2 by glycogen synthase kinase-3 beta (GSK-3β) and mitogen-activating protein kinase (MAPK) [11].

When oxidative stress increases and ROS accumulation occurs, there is a dissociation of the Keap1-Nrf2 complex induced by conformational changes of a Keap1 domain, inhibiting ubiquitination and subsequent degradation of Nrf2. Dissociated from Keap1, Nrf2 is free to translocate into the nucleus and bind specific genomic sequences in order to promote antioxidant enzyme transcription. At the same time, there is a positive Keap1-independent regulation mediated by other kinases. The kinases involved in the phosphorylation and subsequent activation of Nrf2 include protein kinase C (PKC), casein kinase II (CK2), protein kinase R (PKR), c-Jun N-terminal kinase (JNK), and extracellular regulated kinases (ERKs) [12–15]. Finally, brain-derived neurotrophic factor (BDNF) can also promote the activation and subsequent migration of Nrf2 into the nucleus [16].

In addition to its antioxidant activity, Nrf2 is also directly involved in oxidative stress-related pathological processes by regulating their activation. In particular, there is a direct cross-talk between Nrf2 and p62, an autophagy key protein. Nrf2 can promote the expression of genes involved in autophagy, while p62 can compete with Nrf2 in binding to Keap1 in a positive feedback that is associated with cytoprotection [6,17]. In addition, Nrf2 is also involved in a special form of autophagy called mitophagy, the alteration of which allegedly plays an important role in psychiatric disorders. While it preserves mitochondrial integrity, mitophagy entails the elimination of damaged or redundant mitochondria through autophagy [18].

Furthermore, recent studies of Nrf2 found that its inducers promote the suppression of the pro-inflammatory phenotype of microglia through regulating BDNF, the reduction of which is found in many psychiatric disorders characterised by neuroinflammation [16,19–21]. The existence of reciprocal regulation has emerged between BDNF and Nrf2; the latter, in connecting with the exon I promoter of *bdnf*, can activate BDNF; on the other side, BDNF can increase the nuclear translocation of Nrf2, thus promoting its antioxidant activity [16].

Finally, several studies also cast light on the involvement of Nrf2 in ferroptosis, an iron-dependent form of programmed cell death characterised by the accumulation of lipid peroxides (lipids damaged by oxidation). Notably, Nrf2 does not only regulate numerous genes involved in iron metabolism and homoeostasis but also promotes the basal expression of the lipid hydroperoxidase, glutathione peroxidase-4 (GPX4), which converts toxic lipid peroxides to nontoxic lipid alcohols [22].

Impaired response to oxidative stress has been shown in animal models for stress disorders, such as post-traumatic stress disorder (PTSD) [23–26], but only a few studies have focused on Nrf2 [27–29]. PTSD and depression are long considered to represent stress disorders and share common neurobiological patterns [30]. Recently, the Nrf2-depression connection has received attention [2]. We decided to search the literature for studies investigating the ties between depression paradigms in the animal and/or major depressive disorder in humans and Nrf2 as a proxy of a mechanism that counters oxidative stress. Establishing such a relationship would allow us to concentrate on the production of drugs that would promote the search for drugs interfering with intracellular oxidative processes.

2. Methods

To systematically review the ties between depression and Nrf2, we first conducted a PubMed search involving all possible mental and psychiatric disorders and then focused on depression. The inclusion of other than depressive disorders/states was to make sure that depression during the course of other mental disorders was not dealt with. We employed the following search strategy: (psychiatr*[ti] OR schizo*[ti] OR psychot*[ti] OR psychos*[ti] OR depress*[ti] OR MDD[ti] OR BD[ti] OR bipolar[ti] OR Anxiety[ti] OR antidepress*[ti] OR panic[ti] OR obsess*[ti] OR compulsio*[ti] OR "mood disord*"[ti] OR phobi*[ti] OR agoraphob*[ti] OR anorex*[ti] OR anorect*[ti] OR bulimi*[ti] OR "eating disorder*"[ti] OR neurodevelopm*[ti] OR retardation[ti] OR autism[ti] OR autistic[ti] OR ASM[ti] OR adhd[ti] OR "attention-deficit"[ti]) AND nrf2. The choice of the search strategy was based on consultations among all authors.

Following search performance, we characterised the nature of all ensuing records and labelled them accordingly. This resulted in their being either included or excluded. The inclusion/exclusion labelling with the reasons for exclusion is shown in the Online Supplemental material and in Figure 1, where the PRISMA flow diagram is displayed. In carrying out our review, we followed the 2020 Preferred Reporting Items for Systematic reviews and Meta-Analyses (PRISMA) statement [31]. The 2020 PRISMA Checklist may be found in the Online Supplement. We assessed the Risk of bias (RoB) of the included studies with the Cochrane RoB 2.0 tool [32]. We performed an evaluation of the RoB for each included study. The results are shown in the Online Supplement.

Eligibility was based on being an original study on any animal or tissue, including humans, on investigating depression or depression models and providing data on Nrf2 levels. All other studies were excluded. Excluded were case reports, opinion articles, such as editorials, letters to the editor, comments of other work, reviews, and meta-analyses (however, we hand-searched their reference lists to identify other possibly eligible studies), and studies not providing data. Eligibility for each paper was established with the consensus of all authors obtained through Delphi rounds, in which all authors participated, either in-person or online. The same applied to the compilation of the RoB.

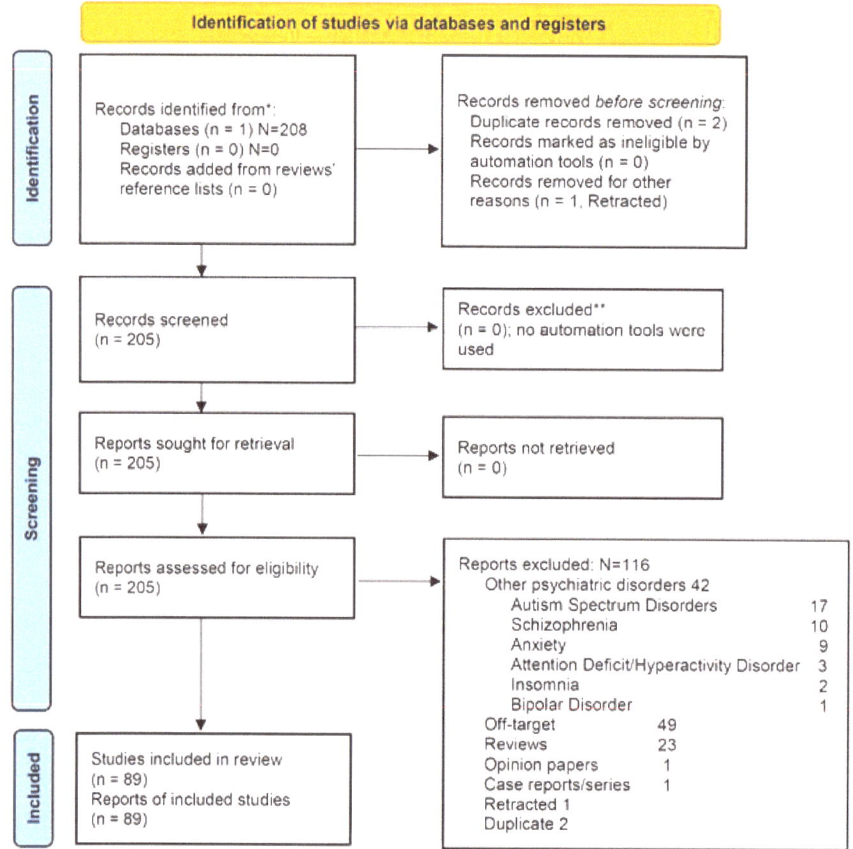

Figure 1. *PRISMA* 2020 flow diagramme (From [31]: Page, M.J.; McKenzie, J.E.; Bossuyt, P.M.; Boutron, I.; Hoffmann, T.C.; Mulrow, C.D.; Shamseer, L.; Tetzlaff, J.M.; Akl, E.A.; Brennan, S.E.; et al. The PRISMA 2020 statement: An updated guideline for reporting systematic reviews. *BMJ* **2021**, *372*, n71. https://doi.org/10.1136/bmj.n71. For more information, visit: http://www.prisma-statement. org/ accessed on 12 February 2023) of the review.

3. Results

Our search, eventually conducted on the 10th of March 2023, yielded 208 results on PubMed, of which 89 studies were eligible, as summarised in Table 1 (human studies, of which two used cell lines *in vitro*) and Table **??** (animal studies). In particular, there were 78 articles labelled Depression and 11 Depression and Anxiety; these amounted to 89 articles. The remaining 116 studies were excluded. Depression-free articles focused on autism (N = 17), schizophrenia (N = 10), anxiety (N = 9), attention-deficit/hyperactivity disorder (N = 3), insomnia (N = 2), and bipolar disorder alone (N = 1). Other articles that did not meet the inclusion criteria were Opinions (N = 1), Case reports (N = 1), and Reviews (N = 23), while many articles were off-target (N = 49) in that they did not focus any psychiatric disorder and were unfocused in their designs or they were unrelated to the subject of our inquiry. Furthermore, there were two duplicates and a retracted paper, but another paper from the same group had not been retracted and dealt with the same issue as the retracted one. Publication dates spanned from 23-March-2006 to 1-March-2023 for the searched papers and 23-April-2013 to 1-March-2023 for the eligible ones. The included and excluded studies with their reasons for exclusion are shown in the Supplement. The selection process and reasons for exclusion are depicted in the PRISMA flowchart (Figure 1).

Of the 89 studies included in this review, most were conducted on mice only (N = 58; 65.17%), 20 were conducted on rats only (22.47%), and three on both rats and mice (3.37%), four on humans (4.49%), two on human cell lines in vitro (2.25%), one on fish and worms each (1.12%). Of the 58 studies carried out on mice only, 50 used only male animals, seven used only female, and one both used animals of both sexes; 31 used C57BL/6 strains; one reported unspecified C57 mice (which were presumably C57BL/6, based on other articles by the same group of authors), 10 used unspecified Swiss strains, five used CD-1, eight used Balb/c, five ICR, four Kunming, one BXD Recombinant Inbred, and one Murphy Roths Large lymphoproliferative Mouse (MRL/lpr); in eight of these studies, investigators used more than one strain. All 20 studies conducted on rats used only male animals, 14 Sprague Dawley and six Wistar, of which two were Hannover and two albino Wistar, while all three studies that employed both mice and rats were conducted using male-only animals, all three used Sprague Dawley rats (1 also Groningen, a strain characterised by high aggression levels [33]) and C57BL/6 mice. The four human studies included patients and matched controls of both sexes, while the only *post mortem* study did not report the sex or the age of included patients. The study that used fish employed the Japanese rice fish, medaka (*Oryzias latipes*), both male and female, and the one that used nematodes used *Caenorhabditis elegans*, while of the two conducted on cell lines, one used the macrophage RAW26.7 line (primary CD14+ monocytes from human donors of both sexes transformed in macrophages through 1-week Colony Stimulation Factor-1 stimulation) and the other used human neuroblastoma SH-SY5Y cells. Of the 85 studies that specified the sex of the animals they used in their experiments, 79 used males (92.94%) and only 12 used females (14.12%). Limiting the sex of animals to the 81 rodent studies, which constituted the bulk of eligible studies included in this review, it results that 73 studies employed male-only animals (90.12%), while only seven studies (9.59%) used female-only animals. This shows a strong bias toward the use of male animals in depression studies of rodent models that cannot be easily translated to humans, given that the majority of people with depression are women [34,35] or female adolescents [36].

Of the eligible studies, most were conducted in China (N = 49, 55.06%; only three were located in Beijing, China's capital, while six studies were conducted in the Guangdong Province, i.e., three in Shenzhen, two in Guangzhou, and one in Zhanjiang), eight in Brazil (8.99%) and seven in Spain, of which one was a multinational study shared with other four countries (6.97%), five in Japan (5.62%), four in Egypt (4.49%), three in the US (3.37%), two in South Korea and France (2.25% each), two in Germany and Poland, with one multinational shared (1.35% each), one each in India, Iran, Italy, Nigeria, Pakistan, Serbia, and Turkey (1.12% each), and one in both Romania and Sweden, sharing the same multinational study as the other three (0.22% each). Of the 89 included studies, 15 did not use any specific drug to identify its effects on Nrf2 but rather focused on the effects of specific animal depression models on the entire antioxidant system. These 15 studies were conducted in China (N = 8), Spain (N = 3), Serbia, the USA, Brazil, and South Korea (N = 1 each). Plant extracts or animal tissue extracts were tested in 29 Chinese studies, three Japanese studies, one Pakistani, one Nigerian, and one South Korean study, for a total of 35 studies, representing 39.33% of all included studies, with China accounting for 82.86% of these studies and plant extract using studies for 59.18% of all Chinese studies.

Table 1. Summary of human studies investigating the role of Nrf2 in depression (*in vivo*, *post mortem*, in vitro on cell lines).

Study	Animal	Paradigm/Model	Location	Design	Results Relating to Nrf2	Conclusions/Observations
Lukic et al., 2014 [37]	Man; 30 patients with MDD (17 ♂; 13 ♂; Ham-D > 14; $\bar{x}$ age 44.77 ± 7.58 yr) vs. 35 HC (19 ♂; 16 ♂; $\bar{x}$ age 39.49 ± 9.64 yr, younger, p = 0.018)	OS	Belgrade, Serbia	PBMC investigated with WB for Nrf2, Keap1, NF-κB, AOEs (MnSOD, CuZnSOD, GPx, GLR) in MDD patients and HC	↑ Nrf2, Keap1, NF-κB, MnSOD, CuZnSOD, and CAT in MDD vs. HC in PBMC cytoplasm; ≈GPx and GLR between MDD and HC, but ↓GLR/GPx in MDD; MnSOD, CuZnSOD, and CAT levels correlated directly with Nrf2 levels, while MnSOD and CuZnSOD correlated with NF-κB levels	Impaired oxidative detoxification capacity in MDD, ↓ capacity of GPx to defend from OS in PBMC of MDD patients; the up-regulation of Nrf2 and NF-κB and their down-stream targets MnSOD, CuZnSOD, and CAT indicate OS status in PBMCs of MDD patients
Mellon et al., 2016 [38]	Man; 20 unmedicated MDD patients (♂, ♂) + 20 HC	MDD human model	San Francisco, CA, USA	I group = 20 unmedicated MDD subjects; II group = 20 age-, sex- and ethnicity-matched HC, before initiation of AD treatment, and in 17/20 of the unmedicated MDD subjects after 8 wk of sertraline treatment→ transcriptome-driven bioinformatic strategy to evaluate the activity of several transcriptional Ctrl pathways	In leukocytes from unmedicated MDD subjects ↑ transcriptional activity of cAMP response element-binding/activating TF (CREB/ATF) and Nrf2. 8 wk sertraline treatment was associated with ↓ in Ham-D scores and ↓ activity of Nrf2, but not in CREB/ATF activity. Several other transcriptional regulation pathways, including the glucocorticoid receptor, NF-κB and (EGR1–4) and interferon-responsive TFs, showed either no significant differences as a function of disease or treatment	CREB/ATF and Nrf2 signalling may contribute to MDD by activating immune cell transcriptome dynamics that ultimately influence central nervous system (CNS) motivational and affective processes via circulating mediators
Martín-Hernández et al., 2018 [39]	*Post-mortem* dlPFC samples of 30 Caucasian pts with MDD, ethnic origin-, gender-, and age-matched to 30 HC (sex not specified, although it was said that men were more, age of death not declared)	NI and OS in MDD *post mortem*	Bilbo, Bizcaia, Euskal Herria, Spain	2 MDD groups: AD-free (N = 15) and AD-treated (N = 15). WB for levels of TLR-4, Hsp60, Hsp70, p-ERK 1/2, p-JNK, p-p38, p38 α/β, I3K, Keap-1, p11,DUSP-2, Nrf-2, NF-κB p65 subunit in cytosol and nucleus of dlPFC neurones	↓ Nrf2 pathway in pts with MDD. AD treatments do not reverse the trend	↑ ERK 1/2 (+22%, t = 2.293, p = 0.03) and JNK (+56%, t = 2.468, p = 0.02) expression in MDD pts, but not p38-MAPK, compared to HC. p-JNK/total JNK and p-p38/total p38 ↑ in MDD > HC. AD-free and AD-treated showed no significant ↑ in Keap-1 expression compared to HC. 21% ↓ of nuclear expression of Nrf2 in MDD pts
Kubick et al., 2020 [40]	*In vitro* cells (macrophage cell line RAW26.7 from 3 ♂ and 4 ♂ human donors, treated with LPS)	Drug repurposing, LPS-induced OS	Hamburg, Germany; Madrid, Spain; Bucharest, Romania; Stockholm, Sweden; Garbatka, Poland	RNA-seq Data Analysis, AI workflow (which drugs activate NRF2?), in vitro cells treated with ZT, Protein Assay (anti-NRF2), Chemiluminescence	RAW264.7 cells treated with ZT (10 μM, 15 h) showed ↑NRF2 levels compared to plac-treated Ctrl cells	Nrf2 pathway is a putative regulator of M1 function in depression; Nrf2 is a potential drug target; ZT activates Nrf2 and its downstream targets

Table 1. *Cont.*

Study	Animal	Paradigm/Model	Location	Design	Results Relating to Nrf2	Conclusions/Observations
Goetzl et al., 2021 [41]	Man; 10 MDD Resp patients (6 ♂; 4 ♂; Ham-D > 14; $\bar{x}$ age 39.0 ± 9.4 yr) 10 MDD NResp patients (6 ♂; 4 ♂; Ham-D > 23; $\bar{x}$ age 41.3 ± 11.6 yr) vs. 10 HC (5 ♂; 5 ♂; $\bar{x}$ age 37.5 ± 10.5 yr)	OS	San Francisco, CA; New Haven, CT, USA	Two groups: MDD Resp → sertraline or escitalopram or fluoxetine × 8 wk at sertraline-equivalent doses; MDD NResp → sertaline or escitalopram, fluoxetine or citalopram at sertraline-equivalent doses; ELISA for Nrf2 and MCh proteins	NDEV levels of NRF2 were statistically ↓ in the NResp at BL and Resp at BL groups than in their Ctrl groups; levels ↑ in the NResp and Resp groups after treatment	NDEV levels of MPs of all functional classes, except complex I-6, NRF2 and PGC-1α, were normalised in MDD participants who responded to SSRI treatment but not in those who failed to respond, as assessed by the psychiatrist; the sample was small
X. Li et al., 2022 [42]	Human neuroblastoma SH-SY5Y cells	H_2O_2 induced SH-SY5Y cell damage	Beijing, China	Human neuroblastoma SH-SY5Y cells used to mimic OS damage *in vitro*. Four groups: untreated Ctrl, H_2O_2-induced injury model, kaempferol treatment, and ginsenoside rh2 treatment. WB to detect Nrf2, Trx, and Akt1. TrxR activity was Measur with the Solarbio thioredoxin reductase activity Kit	Kaempferol and ginsenoside rh2 ↑ the expression of Akt1 and Nrf2, which boosted the targets in the Akt1/Nrf2/Trx pathways cascade working conjointly. Kaempferol works better than ginsenoside rh2 in the Akt1/Nrf2/Trx pathways	Kaempferol and Rh2 could enhance the activity of the Trx system by up-regulating Akt1 to activate Nrf2 *in vitro*

For *abbreviations*, see note to Table ??.

Table 2. Summary of animal studies investigating the role of Nrf2 in depression.

Study	Animal	Paradigm/Model	Location	Design	Results Relating to Nrf2	Conclusions/Observations
Martín-de-Saavedra et al., 2013 [43]	♂WT C517BL/6 mice (Nrf2$^{+/+}$) and Nrf2 KO (Nrf2$^{-/-}$) and ♂Swiss mice (3–4-month-old)	LPS-induced DLB	Madrid, Spain	1. WT (Nrf2^{++}) and KO (Nrf2^{-}) mice were subjected to behav tests (TST, OFT, SPIT)+ biochemical analysis; WT mice received LPS (0.1 mg/kg via IP)+ SFN (1 mg/kg/day via ip for 7 days); 3. KO mice + Rofecoxib (2 mg/kg/day for 7 days)	Nrf2 deletion resulted in DLB (↑ in the immobility time in the TST and by a ↓ in the grooming time in the SPIT); ↓ of Dopa and Ser and ↑ Glu in the PFC;↑ of VEGF and synaptophysin;↑microgliosis. Nrf2 KO mice treatment with rofecoxib reversed their DLB; SFN in LPS-induced depression of WT mice afforded AD-like effects	Inflammation due to a deletion of Nrf2 can lead to a depressive-like phenotype, while the induction of Nrf2 could become a new and interesting target for developing novel AD drugs

Table 2. *Cont.*

Study	Animal	Paradigm/Model	Location	Design	Results Relating to Nrf2	Conclusions/Observations
Mendez-David et al., 2015 [44]	Adult ♂C57BL/6Ntac mice, Nrf2 WT $^{(+/+)}$ and knock-out Nrf2$^{(-/-)}$	Mouse CORT model of DLB	Paris, France	Chronic corticosteroids, chronic fluoxetine 4 wk later; OFT, E + M, NSFT, and SPIT testing during the wk following chronic fluoxetine. Immunoblotting for BDNF, Nrf2, and its downstream targets GCLC, NQO1, and HO-1, in cortical and Hippoc membranes	Chronic fluoxetine restored Nrf2 levels in mouse cortex, as well as GCLC, HO-1, and NQO1 levels that were ↓ by chronic CORT. In the Hippoc, Nrf2 was not ↓, but GCLC, HO-1, and NQO1 levels were ↓; chronic fluoxetine restored GCLC and NQO1 levels, but not HO-1. Chronic fluoxetine ↑ cortical BDNF levels and reversed CORT-induced ↓ in Hippoc BDNF; ↓ cortical and Hippoc BDNF levels in Nrf2$^{(-/-)}$ mice; these were ↑ by chronic fluoxetine	Chronic fluoxetine reverts CORT-induced Nrf2 pathway changes in cortex and Hippoc in a mouse model of depression; Nrf2 enhances BDNF, but fluoxetine enhances cortical and Hippoc BDNF through both Nrf2-dependent and independent pathways
Cunha et al., 2016 [45]	Adult ♂Swiss mice (30–40 g)	Stress-induced DLB	Florianópolis, SC, Brazil	1. I group: creatine or plac + LY294002 icv (PI3K inhibitor), wortmannin icv (PI3K inhibitor), or plac→TST or OFT 2. II group: lithium chloride (nonselective GSK3 inhibitor) or ARA01441 (selective GSK3 inhibitor)→TST or OFT 3. III group: subeffective doses of creatine + subeffective doses of GSK3 inhibitors (ARA01441 or lithium chloride)→TST or OFT 4.1 IV group: CoPP icv (HO-1 activator)→ TST or OFT; creatine or plac + HO-1 inhibitor ZnPP icv or plac→TST and OFT; creatine or plac + CoPP icv or plac→TST or OFT 5. V group: creatine or plac + rapamycin (mTOR inhibitor) or plac→TST and OFT	Treatment with creatine↑ Akt and P70S6K phosphorylation, HO-1/Nrf2, GPx and PSD95 immunocontents. The pretreatment with LY294002 wortmannin, ZnPP (HO-1 inhibitor), or rapamycin (mTOR inhibitor) prevented the AD-like effect of creatine in the TST. Subbeffective dose of either the ARA01441.8, lithium chloride, or the HO-1 inductor CoPP + subeffective dose of creatine ↓ the immobility time in the TST	The AD-like effect of creatine in the TST depends on the activation of Akt, Nrf2/HO-1, GPx, mTOR, and GSK3 inhibition

Table 2. *Cont.*

Study	Animal	Paradigm/Model	Location	Design	Results Relating to Nrf2	Conclusions/Observations
Freitas et al., 2016 [46]	♂, Swiss mice, WT C57BL/6 mice (Nrf2$^{+/+}$) and Nrf2 KO (Nrf2$^{-/-}$) (3–months-old, BW 40–45 g)	Mouse CORT model of DLB	Madrid, Spain	1. Mice assigned to 6 groups (8 mice each): (a) plac, (b) IMI/plac, (c) agmatine/plac as the Ctrl groups, (d) plac/CORST, (e) IMI/CORST, and (f) agmatine/CORST. 2. Mice were assigned to 3 groups (6 mice each): (a) Nrf2$^{+/+}$/plac, (b) Nrf2$^{+/+}$/agmatine, as the Ctrl groups, (c) Nrf2$^{-/-}$/plac, and (d) Nrf2$^{-/-}$/agmatine	Agmatine ↓ CORST-induced DLB, ↑BDNF, synaptotagmin I, Ser and Glut levels; ↓the CORST-induced changes in the morphology of astrocytes and microglia in CA1 subregion of Hippoc; ↑ Nora, Ser, and Dopa levels, CREB phosphorylation, mature BDNF and synaptotagmin I immunocontents, in the Hippoc of Ctrl group. Agmatine's ability to produce an AD-like effect was abolished in Nrf2$^{(-/-)}$ mice	Chronic administration of a low dose of agmatine is able to abolish the behavioural responses in the TST and splash test elicited by the CORST-induced model of depression by a mechanism dependent on the activation of Nrf2 and neuroplasticity-related signalling in mice
Martín-Hernández et al., 2016 [47]	♂outbred Wistar Hannover rats initial BW 200–225 g	CMS-induced DLB	Madrid, Spain	The following groups (n 8 each) were used: (1) Ctrl; (2) Ctrl group +ip injection of sterile plac for 7 days (CT þ Veh), (3) CMS group; (4) a CMS group+ ip plac (CMS þ Veh) group. For experiments requiring the ip injection of AD, 3 other experimental groups were used: (5) CMS group + desipramine; (6) CMS group + escitalopram and (7) CMS group +duloxetine→behav tests + biochemical analysis	In the PFC, CMS ↓Akt and PI3K mRNA expression. Desipramine and duloxetine ↑ CMS-induced Akt levels, but only desipramine restored PI3K levels. CMS ↓Nrf2 mRNA and protein expression levels. Nrf2 inhibitors Keap-1 and p-GSK-3β/GSK-3β ratio ↑ after CMS. Desipramine and duloxetine to CMS rats restored the expression of Nrf2, returned Keap-1 to Ctrl levels and showed a trend towards returning the p-GSK-3b/GSK-3b ratio to its Ctrl levels. CMS ↓ NQO-1, GPx1.AD treatment restored GPx1 levels. Desipramine ↑HO-1. PAR g is modulated by the AD treatments in the PFC	Nrf2 pathway is differentially regulated by AD in the PFC and Hippoc. The Nrf2 pathway is involved in the oxidative/nitrosative damage detected in the PFC, and AD has a therapeutic action through this pathway. It seems that Nrf2 is not involved in the effects caused by CMS in the Hippoc
Martín-Hernández et al., 2016 [48]	♂outbred Wistar Hannover rats; initial BW 200–225 g	CMS-induced DLB	Madrid, Spain	Mice divided into 3 groups (n = 8 each): (1) a Ctrl group; (2) CMS group; (3) CMS group treated with antibiotics (CMS þ ATB)→behav test+ biochemical analysis	CMS protocol ↑intestinal permeability and bacterial translocation. CMS also ↑the expression of the activated form of the MAPK p38 while ↓the expression of Nrf2. The actions of antibiotic administration to prevent bacterial translocation↓MAPK and ↑ Nrf2 pathways	Translocated bacteria could play a role in the pathophysiology of depression through the p38 MAPK pathway, which could aggravate the neuroinflammation and the oxidative/nitrosative damage present in this pathology. Moreover, Nrf2 and its activators may be involved in the consequences of the CMS on the brain

Table 2. *Cont.*

Study	Animal	Paradigm/Model	Location	Design	Results Relating to Nrf2	Conclusions/Observations
Wojnicz et al., 2016 [49]	Adult ♂ Sprague Dawley rats (2.5–3 months old and BW 250–300 g); Nrf2 KO mice	Nrf2 KO mouse model of depression	Madrid, Spain	Seven adult WT rat brain samples→ LC–MS/MS to detect concentrations of neurotransmitters (Adre, Nora, Glu, GABA, DA, 5-HT) and their metabolites (MHPG and 5-HIAA)c. Same procedure in the Hippoc samples of Nrf2 KO mice	LC–MS/MS in adult WT and Nrf2 KO rats showed no significant differences in neurotransmitter values except for GABA, which was strongly ↓ in KO rats	LC–MS/MS method enables rapid quantification of neurotransmitters and their metabolites. It was precise, accurate, sensitive and reproducible. Its application to the mouse model of depression (Nrf2 KO) recorded a↓of Hippoc GABA
Yao et al., 2016 [50]	Adult ♂ C57BL/6 mice, aged 8 wk (BW 20–25 g) and 5 wk; CD-1 mice, aged 14 wk (BW 40–45 g) and ♂ adult Nrf2 KO (Nrf2$^{-/-}$) mice	SDS model of depression	Chiba, Japan	WT and KO mice subjected to SDS to induced DLB (exposed to a different CD1 aggressor mouse each day for 10 min for 10 days)→I group: DLB mice + SFN (10 mg/kg), II group: DLB mice + 7,8-DHF (10 mg/kg); III group: DLB mice + ANA-12 (0.5 mg/kg); IV group: Ctrl→behav tests+ biochemical analysis	↓ Keap1 and Nrf2 in the PFC, CA3 and DG of Hippoc in mice with DLB compared to Ctrl; ↑ serum levels of pro-inflammatory cytokines in Nrf2 KO mice compared to WT mice; ↓ BDNF and TrkB in PFC, CA3 and DG play a role in DLB of Nrf2 KO mice. TrkB agonist, 7,8-DHF, but not antagonist ANA-12, produced AD effects in Nrf2 KO mice. Pretreatment with Nrf2 activator sulforaphane (SFN) prevented the DLB induced after repeated SDS	Keap1-Nrf2 system plays a key role in depression and dietary intake of SFN-rich food during juvenile stages and adolescence can confer stress resilience in adulthood (dietary intake of 0.1% glucoraphanin (a precursor of SFN) containing food during juvenile and adolescent stages also prevented the depression-like phenotype evoked in adulthood, after repeated social defeat stress)
Yao et al., 2016 [51]	Adult ♂ C57BL/6N mice BW 20–26 g	LPS-induced DLB	Chiba, Japan	Mice received ip injection of LPS (0.5 mg/Kg)+ Nrf2 activators TBE-31 or MCE-1→behav tests + ELISA	TBE-31 and MCE-1 ↑ nerve growth factor (NGF)-induced neurite outgrowth in PC12 cells in a concentration-dependent manner. TBE-31 or MCE-1↓ an increase in serum levels of TNF-α after LPS administration. In the TST and FST, TBE-31 or MCE-1 ↑ the mobility time after LPS administration	The Nrf2 activators have AD effects in animal models of depression. The novel Nrf2 activators such as TBE-31 and MCE-1 might be potential therapeutic drugs for inflammation-related depression
Bouvier et al., 2017 [52]	♂ Sprague Dawley rats BW 290–310 g (intruder rats); ♂ WT Groningen rats (WTG, resident rats); C57BL/6J background WT (Nrf2$^{+/+}$) and Nrf2-KO (Nrf2$^{-/-}$) 6-wk-old mice	SDS and CMS-induced DLB	Paris, France	Rats received intense stress first hit produced (SD) + second stressful hit (CMS)→ behav test and biochemical analysis; antioxidants were Tempol, 4-hydroxy-2,2,6,6-tetramethylpiperidine 1-oxyl; 288 µmol/kg^{-1}· day^{-1}, 7,8-DHF on days 5, 7 and 9 after the end of the social defeat protocol, and t-BHQ (continuous infusion during 6–7 days)	Only vulnerable animals developed a DLB after CMS derived from a persistent state of OS and reversed by treatment with antioxidants. This persistent state of OS was due to ↓ BDNF levels. ↓BDNF→↓ nuclear translocation of Nrf2. In Nrf2$^{+/+}$ mice, the activation of Nrf2 translocation restored redox homoeostasis and reversed vulnerability to depression. This mechanism was absent in Nrf2$^{-/-}$ mice	Low BDNF levels in vulnerable animals prevented Nrf2 translocation and consequently prevented the activation of detoxifying/antioxidant enzymes, resulting in the generation of sustained OS

Table 2. *Cont.*

Study	Animal	Paradigm/Model	Location	Design	Results Relating to Nrf2	Conclusions/Observations
Zhang et al., 2017 [53]	♂adult C57BL/6 mice, aged 8 wk (BW 20–25 g)	LPS-induced DLB	Chiba, Japan	Mice received an injection of LPS and SFN→behav test and biochemical analysis. One subgroup of mice received a dietary amount of 0.1% glucoraphanin (a glucosinolate precursor of SFN) at 5 wk→behav test in adulthood (9 wk)	Pretreatment with SFN blocked an ↑ in the serum TNF-α level and an ↑ in microglial activation after LPS administration (0.5 mg/kg); SFN ↑serum IL-10 after LPS administration. In the TST and FST, SFN ↓immobility time after LPS administration, SFN significantly recovered to Ctrl levels for LPS-induced alterations in the proteins such as BDNF, postsynaptic density protein 95 and AMPA receptor 1 (GluA1) and dendritic spine density. Dietary intake of 0.1% glucoraphanin (SFN precursor) food during the juvenile stage and adolescence could prevent the onset of LPS-induced DBS	Dietary intake of SFN-rich broccoli sprouts has prophylactic effects on inflammation-related depressive symptoms→supplementation of SFN-rich broccoli sprouts could be a prophylactic vegetable to prevent or ↓ the relapse by inflammation in the remission state of depressed patients
Zhao et al., 2017 [54]	C57BL/6J ♂mice (adult, 8-wk-old) BW 20–25 g	CRS and ARS-induced depression	Nanjing, Jiangsu, China	Mice subjected to 2 stress paradigms: 8 wk of CRS and 2 h ARS; mice divided into 2 groups: prolonged (4 wk) and short-term (a single inj) Ipt treatment (i.p. 10 mg/kg)→behav tests+ biochemical analysis (ELISA, RT-PCR, WB, IF)	HPA axis was altered after stress, with different responses to CRS (↓r ACTH and CORT, ↑ AVP, but normal CRH) and ARS (↑ CRH, ACTH and CORT, but normal AVP). Prolonged and short-term Ipt treatment normalised stress-induced HPA axis disorders and abnormal behav in mice. CRS and ARS ↑mRNA levels of TNFα, IL-1_, IL-6 and TLR4 and OS molecules (gp91phox, iNOS and Nrf2) in the hypothalamus. IF showed CRS and ARS ↑ microglia activation (CD11b and TNF_) and OS in neurons (NeuN and gp91phox), which were ↓ by Ipt	Activation of ATP-sensitive potassium channel by ipt normalises stress-induced HPA axis disorder and depressive behav by alleviating inflammation and OS in mouse hypothalamus
López-Granero et al., 2017 [55]	BXD RI strains and C57BL/6 WT mice (5–6 wk), ♂, ♂	BXD recombinant inbred mice depression and anxiety model	New York, NY, USA	Two BXD RI mouse strains, BXD21/TyJ RI, BXD84/RwwJ RI and C57BL/6 WT mice were used with 12 animals per strain and 6 animals per sex→behav tests + biochemical analysis	BXD84/RwwJ RI exhibits social avoidance behav and ↓time in elevated open spaces during the EMT. BXD21/TyJ RI ↓immobility time in the and ♂-specific sensitivity is noted; they also ↑Nrf2mRNA levels (no changes in Keap-1). Same cerebral cortex Gpx1 mRNA in BXD21/TyJ RI, BXD84/RwwJ RI and C57BL/6 WT mice. ↑pro-inflammatory response in ♂BXD21/TyJ RI compared to BXD84/RwwJ RI and C57BL/6 WT (↑ IL-6 and TNF mRNA)	BXD84/RwwJ RI strain exhibits anxiety disorders, emotional disorders, anxiety-like behav, and social avoidance-like behavior (2) BXD21/TyJ RI strain shows resistance to depression illness

Table 2. *Cont.*

Study	Animal	Paradigm/Model	Location	Design	Results Relating to Nrf2	Conclusions/Observations
Abueleez and Hendawy, 2018 [56]	Adult ♂ Wistar rats BW 150–200 g	CRS-induced DLB	Cairo, Egypt	Animals were allocated randomly to one of the following 5 groups (n 12 each): non-restrained Ctrl group, CRS group, and 3 other CRS-groups treated with cilostazol, a phosphodiesterase-3 and ROS inhibitor (7.5, 15, 30 mg/kg/day for 4 wk)→SPT, OFT, FST+ Biochemical, RT-PCR, WB analysis	Hippoc cytoplasmic and nuclear Nrf2 expressions were ↓ in CRS-rats, as well as HO-1 and NQO-1mRNA, compared with the Ctrl group. Cilostazol (15 mg/kg/day) prevented ↓nuclear Nrf2, whereas cilostazol (30 mg/kg/day) prevented ↓ in cytoplasmic and nuclear Nrf2 expression. Cilostazol (15 mg/kg/day) prevented ↓ in HO-1, whereas cilostazol (30 mg/kg/day) prevented the decrease in both HO-1 and NQO-1 mRNA	Cilostazol prevented CRS-induced DBL, improving behav tests and hypothalamus–pituitary–adrenal axis hyperactivity. Cilostazol prevented CRS-induced ↑ in Hippoc lipid peroxidation and 8-hydroxy-2′-deoxyguanosine, and a ↓ in antioxidant activities
Omar and Tash, 2017 [57]	50 Adult ♂ Sprague Dawley rats BW 180–220 g	Chronic mild stress model of depression	Cairo, Egypt	Rats divided into the Ctrl (n = 10) and stress (n = 40) groups. Ctrl rats received distilled water. The stress group, subjected to the CMS procedure, was further subdivided into 4 subgroups (n 10 each): I group = distilled water; II group = fluoxetine (10 mg/kg/day); III group = zinc (15 mg/kg/day); IV group = fluoxetine + zinc (treatment for 28 days)→behav tests + biochemical investigations (ELISA, WB, RT-PCR)	Hippoc mRNA and protein levels of Nrf2, HO-1, MTs, GPR39 and BDNF ↑ in response to a combined therapy of fluoxetine and zinc than to either monotherapy. HO-1 and MTs gene expression was correlated with that of Nrf2 in the fluoxetine-only group	Fluoxetine therapy activated the expression of MTs and HO-1 through an Nrf2-dependent pathway. When fluoxetine was escorted by zinc, activated MTs had a positive impact on BDNF through the zinc signalling receptor GPR39, resulting in ↑in neuronal plasticity as well as ↓ of neuronal atrophy and neuronal cell loss
Li et al., 2017 [58]	8- to 10-wk-old ♂ ICR mice	LPS-induced DLB	Ningbo, Zhejiang, China	Mice treated with IL-1β shRNA lentivirus or NS shRNA (Ctrl) lentivirus by DG regions inj + LPS (1 mg/kg, i.p.) or plac→ behav tests (memory deficits with NORT; anxiety-like behaviors with EZM; DLB with SPTand FST). Furthermore, the levels of MDA, SOD, Nrf2, HO-1, TNFα, VGF and BDNF were assayed	IL-1β KO in the Hippoc ↓ the memory ceficits, anxiety- and DLB induced by LPS in mice; it also ameliorated the oxidative and neurcinflammatory responses and abolished the ↓ of VGF and BDNF induced by LPS. Finally, the ↑MDA and ↓SOD, Nrf2 and HO1 induced by LPS were completely prevented with IL-1β shRNA	IL-1β is necessary for the oxidative and neuroinflammatory responses produced by LPS and offers a novel drug target in the IL-1β/oxidative/neuroinflammatory/neurotrophic pathway for treating neuropsychiatric disorders that are closely associated with neuroinflammation, OS and ↓ of VGF and BDNF
Yang et al., 2018 [59]	24 ♂ Sprague Dawley rats, 8-wk-old	LPS-induced DLB	Jining, China	Three groups of 8 rats each: Ctrl, LPS, and LPS + NPB. 24 h after last injection, behav tests + brain tissue analysis	Nrf2 ↓ in LPS group, Nrf, HO1 and NQO-1 levels ↓ in NBP group	Prolonged NBP treatment ameliorated LPS-induced DLB, attenuating LPS-induced NI, and OS

Table 2. *Cont.*

Study	Animal	Paradigm/Model	Location	Design	Results Relating to Nrf2	Conclusions/Observations
Gao et al., 2019 [60]	♂CD-1 mice BW 23–25 g and 8-wk-old ♂C57BL/6 J mice	Effects of allicin on DLB	Yichang, China	Five groups 10 mice each: Ctrl, CSDS, CSDS + allicin (2, 10, or 50 mg/kg). SPT, SIT, and FST → Hippoc tissue collected. Inflam mediator levels assayed through ELISA. Iron concentration and iron-related protein expression Measur by WB. OS and apoptosis markers detected by WB	Allicin ↓ production of ROS, MDA NOX4, and ↑ activities of SOD and Nrf2/HO-1 pathway; CSDS mice performed worse than Ctrl on SPT, SIT, and FST; allicin reversed these impairments, with the highest dose being more effective	Microglia activation and ↑ cytokine in Hippoc of CSDS were ↓ by allicin. Content of iron and protein expression of iron metabolism were aberrant in CSDS mouse Hippoc; allicin improved this phenomenon. It also attenuated enhanced neuronal apoptosis and promoted NLRP3 inflammasome suppression (↓ Hippoc ASC. caspase-1, and IL-1β)
Fan et al., 2018 [61]	72 ♂ 220–240 g Wistar rats	CUMS-induced DLB	Jinan, China	Four groups with N = 18/group: (a) Ctrl (non-CUMS), (b) CUMS, (c) ginsenoside-Rg1 pretreatment (40 mg/kg), (d) ginsenoside-Rg1 pretreatment (40 mg/kg) followed by CUMS. Behav tests + brain removed for immunofluorescence assay, immunohistochemistry and TUNEL staining	Ginsenoside-Rg1 ↓ Nrf2 expression and inhibits p-p38 MAPK and p65 NFκB subunit activation within the vmPFC	Ginsenoside-Rg1 prevented depression-like effects in a rat CUMS model. Chronic ginsenoside-Rg1 pretreatment prior to stress exposure suppressed inflam pathway activity via ↓ proinflam cytokine overexpression and microglial/astrocytic activation; ↓ dendritic spine and synaptic deficits parallel to ↑ synaptic-related proteins in vmPFC. ↓ apoptosis induced by CUMS exposure, ↑ Bcl-2 expression and ↓ cleaved caspase-3 and caspase-9 expression within the vmPFC region
Chu et al., 2019 [62]	24 ♂ pathogen-free Sprague Dawley, 6-wk-old rats + 30 WT and 30 Nrf2$^{-/-}$ KO ♂, 6-wk-old mice	Pollution-induced DLB; tested the Nrf2/NLRP3 pathway in DLB	Shijiazhuang, Hebei Province, China	Twenty-four rats randomised into 3 groups: exposed to FiA, UnA, and CA × 12 wk. 30 WT and 30 Nrf2$^{-/-}$ KO mice randomised into clean air exposure and to UnA × 9 wk. Mice and rats had behav testing. Toxic elements in PFC of rats after PM2.5 exposure were Measur by ICP-MS; neurotransmitter and their metabolites' determination (NA, 5-HIAA, 5-HT, DA, L-Dopa, DOPAC), GSH and GSSG levels in PFC were Measur by HPLC; histopathological changes, neurotrophic factor levels, cytokines, and NLRP3 inflammasome-related protein expression in PFC of rats detected with IHC and WB	CA rats and KO-UA mice displayed depressive-like responses. The NLRP3 signalling pathway was more activated in Nrf2$^{-/-}$ KO than WT mice after PM2.5 exposure × 9 wk	Li, Be, Al, Cr, Co, Ni, Se, Cd, Ba, Ti and Pb were deposited in rat PFC after PM2.5 exposure. Neurotransmitters were significantly altered in PFC of CA rats. The NLRP3 signalling pathway was more activated in Nrf2$^{-/-}$ than WT mice after PM2.5 exposure × 9 wk. The Nrf2/NLRP3 signalling pathway, by modulating inflammation, might play an important role in ambient PM2.5-induced depression

Table 2. *Cont.*

Study	Animal	Paradigm/Model	Location	Design	Results Relating to Nrf2	Conclusions/Observations
Dang et al., 2019 [63]	Adult, 8-wk-old C57BL/6 ♂ mice	PCMS in LPS-induced DLB	Xi'an, 710032, Shaanxi, China	Mice with PCMS (5 min with no mobility × 4 wk) and stress-naïve mice. LPS or plac administered via ip injection → Behav tests (FST, OFT, E + M) → brain removal, analysis through IF to detect IBA-1, IL-1β, Nrf2; WB for NLRP3, ASC, caspase-1, Nrf2, HO-1, NQO-1,TXNIP, Trx and β-actin; Real-time PCR to assess the amount and integrity of total Hippoc RNA	mRNA and protein levels of Nrf2 in stress naïve mice ↓ 26 h post-LPS administration compared with plac-treated mice. Though a significant difference in Nrf2 protein levels was not observed, PCMS mice showed increased gene expression of Nrf2 compared with stress-naïve mice; stress-naïve mice performed worse than PCMS mice on FST, OFT, and E + M	PCMS promotes recovery from LPS-induced behav deficits. Stress naïve mice showed nuclear condensation and acidophilic degeneration after LPS treatment; these neuronal injuries were alleviated in PCMS mice. IF for IBA-1 was used to analyse microglial activation, which was attenuated in PCMS mice. PCMS ameliorated LPS-induced OS, with decreased MDA level, enhanced SOD activities and reduced 8-OHdG. Gene expression of pro-apoptotic Bax was largely ↑ in the Hippoc of stress-naïve mice 26 h post-LPS and ↓ in PCMS mice. PCMS mice showed partially inhibited NLRP3 inflammasome activation (↓ in NLRP3 inflammasome component levels and attenuated IL-1β and TNF-α expression)
Gao et al., 2019 [64]	50 5-wk-old ♂ C57 mice	HFD-induced DLB	Yichang, China	To study OS, MCh function, autophagy, insulin resistance, and NOX/Nrf2 imbalance, mice were randomised into 5 groups of 10 each: Ctrl, HFD, HFD + allicin (50, 100, or 200 mg/kg). After HFD and allicin × 15 wk → behav testing. Blood samples were collected after 12 h fasting periods. All hippocampi were removed for subsequent detection	↑mRNA and protein expressions of NOX2 and NOX4, ↓Nrf2/HO-1 signalling in Hippoc of obese mice. Allicin ↓ NOX2/NOX4 expression and ↑ Nrf2/HO-1 levels	Allicin ↓ weight of obese mice, metabolic indicators, CORST, IR, and corrected HFD-triggered aberrant insulin signalling. HFD induced DLB, which was ameliorated by allicin.↑ ROS, MDA, protein carbonylation triggered by HFD were inhibited by allicin. HFD caused ↑ protein expression of autophagy in the Hippoc, which was reverted by allicin. Allicin ameliorated OS-induced damage through ↑ antioxidant SOD, CAT, GSH, and GPx activity
Arioz et al., 2019 [65]	♂ Balb/c, 12–14-wk-old mice	OS, LPS-DLB	Izmir, Turkey	Effect of MT on NLRP3 inflammasome activation and SIRT1/Nrf2 pathway. Mice randomised into 3 groups: Ctrl, LPS, MT (30 mg/kg × 4) + LPS (5 mg/kg ip). 24 h later, animals performed behav experiments TST, FST → sacrificed. Hippocampi were isolated and used for further analyses; glial cell culture	MT ↑ Nrf2 translocation to nucleus (WB) and Nrf2 target genes HC-1, NQO1, GSTP1, GCLM (qPCR). Cross-talk between Nrf2 and SIRT1 protective pathways: siRNA-mediated Nrf2 knockdown inhibited basal SIRT1 expression; siRNA-mediated SIRT1 knockdown ↓Nrf2 translocation. The beneficial effects of MT on NLRP inflammasome activation were associated with Nrf2 and SIRT1	MT ameliorated LPS-induced behav abnormalities in a mouse model of acute systemic inflammation and depression and decreased NLRP3 inflammasome activation in mice hippocampi (qPCR, WB and IF staining). Beneficial actions of MT are partly and significantly dependent on Nrf2 and SIRT1 activation in LPS and ATP-challenged murine microglia

Table 2. *Cont.*

Study	Animal	Paradigm/Model	Location	Design	Results Relating to Nrf2	Conclusions/Observations
Cigliano et al., 2019 [66]	24 ♂MRL/lpr mice brain samples (8-, 22- or 17-wk-old)	MRL/MpJ-Faslpr lupus-prone depression murine model	Napoli, Italy	To test CLA and FO modulation of the Nrf2 pathway in a mouse depression model, brain samples from 2 groups (n = 8 each), composed of 8- (Young) or 22-wk old (Old) mice were examined to evaluate the age-dependent occurrence of depressive disorder markers (BDNF, TrkB, Synaptophysin, Synapsin I; Synaptotagmin I, PPAR-α, PPAR-γ and the modification of DHA, C18:1, C16:0, and C18:0 content) with rtPCR and WB. 2 additional groups composed of 17-wk-old mice (n = 8 each) were supplemented with FO or CLA × 5 wk, when they reached old age, and were compared with untreated Old mice	FO or CLA ability in modulating Nrf2 pathway was investigated in brain cortex of all experimental groups. Old animals exhibited higher G6PD and GSR activities. Compensatory hyperactivation of GSR and G6PD, as well as ↑GCL and GSRmRNA levels exhibited by Old mice ($p < 0.05$), were ↓by FO and CLA. ↑Nrf2 involvement in the antioxidant activity elicited by FO or CLA (↓ Nrf2 content in nuclear extracts of FO + Old and CLA + Old animals)	Old mice exhibit disrupted Redox homoeostasis, compensatory Nrf2 hyperactivation, ↓ DHA, ↓ BDNF and ↓ of synaptic function proteins (Synaptophysin, Synaptotagmin I, Synapsin I) compared to Young mice. FO and CLA relieve almost all depression markers at a level comparable to Young mice, improving Nrf2-mediated antioxidant defences, ↓ auto-antibody titre and TNF-α concentration, ↑ BDNF and synaptic function proteins (FO > CLA)
Liu et al., 2019 [67]	♂WT C57BL/6 mice (adult, PRMT1[+/+]) BW 22–25 g; PRMT1 KO (PRMT1[−/−]) mice with C57BL/6 background	LPS-induced DLB	Liaocheng, China	PRMT1[+/+] and PRMT1[−/−] mice received plac (10 mL/kg) or LPS (0.5 mg/kg) (ip) → behav testing → sacrifice; Hippoc analysed for total RNA with rt-qPCR. pNF-κB p65, NF-κB, Nrf-2, GFAP, PRMT1 and IBA-1, GAPDH were detected with WB. ROS levels in AST were determined using a specific probe	LPS ↓Nrf-2 expression; PRMT1 deficiency countered this effect. Nrf-2 expression in AST ↓ by ML385, an Nrf-2 inhibitor. PRMT1 KO ↓ expression of IL-1 β and TNF-α in LPS-exposed AST; this was prevented by ML385 pretreatment. PRMT1[−/−] ↓ ROS generation in LPS-exposed cells; levels were restored by ML385 pretreatment	PRMT1[−/−] mice ameliorated LPS-induced DLB and ↑BDNF and PSD-95 expression. PRMT1 deletion alleviates LPS-induced brain injury; down-regulating LPS-promoted expression levels of GFAP and IBA-1 compared with PRMT1[+/+] mice. PRMT1 deficiency ↓IL-1β and TNF-α in Hippoc and PFC of LPS-challenged mice, ↓pNF-κB, ↑SOD and GSH-pX activities in Hippoc and ↑ Nrf-2

Table 2. *Cont.*

Study	Animal	Paradigm/Model	Location	Design	Results Relating to Nrf2	Conclusions/Observations
Rosa et al., 2019 [68]	Adult ♂ Swiss mice (3 months, 30–40 g)	Guanosine AD-like effect via GSK-3β inhibition and MAPK/ERK and Nrf2/HO-1 activation	Florianópolis, Santa Catarina, Brazil	-Effective dose of guanosine (0.05 mg/kg, p.o.)/plac was administered to mice → TST -Sub-effective dose of guanosine (0.01 mg/kg, p.o.)/plac + sub-effective dose of lithium chloride (a non-selective GSK-3β inhibitor, 10 mg/kg, p.o.)/plac → TST, OFT -Sub-effective dose of guanosine/distilled H_2O + sub-effective dose of the selective GSK-3β inhibitor, ARA014418 (0.01 µg/site, icv)/plac → TST and OFT -Effective dose of guanosine→ Hippoc and PFC WB for β-catenin and Nrf2 immunocontent -Effective dose of guanosine/plac + MEK1/2 inhibitor (5 µg/site, icv)/plac → TST -Effective dose of guanosine/+ ZnPP (HO-1 inhibitor, 10 µg/site, icv)/plac→ TST, OFT and WB Hippoc and PFC analysis for HO-1 detection	Guanosine ↓ immobility time on the TST but did not alter OFT parameters. Guanosine ↑ Nrf2 cytosolic fraction immunocontent in Hippoc and PFC, compared to Ctrl. Nrf2 Hippoc nuclear fraction was not altered with guanosine	The combined treatment with sub-effective doses of guanosine (0.01 mg/kg, p.o.) and selective/non-selective GSK-3β inhibitors produced a synergistic AD-like effect in the TST. The AD-like effect of guanosine (0.05 mg/kg, p.o.) was completely prevented by the treatment with MEK1/2 inhibitors, or ZnPP. Guanosine administration (0.05 mg/kg, p.o.) ↑ the immunocontent of β-catenin in the nuclear fraction and Nrf2 in the cytosolic fraction in the Hippoc and PFC. HO-1 immunocontent was also ↑ in the Hippoc and PFC treated with guanosine. Guanosine ↓ depression by ↓ GSK-3β and ↑ MAPK/ERK and Nrf2/HO-1 pathways
Huang et al., 2020 [69]	18 ♂ C57BL/6 mice (7–8-wk-old)	CMS-induced DBL	Shanghai, China	ADSCs were isolated from mouse fat pads and intravenously administered to CMS-exposed C57BL/6 mice at the dose of 1×10^6/wk × 3 wk. Behav test (SPT, FST, TST) + rt-qPCR analysis of brain RNA, ELISA microglia analysis to detect MCP-1, IL-6, TNF-α, and IL-1β; WB with anti-Nrf2, anti-HO-1, anti-NF-κB1, anti-CD29, anti-CD90, anti-CD44, anti-CD105, anti-CD34, anti-vWF, anti-BDNF, anti-TrkB, and anti-GAPDH Abs	CMS promoted DLB, TLR4/NFκB but suppressed Nrf2/HO-1; ADSC treatment had the opposite effect, ↑SPT and ↓ immobility on TST and FST. ADSC and BV2 microglia cell cocultures showed that the ↑ of TLR4 and NFκB induced by LPS was ↓ by treatment with Nrf2-overexpressing vector ADSCs, while the ↓ of Nrf2 decreased the inhibitory effect of ADSCs on LPS-induced TLR4 and NFκB expression. The ↑ of Nrf2 in ADSCs ↓ BV2 LPS-induced inflammatory factor secretion (MCP-1, TNF-α, IL-1β, and IL-6)	ADSC treatment reversed CMS-induced DLB. The BW of the mice in the CMS group slowly ↓ compared to Ctrl, and ADSC treatment restored the CMS-induced BW reduction. ADSC reversed CMS-induced DBS, CMS-induced inflammatory factor expression, and Hippoc microglial polarisation. CMS ↑ MCP-1, TNF-α, IL-1β, and IL-6 expression in serum, but ADSC reversed CMS-induced inflammatory factor production. Immunohisto-chemical detection also showed that the number of apoptotic neuronal cells ↓ with ADSC treatment. BDNF and TrkB ↓ with but ↑with ADSC. CMS induction promoted TLR4/NFκB signalling but suppressed Nrf2/HO-1 signalling, while ADSC treatment had the opposite effect

Table 2. *Cont.*

Study	Animal	Paradigm/Model	Location	Design	Results Relating to Nrf2	Conclusions/Observations
Zborowski et al., 2020 [70]	24 ♂ adult Swiss mice, 60 days old, 25–35 g	DLB in STZ-induced DM mice	Santa Maria, Rio Grande do Sul, Brazil	Animals separated into 4 groups (n = 6 each): Ctrl; STZ-induced DM; (p-IPhSe); DM + (p-ClPhSe)2. Groups II and IV received STZ at a single dose of 200 mg/kg. After 14 days, DM+ mice (blood glucose $\geq$ 200 mg/dL) were enrolled. At day 21, mice performed behav tests (LP, TST, FST). For ex vivo assays, brains were removed, and the samples of the whole cerebral cortex were subjected to WB and OS assays	$\downarrow$ in Keap1, Nrf2 and HO-1 levels in the cerebral cortex of DM mice compared to Ctrl. (p-ClPhSe)2 $\uparrow$ Keap1, Nrf2, and HO-1 levels. A negative correlation was found between glycaemia on the one hand and Keap1, Nrf2 and HO-1 levels on the other	(p-ClPhSe)2 reversed DM+ mice DLB but did not alter mouse spontaneous behaviour; hyperglycaemia; counteracted DM-induced cortical oxidative damage. It did not reverse the $\uparrow$ in adrenal gland weight and the DM-induced decrease in GR content. It modulated the Keap1/Nrf2/HO-1 signalling pathway in DM mice and $\downarrow$ FJC+ cells (a measure of neurodegeneration) in the cerebral cortex of diabetic mice
Casaril et al., 2020 [71]	♂ BALB/c 5–6-wk-old mice	Tumour-induced DLB	Pelotas, Rio Grande do Sul, Brazil	Mice were injected with 50 μL of tumour cell suspension sc; Ctrl mice were injected with PBS. Once tumours became palpable (day 7), tumour size was monitored wkly, and BW and body temperature were recorded. Treatment with CMI (10 mg/kg, i.g.) or canola oil was initiated at day 14 and continued until day 20. 24 h later, mice were submitted to behav tests followed by killing. PFC and Hippoc samples were analysed	4T1 tumour-bearing mice had $\uparrow$ of NFκB, IL-1β, TNF-α, IDO, COX-2, and iNOS and $\downarrow$ of IL-10, Nrf2, and BDNF. CMI treatment $\downarrow$ the expression of inflammatory markers and $\uparrow$ the expression of IL-10, Nrf2, and BDNF	CMI abolished tumour-induced DLB and cognitive impairment; $\downarrow$ tumour-induced NI ($\downarrow$NFκB, IL-1β, TNF-α, IL-10, IDO, and COX-2) and OS (altered expression of iNOS and Nrf2, ROS, NO, lipid peroxidation, and SOD activity) in mouse PFC and Hippoc
Tian et al., 2020 [72]	Adult ♂ Sprague Dawley rats (8–12-wk-old), BW 180–220 g	CUMS-induced DLB	Xi'an, Shaanxi, China	CUMS was used to establish depression and anxiety-like behaviour in rats. The rTMS was performed with a commercially available stimulator for 7 days, and then depression and anxiety-like behav were Measur. Nrf2 expression was Measur by WB and TNF-α, iNOS, IL-1b, IL-6 Measur with ELISA. A small interfering RNA was employed to knockdown Nrf2, after which the neurobehav assessment, Nrf2 nuclear expression, and the amount of inflammation factors were evaluated	CUMS-exposed rats had $\downarrow$ Nrf2 expression compared to Ctrl ($F_{1,8}$ = 2.97, $p < 0.05$). One-wk rTMS treatment $\uparrow$ nuclear Nrf2 protein expression compared to CUMS ($F_{1,18}$ = 3.48, $p < 0.05$)	Application of rTMS exhibited significant AD and anxiolytic-like effects associated with $\uparrow$ Nrf2 nuclear translocation and $\downarrow$ level of TNF-α, iNOS, IL-1β, and IL-6 in the Hippoc. Following Nrf2 silencing, AD and anxiolytic-like effects produced by rTMS were abolished. Moreover, the $\uparrow$ of Nrf2 nuclear translocation, and the $\downarrow$ of TNF-α, iNOS, IL-1β, and IL-6 in Hippoc mediated by rTMS, were reversed by Nrf2 knockdown

Table 2. *Cont.*

Study	Animal	Paradigm/Model	Location	Design	Results Relating to Nrf2	Conclusions/Observations
Li et al., 2020 [73]	6-wk-old ♂ICR mice, 20–22 g	LPS-induced DLB	Liaocheng, ShanDong, China	For the acute inflammation experiment, mice received ip plac or ip Fen (10, 20 and 40 mg/kg) × 7 days prior to LPS injection. After behav tests, all mice were sacrificed. Blood was collected. Brain tissues were isolated for further analysis (siRNA, WB)	Fen dose-dependently ↑ Nrf2 expression from mRNA and Nrf2 protein levels and ↓ Nrf2 ubiquitination. Fen treatment ↑ Nrf2 expression and nuclear translocation in mouse bEnd.3 cells, promoting Nrf2-ARE transcription activity. Nrf2, HO-1, NQO1, and GCLM mRNA; Fen-induced protein expression levels were abolished by Nrf2 knockdown	Fen ↑ antioxidant capacity in bEnd.3 cells after LPS exposure: ↑ SOD, ↑ GPx, ↑ CAT, ↓ ROS, ↓ MDA; ↓ apoptotic rate promoted by LPS; ↓ IL-1β, IL-18, IL-6, TNF-α, and NO; ↓ TNF-κB nuclear expression; ↓ phosphorylation of IKKβ, IκBα and NF-κB. Fen had minimal impact on mouse histological changes and could alleviate symptoms of LPS-induced DLB
Nakayama et al., 2020 [74]	Adult ♂ and ♂ WT Japanese rice fish (medaka, *Oryzias latipes*)	Seasonal changes-induced DLB	Nagoya, Japan	Medaka fish under winter conditions (SC) were divided into 2 groups: one remained in SC with the other transferred to summer-like conditions (LW). 2 wk later, behav tests. Metabolomic and transcriptomic whole brain analyses (microarray analyses, qPCR). Drug screening was conducted to treat winter-induced depression (celastrol)	Inactivation of Nrf2-mediated antioxidant response under winter-like conditions. Celastrol induced Nrf2 expression and NRF2 target genes (GSTω1, GSH, GPx, PG reductase 1, proteasome subunit α type-6,and β type-7 and c-x-c chemokine receptor type 2)	SC in medaka fish: ↓ sociability; ↑ anxiety-like behav; ↓ in circadian clock genes (*PER2, PER3, BMAL1, CLOCK, NPAS2, CRY2*); ↓ GSH, tryptophan, and tyrosine;↑ inflammation markers (IL6, IL10, and BAFF, IL1R2); ↑serotonin levels but ↓ serotonin turnover, ↑glutamate and ↓ taurine; inactivation of RAR and glucocorticoid receptor signalling, with HPA dysregulation. Celastrol activated Nrf2 pathway
Ali et al., 2020 [75]	8–10 wk ♂C57BL/6J mice, divided in 5 groups of 6: normal, plac, LPS (1 mg/kg/day), LPS + MT (10 mg/kg/day), LPS + Fluoxetine (10 mg/kg/day), MT (10 mg/kg/day)	OS; LPS-induced DLB	Shenzhen, Guangdong, China	Open field test, Sucrose preference test, FST, TST, ROS-Measur; ELISA, IF, WB for *ATG* gene products and FOX03a	↑ NF-κB signalling in LPS-treated mice, associated with alterations of redox molecules (Akt, Nrf2, HO-1), which were reversed by MT treatment	MT ↓LPS-induced DLB and autophagy impairment in the brain (via FOX03a signalling), ↓LPS-induced OS and NInfl
Wang et al., 2020 [76]	6 wk ♂C57BL/6J mice, divided in 5 groups of 8: Ctrl; (10 mg/kg) PB; CUMS; CUMS + (10 mg/kg) PB; CUMS + (10 mg/kg) IMI	CUMS-induced DLB mouse model; OS	Nanchang, Jiangxi, China	SPT, OFT, FST, TST, commercial kit for ROS, WB, TUNEL assay	PB treatment → Nrf2 and HO-1 expression, indicating that PB alleviated DLB in mice via activating Nrf2/HO-1 signal pathway	PB alleviated the ↓of sucrose preference and BW, ↓CUMS-induced DLB, ↓ ROS concentrations and inhibited cell apoptosis in Hippoc of CUMS-induced mice. PB ↓ DLB via inhibiting OS and NI, resulting in ↓cell apoptosis in CUMS-induced mice
Liao et al., 2020 [77]	24 ♂Sprague Dawley rats, 3 groups of 8: Ctrl; CMS group (×4 wk); CMS + SalB (30 mg/kg/day)	DLB in CMS-treated rats	Changsha, Hunan, China	WB, PCR (biomarkers of NI); SPT, FST, NSFT	SalB reversed CMS-induced up-regulation of the gene expression of IL-6, IL-1β, and TNF-α in the Hippoc; SalB → anti-inflammatory effect by activating the Nrf2 signal	SalB could alleviate CMS-induced damage to Hippoc neurones. SalB normalised behav changes in CMS rats
Severo et al., 2020 [78]	46 ♂32-day-old Swiss mice in 4 groups: Ctrl; receiving each 4 cncs, 7 cncs; 10 cncs	Protocol of recurrent cncs (4, 7, or 10)	Santa Maria, Rio Grande do Sul, Brazil	TST, HRR, MCh respiration assays, estimation of ROS production and SOD activity, WB	Recurrent cncs did not alter SOD activity, but ↑ expression of NRF2 anc SOD2	Cncs ↓MCh oxygen flux vs. Ctrl; cncs do not induce significant changes in TST

Table 2. *Cont.*

Study	Animal	Paradigm/Model	Location	Design	Results Relating to Nrf2	Conclusions/Observations
Ali et al., 2020 [79]	8 wk C57BL/6J ♂mice in 7groups of 6): normal plac, LPS (1 mg/kg/day), LPS + MT (10 mg/kg/day), LPS + Fluoxetine (10 mg/kg/day), MT (10 mg/kg/day), LPS + MT + luzindole (5 mg/kg/day), LPS – luzindole	LPS induced-DLB	Shenzhen, Guangdong, China	OFT, SPT, FST, TST, Serum ROS Measur, TBARs assay, ELISA (IL-6, IL-1B, TNFa), Immunofluorescence, WB (Nrf2, p-NFkB, NFkB, p-GSK-3β, GSK-3β, Sirt, Ho-1, GAPDH)	MT treatment significantly ↑ Nrf2 and anti-inflammatory protein HO-1 expression which was down-regulated in the presence of MT receptor (MT_1/MT_2) inhibitor, suggesting that MT regulates NF-kB/Nrf2/HO-1 expression in a receptor-dependent manner	MT suppressed LPS-induced DLB, ↓cytokines level, ↓oxidative stress, and normalised LPS-altered Sirt1, Nrf2, and HO-1 expression
Camargo et al., 2020 [80]	♂Swiss mice (30–40 g, 45–60 days of age) divided into 8 groups: plac + plac; plac + ketamine (0.1 mg/kg); plac + guanosine (0.01 mg/kg); (4) plac + ketamine + guanosine; CORT + plac; CORT + ketamine; ketamine+ guanosine; CORT + ketamine + guanosine	CORT-induced animal model of depression	Florianópolis, Santa Catarina, Brazil	Behav tests (TST, OFT, SPIT); WB (GR, NF-κB, IDO-1, GLT-1, Nrf2, HO-1); biochemical analysis (Glutamine synthetase activity, determination of antioxidant enzyme activities and OS markers)	CORT administration ↓ Nrf2 (cytosolic fraction) and HO-1 immunocontent in the Hippoc; a single coadministration with KT + GN could not restore CORT-induced down-regulation on Hippoc Nrf2 and HO-1	Single administration of ketamine (0.1 mg/kg, i.p.) + guanosine (0.01 mg/kg, p.o.) ↓ DLB and Hippoc slice impairments induced by CORT. The behav response obtained by Ketamine + Guanosine was paralleled by the re-establishment of the CORT-induced molecular alterations on Hippoc GR, NF-κB, IDO-1, and GLT-1 immunocontent
Park et al., 2020 [81]	7-wk-old ♂C57BL/6 mice	Reserpine-induced depression and in vitro LPS-stimulated BV2 microglia	Daejeon South Korea	Behav tests (OFT, TST, FST); Electrospray Ionisation Mass Spectrometry; IF (for BDNF, cAMP, CREB); ELISA (IL-6, IL-1b, TNF-a, and IL-10); PCR (Il1b, Il6, TNFα, NOs2, Cox2, Hmox1); WB (iNOS, NF-kB p65, HO-1, Nrf2, p-CREB, CREB, p-p38, p38, p-Erk, Erk, p-JNK, JNK, p-Akt, Akt, and BDNF)	BTS ↑ nuclear translocation of Nrf2 and p-CREB, which act as upstream modulators of HO-1 expression in BV2 microglia	BTS has considerable potential as an anti-NI and AD agent, as it has clear effects on depressive behaviours and associated factors caused by reserpine-induced depression. BDNF and pCREB in the Hippoc ↑ in BTS-treated mice vs. reserpine-treated mice. Il1β, Il6, and TNFα mRNA levels in BTS mice were ↓ vs. reserpine-treated mice.
Zhu et al., 2020 [82]	Adult ♂Sprague Dawley rats BW 200–220 g, 3 groups: Ctrl, low-dose Hsd (50 mg/kg, Hsd-L), high-dose Hsd (50 mg/kg, Hsd-L)	STZ model of type 1 diabetes	Xuzhou, northwestern Jiangsu province, China	Behav test (OFT, TST); ELISA (CORST); Immunohistochemistry (Nrf2)	Hsd caused significant ↑ in Nrf2 levels and up-regulated g-glutamylcysteine synthetase, target gene of Nrf2/ARE signalling	Hsd ameliorate DLB and anxiety-like behaviours of diabetic rats, which are mediated by the enhancement of Glo-1, possibly due to the activation of the Nrf2/ARE pathway

Table 2. *Cont.*

Study	Animal	Paradigm/Model	Location	Design	Results Relating to Nrf2	Conclusions/Observations
Liao et al., 2020 [83]	♂Sprague Dawley rats (BW 180–220 g)	CUMS-induced depression model in rats	Changsha, Hunan, China	Rats randomised into 3 groups of 8: Ctrl, CUMS, CUMS + CUR. After 4 wk: behav tests (SPT, FST, OFT, NSFT); Determination of serum CORST; Hippoc: WB (NOX2, 4-HNE, Nrf2, pCREB, CREB, PSD-95, synaptophysin, PCNA), PCR (Nrf2, NQO-1, HO-1); Immunohistochemical staining	Nrf2 signal pathway was inhibited under CUMS, and chronic administration of CUR enhanced Nrf2 translocation from cytoplasm to nucleus and ↑ expression of antioxidant enzymes through Nrf2 signal pathway, thereby protecting the brain against CUMS-induced depression	CUR relieves depressive-like state through the mitigation of OS and the activation of Nrf2-ARE signalling pathway. DLB in CUMS-treated rats successfully corrected after CUR; CUR could effectively ↓ protein expression of OS markers (NOX2, 4-HNE, and MDA) and ↑ the activity of CAT; CUR also reversed CUMS-induced inhibition of Nrf2-ARE signalling pathway along with ↑ the mRNA expression of NQO-1 and HO-1; CUR also ↑ the ratio of pCREB/CREB and synaptic-related protein (BDNF, PSD-95, and synaptophysin); CUR could effectively reverse CUMS-induced reduction in spine density and total dendritic length.
Qu et al., 2021 [84]	♂adult C57BL/6 mice and ♂adult Nrf2 KO mice (Nrf2 $^{-/-}$) mice	Nrf2 KO mice depression-like phenotypes	Chiba, Japan	Behav tests (LMT, TST, FST, SPT); brain mPFC homogenates: WB for GluA1 and PSD-95	(R)-ketamine(10 mg/kg) could produce rapid-acting and long-lasting AD-like effects in Nrf2 KO mice via the BDNF-TrkB signalling pathway	(R)-KT can produce rapid and long-lasting AD-like actions in Nrf2 KO mice via TrkB signalling; (R)-KT significantly attenuated TST and FST ↑ immobility in Nrf2 KO mice; on the SPT, (R)-KT significantly ameliorated ↓ SPT preference. ↓ expression of GluA1 and PSD-95 in the mPFC of Nrf2 KO mice was significantly improved after a single (R)-KT injection and pretreatment with the TrkB antagonist ANA-12 (0.5 mg/kg) blocked the rapid and long-lasting AD-like effects of (R)-KT. ANA-12 significantly antagonised the beneficial effects of (R)-KT on ↓ expression of synaptic proteins in the mPFC

Table 2. *Cont.*

Study	Animal	Paradigm/Model	Location	Design	Results Relating to Nrf2	Conclusions/Observations
Li et al., 2021 [85]	Adult C57BL/6J ♂ mice BW 25–30 g	LPS-induced DLB	Shenzhen, Guangdong, China	Animals were divided into four groups of 10: Ctrl, LPS (2 mg/kg/day), LPS + Ibrutinib (50 mg/kg/day), and Ibrutinib (50 mg/kg/day). Behav tests (OFT, SPT, FST); ROS, NO, H_2O_2; TBAR Assay, ELISA, IF, Golgi staining, WB (BDNF, Nrf-2, NF-κB, HO-1)	Ibrutinib alleviated redox signalling changes, including altered LPS-induced Nrf2, HO-1, and SOD2 expression; ibrutinib, in the presence of LPS, ↑ the expression of Nrf2 and its target proteins, including HO-1 and SOD2	Ibrutinib ↓ LPS-induced DLB and NI by inhibiting NF-κB activation, ↓ pro-inflammatory cytokine levels, normalising redox signalling and its downstream components, including Nrf2, HO-1, and SOD2, and glial cell activation markers, such as IBA-1 and GFAP; ibrutinib ↓ LPS-activated inflammasome activation by targeting NLRP3/P38/Caspase-1 signalling. LPS ↓ the number of dendritic spines and expression of BDNF, and synaptic-related markers, including PSD95, SNAP25, and synaptophysin, were ↑ by ibrutinib in mouse Hippoc
Yan et al., 2021 [86]	♂ Kunming mice BW 18–22 g	D-GalN-induced animal model	Shenyang, Lioning, China	Four groups: Ctrl, D-GalN, NKT (5 mg/kg), NKT (10 mg/kg). Behav tests (SPT, FST, TST, NFT), WB (Ho-1, Nrf-2)	NKT can effectively ↓ OS in the model group, which may be caused by activating the Nrf2/HO-1/NQO1 signalling pathway, promoting the nuclear translocation of Nrf2, and ↑ the expression of downstream antioxidant protein HO-1 and NQO1 to weaken OS	NKT (5 mg/kg) co-treatment remarkably ameliorates D-GalN-induced anxiety- and depression-like behaviours. NKT could ↑ serum alanine transaminase and aspartate transaminase levels, alleviate hyperammonaemia-induced OS by activating Keap1/Nrf2/HO-1 antioxidant pathways, ↓ the expression of inducible NOs and NOX2 in Hippoc and prefrontal cortex, ↑ the vitality of SOD, ↑ catalase and GSH levels in serum, liver, and brain, and significantly ↓ the generation of MDA. NKT also ↓ the level of ammonia in serum and brain and ↑ the activity of glutamine synthase in the Hippoc and prefrontal cortex

Table 2. *Cont.*

Study	Animal	Paradigm/Model	Location	Design	Results Relating to Nrf2	Conclusions/Observations
Herbet et al., 2021 [87]	♂ adult Albino Swiss mice BW 25–35 g	Mouse CORT model of DLB	Lublin, Poland	Five groups of 8: (1) Ctrl; (2) stress Ctrl or positive Ctrl of depression: CORST (20 mg/kg) for 21 days; (3) fluoxetine (10 mg/kg) and CORST for 21 days; (4) Mito-TEMPO (1 mg/kg) and CORST; (5) fluoxetine, Mito-TEMPO and CORST. Behav tests (FST, TST). Evaluation of the level of mRNA expression of *Adora1, Ogg1, Msra, Nrf2* and *Tfam* in mouse Hippoc	↑ of *Ogg1. Adora1* and *Nrf2* in the Hippoc of mice receiving CORST and Mito-TEMPO as compared to the CORST Ctrl group	Behavioural research data showed the AD effect of fluoxetine and Mito-TEMPO administered to mice alone and in combination. Molecular findings indicate a significant impact of chronic stress on the oxidation-reduction balance and an antioxidant effect of Mito-TEMPO. The results obtained in the study suggest that Mito-TEMPO protects DNA against oxidative damage and may be beneficial in the way of cellular function improvement under the conditions of chronic stress. *Adora1, Msra, Nrf2* and *Tfam* genes may be involved in mediating the antioxidant effect of the combined fluoxetine–Mito-TEMPO treatment
Yao et al., 2021 [88]	♂ adult C57BL/6 mice (8-wk-old, 20–25 g BW), CD-1 mice (14-wk-old, 40–45 g BW), and ♂ adult Nrf2 homozygous KO mice (Nrf2 $^{-/-}$)	LPS-induced and CSDS models of depression; Nrf2 KO mice	Guangzhou, Guangdong, China	PCR for *Nrf2, Bdnf, Gapdh* genes, and BDNF; WB; Behav tests (locomotion, TST, FST)	Nrf2 activator SFN showed AD-like effects in the LPS-induced and CSDS models of depression by ↑ the expression of BDNF	Activation of Nrf2 by SFN showed fast-acting AD-like effects in mice by activating BDNF, ↓ expression of its transcriptional co-repressors (*HDAC2, mSin3A*, and *MeCP2*), and restoring normal synaptic transmission; in contrast, SFN did not affect the protein expression of BDNF and its transcriptional repressor proteins in mPFC and Hippoc, nor did it ↓ DLB and abnormal synaptic transmission in Nrf2 KO mice. In the CSDS mouse model, Nrf2 and BDNF protein levels in mPFC and Hippoc were ↓ compared to Ctrl and CSDS-resilient mice; in contrast, protein levels of BDNF transcriptional repressors in the CSDS-susceptible mice were ↑ than those of Ctrl and CSDS-resilient mice

Table 2. *Cont.*

Study	Animal	Paradigm/Model	Location	Design	Results Relating to Nrf2	Conclusions/Observations
Salama et al., 2021 [89]	Adult ♂ Wistar albino rats, BW 150 ± 20 g	Ciprofloxacin-induced depression	Cairo, Egypt	Tested camphor as AD. 5 groups. I (normal Ctrl): normal plac. II: camphor (10 mg/kg; i.p.) × 21 days. Group III (depression Ctrl): ciprofloxacin only. Groups IV and V: ciprofloxacin + camphor (5 and 10 mg/kg; i.p.) × 21 days. Behav tests (FST, activity cage, and Rotarod). Measur of OS and antioxidant biomarkers (MDA, NO, Nrf2), inflammatory biomarkers (TLR4, TNF-α), neurotransmitters; histopathology	Camphor $\uparrow$ catalase and Nrf-2 activities, $\downarrow$ NO, MDA, TNF-α, TLR4 serum levels, and $\uparrow$ brain contents of 5-HT, DA, GABA, and P190-RHO GTP protein, normalising fronto-cortical neuronal cell structure and function	The beneficial effect of camphor as AD could be mainly attributed to its antioxidant and anti-inflammatory abilities that $\uparrow$ catalase, Nrf-2 expression, and $\downarrow$ NO, MDA, TNF-α, and TLR4 production. In addition, it up-regulated P190-RHO GTP protein, an actin reorganiser, thus improving locomotor activity and restoring neurotransmitter function and structure, countering histopathological changes; hence, it may be beneficial in $\downarrow$ ciprofloxacin-induced depression
Naß et al., 2021 [90]	*C. elegans* strains (N2 WT, QV225 *skn-1* deficient, and VC289 *prdx2* deficient)	*Skn-1* (which corresponds to the human Nrf2) and *prdx2*-deficient mutants of *C. elegans*	Mainz, Germany	Examination of the antioxidant activity of UA compared to fluoxetine in *C. elegans* WT and *skn-1*- and *prdx2*-deficient strains through H$_2$DCF-DA and jugl., and osmotic and heat-stress assays. Analysis of the binding of UA to human PRDX2 and Skn-1 proteins by molecular docking and microscale thermophoresis	UA exerted stronger antioxidant activities than fluoxetine. Additionally, induction of stress resistance towards osmotic and heat stress was observed. qRT-PCR showed UA to up-regulate *skn-1* and *prdx2* expression	UA exerted antioxidant effects and induced stress resistance through *Prdx2* and *Skn-1*. Additionally, it $\uparrow$ the expression of antioxidant genes and prolonged lifespan. In many of these experiments, UA outperformed fluoxetine
Yang et al., 2021 [91]	Adult ♂ BALB/c (10-wk) mice, BW 20–22 g	Cancer-related fatigue model of depression	Changsha, Hunan, China	Test effect of Chinese herb couple Fuzi and Ganjiang (*Aconitum carmichaelii* Debx and *Zingiber officinale* Rosc) on NI (tested on cultured BV2 microglial cells; tested viability, LPS-induction), which in turn induces cancer-related fatigue (tumour inoculation). NO detected through NO$_2^-$, ROS, ELISA, IF, Nrf2 siRNA transfection of BV2 cells, WB, immunohistochemistry, Hippoc and cortex; 7 days post-inoculation, mice randomised into Ctrl, tumour-model, minocycline, low-, intermediate- and high-dose Fuzi and Ganjiang $\rightarrow$ FST, OFT, TST, and E + M $\rightarrow$ sacrifice	Fuzi and Ganjiang did not affect BV2 viability, $\downarrow$ TNF-α, $\downarrow$ IL-6, and $\downarrow$ ROS production, abolished iNOS-mediated NO, $\downarrow$ COX2-mediated prostaglandin E$_2$ and $\downarrow$ NF-κB in LPS-induced BV2 microglia; in the same cells, Fuzi and Ganjiang $\uparrow$ Nrf2/HO-1 signalling pathway; low and high doses $\downarrow$ immobility in the TST, high dose $\uparrow$ open arm time in the E + M and $\downarrow$ immobility in the FST in tumour-model mice, $\downarrow$ iNOS and COX2 in PFC and Hippoc of tumour model mice with cancer-related fatigue–induced depression	Fuzi and Ganjiang counteracted NI and related depression by activating the Nrf2 pathway
Zhu et al., 2021 [92]	♂ Sprague Dawley rats (BW 200–220 g)	DM-associated DLB	Xuzhou northwestern Jiangsu province, China	Rats received 60 mg/kg, ip STZ injection and were divided into 3 groups of 10 each: DM model group; low dose hesperetin-treated DM group (50 mg/kg), high-dose hesperetin-treated DM group (150 mg/kg) + normal Ctrl $\rightarrow$ behav tests, biochemical analysis	Hesperetin $\uparrow$ Nrf2 and its related genes and proteins (Glo-1 and γ-GCS); $\downarrow$ high glucose-induced neuronal damage through the activation of the Nrf2/ARE pathway in SH-SY5Y cells	Hesperetin ameliorated DM-associated anxiety and DLB in rats ($\uparrow$OFT, SPT, and FST performance) and $\uparrow$ Nrf2/ARE pathway activation

Table 2. *Cont.*

Study	Animal	Paradigm/Model	Location	Design	Results Relating to Nrf2	Conclusions/Observations
Wang et al., 2021 [93]	♂Sprague Dawley rats (BW 160–180 g)	CUMS-induced depression	Zhengzhou, Henan, China	Rats divided into 5 groups (10 rats each): Ctrl, catalpol; CUMS model; CUMS + catalpol; fluoxetine +CUMS→ behav tests (OFT, SPT, and FST) before and after stress/drug; hippoc for histological and biochemical analysis	CUMS caused ↓ mRNA and protein expression of Nrf2 and HO-1 in rat hippoc, whereas separate administration of both catalpol and fluoxetine reversed CUMS-induced Nrf2 and HO-1 abnormalities	Catalpol improved OFT, FST CUMS-induced abnormalities; ↑ Hippoc PI3K, Akt, Nrf2, HO-1, TrkB, and BDNF (↓ in CUMS-rats); ↑ the Hippoc SOD, catalase, GPX, GSTs; ↑ glutathione levels, ↓ in thiobarbituric acid reactive substances level in CUMS-induced depression
Wu et al., 2021 [94]	Adult ♂Kunming mice (BW 18–22 g, 3–4-wk-old)	Hyperglycaemia-induced DLB	Zhengzhou, Henan, China	Ten mice selected as the Ctrl group. 80 mice received STZ (150 mg/kg once, ip) injection; 50 mice selected (blood glucose >200 mg/dL) and divided into 5 groups (10 mice each):the plac group; the catalpol (5–10–20 mg/kg)) group; the fluoxetine (20 mg/kg)+ metformin (100 mg/kg) group→ behav tests+ biochemical analysis on brain tissues	The levels of Nrf2 and HO-1 in hippoc and frontal cortex of STZ-induced hyperglycaemic mice significantly ↓, while 20 mg/kg catalpol reversed the abnormal Nrf2 and HO-1 protein levels	Catalpol reversed TST, FST, and OFT abnormalities and abnormal PI3K and Akt phosphorylation; ↑Nrf2-HO1, SOD, GPX and GSTs; ↓ GSH and MDA in hippoc and frontal cortex of STZ-induced hyperglycaemic mice with DLB
Rahman et al., 2021 [95]	Adult BALB/c ♂mice BW 25–30 g (7–8-wk-old)	LPS-induced DLB	Zhengzhou, Henan, China	Mice divided into 5 groups of 10: (1) plac Ctrl; (2) Plac-LPS (LPS); (3) Xn-LPS (Xn = 10 mg/kg BW, *i.g.*) (4) Xn-LPS (Xn = 20 mg/kg BW, *i.g.*); (5) Fluoxetine-LPS→ behav tests; blood collected brain tissues collected for biochemical analysis	Xn significantly ($p < 0.001$) ↑ Nrf2 and HO-1 expression in the Hippoc, ↓ OS	Pretreatment with Xn (10 and 20 mg/kg, *i.g.*) reversed the behav impairments (FST and TST) with no effect on Locomotion; 20 mg improved anhedonic behavior (SPT). Xn dose-dependently prevented the LPS-induced NI, OS and nitrosative stress; ↓ activated gliosis via ↓ of Iba-1 and GFAP in hippoc; ↓ the expression of p-NF-κB and cleaved caspase-3
Tao et al., 2021 [96]	60 ♂ C57BL/6J mice BW 18–22 g	CUMS induced DLB	Nanjing 210023, China	Four groups: Ctrl, CUMS, CUMS + Magnolol (50 mg/kg, i.g. × 3 wk), and CUMS + Magnolol (100 mg/kg, MA-H i.g. × 3 wk). → behav tests (SPT,OFT,SFT,TST) → sacrificed → Hippoc tissue collected; ELISA for TNF-α, IL-1β, IL-6 IL-4, IL-10; PCR for *Arg1, Ym1, Fizz1* and *Klf4*; Flow cytometry for ROS. Immunofluorescence for Iba-1 + CD16/32+ and Iba-1 + CD206+; WB for Nrf2, HO-1, NLRP3, caspase-1 p20 and IL-1β	Magnolol ↑ Nrf2, HO-1; ↓ NLRP3, caspase-1 p20, IL-1β both *in vivo* and *in vitro*. Magnolol ↓ ROS concentration, promoted Nrf2 nucleus translocation, and prevented Nrf2 ubiquitination. Nrf2 knockdown abolished the Magnolol-mediated microglial polarisation	Magnolol attenuated CUMS-stimulated depression by inhibiting M1 polarization and inducing M2 polarisation via Nrf2/HO-1/NLRP3 signalling

Table 2. *Cont.*

Study	Animal	Paradigm/Model	Location	Design	Results Relating to Nrf2	Conclusions/Observations
Huang et al., 2021 [97]	♂Sprague Dawley rats BW 180–220 g	CUMS induced depression model	Wenzhou Zhejiang, China	Rats divided into 4 groups of 8: Ctrl, CUMS, CUMS+ NC (Hippoc injection LV-pCDH-Nrf2-NC) CUMS+ Nrf2 (Hippoc injection LV-pCDH-Nrf2) → behave tests (SPT, EMZ, OFT, FST, MWM) → sacrifice → Hippoc tissue collected →ELISA for TNF-α, IL-1β, IL-10; WB for Nrf2/β-actin/Wfs1; PCR for miR-17-5P/Nrf2/Wfs1	Nrf2 weakly expressed in CUMS-treated rats. Nrf2 ↓ cognitive dysfunction and inflammatory brain injury. Nrf2 ↓ in CUMS treated rats. Nrf2 up-regulation reversed the trends in behav tests and the changes inflammation-related cytokine levels in CUMS-treated rats. Nrf2 inhibited miR-17-5p → limit Wfs1 transcription. miR-17–5p ↑ or Wfs1 ↓ reversed the role of Nrf2 in reliving inflammatory injury of murine Hippoc neurones	CUMS group ↓ performance in behave tests. CUMS treated group ↑ TNF-α, IL-1β and ↓ IL-10
Song et al., 2021 [98]	Adult ♂Kunming mice BW 16–18 g	CORST-induced depression model	Zhengzhou, Henan, China	Four groups of 11: Ctrl group, CORST group, the CORST + catalpol (20 mg/kg) and CORST + fluoxetine (20 mg/kg i-gastr) 21 days → behav test (FST,OFT,TST) → sacrifice → Hippoc tissue, cortex and serum collected → WB for NF-κB/Nrf2/HO-1; ELISA for IL-1β, TNF-α, iNOS, and NO. Cortical tissue ELISA for IL-1β, TNF-α, iNOS, and NO. Detected in the serum level of CORST, ACH and CRH	Catalpol ↑ Nrf2 and antioxidant defence (GSH/GST) and ↓ oxidative damage	CORST → DLB in mice in behav tests, ↑ serum CORT/CRH/ACTH, ↑ NF-κB in the Hippoc and frontal cortex, and ↓ Nrf2. CORST ↑ IL-1β, TNF-α, iNOS NO and MDA while ↓ GSH and GST in Hippoc and frontal cortex. Catalpol administration suppressed the abnormalities of the above indicators
Guan et al., 2021 [99]	Adult ♂Kunming mice BW 18–22 g	CUMS- induced depression model	Zhengzhou, Henan, China	Six group of 10: Ctrl group, CUMS group (CUMS + double distilled water i-gastr), CUMS + quercetin (10 mg/kg i-gastr), CUMS + quercetin (20 mg/kg i-gastr), CUMS + quercetin (40 mg/kg i-gastr), CUMS + FH (20 mg/kg i-gastr) for 21 days → behav test (SPT,OFT,FST) → sacrifice → Hippoc tissue collected → WB for PI3K/Akt/Nrf2/HO-1 detection kits for iNOS/NO/SOD/GST/ GPx/GSH/MD	CUMS ↓ PI3K/Akt, Nrf2/HO-1in the Hippoc of mice. (all $p < 0.01$); Quercetin (40 mg/Kg) ↑ PI3K/Akt, Nrf2/HO-1	CUMS for 21 days ↓ performance in behav tests (SPT,OFT,FST) quercetin(40 mg/kg) ↑ (SPT,OFT,FST) performance; CUMS ↓ SOD GST (both $p < 0.01$),quercetin at 40 mg/kg ↑ SOD,GST. CUMS ↑MDA, NO, iNOS while quercetin 40 mg/kg ↓ them; quercetin (20 mg/kg) ↑ performance (SPT,OFT,FST) and Akt, SOD, HO-1 and ↓ iNOS/MDA; quercetin (10 mg/kg) no reversal effect on the above indicators

Table 2. *Cont.*

Study	Animal	Paradigm/Model	Location	Design	Results Relating to Nrf2	Conclusions/Observations
D.-c. Sun et al., 2022 [100]	5–6-wk-old ♂ICR mice BW 18–22 g	CUMS-induced depression; literature-based prediction	Tianjin, China	Animals were divided into 6 groups (n = 10) including the "normal" (non-stressed mice), the CUMS (CUMS mice administered with distilled water), the venlafaxine (CUMS + venlafaxine), *SE* high-, medium-, and low-dose groups (CUMS + *SE* 1.8, 1.35 and 0.9 g/kg). → FST, TST, and OFT. DA in Hippoc and cerebral cortex, IL-2 and CORST in blood, Nrf2, Keap1, NADP, NQO1 and HO-1 in mice were Measur by ELISA and WB	In the *SE* group: up-regulation of Nrf2, Keap1, NQO1 and HO-1 vs. Ctrl. *SE* (1.8, 1.35 g/kg) ↓ immobility in FST and TST, while *SE* exhibited no significan: effect in OFT. Compared with the CUMS group, the *SE* group (0.9 g/kg) showed significant differences in Hippoc and cortical DA levels. The *SE* (1.8 g/kg) group significantly ↓ the activities of CORST ($p < 0.05$) but not serum IL-2 ($p > 0.05$). WB showed ↑ Nrf2, Keap1, NQO1, and HO-1 in the *SE* group (1.8 g/kg)	Compared with the CUMS group, the protein contents of Nrf2, Keap1, NQO1, HO-1 ↑ in the *SE* group; *SE* may enhance antioxidant effects through regulating Nrf2-ARE signalling pathways. *SE* ↑ Nrf2 and downstream anti-OS targets
Cheng et al., 2022 [101]	♂Sprague Dawley rats; BW 130–150 g	CUMS-induced depression model in rats	Xiamen, China	Rats divided into 4 groups of 10: CUMS group, acupuncture group (acupoints in the skull) fluoxetine group (2.1 mg/kg, i-gastr for 28 days) and Ctrl group. Behav test (OFT, SPT, FST) → sacrificed → Hippoc tissue collected → ELISA for ROS, WB for Nrf2/HO-1, Immunohistochemical for Bcl-2/caspase-3	CUMS↓ Nrf2/HO-1. Acupuncture pretreatment ↑ antioxidant enzymes in the Nrf2 pathway	Acupuncture/fluoxetine ↑ sugar preference SPT and ↓ immobility time in Behav test. Acupuncture ↓ Bax2/caspase-3 and ↑ Bcl-2. Acupuncture improved DLB of CUMS rats. Acupuncture showed AD effects in ↓oxidative stress products regulating the Nrf2/HO-1 pathway
Dang et al., 2021 [102]	♂C57BL/6J mice (aged 7–8 wk) and retired ♂CD-1 mice (aged 16–20 wk)	CSDS depression model	Yuzhong District, Chongqing, 400016, China	Behav tests (SI, SPT, OFT, EPM, NOR, TST, FST). Hippoc and mPFC tissues for Nissl staining, immunofluorescence, targeted energy metabolomics analysis, ELISA, Measur of MDA, SOD, GSH, GSH-PX, T-AOC. WB and PCR for Sirt1/Nrf2/HO-1/Gpx4 signalling pathway. EX527, a Sirt1 inhibitor and ML385, an Nrf2 inhibitor 30 min before EDA injection daily	EDA (in CSDS model) ↑ expressions of Sirt1, Nrf2, HO-1 and Gpx4 in the Hip. EX527 and ML385 reversed the effect of EDA	EDA ameliorated CSDS-induced depressive and anxiety-like behaviours (↓ of neuronal loss, microglial activation, astrocyte activation in the Hip and mPFC)
Xia et al., 2022 [103]	♂C57BL/6 mice (8-wk-old)	DSS-induced IBF-associated depression and anxiety	Yangling, China	Test the effect of 6-wk 100 mg/kg BW/day sesamol for DSS-induced mice who developed IBF after DSS induction. Behav F + M (time spent and % entries in open arms), MBT (total and % travelled distance); TST (immobility time); Tissue Claudin 1, TNF-α, IL-1β, Iba1, and GFAP; Cortical mRNA expression of TLR-4, iNOS, COX-2, TNF-α, IL-1β, and IL-6; MDA, SOD, and GSH content in serum and cortex; Nrf2–HO-1, NQO1, HO-1 and their mRNA expression in cortex; BDNF in Hippoc CA3, mRNA expression of BDNF in cortex and Hippoc, p-TrkB/TrkB, p-CREB/CREB, PSD length and width, PSD-95 in cortex and Hippoc CA3; cortical NA and 5-HT content and mRNA expression of Htr1a and Htr2a, and PSD-95 mRNA expression in Hippoc and cortex; WB for 5-HT1AR F and 5-HT2AR F, Hippoc PSD-95, TLR-4, p-NF-κB/NF-κB, iNOS, COX-2, and p-IKBα/IKBα	Sesamol ↓ TLR-4/NF-κB, ↓ OS, ↑ Nrf2-HC-1 pathway, ↑ BDNF, ↑ BDNF/TrkB/CREB signalling pathway, and ↑ NA and 5-HT levels; ↑ Claudin 1 levels, thus reducing epithelial junction dysfunction, and improved behav test performance. It ↓ TNF-α, IL-1β, ↑ NQO1, ↑ Hippoc, cortical BDNF	Sesamol ↓ inflam, epithelial barrier dysfunction, OS damage, energy metabolism and anxiety-like behaviours via the gut-brain axis; it restored pro-inflammatory cytokines synaptic impairment in a DSS-induced IBF mouse model of depression and anxiety and ↓ DLB and anxiety

Table 2. *Cont.*

Study	Animal	Paradigm/Model	Location	Design	Results Relating to Nrf2	Conclusions/Observations

Table 2. *Cont.*

Study	Animal	Paradigm/Model	Location	Design	Results Relating to Nrf2	Conclusions/Observations
Li J. et al., 2022 [104]	♂Sprague Dawley rats; BW 200–250 g	CRS-induced DLB	Beijing, China	Experiment 1: rats divided into 4 groups: CRS, Ctrl, CRS + Escitalopram 1 mg/kg/day i.g. × 28 days, CRS+ Rg1 20 mg/kg/day i.g. × 28 days. → behav tests (SPT, OFT, FST) → sacrifice → Hippoc tissue and serum collected. Experiment 2: rats divided into 4 groups: Ctrl, CRS, CRS + lenti + shNC (injection *in vivo* with lenti-shRNA into lateral ventricle), CRS + lenti + shGAS5 (injection *in vivo* with lenti-shGAS5 into lateral ventricle) → 48h CRS × 28 days → behav tests (SPT, OFT, FST) → sacrifice → Hippoc tissue and serum collected → qPCR for GAS5. Experiment 3: rats divided into 4 group: Ctrl, CRS, CRS + Rg1 (20 mg/Kg/day i.g. 1 h after CRS stimulation) +lenti vector (*in vivo* injection with lenti-shRNA into lateral ventricle), CRS+ Rg1(20/mg/Kg/day i-gastr 1 h after CRS stimulation) + lenti-GAS5 (*in vivo* injection with lenti-shGAS5 into lateral ventricle) → 48 h CRS × 28 days → behav tests (SPT, OFT, FST). → sacrificed → Hippoc tissue collected → qPCR for GAS-5; IF for IBA-1. Experiment 4: Rat microglia cell line HAPI → with LPS, (1 μM) or solution × 2 h → treated with Rg1 (5 μM, 10 μM, 20 μM) for 24 h → HAPI WB for COX-2/iNOS/SOCS3. PC-12cell → with CORST (400 μM) or solution for 2 h → treated with Rg1 (5 μM/10 μM/20 μM) for 24 h. Experiment 5: HAPI → GAS5 shRNA after 48 h → LPS (1 μM) × 24 h. HAPI qPCR GAS5, ELISA for TNF-α, IL-1β, IL-6, WB for COX-2/iNOS/SOCS3. PC-12 → GAS5 shRNA after 48 h → CORST (400 μM) × 24 h. PC-12 qPCR for GAS5. Experiment 6: GAS5 KO Nrf2/SOCS3. HAPI/PC-12 → GAS5 shRNA. 48 after → HAPI qPCR, WB for SOCS3. PC-12 qPCR,	SOCS3/Nrf2 ↓ in CRS group compared to CRS+ Rg1, CRS + Escitalopram and Ctrl. CRS rats showed ↑ EZH2 in Hippoc, ↓ SOCS3/Nrf2, ↑ EZH2 than other groups. GAS5 → ↑ EZH2, ↑ GAS5 in CRS rats compared to other groups. CRS + lenti + shGAS5 → ↑ Nrf2 compared to CRS + lenti + shNC and CRS; Rg1 ↑ ATP, SOCS3/Nrf2 ↓ GAS5 overexpression-related ATP, SOCS3, and Nrf2 reduction. CORST ↓ Nrf2/HO-1; Rg1 treatment ↑ Nrf2/HO-1. CORST stimulation ↓ Nrf2 and HO-1, GAS5 knockdown-related Nrf2 and HO-1 increase. GAS5 knockdown ↓ EZH2 and H3K27me3 enrichment in Nrf2 promoter region in PC-12 cells. GAS5 knockdown ↑ Nrf2 in PC12 and HAPI. In PC-12, Nrf2 knockdown reversed the protective effect of Rg1 on ROS production and on MCh membrane potential	CRS ↓ behav test performance, ↑ Hippoc IBA-1, ↑ Hippoc TNF-α/IL-1β/IL-6, ↓ Hippoc ATP and ↑ MCh injury compared to CRS + Rg1, CRS + Escitalopram and Ctrl. CRS, CRS + lenti + shNC ↑ Hippoc GAS5 compared to CRS + lenti + shGAS5 and Ctrl. ↑ performance on OFT and FST in CRS + lenti + shGAS5 and Ctrl compared to other groups. SFT ↑ performance in CRS + lenti + shGAS5 and Ctrl > than in other groups. GAS5 knockdown ↓ CRS-induced DLB. CRS + lenti + shGAS5 ↓ Hippoc IBA-1 compared to other groups. ↓ serum and Hippoc TNF-α, IL-1β, and IL-6 levels in Ctrl and CRS + lenti + shGAS5 groups compared to other groups. ↓ Hippoc ATP in CRS than in CRS + lenti + shNC, CRS + lenti + shGAS5, and Ctrl. ↑ MCh injury in CRS compared to CRS + lenti + shGAS5; GAS5 knockdown did not affect EZH2 expression. ↑ Hippoc GAS5 5 in CRS than CRS + Rg1 + lenti; Rg1 ↓ DLB, ↓ IBA-1, ↓ microglial activation, and ↓ MCh dysfunction; ↑ GAS5 overexpression → partly reversed the effect of Rg1 on DLB, on microglial activation, on IBA-1, and on MCh dysfunction. GAS5 overexpression → ↑ IBA-1, LPS ↑ TNF-α, IL-1β, and IL-6; Rg1 treatment ↓ TNF-α, IL-1β, no significant effect on IL-6. LPS ↑ COX-2/iNOS and ↓ SOCS3 in HAPI. Rg1 treatment ↓ MCh dysfunction. GAS5 inhibition exhibited a protective effect, similar to that of Rg1 treatment

Table 2. *Cont.*

Study	Animal	Paradigm/Model	Location	Design	Results Relating to Nrf2	Conclusions/Observations
Li J. et al., 2022 [104]	♂Sprague Dawley rats; BW 200–250 g	CRS-induced DLB	Beijing, China	WB for Nrf2. RIP and RNA pull-down assay → validate binding between GAS5/EZH2. ChIP-qPCR for EZH2/H3K27me3 in the promoter region of SOCS3 in HAPI. **Experiment 7:** SOCS3/NRF2 KO ↓ protective effects of Rg1 in HAPI7 PC-12 *in vitro* model. HAPI → SOCS3 shRNA. 48 h after → LPS (1 μM) for 2 h, → Rg1 (5 μM, 10 μM, 20 μM) × 24 h. ELISA for TNF-α, IL-1β, IL-6 WB for COX2/iNOS/SOCS3. PC-12 → NRF2 shRNA 48 h after → CORST (400 μM) × 2 h →Rg1 (5 μM, 10 μM, 20 μM) × 24 h. IF for IBA-1, WB for Nrf2/SOCS3/EZH2; ELISA for TNF-α/IL-1β/IL-6, TEM for MCh morphology, mitoSOX kit for MCh ROS detection, JC-1 assay to assess MCh membrane potential; ATP assay kit for ATP; q PCR for GAS5. ChIP-qPCR for EZH2/H3K27me3 in Nrf2 promoter region in PC-12		
Muhammad et al., 2022 [105]	♂Sprague Dawley rats; BW 180–200 g	LPS-induced DLB	Islamabad, Pakistan	Rats divided into 2 groups of 10, one treated with fluoxetine (5 mg/kg) + LPS, CAR20 or CAR50 + LPS → single dose of each × 5 days, ip + LPS → or single dose or after CAR20/50 (3rd, 4th day 1 mg/kg), and another, divided into 3 subgroupings (10 animals/group): ATRA + LPS, ATRA + LPS + CAR20/50, ATRA + LPS+ fluoxetine. ATRA ip injection 30 min before LPS. At Day 2, → behav tests (FST, LDB, E + M, sucrose SPIT) → sacrificed → Hippoc and cortical tissue collected and processed to RT-PCR and ELISA for Nrf2/HO-1. Immunohistochemical analysis for OS-related molecules	LPS ↓ Nrf2/HO-1 expression in cortex and Hippoc compared to Ctrl ($p < 0.05$). CAR20/50 + LPS ↑ Nrf2 and HO-1 compared to the LPS group ($p < 0.01$)	LPS ↑ ROS and DLB. CAR20/50 ↓ DLB and↑ Nrf2/HO-1; no differences between CAR20/50 + ATRA + LPS and ATRA + LPS

Table 2. *Cont.*

Study	Animal	Paradigm/Model	Location	Design	Results Relating to Nrf2	Conclusions/Observations
Shen et al., 2022 [106]	♂C57BL/6J mice (2 months old, BW 20–25 g)	LPS-induced CUMS and DLB	Hefei, Anhui, China	Experiment 1, mice randomised into 5 groups of 6: Ctrl, 2 h post-LPS, 6 h post-LPS, 12 h post-LPS, and 24 h post-LPS. After LPS injection → mice sacrificed → Hippoc tissue collected. Experiment 2, mice randomised into 6 groups of 6: Ctrl, LPS, LPS + PSP, LPS + fluoxetine, (30 mg/kg ip), LPS + calpeptin (2 mg/kg calpeptin ip), and LPS+ MCC (50 mg/kg ip). 24 h after LPS injection → sacrificed → Hippoc tissue collected. Experiment 3, mice randomised into 3 groups of 6: Ctrl, LPS, and LPS + NAC. 24 h after LPS injection → sacrificed → Hippoc tissue collected. Mice further randomised into 4 groups: Ctrl, CUMS, CUMS + PSP (i-gastr × 21 days 400 mg/kg) and CUMS + MCC (5 mg/kg ip × 7 days) After 21 day → behav tests (TST, SPT, FST, OFT) → sacrificed → Hippoc tissue collected. Hippoc tissue WB for OS-related molecules and IF for Nrf2	LPS ↓ Nrf2 expression compared to Ctrl in Hippoc CA1 region ($p < 0.05$). Treatment with PSP improved LPS-mediated ↓ in Nrf2 expression in CA1 ($p < 0.05$). LPS+ PSP ↑ Nrf2 compared to Ctrl ($p < 0.05$)	LPS treatment ↑ calpain-1 expression and substrate degradation, activated the NLRP3 inflammasome signalling pathway and glial cells, and Nrf2 and calpastatin expression. Treatment with PSP prevented these LPS-induced changes. Nrf2 expression was ↓ in the hippocampi of animals treated with LPS or subjected to CUMS, and this decrease was prevented by PSP
Jiang et al., 2022 [107]	6–8–wk-old ♂C57BL/6J mice ($n = 60$, BW 20–22 g); and 12-month-old retired breeders ♂CD-1 mice ($n = 70$, BW 20–22 g)	CSDS-induced DLB	Beijing, Changsha, Luzhou, China	After a 1-wk adaptation, mice randomised into Ctrl, model, and treatment groups IMI (15 mg/kg ip), and Rb1 (35 and 70 mg/kg ip). Except for Ctrl, all animals subjected to CSDS treatment → behav testing (SIT, OFT, TST, FST) over the next 4 wk. IMI p.o. × 32 days. Hippoc protein concentration, CAT, SOD, and LPO Measur with specific kits. Pro-inflammatory mediator (TNF-α, IL-18, IL-1β) serum levels Measur with ELISA. WB for SIRT1, NLRP3, cleaved Caspase-1, ASC, IL-1β, HO-1, Nrf2, and β-actin	Nrf2 and HO-1 expression substantially ↓ in rat CSDS Hippoc compared to Ctrl. 70 mg/kg Rb1 ↑ Nrf2 expression	Rb1 treatment can rescue DLB-like social avoidance and behav despair of CSDS-induced mice. Rb1 attenuates pro-inflammatory cytokine production and inhibits the activation of NLRP3 inflammasome. Furthermore, Rb1 normalised OS following Nrf2/HO-1 and SIRT1 activation. Findings show that Rb1 attenuates NI

Table 2. *Cont.*

Study	Animal	Paradigm/Model	Location	Design	Results Relating to Nrf2	Conclusions/Observations
de Souza et al., 2022 [108]	♂ adult Swiss mice BW 25–30 g	CUMS-induced depression	Fortaleza, Ceará, Brazil	Mice were exposed to a variety of stressful events (restraint, tilted cage, intermittent circle between lights on and off, constant light, water deprivation, wet cage, electric shock) × 28 days; from the 14th day they received DMF 50 and 100 mg/kg or fluoxetine 10 mg/kg or plac. On the 29th day → behav tests (OFT, FST, SPT, NOR). Mice divided into two groups, Ctrl and CUMS, which were subdivided into plac (carboxymethyl cellulose 0.5%), DMF50, DMF100 or fluoxetine 10 mg/kg (FLU) groups. 14 days after the beginning of the procedure, treatments were started in all groups, 30 min before daily stress application in CUMS groups. Simultaneously, Ctrl received the same treatments. Immunoenzymatic assay for TNF-α and IL-1β concentrations in Hippoc. IF to analyse subfields CA1, CA3, and DG	MMF (active metabolite of DMF) showed an efficient binding interaction with the Keap1 protein, leading to Nrf2 activation	DMF is a promising prodrug for the treatment of MDD and for comorbidities such as memory deficit. DMF modulated the neuroinflammatory pathway, ↑ astrocyte expression, ↓ microglia expression and the production of the pro-inflammatory cytokines IL-1β and TNF-α. MMF bound the Keap1 protein → activated Nrf2; it also bound HCAR2 protein, leading to a complex signalling cascade cross-talking with the Nrf2 pathway. It is suggested that one of the mechanisms involved in the AD effect of DMF is NI suppression, triggered by the binding of its metabolite MMF to the HCAR2 protein
Fasakin et al., 2022 [109]	Adult ♂ Wistar Albino rats, BW 181 ± 13 g	Depression and anxiety linked to NI and OS.	Akure, Nigeria	Six rats randomly housed in each cage: Group 1 = normal Ctrl. Groups 2 to 5 = CSAE once daily (doses of 5, 50, 500, and 2000 mg/kg BW, respectively). Groups 6 to 9 = DSAE once daily (doses of 5, 50, 500, and 2000 mg/kg BW, respectively). Groups 10 to 13 = NTAE once daily (doses of 5, 50, 500, and 2000 mg/kg BW, respectively). Groups 14 to 17 = CMAE once daily (doses 5, 50, 500, and 2000 mg/kg BW, respectively). Behav Tests: E + M, FST, TST. 91 days after alkaloid administration, rt-qPCR on Hippoc tissue. Biochemical analyses were performed to assess MAO activity, dopamine concentration, ACE activity, AChE activity, GDH activity, ROS concentrations. Concentrations of TNF-α, IL-1β, and IL-10 were evaluated using ELISA. Gas chromatograph–mass spectrometry analysis of alkaloid-rich extracts was performed	CSAE, NTAE, and CMAE ↑ 5-HT, BDNF, CREB and Nrf2 while ↓ NF-κB, GSK3β JNK3 and nesfatin-1 at 5, 50 and 500 mg/kg BW sub-chronic exposure. However, DSAE sub-chronic administration ↓ serotonin, BDNF, CREB and Nrf2 while ↑ NF-κB, GSK3β and JNK3	This study established the AD and anxiolytic potentials of CSAE, CMAE, and NTAE via improving monoaminergic bioavailability, activation of neurotrophic signalling cascades and de-hyper-activation of hypothalamic–pituitary–adrenal axis. The difference between the effects of CSAE, CMAE, and NTAE at low and high doses may be the underlying factor in the discrepancies between the results of studies that have evaluated depression and psychoactive substance use

Table 2. *Cont.*

Study	Animal	Paradigm/Model	Location	Design	Results Relating to Nrf2	Conclusions/Observations
Tan et al., 2022 [110]	Adult ♂C57BL/6 mice (BW 20 ± 2 g)	Behav despair mouse model *in vivo* and H_2O_2-induced PC12 cell model *in vitro*	Shenyang, 110016, China	Mice randomised into 4 groups of 10: plac, Fluoxetine 15 mg/kg, i.g.), SEO 250 group (250 mg/kg, i.g.) and SEO 750 group (750 mg/kg, i.g.). Flu and SEO suspended in 0.5% CMC-Na, plac 0.5% CMC-Na. Treatment once daily × 9 consecutive days. Behav tests (FST, TST). Serum, liver, brain, and cell supernatant MDA, SOD, CAT, and GSH levels detected through ELISA. Hippoc, cortex, and cell supernatant processed through WB (Nrf2, HO-1,PI3K,p-PI3K,p-Akt, GSK3β, p-GSK3β, LaminB, β-actin). Hippoc and cortical histopathology. PC12 cell culture. Groups: Ctrl, plac (300 μmol/L H_2O_2), SEO (0.78–25 μg/mL + 300 μmol/L H_2O_2). IF for PC12 cells with anti-Nrf2. 18 batches of SEO analysed by GC-MS. DPPH and ABTS assay	After FST and TST, expressions of nuclear Nrf2 in Hippoc and cortex significantly ↓. SEO treatment could reverse these changes, which indicates that SEO might promote Nrf2 translocation to reduce OS	SEO showed a promising AD-like effect in mice; the mechanisms of this effect might contribute to the antioxidant activities of SEO. SEO could ↑ Nrf2/HO-1 pathway to improve OS status and exert AD-like effects. PI3K/Akt/GSK3β signalling pathway might also be involved
Wu et al., 2022 [111]	Adult ♂Sprague Dawley rats (BW 230–250 g, 6–7-wk-old)	CUMS-induced depression	Zhengzhou, Henan, China	After 7 days of adaptive life, all rats were evaluated by behav tests (OFT, FST and SPT. On the next day, rats were randomly divided into five groups: (1) Ctrl group (n = 10); (2) CUMS group (n = 10); (3) CUMS plus Catalpol group (n = 10); (4) CUMS plus ZnPP group (n = 6); (5) CUMS plus ZnPP + Catalpol group (n = 6). CUMS lasted for 3 wk. Catalpol was administered orally (dose of 10 mg/kg and volume of 20 mL/kg) for 3 wk. ZnPP injected into the intracerebroventricular (dose of 10 μg/rat and volume of 5 μL/rat) for 3 wk. 24 h after the last drug intervention, behav. tests were performed. Biochemical analyses were performed to determine Hippoc total protein and the levels of iNOS, GPX and GST in the Hippoc (using catalytic reaction plus chemical colourimetry method and the levels of NO, MDA, SOD, and GSH (nitrate reductase method, barbituric acid sulfate method, hydroxylamine method, and dithiodinitrobenzoic acid method, respectively)	CUMS exposure ↓ the phosphorylation level of ERK1/2, the nuclear expression level of Nrf2, and the total intracellular levels of HO-1, SOD, GPX, GST, and GSH and ↑ the level of peroxide MDA. Before intervention by HO-1's antagonist ZnPP, catalpol significantly reversed these abnormalities. However, after intervention by HO-1's antagonist ZnPP, the reversal effects of oral administration catalpol on the phosphorylation level of ERK1/2, the nuclear expression level of Nrf2, and the total intracellular levels of HO-1, SOD, GPX, GST, GSH, and MDA were completely abolished, which implied that HO-1's antagonist ZnPP may abolish the regulation of catalpol on antioxidant and peroxide related factors in the Hippoc of CUMS-induced DLB rats	The AD-like effects of catalpol might be due to HO-1 that up-regulates the ERK1/2/Nrf2/HO-1 pathway-related factors to enhance the antioxidant defence, triggering the down-regulation of the COX-2/iNOS/NO pathway-related factors to inhibit NI, and the up-regulation of the BDNF/TrkB pathway to enhance neurotrophic effects

Table 2. *Cont.*

Study	Animal	Paradigm/Model	Location	Design	Results Relating to Nrf2	Conclusions/Observations
Zhao et al., 2022 [112]	♂C57BL/6 mice (8-wk-old)	CP-induced anxiety and DBL	Zhanjiang, Guangdong, China	Mice received CP ip × 2 wk and TSP × 4 wk → behav, biochemical, and molecular tests conducted	Protein expression of Keap1 ↑, Nrf2 and HO-1 ↓ in CP compared with Ctrl. TSP ↑ Nrf2 and HO-1	TSP ↓ behav test abnormalities, ↓ SOD, MDA, and GSP abnormalities; ↓ expression of IBA-1, TNF-α and IL-1β, ↓ Hippoc apoptosis in CA1 and CA3 subfields; ↑ Nrf2/HO-1 signalling pathway; ↑ BDNF/TrkB/CREB signalling pathway, and ↓ the Bcl-2/BAX/caspase-3 apoptosis pathway
Bansal et al., 2022 [113]	3-month-old ♂BALB/c mice (BW 25–30 g)	OBX-induced depression	Chandigarh, Panjab, India	Thirty mice randomised in 3 groups of 10. I group = sham; II = OBX + plac plac; III and IV groups = OBX + Ro61-8048 (0.2 and 0.4 mg/kg) × 14 days → behav tests (OFT, SFT, SPT), biochemical analysis, ELISA (Nrf2, BDNF, NF-κB, TNF-α and IFN-γ), HPLC (5-HT, 5-HIAA, TRP, and KYN levels), SF(QA) and qPCR (Hippoc and cortex based gene expression analysis) were performed	KMO inhibition with higher dose of Ro61-8048 (0.4 mg/kg) in OBX mice showed: ↓ CORST levels; ↑ Nrf2 and GCLC with no significant effect on HO-1, ↑ PI3K and Akt, ↓ GSK3β gene expression and ↑ GFAP expression, ↓ MDA and nitrite, ↓ NF-κB and IFN-γ levels with no effect on TNF-α, ↑ BDNF, ↑ TRP, 5-HT and 5-HIAA, ↓ QA. The lower dose of Ro61-8048 (0.2 mg/kg) showed no significant effect	OBX ↑ OS in the brain; Ro61-8048 ↓ the effect of KMO and ↓ OS. KMO inhibition and QA-associated down-regulation ↓ Keap-1 and ↑ Nrf2 expression
Yu et al., 2022 [114]	♂C57BL/6 mice (6–8 wk, BW 20 ± 2 g)	LPS-induced DLB	Fuzhou, Fujian, China	Mice were divided into 4 groups (9 each): CON, the Ctrl group; CON + RA (80 mg/kg); LPS (0.1 mg/kg); LPS + RA (80 mg/kg). Behav tests were carried out 24 h later → mice sacrificed, and brains used for IF, histopathological analysis, WB, and rt-qPCR	Nrf2 ↓ after LPS injection compared to Ctrl, RA administration attenuated the ↓ of Nrf2	RA improved behav test performances and BW; ↓ histologic brain damage in LPS-exposed mice; blocked the ↓ of BDNF (rt-qPCR and WB) and of Nrf2 (WB). LPS ↓ LC3 (IF), ↓ p21 and p62, the levels of which were restored by RA. ME1, IDH1, and 6-PGDH mRNA were ↓ in the LPS group and ↑ after RA. RA inhibited the LPS-induced ↑of CD44, iNOS, TNFα, and IL-1β mRNA expression
Kang et al., 2022 [115]	♂C57BL/6J mice (8-wk-old, BW 21–23 g, n = 36)	CRS-induced DLB	Daejeon, South Korea	Mice divided into 4 groups of 9: normal; CRS (stress and distilled water), PSE (stress and 100 mg/kg PSE), and positive Ctrl (stress and 100 mg/kg NAC). Behav tests (OFT, FST, TST); murine microglial cells (BV2) and Hippoc neuronal cells (HT22) were cultured for cytotoxicity, cytokines, and phagocytosis assays. IF and WB were also performed	p-Nrf2 ↑ with PSE ($p < 0.05$ for 40 μg/mL). PSE pretreatment increased the level of HO-1 protein expression ($p < 0.05$ for the 10 and 20 μg/mL doses). NAC exerted similar effects as PSE	PSE (10, 20, and 40 μg/mL) attenuated the ↑ in inflammatory factors (NO and TNF-α), translocation of NF-κB, and phenotypic transformations in a dose-dependent manner. These inhibitory effects of PSE on microglia were supported by its regulatory effects on the CX$_3$CR1/Nrf2 pathway. PSE (100 mg/kg) ↓ sickness, anxiety, and DLB in mice subjected to CRS and improved behav test performance

Table 2. *Cont.*

Study	Animal	Paradigm/Model	Location	Design	Results Relating to Nrf2	Conclusions/Observations
Wang et al., 2022 [116]	40 ♂ ICR 7-wk-old mice	PTZ-induced seizures and depression	Liaocheng, western Shandong, China	Mice divided into 4 groups of 10: Ctrl, PTZ, PTZ + EPA, and PTZ + DHA groups. Behav tests (OFT; TST; FST). Hippoc DHA, EPA and neurotransmitter levels were detected by GC. Histopathology of Hippoc was conducted through H&E-staining, FJB-staining and IF. Hippoc proteins were analysed by WB	The protein expressions of p-Nrf2 and HO-1 were significantly ↓ in the PTZ group compared with Ctrl. EPA and DHA significantly ↑ p-Nrf2 and HO-1 (higher levels in EPA) group	↓ GSH, ↓ SOD activity, ↑ MDA, ↓ GPx4, xCT, HO-1, and p-Nrf2 in the PTZ group, reversed by EPA and DHA. EPA and DHA improved behav test results; ↑ GABA and GABARA1A levels without changing Glu levels; improved the morphology of cell damage (EPA repaired > than DHA); ↓ Hippoc total Fe (which was ↑ in the PTZ group) and ameliorated iron-homoeostasis-related proteins; ↓ pro-IL-1β, IL-1β, and TNFα (which were ↑ in the PTZ group); EPA inhibited NLRP3 and ↓ Caspase-1 levels; DHA ↓ ASC, pro-Caspase-1 and Caspase-1
Nasehi et al., 2022 [117]	Adult ♂ mice, sons of pregnant mothers from Pasteur Institute, Tehran, Iran (Balb/c?), fed ad libitum and kept in 12 h/12 h dark/light cycles	Maternal separation-induced depression associated with CVD	Zanjan, Iran	Depression was induced with 3 h daily separation from mothers while weaning; mice with maternal separation-induced depression and CV comorbidities treated with fluoxetine or SAHA + fluoxetine ip →behav tests (OFT [horizontal and vertical locomotion], E + M (Y-maze), TST, FST, SPIT [grooming]) + gene analysis with RT- qPCR; content and gene expression of PPAR-α, NOX-2, -4, PGC-1α, and Nrf2	SAHA ↓the *NOX-4* gene expression level in mice treated with SAHA + fluoxetine without significantly changing *NOX-2* expression. SAHA ↑ PGC-1α and Nrf2 in heart tissues of maternally separated mice, but had no effect on *PPAR-α* gene expression. Maternal separation induced depressive behaviour on all behav tests. SAHA and fluoxetine had no effect on OFT locomotion, but ↑ time spent in open arm of the Y-maze, and ↓ immobility time on FST and TST	SAHA reversed DLB and restored mitochondrial function in the heart via an effect on NOX-4 and mitochondrial biogenesis gene (PGC-1α and Nrf-2) expression levels. The authors speculated that when fluoxetine lacks effectiveness on cardiac mitochondrial biogenesis and cardiac *NOX-4* gene expression, patients with depression may develop a recurrence of their depression
Pei et al., 2022 [118]	50 ♂ Balb/c mice (BW = 25 g)	LPS-induced depression	Changchun, Jilin, China	Mice were divided into 5 groups (n = 10 each): Ctrl, LPS-induced, palmatine (50 and 100 mg/kg) group, and fluoxetine groups. Each group except Ctrl received LPS (5 g/L, 3 μL/mice). OFT and EPM were performed; SOD, TNFα, IL1β, and IL6 detected with ELISA; apoptosis and ROS with flow cytometry; total Hippoc proteins analysed with WB (BAX, Bcl-2, Nrf2, HO-1, β-actin)	Nrf2 and HO-1 in the LPS-induced group were ↓ compared with the Ctrl group. After palmatine treatment, their protein expression levels were significantly ↑	Palmitine ameliorated mitochondrial damage, improved behav tests, ↓ levels of SOD, TNF-α, IL-1β, and IL-6. It also ↓ neuronal apoptosis in the Hippoc, and depression through BAX/Bcl-2 and Nrf2/HO-1 signalling pathways

Table 2. *Cont.*

Study	Animal	Paradigm/Model	Location	Design	Results Relating to Nrf2	Conclusions/Observations
J.Y. Sun et al., 2022 [119]	60 ICR ♂ mice (6–8-wk, BW 18–22 g)	CORT-induced depression	Chengdu, Sichuan, China	All mice were randomly divided into 6 groups (n = 6): Ctrl group, CORT group, Fluoxetine (3 mg/kg), EOP low-dose group (1 mg/kg), EOP medium-dose group (2 mg/kg) and EOP high-dose group (3 mg/kg). Except for the Ctrl group, mice in other groups were injected with CORT 20 mg/kg ip daily for 4 consecutive wk→behav tests (TST, FST, SPT, OFT)+ blood samples and brain tissues for analysis	EOP ↓ CORT-induced DLB in mice (by improving behav tests), ↓ CRH, ACTH and cortisol in the brain tissues, ↑the phosphorylation of PI3K and Akt in Hippoc neurons and PC12 cells, ↑ Nrf2 and ↓ the CORT-induced OS. It also ↓ CORT-induced Hippoc neuron injury and apoptosis and ↑ the proliferation ability and cell viability of PC12 cells	EOP had a significant AD effect on the symptoms of CORT-induced depression in mice. EOP exerted anti-apoptotic effects on Hippoc neurons through PI3K/Akt/Nrf2 signalling pathway
He et al., 2022 [120]	♂ adult C57BL/6 mice (8-wk-old, BW 20–25 g), CD-1 mice (14-wk-old, BW 40–45 g), ♂ adult YFP-Nrf2 WT mice, and YFP-Nrf2 KO mice	CSDS-induced DLB	Guangzhou, Guangdong, China	For the CSDS procedure, C57BL/6 mice or YFP mice were defeated by CD1 mice (10 min × 10 days). SIT performed to select susceptible mice (70%) → locomotion test, FST, and SPT. TREM2-HDO (icv) was injected on day 0, SFN (ip) 30 min before the CSDS and LPS. CSDS-susceptible mice were used for luciferase assay (lateral habenular BV2 cultured microglia cells treated with SFN or siRNA-Nrf2), ChIP assay (with Nrf2 antibody), qPCR (levels of TREM2 promoter mRNA and Trem2 mRNA); ImBl (TREM2, Nrf2, BDNF, arginase, phospho-TrkB, PSD-95, TrkB, β-actin antibody, β-tubulin antibody, GAPDH), IF (anti-TREM2, anti-IBA-1, anti-arginase1 [microglial anti-inflammatory phenotype marker]), dendritic spine analysis	SFN → ↑ Nrf2 → ↑ microglial arginase 1+ phenotype by initiating TREM2 transcription in mPFC, improving DLB in CSDS mice. Nrf2 KO ↓ TREM2 and microglial arginase 1+; both → DBL. Nrf2 KO and ↓ TREM2 expression associated with ↓ BDNF/TrkB signalling pathway	Luciferase assay showed that Nrf2 regulates TREM2 transcription; qPCR showed that ↑ SFN and ↑ Nrf2 → ↑ TREM2 promoter activity and ↑ TREM2 mRNA (which were ↓ in CSDS-mouse mPFC). ImBl and qPCR → TREM2-HDO dose-dependently decreased mRNA and protein expression of both TREM2 and Nrf2. IF showed TREM2 to colocalise with IBA1 and arginase 1+ in microglial cells. LPS ↓ TREM2 and arginase 1 and ↑ IBA1; these effects were reversed by SFN. qPCR → reversal by SFN of the reduction in the anti-inflammatory cytokines IL-4 and IL-10 in the mPFC of CSDS mice. ImBl → reversal by SFN of the reduction in Nrf2, TREM2, and arginase 1 expression in the mPFC of CSDS mice. Nrf2 KO ↓ TREM2 and arginase 1+ microglial phenotypes in mPFC

Table 2. *Cont.*

Study	Animal	Paradigm/Model	Location	Design	Results Relating to Nrf2	Conclusions/Observations
Samy et al., 2022 [121]	10 wk-old ♂ Balb/c mice (BW = 25 g)	Post-OVX depression	Alexandria, Egypt	Eighty mice assigned to 5 groups (n = 16 each) receiving treatments or plac × 3 wk: Ctrl (sham-operated + plac); OVX mice + CarA (20 mg/kg/day, p.o.); OVX+ SnPP-IX (50 μmol/kg/day i.p); combined-treatment OVX + CarA + SnPP-IX received CarA; Behav tests. HO-1 activity and brain oxidative and antioxidant markers were determined. Isolation of Hippoc and cortex RNA for RT-qPCR; protein extraction for WB; determination of brain 5-HT levels through HPLC	↓ Nrf2, Trx-1, and BDNF in OVX mouse brains compared with Ctrl. ↑ Nrf2, Trx-1, and BDNF with CarA either solely or combined with SnPP-IX (except for BDNF, which ↓ in the treatment combination group). Mice treated with SnPP-IX alone did not show any significant impact on Hippoc or cortical Nrf2, Trx-1, or BDNF	Improvement of FST, TST, SPT with CarA and partial improvement with CarA + SnPP-IX. ↓ Nrf2, Trx-1, and BDNF in OVX mouse brain, ↑ by CarA. ↑ in MDA and ↓ in GSH and SOD in OVX mice. ↑ GSH, SOD activity and ↓ MDA in CarA group (antioxidant effect attenuated in CarA + SnPP-IX). ↑ OS, ↑ MDA, ↓ GSH and SOD with SnPPIX alone. CarA ↑ 5-HT level in mouse Hippoc and PFC. Significant thinning of the pyramidal and granule cell layers of the HippP and DG of OVX mice, which was restored to the original thickness by CarA
Si et al., 2023 [122]	♂ 6–8-wk-old Sprague Dawley rats (BW 160–180 g, n = 70)	CUMS-induced DLB	Wuhan, Hubei, China	Rats sorted into 2 groups of 35, Ctrl and CUMS. Behav tests performed to select CUMS-susceptible rats (SUS). SUS randomly divided into 2 groups: CUMS-SS (1 SUS + 3 Ctrl) and CUMS-SI (1 SUS). After 4 wk of SI or SS, the rats in different groups underwent behav tests. Total RNA was isolated for RT-PCR and total protein for WB. IF performed to observe Hippoc astrocyte changes	mRNA expression of Nrf2, HO-1 and NQO1 ↓in CUMS-SI and significantly lower in the CUMS-SI group compared to CUMS-SS	SUS exhibited DLB, memory deficits and social withdrawal. qRT-PCR and WB analysis showed ↓ Nrf2, HO-1, NQO1 and p-ERK in SUS and ↑ Keap1. IF showed ↓GFAP+ astrocytes in SUS. SI after CUMS perpetuated DLB, memory deficits and social withdrawal (overall worsening in behav tests and no BW ↑); ERK/Keap1/Nrf2 signalling was suppressed in CUMS-SI (↓ Nrf2, ERK, and ↑ Keap1 compared to CUMS-SS)
Wang et al., 2023 [123]	♂ 8–10-wk-old adult C57BL6 mice	CUMS-induced DLB	Cangzhou, Hebei, China	13 mice per group: Ctrl, exposed to CUMS + plac, CUMS + PD 100 mg/kg/day delivered by oral gavage, CUMS + PD 200 mg/kg/day oral gavage. → SPT, FST, TST, SIT for depression and MBI, E + M for anxiety → killing → Hippoc neuronal culture and staining, IF for Iba-1 and GFAP in Hippoc CA1, CA3, and DG, rt-qPCR for *Iba-1, Gfap, Gapdh, Hmox1, Nqo1,* and *Nrf2* forward and reverse primers, WB for synaptophysin and PSD-95, ELISA for DA, 5-HT, IL-1β, IL-6, and TNF-α	CUMS → ↓ BW, in SPT ↓ sucrose preference, in FST and TST ↑ immobility time, in SIT ↓ affiliation, sociability, memory, and novelty; all these effects were counteracted by oral PD 200 mg/kg/day; CUMS ↑ marble burying in the MBI and time on closed arms in E + M, both counteracted by PD 200 mg/kg/day. PD ↑ PSD-95 and synaptophysin content in Hippoc, dendrite length and number. CUMS ↑ NI markers Iba-1 and GFAP, IL-1β, IL-6, and TNFα in Hippoc, ↑ release and nuclear translocation of NFκB, ↑ ROS, SOD, GPx, ↓ Hmox1, Nqo1, and Nrf2 levels; all these CUMS effects were counteracted by PD	PD reverts changes induced by CUMS on depressive and anxiety-related behaviour; PD restored all changes induced by CUMS and hippocampal neuronal function

Table 2. *Cont.*

Study	Animal	Paradigm/Model	Location	Design	Results Relating to Nrf2	Conclusions/Observations
Smanietto et al., 2023 [124]	*Swiss* ♂ mice (BW 25–35 g)	CUMS-induced DLB	Pelotas, Rio Grande do Sul, Brazil	CUMS × 28 days → intranasal IL-4 vs. plac; 8 mice per group, 4 groups (Ctrl, Ctrl–IL-4, CUMS, CUMS–IL-4) subjected to water/food/movement restriction, tail pressure, and inclined box with wet bedding; → OFT, TST, SPiT → killing, tissue extraction and analyses (PFC and Hippoc, adrenals, lymph nodes, thymus, spleen [organs], blood to determine ROS, IDO, CAT, SOD, LPO, MDA, NO metabolites, gene expression, CORST levels)	In OFT no differences in crossings and rearings among groups; in TST, CUMS ↑ immobility and IL-4 reversed it; in the SPiT, CUMS ↓ grooming and IL-4 reversed it. CUMS → ↑ CORST and lymph organs, ↑ NF-κB expression, IL-4 → ↓ CORST and organ weight; ↓ NF-κB expression; in both PFC and Hippoc, CUMS → ↑ IDO, IL-1β, NF-κB, Nrf2, ROS, SOD, and CAT and ↓ BDNF (no effect on LPO in Hippoc and ↑ in PFC; similar effects for MDA, but no effect on NO metabolites in both PFC and Hippoc); all CUMS effects were reversed by IL-4	IL-4 counteracts CUMS-induced DLB; the effects of IL-4 are most evident in the PFC. IL-4 protects against chronic stress-induced depressive behaviour through ↓ of NI and OS
Zhang et al., 2023 [125]	♂7–8-wk-old adult C57BL6 mice	LPS-induced depression and anxiety	Guiyang, Guizhou, China	Mice randomised into Plac (Ctrl), to the phenolic glycoside gastrodin (major component of *Gastrodia elata*) 25, 50, or 100 mg/kg/day) or minocycline (microglial inhibitor) 50 mg/kg/day i.p. × 3 days → Plac (Ctrl), gastrodin 25, 50, or 100 mg/kg/day + Plac or LPS 0.25 mg/day, or minocycline 50 mg/kg/day + Plac or LPS 0.25 mg/day i.p. × 10 days; gastrodin-receiving animals were also treated with Nrf2 antagonist ML385 30 mg/kg/day i.p. × 10 days. Network pharmacology analysis, molecular docking on PPAR-γ, STAT6, and Nrf2; → OFT, SPT, E + M, FST → killing, cultured neonatal microglia, Hippoc microglia, rt-qPCR for *Iba-1, β-actin, Nlrp3, iNOS, IL-1β, IL-10, TNF-α, Cd11b, Cd68*, and *Arg-1⁺*; ELISA for IL-1β, IL-6, IL-10, TNF-α, and NO; WB for Nrf2, pNrf2, and GAPDH; IF for Hippoc Iba-1, iNOS, Arg-1, Nrf2	LPS ↓ sucrose preference in SPT, ↑ immobility time in FST, ↓ number of entries or time of stay in open arms in E + M, ↓ time of stay in central area in OFT; all DLB and anxiety behav reversed by higher doses of gastrodin and minocycline; LPS ↑ Iba-1 staining in Hippoc microglia, thickened cell-body and shortened cell processes; LPS ↑ CD11b, CD86, NLRP3, iNOS; gastrodin 100 mg/kg/day and minocycline reversed NI-induced changes. Gastrodin 100 mg/kg/day or minocycline ↑ the microglial anti-inflammatory marker Arg-1 and Hippoc Arg-1⁺ microglia in LPS-treated, while they ↓ IL-1β and IL-6 and ↑ IL-10. ML385 blocked the effect of gastrodin on ↑ Arg-1⁺ cell proportion in Hippoc microglia and the subsequent ↑ of Nrf2, pNrf2 and nuclear Nrf2 translocation	Gastrodin through Nrf2 prevents depressive and anxiety behaviours through counteracting NI actions. Gastrodin could represent a useful addition to the treatment of depressive and anxiety disorders, as it promptly crosses the BBB

Note. *Abbreviations:* Ab(s), antibody(ies); ABTS, 2,2′-azino-bis-3-ethylbenzothiazoline-6-sulfonic acid; ACE Angiotensin-converting enzyme; AChE, Acetyl-cholinesterase; ACMS, adipose-derived mesenchymal stem cell; AD, antidepressant; ADSC, adipose-derived mesenchymal stem cells; AI, Artificial Intelligence, Akt, Protein kinase B; AOEs, antioxidant enzymes; ARE, antioxidant response elements; ARS, acute restraint stress ASC, apoptosis-associated speck-like; AST, astrocytes; ATP, adenosine triphosphate; ATRA, all-trans retinoic acids; BAFF, B-cell-activating factor of the tumour-necrosis-factor family, B-cell activating factor; BAX, Bcl-2-associated X protein, bcl-2-like protein 4; BBB, blood–brain barrier; Bcl-2, Bcl-2, B-cell lymphoma 2; BDNF, brain-derived neurotrophic factor; behav, behavioural; bEnd.3, brain endothelial cell line 3; Bilat, bilateral;BL, baseline; BTS, Bangpungtongsung-san; BW, body weight; *C. elegans, Caenorhabditis elegans*, a common worm (nematode); CA, concentrated PM2.5 air; cAMP, cyclic adenosine monophosphate; CarA, carnosic acid; CAR20, carveol 20 mg/kg; CAR50, carveol 50 mg/kg; CAT, catalase; ChIP, Chromatin immunoprecipitation; CLA, conjugated linoleic acid; CMAE, *Carica papaya*; CMC-Na, Sodium carboxymethylcellulose; CMI, 3-[(4-chlorophenyl) selanyl]-1-methyl-1H-indole; CMS, Chronic mild stress; cncs, concussions; CORT model, elevated corticosteroid anxiety/depression mouse model; CORST, corticosterone; CP, cyclophosphamide; CRS, chronic restraint stress; CREB, cyclic AMP responsive element binding protein; CRS, chronic restraint stress; CSAE, Cannabis sativa; CSDS, chronic social defeat stress; Ctrl, control(s); CUMS, chronic unpredictable mild stress; CUR, Curcumine; CuZnSOD, copper zinc superoxide dismutase; CVD, cardiovascular disease; CX₃CR1, CX₃C chemokine receptor 1; DA, dopamine; DG, dentate gyrus; D-GalN,

D-galactosamine; DHA, docosaheaxaenoic acid; 7,8-DHF, 7,8-Dihydroxyflavone; DLB, depressive-like behaviour; DM, diabetes mellitus; DMF, Dimethyl fumarate; DOPAC, 3,4-dihydroxyphenylacetic acid; DPFH, 2,2-diphenyl-1-picrylhydrazyl; DSAE, Datura stramonium; DSS, dextran sulfate sodium; EGR1–4, early growth response proteins 1–4; ELISA, enzyme-linked immune sorbent test; EPA, eicosapentaenoic acid; ERK, extracellular signal-regulated protein kinase; EZH2, Enhancer of zeste homolog 2; E + M, elevated plus (or X or Y) maze; EZM, elevated zero maze; Fen, fenretinide; FiA, filtered air; FJB, Fluoro-Jade B; FJC+, Fluoro-Jade C-positive cells; FO, fish oil; FST, forced swimming test (Porsolt); GABA, γ-hydroxybutyric acid; GABARA1A, the α1 subunit of the GABA$_A$ receptor; GAPDH, glyceraldehyde-3-dehydrogenase; GAS5, long non-coding RNA growth arresting-specifc 5; GC, gas chromatography; GC-MS, gas chromatography–mass spectrometry; GCLC, glutamate-cysteine ligase catalytic subunit; GCLM, glutamate-cysteine ligase modifier subunit; GDH, Glutamate dehydrogenase; GFAP, glial fibrillary acid protein; Glo-1, glyoxalase 1; GLR, glutathione reductase; Glu, glutamate; GPx, glutathione peroxidase; GR, glucocorticoid receptor; GSH, glutathione; GSK-3β, glycogen synthase kinase 3-beta; GST, Glutathione S transferase; GSTP1, Glutathione S-transferase Pi; GSTω1, Glutathione S-transferase omega-1; GSSG, glutathione disulphide; G6PD, glucose-6-phosphodehydrogenase; h, hours; Ham-D, Hamilton Depression Rating Scale; HAPI, microglial cell line Highly Aggressively Proliferating Immortalised; HCAR2, Hydroxycarboxylic Acid Receptor 2; HC, healthy control(s); HDAC2, Histone Deacetylase 2 gene; HDOs, heteroduplex oligonucleotides; HFD, high fat diet; Hippoc, hippocampus, hippocampal; HippP, hippocampus proper; HO-1, haemoxygenase; HPLC, high-performance liquid chromatography; HRR, High resolution respirometry; Hsd, hesperidin; hsp-60, -70, 60 and 70 kDa heat shock proteins; Hst, hesperetin, i.e., the aglycone of the flavonoid hesperidin; H&E, Haematoxylin-Eosin; H$_2$DCF-DA, 2′,7′ -dichlorodihydrofluorescein diacetate; H$_2$O, water; H3K27me3, Methylation of histone 3 on lysine 27; IBA-1, ionised calcium-binding adapter molecular 1; IBF, inflammatory bowel disease; ICP-MS, inductively coupled plasma mass spectrometry; ICR, imprinting control region (mice); icv, intracerebroventricular; IDO-1, indoleamine-2,3-dioxygenase 1; IF, immunofluorescence; i.g., intragastric gavage; IKK-β, inhibitor of nuclear factor kappa-B kinase subunit beta; IL-1 . . . 10, Interleukin 1, 2, 3, . . . -10; IL-1β, Interleukin 1-beta; inflam, inflammation, inflammatory; ImBl, immunoblotting; IMI, Imipramine Hydrocloride; iNOS, inducible nitric oxide synthase; ip, intraperitoneal; Ipt, iptakalim IκBα, nuclear factor of kappa light polypeptide gene enhancer in B-cells inhibitor, alpha; i-gastr, intragastrically; JNK, c-Jun N-terminal kinase; jugl., juglone, a naphthoquinone from walnut; Keap1, kelch-like ECH-associated protein 1; KMO, kynurenine monooxygenase; KO, knock-out; KT, ketamine; L-dopa, levodopa, L-3,4-dihydroxyphenylalanine; LDB, light-dark box; lenti, lentivirus; Lenti-GAS5, lentivirus-packaged GAS5 overexpressing plasmid; LPO, lipid peroxidase; LPS, lipopolysaccharide; LW, long day and warm temperature; M1, Macrophage M1 phenotype, MAO, Monoamine oxidase; MAPK, mitogen-activated protein kinase; MBT, marble burying test; MCE-1, ($\pm$)-3-ethynyl-3-methyl-6-oxocyclohexa-1,4-dienecarbonitrile, Nrf2 activator; MCC, MCC950; MCh, mitochondrium, mitochondrial; MDA, malondialdehyde; MDD, major depressive disorder; Measur, measurement, measured; *MeCP2*, methyl CpG binding protein 2; MEK1/2, mitogen-activated protein kinase kinases 1 and 2; Mito–TEMPO, 2-(2,2,6,6-Tetramethylpiperidin-1-oxyl-4-ylamino)-2-oxoethyl) triphenylphosphonium chloride; MMF, Monomethyl fumarate; MnSOD, manganese superoxide dismutase; MT, melatonin; NA, noradrenaline; NAC, *N*-acetyl-Lcysteine; NADP, Nicotinamide adenine dinucleotide phosphate; NBP, Dl-3-n-Butylphthalide; NBT, nesting building test; NDEVs, Enrichment of plasma neuron-derived extracellular vesicles; NF-κB, nuclear factor kappa B; NI, neuroinflammation; NKT, Nootkatone; NLRP3, Nucleotide-binding oligomerization domain containing 3 inflammasome; NO, nitric oxide; NORT, Novel Object Recognition Test; NOs, nitric oxide synthase; NOX, NADP oxidase; NQO-1, NAD(P)H Quinone Dehydrogenase (oxidoreductase) 1; NResp, non-responder(s); Nrf2, recox-sensing nuclear factor (erythroid-derived 2)-like 2; NS, non-silencing; NSFT, Novelty-suppressed feeding test; NTAE, Nicotiana tabacum; OBX, olfactory bulbectomy; OFT, open-field test; OS, oxidative stress; OVX, ovariectomised, ovariectomy; p, phospho, phosphorylated; *p*, statistical significance probability; PB, Pinocembrin; PBMC, peripheral blood mononuclear cells; PBS, phosphate-buffered saline; (p-ClPhSe)2, *p*-chlorodiphenyl diselenide; PCMS, predictable, chronic mild stress; PD, Polydatin, a resveratrol derivative of the herbal medicine *Polygonum cuspidatum*; PFC, prefrontal cortex, dl, dorsolateral, vm, ventromedial; PI3K, phosphatidylinositol 3-kinase; plac, placebo or vehicle or saline; p.o., *per os*, orally; PG, prostaglandin; PGC-1α, peroxisome proliferator-activated receptor-gamma coactivator; PPAR-α, -γ, Peroxisome Proliferator Receptor-alpha, -gamma; PRC2, Polycomb repressive complex 2; PRMT1, protein arginine methyltransferase; PSD-95, 95 kD post-synaptic density protein; pts, patients; PSE, pinus spp. succinum extract;. PSP, Polysaccharides from *Polygonatum cyrtonema*; PTZ, pentylenetetrazol; QA quinolinic acid; qPCR, quantitative polymerase chain reaction; RA, rosmarinic acid; Rb1, Retinoblastoma protein-1; Resp, responder(s); Rg1, Ginsenoside Rg1, the main active compound of ginseng; RI, recombinant inbred; RIP, RNA-binding protein immunoprecipitation; ROS, reactive oxygen species; Ro61-8048, 3,4-dimethoxy-N-[4-(3-nitrophenyl)thiazol-2-yl]benzenesulfonamide, high-affinity kynurenine 3-hydroxylase (KMO) inhibitor; rt-qPCR, real-time quantitative polymerase chain reaction; SAHA, suberoyanilide hydroxamic acid, vorinostat, a histone deacetylase inhibitor; SalB, Salvianolic Acid; SC, short day and cool temperature; SDS, social defeat stress; *SE, Salicornia europaea*; SEO, *Schisandra Chinensis* oil; SF spectrofluorometer; SFN, sulforaphane; shGAS5, lentivirus-packaged GAS5 shRNA; shNC, shRNA-containing lentivirus, negative-control; shRNA, short (small) hairpin RNA to silence specific genes, e.g., lentivirus-packaged GAS5 shRNA); SI, social isolation; siRNA, small interfering ribonucleic acid; SIRT1, sirtuin-1, silent information regulator 2 homolog 1; SIT, social interaction test; SNAP25, 25kD synaptosomal associated protein; SnPP-IX, tin protoporphyrin IX; SOCS3, Suppressor of cytokine signalling 3; SODs, superoxide dismutases; SPlT, splash test; SPT, sucrose preference test; SS, social support; STAT6, signal transducer and activator of transcription 6; STZ, streptozotocin; SUS, CUMS-susceptible; TBAR, Thiobarbituric acid reactive substance; TBE-31, ($\pm$)-(4bS,8aR,10aS)-10a-ethynyl-4b,8,8-trimethyl-3,7-dioxo-3,4b,7,8,8a,9,10,10a-octahydrophenanthrene-2,6-dicarbonitrile, Nrf2 activator; *t*-BHQ, *tert*-Butylhydroquinone, Nrf2 nuclear translocator; TEM, transmission electron microscopy; Tempol, 4-hydroxy-2,2,6,6-tetramethylpiperidine 1-oxyl;

1-oxyl; 288 μmol/kg-1·day-1, an antioxidant; TLR4, Toll-like receptor 4; TMS, transcranial magnetic stimulation, d, deep, r, repetitive; TNF-α, tumour necrosis factor-alpha; TREM2, Triggering receptor expressed on myeloid cell-2; TrkB, tropomyosin receptor kinase B; Trx, thioredoxin; TrxR, thioredoxin reductase; TSP, Tilapia Skin Peptides; TST, tail suspension test; TUNEL, Terminal-deoxynucleoitidyl Transferase mediated Nick end labelling; TXNIP, thioredoxin-interacting protein; UA, ursolic acid; UnA, unfiltered air; WB, Western blot(ting); Wfs1, Wolfram syndrome 1; wk, week(s); WT, wild type; $\bar{x}$, mean; xCT, solute carrier family 7 member 11, SLC7A11; YFP, Thy1-yellow fluorescent protein; Xn, Xanthohumol; yr(s), year(s); ZnPP, zinc protoporphyrin IX; ZT, Zileuton; γ-GCS, γ-glutamylcysteine synthetase; 4-HNE, 4-hydroxy-2-nonenal; 5-HIAA, 5-hydroxyindoleacetic acid; 5-HT, 5-hydroxytryptamine, serotonin; 8-OHdG, 8-hydroxy-2′-deoxyguanosine; ±, SD, standard deviation; ×, for, per; ≈, about equal, not different; ♀, females; ♂, males; ↓, decreased, lower; ↑, increased, higher, →, induced, followed.

4. Discussion

In this review, we collected data on animal and human studies on depression and the role of Nrf2 as a probe of the antioxidant system function and found an antidepressant effect to be the consequence of the activation of the Nrf2-HO-1 pathway, with no studies pointing to the opposite direction. In this context, the role of neuroinflammation in being associated with depressive-like states clearly emerges, as well as the role of antioxidant systems in countering neuroinflammation and its markers, with Nrf2 standing at its crossroad. We may thus suppose that depression involves multiple systems and the antioxidant system, and Nrf2, in particular, counteract many actions of the failing multi-system state associated with depressive behaviour. Many of the substances that help overcome the alterations caused in the organism by depression are of natural origin and constitute long-used remedies for many other conditions. The antioxidant research line in depression that focuses on the actions of Nrf2 is relatively recent, dating back to 2013 [43], although the substance was isolated in 1994 [126], almost three decades ago.

The rate of publications on this issue has considerably increased across the years, with papers focusing on psychiatric disorders increasing almost exponentially from 26 in 2020 to 37 in 2021 and to 58 in 2022, witnessing increased interest and awareness of its importance in the pathophysiology of psychiatric disorders. Studies on depression reflect a similar trend. From eight studies in 2019, which increased to 16 in 2020, 18 in 2021, and 23 in 2022 to the beginning of 2023.

Most of the studies were conducted on rodents, mainly mice and, most importantly, male animals. This limits the extension of results to female animals, but it should be noted that findings in females were not dissimilar from those in males.

The studies included here were based on different animal depression paradigms, with many studies using stress-induced DLB, mostly elicited by LPS [40,43,51,53,58,59,63,65,67,73,75,79,85,88,95,105,106,125], or through corticosterone or cortisol [44,46,80,87,98], while other studies used chronic unpredictable mild stress [61,72,76,83,93,96,97,99–101,106,108,111,122,124] or social defeat stress [50,52,88,102,107,120]. Two studies used tumour-related depression models [71,91], while just one study each used post-ovariectomised depression [121], pentylenetetrazol-induced seizures [116], maternal separation [117], and olfactory bulbectomy [113] models (Table 1). Various studies used Nrf2 knock-out animals [49,62,84,88,120] or knock-out of the inflammation and histone methylation modulator protein arginine methyltransferase [67]. Nrf2 knock-out or knock-down apparently impairs neurochemical indexes and behavioural test performance, while protein arginine methyltransferase knock-out decreases depressive indexes [67]. Despite heterogeneity, findings all pointed to the Nrf2-antioxidant pathway restoring impaired neurochemistry and behavioural test performance.

Anti-inflammatory or anti-neuroinflammatory effects of employed antidepressants and putative antidepressant substances, such as the many plant extracts used in the studies we considered here, were shown to be related to various improvements in neurochemical indexes and behavioural tests [43,48,59,71,76,81,85,91,95,108,111,123,125].

Studies were consistent in measuring cyclic AMP responsive element binding protein (CREB) activity, in that increased phospho-CREB or CREB levels were associated with an antidepressant response [38,46,81,83,103,109,112]; remarkably, studies on peripheral tissues in man [38] and hippocampal content in mice [81] pointed to the same direction. Additionally, studies of BDNF showed that increased BDNF activity was associated with antidepressant effects and that animals displaying depressive features have low BDNF contents in the periphery and the brain [44,52,57,57,58,66,67,69,81,83,85,88,93,111,112,114,121]. It has recently been shown that most, if not all, antidepressant drugs bring about their antidepressant effects by binding the tyrosine kinase receptor 2 (TrkB), a molecule tied to the action of BDNF [127,128]. The findings of the studies reviewed here quite match this concept; increased TrkB/BDNF signalling was related to antidepressant effects, and low levels were associated with depressive behaviour [11,50,66,69,84,93,103,111,120]. Increased activity also of protein kinase B (Akt-1), which is an antioxidant in the wingless-GSK pathway and carries on some of the biological [128–130] and clinical [131] actions of lithium, which also inhibits GSK-3β [132–134], was found to be associated with antidepressant effects in the studies included in this review [42,45,47,74,75,93,94,110,119]. On the opposite side are found nuclear factor kappa B (NF-κB) and glycogen synthase kinase (GSK) pathways; the higher the NF-κB [38,39,67,69,73,75,80,85,91,98,100,103,113,115] and GSK3β levels [45,79,109], the worse the depressive indexes. Therefore, it appears that on one side, there are antioxidant systems and growth factors promoting plasticity, such as Nrf2–HO-1, BDNF, Akt, TrkB, and CREB, which are protective from depression and, on the other side, GSK-3β and NF-κB, which promote depressive behaviour (Figure 2). Therefore, we may suppose that new antidepressant drug discovery could involve promoting drugs that increase Nrf2–HO-1, BDNF, Akt, TrkB, and CREB activity, while downplaying GSK-3β and NF-κB. However, it is not so simple to identify the characteristics of drugs that act only where they will wishfully carry on the desired effect because drugs acting within cells must cross plasma membranes selectively, reach their intracellular target in those cells needing their actions, and not everywhere. Depression is a multi-system and multi-organ condition but not an all-system derangement, so the presence of an extraordinarily penetrating drug in some cells could be related to undesirable or adverse effects. Furthermore, not all actions of Nrf2 are good for the body. Nrf2 may promote tumours [135,136] and atherogenesis [137,138], so any drug acting on Nrf2 must avoid these two and maybe other unknown potentially harmful actions.

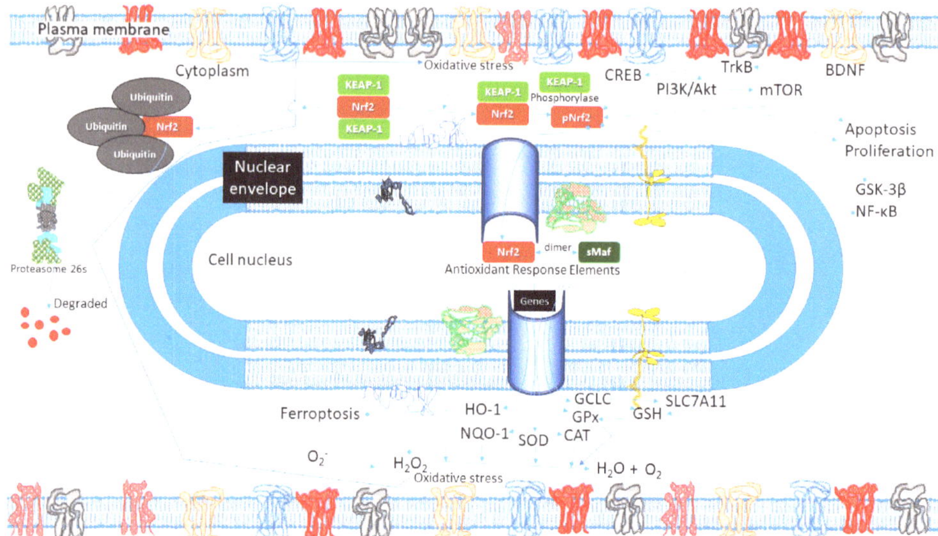

Figure 2. Simplified depiction of the mechanism of action of Nrf2 and the pathways it interacts with. Arrows, activation; broken lines with rhomboid ending, inhibition. *Abbreviations.* Akt, Protein kinase B; BDNF, brain-derived neurotrophic factor; CREB, cyclic AMP responsive element binding protein; GCLC, Glutamate-cysteine ligase; GPx, glutathione peroxidase; GSH, glutathione; GSK-3β, glycogen synthase kinase 3-beta; HO-1, haem oxygenase 1; mTOR, mammalian target of rapamycin; NF-κB, nuclear factor kappa B; NQO-1, NAD(P)H quinone dehydrogenase 1; PI3K, phosphatidylinositol-3-kinase; SLC7A11, Solute Carrier Family 7 Member 11 (cystine/glutamate transporter); sMAF, small musculoaponeurotic fibrosarcoma proteins; SOD, superoxide dismutase; TrkB, tropomyosin receptor kinase B.

5. Limitations

This review has several limitations in that the animal studies focused mostly on rodents (mice and rats), and few focused on other animals (fish and nematodes) or cell lines; therefore, the results cannot extend to other animal species. Human studies were only four (one *post mortem*), and their methodologies were too different from animal protocols; hence the results from the latter are not translatable to the former. Furthermore, a sex bias exists in animal studies (mostly males) and not in the few human studies. Depression is overrepresented in women in humans, so it might be that testing it in male animals is not a good idea, but female rodents pose significant problems in conducting scientific experiments using the paradigms of the studies included in this review. The animal models of depression are not easy to translate to humans and into clinical practice.

6. Conclusions

Nrf2 is at the crossroad of many cellular actions. It directs antioxidant pathways and receives many intracellular signals to which it attempts to respond in a balanced way. Its function appears to be impaired in depression, and it is also possible that its manipulation could prove to be beneficial in human depression; however, there is much to discover. There is hope to discover drugs that cross the blood–brain barrier, or use some phytopharmaca or their derivatives, such as gastrodin, or use drugs stimulating anti-inflammatory cytokines that could counteract neuroinflammation. From the studies included in this review, depression appears to be strictly tied to inflammatory mechanisms. Provided that new drugs acting on the Nrf2 antioxidant pathway can avoid tumorigenesis and atherogenesis, Nrf2 can be a useful target for novel drug development.

Supplementary Materials: The following supporting information can be downloaded at: https://www.mdpi.com/article/10.3390/antiox12040817/s1, Online Supplement.

Author Contributions: G.S., G.D.K., M.M., D.P.R.C. and L.M. (Laura Monti) conceived the review; S.M., A.B. and G.D.K. wrote the introduction; S.M., A.B., O.M.F., L.M. (Lorenzo Moccia), E.B., A.S., D.P.R.C. and G.D.K. wrote the methods, decided eligibility along all others and wrote results and discussion; L.M. (Laura Monti), L.M. (Lorenzo Moccia), M.M., D.J., G.D.K., L.J., S.M., A.B., D.P.R.C. and G.D.K. performed literature searches; S.M., A.B., L.M. (Laura Monti), L.M. (Lorenzo Moccia), M.M. and G.D.K. provided the first draft; A.S., D.J., L.J., L.M. (Lorenzo Moccia), D.P.R.C., G.D.K. and G.S. supervised the writing of the manuscript, wrote the conclusions, and edited the final form. All authors have read and agreed to the published version of the manuscript.

Funding: This research received no external funding.

Institutional Review Board Statement: Not applicable.

Informed Consent Statement: Not applicable.

Data Availability Statement: The data presented in this study are available in the article and Supplementary Materials.

Acknowledgments: The authors are grateful to the Scientific Administration of the Bibliographic and Bibliometric Support Service, Fondazione Policlinico A. Gemelli IRCCS, in particular, Maria Pattuglia, and Mimma Ariano, Ales Casciaro, Teresa Prioreschi, and Susanna Rospo, Librarians of the Sant'Andrea Hospital, Faculty of Medicine and Psychology, Sapienza University of Rome, for rendering precious bibliographical material accessible. They all have consented to the acknowledgement.

Conflicts of Interest: The authors declare no conflict of interest.

References

1. Cao, H.; Zuo, C.; Gu, Z.; Huang, Y.; Yang, Y.; Zhu, L.; Jiang, Y.; Wang, F. High frequency repetitive transcranial magnetic stimulation alleviates cognitive deficits in 3xTg-AD mice by modulating the PI3K/Akt/GLT-1 axis. *Redox Biol.* **2022**, *54*, 102354. [CrossRef]
2. Zuo, C.; Cao, H.; Song, Y.; Gu, Z.; Huang, Y.; Yang, Y.; Miao, J.; Zhu, L.; Chen, J.; Jiang, Y.; et al. Nrf2: An all-rounder in depression. *Redox Biol.* **2022**, *58*, 102522. [CrossRef]
3. Betteridge, D.J. What is oxidative stress? *Metabolism* **2000**, *49* (Suppl. 1), 3–8. [CrossRef]

4. Francisqueti-Ferron, F.V.; Ferron, A.J.T.; Garcia, J.L.; Silva, C.C.V.A.; Costa, M.R.; Gregolin, C.S.; Moreto, F.; Ferreira, A.L.A.; Minatel, I.O.; Correa, C.R. Basic concepts on the role of nuclear factor erythroid-derived 2-like 2 (Nrf2) in age-related diseases. *Int. J. Mol. Sci.* **2019**, *20*, 3208. [CrossRef]

5. Saha, S.; Buttari, B.; Panieri, E.; Profumo, E.; Saso, L. An overview of Nrf2 signaling pathway and its role in inflammation. *Molecules* **2020**, *25*, 5474. [CrossRef]

6. Zhang, W.; Feng, C.; Jiang, H. Novel target for treating Alzheimer's Diseases: Crosstalk between the Nrf2 pathway and autophagy. *Ageing Res. Rev.* **2021**, *65*, 101207. [CrossRef]

7. Sivandzade, F.; Prasad, S.; Bhalerao, A.; Cucullo, L. NRF2 and NF-κB interplay in cerebrovascular and neurodegenerative disorders: Molecular mechanisms and possible therapeutic approaches. *Redox Biol.* **2019**, *21*, 101059. [CrossRef] [PubMed]

8. Osama, A.; Zhang, J.; Yao, J.; Yao, X.; Fang, J. Nrf2: A dark horse in Alzheimer's disease treatment. *Ageing Res. Rev.* **2020**, *64*, 101206. [CrossRef]

9. Tong, K.I.; Padmanabhan, B.; Kobayashi, A.; Shang, C.; Hirotsu, Y.; Yokoyama, S.; Yamamoto, M. Different electrostatic potentials define ETGE and DLG motifs as hinge and latch in oxidative stress response. *Mol. Cell. Biol.* **2007**, *27*, 7511–7521. [CrossRef] [PubMed]

10. Mohan, S.; Gupta, D. Crosstalk of toll-like receptors signaling and Nrf2 pathway for regulation of inflammation. *Biomed. Pharmacother.* **2018**, *108*, 1866–1878. [CrossRef] [PubMed]

11. Chowdhry, S.; Zhang, Y.; McMahon, M.; Sutherland, C.; Cuadrado, A.; Hayes, J.D. Nrf2 is controlled by two distinct β-TrCP recognition motifs in its Neh6 domain, one of which can be modulated by GSK-3 activity. *Oncogene* **2013**, *32*, 3765–3781. [CrossRef] [PubMed]

12. Huang, H.-C.; Nguyen, T.; Pickett, C.B. Phosphorylation of Nrf2 at Ser-40 by protein kinase C regulates antioxidant response element-mediated transcription. *J. Biol. Chem.* **2002**, *277*, 42769–42774. [CrossRef] [PubMed]

13. Cullinan, S.B.; Diehl, J.A. PERK-dependent activation of Nrf2 contributes to redox homeostasis and cell survival following endoplasmic reticulum stress. *J. Biol. Chem.* **2004**, *279*, 20108–20117. [CrossRef] [PubMed]

14. Keum, Y.-S.; Yu, S.; Chang, P.P.-J.; Yuan, X.; Kim, J.-H.; Xu, C.; Han, J.; Agarwal, A.; Kong, A.-N.T. Mechanism of action of sulforaphane: Inhibition of p38 mitogen-activated protein kinase isoforms contributing to the induction of antioxidant response element-mediated heme oxygenase-1 in human hepatoma HepG2 cells. *Cancer Res.* **2006**, *66*, 8804–8813. [CrossRef] [PubMed]

15. Pi, J.; Bai, Y.; Reece, J.M.; Williams, J.; Liu, D.; Freeman, M.L.; Fahl, W.E.; Shugar, D.; Liu, J.; Qu, W.; et al. Molecular mechanism of human Nrf2 activation and degradation: Role of sequential phosphorylation by protein kinase CK2. *Free Radic. Biol. Med.* **2007**, *42*, 1797–1806. [CrossRef] [PubMed]

16. Nguyen, C.D.; Yoo, J.; Hwang, S.Y.; Cho, S.Y.; Kim, M.; Jang, H.; No, K.O.; Shin, J.C.; Kim, J.H.; Lee, G. Bee venom activates the Nrf2/HO-1 and TrkB/CREB/BDNF pathways in neuronal cell responses against oxidative stress induced by Aβ1-42. *Int. J. Mol. Sci.* **2022**, *23*, 1193. [CrossRef]

17. Lau, A.; Wang, X.-J.; Zhao, F.; Villeneuve, N.F.; Wu, T.; Jiang, T.; Sun, Z.; White, E.; Zhang, D.D. A noncanonical mechanism of Nrf2 activation by autophagy deficiency: Direct interaction between Keap1 and p62. *Mol. Cell. Biol.* **2010**, *30*, 3275–3285. [CrossRef]

18. Guo, L.; Jiang, Z.-M.; Sun, R.-X.; Pang, W.; Zhou, X.; Du, M.-L.; Chen, M.-X.; Lv, X.; Wang, J.-T. Repeated social defeat stress inhibits development of hippocampus neurons through mitophagy and autophagy. *Brain Res. Bull.* **2022**, *182*, 111–117. [CrossRef]

19. Hashimoto, K. Essential role of Keap1-Nrf2 signaling in mood disorders: Overview and future perspective. *Front. Pharmacol.* **2018**, *9*, 1182. [CrossRef]

20. Björkholm, C.; Monteggia, L.M. BDNF a key transducer of antidepressant effects. *Neuropharmacology* **2016**, *102*, 72–79. [CrossRef]

21. Tang, R.; Cao, Q.-Q.; Hu, S.-W.; He, L.-J.; Du, P.-F.; Chen, G.; Fu, R.; Xiao, F.; Sun, Y.-R.; Zhang, J.-C.; et al. Sulforaphane activates anti-inflammatory microglia, modulating stress resilience associated with BDNF transcription. *Acta Pharmacol. Sin.* **2022**, *43*, 829–839. [CrossRef] [PubMed]

22. Abdalkader, M.; Lampinen, R.; Kanninen, K.M.; Malm, T.M.; Liddell, J.R. Targeting Nrf2 to suppress ferroptosis and mitochondrial dysfunction in neurodegeneration. *Front. Neurosci.* **2018**, *12*, 466. [CrossRef] [PubMed]

23. Wilson, C.B.; McLaughlin, L.D.; Nair, A.; Ebenezer, P.J.; Dange, R.; Francis, J. Inflammation and oxidative stress are elevated in the brain, blood, and adrenal glands during the progression of post-traumatic stress disorder in a predator exposure animal model. *PLoS ONE* **2013**, *8*, e76146. [CrossRef]

24. Sun, X.R.; Zhang, H.; Zhao, H.T.; Ji, M.H.; Li, H.H.; Wu, J.; Li, K.Y.; Yang, J.J. Amelioration of oxidative stress-induced phenotype loss of parvalbumin interneurons might contribute to the beneficial effects of environmental enrichment in a rat model of post-traumatic stress disorder. *Behav. Brain Res.* **2016**, *312*, 84–92. [CrossRef] [PubMed]

25. Alzoubi, K.H.; Al Subeh, Z.Y.; Khabour, O.F. Molecular targets for the interactive effect of etazolate during post-traumatic stress disorder: Role of oxidative stress, BDNF and histones. *Behav. Brain Res.* **2019**, *369*, 111930. [CrossRef]

26. Li, H.; Tofigh, A.M.; Amirfakhraei, A.; Chen, X.; Tajik, M.; Xu, D.; Motevalli, S. Modulation of astrocyte activity and improvement of oxidative stress through blockage of NO/NMDAR pathway improve posttraumatic stress disorder (PTSD)-like behavior induced by social isolation stress. *Brain Behav.* **2022**, *12*, e2620. [CrossRef]

27. Hong, C.; Cao, J.; Wu, C.F.; Kadioglu, O.; Schüffler, A.; Kauhl, U.; Klauck, S.M.; Opatz, T.; Thines, E.; Paul, N.W.; et al. The Chinese herbal formula Free and Easy Wanderer ameliorates oxidative stress through KEAP1-NRF2/HO-1 pathway. *Sci. Rep.* **2017**, *7*, 11551. [CrossRef]

28. Zhou, C.H.; Xue, F.; Xue, S.S.; Sang, H.F.; Liu, L.; Wang, Y.; Cai, M.; Zhang, Z.J.; Tan, Q.R.; Wang, H.N.; et al. Electroacupuncture pretreatment ameliorates PTSD-like behaviors in rats by enhancing hippocampal neurogenesis via the Keap1/Nrf2 antioxidant signaling pathway. *Front. Cell. Neurosci.* **2019**, *13*, 275. [CrossRef]

29. Recinella, L.; Chiavaroli, A.; Orlando, G.; Ferrante, C.; Veschi, S.; Cama, A.; Marconi, G.D.; Diomede, F.; Gesmundo, I.; Granata, R.; et al. Effects of growth hormone-releasing hormone receptor antagonist MIA-602 in mice with emotional disorders: A potential treatment for PTSD. *Mol. Psychiatry* **2021**, *26*, 7465–7474. [CrossRef]

30. McEwen, B.S. Mood disorders and allostatic load. *Biol. Psychiatry* **2003**, *54*, 200–207. [CrossRef]

31. Page, M.J.; McKenzie, J.E.; Bossuyt, P.M.; Boutron, I.; Hoffmann, T.C.; Mulrow, C.D.; Shamseer, L.; Tetzlaff, J.M.; Akl, E.A.; Brennan, S.E.; et al. The PRISMA 2020 statement: An updated guideline for reporting systematic reviews. *BMJ* **2021**, *372*, n71. [CrossRef] [PubMed]

32. Higgins, J.P.T.; Savović, J.; Page, M.J.; Elbers, R.G.; Sterne, J.A.C. Chapter 8: Assessing risk of bias in a randomized trial. In *Cochrane Handbook for Systematic Reviews of Interventions*; Higgins, J.P.T., Thomas, J., Chandler, J., Cumpston, M., Li, T., Page, M.J., Welch, V.A., Eds.; Version 6.3, updated February 2022; Cochrane: London, UK, 2022; Available online: www.training.cochrane.org/handbook (accessed on 12 February 2023).

33. de Boer, S.F.; van der Vegt, B.J.; Koolhaas, J.M. Individual variation in aggression of feral rodent strains: A standard for the genetics of aggression and violence? *Behav. Genet.* **2003**, *33*, 485–501. [CrossRef]

34. Albert, P.R. Why is depression more prevalent in women? *J. Psychiatry Neurosci.* **2015**, *40*, 219–221. [CrossRef]

35. Kuehner, C. Why is depression more common among women than among men? *Lancet Psychiatry* **2017**, *4*, 146–158. [CrossRef] [PubMed]

36. Tang, J.; Zhang, T. Causes of the male-female ratio of depression based on the psychosocial factors. *Front. Psychol.* **2022**, *13*, 1052702. [CrossRef]

37. Lukic, I.; Mitic, M.; Djordjevic, J.; Tatalovic, N.; Bozovic, N.; Soldatovic, I.; Mihaljevic, M.; Pavlovic, Z.; Radojcic, M.B.; Maric, N.P.; et al. Lymphocyte levels of redox-sensitive transcription factors and antioxidative enzymes as indicators of pro-oxidative state in depressive patients. *Neuropsychobiology* **2014**, *70*, 1–9. [CrossRef]

38. Mellon, S.H.; Wolkowitz, O.M.; Schonemann, M.D.; Epel, E.S.; Rosser, R.; Burke, H.B.; Mahan, L.; Reus, V.I.; Stamatiou, D.; Liew, C.C.; et al. Alterations in leukocyte transcriptional control pathway activity associated with major depressive disorder and antidepressant treatment. *Transl. Psychiatry* **2016**, *6*, e821. [CrossRef] [PubMed]

39. Martín-Hernández, D.; Caso, J.R.; Javier, M.J.; Callado, L.F.; Madrigal, J.L.M.; García-Bueno, B.; Leza, J.C. Intracellular inflammatory and antioxidant pathways in postmortem frontal cortex of subjects with major depression: Effect of antidepressants. *J. Neuroinflamm.* **2018**, *15*, 251. [CrossRef]

40. Kubick, N.; Pajares, M.; Enache, I.; Manda, G.; Mickael, M.E. Repurposing zileuton as a depression drug using an AI and in vitro approach. *Molecules* **2020**, *25*, 2155. [CrossRef]

41. Goetzl, E.J.; Wolkowitz, O.M.; Srihari, V.H.; Reus, V.I.; Goetzl, L.; Kapogiannis, D.; Heninger, G.R.; Mellon, S.H. Abnormal levels of mitochondrial proteins in plasma neuronal extracellular vesicles in major depressive disorder. *Mol. Psychiatry* **2021**, *26*, 7355–7362. [CrossRef] [PubMed]

42. Li, X.; Wang, Y.B.; Wang, C.C.; Jing, R.; Mu, L.H.; Liu, P.; Hu, Y. Antidepressant mechanism of kaixinsan and its active compounds based on upregulation of antioxidant thioredoxin. *Evid.-Based Complement. Altern. Med.* **2022**, *2022*, 7302442. [CrossRef] [PubMed]

43. Martín-de-Saavedra, M.D.; Budni, J.; Cunha, M.P.; Gómez-Rangel, V.; Lorrio, S.; Del Barrio, L.; Lastres-Becker, I.; Parada, E.; Tordera, R.M.; Rodrigues, A.L.S.; et al. Nrf2 participates in depressive disorders through an anti-inflammatory mechanism. *Psychoneuroendocrinology* **2013**, *38*, 2010–2022. [CrossRef] [PubMed]

44. Mendez-David, I.; Tritschler, L.; Ali, Z.E.; Damiens, M.H.; Pallardy, M.; David, D.J.; Kerdine-Römer, S.; Gardier, A.M. Nrf2-signaling and BDNF: A new target for the antidepressant-like activity of chronic fluoxetine treatment in a mouse model of anxiety/depression. *Neurosci. Lett.* **2015**, *597*, 121–126. [CrossRef] [PubMed]

45. Cunha, M.P.; Budni, J.; Ludka, F.K.; Pazini, F.L.; Rosa, J.M.; Oliveira, Á.; Lopes, M.W.; Tasca, C.I.; Leal, R.B.; Rodrigues, A.L.S. Involvement of PI3K/Akt signaling pathway and its downstream intracellular targets in the antidepressant-like effect of creatine. *Mol. Neurobiol.* **2016**, *53*, 2954–2968. [CrossRef]

46. Freitas, A.E.; Egea, J.; Buendia, I.; Gómez-Rangel, V.; Parada, E.; Navarro, E.; Casas, A.I.; Wojnicz, A.; Ortiz, J.A.; Cuadrado, A.; et al. Agmatine, by improving neuroplasticity markers and inducing Nrf2, prevents corticosterone-induced depressive-like behavior in mice. *Mol. Neurobiol.* **2016**, *53*, 3030–3045. [CrossRef] [PubMed]

47. Martín-Hernández, D.; Bris, Á.G.; MacDowell, K.S.; García-Bueno, B.; Madrigal, J.L.; Leza, J.C.; Caso, J.R. Modulation of the antioxidant nuclear factor (erythroid 2-derived)-like 2 pathway by antidepressants in rats. *Neuropharmacology* **2016**, *103*, 79–91. [CrossRef] [PubMed]

48. Martín-Hernández, D.; Caso, J.R.; Bris, Á.G.; Maus, S.R.; Madrigal, J.L.; García-Bueno, B.; MacDowell, K.S.; Alou, L.; Gómez-Lus, M.L.; Leza, J.C. Bacterial translocation affects intracellular neuroinflammatory pathways in a depression-like model in rats. *Neuropharmacology* **2016**, *103*, 122–133. [CrossRef] [PubMed]

49. Wojnicz, A.; Avendaño Ortiz, J.; Casas, A.I.; Freitas, A.E.; López, M.G.; Ruiz-Nuño, A. Simultaneous determination of 8 neurotransmitters and their metabolite levels in rat brain using liquid chromatography in tandem with mass spectrometry: Application to the murine Nrf2 model of depression. *Clin. Chim. Acta* **2016**, *453*, 174–181. [CrossRef]

50. Yao, W.; Zhang, J.-c.; Ishima, T.; Dong, C.; Yang, C.; Ren, Q.; Ma, M.; Han, M.; Wu, J.; Suganuma, H.; et al. Role of Keap1-Nrf2 signaling in depression and dietary intake of glucoraphanin confers stress resilience in mice. *Sci. Rep.* **2016**, *6*, 30659. [CrossRef]
51. Yao, W.; Zhang, J.-C.; Ishima, T.; Ren, Q.; Yang, C.; Dong, C.; Ma, M.; Saito, A.; Honda, T.; Hashimoto, K. Antidepressant effects of TBE-31 and MCE-1, the novel Nrf2 activators, in an inflammation model of depression. *Eur. J. Pharmacol.* **2016**, *793*, 21–27. [CrossRef]
52. Bouvier, E.; Brouillard, F.; Molet, J.; Claverie, D.; Cabungcal, J.H.; Cresto, N.; Doligez, N.; Rivat, C.; Do, K.Q.; Bernard, C.; et al. Nrf2-dependent persistent oxidative stress results in stress-induced vulnerability to depression. *Mol. Psychiatry* **2017**, *22*, 1701–1713, Erratum in *Mol. Psychiatry* **2017**, *22*, 1795. [CrossRef] [PubMed]
53. Zhang, J.-c.; Yao, W.; Dong, C.; Yang, C.; Ren, Q.; Ma, M.; Han, M.; Wu, J.; Ushida, Y.; Suganuma, H.; et al. Prophylactic effects of sulforaphane on depression-like behavior and dendritic changes in mice after inflammation. *J. Nutr. Biochem.* **2017**, *39*, 134–144, Erratum in *J. Nutr. Biochem.* **2021**, *88*, 108550; *J. Nutr. Biochem.* **2021**, *89*, 108562. [CrossRef]
54. Zhao, X.-J.; Zhao, Z.; Yang, D.-D.; Cao, L.-L.; Zhang, L.; Ji, J.; Gu, J.; Huang, J.-Y.; Sun, X.-L. Activation of ATP-sensitive potassium channel by iptakalim normalizes stress- induced HPA axis disorder and depressive behaviour by alleviating inflammation and oxidative stress in mouse hypothalamus. *Brain Res. Bull.* **2017**, *130*, 146–155. [CrossRef] [PubMed]
55. López-Granero, C.; Antunes Dos Santos, A.; Ferrer, B.; Culbreth, M.; Chakraborty, S.; Barrasa, A.; Gulinello, M.; Bowman, A.B.; Aschner, M. BXD recombinant inbred strains participate in social preference, anxiety and depression behaviors along sex-differences in cytokines and tactile allodynia. *Psychoneuroendocrinology* **2017**, *80*, 92–98. [CrossRef] [PubMed]
56. Abuelezz, S.A.; Hendawy, N. Insights into the potential antidepressant mechanisms of cilostazol in chronically restraint rats: Impact on the Nrf2 pathway. *Behav. Pharmacol.* **2018**, *29*, 28–40. [CrossRef]
57. Omar, N.N.; Tash, R.F. Fluoxetine coupled with zinc in a chronic mild stress model of depression: Providing a reservoir for optimum zinc signaling and neuronal remodeling. *Pharmacol. Biochem. Behav.* **2017**, *160*, 30–38. [CrossRef]
58. Li, M.; Li, C.; Yu, H.; Cai, X.; Shen, X.; Sun, X.; Wang, J.; Zhang, Y.; Wang, C. Lentivirus-mediated interleukin-1β (IL-1β) knock-down in the hippocampus alleviates lipopolysaccharide (LPS)-induced memory deficits and anxiety- and depression-like behaviors in mice. *J. Neuroinflamm.* **2017**, *14*, 190. [CrossRef]
59. Yang, M.; Dang, R.; Xu, P.; Guo, Y.; Han, W.; Liao, D.; Jiang, P. Dl-3-n-Butylphthalide improves lipopolysaccharide-induced depressive-like behavior in rats: Involvement of Nrf2 and NF-κB pathways. *Psychopharmacology* **2018**, *235*, 2573–2585. [CrossRef]
60. Gao, W.; Wang, W.; Liu, G.; Zhang, J.; Yang, J.; Deng, Z. Allicin attenuated chronic social defeat stress induced depressive-like behaviors through suppression of NLRP3 inflammasome. *Metab. Brain Dis.* **2019**, *34*, 319–329. [CrossRef]
61. Fan, C.; Song, Q.; Wang, P.; Li, Y.; Yang, M.; Yu, S.Y. Neuroprotective effects of ginsenoside-Rg1 against depression-like behaviors via suppressing glial activation, synaptic deficits, and neuronal apoptosis in rats. *Front. Immunol.* **2018**, *9*, 2889. [CrossRef]
62. Chu, C.; Zhang, H.; Cui, S.; Han, B.; Zhou, L.; Zhang, N.; Su, X.; Niu, Y.; Chen, W.; Chen, R.; et al. Ambient PM2.5 caused depressive-like responses through Nrf2/NLRP3 signaling pathway modulating inflammation. *J. Hazard. Mater.* **2019**, *369*, 180–190. [CrossRef]
63. Dang, R.; Guo, Y.-Y.; Zhang, K.; Jiang, P.; Zhao, M.-G. Predictable chronic mild stress promotes recovery from LPS-induced depression. *Mol. Brain* **2019**, *12*, 42. [CrossRef]
64. Gao, W.; Wang, W.; Zhang, J.; Deng, P.; Hu, J.; Yang, J.; Deng, Z. Allicin ameliorates obesity comorbid depressive-like behaviors: Involvement of the oxidative stress, mitochondrial function, autophagy, insulin resistance and NOX/Nrf2 imbalance in mice. *Metab. Brain Dis.* **2019**, *34*, 1267–1280. [CrossRef]
65. Arioz, B.I.; Tastan, B.; Tarakcioglu, E.; Tufekci, K.U.; Olcum, M.; Ersoy, N.; Bagriyanik, A.; Genc, K.; Genc, S. Melatonin attenuates LPS-induced acute depressive-like behaviors and microglial NLRP3 inflammasome activation through the SIRT1/Nrf2 pathway. *Front. Immunol.* **2019**, *10*, 1511. [CrossRef] [PubMed]
66. Cigliano, L.; Spagnuolo, M.S.; Boscaino, F.; Ferrandino, I.; Monaco, A.; Capriello, T.; Cocca, E.; Iannotta, L.; Treppiccione, L.; Luongo, D.; et al. Dietary supplementation with fish oil or conjugated linoleic acid relieves depression markers in mice by modulation of the Nrf2 pathway. *Mol. Nutr. Food Res.* **2019**, *63*, e1900243. [CrossRef] [PubMed]
67. Liu, H.; Jiang, J.; Zhao, L. Protein arginine methyltransferase-1 deficiency restrains depression-like behavior of mice by inhibiting inflammation and oxidative stress via Nrf-2. *Biochem. Biophys. Res. Commun.* **2019**, *518*, 430–437. [CrossRef] [PubMed]
68. Rosa, P.B.; Bettio, L.E.B.; Neis, V.B.; Moretti, M.; Werle, I.; Leal, R.B.; Rodrigues, A.L.S. The antidepressant-like effect of guanosine is dependent on GSK-3β inhibition and activation of MAPK/ERK and Nrf2/heme oxygenase-1 signaling pathways. *Purinergic Signal.* **2019**, *15*, 491–504. [CrossRef] [PubMed]
69. Huang, X.; Fei, G.-Q.; Liu, W.-J.; Ding, J.; Wang, Y.; Wang, H.; Ji, J.-L.; Wang, X. Adipose-derived mesenchymal stem cells protect against CMS-induced depression-like behaviors in mice via regulating the Nrf2/HO-1 and TLR4/NF-κB signaling pathways. *Acta Pharmacol. Sin.* **2020**, *41*, 612–619. [CrossRef]
70. Zborowski, V.A.; Heck, S.O.; Vencato, M.; Pinton, S.; Marques, L.S.; Nogueira, C.W. Keap1/Nrf2/HO-1 signaling pathway contributes to p-chlorodiphenyl diselenide antidepressant-like action in diabetic mice. *Psychopharmacology* **2020**, *237*, 363–374. [CrossRef]
71. Casaril, A.M.; Domingues, M.; Bampi, S.R.; Lourenço, D.A.; Smaniotto, T.Â.; Segatto, N.; Vieira, B.; Seixas, F.K.; Collares, T.; Lenardão, E.J.; et al. The antioxidant and immunomodulatory compound 3-[(4-chlorophenyl)selanyl]-1-methyl-1H-indole attenuates depression-like behavior and cognitive impairment developed in a mouse model of breast tumor. *Brain Behav. Immun.* **2020**, *84*, 229–241. [CrossRef]

72. Tian, L.; Sun, S.S.; Cui, L.B.; Wang, S.Q.; Peng, Z.W.; Tan, Q.R.; Hou, W.G.; Cai, M. Repetitive transcranial magnetic stimulation elicits antidepressant- and anxiolytic-like effect via nuclear factor-E2-related factor 2-mediated anti-inflammation mechanism in rats. *Neuroscience* **2020**, *429*, 119–133. [CrossRef]

73. Li, T.; Zheng, L.N.; Han, X.H. Fenretinide attenuates lipopolysaccharide (LPS)-induced blood-brain barrier (BBB) and depressive-like behavior in mice by targeting Nrf-2 signaling. *Biomed. Pharmacother.* **2020**, *125*, 109680. [CrossRef]

74. Nakayama, T.; Okimura, K.; Shen, J.; Guh, Y.J.; Tamai, T.K.; Shimada, A.; Minou, S.; Okushi, Y.; Shimmura, T.; Furukawa, Y.; et al. Seasonal changes in NRF2 antioxidant pathway regulates winter depression-like behavior. *Proc. Natl. Acad. Sci. USA* **2020**, *117*, 9594–9603. [CrossRef] [PubMed]

75. Ali, T.; Rahman, S.U.; Hao, Q.; Li, W.; Liu, Z.; Ali Shah, F.; Murtaza, I.; Zhang, Z.; Yang, X.; Liu, G.; et al. Melatonin prevents neuroinflammation and relieves depression by attenuating autophagy impairment through FOXO3a regulation. *J. Pineal Res.* **2020**, *69*, e12667. [CrossRef] [PubMed]

76. Wang, W.; Zheng, L.; Xu, L.; Tu, J.; Gu, X. Pinocembrin mitigates depressive-like behaviors induced by chronic unpredictable mild stress through ameliorating neuroinflammation and apoptosis. *Mol. Med.* **2020**, *26*, 53. [CrossRef]

77. Liao, D.; Chen, Y.; Guo, Y.; Wang, C.; Liu, N.; Gong, Q.; Fu, Y.; Fu, Y.; Cao, L.; Yao, D.; et al. Salvianolic acid B improves chronic mild stress-induced depressive behaviors in rats: Involvement of AMPK/SIRT1 signaling pathway. *J. Inflamm. Res.* **2020**, *13*, 195–206. [CrossRef] [PubMed]

78. Severo, L.; Godinho, D.; Machado, F.; Hartmann, D.; Fighera, M.R.; Soares, F.A.; Furian, A.F.; Oliveira, M.S.; Royes, L.F. The role of mitochondrial bioenergetics and oxidative stress in depressive behavior in recurrent concussion model in mice. *Life Sci.* **2020**, *257*, 117991. [CrossRef] [PubMed]

79. Ali, T.; Hao, Q.; Ullah, N.; Rahman, S.U.; Shah, F.A.; He, K.; Zheng, C.; Li, W.; Murtaza, I.; Li, Y.; et al. Melatonin act as an antidepressant via attenuation of neuroinflammation by targeting Sirt1/Nrf2/HO-1 signaling. *Front. Mol. Neurosci.* **2020**, *13*, 96. [CrossRef] [PubMed]

80. Camargo, A.; Dalmagro, A.P.; Rosa, J.M.; Zeni, A.L.B.; Kaster, M.P.; Tasca, C.I.; Rodrigues, A.L.S. Subthreshold doses of guanosine plus ketamine elicit antidepressant-like effect in a mouse model of depression induced by corticosterone: Role of GR/NF-κB/IDO-1 signaling. *Neurochem. Int.* **2020**, *139*, 104797. [CrossRef]

81. Park, B.-K.; Kim, N.-S.; Kim, Y.-R.; Yang, C.; Jung, I.-C.; Jang, I.-S.; Seo, C.-S.; Choi, J.-J.; Lee, M.-Y. Antidepressant and anti-neuroinflammatory effects of bangpungtongsung-san. *Front. Pharmacol.* **2020**, *11*, 958. [CrossRef]

82. Zhu, X.; Liu, H.; Liu, Y.; Chen, Y.; Liu, Y.; Yin, X. The antidepressant-like effects of hesperidin in streptozotocin-induced diabetic rats by activating Nrf2/ARE/Glyoxalase 1 pathway. *Front. Pharmacol.* **2020**, *11*, 1325. [CrossRef] [PubMed]

83. Liao, D.; Lv, C.; Cao, L.; Yao, D.; Wu, Y.; Long, M.; Liu, N.; Jiang, P. Curcumin attenuates chronic unpredictable mild stress-induced depressive-like behaviors via restoring changes in oxidative stress and the activation of Nrf2 signaling pathway in rats. *Oxid. Med. Cell. Longev.* **2020**, *2020*, 9268083. [CrossRef]

84. Qu, Y.; Shan, J.; Wang, S.; Chang, L.; Pu, Y.; Wang, X.; Tan, Y.; Yamamoto, M.; Hashimoto, K. Rapid-acting and long-lasting antidepressant-like action of (R)-ketamine in Nrf2 knock-out mice: A role of TrkB signaling. *Eur. Arch. Psychiatry Clin. Neurosci.* **2021**, *271*, 439–446. [CrossRef] [PubMed]

85. Li, W.; Ali, T.; He, K.; Liu, Z.; Shah, F.A.; Ren, Q.; Liu, Y.; Jiang, A.; Li, S. Ibrutinib alleviates LPS-induced neuroinflammation and synaptic defects in a mouse model of depression. *Brain Behav. Immun.* **2021**, *92*, 10–24. [CrossRef] [PubMed]

86. Yan, T.; Li, F.; Xiong, W.; Wu, B.; Xiao, F.; He, B.; Jia, Y. Nootkatone improves anxiety- and depression-like behavior by targeting hyperammonemia-induced oxidative stress in D-galactosamine model of liver injury. *Environ. Toxicol.* **2021**, *36*, 694–706. [CrossRef]

87. Herbet, M.; Szumełda, I.; Piątkowska-Chmiel, I.; Gawrońska-Grzywacz, M.; Dudka, J. Beneficial effects of combined administration of fluoxetine and mitochondria- targeted antioxidant at in behavioural and molecular studies in mice model of depression. *Behav. Brain Res.* **2021**, *405*, 113185. [CrossRef]

88. Yao, W.; Lin, S.; Su, J.; Cao, Q.; Chen, Y.; Chen, J.; Zhang, Z.; Hashimoto, K.; Qi, Q.; Zhang, J.-C. Activation of BDNF by transcription factor Nrf2 contributes to antidepressant-like actions in rodents. *Transl. Psychiatry* **2021**, *11*, 140. [CrossRef]

89. Salama, A.; Mahmoud, H.A.; Kandeil, M.A.; Khalaf, M.M. Neuroprotective role of camphor against ciprofloxacin induced depression in rats: Modulation of Nrf-2 and TLR4. *Immunopharmacol. Immunotoxicol.* **2021**, *43*, 309–318. [CrossRef]

90. Naß, J.; Abdelfatah, S.; Efferth, T. The triterpenoid ursolic acid ameliorates stress in *Caenorhabditis elegans* by affecting the depression-associated genes *skn-1* and *prdx2*. *Phytomedicine* **2021**, *88*, 153598. [CrossRef]

91. Yang, S.; Yang, Y.; Chen, C.; Wang, H.; Ai, Q.; Lin, M.; Zeng, Q.; Zhang, Y.; Gao, Y.; Li, X.; et al. The anti-neuroinflammatory effect of fuzi and ganjiang extraction on LPS-induced BV2 microglia and its intervention function on depression-like behavior of cancer-related fatigue model mice. *Front. Pharmacol.* **2021**, *12*, 670586. [CrossRef]

92. Zhu, X.; Zhang, Y.-M.; Zhang, M.-Y.; Chen, Y.-J.; Liu, Y.-W. Hesperetin ameliorates diabetes- associated anxiety and depression-like behaviors in rats via activating Nrf2/ARE pathway. *Metab. Brain Dis.* **2021**, *36*, 1969–1983. [CrossRef] [PubMed]

93. Wang, J.; Chen, R.; Liu, C.; Wu, X.; Zhang, Y. Antidepressant mechanism of catalpol: Involvement of the PI3K/Akt/Nrf2/HO-1 signaling pathway in rat hippocampus. *Eur. J. Pharmacol.* **2021**, *909*, 174396. [CrossRef] [PubMed]

94. Wu, X.; Wang, J.; Song, L.; Guan, Y.; Cao, C.; Cui, Y.; Zhang, Y.; Liu, C. Catalpol weakens depressive-like behavior in mice with streptozotocin-induced hyperglycemia via PI3K/AKT/Nrf2/HO-1 signaling pathway. *Neuroscience* **2021**, *473*, 102–118. [CrossRef]

95. Rahman, S.U.; Ali, T.; Hao, Q.; He, K.; Li, W.; Ullah, N.; Zhang, Z.; Jiang, Y.; Li, S. Xanthohumol attenuates lipopolysaccharide-induced depressive like behavior in mice: Involvement of NF-κB/Nrf2 signaling pathways. *Neurochem. Res.* **2021**, *46*, 3135–3148. [CrossRef]

96. Tao, W.; Hu, Y.; Chen, Z.; Dai, Y.; Hu, Y.; Qi, M. Magnolol attenuates depressive-like behaviors by polarizing microglia towards the M2 phenotype through the regulation of Nrf2/HO-1/NLRP3 signaling pathway. *Phytomedicine* **2021**, *91*, 153692. [CrossRef]

97. Huang, X.; Qu, Q.; Ren, D. Nrf2 alleviates cognitive dysfunction and brain inflammatory injury via mediating Wfs1 in rats with depression-like behaviors. *Inflammation* **2022**, *45*, 399–413. [CrossRef]

98. Song, L.; Wu, X.; Wang, J.; Guan, Y.; Zhang, Y.; Gong, M.; Wang, Y.; Li, B. Antidepressant effect of catalpol on corticosterone-induced depressive-like behavior involves the inhibition of HPA axis hyperactivity, central inflammation and oxidative damage probably via dual regulation of NF-κB and Nrf2. *Brain Res. Bull.* **2021**, *177*, 81–91. [CrossRef] [PubMed]

99. Guan, Y.; Wang, J.; Wu, X.; Song, L.; Wang, Y.; Gong, M.; Li, B. Quercetin reverses chronic unpredictable mild stress-induced depression-like behavior in vivo by involving nuclear factor-E2-related factor 2. *Brain Res.* **2021**, *1772*, 147661. [CrossRef]

100. Sun, D.-C.; Wang, R.-R.; Xu, H.; Zhu, X.-H.; Sun, Y.; Qiao, S.-H.; Qiao, W. A network pharmacology-based study on antidepressant effect of *Salicornia europaea* L. extract with experimental support in chronic unpredictable mild stress model mice. *Chin. J. Integr. Med.* **2022**, *28*, 339–348. [CrossRef]

101. Cheng, W.-J.; Li, P.; Huang, W.-Y.; Huang, Y.; Chen, W.-J.; Chen, Y.-P.; Shen, J.-L.; Chen, J.-K.; Long, N.-S.; Meng, X.-J. Acupuncture relieves stress-induced depressive behavior by reducing oxidative stress and neuroapoptosis in rats. *Front. Behav. Neurosci.* **2022**, *15*, 783056. [CrossRef]

102. Dang, R.; Wang, M.; Li, X.; Wang, H.; Liu, L.; Wu, Q.; Zhao, J.; Ji, P.; Zhong, L.; Licinio, J.; et al. Edaravone ameliorates depressive and anxiety-like behaviors via Sirt1/Nrf2/HO-1/Gpx4 pathway. *J. Neuroinflamm.* **2022**, *19*, 41. [CrossRef] [PubMed]

103. Xia, B.; Liu, X.; Li, X.; Wang, Y.; Wang, D.; Kou, R.; Zhang, L.; Shi, R.; Ye, J.; Bo, X.; et al. Sesamol ameliorates dextran sulfate sodium-induced depression- like and anxiety-like behaviors in colitis mice: The potential involvement of the gut-brain axis. *Food Funct.* **2022**, *13*, 2865–2883. [CrossRef] [PubMed]

104. Li, J.; Gao, W.; Zhao, Z.; Li, Y.; Yang, L.; Wei, W.; Ren, F.; Li, Y.; Yu, Y.; Duan, W.; et al. Ginsenoside Rg1 reduced microglial activation and mitochondrial dysfunction to alleviate depression-like behaviour via the GAS5/EZH2/SOCS3/NRF2 axis. *Mol. Neurobiol.* **2022**, *59*, 2855–2873. [CrossRef] [PubMed]

105. Muhammad, A.J.; Hao, L.; Al Kury, L.T.; Rehman, N.U.; Alvi, A.M.; Badshah, H.; Ullah, I.; Shah, F.A.; Li, S. Carveol promotes Nrf2 contribution in depressive disorders through an anti-inflammatory mechanism. *Oxid. Med. Cell. Longev.* **2022**, *2022*, 4509204. [CrossRef] [PubMed]

106. Shen, F.; Xie, P.; Li, C.; Bian, Z.; Wang, X.; Peng, D.; Zhu, G. Polysaccharides from *Polygonatum cyrtonema* Hua reduce depression-like behavior in mice by inhibiting oxidative stress-calpain-1-NLRP3 signaling axis. *Oxid. Med. Cell. Longev.* **2022**, *2022*, 2566917. [CrossRef]

107. Jiang, N.; Zhang, Y.; Yao, C.; Huang, H.; Wang, Q.; Huang, S.; He, Q.; Liu, X. Ginsenosides Rb1 attenuates chronic social defeat stress-induced depressive behavior via regulation of SIRT1-NLRP3/Nrf2 pathways. *Front. Nutr.* **2022**, *9*, 868833. [CrossRef]

108. de Souza, A.G.; Lopes, I.S.; Filho, A.J.M.C.; Cavalcante, T.M.B.; Oliveira, J.V.S.; de Carvalho, M.A.J.; de Lima, K.A.; Jucá, P.M.; Mendonça, S.S.; Mottin, M.; et al. Neuroprotective effects of dimethyl fumarate against depression-like behaviors via astrocytes and microglia modulation in mice: Possible involvement of the HCAR2/Nrf2 signaling pathway. *Naunyn Schmiedebergs Arch. Pharmacol.* **2022**, *395*, 1029–1045. [CrossRef]

109. Fasakin, O.W.; Oboh, G.; Ademosun, A.O.; Lawal, A.O. The modulatory effects of alkaloid extracts of *Cannabis sativa, Datura stramonium, Nicotiana tabacum* and male *Carica papaya* on neurotransmitter, neurotrophic and neuroinflammatory systems linked to anxiety and depression. *Inflammopharmacology* **2022**, *30*, 2447–2476. [CrossRef]

110. Tan, L.; Yang, Y.; Peng, J.; Zhang, Y.; Wu, B.; He, B.; Jia, Y.; Yan, T. *Schisandra chinensis* (Turcz.) Baill. essential oil exhibits antidepressant-like effects and against brain oxidative stress through Nrf2/HO-1 pathway activation. *Metab. Brain Dis.* **2022**, *37*, 2261–2275. [CrossRef]

111. Wu, X.; Liu, C.; Wang, J.; Guan, Y.; Song, L.; Chen, R.; Gong, M. Catalpol exerts antidepressant-like effects by enhancing anti-oxidation and neurotrophy and inhibiting neuroinflammation via activation of HO-1. *Neurochem. Res.* **2022**, *47*, 2975–2991. [CrossRef]

112. Zhao, Y.-T.; Yin, H.; Hu, C.; Zeng, J.; Zhang, S.; Chen, S.; Zheng, W.; Li, M.; Jin, L.; Liu, Y.; et al. Tilapia skin peptides ameliorate cyclophosphamide-induced anxiety- and depression-like behavior via improving oxidative stress, neuroinflammation, neuron apoptosis, and neurogenesis in mice. *Front. Nutr.* **2022**, *9*, 882175. [CrossRef]

113. Bansal, Y.; Singh, R.; Sodhi, R.K.; Khare, P.; Dhingra, R.; Dhingra, N.; Bishnoi, M.; Kondepudi, K.K.; Kuhad, A. Kynurenine monooxygenase inhibition and associated reduced quinolinic acid reverses depression-like behaviour by upregulating Nrf2/ARE pathway in mouse model of depression: *In-vivo* and *In-silico* studies. *Neuropharmacology* **2022**, *215*, 109169. [CrossRef] [PubMed]

114. Yu, Y.; Li, Y.; Qi, K.; Xu, W.; Wei, Y. Rosmarinic acid relieves LPS-induced sickness and depressive-like behaviors in mice by activating the BDNF/Nrf2 signaling and autophagy pathway. *Behav. Brain Res.* **2022**, *433*, 114006. [CrossRef] [PubMed]

115. Kang, J.-Y.; Baek, D.-C.; Son, C.-G.; Lee, J.-S. Succinum extracts inhibit microglial- derived neuroinflammation and depressive-like behaviors. *Front. Pharmacol.* **2022**, *13*, 991243. [CrossRef] [PubMed]

116. Wang, X.; Xiao, A.; Yang, Y.; Zhao, Y.; Wang, C.C.; Wang, Y.; Han, J.; Wang, Z.; Wen, M. DHA and EPA prevent seizure and depression-like behavior by inhibiting ferroptosis and neuroinflammation via different mode-of-actions in a pentylenetetrazole-induced kindling model in mice. *Mol. Nutr. Food Res.* **2022**, *66*, e2200275. [CrossRef]

117. Nasehi, L.; Morassaci, B.; Ghaffari, M.; Sharafi, A.; Dehpour, A.R.; Hosseini, M.-J. The impacts of vorinostat on NADPH oxidase and mitochondrial biogenesis gene expression in the heart of mice model of depression. *Can. J. Physiol. Pharmacol.* **2022**, *100*, 1077–1085. [CrossRef]

118. Pei, H.; Zeng, J.; He, Z.; Zong, Y.; Zhao, Y.; Li, J.; Chen, W.; Du, R. Palmatine ameliorates LPS-induced HT-22 cells and mouse models of depression by regulating apoptosis and oxidative stress. *J. Biochem. Mol. Toxicol.* **2022**, *37*, e23225. [CrossRef]

119. Sun, J.Y.; Liu, Y.T.; Jiang, S.N.; Guo, P.M.; Wu, X.Y.; Yu, J. Essential oil from the roots of *Paeonia lactiflora pall.* has protective effect against corticosterone-induced depression in mice via modulation of PI3K/Akt signaling pathway. *Front. Pharmacol.* **2022**, *13*, 999712. [CrossRef]

120. He, L.; Zheng, Y.; Huang, L.; Ye, J.; Ye, Y.; Luo, H.; Chen, X.; Yao, W.; Chen, J.; Zhang, J.C. Nrf2 regulates the arginase 1+ microglia phenotype through the initiation of TREM2 transcription, ameliorating depression-like behavior in mice. *Transl. Psychiatry* **2022**, *12*, 459. [CrossRef]

121. Samy, D.M.; Mostafa, D.K.; Saleh, S.R.; Hassaan, P.S.; Zeitoun, T.M.; Ammar, G.A.G.; Elsokkary, N.H. Carnosic acid mitigates depression-like behavior in ovariectomized mice via activation of Nrf2/HO-1 pathway. *Mol. Neurobiol.* **2022**, *60*, 610–628. [CrossRef]

122. Si, L.; Xiao, L.; Xie, Y.; Xu, H.; Yuan, G.; Xu, W.; Wang, G. Social isolation after chronic unpredictable mild stress perpetuates depressive-like behaviors, memory deficits and social withdrawal via inhibiting ERK/KEAP1/NRF2 signaling. *J. Affect. Disord.* **2023**, *324*, 576–588. [CrossRef] [PubMed]

123. Wang, J.; Men, Y.; Wang, Z. Polydatin alleviates chronic stress-induced depressive and anxiety-like behaviors in a mouse model. *ACS Chem. Neurosci.* **2023**, *14*, 977–987. [CrossRef] [PubMed]

124. Smaniotto, T.Â.; Casaril, A.M.; de Andrade Lourenço, D.; Sousa, F.S.; Seixas, F.K.; Collares, T.; Woloski, R.; da Silva Pinto, L.; Alves, D.; Savegnago, L. Intranasal administration of interleukin-4 ameliorates depression-like behavior and biochemical alterations in mouse submitted to the chronic unpredictable mild stress: Modulation of neuroinflammation and oxidative stress. *Psychopharmacology* **2023**, *240*, 935–950. [CrossRef] [PubMed]

125. Zhang, J.; Li, L.; Liu, Q.; Zhao, Z.; Su, D.; Xiao, C.; Jin, T.; Chen, L.; Xu, C.; You, Z.; et al. Gastrodin programs an Arg-1$^+$ microglial phenotype in hippocampus to ameliorate depression- and anxiety-like behaviors via the Nrf2 pathway in mice. *Phytomedicine* **2023**, *113*, 154725. [CrossRef] [PubMed]

126. Moi, P.; Chan, K.; Asunis, I.; Cao, A.; Kan, Y.W. Isolation of NF-E2-related factor 2 (Nrf2), a NF-E2-like basic leucine zipper transcriptional activator that binds to the tandem NF-E2/AP1 repeat of the beta-globin locus control region. *Proc. Natl. Acad. Sci. USA* **1994**, *91*, 9926–9930. [CrossRef]

127. Casarotto, P.C.; Girych, M.; Fred, S.M.; Kovaleva, V.; Moliner, R.; Enkavi, G.; Biojone, C.; Cannarozzo, C.; Sahu, M.P.; Kaurinkoski, K.; et al. Antidepressant drugs act by directly binding to TRKB neurotrophin receptors. *Cell* **2021**, *184*, 1299–1313.e19. [CrossRef]

128. Casarotto, P.; Umemori, J.; Castrén, E. BDNF receptor TrkB as the mediator of the antidepressant drug action. *Front. Mol. Neurosci.* **2022**, *15*, 1032224. [CrossRef]

129. Meffre, D.; Massaad, C.; Grenier, J. Lithium chloride stimulates PLP and MBP expression in oligodendrocytes via Wnt/β-catenin and Akt/CREB pathways. *Neuroscience* **2015**, *284*, 962–971. [CrossRef]

130. Mehrafza, S.; Kermanshahi, S.; Mostafidi, S.; Motaghinejad, M.; Motevalian, M.; Fatima, S. Pharmacological evidence for lithium-induced neuroprotection against methamphetamine-induced neurodegeneration via Akt-1/GSK3 and CREB-BDNF signaling pathways. *Iran. J. Basic Med. Sci.* **2019**, *22*, 856–865. [CrossRef]

131. Sani, G.; Perugi, G.; Tondo, L. Treatment of Bipolar Disorder in a Lifetime Perspective: Is Lithium Still the Best Choice? *Clin. Drug Investig.* **2017**, *37*, 713–727. [CrossRef]

132. Rahimi, H.R.; Ghahremani, M.H.; Dehpour, A.R.; Sharifzadeh, M.; Ejtemaei-Mehr, S.; Razmi, A.; Ostad, S.N. The neuroprotective effect of lithium in cannabinoid dependence is mediated through modulation of cyclic AMP, ERK1/2 and GSK-3β phosphorylation in cerebellar granular neurons of rat. *Iran. J. Pharm. Res.* **2015**, *14*, 1123–1135. [PubMed]

133. Li, B.; Ren, J.; Yang, L.; Li, X.; Sun, G.; Xia, M. Lithium Inhibits GSK3β Activity via Two Different Signaling Pathways in Neurons After Spinal Cord Injury. *Neurochem. Res.* **2018**, *43*, 848–856. [CrossRef] [PubMed]

134. Monaco, S.A.; Ferguson, B.R.; Gao, W.J. Lithium inhibits GSK3β and augments GluN2A receptor expression in the prefrontal cortex. *Front. Cell. Neurosci.* **2018**, *12*, 16. [CrossRef] [PubMed]

135. De Nicola, G.M.; Karreth, F.A.; Humpton, T.J.; Gopinathan, A.; Wei, C.; Frese, K.; Mangal, D.; Yu, K.H.; Yeo, C.J.; Calhoun, E.S.; et al. Oncogene-induced Nrf2 transcription promotes ROS detoxification and tumorigenesis. *Nature* **2011**, *475*, 106–109. [CrossRef]

136. Cirone, M.; D'Orazi, G. NRF2 in cancer: Cross-talk with oncogenic pathways and involvement in gammaherpesvirus-driven carcinogenesis. *Int. J. Mol. Sci.* **2022**, *24*, 595. [CrossRef] [PubMed]

137. Barajas, B.; Che, N.; Yin, F.; Rowshanrad, A.; Orozco, L.D.; Gong, K.W.; Wang, X.; Castellani, L.W.; Reue, K.; Lusis, A.J.; et al. NF-E2-related factor 2 promotes atherosclerosis by effects on plasma lipoproteins and cholesterol transport that overshadow antioxidant protection. *Arterioscler. Thromb. Vasc. Biol.* **2011**, *31*, 58–66. [CrossRef]

138. Wu, W.; Hendrix, A.; Nair, S.; Cui, T. Nrf2-mediated dichotomy in the vascular system: Mechanistic and therapeutic perspective. *Cells* **2022**, *11*, 3042. [CrossRef]

 *antioxidants*

MDPI

Article

Rutin Protects Fibroblasts from UVA Radiation through Stimulation of Nrf2 Pathway

Elisabetta Tabolacci [1,2,†], Giuseppe Tringali [2,3,*,†], Veronica Nobile [1,2], Sara Duca [4], Michela Pizzoferrato [2,3], Patrizia Bottoni [5] and Maria Elisabetta Clementi [6,*]

1 Dipartimento di Sanità Pubblica e Scienze della Vita, Sezione di Medicina Genomica, Università Cattolica del Sacro Cuore, Largo F. Vito 1, 00168 Rome, Italy; elisabetta.tabolacci@unicatt.it (E.T.)
2 Fondazione Policlinico Universitario A. Gemelli IRCSS, 00168 Rome, Italy
3 Pharmacology Section, Department of Health Care Surveillance and Bioethics, Università Cattolica del Sacro Cuore, Largo F. Vito 1, 00168 Rome, Italy
4 UOC Genetica Medica, Fondazione Policlinico Universitario A. Gemelli IRCCS, 00168 Rome, Italy
5 Dipartimento di Scienze Biotecnologiche di Base, Cliniche Intensivologiche e Perioperatorie, Università Cattolica del Sacro Cuore, Largo F. Vito 1, 00168 Rome, Italy
6 Istituto di Scienze e Tecnologie Chimiche "Giulio Natta" SCITEC-CNR, Largo Francesco Vito 1, 00168 Rome, Italy
* Correspondence: giuseppe.tringali@unicatt.it (G.T.); elisabetta.clementi@scitec.cnr.it (M.E.C.); Tel.: +39-06-30154215 (M.E.C.)
† These authors contributed equally to this work.

Abstract: This study explores the photoprotective effects of rutin, a bioflavonoid found in some vegetables and fruits, against UVA-induced damage in human skin fibroblasts. Our results show that rutin increases cell viability and reduces the high levels of ROS generated by photo-oxidative stress (1 and 2 h of UVA exposure). These effects are related to rutin's ability to modulate the Nrf2 transcriptional pathway. Interestingly, activation of the Nrf2 signaling pathway results in an increase in reduced glutathione and Bcl2/Bax ratio, and the subsequent protection of mitochondrial respiratory capacity. These results demonstrate how rutin may play a potentially cytoprotective role against UVA-induced skin damage through a purely antiapoptotic mechanism.

Keywords: rutin; UVA-induced damage; Nrf2 pathway; oxidative stress; mitochondrial function

Citation: Tabolacci, E.; Tringali, G.; Nobile, V.; Duca, S.; Pizzoferrato, M.; Bottoni, P.; Clementi, M.E. Rutin Protects Fibroblasts from UVA Radiation through Stimulation of Nrf2 Pathway. *Antioxidants* 2023, 12, 820. https://doi.org/10.3390/antiox12040820

Academic Editor: Marcel Bonay

Received: 11 March 2023
Revised: 21 March 2023
Accepted: 24 March 2023
Published: 28 March 2023

1. Introduction

The risks and damages induced by excessive exposure to UV radiation, especially the solar one, have been the object of careful evaluations and studies, especially in recent decades. Prolonged exposure to solar UV radiation can cause acute and chronic harmful effects to the skin, eyes, connective tissue, blood vessels, and, in the most serious cases, skin tumors. In this regard, among the chronic effects of major health relevance, there are some neoplastic forms such as melanoma, basal cell carcinoma (BCC), and squamous cell carcinoma (SCC). These types of cancers often appear on areas of the skin that are not adequately protected from solar radiation [1–4].

There are different types of ultraviolet radiation, which have different effects depending on the wavelength: the longer the wavelength is, the greater is the ability of the rays to penetrate the atmosphere and reach the deep layers of the skin tissue [5]. In this context, UVA rays (315–400 nm), characterized by a constant intensity throughout the year, represent around 95% of the UV radiation capable of reaching dermal fibroblasts, also causing long-term damage [6,7]. Exposure to UVA rays leads to (i) an increased oxidative stress in fibroblasts, (ii) a decreased secretion and synthesis of matrix proteins, (iii) cellular dysfunction, and, finally, (iv) apoptosis: these processes first promote skin aging and later promote the predisposition to skin diseases [7–11]. The reactive oxygen species (ROS) produced (after UVA exposure) interact with lipid-rich membranes, enzymes, and nuclear

DNA by changing their structures and interfering with their functions. In addition, the accumulation of ROS in cells leads to the initiation of cellular reactions with dysfunctions of mitochondria activities. Therefore, ROS accumulation in skin cells represents the main cause of photodamage, photoaging, and photocarcinogenesis [12–15]. At the molecular level, it is also known that UVA-generated ROS lead to a decreased expression of erythroid nuclear factor 2-like 2 (Nrf2), a marker of oxidative stress [16–18]. Nrf2 is a key regulator in the protection of skin cells from oxidative stress and UVA radiation [19]. In fact, increasing intracellular Nrf2 leads to the maintenance of the cell's reducing potential, the decrease in apoptotic genes, and the maintenance of mitochondrial integrity resulting in improved cell viability [20,21].

The use of natural substances that can protect cells from oxidative stress has now become the subject of numerous studies, either in the form of nutraceuticals or used topically as protectants. In our study, we planned to analyze the effect of rutin, a flavonoid naturally occurring in many foods, especially buckwheat, apricots, cherries, grapefruit, plums, and oranges, and it was found in high amounts in the waste from wine production from grapes. It is often used in patients with capillary fragility, varicose veins, and hematomas, but recent papers have shown an important antioxidant effect [22–24]. Therefore, our experimental design aimed at studying the effect of rutin on fibroblast cultures exposed to UVA radiation, focusing our preliminary/exploratory study on oxidative stress and its mechanisms.

2. Materials and Methods

2.1. Cells and Treatments

Fibroblasts derived from a healthy control male (CTRL, coded CTRL1) of 31 years old were employed. A fibroblast cell culture was established after around 1 week from the date of biopsy. Skin biopsy was performed at the left leg (around 10 cm from the ankle) and was obtained after a signed Informed Consent (prot. N. 9917/15 and prot.cm 10/15 of Ethics Committee at the Catholic University of Rome). The cell culture was grown in DMEM medium (Sigma Aldrich, St. Louis, MO, USA), supplemented with 10% fetal bovine serum (FBS), 1% penicillin/streptomycin, and 2.5% L-glutamine at 37 °C with 5% CO_2. For the subsequent experiments, cells were seeded with 80% of confluency. UVA exposure was produced using a lamp (Vilber Lourmat VL-62C Power 6W; Vilber Lourmat Deutschland GmbH, Eberhardzell, Germany) at 365 nm, and was placed 10 cm from the source for 1, 2, 3, 4, and 5 h at an intensity of ~0.06 J/cm^2/s. To minimize radiation uptake by the medium, the cells were kept in PBS during exposure, and immediately after exposure, the culture medium was replaced and the cells were put in an incubator for 24 h before proceeding to the different assays. Control cells (unirradiated) were maintained for the same period under the same experimental condition.

Considering the results obtained from the cell viability curve, we chose to continue the subsequent experiments by exposing the cells to UVA radiation for 1 and 2 h.

Rutin (purchased from Sigma) was dissolved in DMSO to a final concentration of 10 mM and then, before use, the solution was diluted in PBS to the desired concentrations. Rutin was added to the cells 24 h before UVA exposure. To determine a dose–response curve, we evaluated the cell viability after UVA exposure for 1 and 2 h, using rutin at 5, 10, 20, and 25 μM. Based on the data obtained, we chose 10 μM as the optimal concentration.

All experiments were also conducted in the presence of DMSO alone at the highest concentration used to dissolve 25 μM rutin. No difference was found compared with control cells.

2.2. Cell Viability

Cell viability was assessed with the MTS assay (Promega srl, Padua, Italy), according to the manufacturer's instructions. Briefly, after UVA exposure and various experimental treatments, MTS reagent was added to cells and plated in 96-well plates at a cell density of 2×10^4 cells/well, reaching around 80% of confluency. The assay provides a sensitive measurement of the normal metabolic state of cells, reflecting early changes in cellular redox homeostasis. Intracellular soluble formazan produced by the cell reduction of MTS is proportional to the number of live cells, and it was measured by recording the absorbance of each well of the plate with a plate reader at 490 nm. The cellular viability was expressed as % compared with control cells.

2.3. ROS Measurement

An intracellular ROS detection kit containing $2',7'$-dichlorofluorescein diacetate (DCF-DA) was used to measure the formation of intracellular ROS (Abcam, Cambridge, UK). Briefly, DCF-DA was applied in accordance with the manufacturer's instructions to fibroblasts grown in 96-well microplates (20,000 cells per well) and subjected to various experimental settings. When ROS are present, DCF-DA, an originally non-fluorescent molecule, is oxidized to DCF, a highly fluorescent chemical. A CytoFluor multiwell plate reader (Victor3-Wallac-1420; PerkinElmer, Waltham, MA, USA) was used to measure the fluorescence at 485/538 nm of excitation/emission. The amount of ROS produced was proportional to the emitted fluorescence and measured by fluorescence intensity.

2.4. Nrf2 Detection Assay

A cell-based colorimetric ELISA kit was used to measure the intracellular levels of Nrf2 (LSBio, LifeSpan Biosciences; Seattle, WA, USA). In a 96-well plate, fibroblasts were seeded at a density of 20,000 cells per well. Following the UVA exposure and treatment, cells were fixed with 4% formaldehyde. Finally, the plate was incubated at 4 °C overnight with the addition of quenching solution, blocking buffer, and the primary antibodies (against-Nrf2—a rabbit polyclonal antibody; against-GAPDH—a mouse monoclonal antibody, employed as an internal positive control and used for normalization). The samples were examined with a microplate reader at a wavelength of 450 nm after the addition of the peroxidase-conjugated secondary antibody. The colorimetric results were provided as a percentage compared with the untreated control and normalized as the OD450 of the Nrf2/GAPDH ratio.

2.5. Total and Reduced Glutathione Assay

Total (GSH + GSSG) and reduced glutathione (GSH) levels were assessed using a colorimetric Assay Kit (ab239709, Abcam, Cambridge, UK). Briefly, fibroblasts seeded in Petri dishes at 80% confluence were pre-treated with 10 µM rutin for 24 h and then exposed to UVA irradiation for 1 and 2 h. After irradiation, cells were lysed by adding the lysis buffer provided in the kit and centrifuged at $14,000 \times g$ for 10 min. Total glutathione and reduced fraction were measured in the supernatants after normalization for mg/protein (see below). The glutathione assay is based on the reaction between DTNB (glutathione substrate) and glutathione that generates 2-nitro-5-thiobenzoic acid, which has a yellow color, and therefore, the concentration of GSH is determined by measuring the absorbance at 412 nm. To detect only the reduced form of glutathione, glutathione reductase is omitted from the assay. Standard curves (substrates provided by the kit) were used to calculate the absolute value of total glutathione (ng/mL) and glutathione in reduced form (µg/mL).

Protein concentration was determined with a Protein Assay (Bio-Rad, Hercules, CA, USA) in 96-well microplates using a calibration curve for BSA.

2.6. Measurement of Iron Level

Total iron levels in the cytoplasm of fibroblasts treated according to the experimental design previously described were assessed with the Iron Assay kit (Abcam, ab83366), according to the manufacturer's instructions. Briefly, 2×10^6 cells were lysed in 250 µL of assay buffer provided by the kit, and 50 µL of lysates, at the same protein concentration, was placed in 96 wells plate. Calibration standard curves were placed in the same plate. For the determination of total iron (II + III), an iron reducer was added to the wells so that all ferric iron was reduced to ferrous. The plate was incubated at 37 °C for 30 min, then the Iron Probe was added and incubated for an additional 60 min. Free iron (II) reacts with the Iron Probe to produce a stable-colored complex, and it was evaluated immediately on a colorimetric microplate reader (OD = 593 nm). The iron(II) and total iron content of the test samples can be directly determined from the standard curve (standard provided by the kit). Iron(III) was calculated as total iron-iron(II). Values were presented as nmoles present in the 50 µL of the cellular lysates.

2.7. Bax and Bcl2 Detection Assay

Bax and Bcl2 proteins were used to assess the degree of apoptosis of fibroblasts exposed to UVA radiation for 1 and 2 h with and without pretreatment with 10 µM rutin. Intracellular concentrations of Bax and Bcl2 proteins were determined using a colorimetric cell ELISA kit from Assay Biotechnology (Sunnyvale, CA, USA). Cells were seeded in a 96-well plate at a density of 20,000 cells/well and treated as previously reported. Twenty-four hours after treatment, primary antibodies (rabbit polyclonal against Bax, rabbit polyclonal against Bcl2, and mouse monoclonal against GADPH, the latter used as housekeeping) were added to the cells, followed by peroxidase-coupled secondary antibodies.

A microplate reader was used to measure the samples at 450 nm, and the ODs of Bax and Bcl2 were normalized for housekeeping. To establish an arbitrary anti-apoptotic index, we expressed the ratio of OD of Bcl2 (anti-apoptotic protein) vs. OD of Bax (apoptotic protein).

2.8. High-Resolution Respirometry (HRR)

Respiration in intact fibroblasts was monitored with high-resolution respirometry (Oroboros Oxygraph-2k, Innsbruck, Austria) operating at 37 °C with a 2 mL chamber volume [25]. Cellular respiration experiments were carried out in two O2k chambers operated in parallel after calibration of the oxygen sensors at air saturation and an instrumental background correction. Calibration with air-saturated medium was performed immediately before the oxygen flux measurement was taken. The data acquisition and analysis were carried out using DatLab software (Oroboros Instruments). Fibroblasts were seeded in a Petri dish with 80% of confluency. After seeding, cells were pre-treated with 10 µM of rutin for 24 h and then were exposed to UVA for 1 and 2 h. After irradiation, fibroblasts were trypsinized, counted, resuspended in supplemented DMEM medium to a final concentration of 1×10^6 cells/mL, added to each Oxygraph Chamber (Chamber A and Chamber B), and thereafter investigated using a phosphorylation control protocol [26]. The experiments began with the measurement of the basal oxygen consumption rate (OCR), followed for about 10 min, until a steady state level was obtained (Basal OCR). ATP synthase was inhibited by the addition of oligomycin (2 µg/mL) at each chamber to detect the OCR from Proton Leak. The maximal respiration capacity (Maximal OCR) was obtained by the addition of small volumes of the uncoupler carbonyl cyanide-4-(trifluoromethoxy) phenylhydrazone (FCCP, 0.25 µM FCCP/step) and by the instantaneous observation of its effect on cellular respiration in the uncoupled state. Cell respiration was then measured in the presence of 0.5 µM rotenone, which selectively inhibits Complex I, and then in the presence of 2.5 µM antimycin A, which inhibits Complex III, to estimate Residual OCR. In addition to instrumental background, the mitochondrial respiration was corrected for the oxygen flux due to residual OCR.

2.9. Caspase-3 Activity

The supernatants obtained, as reported for the GSH assay, were also used to evaluate the caspase-3 activity, quantified using a specific colorimetric kit (Caspase-3 Assay Kit Abcam ab39401). Briefly, cell lysates, at the same protein concentration, were mixed with an equal volume of reaction buffer containing 10 mM DTT and 200 µM DEVD-p-NA substrate. The mixture was incubated at 37 °C for 2 h. Optical density was measured at 400 nm with a microplate reader. The number-fold increase in caspase-3 activity was determined by comparing the results of the samples (irradiated and/or treated cells) with the level of the untreated and unirradiated control. Data were expressed as the percentage of activity compared with the control cells.

2.10. Statistical Analysis

Each experiment was replicated at least three times with up to eight replicates per group. Results were displayed as means ± SEM. Data were analyzed by one-way ANOVA with the Newman–Keuls post hoc test by using PrismTM software (GraphPad, San Diego, CA, USA). Statistical analysis for HHR was performed using the Newman–Keuls Multiple Comparison Test. The level of significance was set at $p \leq 0.05$.

3. Results

We first tested the response of fibroblasts exposed to different times of UVA irradiation by evaluating their viability. After a time titration curve (1, 2, 3, 4, and 5 h) of UVA exposure, fibroblasts showed a significant loss of viability of 35 and 50% after 1 and 2 h, respectively (Figure 1), compared with control cells (blue line), which were not treated with UVA, but kept under a hood outside the incubator. From the 3rd hour onward, cell death was also observed in control cells and therefore not attributable exclusively to the action of UVA. In light of these results, we chose 1 and 2 h stay under the lamp as the optimal time of UVA exposure.

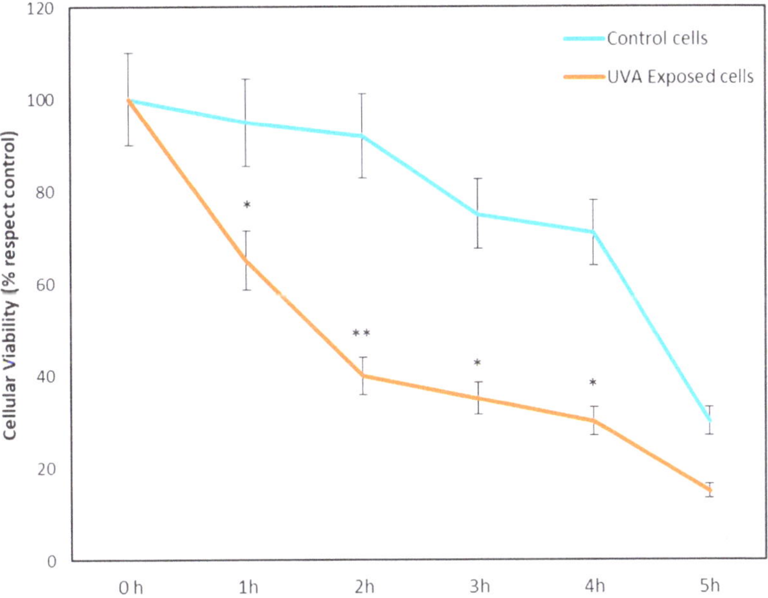

Figure 1. Cell viability after 1, 2, 3, 4, and 5 h of UVA exposure compared with control. Cell viability is expressed as % relative to control (unexposed cells, blue line), which was held for the same time under the same experimental conditions (see details in Section 2). Compared with control fibroblasts, UVA exposure (orange line) produced a significant reduction in cell viability as early as 1 and 2 h. * $p < 0.05$ and ** $p < 0.01$ exposed vs. unexposed.

After evaluating the UVA exposure conditions, we sought to identify the optimal concentration of rutin, the molecule we chose as a potential protective agent. We then pretreated the fibroblasts with different concentrations of rutin (5, 10, 15, and 25 μM) and, after 24 h, the cells were subjected to UVA irradiation for 1 and 2 h. As shown in Figure 2, which reports cell viability, pretreatment with rutin had no effect on cells not exposed to UVA (blue bars), while it protected cells from UVA-dependent cytotoxicity. This result was evident even at low concentrations (5 µM): after exposure, there were 15% and 32% increases in viability at 1 h and 2 h, respectively, compared with the control (UVA-exposed cells without rutin). The result was even more pronounced at 10 µM, when the increase in cell viability after 1 h of exposure was 20% and 45% compared with controls. These data demonstrated the protective effect of rutin on UVA-exposed cells, and in order to identify the molecular mechanisms underlying this phenomenon, we chose a concentration of 10 µM for subsequent experiments.

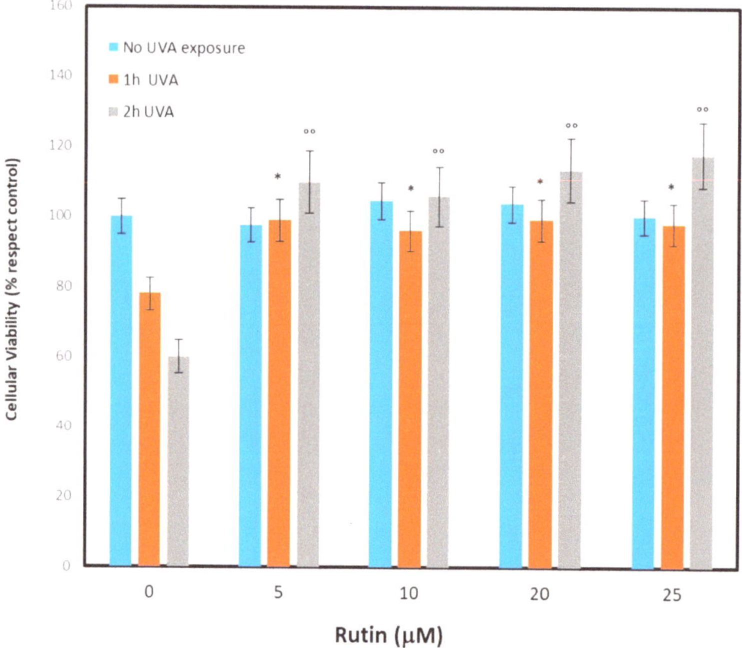

Figure 2. Cell viability in fibroblasts unexposed and exposed to UVA for 1 and 2 h before and after pre-treatment with different concentrations of rutin. Cell viability is expressed as % vs. control (unexposed and untreated cells). Different concentrations of rutin were employed as follows: 0 (absence of pre-treatment), 5, 10, 20, and 25 µM of rutin. Statistical analysis was performed by comparing data under the same experimental conditions (0, 1, and 2 h of UVA exposure). * $p < 0.05$ for 1 h exposure; °° $p < 0.01$ for 2 h exposure.

To investigate the molecular mechanisms underlying UVA damage and the protective effect of rutin, we measured the presence of ROS after administration of 10 µM rutin and the subsequent exposure to UVA radiation for 1 and 2 h, respectively (Figure 3). ROS were measured as fluorescence intensity and expressed as a percentage compared with unirradiated, non-rutin-treated cells (arbitrarily set as 100%). As can be seen, in cells without rutin, the increase in ROS was 20% after 1 h of exposure and tripled after 2 h. In contrast, in the rutin-pretreated cells, the oxidation levels were similar to those of the unexposed cells, showing a significant reduction in ROS compared with the untreated cells.

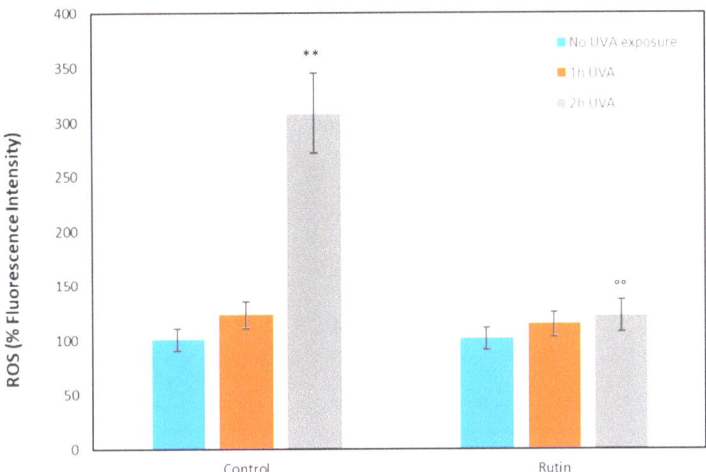

Figure 3. Reactive oxygen species (ROS) for fibroblasts unexposed (blue bars) and exposed to UVA for 1 h (orange bars) and 2 h (gray bars) in the absence (control) and pre-treated with 10 μM of rutin. ROS are expressed as percentage of fluorescence intensity compared with the unexposed and untreated control, arbitrarily considered 100%. Statistical analysis was performed by comparing data between unexposed vs. exposed and between untreated vs. treated fibroblasts. ** $p < 0.01$ unexposed vs. exposed; °° $p < 0.01$ untreated vs. treated.

As it is known that increased ROS within cells lead to an immediate increase in Nrf2, which seeks to counteract cellular oxidation, we measured this protein in fibroblasts treated according to the previous experimental protocol. In Figure 4, we report Nrf2 values in cells unexposed and in cells exposed for 1 and 2 h to UVA irradiation, with and without pre-treatment with 10 μM of rutin. Even in cells not exposed to radiation, rutin evidently resulted in increased Nrf2 levels (reaching threefold values compared with control) and, after UVA exposure, a 3.5- and 4-fold increase at 1 and 2 h of radiation exposure, respectively. Note that in fibroblasts not pre-treated with rutin, the Nrf2 level did not significantly change after 1 h of exposure, whereas a slight decrease occurred after 2 h of UVA irradiation without reaching statistical significance.

As Nrf2 is known to induce an increase in the antioxidant defenses of cells through several mechanisms among which the increase in the antioxidant potential of cells emerges, the next step was to assess the glutathione in the cytosol of treated fibroblasts according to our experimental design. Table 1 shows the values of total glutathione expressed in ng/μL and reduced glutathione in μg/μL. As can be seen, there were no significant changes in total glutathione in the various samples, demonstrating that both UVA and rutin treatment did not change glutathione expression. In contrast, the reduced form of glutathione (GSH) was significantly decreased following UVA treatment and significantly increased in cells pre-treated with rutin. These data indicate that fibroblasts, in order to counteract the UVA-induced increase in ROS, consumed the cellular antioxidant stores resulting in GSH depletion, which decreased from 2.30 (μg/μL) to 1.36 and 1.07 after 1 and 2 h of irradiation, respectively. In contrast, pre-treatment with rutin doubled the amount of GSH before radiation exposure (which increased from 2.20 μg/μL to 5.20 μg/μL) and, following UVA rays, kept the antioxidant potential of the cells high by counteracting ROS formation, as seen.

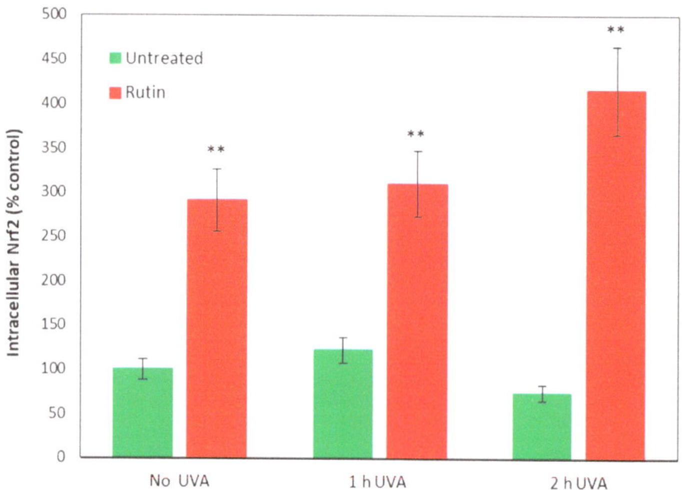

Figure 4. Nrf2 dosage after 1 and 2 h of UVA irradiation with and without 10 μM of rutin. Results are presented as percentage of protein compared with untreated and unexposed control (arbitrarily set as 100%). Rutin induced an increase in Nrf2 levels in unexposed and exposed fibroblasts at both 1 and 2 h of UVA irradiation. ** $p < 0.01$ untreated vs. treated.

Table 1. Total Glutathione (expressed as ng/μL) and Reduced Glutathione (expressed as μg/μL) in fibroblasts in different experimental conditions. The values were calculated as reported in the Section 2. * $p < 0.05$ unirradiated vs. UVA-treated; °° $p < 0.01$ untreated vs. rutin.

	Total Glutathione (ng/μL)		Reduced Glutathione (μg/μL)	
	Untreated	*Rutin*	*Untreated*	*Rutin*
Unirradiated	57.4 ± 4.5	59.5 ± 7.1	2.30 ± 0.50	$5.20^{°°} \pm 0.62$
UVA 1 h	57.6 ± 5.5	55.3 ± 4.6	$1.36^* \pm 0.22$	$4.63^{°°} \pm 0.60$
UVA 2 h	53.3 ± 4.2	59.5 ± 6.5	$1.07^* \pm 0.20$	$4.06^{°°} \pm 0.65$

The increase in Nrf2 and GSH is thus a cellular defense mechanism that rutin induces to counteract UVA damage. As recent work has shown that rutin acts as a cytoprotective agent through the induction of Nrf2 by both preventing ferroptosis and apoptosis, we further investigated the molecular basis of the phenomenon observed in our experimental system. First, we assessed intracellular levels of total iron (as a marker of ferroptosis) and of the two forms, oxidized Fe(II) and reduced Fe(III), in fibroblasts irradiated with UVA for 1 and 2 h in the presence and absence of 10 μM of rutin. As can be seen from Figure 5, no significant changes in total iron and ferrous and ferric iron were observed in fibroblasts exposed to UVA or treated with rutin: this excludes that the cytotoxicity shown by UVA was induced by ferroptosis.

Next, to assess the presence of apoptotic mechanisms, we measured the amount of Bcl2 and Bax in the cells and reproduced the value of the "antiapoptotic index" by expressing the ratio of Bcl2 (antiapoptotic) to Bax (pro-apoptotic) in fibroblasts exposed to UVA radiation (for 1 and 2 h) with and without pre-treatment with rutin (Figure 6). As evident in the radiation-exposed cells, Bcl2/Bax decreased by 20% (UVA 1 h) and 25% (UVA 2 h), highlighting a prevalence of pro-apoptotic mechanisms, whereas the presence of rutin practically maintained the ratio as in cells not exposed to UVA.

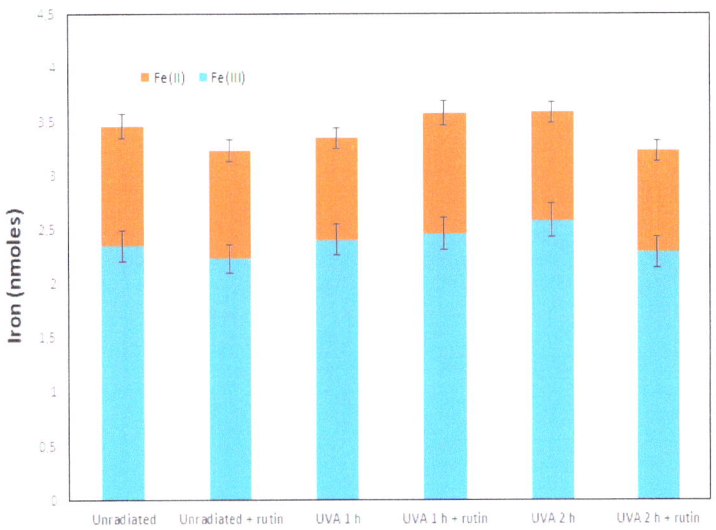

Figure 5. Total iron and oxidized Fe(II) and reduced Fe(III) in the cytosol of fibroblasts unexposed and exposed to UVA and untreated and pre-treated with rutin. Values of iron and its forms are expressed in nmoles and were calculated as reported in the Section 2.

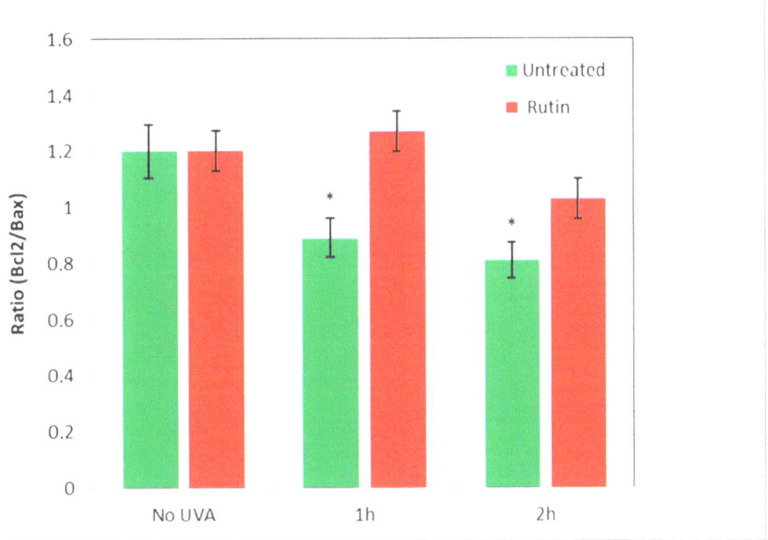

Figure 6. Bcl2/Bax ratio after 1 and 2 h of UVA irradiation with and without 10 µM of rutin. Anti-apoptotic index (expressed as Bcl2/Bax ratio) was calculated in untreated fibroblasts (green bars) and fibroblasts treated with 10 µM of rutin (red bars) unexposed (No-UVA) and exposed for 1 and 2 h of UVA radiation. Pre-treatment with rutin prevented the decrease in the ratio in UVA-exposed fibroblasts. * $p < 0.05$ untreated vs. treated.

To assess the protective effects of rutin on the respiratory capacity of human dermal fibroblasts, HRR measurements were performed on cells exposed to UVA radiation (2 h) with and without pre-treatment with 10 µM of rutin. HRR measures, displayed as oxygen flux per cell number, revealed that rutin induced a statistically significant increase in basal respiration of treated fibroblasts compared to control cells. Similarly, a significant growth in basal OCR was measured in fibroblasts exposed to UVA and pretreated with rutin vs. UVA exposed (Figure 7A). Oligomycin-sensitive respiration (Proton Leak), which is induced by inhibiting ATP synthase with oligomycin and corresponds to resting, non-phosphorylating electron transfer, showed a slight trend to increase following rutin treatment, without reaching statistical significance (Figure 7B). Similar to basal OCR, the level of maximal uncoupled respiratory activity (Maximal OCR), recorded in the presence of optimal uncoupler (FCCP) concentrations, was positively influenced by rutin. Notably, maximal OCR, which is a measure of functionality of the mitochondrial respiratory system independently of the cellular energy demand, was significantly increased by about 40% in UVA exposed and rutin-pre-treated cells compared to UVA-exposed cells (Figure 7C). Lastly, the residual OCR, measured upon addition of rotenone and antimycin A, revealed among samples several significant differences in non-mitochondrial oxygen-consuming processes as shown in Figure 7D.

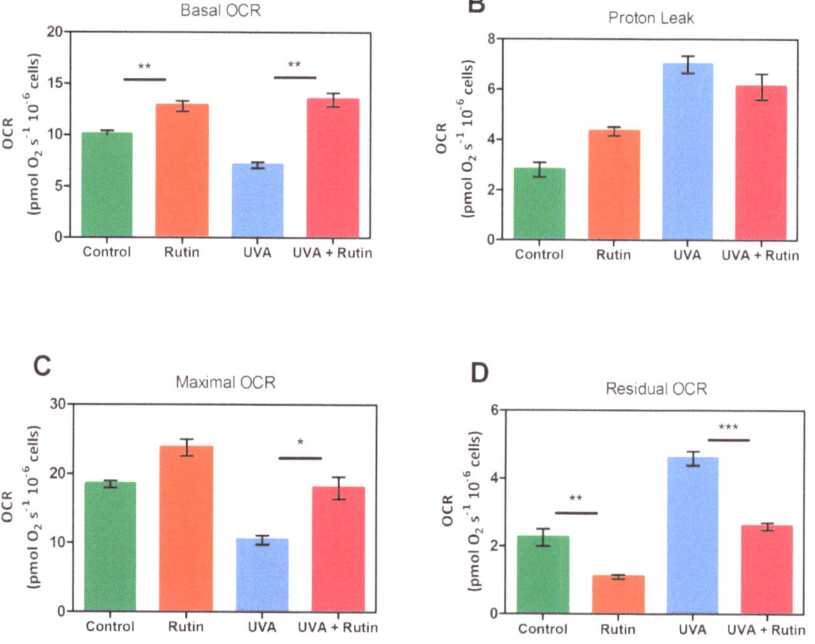

Figure 7. High-resolution respirometry measurements after 2 h of UVA exposure with and without 10 µM of rutin pre-treatment. Measurements were carried out on untreated and unexposed fibroblasts (green bars), fibroblasts treated with 10 µM of rutin (orange bars), those exposed for 2 h to UVA radiation, and those with (red bars) and without (blue bars) pre-treatment with rutin. (**A**) Basal oxygen consumption rate (Basal OCR), (**B**) Proton Leak, (**C**) Maximal OCR, and (**D**) non-mitochondrial respiration (Residual OCR) are expressed as (pmol/(s × 10^6 cells)) and are average values ± SD of three independent experiments performed in duplicate. The oxygen consumption rate obtained after addition of 0.5 µM rotenone and 2.5 µM antimycin A (Residual OCR) was subtracted from all other OCRs. * $p \leq 0.05$, ** $p \leq 0.01$, *** $p \leq 0.001$.

Finally, because all data converged to indicate an inhibition of apoptotic mechanisms by rutin on UVA-exposed cells, caspase-3 activity was measured. It was assumed [27] that caspases regulate the final stages of apoptosis and, in particular, caspase-3 is an "executor" of cell death. Figure 8 shows the values of caspase-3 activity in fibroblasts exposed to UVA radiation with and without rutin pretreatment. As can be seen, exposure to 1 and 2 h of radiation caused an increase by 50 and 100%, respectively, of caspase activity in untreated cells. Pretreatment with rutin, consistent with previous data, did not change caspase activity compared with unirradiated cells.

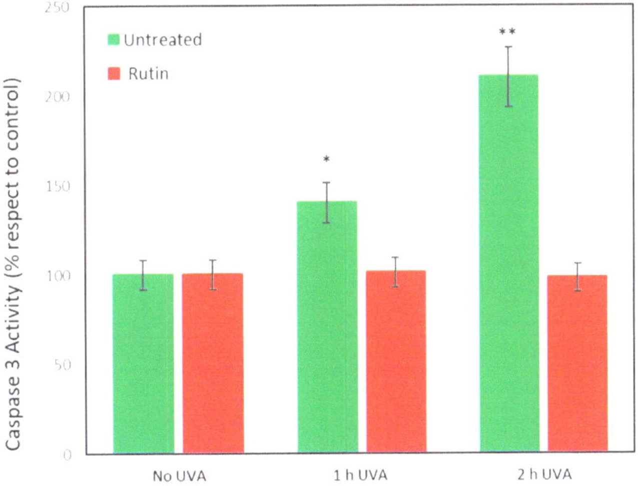

Figure 8. Caspase-3 activity in unexposed and exposed fibroblasts with and without rutin pretreatment. Values are expressed as percentage compared with unirradiated cells (assumed as 100%) in fibroblasts untreated (green bars) and pretreated with rutin 10 μM (red bars) exposed for 1 and 2 h to UVA radiation. Note the protective effect of rutin in irradiated cells with practically unchanged levels of caspase-3 activity. * $p < 0.05$ for 1 h of exposure vs. unexposed and untreated; ** $p < 0.01$ for 2 h of exposure vs. unexposed and untreated.

In conclusion, our results show that fibroblasts irradiated with UVA for 1 and 2 h reach cell death by triggering apoptotic-type mechanisms, and pretreatment with rutin (24 h before exposure) preserves cells from this cell death phenomenon.

4. Discussion

Rutin, a naturally occurring flavonoid glycoside in many plant species and/or in the waste of their production cycle (e.g., from the wine production chain), has been shown to exhibit several biological activities, such as antimicrobial, anticarcinogenic, antithrombotic, cardioprotective, and neuroprotective [23,28,29]. Biological actions would seem to be due to its antioxidant, anti-inflammatory, and antiapoptotic properties, capable of protecting the cells from the harmful effects, especially free radicals [30]. For this reason, rutin effects have also been studied on several neurodegenerative diseases, including Alzheimer's disease (AD), Parkinson's disease (PD), and Huntington's disease (HD) [28,30,31]. Furthermore, the rutin in the phospholipid complex is better soluble and permeable than the free rutin, thus allowing its delivery via the skin for the treatment of acute and chronic inflammatory diseases in vivo [32].

In this exploratory work, we exposed to UVA radiation a human fibroblast cell line to obtain an experimental model of photo-induced damage. Ultraviolet radiation triggers a series of chain reactions in exposed cells, that initially determine an increase in intracellular ROS levels, which favor apoptotic processes and consequently cell death. Our previous works, conducted on different cellular models, showed that the increase in ROS levels, consequent to photo-oxidative damage, are linked to the activation of the signaling pathways of the transcription factor Nrf2 [33–35]. The results exposed here follow our data previously obtained. In fact, fibroblasts exposed to UVA rays, especially after two hours, show an increase of intracellular ROS levels and of the associated apoptotic molecular events (increase in the Bcl2/Bax ratio), involving the Nrf2 signal transduction pathways [36–42].

Interestingly, the increase in ROS observed during the two-hour UVA exposure is not associated with an increase in Nrf-2 nuclear levels. Usually, the increase in ROS levels, induced by pro-oxidative stimuli, causes a ready activation of the Nrf-2 pathway, but in the case of UVA rays, the increase in ROS levels is not associated with an immediate increase in Nrf2 nuclear levels. In fact, in the case of photo-oxidative stimuli, intranuclear levels of Nrf2 remain unvaried in the first hours and increase after 3 h of exposure, as already observed in our work [33,43,44] and reported in the literature [45]. The cellular response and nuclear translocation of Nrf2 probably follow different times and modes depending on the chemical and physical properties of the noxious stimuli. Based on the results obtained in our study, it can be hypothesized that GSH, the most important antioxidant in cells, directly interacts with ROS to form oxidized glutathione (GSSG) and that regeneration of oxidized GSH (GSSG) is mediated by an enzyme, glutathione reductase (GSR), which is stimulated by Nrf2. In our experimental system, we see that reduced GSH decreases significantly upon exposure to UVA, a sign of an immediate attempt by the cells to defend themselves against oxidative stress. Subsequently, the decrease in GSH concentration due to UVA-stimulated ROS production causes an alteration of redox homeostasis in cells that can no longer defend themselves against reactive oxygen species.

The mechanism of the short-term activation of the Nrf2 signaling pathway is an important pathway for protection from oxidative damage in skin cells, but in our experimental system, it fails to intervene.

Pretreatment with rutin therefore seems to predispose cells to be more effectively equipped to deal with the damages of photo-exposure and the consequent increase in ROS, given that Nrf2 already increases after 24 h in unexposed cells. This increase also leads to a greater capacity of cells to be reactive, in the short term, against oxidative stress, as demonstrated by the values of reduced GSH. The increase in Nrf2 therefore provides cells with antioxidant potential that facilitates the disposal of reactive oxygen species after an acute insult such as UVA exposure.

Within this framework, it is interesting to underline that rutin can protect fibroblasts from UVA damage both directly as an ROS scavenger and indirectly in the form of rutin "quinone", stimulating endogenous antioxidant systems mediated by Nrf2 [46]. In fact, rutin is an antioxidant flavonoid, which is quickly oxidized and converted to quinone during its action as an ROS scavenger. This is possible because the chemical structure of rutin presents some isomers Orto and Para-diphenols, which, in conditions of oxidative stress, are converted into the correspondent quinones. The latter easily bind to the thiol-groups present on the Keap-1 protein, facilitating the nuclear translocation of Nrf2 [47,48].

However, whatever the mechanism, activation of the Keap1/Nrf2/Are signaling pathway clearly plays a central role in mediating the cytoprotective/antioxidative response of rutin in the presence of photooxidative stress. This hypothesis is reinforced by the results obtained in the successive experiments.

As the recent scientific paper [49] demonstrated that the effect of intense UVA radiation (4 kJ/m^2) for 1–10 min increases cytosolic iron in skin fibroblasts—an event that rapidly leads to cell death by ferroptosis—it seemed important to us to evaluate the involvement of these mechanisms in our experimental system. We observed that a lower radiation supplied

$(0.06 \text{ J}/\text{cm}^2)$, such as the one we used, does not modify the amount of cytoplasmic Fe(II) and Fe(III); on the contrary, it brings out mechanisms of apoptotic type at long-term (2 h).

Therefore, the mechanism we hypothesize about the role of rutin is to activate Nrf2, which not only controls genes for antioxidant enzymes but also prevents apoptosis. Indeed, it has been shown that UVA radiation induces apoptosis through the intrinsic pathway [50], of which Bax is the most important regulatory factor leading to the formation of macropores in the outer mitochondrial membrane and the activation of Caspase-3, the true effector of cell death. Rutin, by increasing Bcl2 and decreasing Bax, leads to an arrest of the apoptotic cascade.

Moreover, Nrf2 is also able to directly modulate mitochondrial functions, including mitophagy and mitochondrial respiration, by forming on the mitochondrial membrane the complex KEAP1/NRF2/PGAM5 (phosphoglycerate mutase 5). In the presence of redox imbalance, the complex on the outer membrane of mitochondria degrades and causes the dissociation of Nrf2. Free Nrf2 enters the nucleus and activates antioxidant and antiapoptotic gene expression to mitigate oxidative mitochondrial stress [51–53]. In confirmation of this, experimental evidence shows that cells of Nrf2-KO animals are highly sensitive to chemical-induced mitochondrial damage, whereas chemo-preventive agents protect against mitochondrial damage [54,55]. The results of our study confirm this axiom; in fact, HRR measurements obtained on intact fibroblasts under physiological feeding conditions with a mating control protocol [56,57] revealed that the antioxidant activity exerted by rutin and its ability to activate Nrf2 is also significantly displayed at the mitochondrial level. Indeed, cellular respiration was found to be strongly inhibited in UVA-exposed fibroblasts and maintained at levels comparable to those of the control in rutin-treated cells.

In conclusion, the present study extends our knowledge on rutin's nutraceutical properties. In fact, although preliminary and worth further analysis and validation, these data demonstrate that rutin is a substance with high photoprotective properties, capable of protecting fibroblasts' UVA-induced oxidative damage, via the Nrf2 signaling pathway. Nrf2, a transcription factor that regulates the gene expression of a large variety of antioxidants and phase II detoxification cytoprotective enzymes, when upregulated by rutin, maintains both redox homeostasis and the anti-apoptotic response in fibroblasts exposed to UVA rays. Therefore, based on our results, one could hypothesize the use of the rutin as a natural remedy, in food supplements or cosmetic preparations, to protect the skin from the sun, also in the light of issues of environmental sustainability and the circular economy.

Author Contributions: Conceptualization, E.T., G.T. and M.E.C.; methodology, V.N., S.D. and M.P.; formal analysis, P.B.; investigation, P.B., V.N. and S.D.; data curation, M.P.; writing—original draft preparation, E.T., G.T., P.B. and M.E.C.; writing—review and editing, E.T., G.T., P.B. and M.E.C. All authors have read and agreed to the published version of the manuscript.

Funding: This research was funded by CNR institutional funding (to M.E.C), by the internal funding of UCSC (Fondi di Ateneo D1: 2021–2022 to G.T).

Institutional Review Board Statement: The study was conducted in accordance with the Declaration of Helsinki and approved by the Ethics Committee of Catholic University (prot. N. 9917/15 and prot.cm 10/15, 8 April 2021) to perform human skin biopsies.

Informed Consent Statement: Informed consent was obtained from individuals involved in this study.

Data Availability Statement: All data supporting the findings of this study are available within the article.

Conflicts of Interest: The authors declare no conflict of interest.

References

1. Bosch, R.; Philips, N.; Suárez-Pérez, J.A.; Juarranz, A.; Devmurari, A.; Chalensouk-Khaosaat, J.; González, S. Mechanisms of photoaging and cutaneous photocarcinogenesis, and photoprotective strategies with phytochemicals. *Antioxidants* **2015**, *2*, 248–268. [CrossRef] [PubMed]
2. Cadet, J.; Douki, T. Formation of UV-induced DNA damage contributing to skin cancer development. *Photochem. Photobiol. Sci.* **2018**, *12*, 1816–1841. [CrossRef] [PubMed]
3. Salminen, A.; Kaarniranta, K.; Kauppinen, A. Photoaging: UV radiation-induced inflammation and immunosuppression accelerate the aging process in the skin. *Inflamm. Res.* **2022**, *71*, 817–831. [CrossRef] [PubMed]
4. Jin, S.G.; Padron, F.; Pfeifer, G.P. UVA Radiation, DNA Damage, and Melanoma. *ACS Omega* **2022**, *7*, 32936–32948. [CrossRef]
5. National Toxicology Program. Ultraviolet radiation related exposures: Broad-spectrum ultraviolet (UV) radiation, UVA, UVB, UVC, solar radiation, and exposure to sunlamps and sunbeds. *Rep. Carcinog. Carcinog. Profiles* **2002**, *10*, 250–254.
6. Krutmann, J. Ultraviolet A radiation-induced biological effects in human skin: Relevance for photoaging and photodermatosis. *J. Dermatol. Sci.* **2000**, *23*, S22–S26. [CrossRef]
7. Negre-Salvayre, A.; Salvayre, R. Post-Translational Modifications Evoked by Reactive Carbonyl Species in Ultraviolet-A-Exposed Skin: Implication in Fibroblast Senescence and Skin Photoaging. *Antioxidants* **2022**, *11*, 2281. [CrossRef]
8. von Thaler, A.K.; Kamenisch, Y.; Berneburg, M. The role of ultraviolet radiation in melanomagenesis. *Exp. Dermatol.* **2010**, *19*, 81–88. [CrossRef]
9. Mullenders, L.H.F. Solar UV damage to cellular DNA: From mechanisms to biological effects. *Photochem. Photobiol. Sci.* **2018**, *17*, 1842–1852. [CrossRef]
10. Ivanov, I.V.; Mappes, T.; Schaupp, P.; Lappe, C.; Wahl, S. Ultraviolet radiation oxidative stress affects eye health. *J. Biophotonics* **2018**, *11*, e201700377. [CrossRef]
11. Bernerd, F.; Passeron, T.; Castiel, I.; Marionnet, C. The Damaging Effects of Long UVA (UVA1) Rays: A Major Challenge to Preserve Skin Health and Integrity. *Int. J. Mol. Sci.* **2022**, *23*, 8243. [CrossRef]
12. Gilchrest, B.A. Photoaging. *J. Investig. Dermatol.* **2013**, *133*, E2–E6. [CrossRef] [PubMed]
13. Kammeyer, A.; Luiten, R.M. Oxidation events and skin aging. *Ageing Res. Rev.* **2015**, *21*, 16–29. [CrossRef] [PubMed]
14. Xian, D.; Lai, R.; Song, J.; Xiong, X.; Zhong, J. Emerging Perspective: Role of Increased ROS and Redox Imbalance in Skin Carcinogenesis. *Oxidative Med. Cell. Longev.* **2019**, *2019*, 8127362. [CrossRef] [PubMed]
15. Garg, C.; Sharma, H.; Garg, M. Skin photo-protection with phytochemicals against photo-oxidative stress, photo-carcinogenesis, signal transduction pathways and extracellular matrix remodeling-An overview. *Ageing Res. Rev.* **2020**, *62*, 101127. [CrossRef] [PubMed]
16. Kulms, D.; Schwarz, T. Mechanisms of UV-induced signal transduction. *J. Dermatol.* **2002**, *29*, 189–196. [CrossRef]
17. Ikehata, H.; Yamamoto, M. Roles of the KEAP1-NRF2 system in mammalian skin exposed to UV radiation. *Toxicol. Appl. Pharmacol.* **2018**, *360*, 69–77. [CrossRef]
18. Ryšavá, A.; Vostálová, J.; Rajnochová Svobodová, A. Effect of ultraviolet radiation on the Nrf2 signaling pathway in skin cells. *Int. J. Radiat. Biol.* **2021**, *97*, 1383–1403. [CrossRef]
19. Kahremany, S.; Hofmann, L.; Gruzman, A.; Dinkova-Kostova, A.T.; Cohen, G. NRF2 in dermatological disorders: Pharmacological activation for protection against cutaneous photodamage and photodermatosis. *Free Radic. Biol. Med.* **2022**, *188*, 262–276. [CrossRef]
20. Tulah, A.S.; Birch-Machin, M.A. Stressed out mitochondria: The role of mitochondria in ageing and cancer focussing on strategies and opportunities in human skin. *Mitochondrion* **2013**, *13*, 444–453. [CrossRef]
21. Süntar, I.; Çetinkaya, S.; Panieri, E.; Saha, S.; Buttari, B.; Profumo, E.; Saso, L. Regulatory Role of Nrf2 Signaling Pathway in Wound Healing Process. *Molecules* **2021**, *26*, 2424. [CrossRef]
22. Gęgotek, A.; Domingues, P.; Skrzydlewska, E. Natural Exogenous Antioxidant Defense against Changes in Human Skin Fibroblast Proteome Disturbed by UVA Radiation. *Oxidative Med. Cell. Longev.* **2020**, *2020*, 3216415. [CrossRef] [PubMed]
23. Rahmani, S.; Naraki, K.; Roohbakhsh, A.; Hayes, A.W.; Karimi, G. The protective effects of Rutin on the liver, kidneys, and heart by counteracting organ toxicity caused by synthetic and natural compounds. *Food Sci. Nutr.* **2022**, *11*, 39–56. [CrossRef] [PubMed]
24. Rotimi, D.E.; Elebiyo, T.C.; Ojo, O.A. Therapeutic potential of Rutin in male infertility: A mini review. *J. Integr. Med.* **2023**, *in press.* [CrossRef]
25. Pesta, D.; Gnaiger, E. High-resolution respirometry: OXPHOS protocols for human cells and permeabilized fibers from small biopsies of human muscle. *Mitochondrial Bioenerg. Methods Protoc.* **2012**, *810*, 25–58.
26. Gnaiger, E. Mitochondrial Pathways and Respiratory Control: An Introduction to OXPHOS Analysis. In *Mitochondr Physiol Network 19.12*, 4th ed.; Oroboros MiPNet Publications: Innsbruck, Austria, 2014; p. 80.
27. Fuchs, Y.; Steller, H. Programmed cell death in animal development and disease. *Cell* **2011**, *147*, 742–758. [CrossRef]
28. Enogieru, A.B.; Haylett, W.; Hiss, D.C.; Bardien, S.; Ekpo, O.E. Rutin as a Potent Antioxidant: Implications for Neurodegenerative Disorders. *Oxidative Med. Cell. Longev.* **2018**, *2018*, 6241017. [CrossRef]
29. Stoyanova, N.; Spasova, M.; Manolova, N.; Rashkov, I.; Georgieva, A.; Toshkova, R. Quercetin- and Rutin-Containing Electrospun Cellulose Acetate and Polyethylene Glycol Fibers with Antioxidant and Anticancer Properties. *Polymers* **2022**, *14*, 5380. [CrossRef] [PubMed]

30. Cordeiro, L.M.; Soares, M.V.; da Silva, A.F.; Machado, M.L.; Bicca Obetine Baptista, F.; da Silveira, T.L.; Arantes, L.P.; Soares, F.A.A. Neuroprotective effects of Rutin on ASH neurons in Caenorhabditis elegans model of Huntington's disease. *Nutr. Neurosci.* **2022**, *25*, 2288–2301. [CrossRef]

31. Goyal, J.; Verma, P.K. An Overview of Biosynthetic Pathway and Therapeutic Potential of Rutin. *Mini Rev. Med. Chem.* **2023**. [CrossRef]

32. Kalita, B.; Das, M.K. Rutin-phospholipid complex in polymer matrix for long-term delivery of rutin via skin for the treatment of inflammatory diseases. *Artif. Cells Nanomed. Biotechnol.* **2018**, *46*, 41–56. [CrossRef] [PubMed]

33. Clementi, M.E.; Sampaolese, B.; Sciandra, F.; Tringali, G. Punicalagin Protects Human Retinal Pigment Epithelium Cells from Ultraviolet Radiation-Induced Oxidative Damage by Activating Nrf2/HO-1 Signaling Pathway and Reducing Apoptosis. *Antioxidants* **2020**, *9*, 473. [CrossRef] [PubMed]

34. Clementi, M.E.; Pizzoferrato, M.; Bianchetti, G.; Brancato, A.; Sampaolese, B.; Maulucci, G.; Tringali, G. Cytoprotective Effect of Idebenone through Modulation of the Intrinsic Mitochondrial Pathway of Apoptosis in Human Retinal Pigment Epithelial Cells Exposed to Oxidative Stress Induced by Hydrogen Peroxide. *Biomedicines* **2022**, *10*, 503. [CrossRef] [PubMed]

35. Bianchetti, G.; Clementi, M.E.; Sampaolese, B.; Serantoni, C.; Abeltino, A.; De Spirito, M.; Sasson, S.; Maulucci, G. Investigation of DHA-Induced Regulation of Redox Homeostasis in Retinal Pigment Epithelium Cells through the Combination of Metabolic Imaging and Molecular Biology. *Antioxidants* **2022**, *11*, 1072. [CrossRef] [PubMed]

36. Kansanen, E.; Kuosmanen, S.M.; Leinonen, H.; Levonen, A.L. The Keap1-Nrf2 pathway: Mechanisms of activation and dysregulation in cancer. *Redox Biol.* **2013**, *1*, 45–49. [CrossRef]

37. Niture, S.K.; Jaiswal, A.K. Inhibitor of Nrf2 (INrf2 or Keap1) protein degrades Bcl-xL via phosphoglycerate mutase 5 and controls cellular apoptosis. *J. Biol. Chem.* **2014**, *289*, 22019. [CrossRef]

38. Otsuki, A.; Yamamoto, M. Cis-element architecture of Nrf2-sMaf heterodimer binding sites and its relation to diseases. *Arch. Pharm. Res.* **2020**, *43*, 275–285. [CrossRef]

39. Shilovsky, G.A. Lability of the Nrf2/Keap/ARE Cell Defense System in Different Models of Cell Aging and Age-Related Pathologies. *Biochemistry* **2022**, *87*, 70–85. [CrossRef]

40. Zhang, Q.; Liu, J.; Duan, H.; Li, R.; Peng, W.; Wu, C. Activation of Nrf2/HO-1 signaling: An important molecular mechanism of herbal medicine in the treatment of atherosclerosis via the protection of vascular endothelial cells from oxidative stress. *J. Adv. Res.* **2021**, *34*, 43–63. [CrossRef]

41. Badibostan, H.; Eizadi-Mood, N.; Hayes, A.W.; Karimi, G. Protective effects of natural compounds against paraquat-induced pulmonary toxicity: The role of the Nrf2/ARE signaling pathway. *Int. J. Environ. Health Res.* **2023**, 1–14. [CrossRef]

42. Moghaddam, A.H.; Eslami, A.; Jelodar, S.K.; Ranjbar, M.; Hasantabar, V. Preventive effect of quercetin-Loaded nanophytosome against autistic-like damage in maternal separation model: The possible role of Caspase-3, Bax/Bcl-2 and Nrf2. *Behav. Brain Res.* **2023**, *441*, 114300. [CrossRef] [PubMed]

43. Tringali, G.; Sampaolese, B.; Clementi, M.E. Expression of early and late cellular damage markers by ARPE-19 cells following prolonged treatment with UV-A radiation. *Mol. Med. Rep.* **2016**, *14*, 3485–3489. [CrossRef] [PubMed]

44. Clementi, M.E.; Sampaolese, B.; Lazzarino, G.; Tringali, G. Ultraviolet A radiation induces cortistatin overexpression and activation of somatostatin receptors in ARPE-19 cells. *Mol. Med. Rep.* **2018**, *17*, 5538–5543. [CrossRef] [PubMed]

45. Ryšavá, A.; Čížková, K.; Franková, J.; Roubalová, L.; Ulrichová, J.; Vostálová, J.; Vrba, J.; Zálešák, B.; Rajnochová Svobodová, A. Effect of UVA radiation on the Nrf2 signalling pathway in human skin cells. *J. Photochem. Photobiol. B Biol.* **2020**, *209*, 111948. [CrossRef] [PubMed]

46. Suraweera, T.L.; Rupasinghe, H.V.; Dellaire, G.; Xu, Z. Regulation of Nrf2/ARE Pathway by Dietary Flavonoids: A Friend or Foe for Cancer Management? *Antioxidants* **2020**, *9*, 973. [CrossRef] [PubMed]

47. Dinkova-Kostova, A.T.; Wang, X.J. Induction of the Keap1/Nrf2/ARE pathway by oxidizable diphenols. *Chem.-Biol. Interact.* **2011**, *192*, 101–106. [CrossRef]

48. Sthijns, M.M.; Schiffers, P.M.; Janssen, G.M.; Lemmens, K.J.; Ides, B.; Vangrieken, P.; Bouwman, F.G.; Mariman, E.C.; Pader, I.; Arnér, E.S.; et al. Rutin protects against H_2O_2-triggered impaired relaxation of placental arterioles and induces Nrf2-mediated adaptation in Human Umbilical Vein Endothelial Cells exposed to oxidative stress. *Biochim. Biophys. Acta (BBA)-Gen. Subj.* **2017**, *1861*, 1177–1189. [CrossRef]

49. Smith, M.J.; Fowler, M.; Naftalin, R.J.; Siow, R.C.M. UVA irradiation increases ferrous iron release from human skin fibroblast and endothelial cell ferritin: Consequences for cell senescence and aging. *Free Radic. Biol. Med.* **2020**, *155*, 49–57. [CrossRef]

50. Hussar, P. Apoptosis Regulators Bcl-2 and Caspase-3. *Encyclopedia* **2022**, *2*, 1624–1636. [CrossRef]

51. Gado, F.; Ferrario, G.; Della Vedova, L.; Zoanni, B.; Altomare, A.; Carini, M.; Aldini, G.; D'Amato, A.; Baron, G. Targeting Nrf2 and NF-κB Signaling Pathways in Cancer Prevention: The Role of Apple Phytochemicals. *Molecules* **2023**, *28*, 1356. [CrossRef]

52. Lo, S.C.; Hannink, M. PGAM5 tethers a ternary complex containing Keap1 and Nrf2 to mitochondria. *Exp. Cell Res.* **2008**, *314*, 1789–1803. [CrossRef]

53. Zeb, A.; Choubey, V.; Gupta, R.; Kuum, M.; Safiulina, D.; Vaarmann, A.; Gogichaishvili, N.; Liiv, M.; Ilves, I.; Tämm, K.; et al. A novel role of KEAP1/PGAM5 complex: ROS sensor for inducing mitophagy. *Redox Biol.* **2021**, *48*, 102186. [CrossRef]

54. Esteras, N.; Abramov, A.Y. Nrf2 as a regulator of mitochondrial function: Energy metabolism and beyond. *Free Radic. Biol. Med.* **2022**, *189*, 136–153. [CrossRef]

55. Inoue-Yanagimachi, M.; Himori, N.; Uchida, K.; Tawarayama, H.; Sato, K.; Yamamoto, M.; Namekata, K.; Harada, T.; Nakazawa, T. Changes in glial cells and neurotrophic factors due to rotenone-induced oxidative stress in Nrf2 knockout mice. *Exp. Eye Res.* **2023**, *226*, 109314. [CrossRef] [PubMed]
56. Brand, M.D.; Nicholls, D.G. Assessing mitochondrial dysfunction in cells. *Biochem. J.* **2011**, *435*, 297–312. [CrossRef] [PubMed]
57. Bottoni, P.; Pontoglio, A.; Scarà, S.; Pieroni, L.; Urbani, A.; Scatena, R. Mitochondrial Respiratory Complexes as Targets of Drugs: The PPAR Agonist Example. *Cells* **2022**, *11*, 1169. [CrossRef] [PubMed]

 *antioxidants*

 **MDPI**

Review

Nrf2-Mediated Antioxidant Defense and Thyroid Hormone Signaling: A Focus on Cardioprotective Effects

Laura Sabatino

Institute of Clinical Physiology, National Research Council, Via Moruzzi 1, 56124 Pisa, Italy; laura.sabatino@cnr.it; Tel.: +39-050-315-2659

Abstract: Thyroid hormones (TH) perform a plethora of actions in numerous tissues and induce an overall increase in metabolism, with an augmentation in energy demand and oxygen expenditure. Oxidants are required for normal thyroid-cell proliferation, as well as for the synthesis of the main hormones secreted by the thyroid gland, triiodothyronine (T3) and thyroxine (T4). However, an uncontrolled excess of oxidants can cause oxidative stress, a major trigger in the pathogenesis of a broad spectrum of diseases, including inflammation and cancer. In particular, oxidative stress is implicated in both hypo- and hyper-thyroid diseases. Furthermore, it is important for the TH system to rely on efficient antioxidant defense, to maintain balance, despite sustained tissue exposure to oxidants. One of the main endogenous antioxidant responses is the pathway centered on the nuclear factor erythroid 2-related factor (Nrf2). The aim of the present review is to explore the multiple links between Nrf2-related pathways and various TH-associated conditions. The main aspect of TH signaling is described and the role of Nrf2 in oxidant–antioxidant homeostasis in the TH system is evaluated. Next, the antioxidant function of Nrf2 associated with oxidative stress induced by TH pathological excess is discussed and, subsequently, particular attention is given to the cardioprotective role of TH, which also acts through the mediation of Nrf2. In conclusion, the interaction between Nrf2 and most common natural antioxidant agents in altered states of TH is briefly evaluated.

Keywords: thyroid hormones; oxidative stress; antioxidants; Nrf2; cardioprotection

Citation: Sabatino, L. Nrf2-Mediated Antioxidant Defense and Thyroid Hormone Signaling: A Focus on Cardioprotective Effects. *Antioxidants* **2023**, *12*, 1177. https://doi.org/10.3390/antiox12061177

Academic Editor: Paola Venditti

Received: 30 April 2023
Revised: 24 May 2023
Accepted: 28 May 2023
Published: 30 May 2023

1. Molecular Aspects of Thyroid-Hormone Signaling: An Overview

The thyroid hormones (THs) include the prohormone thyroxine (T4) and the biologically active form triiodothyronine (T3) and regulate a wide range of genes, intervening in many physiological processes, such as cell growth, development, differentiation, and survival [1]. They are synthesized in the thyroid follicles after the iodination of thyroglobulin (Tg) by thyroid peroxidase (TPO) [2].

Largely in the form of T4, THs are released in the circulation, where they are mostly bound to transport proteins and reach the peripheral tissues, where the 5′-monodeiodinases (DIO1 and DIO2) catalyze T4 to T3 activation [2]. A third monodeiodinase (DIO3) has been described in the cells and mediates T4 conversion to metabolically inactive reverse T3 (rT3) [2].

The signaling of TH in the target cells is highly complex and finely regulated [3]. Transporters of TH mediate the uptake of TH and, once inside the cell, TH can mediate genomic effects in the nucleus, binding to specific receptors (TRs), which directly interact with responding elements (TREs) in the target promoters, thus regulating the transcription of specific genes (Figure 1) [4]. Two genes encoding TRs have been described, THRA and THRB, codifying for TRα and TRβ, respectively, and these two isoforms are differently expressed during embryonic development and in adult life [1,5].

Furthermore, TH actions can also be exerted by the so-called non-genomic mechanisms, which occur in a short time (from seconds to minutes) and do not require the direct interaction of TH with TRs and DNA, but act trough intracellular signaling pathways, which indirectly regulate gene transcription [6]. Most of these effects, observed in different tissues, begin with TH interactions with receptors located in the plasma membrane, mitochondria, or cytoplasm [7–9]. The receptors involved in non-classical actions may or may not have structural homologies with TRs, and some actions that initiate at the plasma membrane may also regulate the fate and function of nuclear TRs [6].

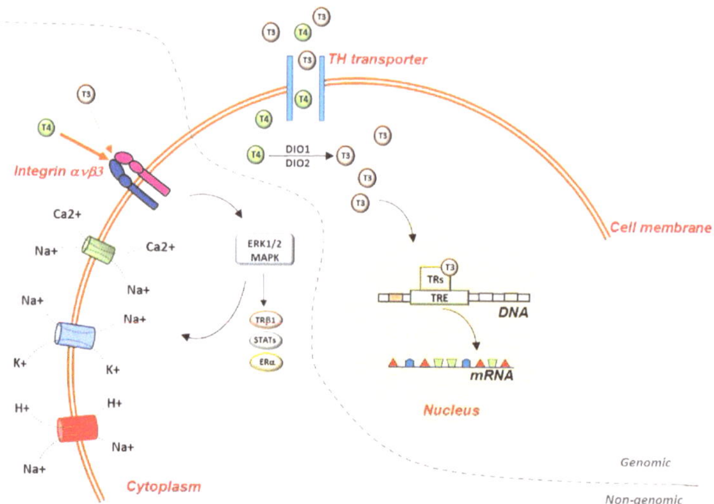

Figure 1. Representation of genomic and non-genomic actions of TH in the cell. Genomic actions begin at the plasma membrane and THs enter the cell through specific TH transporters. Once in the cell, T4 is converted to T3 by D1 and D2 deiodinases and T3 enters the nucleus, where it binds to specific receptors, which mediate the interaction with the DNA. Non-genomic mechanisms require the mediation of integrin αvβ3, which has a higher binding affinity for T4 than T3. Once in the cell, THs activate several MAPK-mediated signaling pathways. At the plasma-membrane level, TH regulate glucose transporter, Na^+/K^+-ATPase, Na^+/H^+-exchanger, Ca^{2+}-ATPase, and the Na^+-sensitive amino-acid transporter. TH: thyroid hormones; T3: triiodothyronine; T4: thyroxine; TRs: thyroid-hormone receptors; TRE: thyroid-responsive elements; DIO1: deiodinase 1; DIO2: deiodinase 2; STATs: signal transducer and activator of transcription 1α and 3; ERα: estrogen receptor α.

In the plasma membrane, the heterodimer protein integrin αvβ3 has been demonstrated to be of central importance in the mediation of TH effects on cell angiogenesis and proliferation; in fact, high concentrations of integrin are detectable on vascular and tumor cells. Integrin αvβ3 interacts with many structural proteins in the extracellular matrix, playing a crucial role in transducing important signals either from the outside into the cells or from the intracellular to the extracellular compartment [9]. The main intracellular signaling cascade triggered by TH-integrin αvβ3 interaction is the mitogen-activated protein kinase (MAPK; ERK1/2) via phospholipase C (PLC) and protein kinase Cα (PKCα). The TH-activated MAPK mediates the serine phosphorylation of several nuclear trans-activator proteins, such as the signal transducer and activator of transcription 1α and 3 (STAT1α and STAT3), TRβ1, estrogen receptor α (ERα), etc. [10,11]. Furthermore, at the plasma-membrane level, TH effects have been associated with the regulation of essential membrane-transport systems, such as glucose transporter, Na^+/K^+-ATPase, Na^+/H^+-exchanger, Ca^{2+}-ATPase, and the Na^+-sensitive amino-acid transporter (Figure 1) [12–14].

Moreover, acting on a cytoplasmic truncated form of nuclear TRα1 (TRΔα1), TH can regulate dynamic and structural changes in the cellular architecture, through the conversion of soluble to fibrous actin, which is important for cells' motility and interaction with the environment (for example, in glia and neurons). In vitro studies showed that the ability of astrocytes to adhere to the culture dish was associated with the presence of TH, and that the deprivation of TH in the culture medium induces the loss of the major actin filaments in the cells. It was demonstrated that, upon the administration of TH to the culture medium, this effect can be rapidly reversed in a few minutes by the mediation of truncated delta TRs, TRΔα1 and TRΔα2 [15].

Of the principal extra-nuclear activities, the regulatory action of TH on mitochondria is of central importance in cell metabolism and requires the presence of truncated TRα isoforms. More specifically, truncated forms of nuclear TRα1, with molecular weights of 43 kDa (p43) and 28 kDa (p28), have been described in the matrix and the inner membranes of mitochondria, respectively [16–18]. These forms lack of DNA-binding domain of TRα1, and the binding affinity of p28 for T3 is higher than that of p43 [15]. Furthermore, both receptors are targeted to mitochondria, but only p28 was demonstrated to enter the organelle in a T3-dependent way [17]. In the last decade, many other forms of nuclear-receptor superfamily have been described in the mitochondria, suggesting that the physiological importance of the interaction between TH and the organelles may be more complex than initially thought [18].

It is well known that TRs and steroid receptors have similar structural characteristics and conserved domains, and that they play a major role in the mediation of TH and steroid-hormone regulatory activity in the mitochondrial enzymes of oxidative phosphorylation (OXPHOS) [19]. It is believed that TH and steroid hormones can coordinate OXPHOS nuclear and mitochondrial gene expression and protein biosynthesis, in both normal and pathological contexts, in order to regulate energy metabolism [20]. The finding in the mitochondrial genome of sequences similar to nuclear hormone-response elements (HREs) and the presence of TRs and steroid receptors in mitochondria suggest the possibility of parallel hormonal regulation at the nuclear and mitochondrial level [21]. Interestingly, in the nucleus, the TH and steroid-hormone activation of OXPHOS genes can occur through both direct and indirect interaction with HRE-containing genes encoding for specific transcription factors, such as nuclear respiratory factors 1 and 2 (NRF1 and NRF2) and peroxisome proliferator-activated receptor γ coactivator 1α (PGC-1 α), which, in turn, can induce genes encoding mitochondrial transcription factors, such as TFAM, TFB1M,s and TFB2M, which activate mitochondrial OXPHOS gene expression [21].

2. Oxidant–Antioxidant Homeostasis: Nrf2 and Thyroid Protection

Oxidants are required for regular activities in cells and are continuously produced by endogenous processes or obtained from exogenous sources. The basal level of oxidants is maintained by active oxygen scavenging through enzymatic superoxide dismutase (SOD), catalase (CAT), glutathione peroxidase (GPx), and glutathione reductase (GR), as well as non-enzymatic antioxidant molecules, such as reduced glutathione (GSH) [22].

Several organelles, including mitochondria, the endoplasmic reticulum (ER), and peroxisomes, as well as enzymatic systems, such as xanthine oxidase, lipoxygenase, and nitric oxide synthase, are important sources of oxidants in mammalian cells [23]. In normal cells, oxidants are mainly generated by mitochondrial oxidative phosphorylation, and moderate amounts of oxidants have positive effects as regulators of inflammation, immune activity, and stress response, as protectors against invading harmful pathogens and as mediators of healing and repairing processes [22]. In addition to mitochondria, other organelles, such as the endoplasmic reticulum and peroxisomes, can produce oxidants, and their relative contributions vary according to the cell type [23]. Oxidative stress is an effect of a redox imbalance between oxidants and antioxidant defense. It can be induced by the excessive production of oxidants and/or reduced antioxidant capacity, thus provoking molecular damage [24]. Furthermore, oxidants produced in different cellular compartments determine a positive feedback circuit, supporting pathological conditions associated with oxidative stress [25,26].

More specifically, in the thyrocytes, H_2O_2 is the primary oxidative agent required by the TPO enzyme for regular hormonogenesis; hence, oxidants are continuously produced, even in physiological conditions [27]. However, since an uncontrolled excess of oxidants can rapidly cause oxidative stress, follicular cells have to guarantee the presence of efficient protective mechanisms. Recently, the antioxidant pathway centered on the nuclear factor erythroid 2-related transcription factor 2 (Nrf2) and its cytoplasmic inhibitor, Kelch-like ECH-associated protein (Keap1) has gained increasing relevance as an efficient antioxidant system in the thyroid [28,29].

In normal conditions, Keap1 acts as an adaptor targeting Nrf2 for poly-ubiquination by Cullin 3-based ubiquitin E3 ligase (Cul3-Rbx1) and proteasomal degradation. In the presence of redox-disrupting stimuli, Keap1 thiol groups react with oxidants, such as H_2O_2, leading to the inactivation of the Keap1 stabilizing function, inducing the impairment of Nrf2 poly-ubiquination and the accumulation of Nrf2 in the nucleus. At the nuclear level, Nrf2 acts as a transcription factor, interacting with the so-called antioxidant response elements (AREs) in the promoters of numerous target genes, encoding antioxidant enzymes and other cytoprotective molecules (Figure 2) [30]. Furthermore, Nrf2 is believed to control the basal and inducible expression of over 1000 genes involved in antioxidant defense, detoxification, inflammatory response, and proteasomal and autophagic degradation and metabolism. The role of Nrf2 was extensively studied in Nrf2 knockout mice, in which the expression level of antioxidant and cytoprotective genes decreased, whereas a higher level of oxidative damage was augmented [31]. Although Nrf2 is ubiquitously expressed, its role as a multiple-organ protector is not only due to the regulation of ubiquitous cytoprotective genes, but also to the regulation of tissue-specific genes involved in highly specialized functions in different tissues [32].

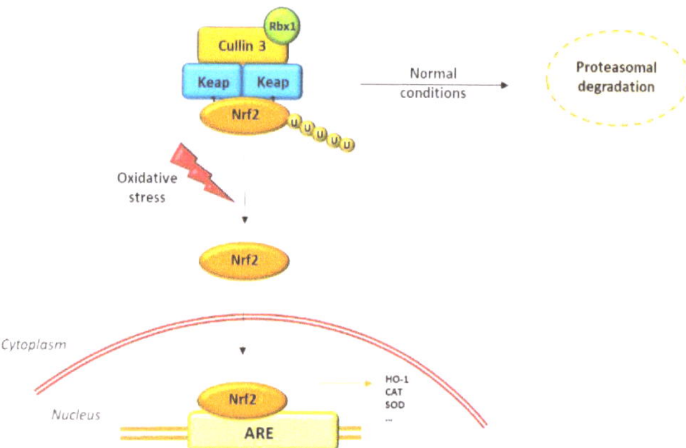

Figure 2. Representation of Nrf2–ARE pathway activation by oxidants. In normal conditions, Keap1 acts as an adaptor targeting Nrf2 and inducing proteasomal degradation. In presence of excess oxidant stimuli, Keap1's stabilizing function is inactivated and Nrf2 accumulates in the nucleus. At the nuclear level, Nrf2 acts as a transcription factor interacting with the ARE sequences in the promoters of numerous target genes encoding antioxidant enzymes and other cytoprotective molecules.

While a minimal oxidative load is required for normal thyroid-gland function, antioxidant protection is activated when oxidative stress occurs, in conditions of iodine overload; Nrf2, in association with its cytoplasmic inhibitor Keap1, is considered the main mediator of the antioxidant response [33]. Experimental studies evidenced that Nrf2 promotes antioxidant defense in the thyroid gland by stimulating the expression of cytoprotective molecules, such as GPx2, GR1, thioredoxin 1 (TXN1), thioredoxin reductase 1 (TXNRD1), sulfiredoxin 1 (SRXN1), and NAD(P)H quinone dehydrogenase (NQO1), which are known to have determining roles in regular thyroid activity [34–36]. In a study on a Nrf2-KO mouse model, it was shown that the cytoprotective activity was dramatically lost after iodine overload, whereas in wild-type mice, despite the excessive iodine exposure, no augmented protein or lipid levels were observed, suggesting that Nrf2-dependent antioxidant machinery was efficiently activated to neutralize oxidative-stress onset [33]. Furthermore, in the same study, it was observed that Nrf2, in addition to maintaining thyroid homeostasis, plays an important role in the regulation of Tg synthesis through the direct regulation of ARE sequences present in the Tg gene, both in rodents and in humans. Interestingly, the fact that in Nrf2-KO mice, a relevant increase in iodinated Tg was observed indicates that Nrf2 is involved in the regulation of both the synthesis and the iodination of Tg [33].

The main Nrf2 signaling aspects in thyroidal pathological contexts, discussed in the present review, are summarized in Figure 3.

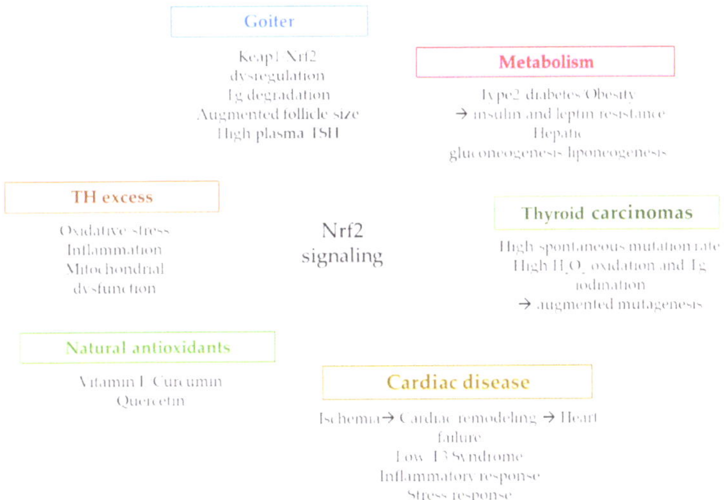

Figure 3. Nrf2 signaling in different pathological contexts.

Some reports on mice and humans showed the involvement of the Keap1/Nrf2 system in goiter formation and that the gene-based over-activation of Nrf2 in the thyroid can induce goiter formation with variable phenotypic characteristics [33,37]. Germline mutations of Keap1 are very rare and usually associated with non-toxic multinodular goiters, characterized by the nodular enlargement of the thyroid without thyroid dysfunction or inflammation [38]. It is reasonable to assume that the germline Keap1 mutation may determine Nrf2 activation in the thyroid, as in all other tissues; however, no diseases other than goiters were described in the patient affected, and this intriguing aspect needs to be investigated further [32]. The most frequently used experimental model for Nrf2-pathway activation is the Keap1 knockdown (Keap1[KD]) mouse, also referred as Keap1 hypomorphic, which has very low expression levels of the Keap1 gene [39]. Keap1[KD] mice have enlarged goiters, with dilated follicles, the absence of nodules and hyperplasia, and decreased plasma levels of T4, which are normalized in adult life by the activity of stimulating hormone (TSH), the pituitary hormone that stimulates follicular thyroid-cell growth and function [40]. Furthermore, the finding in Keap1[KD]-mouse thyroids of high concentrations of Tg-degrading cathepsin enzymes indicates that the lysosomal degradation of Tg may offer important support for the pathogenesis of goiters in Keap1[KD] mice [28,40].

The Keap1[KD] mouse model was also used in studies of metabolic diseases, especially in the context of type 2 diabetes and obesity, where, in addition to its protective role against oxidative-stress damage, Nrf2 can interact with pathways not directly associated with cytoprotection, such as in the hypothalamus, where Nrf2 improves insulin and leptin resistance, preventing the progression of diabetes mellitus [41], or in the liver, where Nrf2 has been described as a potential repressor of hepatic gluconeogenesis and lipogenesis [42].

Unfortunately, no data on thyroids were reported in these studies, which might be an interesting starting point for future studies.

The signaling of Nrf2 was also found to be activated in thyroid carcinomas, where it exerts a dual role, since, in addition to conferring protection against oxidative stress, it also promotes drug resistance to malignant cells [32,43]. Thyroid tumors are quite frequent in the population and the spontaneous mutation rate (preferentially single-base modifications) in the thyroid is higher than in other tissues. Furthermore, the presence of H_2O_2 for iodine oxidation and Tg iodination might account for the high mutagenesis rate in the thyroid [44]. In the attempt to better define the molecular mechanisms leading to Nrf2 activation in thyroid cancer, several Nrf2-gain-of-function and Keap1-loss-of-function somatic mutations were described in many human cancers in different tissues, including the liver, kidneys, lungs, and others [43,45]. The Nrf2 appears to be able to function not only as a tumor suppressor but also as an oncogene. In fact, while Nrf2 initially acts as a cancer-preventive factor, protecting cells from carcinogens and oxidative stress, the persistent activation of Nrf2 activates its oncogene properties and reduces radiotherapy- and chemotherapy-induced cytotoxicity, enhancing drug resistance in cancer cells [46].

Therefore, the inhibition of Nrf2 signaling is increasingly considered a potential target to overcome drug resistance and provides a novel strategy through which to increase the efficacy of traditional treatments.

3. Thyroid Hormone Excess, Oxidative Stress, and Nrf2 Activation

Diseases of THs are strongly associated with oxidative stress, and while, on one hand, oxidants interfere with the synthesis, activity, and metabolism of hormones, the reverse condition is also possible, and TH can regulate the antioxidant levels in cells.

Depending on the tissue demand, in normal conditions, a baseline level of oxidants is necessary to preserve cell homeostasis, and this number of oxidants is generally low in most tissues. When the oxidants exceed the ability of the cells to remove the oxidant surplus, oxidative stress arises. Thus, the role of oxidants in cells depends mainly on their initial concentrations, which determine the downstream cellular responses.

The THs are key regulators of cellular metabolism, and several studies found that in hyperthyroidism, the augmented metabolic demand promotes the synthesis of chemical energy by mitochondrial oxidation-reduction reactions, thus increasing the oxidant levels in the cell and inducing lower antioxidant ability [47,48].

The available data indicate that TH administration increases H_2O_2 generation by mitochondria in rat tissues, and this event is often associated with increased rates of oxygen consumption in target tissues, such as the liver, kidneys, heart, and skeletal muscles, where the need for metabolic capacity is higher [49,50]. Variability in the antioxidant response leading to an imbalance in oxidant clearance was observed in the tissues of hyperthyroid-induced animals, and further variations were appreciable, according to the age and the characteristics of the animals undergoing TH treatment [51].

Hyperthyroidism and thyrotoxicosis have been associated with the activation of Nrf2 signaling in TH target tissues. More specifically, in rat livers, T3 administration led to a rapid and transient cytosol-to-nuclear translocation of Nrf2, and it was hypothesized that the increase in oxidative status induced by T3 administration may inactivate Keap1-mediated ubiquination/degradation and expand Nrf2 nuclear-pool availability [52].

Several studies hypothesized that Nrf2 activation is triggered by mitogen-activated protein kinases (MAPKs) produced by T3-induced oxidants; however, the exact role of MAPKs and the underlying molecular mechanism remain poorly defined [52]. On the other hand, some other studies evaluated the hypothesis that the direct phosphorylation of Nrf2 by MAPKs contributes little to the modulation of Nrf2 activity and suggested that MAPKs mainly regulate the Nrf2 signaling pathway through indirect mechanisms [53].

In the last decades, several studies have provided new approaches to detailing the interactions between the TH system and mitochondrial compartments and to elucidating the effects of TH on electron-transport complexes and the existing relationship with oxidative metabolism [54]. Recently, it was demonstrated that respiratory complexes are organized in higher-order structures, called supercomplexes, which guarantee the major stabilization of the assembly and better control over oxidant production in the electron-transport chain, thanks to the better accessibility of substrates necessary for enzymatic reactions [55,56]. Moreover, the discovery of supercomplexes represents an important step forward in the study of the functional and structural properties of the mitochondrial respiratory chain, even though their functional advantages and their possible pathophysiological involvement in TH disease are far from being fully understood.

In an experimental model of hyperthyroidism, it was found that more than 58% of mitochondria were swollen, and that their cristae were radially oriented towards the center of organelles [57]. Alterations in mitochondrial morphology can slightly reduce the efficiency of phosphorylation, whereas the TH-induced increase in mitochondrial respiratory complexes explains the increase in respiratory rate [58].

In normal conditions, Nrf2 affects the mitochondrial membrane potential, fatty-acid oxidation, the availability of substrates for respiration (NADH and FADH2/succinate), and ATP synthesis. In conditions of stress, Nrf2 activation counteracts oxidant production in mitochondria via the transcriptional upregulation of uncoupling protein 3 and influences mitochondrial biogenesis by maintaining adequate levels of NRF1 and PGC-1α, as well as by promoting purine-nucleotide biosynthesis in rapidly growing cells [59].

The Nrf2 plays an important role in the maintenance of mitochondrial homeostasis and structural integrity. This is especially true in conditions of oxidative, electrophilic, and inflammatory stress, when the request for cytoprotective responses is crucial for the survival of the cell and the organism. The effects on mitochondria are among the principal protective mechanisms mediated by Nrf2. Diseases of the THs, analogously to many other pathological conditions, are characterized by oxidative stress, inflammation, and mitochondrial dysfunction as essential components of their pathogenesis. Therefore, Nrf2's possible involvement holds promise for disease prevention and treatment.

4. Thyroid Hormones and Their Antioxidant Role in Cardioprotection: Nrf2 Mediation

Experimental studies showed the negative effects of TH-altered metabolism on cardiac function, cell protection, and mitochondrial function, whereas the reversibility of these conditions restores the euthyroid state, suggesting that TH exert an important cardioprotective role [60].

Oxidative stress is a determining factor in the pathological progression of cardiac diseases, and excess of oxygen species may occur when oxygen supply is limited, such as during cardiac ischemia. In these conditions, oxidants can provoke irreversible damage by oxidation-membrane phospholipids, proteins, and DNA [61]. Subsequently, the heart reacts with a remodeling process that starts as a compensatory event characterized by the hypertrophy of surviving myocytes and the fibrosis of non-myocyte components, but soon involves the activation of the neuroendocrine and inflammatory systems and leads to decompensation and heart failure [62]. In particular, the progression to heart failure is associated with a progressive compromise of mitochondrial respiratory activity and a reduction in its capacity to produce ATP, which, in turn, leads to secondary dysregulation and altered Ca^{2+} handling and energy deficiency [63].

In both clinical settings and experimental studies of acute myocardial infarction, the reduction in circulating T3 levels (low-T3 syndrome) is one of the principal alterations observed and correlates with intense pro-inflammatory and stress responses [64]. The low-T3 state induces several important molecular, biochemical, and histological changes in the myocardium [65] and, for a long time, low T3 has been considered part of a beneficial adaptive mechanism aiming to reduce cardiac energy expenditure. However, clinical and experimental data demonstrated that low T3 is a strong prognostic predictor of short-term and long-term mortality [66,67], and that constant and low-level T3 administration allow the normalization of the hormone in the serum, attenuate myocardial damage, reduce remodeling, and prevent oxidative stress, with the final effect of improving cardiac function [68,69].

Many Nrf2-regulated enzymes are involved in the pathogenesis of cardiovascular diseases and may act as specific markers of the progression towards heart failure. These genes include antioxidant-related genes [70], stress-response genes [71], and genes limiting the inflammatory processes and conferring protection against ischemia/reperfusion events [72].

Coronary artery disease and ischemic heart disease are the most prevalent causes of mortality worldwide, and post-myocardial infarction hypertrophy, fibrosis, and apoptosis are the major events driving the progression towards heart failure. Coronary interventions and revascularization initially provide benefits after acute myocardial infarction; however, ischemia/reperfusion injury occurring during revascularization may worsen general cardiac conditions due to oxidant formation and inflammatory infiltration [72–75]. In this context, Nrf2 has been demonstrated to play a central role in cardiac protection, through the regulation of a broad spectrum of target genes [76,77].

Mouse models of Nrf2 overexpression or Nrf2 knockouts have been widely used to characterize Nrf2's role in cardiac pathological contexts. In mice with constitutively active Nrf2 cardiac overexpression, beyond the increased expression of antioxidant genes, some hypertrophic genes (i.e., genes for natriuretic peptides A and B) are also stimulated, increasing the risk of developing pathological cardiac remodeling [78]. By contrast, Nrf2-KO mice have a marked exposure to oxidative insult and oxidative-stress-associated pathologies [79]. Although some antioxidant gene expression is still appreciable in Nrf2-KO mice, it is not sufficient to compensate for oxidative stress and cardiac hypertrophy due to acute exercise stress, leading to cardiac dysfunction [79,80]. Figure 4 reports a schematic representation of the main cardiac phenotypes associated with Nrf2 over expression or lack of expression.

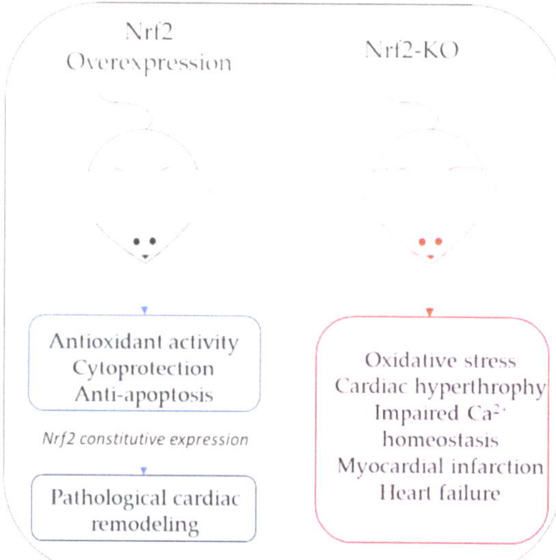

Figure 4. Schematic representation of main cardiac phenotypes associated with Nrf2 overexpression or lack of expression (Nrf-KO mice).

Several antioxidant agents exert their protective effects on ischemia/reperfusion injury through the induction of Nrf2-regulated pathways [81–85]. Some examples of Nrf2's effects in myocardial ischemia/reperfusion experimental models are reported in Table 1.

Table 1. Examples of experimental approaches in the study of Nrf2-mediated cardioprotection.

I/R Model	Nrf2 Activation Effects	Downstream Targets	Ref.
Nrf2-KO, C57BL/6J mice	Attenuation of MI size, decreased cardiomyocyte apoptosis	GPx, HO-1, NQO1, TXN1, CYBRD, ALDOSE REDUCTASE	[77]
Fh1-KO mice	Cardioprotection	HO-1, NQO1, MTHFD2, GSTA1	[82]
Sprague-Dawley rats (LA pretreatment)	Reduction of cardiomyocyte necrosis, apoptosis and inflammation	HO-1, GST, SOD, NADPH-regenerating enzymes	[81]
Nrf2-KO, C57BL/6J mice (PGD2 pretreatment)	Cardioprotection	GCLC, GR, G6PD, HO-1, SLC7A11, GSTA2	[83]
C57BL/6J mice (Bortezomib pretreatment)	Redox homeostasis, preservation of cardiac systolic function, reduction of MI size	HO-1, GSH, SOD1, CAT	[84]

GPx: glutathione peroxidase; HO-1: heme oxygenase-1; NQO1: NAD(P)H quinone dehydrogenase; TXN1: thioredoxin 1; CYBRD: Cytochrome b reductase; Fh1: fumarate hydratase 1; MTHFD2: Methylenetetrahydrofolate dehydrogenase (NADP+ dependent) 2; GSTA: glutathione-S transferase; LA: a-lipoic acid; SOD: superoxide dismutase; PGD2: prostaglandin D2; GCLC: glutam Fh1: ate-cysteine ligase catalytic; SLC7A11: solute carrier family 7 member 11; GSH: reduced glutathione; CAT: catalase.

The Nrf2 affects cell survival through some mediators, such as the anti-apoptotic proteins Bcl-2 and the heme oxygenase-1 (HO-1), a stress protein with antioxidant, anti-apoptotic, anti-thrombotic, and anti-inflammatory properties [86], and it is therefore considered a reliable marker of oxidative stress [87]. During ischemia/reperfusion, Nrf2's dissociation from Keap1 is encouraged and Nrf2 translocation to the cardiomyocyte nucleus increases, thus increasing antioxidant responses [88]. The stimulation of Nrf2 in cardioprotection is associated with the activation of the pro-survival pathway phosphoinositide 3-kinase (PI3K)/Akt kinase, which is considered a key factor in many aspects of cardiac physiology, such as cell survival, contractility, and electrophysiology [89]. Moreover, the PI3K/Akt pathway is considered to be involved in T3 protection against ischemic injury, both in vivo and in vitro. In fact, in H_2O_2-treated cardiomyocytes, pre-treatment with T3 stimulates PI3K and Akt signaling through their phosphorylation [88,89], and in a mouse experimental model, TH-replacement therapy restores myocardial function after ischemia/reperfusion injury [90]. Th levels of Nrf2 increase in response to T3 treatment, suggesting the pivotal role of this factor in the mediation of T3's protective function in cardiomyocytes [88]. Moreover, HO-1, which is regulated by Nrf2, is also augmented after T3 treatment in vitro, supporting resistance to oxidative stress and mitochondrial biogenesis [59].

5. Natural Antioxidants, TH Signaling, and Nrf2 Mediation

Several conditions and chemical substances can interfere with thyroid function and affect the secretion of thyroid hormones and their availability to target tissues. Altered TH levels can cause relevant changes in the ratio of antioxidant enzymes leading to imbalances in the clearance of oxidants, leading to the deterioration of cellular proteins, lipids, and DNA.

In recent years, many efforts have been made to individuate antioxidant molecules commonly present in nature as therapeutic agents to counter the effects of excessive TH-induced oxidant production [91]. In the present paper, the effects of some natural substances and their modulation of Nrf2 signaling in the antioxidant response to altered TH levels will be discussed.

Vitamin E and curcumin, low-molecular-mass antioxidants, are considered among the most effective natural protective agents against the oxidative stress occurring in hyperthyroidism. They both have a potent oxidant-scavenging activity, but their methods of action are quite different [91]. Vitamin E is lipid-soluble and, once incorporated in the membrane bilayer, interferes with the synthesis of lipid peroxides and carbonyl groups, acting as a membrane stabilizer and limiting lipid-chain peroxidation [92]. Curcumin is a phenolic compound isolated from the rhizome of turmeric (Curcuma longa), which is commonly employed as a spice and food colorant and used in traditional Indian and Chinese medicine for its antioxidant, anti-inflammatory, and anti-carcinogenic properties [93]. Curcumin exerts scavenging effects, directly quenching free radicals or intervening in the oxidative cascade, preventing oxidant formation [94]. Furthermore, it was found that curcumin may inhibit the oxidation of low-density lipoproteins (LDL), thus playing an important role in cardiovascular protection [95]. In addition, curcumin may induce conformational changes in sarcoplasmic reticulum Ca^{2+}-ATPase (SERCA), preventing the enzyme from interacting with ATP and blocking Ca^{2+} from entering the sarcoplasmic reticulum [96].

A recent study on rat hearts supported the idea that Nrf2 may be activated after vitamin E/curcumin administration and that it neutralizes altered TH-induced oxidative stress [91]. In oxidant-mediated oxidative damage, the regulation of antioxidant enzymes depends on the TH levels. Hyperthyroidism upregulates SOD and GR and downregulates CAT and GPx, whereas all enzymes are downregulated in hypothyroidism. However, in experimental hyper- and hypo-thyroid animals, in response to vitamin E and curcumin administration, a different regulation of antioxidant enzymes was described. Curcumin alone ameliorated SOD and CAT activities in both TH-altered states, whereas vitamin E alone stabilized SOD and CAT only in hyperthyroid conditions, and no response was observed in the hypothyroid rats. Interestingly, the combined administration of the two compounds normalized the GPx and GR activities compared to the administration of each compound alone. In this study, the optimal in silico interaction observed between vitamin E and curcumin with Keap1 factor strongly suggested that the antioxidant effects of both compounds were mediated by the Keap1-Nrf2 system, Nrf2 release to the nucleus, and ARE-sequence activation [91].

Quercetin is also a flavonoid found in some vegetables and fruits, and it showed antioxidant, anti-inflammatory. and anti-proliferative properties, as well as the ability to suppress lipid peroxidation [97]. In experimental models of hyperthyroidism, it was demonstrated that the administration of quercetin can protect liver functions from oxidative stress [97], and that this effect is mediated by Nrf2 activation and subsequent increases in HO-1 and NQO-1 production [98,99]. Recently, it was observed that quercetin may downregulate the gene expression of thyroid-restricted genes (sodium/iodide symporter, thyroid peroxidase, Tg, and the thyrotropin receptor), indicating that this compound has important inhibitory effects on TH metabolism [100]. However, quercetin might carry a potential toxicity risk if it is administered in excess, since it can inhibit thyroid-cell growth and iodide uptake, thereby negatively interfering with thyroid function. In general, caution should always be exercised in drawing conclusions as to the potential beneficial (or harmful) effects of natural compounds on the metabolism of TH, and a careful toxicological evaluation should be performed in view of their use for therapeutic purposes.

6. Conclusions

In summary, Nrf2 can be considered to play a key role in cellular homeostasis, with important regulatory effects on a large battery of genes with cytoprotective actions, exerted in response to different stimuli, such as oxidative stress, inflammation, cell growth, and energy supply. Emerging evidence indicates that Nrf2 plays a critical role in restoring redox balance and metabolic homeostasis in conditions of TH alterations. Ongoing research, including studies on cell cultures and experimental animal models, provides a better understanding of Nrf2 signaling in normal and pathological settings in which TH are involved. While the beneficial role of Nrf2 is well defined, less established is how the regulation of this pathway can be successfully used for disease prevention. First, this difficulty must be ascribed to the fact that developments in research continuously identify the wide range of pathological contexts in which Nrf2 plays a role. Second, the dual function of Nrf2 in disease underlines the need to accurately evaluate the clinical context in developing the most suitable and targeted Nrf2-based therapy.

Funding: This research received no external funding.

Institutional Review Board Statement: Not applicable.

Informed Consent Statement: Not applicable.

Data Availability Statement: Not applicable.

Conflicts of Interest: The authors declare no conflict of interest.

References

1. Oetting, A.; Yen, P.M. New insights into thyroid hormone action. *Best. Pract. Res. Clin. Endocrinol. Metab.* **2007**, *21*, 193–208. [CrossRef] [PubMed]
2. Citterio, C.E.; Targovnik, H.M.; Arvan, P. The role of thyroglobulin in thyroid hormonogenesis. *Nat. Rev. Endocrinol.* **2019**, *15*, 323–338. [CrossRef]
3. Cheng, S.-Y.; Leonard, J.L.; Davis, P.J. Molecular aspects of thyroid hormone actions. *Endocr. Rev.* **2010**, *31*, 139–170. [CrossRef]
4. Lazar, M.A. Thyroid hormone action: A binding contract. *J. Clin. Investig.* **2003**, *112*, 497–499. [CrossRef]
5. Brent, G.A. Mechanisms of thyroid hormoneaction. *J. Clin. Investig.* **2012**, *122*, 3035–3043. [CrossRef]
6. Davis, P.J.; Davis, F.B.; Mousa, S.A.; Luidens, M.K.; Lin, H.Y. Membrane receptor for thyroid hormone: Physiologic and pharmacologic implications. *Annu. Rev. Pharmacol. Toxicol.* **2011**, *51*, 99–115. [CrossRef] [PubMed]
7. Davis, P.J.; Goglia, F.; Leonard, J.L. Nongenomic actions of thyroid hormone. *Nat. Rev. Endocrinol.* **2016**, *12*, 111–121. [CrossRef] [PubMed]
8. Davis, P.J.; Leonard, J.L.; Lin, H.Y.; Leinung, M.; Mousa, S.A. Molecular Basis of Nongenomic Actions of Thyroid Hormone. *Vitam. Horm.* **2018**, *106*, 67–96. [CrossRef]
9. Davis, P.J.; Mousa, S.A.; Lin, H.Y. Nongenomic Actions of Thyroid Hormone: The Integrin Component. *Physiol. Rev.* **2021**, *101*, 319–352. [CrossRef] [PubMed]
10. Davis, P.J.; Shih, A.; Lin, H.-Y.; Martino, L.J.; Davis, F.B. Thyroxine promotes association of mitogen-activated protein kinase (MAPK) and nuclear thyroid hormone receptor (TR) and causes serine phosphorylation of TR. *J. Biol. Chem.* **2000**, *275*, 38032–38039. [CrossRef]
11. Tang, H.Y.; Lin, H.Y.; Zhang, S.; Davis, F.B.; Davis, P.J. Thyroid hormone causes mitogen-activated protein kinase-dependent phosphorylation of the nuclear estrogen receptor. *Endocrinology* **2004**, *145*, 3265–3272. [CrossRef] [PubMed]
12. Chi, H.-C.; Tsai, C.-Y.; Tsai, M.-M.; Yeh, C.-T.; Lin, K.-H. Molecular functions and clinical impact of thyroid hormone-triggered autophagy in liver-related diseases. *J. Biomed. Sci.* **2019**, *26*, 24. [CrossRef] [PubMed]
13. Tedeschi, L.; Vassalle, C.; Iervasi, G.; Sabatino, L. Main Factors Involved in Thyroid Hormone Action. *Molecules* **2021**, *26*, 7337. [CrossRef] [PubMed]
14. Incerpi, S.; Gionfra, F.; De Luca, R.; Candelotti, E.; De Vito, P.; Percario, Z.A.; Leone, S.; Gnocchi, D.; Rossi, M.; Caruso, F.; et al. Extranuclear effects of thyroid hormones and analogs during development: An old mechanism with emerging roles. *Front. Endocrinol.* **2022**, *13*, 961744. [CrossRef]
15. Leonard, J.L.; Farwell, A.P.; Yen, P.M.; Chin, W.W.; Stula, M. Differential expression of thyroid hormone receptor isoforms in neurons and astroglial cells. *Endocrinology* **1994**, *135*, 548–555. [CrossRef]
16. Wrutniak-Cabello, C.; Casas, F.; Cabello, G. Thyroid hormone action in mitochondria. *J. Mol. Endocrinol.* **2001**, *26*, 67–77. [CrossRef]
17. Pessemesse, L.; Lepourry, L.; Bouton, K.; Levin, J.; Cabello, G.; Wrutniak-Cabello, C.; Casas, F. p28, a truncated form of TRα1 regulates mitochondrial physiology. *FEBS Lett.* **2014**, *588*, 4037–4043. [CrossRef]
18. Wrutniak-Cabello, C.; Casas, F.; Cabello, G. Mitochondrial T3 receptor and targets. *Mol. Cell. Endocrinol.* **2017**, *458*, 112–120. [CrossRef]
19. Psarra, A.M.; Solakidi, S.; Sekeris, C.E. The mitochondrion as a primary site of action of steroid and thyroid hormones: Presence and action of steroid and thyroid hormone receptors in mitochondria of animal cells. *Mol. Cell. Endocrinol.* **2006**, *246*, 21–33. [CrossRef]
20. Scheller, K.; Seibel, P.; Sekeris, C.E. Glucocorticoid and thyroid hormone receptors in mitochondria of animal cells. *Int. Rev. Cytol.* **2003**, *222*, 1–61. [CrossRef]
21. Psarra, A.M.; Sekeris, C.E. Steroid and thyroid hormone receptors in mitochondria. *IUBMB Life* **2008**, *60*, 210–223. [CrossRef] [PubMed]
22. Finkel, T. Signal transduction by reactive oxygen species. *J. Cell Biol.* **2011**, *194*, 7–15. [CrossRef]
23. Holmström, K.M.; Finkel, T. Cellular mechanisms and physiological consequences of redox-dependent signalling. *Nat. Rev. Mol. Cell Biol.* **2014**, *15*, 411–421. [CrossRef] [PubMed]
24. Sies, H.; Jones, D.P. Reactive Oxygen Species (ROS) as Pleiotropic Physiological Signalling Agents. *Nat. Rev. Mol. Cell Biol.* **2020**, *21*, 363–383. [CrossRef] [PubMed]
25. Fukai, T.; Ushio-Fukai, M. Cross-Talk between NADPH Oxidase and Mitochondria: Role in ROS Signaling and Angiogenesis. *Cells* **2020**, *9*, 1849. [CrossRef]
26. Irazabal, M.V.; Torres, V.E. Reactive Oxygen Species and Redox Signaling in Chronic Kidney Disease. *Cells* **2020**, *9*, 1342. [CrossRef]

27. Poncin, S.; Colin, I.M.; Gérard, A.C. Minimal oxidative load: A prerequisite for thyroid cell function. *J. Endocrinol.* **2009**, *201*, 161–167. [CrossRef]

28. Chartoumpekis, D.V.; Ziros, P.G.; Habeos, I.G.; Sykiotis, G.P. Emerging roles of Keap1/Nrf2 signaling in the thyroid gland and perspectives for bench-to-bedside translation. *Free Radic. Biol. Med.* **2022**, *190*, 276–283. [CrossRef]

29. Thanas, C.; Ziros, P.G.; Chartoumpekis, D.V.; Renaud, C.O.; Sykiotis, G.P. The Keap1/Nrf2 Signaling Pathway in the Thyroid-2020 Update. *Antioxidants* **2020**, *9*, 1082. [CrossRef]

30. Kobayashi, A.; Kang, M.I.; Okawa, H.; Ohtsuji, M.; Zenke, Y.; Chiba, T.; Igarashi, K.; Yamamoto, M. Oxidative stress sensor Keap1 functions as an adaptor for Cul3-based E3 ligase to regulate proteasomal degradation of Nrf2. *Mol. Cell. Biol.* **2004**, *24*, 7130–7139. [CrossRef]

31. Li, J.; Stein, T.D.; Johnson, J.A. Genetic dissection of systemic autoimmune disease in Nrf2-deficient mice. *Physiol. Genom.* **2004**, *18*, 261–272. [CrossRef]

32. Renaud, C.O.; Ziros, P.G.; Chartoumpekis, D.V.; Bongiovanni, M.; Sykiotis, G.P. Keap1/Nrf2 Signaling: A New Player in Thyroid Pathophysiology and Thyroid Cancer. *Front. Endocrinol.* **2019**, *10*, 510. [CrossRef] [PubMed]

33. Ziros, P.G.; Habeos, I.G.; Chartoumpekis, D.V.; Ntalampyra, E.; Somm, E.; Renaud, C.O.; Bongiovanni, M.; Trougakos, I.P.; Yamamoto, M.; Kensler, T.W.; et al. NFE2-Related Transcription Factor 2 Coordinates Antioxidant Defense with Thyroglobulin Production and Iodination in the Thyroid Gland. *Thyroid* **2018**, *28*, 780–798. [CrossRef] [PubMed]

34. Leoni, S.G.; Kimura, E.T.; Santisteban, P.; De la Vieja, A. Regulation of thyroid oxidative state by thioredoxin reductase has a crucial role in thyroid responses to iodide excess. *Mol. Endocrinol.* **2011**, *25*, 1924–1935. [CrossRef]

35. Poncin, S.; Gerard, A.C.; Boucquey, M.; Senou, M.; Calderon, P.B.; Knoops, B.; Lengele, B.; Many, M.C.; Colin, I.M. Oxidative stress in the thyroid gland: From harmlessness to hazard depending on the iodine content. *Endocrinology* **2008**, *149*, 424–433. [CrossRef] [PubMed]

36. Hayes, J.D.; Dinkova-Kostova, A.T. The Nrf2 regulatory network provides an interface between redox and intermediary metabolism. *Trends Biochem. Sci.* **2014**, *39*, 199–218. [CrossRef]

37. Teshiba, R.; Tajiri, T.; Sumitomo, K.; Masumoto, K.; Taguchi, T.; Yamamoto, K. Identification of a KEAP1 germline mutation in a family with multinodular goitre. *PLoS ONE* **2013**, *8*, e65141. [CrossRef]

38. Nishihara, E.; Hishinuma, A.; Kogai, T.; Takada, N.; Hirokawa, M.; Fukata, S.; Ito, M.; Yabuta, T.; Nishikawa, M.; Nakamura, H.; et al. A Novel Germline Mutation of KEAP1 (R483H) Associated with a Non-Toxic Multinodular Goiter. *Front. Endocrinol.* **2016**, *7*, 131. [CrossRef]

39. Taguchi, K.; Maher, J.M.; Suzuki, T.; Kawatani, Y.; Motohashi, H.; Yamamoto, M. Genetic analysis of cytoprotective functions supported by graded expression of Keap1. *Mol. Cell. Biol.* **2010**, *30*, 3016–3326. [CrossRef]

40. Ziros, P.G.; Renaud, C.O.; Chartoumpekis, D.V.; Bongiovanni, M.; Habeos, I.G.; Liao, X.H.; Refetoff, S.; Kopp, P.A.; Brix, K.; Sykiotis, G.P. Mice Hypomorphic for Keap1, a Negative Regulator of the Nrf2 Antioxidant Response, Show Age-Dependent Diffuse Goiter with Elevated Thyrotropin Levels. *Thyroid* **2021**, *31*, 23–35. [CrossRef]

41. Yagishita, Y.; Uruno, A.; Fukutomi, T.; Saito, R.; Saigusa, D.; Pi, J.; Fukamizu, A.; Sugiyama, F.; Takahashi, S.; Yamamoto, M. Nrf2 Improves Leptin and Insulin Resistance Provoked by Hypothalamic Oxidative Stress. *Cell Rep.* **2017**, *18*, 2030–2044. [CrossRef] [PubMed]

42. Slocum, S.L.; Skoko, J.J.; Wakabayashi, N.; Aja, S.; Yamamoto, M.; Kensler, T.W.; Chartoumpekis, D.V. Keap1/Nrf2 pathway activation leads to a repressed hepatic gluconeogenic and lipogenic program in mice on a high-fat diet. *Arch. Biochem. Biophys.* **2016**, *591*, 57–65. [CrossRef] [PubMed]

43. Lau, A.; Villeneuve, N.F.; Sun, Z.; Wong, P.K.; Zhang, D.D. Dual roles of Nrf2 in cancer. *Pharmacol. Res.* **2008**, *58*, 262–270. [CrossRef] [PubMed]

44. Maier, J.; van Steeg, H.; van Oostrom, C.; Karger, S.; Paschke, R.; Krohn, K. Deoxyribonucleic acid damage and spontaneous mutagenesis in the thyroid gland of rats and mice. *Endocrinology* **2006**, *147*, 3391–3397. [CrossRef] [PubMed]

45. Menegon, S.; Columbano, A.; Giordano, S. The dual roles of NRF2 in cancer. *Trends Mol. Med.* **2016**, *22*, 578–593. [CrossRef]

46. Jeddi, F.; Soozangar, N.; Sadeghi, M.R.; Somi, M.H.; Samadi, N. Contradictory roles of Nrf2/Keap1 signaling pathway in cancer prevention/promotion and chemoresistance. *DNA Repair* **2017**, *54*, 13–21. [CrossRef]

47. Mancini, A.; Di Segni, C.; Raimondo, S.; Olivieri, G.; Silvestrini, A.; Meucci, E.; Currò, D. Thyroid Hormones, Oxidative Stress, and Inflammation. *Mediat. Inflamm.* **2016**, *2016*, 6757154. [CrossRef]

48. He, L.; He, T.; Farrar, S.; Ji, L.; Liu, T.; Ma, X. Antioxidants Maintain Cellular Redox Homeostasis by Elimination of Reactive Oxygen Species. *Cell. Physiol. Biochem.* **2017**, *44*, 532–553. [CrossRef]

49. Asayama, K.; Kato, K. Oxidative muscular injury and its relevance to hyperthyroidism. *Free Radic. Biol. Med.* **1990**, *8*, 293–303. [CrossRef]

50. Dekkers, J.C.; van Doornen, L.J.; Kemper, H.C. The role of antioxidant vitamins and enzymes in the prevention of exercise-induced muscle damage. *Sports Med.* **1996**, *21*, 213–238. [CrossRef]

51. Venditti, P.; Di Meo, S. Thyroid hormone-induced oxidative stress. *Cell. Mol. Life Sci.* **2006**, *63*, 414–434. [CrossRef]

52. Romanque, P.; Cornejo, P.; Valdés, S.; Videla, L.A. Thyroid hormone administration induces rat liver Nrf2 activation: Suppression by N-acetylcysteine pretreatment. *Thyroid* **2011**, *21*, 655–662. [CrossRef]

53. Sun, Z.; Huang, Z.; Zhang, D.D. Phosphorylation of Nrf2 at multiple sites by MAP kinases has a limited contribution in modulating the Nrf2-dependent antioxidant response. *PLoS ONE* **2009**, *4*, e6588. [CrossRef]

54. Venediktova, N.; Solomadin, I.; Nikiforova, A.; Starinets, V.; Mironova, G. Functional State of Rat Heart Mitochondria in Experimental Hyperthyroidism. *Int. J. Mol. Sci.* **2021**, *22*, 11744. [CrossRef]

55. Letts, J.A.; Sazanov, L.A. Clarifying the supercomplex: The higher-order organization of the mitochondrial electron transport chain. *Nat. Struct. Mol. Biol.* **2017**, *24*, 800–808. [CrossRef]

56. Lobo-Jarne, T.; Ugalde, C. Respiratory chain supercomplexes: Structures, function and biogenesis. *Semin. Cell. Dev. Biol.* **2018**, *76*, 179–190. [CrossRef] [PubMed]

57. Venediktova, N.I.; Mashchenko, O.V.; Talanov, E.Y.; Belosludtseva, N.V.; Mironova, G.D. Energy metabolism and oxidative status of rat liver mitochondria in conditions of experimentally induced hyperthyroidism. *Mitochondrion* **2020**, *52*, 190–196. [CrossRef] [PubMed]

58. Venediktova, N.I.; Pavlik, L.L.; Belosludtseva, N.V.; Khmil, N.V.; Murzaeva, S.V.; Mironova, G.D. Formation of lamellar bodies in rat liver mitochondria in hyperthyroidism. *J. Bioenerg. Biomembr.* **2018**, *50*, 289–295. [CrossRef]

59. Piantadosi, C.A.; Carraway, M.S.; Babiker, A.; Suliman, H.B. Heme oxygenase-1 regulates cardiac mitochondrial biogenesis via Nrf2-mediated transcriptional control of nuclear respiratory factor-1. *Circ. Res.* **2008**, *103*, 1232–1240. [CrossRef] [PubMed]

60. Mastorci, F.; Sabatino, L.; Vassalle, C.; Pingitore, A. Cardioprotection and Thyroid Hormones in the Clinical Setting of Heart Failure. *Front. Endocrinol.* **2020**, *10*, 927. [CrossRef] [PubMed]

61. Kutala, V.K.; Khan, M.; Angelos, M.G.; Kuppusamy, P. Role of oxygen inpostischemic myocardial injury. *Antioxid. Redox Signal.* **2007**, *9*, 1193–1206. [CrossRef] [PubMed]

62. Hill, M.F.; Singal, P.K. Antioxidant and oxidative stress changes during heart failure subsequent to myocardial infarction in rats. *Am. J. Pathol.* **1996**, *148*, 291–300. [PubMed]

63. Huss, J.M.; Kelly, D.P. Mitochondrial energy metabolism in heart failure: A question of balance. *J. Clin. Investig.* **2005**, *115*, 547–555. [CrossRef]

64. Kimur, T.; Kotajima, N.; Kanda, T.; Kuwabara, A.; Fukumura, Y.; Kobayashi, I. Correlation of circulating interleukin-10 with thyroid hormone in acute myocardial infarction. *Res. Commun. Mol. Pathol. Pharmacol.* **2001**, *110*, 53–58.

65. Pantos, C.; Mourouzis, I.; Saranteas, T.; Brozou, V.; Galanopoulos, G.; Kostopanagiotou, G.; Cokkinos, D.V. Acute T3 treatment protects the heart against ischemia-reperfusion injury via TRα1 receptor. *Mol. Cell. Biochem.* **2011**, *353*, 235–241. [CrossRef]

66. Iervasi, G.; Pingitore, A.; Landi, P.; Raciti, M.; Ripoli, A.; Scarlattini, M.; L'Abbate, A.; Donato, L. Low-T3 syndrome: A strong prognostic predictor of death in patients with heart disease. *Circulation* **2003**, *107*, 708–713. [CrossRef]

67. Pantos, C.; Mourouzis, I.; Markakis, K.; Tsagoulis, N.; Panagiotou, M.; Cokkinos, D.V. Long-term thyroid hormone administration reshapes left ventricular chamber and improves cardiac function after myocardial infarction in rats. *Basic Res. Cardiol.* **2008**, *103*, 308–318. [CrossRef]

68. Pantos, C.; Mourouzis, I.; Tsagoulis, N.; Markakis, K.; Galanopoulos, G.; Roukounakis, N.; Perimenis, P.; Liappas, A.; Cokkinos, D.V. Thyroid hormone at supra-physiological dose optimizes cardiac geometry and improves cardiac function in rats with old myocardial infarction. *J. Physiol. Pharmacol.* **2009**, *60*, 49–56. [PubMed]

69. de Castro, A.L.; Tavares, A.V.; Campos, C.; Fernandes, R.O.; Siqueira, R.; Conzatti, A.; Bicca, A.M.; Fernandes, T.R.; Sartório, C.L.; Schenkel, P.C.; et al. Cardioprotective effects of thyroid hormones in a rat model of myocardial infarction are associated with oxidative stress reduction. *Mol. Cell. Endocrinol.* **2014**, *391*, 22–29. [CrossRef] [PubMed]

70. Jaiswal, A.K. Nrf2 signaling in coordinated activation of antioxidant gene expression. *Free Radic. Biol. Med.* **2004**, *36*, 1199–1207. [CrossRef]

71. Itoh, K.; Ishii, T.; Wakabayashi, N.; Yamamoto, M. Regulatory mechanisms of cellular response to oxidative stress. *Free Radic. Res.* **1999**, *31*, 319–324. [CrossRef] [PubMed]

72. Zhang, H.; Liu, Y.; Cao, X.; Wang, W.; Cui, X.; Yang, X.; Wang, Y.; Shi, J. Nrf2 Promotes Inflammation in Early Myocardial Ischemia-Reperfusion via Recruitment and Activation of Macrophages. *Front. Immunol.* **2021**, *12*, 763760. [CrossRef] [PubMed]

73. Takemura, G.; Nakagawa, M.; Kanamori, H.; Minatoguchi, S.; Fujiwara, H. Benefits of reperfusion beyond infarct size limitation. *Cardiovasc. Res.* **2009**, *83*, 269–276. [CrossRef]

74. Eefting, F.; Rensing, B.; Wigman, J.; Pannekoek, W.J.; Liu, W.M.; Cramer, M.J.; Lips, D.J.; Doevendans, P.A. Role of apoptosis in reperfusion injury. *Cardiovasc. Res.* **2004**, *61*, 414–426. [CrossRef] [PubMed]

75. Sun, Y. Myocardial repair/remodelling following infarction: Roles of local factors. *Cardiovasc. Res.* **2009**, *81*, 482–490. [CrossRef]

76. Zhou, S.; Sun, W.; Zhang, Z.; Zheng, Y. The role of Nrf2-mediated pathway in cardiac remodeling and heart failure. *Oxid. Med. Cell. Longev.* **2014**, *2014*, 260429. [CrossRef]

77. Xu, B.; Zhang, J.; Strom, J.; Lee, S.; Chen, Q.M. Myocardial ischemic reperfusion induces de novo Nrf2 protein translation. *Biochim. Biophys. Acta* **2014**, *1842*, 1638–1647. [CrossRef] [PubMed]

78. Shanmugam, G.; Narasimhan, M.; Tamowski, S.; Darley-Usmar, V.; Rajasekaran, N.S. Constitutive activation of Nrf2 induces a stable reductive state in the mouse myocardium. *Redox Biol.* **2017**, *12*, 937–945. [CrossRef]

79. Muthusamy, V.R.; Kannan, S.; Sadhaasivam, K.; Gounder, S.S.; Davidson, C.J.; Boeheme, C.; Hoidal, J.R.; Wang, L.; Rajasekaran, N.S. Acute exercise stress activates Nrf2/ARE signaling and promotes antioxidant mechanisms in the myocardium. *Free Radic. Biol. Med.* **2012**, *52*, 366–376. [CrossRef]

80. Gutiérrez-Cuevas, J.; Galicia-Moreno, M.; Monroy-Ramírez, H.C.; Sandoval-Rodriguez, A.; García-Bañuelos, J.; Santos, A.; Armendariz-Borunda, J. The Role of NRF2 in Obesity-Associated Cardiovascular Risk Factors. *Antioxidants* **2022**, *11*, 235. [CrossRef]

81. Deng, C.; Sun, Z.; Tong, G.; Yi, W.; Ma, L.; Zhao, B.; Cheng, L.; Zhang, J.; Cao, F.; Yi, D. α-Lipoic acid reduces infarct size and preserves cardiac function in rat myocardial ischemia/reperfusion injury through activation of PI3K/Akt/Nrf2 pathway. *PLoS ONE* **2013**, *8*, e58371. [CrossRef] [PubMed]

82. Ashrafian, H.; Czibik, G.; Bellahcene, M.; Aksentijević, D.; Smith, A.C.; Mitchell, S.J.; Dodd, M.S.; Kirwan, J.; Byrne, J.J.; Ludwig, C.; et al. Fumarate is cardioprotective via activation of the Nrf2 antioxidant pathway. *Cell Metab.* **2012**, *15*, 361–371. [CrossRef] [PubMed]

83. Katsumata, Y.; Shinmura, K.; Sugiura, Y.; Tohyama, S.; Matsuhashi, T.; Ito, H.; Yan, X.; Ito, K.; Yuasa, S.; Ieda, M.; et al. Endogenous prostaglandin D2 and its metabolites protect the heart against ischemia-reperfusion injury by activating Nrf2. *Hypertension* **2014**, *63*, 80–87. [CrossRef]

84. Liu, C.; Zhou, J.; Wang, B.; Zheng, Y.; Liu, S.; Yang, W.; Li, D.; He, S.; Lin, J. Bortezomib alleviates myocardial ischemia reperfusion injury via enhancing of Nrf2/HO-1 signaling pathway. *Biochem. Biophys. Res. Commun.* **2021**, *556*, 207–214. [CrossRef]

85. Liu, Z.; Zhang, F.; Zhao, L.; Zhang, X.; Li, Y.; Liu, L. Protective Effect of Pravastatin on Myocardial Ischemia Reperfusion Injury by Regulation of the miR-93/Nrf2/ARE Signal Pathway. *Drug Des. Dev. Ther.* **2020**, *14*, 3853–3864. [CrossRef] [PubMed]

86. Loboda, A.; Jazwa, A.; Grochot-Przeczek, A.; Rutkowski, A.J.; Cisowski, J.; Agarwal, A.; Jozkowicz, A.; Dulak, J. Heme oxygenase-1 and the vascular bed: From molecular mechanisms to therapeutic opportunities. *Antioxid. Redox Signal.* **2008**, *10*, 1767–1812. [CrossRef]

87. Das, A.; Gopalakrishnan, B.; Voss, O.H.; Doseff, A.I.; Villamena, F.A. Inhibition of ROS-induced apoptosis in endothelial cells by nitrone spin traps via induction of phase II enzymes and suppression of mitochondria-dependent pro-apoptotic signaling. *Biochem. Pharmacol.* **2012**, *84*, 486–497. [CrossRef] [PubMed]

88. Zeng, B.; Liu, L.; Liao, X.; Zhang, C.; Ruan, H. Thyroid hormone protects cardiomyocytes from H2O2-induced oxidative stress via the PI3K-AKT signaling pathway. *Exp. Cell Res.* **2019**, *380*, 205–215. [CrossRef]

89. Rota, M.; Boni, A.; Urbanek, K.; Padin-Iruegas, M.E.; Kajstura, T.J.; Fiore, G.; Kubo, H.; Sonnenblick, E.H.; Musso, E.; Houser, S.R.; et al. Nuclear targeting of Akt enhances ventricular function and myocyte contractility. *Circ. Res.* **2005**, *97*, 1332–1341. [CrossRef]

90. Mourouzis, I.; Mantzouratou, P.; Galanopoulos, G.; Kostakou, E.; Roukounakis, N.; Kokkinos, A.D.; Pantos, C. Dose-dependent effects of thyroid hormone on post-ischemic cardiac performance: Potential involvement of Akt and ERK signalings. *Mol. Cell. Biochem.* **2012**, *363*, 235–243. [CrossRef]

91. Mishra, P.; Paital, B.; Jena, S.; Swain, S.S.; Kumar, S.; Yadav, M.K.; Chainy, G.B.N.; Samanta, L. Possible activation of NRF2 by Vitamin E/Curcumin against altered thyroid hormone induced oxidative stress via NFκB/AKT/mTOR/KEAP1 signalling in rat heart. *Sci. Rep.* **2019**, *9*, 7408. [CrossRef] [PubMed]

92. Subudhi, U.; Das, K.; Paital, B.; Bhanja, S.; Chainy, G.B. Alleviation of enhanced oxidative stress and oxygen consumption of L-thyroxine induced hyperthyroid rat liver mitochondria by vitamin E and curcumin. *Chem. Biol. Interact.* **2008**, *173*, 105–114. [CrossRef] [PubMed]

93. Maheshwari, R.K.; Singh, A.K.; Gaddipati, J.; Srimal, R.C. Multiple biological activities of curcumin: A short review. *Life Sci.* **2006**, *78*, 2081–2087. [CrossRef]

94. Priyadarsini, K.I.; Maity, D.K.; Naik, G.H.; Kumar, M.S.; Unnikrishnan, M.K.; Satav, J.G.; Mohan, H. Role of phenolic O-H and methylene hydrogen on the free radical reactions and antioxidant activity of curcumin. *Free Radic. Biol. Med.* **2003**, *35*, 475–484. [CrossRef] [PubMed]

95. Witztum, J.L.; Steinberg, D. Role of oxidized low density lipoprotein in atherogenesis. *J. Clin. Investig.* **1991**, *88*, 1785–1792. [CrossRef] [PubMed]

96. Bilmen, J.G.; Khan, S.Z.; Javed, M.H.; Michelangeli, F. Inhibition of the SERCA Ca^{2+} pumps by curcumin. Curcumin putatively stabilizes the interaction between the nucleotide-binding and phosphorylation domains in the absence of ATP. *Eur. J. Biochem.* **2001**, *268*, 6318–6327. [CrossRef]

97. Jiang, Y.; Xie, G.; Alimujiang, A.; Xie, H.; Yang, W.; Yin, F.; Huang, D. Protective Effects of Querectin against MPP+-Induced Dopaminergic Neurons Injury via the Nrf2 Signaling Pathway. *Front. Biosci.* **2023**, *28*, 42. [CrossRef]

98. Panda, S.; Kar, A. Annona squamosa seed extract in the regulation of hyperthyroidism and lipid-peroxidation in mice: Possible involvement of quercetin. *Phytomedicine* **2007**, *14*, 799–805. [CrossRef]

99. Zhao, P.; Hu, Z.; Ma, W.; Zang, L.; Tian, Z.; Hou, Q. Quercetin alleviates hyperthyroidism-induced liver damage via Nrf2 signaling pathway. *Biofactors* **2020**, *46*, 608–619. [CrossRef] [PubMed]

100. Giuliani, C.; Bucci, I.; Di Santo, S.; Rossi, C.; Grassadonia, A.; Piantelli, M.; Monaco, F.; Napolitano, G. The flavonoid quercetin inhibits thyroid-restricted genes expression and thyroid function. *Food Chem. Toxicol.* **2014**, *66*, 23–29. [CrossRef]

Article

High-Frequency Repetitive Magnetic Stimulation Activates Bactericidal Activity of Macrophages via Modulation of p62/Keap1/Nrf2 and p38 MAPK Pathways

Therese B. Deramaudt [1,*], Ahmad Chehaitly [1], Théo Charrière [1], Julie Arnaud [1] and Marcel Bonay [1,2]

[1] U1179 INSERM, END-ICAP, UFR des Sciences de la Santé-Simone Veil, Université de Versailles Saint-Quentin-en-Yvelines, 78180 Montigny-le-Bretonneux, France; marcel.bonay@aphp.fr (M.B.)
[2] Service de Physiologie-Explorations Fonctionnelles, Hôpital Ambroise Paré, Assistance Publique-Hôpitaux de Paris, 92100 Boulogne-Billancourt, France
* Correspondence: therese.deramaudt@uvsq.fr; Tel.: +33-1704-29415

Abstract: The effects of repetitive magnetic stimulation (rMS) have predominantly been studied in excitable cells, with limited research in non-excitable cells. This study aimed to investigate the impact of rMS on macrophages, which are crucial cells in the innate immune defense. THP-1-derived macrophages subjected to a 5 min session of 10 Hz rMS exhibited increased Nrf2 activation and decreased Keap1 expression. We found that activation of the Nrf2 signaling pathway relied on rMS-induced phosphorylation of p62. Notably, rMS reduced the intracellular survival of *Staphylococcus aureus* in macrophages. Silencing Nrf2 using siRNA in THP-1-derived macrophages or utilizing Nrf2 knockout in alveolar macrophages abolished this effect. Additionally, rMS attenuated the expression of IL-1β and TNF-α inflammatory genes by *S. aureus* and inhibited p38 MAPK activation. These findings highlight the capacity of rMS to activate the non-canonical Nrf2 pathway, modulate macrophage function, and enhance the host's defense against bacterial infection.

Keywords: Nrf2; THP-1-derived macrophages; alveolar macrophages; repetitive magnetic stimulation; p38 MAPK; p62; Nrf2 knockout mice; Keap1; *Staphylococcus aureus*

Citation: Deramaudt, T.B.; Chehaitly, A.; Charrière, T.; Arnaud, J.; Bonay, M. High-Frequency Repetitive Magnetic Stimulation Activates Bactericidal Activity of Macrophages via Modulation of p62/Keap1/Nrf2 and p38 MAPK Pathways. *Antioxidants* **2023**, *12*, 1695. https://doi.org/10.3390/antiox12091695

Academic Editor: Reto Asmis

Received: 21 July 2023
Revised: 9 August 2023
Accepted: 16 August 2023
Published: 30 August 2023

1. Introduction

Macrophages, as sentinel cells of the innate immune system, play a crucial role in innate immunity by recognizing and eliminating foreign invaders. Their ability to rapidly respond to microbial challenges and modulate immune responses is essential for host defense [1]. Macrophages migrate to sites of aggression, phagocytize external particles including microorganisms and cell debris, and initiate rapid responses for their elimination. While phagocytosis is essential for protection against bacterial infection, it can also serve as a pathway for bacteria, such as *Staphylococcus aureus*, to enter macrophages. These intracellular pathogens have developed evasion mechanisms to survive, escape, and disseminate within the host.

S. aureus is a Gram-positive bacterium that has colonized the human skin and mucous membranes. A versatile and commensal opportunistic pathogen, *S. aureus* can cause various infections in humans, including skin and soft tissue infections, pneumonia, osteomyelitis, and endocarditis [2]. Studies in a murine model of airway infection have highlighted the active contribution of macrophages in clearing methicillin-resistant *S. aureus* from the lungs. Depletion of alveolar macrophages in this mouse model resulted in increased mortality [3]. The prevalence of antibiotic-resistant *S. aureus* strains presents significant challenges, leading to higher morbidity, mortality, and healthcare costs worldwide [4]. Therefore, it is crucial to develop novel therapeutic strategies that can suppress the intracellular load of *S. aureus* while minimizing potential side effects to the host. *S. aureus* has the ability to manipulate the MAPK signaling pathway and modulate host cell responses for its benefit.

Activation of specific MAPK pathways, including ERK, JNK and p38, has been implicated in promoting cell processes such as internalization, apoptosis and the expression of inflammatory cytokines and chemokines [5–8].

Our previous studies have confirmed the significant role of Nrf2 activity in influencing macrophage bactericidal function and immune responses [5,9]. Nrf2, a master transcriptional factor, acts as a key regulator in the adaptive responses to oxidative stress by regulating the expression of genes involved in antioxidant and cytoprotective mechanisms. Activation of Nrf2 leads to the transcriptional upregulation of cytoprotective genes that encode molecules involved in detoxifying, antioxidant and anti-inflammatory pathways [10].

Interestingly, Nrf2 can be activated through both canonical and non-canonical mechanisms [11]. The canonical mechanism involves the activation of Nrf2 through the Nrf2/Keap1-Cullin3-E3 ubiquitin ligase complex. Under homeostatic conditions, Nrf2 is sequestered in the cytoplasm by Keap1, which promotes its ubiquitination and subsequent degradation by the proteasome. However, in response to oxidative stress or electrophilic insults, Keap1 undergoes a conformational change, leading to the release of Nrf2 and its translocation into the nucleus. Nuclear Nrf2 then heterodimerizes with small Maf proteins to form a cofactor complex that binds to antioxidant responsive element (ARE) sequences, resulting in the upregulation of genes coding for proteins involved in antioxidant and anti-inflammatory responses.

Several non-canonical mechanisms of Nrf2 activation have also been described and involved direct interactions between Keap1 and various proteins, such as the scaffold protein p62/SQSTM1 [12], the dipeptidyl peptidase III enzyme [13], the nuclear protein prothymosin α [14], and the nuclear protein partner and localizer of BRCA2 (PALB2) [15]. These interactions promote the nuclear accumulation of Nrf2 and the transcriptional regulation of Nrf2-targeted genes.

Repetitive magnetic stimulation (rMS) is a non-invasive technique primarily used in the treatment of neurological disorders, with potential therapeutic applications in neuroregeneration and wound healing [16,17]. While the cellular responses elicited by rMS have been extensively documented in excitable cells, limited studies have investigated the impact of rMS on non-excitable cells. A recent report has highlighted the ability of microglia, the resident immune cells of the central nervous system, to modulate neural excitability and plasticity through the release of cytokines following high-frequency repetitive transcranial magnetic stimulation (rTMS) in mice [18]. Additionally, high-frequency rTMS treatment of bone mesenchymal stromal cells has been shown to induce autophagy through the NMDAR-Ca^{2+}-ERK-mTOR signaling pathway [19]. Still, it is important to note that the effects of rMS on cellular processes can vary depending on the cell type and are influenced by various stimulatory parameters, including the intensity, frequency, and duration of the magnetic pulses. Nonetheless, the molecular and cellular mechanisms underlying the specific effects of rMS on cell types outside the nervous system remain poorly understood.

This study aims to explore the effect of rMS on the intracellular survival of *S. aureus* in THP-1-derived macrophages and murine primary alveolar macrophages. We demonstrate that the induction of the Nrf2/Keap1/p62 signaling pathway by rMS is critical for controlling the intracellular load of *S. aureus*. Furthermore, we established that rMS treatment represses the *S. aureus*-mediated activation of p38 MAPK. To the best of our knowledge, this is the first study to provide insights into the molecular and cellular mechanisms of a single 5 min session of high-frequency rMS in in vitro THP-1-derived macrophages and ex vivo murine primary alveolar macrophages.

2. Materials and Methods

2.1. Antibodies and Reagents

A Maxima First Strand cDNA synthesis kit, Lipofectamine® LTX & PLUS™ Reagent, Trizol, CellROX™ Green Oxidative Stress Reagent, Fluoromount-G™ (EMS), RIPA Lysis and Extraction Buffer, sodium pyruvate, Pierce protease and phosphatase inhibitor cocktail,

SYTO9^{TM} dye from the LIVE/DEAD BacLight bacterial counting kit, anti-phospho-p62 antibody (pS349, Invitrogen), 3-(4,5-dimethylthiazol-2-yl)-2,5-diphenyltetrazolium bromide (MTT) cell viability assay kit (Biotium Inc.), secondary antibodies donkey anti-mouse Alexa Fluor 488, donkey anti-rabbit Alexa Fluor 594, and an iBind Western blot system were purchased from Fisher Scientific (Illkirch, France). A nuclear extraction kit, phorbol 12-myristate 13-acetate (PMA), dimethyl sulfoxide (DMSO) $\geq$ 99.7%, β-mercaptoethanol, paraformaldehyde, H_2O_2 30% in H_2O, rabbit anti-phosphorylated ERK (pT185/pY187) antibody, rabbit anti-ERK antibody, mouse anti-phospho-p38 (pT180/pY182) antibody, rabbit-anti-p62 antibody, rabbit anti-Keap1 antibody, and mouse anti-GAPDH antibody were obtained from Merck (Saint-Quentin-Fallavier, France). iTaq SYBRgreen qPCR Supermix, a DC protein assay kit, and 4–20% mini-Protean precast protein gels were purchased from BioRad (Marnes-la-Coquette, France). RPMI (Roswell Park Memorial Institute) 1640 culture medium, phosphate-buffered saline (PBS), gentamicin, penicillin/streptomycin mixture, Hoechst 33342, and heat-inactivated fetal bovine serum were obtained from Eurobio Scientific (Les Ulis, France). CO_2 $\geq$ 99.99% was purchased from Messer (France). Annexin V-fluorescein isothiocyanate (FITC) apoptosis detection reagent, mouse anti-HO-1 antibody, rabbit anti-NQO1 antibody, and rabbit anti-Lamin A/C antibody were purchased from Abcam (Paris, France). Mouse anti-phospho-JNK (pT183/pY185) antibody, mouse anti-JNK antibody, mouse anti-p38 antibody were acquired from BD Biosciences (Le Pont de Claix, France). Anti-Nrf2 antibody was obtained from ProteinTech (Manchester, UK). Tryptic soy broth (TSB) and agar-supplemented TSB (TSB agar) were obtained from Conda laboratories (Dutscher, Bernolsheim, France). IRDye$^{®}$ 680CW goat anti-mouse IgG, and IRDye$^{®}$ 800CW Goat anti-Rabbit IgG secondary antibodies were purchased from LI-COR$^{®}$ Biosciences (Bad Homburg, Germany).

2.2. Repetitive Magnetic Stimulation (rMS)

Repetitive magnetic stimulation was delivered using a B65 refrigerated butterfly coil connected to a MagPro R30 magnetic stimulator (Magventure, Denmark) purchased from Mag2Health (Villennes-sur-Seine, France). Cells were placed on the cool coil and were subjected to one session of 10 series of 100 biphasic pulses with a 25 s interval between series, resulting in a total of 1000 pulses (10 Hz at 80% of maximum output). In parallel, control cells were exposed to the same environment for the same duration as the stimulated cells but without stimulation. rMS at 5 Hz and 15 Hz were also tested and showed no effect on HO-1 and NQO1 mRNA expressions (Supplementary Figure S1).

2.3. Bacterial Strain, Growth Culture, and Fluorescence Labeling

The Gram-positive bacteria *Staphylococcus aureus* strain (ATCC 25923) was grown aerobically in Trypticase soy broth to the optical density of 1 at 37 °C under agitation. Glycerol stocks of *S. aureus* were prepared and aliquoted. When required, frozen stocks were thawed and bacteria were diluted in sterile PBS at the appropriate concentration.

For phagocytosis assay, *S. aureus* was incubated with SYTO9 dye for 15 min in the dark following the manufacturer's instructions. For immunofluorescence microscopy, *S. aureus* was incubated with 10 µg/mL Hoechst 33342 in the dark. After 1 h, Hoechst 33342-stained bacteria were washed twice with PBS and resuspended in PBS.

2.4. Cell Culture, Cell Differentiation, and Cell Transfection

Human monocytic THP-1 cell line (ATCC$^{®}$ TIB-202^{TM}) was maintained in RPMI 1640 Glutabio medium supplemented with 10% heat-inactivated fetal bovine serum, 10 mM HEPES buffer, 1 mM sodium pyruvate, and 50 µM β-mercaptoethanol in a humidified atmosphere at 37 °C and 5% CO_2. Terminal differentiation of THP-1 to macrophages was obtained by rinsing cells twice with PBS prior to incubation with 50 nM PMA in β-mercaptoethanol-free complete RPMI 1640 medium for 48 h.

S. aureus was added to the cell culture at a multiplicity of infection (MOI) of 10. For experiments requiring an incubation time longer than 3 h, 10 µg/mL gentamicin was added to the cell culture medium to inhibit growth of extracellular bacteria.

In order to knockdown Nrf2, THP-1-derived macrophages were transfected with a pre-designed siRNA targeting Nrf2 (s9492; ThermoFisher Scientific) or a universal negative control siRNA (1027310; Qiagen) using Lipofectamine LTX following the manufacturer's recommended protocol.

2.5. Animals and Isolation of Primary Mouse Alveolar Macrophages

Breeding of the C57BL/6J Nrf2 knockout (Nrf2$^{-/-}$) mouse strain, provided by Dr Yamamoto (Tohoku University) and purchased from Riken BRC [20], was conducted according to National and European legislation and institutional guidelines for the care and use of laboratory animals approved by the Animal Ethics Committee CEEA47 PELVIPHARM (2Care animal facility, University of Versailles Saint-Quentin-en-Yvelines, UFR Santé Simone Veil) and the Ministry of Higher Education, Research and Innovation APAFIS#8785-201610191723731v3. Mice were maintained in a standard 12 h light/12 h dark cycle with access to food and water ad libitum.

Bronchoalveolar lavage was performed following carbon dioxide euthanasia of mice and alveolar macrophages were collected via repeated lavages with 1 mL ice-cold PBS. After cell resuspension, total cell count was obtained using Countess automated Cell counter (Invitrogen, ThermoFisher Scientific, Waltham, MA, USA). Primary alveolar macrophages were maintained in complete RPMI 1640 medium complemented with penicillin/streptomycin mixture overnight. The antibiotics were removed prior to performing ex vivo experiments by washing cells twice with PBS, followed by the addition of antibiotics-free complete RPMI 1640 medium.

2.6. Bacteria Intracellular Survival Assay

THP-1-derived macrophages, seeded in 24-well plates at 2.5×10^5 cells/well, were treated with rMS for 3 h prior to infection with *S. aureus* at an MOI of 10. Gentamicin was added to the cell culture medium 1 h after infection to inhibit growth of extracellular bacteria, and macrophages were incubated for a total of 24 h before the colony-forming unit (CFU) assay. Briefly, macrophages were washed twice with ice-cold PBS followed by 20 min incubation in 1 mL ice-cold sterile water to lyse cells. Numeration of intracellular bacteria was obtained by plating 5-fold serial dilutions on TSB solidified with 1.5% agar and incubating at 37 °C for 24 h.

2.7. Subcellular Fractionation and Immunoblotting

For the subcellular fractionation, a nuclear extraction kit was used according to the manufacturer's protocol. Briefly, following treatment of THP-1-derived macrophages, 3×10^6 cells per condition were collected and washed with PBS. After centrifugation, cell pellets were incubated for 15 min with ice-cold cytoplasmic lysis buffer supplemented with a protease and phosphatase inhibitor cocktail. Cell membranes were then disrupted via repeated passages through a 27G needle. After 20 min centrifugation at $8000 \times g$ at 4 °C, the cytoplasmic fractions were collected and the remaining cell pellets were resuspended in ice-cold nuclear extraction buffer supplemented with protease and phosphatase inhibitor cocktail. Nuclear membranes were disrupted via repeated passages through a 27G needle, followed by incubation of the nuclear suspensions at 4 °C on an orbital shaker. After 1 h, nuclear fractions were collected by centrifugation at $16{,}000 \times g$ for 5 min at 4 °C.

For extraction of total protein lysates, 1×10^6 THP-1-derived macrophages were rinsed once with PBS and lysed in ice-cold RIPA buffer supplemented with a protease and phosphatase inhibitor cocktail. After 30 min incubation on ice, cell lysates were centrifuged at $12{,}000 \times g$ for 5 min at 4 °C. The supernatants containing protein extracts were collected.

Protein concentrations for total protein extracts and subcellular fractions were measured using a DC protein assay kit. Protein extracts were resolved by SDS-PAGE and transferred to a polyvinylidene difluoride membrane (Immobilon-FL, Merck, Darmstadt, Germany). Immunoblotting was performed using the iBind flex western system according to the manufacturer's instructions. Briefly, primary antibodies targeting Nrf2, HO-1, NQO1, p62, phospho-p62, p38, phospho-p38, ERK, phospho-ERK, JNK, phospho-JNK, GAPDH, and secondary antibodies, IRDye680RD and IRDye800RD, were diluted in iBind Flex FD solution. Fluorescence signals were acquired using Odyssey CLx imaging system (LI-COR) and densitometric analysis was achieved using Image Studio Lite v4.0.

2.8. Total RNA Extraction and Real-Time Quantitative PCR Analysis

Total RNA was isolated from treated THP-1-derived macrophages using Trizol reagent and chloroform extraction technique following the manufacturer's instructions. RNA concentrations were determined using the NanoPhotometer® N120 (Implen; München, Germany). First, 1 µg of total RNA was reverse transcribed to cDNA using a Maxima First strand cDNA synthesis kit. Quantitative analysis was achieved using real-time quantitative PCR (RT-qPCR), with each cDNA sample performed in triplicate. qPCR was realized using the BioRad CFX384 Touch Real-Time PCR Detection system and iTaq SYBRgreen qPCR mix. Table 1 lists the specific primers used for qPCR, Nrf2, HO-1, NQO1, IL-6, IL-1β, TNF-α, and 18S rRNA, and synthetized by Eurogentec (Seraing, Belgium). The cycling parameters for qPCR were 95 °C for 3 min, 40 cycles of 95 °C for 5 s and 60 °C for 20 s, with a melting curve from 65 °C to 95 °C. The cycle threshold (Ct) values of each target genes were first normalized to that of the reference gene 18S rRNA (ΔCt) then the final values (ΔΔCt values) were expressed as folds of control. Data were analyzed on the BioRad CFX manager 3.1 using the ΔΔCt method.

Table 1. List of human primer sequences for RT-qPCR.

Gene	Forward Primer	Reverse Primer	Gene Accession Code
Nrf2	5'-AGCGACGGAAAGAGTATGAG-3'	5'-GTTGGCAGATCCACTGGTTT-3'	NM_001145413.2
HO-1	5'-TCCGATGGGTCCTTACACTC-3'	5'-TAACCAACCCAGCCAAGAGA 3'	NM_002133.2
NQO1	5'-CAGACGCCCGAATTCAAATC-3'	5'-AGGCTGCTTGGAGCAAAATACA-3'	NM_000903.2
IL-1β	5'-AATGATGGCTTATTACAGTGGCA-3'	5'-GTCGGAGATTCGTAGCTGGA-3'	NM_000576.3
IL-6	5'-GTAGCCGCCCCACACAGA-3'	5'-CATGTCTCCTTTCTCAGGGCTG-3'	NM_000600.5
TNF-α	5'-GGAGAAGGGTGACCGACTC-3'	5'-TGGGAAGGTTGGATGTTCGT-3'	NM_000594
18S rRNA	5'-GATAGCTCTTTCTCGATTCCG-3'	5'-TAGTTAGCATGCCAGAGTC-3'	M10098.1

2.9. Analysis of Cell Phagocytosis and Cell Apoptosis via Flow Cytometry

For the assessment of cell phagocytosis, 1×10^6 cells/well THP-1-derived macrophages were treated with rMS 3 h prior to infection with SYTO9-labeled *S. aureus*. After 1 h infection, macrophages were scraped, washed with PBS, and collected for flow cytometry analysis.

To assess cell apoptosis 24 h after rMS treatment, 1×10^6 cells/well THP-1-derived macrophages were stained for 10 min in the dark with Annexin V-FITC reagent. Non-adherent and adherent cells were collected and after 3 washes with PBS, cells were analyzed via flow cytometry.

All flow cytometric analysis was immediately realized on the BD LSRFortessa™ using FACSDiva 7.0 software (BD Biosciences, Franklin Lakes, NJ, USA).

2.10. Quantification of Intracellular ROS Production

Intracellular ROS was measured in live cells using the CellROX™ green fluorogenic probe. Briefly, 2.5×10^5 THP-1-derived macrophages grown on 12 mm-diameter coverslips were treated with rMS. After 6 h incubation, CellROX reagent was added to the cell culture medium. After an additional 30 min, cells were washed with PBS, then fixed in 4% paraformaldehyde (PFA) in PBS. Nuclei were counterstained with 4',6-diamidino-2-phenylindole (DAPI), and macrophages were mounted on glass slides using Fluoromount g aqueous mounting medium.

Positive control for ROS production consisted of treatment of macrophages with H_2O_2. Confocal fluorescent images were taken using Leica SP8 confocal microscope (Leica Microsystems, Wetzlar, Germany) and quantitative analysis of fluorescent signals was performed for 7 fields per treatment using Image J v1.53k (National Institutes of Health, Rockville, MD, USA).

2.11. Cell Viability Assay

Cell viability was assessed using the MTT assay. Briefly, 5×10^4 cells/well THP-1-derived macrophages seeded in a 96-well plate were treated for 24 h before measuring the cellular metabolic activity following the manufacturer's protocol. Absorbance values were obtained at 550 nm and 600 nm using the FLUOstar Omega microplate reader (BMG Labtech).

2.12. Immunofluorescence Microscopy

THP-1-derived macrophages were obtained by seeding 5×10^4 THP-1 cells on glass coverslips in 24-well plates followed by differentiation with PMA for 48 h. After rMS treatment and/or *S. aureus* infection, macrophages were fixed in 4% PFA in PBS, permeabilized with 0.1% Triton X-100 for 5 min, blocked with 1% BSA in PBS for 1 h, and then incubated overnight at 4 °C with the primary antibodies targeting p62 or phospho-p38 diluted in 1% BSA in PBS. After being washed with PBS, cells were incubated with secondary antibodies diluted in 1% BSA for 1 h at room temperature. Nuclei were counterstained with Hoechst 33342. Coverslips were mounted on slides using Fluoromount-G. Confocal images of p62/SQSTM1 punctate structures were acquired using a Leica SP8 confocal microscope, 40× magnification. Quantitative analyses of fluorescent signals were performed on 10 fields per treatment using Image J v1.53k (National Institutes of Health, Bethesda, MD, USA).

2.13. Statistical Analyses

Values are presented as means $\pm$ standard errors of mean (SEM). The imaging flow cytometry results presented are means $\pm$ SEM of 3 independent experiments of 100,000 events. Statistical comparisons were performed using a two-tailed unpaired t-test for comparisons between 2 groups, or one-way analysis of variance with Tukey's multiple group comparison test for multiple comparisons. Data analyses were performed with GraphPad Prism statistical software (v8.0.2). Differences were considered to be statistically significant at $p < 0.05$.

3. Results

3.1. High-Frequency rMS Activates Nrf2 in THP-1-Derived Macrophages

In order to determine the effect of repetitive magnetic stimulation on macrophages, THP-1-derived macrophages were treated with rMS. After the indicated incubation time, RT-qPCR analysis showed a significant increase in Nrf2 transcriptional expression at 4 h and 6 h after rMS treatment (Figure 1A). At 24 h after rMS treatment, Nrf2 protein expression was measured via Western blot, and was found to be significantly increased compared to control (Figure 1B). Confocal images were taken of THP-1-derived macrophages treated with rMS and immunolabeled with a specific antibody against Nrf2, with the nucleus counterstained with DAPI. An increasing amount of Nrf2 localized within the nucleus was observed compared to control (Figure 1C). To quantify the amount of Nrf2 localized in the nucleus, nuclear extracts were examined via Western blot. A significant increase in nuclear Nrf2 was observed as early as 30 min after rMS treatment (Figure 1D). Moreover, we found that the expression of Keap1 protein was significantly decreased in THP-1-derived macrophages 3 h after rMS treatment (Figure 1E). These results suggest that rMS treatment is able to modulate the activation of Nrf2 and Keap1 by affecting their expression and/or localization in macrophages.

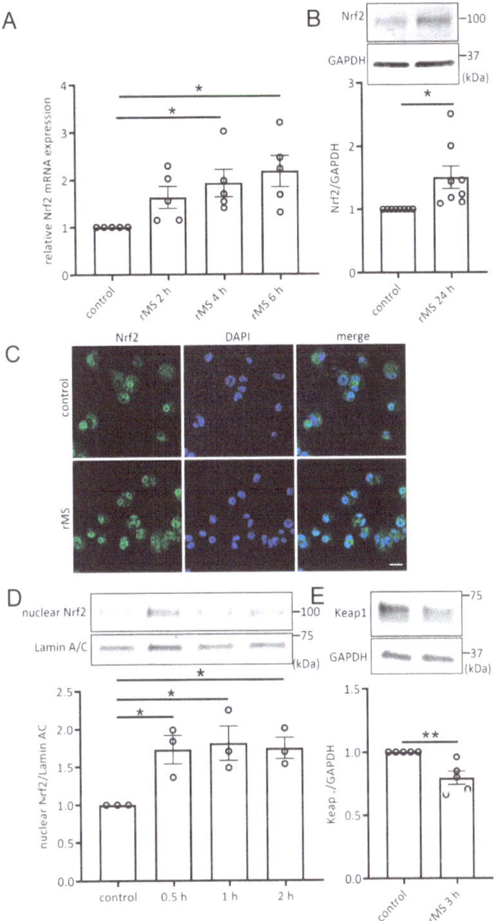

Figure 1. rMS activates Nrf2 in THP-1-derived macrophages. (**A**) THP-1-derived macrophages subjected to rMS treatment (10 Hz). Nrf2 mRNA expression was determined via RT-qPCR. Data were analyzed using a one-way ANOVA followed by Dunnett's multiple comparisons. (**B**) Nrf2 protein expression was detected via Western blotting 24 h after rMS treatment and analyzed using densitometric analysis. Statistical analysis was performed using two-tailed unpaired *t*-test. (**C**) Immunofluorescence microscopy. Nuclear translocation of Nrf2 1 h after rMS treatment. THP-1-derived macrophages were immunostained with specific Nrf2 antibody; nuclei were conterstained with DAPI. Bars: 20 μm. (**D**) Nuclear Nrf2 protein expression was determined at the indicated time point using Western blotting and analyzed using densitometry analysis. Data are expressed as Nrf2 protein expression normalized to lamin A/C control values. Statistical analysis was performed using one-way ANOVA followed by Dunnett's multiple comparisons. (**E**) Keap1 protein expression was detected 3 h after rMS treatment and analyzed using densitometric analysis. Statistical analysis was performed using two-tailed unpaired *t*-test. Data are shown as mean ± SEM * $p < 0.05$, ** $p < 0.01$.

3.2. rMS Activates Nrf2 Signaling Pathway

To determine whether the nuclear translocation of Nrf2 leads to the activation of the Nrf2 signaling pathway, we examined the transcriptional and protein expressions of HO-1 and NQO1, two downstream targets of Nrf2. RT-qPCR analyses were performed on total mRNA extracted from THP-1-derived macrophages at 2, 4 and 6 h after rMS treatment. RT-qPCR showed a significant increase in HO-1 mRNA expression at 6 h (Figure 2A).

Immunoblotting analysis conducted on protein lysates extracted from THP-1-derived macrophages 24 h after rMS treatment revealed a significant increase in HO-1 protein expression (Figure 2A). Additionally, a significant increase in NQO1 mRNA expression level was observed at 4 h and 6 h after rMS treatment and NQO1 protein expression level exhibited a significant increase at 24 h after rMS treatment (Figure 2B). These results confirmed the activation of the Nrf2 signaling pathway by rMS.

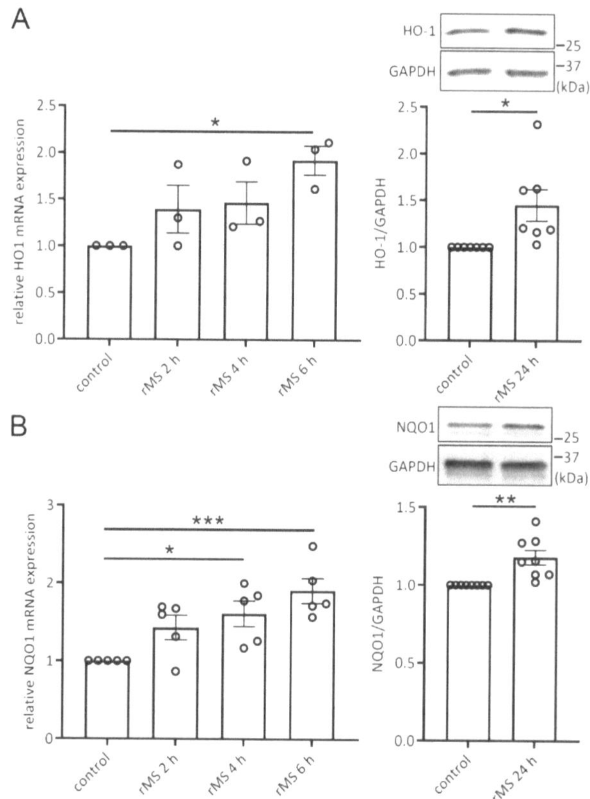

Figure 2. rMS increases HO-1 and NQO1 mRNA and protein expression levels. (**A**) Transcriptional expression of HO-1 at the indicated time points was determined via SYBRgreen RT-qPCR. Data shown are mean ± SEM from independent experiments. Statistical analyses were performed using one-way ANOVA followed by Dunnett's multiple comparisons. HO-1 protein expression was determined at 24 h after rMS treatment. (**B**) NQO1 mRNA and protein expressions. * $p < 0.05$; ** $p < 0.01$; *** $p < 0.005$.

3.3. rMS Increased Phosphorylated p62 and Total p62 Protein Expressions

Next, we hypothesized that the observed decrease in Keap1 expression and increase in Nrf2 expression levels resulted from rMS modulation of p62 phosphorylation and p62 expression levels. Total protein extracts were collected from THP-1-derived macrophages 1 h after rMS treatment for phosphorylated p62 immunoblotting analysis and 3 h after rMS treatment for p62 immunoblotting analysis. Data analysis showed a significant increase in p62 protein expression, as well as a significant increase in phosphorylated p62 (Figure 3A,B). Furthermore, confocal images were taken from THP-1-derived macrophages treated with rMS for 3 h. The macrophages were then immunolabeled with a specific p62 antibody, and the nucleus was counterstained with Hoechst33342 (Figure 3C). Using Image J software v1.53k, the number of p62 puncta and the fluorescence intensity of p62 in the puncta were quantified. Our data showed a decrease in the number of p62 puncta per cell in rMS-treated macrophages, along with a decrease in p62 fluorescent intensity within the puncta of each cell (Figure 3D,E). These results suggest that rMS induces the phosphorylation of p62, which enables the binding of phosphorylated p62 with Keap1, and consequently allows the released Nrf2 to translocate into the nucleus. The phosphorylated p62-Keap1 complex subsequently undergoes p62-dependent autophagy degradation [21].

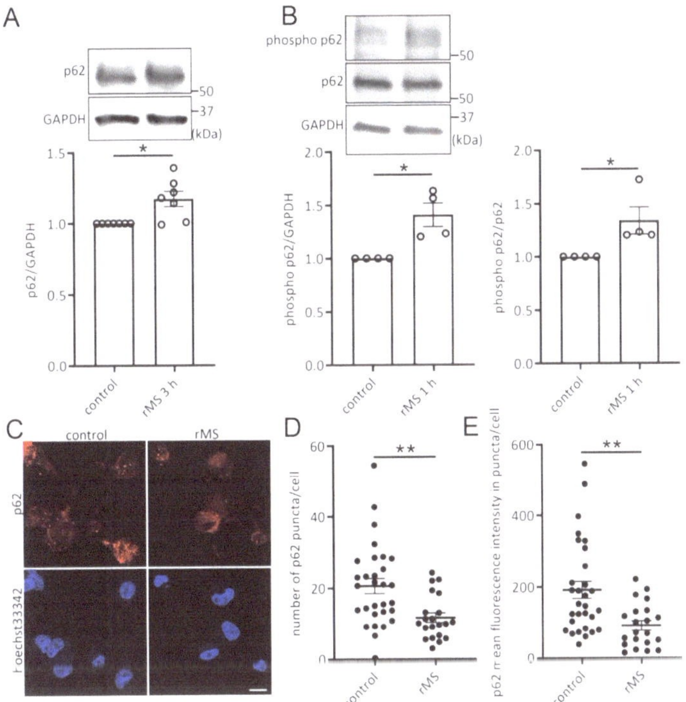

Figure 3. rMS activates p62 and increases p62 protein expression level. (**A,B**) p62 and phosphorylated p62 protein expressions were determined via Western blotting 3 h and 1 h after rMS treatment, respectively. (**C**) Confocal microscopy images of THP-1-derived macrophages treated with rMS. Immunofluorescence labeling with p62 and counterstained with Hoechst33342 was performed. Bars: 10 μm. (**D,E**) Quantitative analysis of p62 puncta and p62 fluorescence intensity in the puncta using Image J. Mean ± SEM were obtained from analysis of n = 100 cells from 2 independent experiments. * $p < 0.05$, ** $p < 0.01$. Statistical analyses were performed using two-tailed unpaired *t*-test.

3.4. Effects of rMS on Cell Viability, Oxidative Stress, Inflammation, and Macrophage Bactericidal Function

To further investigate the potential impact of rMS treatment on cellular processes, THP-1-derived macrophages were subjected to rMS and incubated for the indicated time before analysis. Firstly, we assessed the effect of rMS on cell viability using the MTT assay and cell apoptosis by staining with Annexin V. Flow cytometry analysis indicated that rMS treatment of THP-1-derived macrophages had no significant impact on cell viability or cell apoptosis (Figure 4A,B). We also performed RT-qPCR on mRNA extracted from THP-1-derived macrophages 6 h after rMS treatment to evaluate the transcriptional expression of genes coding for the inflammatory markers IL-1β, IL-6, and TNF-α. No significant changes were observed in their mRNA expressions (Figure 4C). Next, we investigated whether rMS affected oxidative stress by staining rMS-treated THP-1-derived macrophages with CellROX fluorogenic probe, which detected ROS in live cells. Data analysis showed no significant effect of rMS on oxidative stress (Figure 4D). We then examined the impact of rMS on macrophage phagocytic activity. THP-1-derived macrophages were treated with rMS and incubated for 3 h prior to infection with SYTO9-labeled *S. aureus* for 1 h. Flow cytometry analysis revealed no significant difference in macrophage phagocytosis activity after rMS as compared to control (Figure 4E). Furthermore, we assessed the bactericidal activity of rMS-treated macrophages by determining the intracellular survival of *S. aureus* 24 h after infection. Gentamicin was added to the cell culture medium 1 h after infection to eliminate extracellular bacteria. CFU analysis demonstrated a significant decrease in intracellular survival of *S. aureus* in rMS-treated THP-1-derived macrophages (Figure 4F). Collectively, these findings indicate that 5 min magnetic stimulation does not affect the cell viability, cell apoptosis, inflammation, or oxidative stress of THP-1-derived macrophages. Notably, rMS appears to impact the intracellular survival of *S. aureus*.

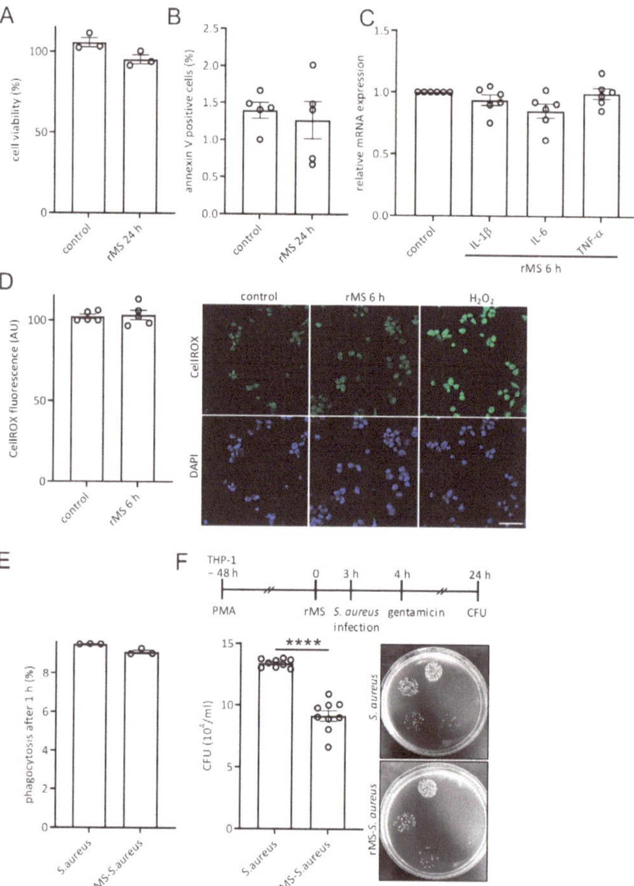

Figure 4. Effects of rMS on cell viability, oxidative stress, inflammation, and macrophage bactericidal function. (**A**) Cell viability measured via MTT assay 24 h after rMS treatment. (**B**) Cell apoptosis was determined using the Annexin V staining. (**C**) Inflammation was determined by quantifying IL-6, IL-1β, and TNF-α mRNA expressions 6 h after rMS treatment using RT-qPCR. (**D**) Oxidative stress was measured using a CellROX fluorogenic probe in live cells to detect ROS. Quantification of CellROX green fluorescence involved Image J. H₂O₂-treated cells were used as positive control. Scale bar: 50 μm. (**E**) THP-1-derived macrophages were infected with SYTO9-labeled *S. aureus*. After 1 h infection, phagocytosis was determined via FACS. (**F**) *S. aureus* intracellular survival using CFU assay. Three hours after rMS treatment, THP-1-derived macrophages were infected with *S. aureus*. Extracellular bacteria were eliminated via addition of gentamicin 1 h after infection. Cells were incubated for 24 h and CFU was determined. Data are mean ± SEM. Statistical analysis was performed using two-tailed unpaired *t*-test **** *p* < 0.0001.

3.5. rMS-Mediated Decrease in S. aureus Intracellular Survival Is Nrf2 Dependent

In order to investigate whether the bactericidal activity of rMS-treated THP-1-derived macrophages resulted from Nrf2 activity, macrophages were transfected with either scrambled siRNA or siRNA specifically targeting Nrf2. Firstly, Western blot analysis confirmed the reduced expression level of Nrf2 and HO-1 proteins in Nrf2-targeted siRNA-transfected macrophages compared to those transfected with scrambled siRNA (Figure 5A). Next, we examined the intracellular survival of *S. aureus* in scrambled or Nrf2 siRNA-transfected macrophages. As expected, CFU assay demonstrated a significant decrease in intracellular survival of *S. aureus* in scrambled siRNA-transfected THP-1-derived macrophages treated with rMS compared to untreated cells. Interestingly, THP-1-derived macrophages transfected with Nrf2-targeted siRNA showed no rMS-mediated decrease in intracellular survival of *S. aureus* compared to untreated cells (Figure 5B).

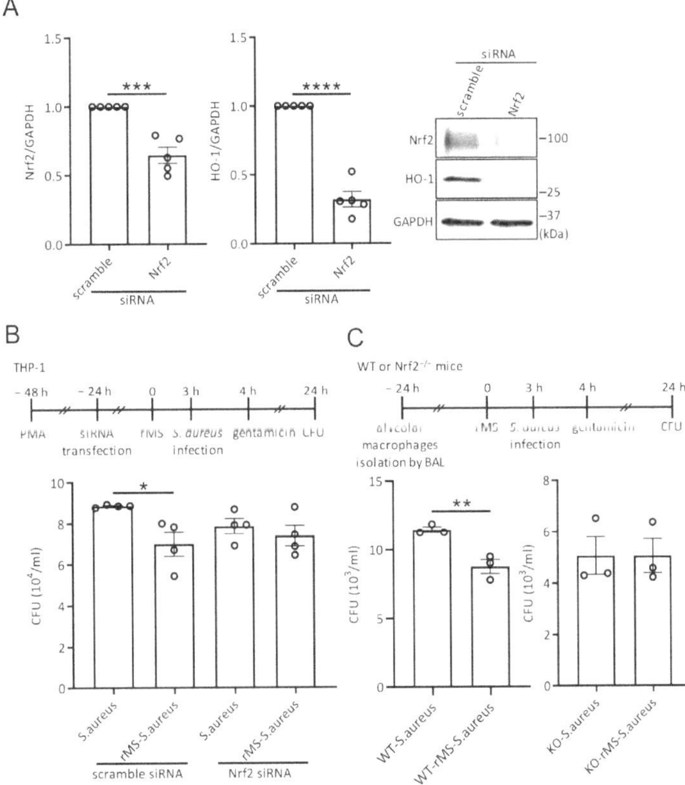

Figure 5. rMS-mediated decrease in *S. aureus* intracellular survival is Nrf2-dependent. (**A**) THP-1 derived macrophages were transfected with scramble or Nrf2-targeted siRNA 24 h prior to rMS treatment. Then, 24 h after rMS treatment, protein lysates were analyzed Nrf2 and HO-1 protein expressions were determined via Western blotting. Statistical analysis was performed using a two-tailed *t*-test *** $p < 0.0003$, **** $p < 0.0001$. (**B**) CFU with transfected THP-1 derived macrophages. (**C**) Primary alveolar macrophages were isolated from WT and Nrf2 knockout mice and used in bacteria intracellular survival assay. Data are means ± SEM. * $p < 0.05$, ** $p < 0.01$ as determined by one-way ANOVA with Tukey's multiple comparisons.

Additionally, we evaluated the intracellular survival of *S. aureus* in primary alveolar macrophages isolated from Nrf2 knockout (KO) mice and their wild-type (WT) littermates (Figure 5C). Alveolar macrophages were isolated from bronchoaleolar lavages, cultured for 24 h, and then subjected to rMS treatment and *S. aureus* infection. Gentamicin was added to the cell culture medium 1 h after infection to inhibit the growth of extracellular bacteria. After 24 h, CFU numeration was determined. Quantitative analysis revealed that the rMS-mediated decrease in intracellular survival of *S. aureus* observed in alveolar macrophages from WT mice did not persist in rMS-treated alveolar macrophages isolated from Nrf2 KO mice. These results suggest the involvement of Nrf2 in the effect of rMS on the bactericidal activity of macrophages.

3.6. rMS Decreases IL-1β and TNF-α mRNAs Expression Levels and p62 Protein Expression Level in S. aureus-Infected THP-1-Derived Macrophages

Upon infection, *S. aureus* induces the expression of genes encoding pro-inflammatory cytokines, including IL-6, IL-1β, and TNF-α [5]. In THP-1-derived macrophages treated with rMS for 3 h prior to *S. aureus* infection, RT-qPCR analysis revealed a significant reduction in the transcriptional expression of IL-1β and TNF-α (Figure 6A). However, no significant change was observed for IL-6 mRNA expression.

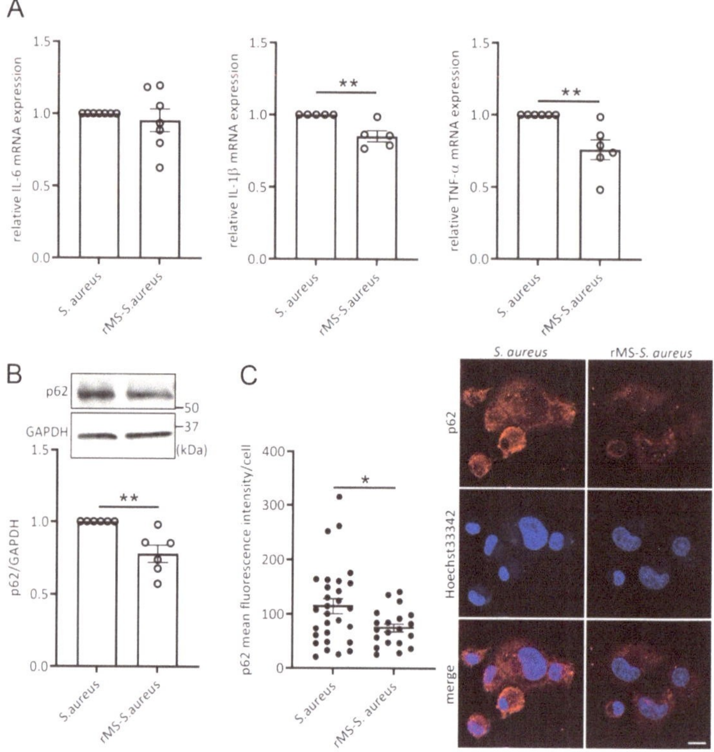

Figure 6. rMS decreases IL-1β and TNF-α mRNA expression levels and p62 protein levels in *S. aureus*-infected THP-1-derived macrophages. (**A**) rMS-treated THP-1-derived macrophages were infected with *S. aureus*. Total RNAs were isolated 3 h after infection. IL-6, IL-1β, and TNF-α mRNA expressions were quantified by RT-qPCR. (**B**) Immunoblot analysis. Protein lysates were extracted from THP-1-derived macrophages after 3 h treatment with rMS and infected with *S. aureus* for 3 h before cell lysis. Immunoblots using the indicated antibodies are representative of 6 independent experiments. (**C**) Immunofluorescence microscopy. THP-1-derived macrophages were treated with rMS then infected with Hoechst 33342-stained *S. aureus*. Cells were fixed with PFA 4% and were immunostained with p62 antibody. Nucleus was stained with Hoechst 33342. Bar scale: 10 μm. p62 mean fluorescence intensity in puncta by cell was quantified using Image J v1.53k. * $p < 0.05$, ** $p < 0.01$. Statistical analyses were performed using a two-tailed unpaired *t*-test.

Moreover, THP-1-derived macrophages pretreated with rMS for 3 h and subsequently infected with *S. aureus* for an additional 3 h exhibited a decrease in p62 protein expression (Figure 6B). Analysis of confocal images using Image J v1.53k demonstrated a similar decrease in p62 fluorescent intensity in the puncta of cells (Figure 6C). These results suggest the potential involvement of autophagy in the intracellular survival of *S. aureus*.

3.7. rMS Decreases S. aureus-Mediated Phosphorylation of p38 MAPK and Its Nuclear Localization

Since previous studies have shown that *S. aureus* is able to modulate the JNK MAPK signaling pathway in RAW264.7 macrophages [22], our aim was to investigate the involvement of the MAPK signaling pathway in THP-1-derived macrophages treated with rMS and subsequently infected with *S. aureus*. Protein lysates extracted from THP-1-derived macrophages stimulated with rMS for 3 h and then infected with *S. aureus* for an additional 1 h were analyzed via Western blot. The results showed that rMS treatment alone had no significant effect on the activation of p38, JNK, and ERK MAPK signaling (Figure 7A). In contrast, infection of THP-1-derived macrophages with *S. aureus* strongly induced phosphorylation of p38 and JNK, while ERK phosphorylation remained unchanged. Interestingly, in *S. aureus*-infected cells, rMS treatment specifically inhibited the phosphorylation of p38 compared to control cells.

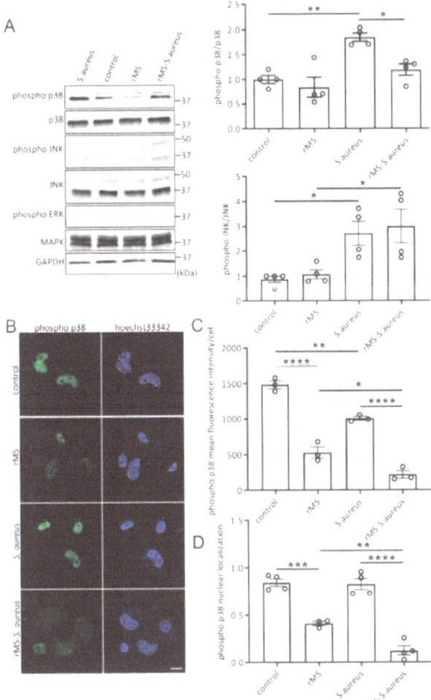

Figure 7. rMS decreases p38 MAPK phosphorylation and nuclear localization. (**A**) THP-1-derived macrophages were treated with rMS 3 h prior to infection with *S. aureus*. Cell lysis was performed 1 h after infection using RIPA supplemented with a protease and phosphatase inhibitors cocktail. The immunoblots shown here are representative of 4 independent experiments. Densitometric analysis was performed using Image Studio Lite. (**B**) Immunofluorescence confocal microscopy. Confocal images of THP-1-derived macrophages treated with rMS and infected with Hoechst-labeled *S. aureus*. (**C**) Quantitative analysis of phospho-p38 fluorescence intensity per cell. (**D**) Quantification of the phospho-p38 localization in the nucleus in THP-1-derived macrophages treated with rMS and infected with *S. aureus*. * $p < 0.05$, ** $p < 0.01$, *** $p < 0.0005$, **** $p < 0.0001$ were determined via one-way ANOVA with Tukey's multiple comparisons.

To further validate the findings from the Western blot analysis, we acquired confocal images of THP-1-derived macrophages labeled with a specific phospho-p38 MAPK antibody, counterstained the nuclei with Hoechst 33342, and conducted quantitative analysis using Image J software v1.53k (Figure 7B). rMS-treated THP-1-derived macrophages exhibited a significant decrease in the fluorescence intensity of phospho-p38 MAPK compared to control cells or *S. aureus*-infected cells (Figure 7C). Additionally, analysis of confocal images revealed a reduction in the nuclear localization of phospho-p38 MAPK in rMS-treated macrophages compared to control macrophages or *S. aureus*-infected macrophages (Figure 7D).

4. Discussion

In this study, we demonstrated that the use of a single 5 min session of high-frequency repetitive magnetic stimulation is able to decrease the intracellular load of *S. aureus* in THP-1-derived macrophages and in primary alveolar macrophages. We determined that rMS treatment activated the Nrf2/Keap1 signaling pathway through a non-canonical mechanism. rMS increased p62 protein expression and p62 phosphorylation, which increased its affinity binding to Keap1, leading to the release of Nrf2 and its translocation to the nucleus. We demonstrated that Nrf2 was necessary to the bactericidal activity of macrophages since Nrf2 knockdown, using siRNA in THP-1-derived macrophages or alveolar macrophages isolated from Nrf2 knockout mice, led to an abolition of the rMS-mediated bactericidal activity. Furthermore, we showed that *S. aureus*-induced activation of p38 MAPK was repressed by rMS treatment. rMS-mediated repression of p38 MAPK may explain the decrease in *S. aureus*-induced expression of genes coding for the pro-inflammatory cytokines IL-1β and TNF-α. To the best of our knowledge, this study is the first report to link macrophage biological activity and the molecular mechanisms elicited by rMS treatment.

rMS has become an increasingly popular non-invasive technique for modulating neural excitability and plasticity in clinical and preclinical models, as well as in cultured neurons. However, the effect of rMS on non-excitable cells should not be overlooked. Recent reports have described the significant role of rMS in various cell types, including microglia, mesenchymal stromal cells, chondrocytes, and endothelial cells, despite the discrepancies in the stimulatory parameters, which include variations in intensity, frequency, and duration [18,19,23–25].

One recent report conducted on lipopolysaccharide-stimulated microglia isolated from newborn Wistar rat brains and treated twice with 10 Hz rTMS showed no effect of rTMS on IL-1β and TNF-α secretions while increasing the production of IL-4 and IL-10. The authors stated that rTMS treatment could promote the anti-inflammatory polarization of microglia [25]. Similarly, another recent study demonstrated the role played by microglia in inducing synaptic plasticity through the release of cytokines upon rTMS treatment in a mouse model of organotypic brain tissue. Furthermore, in vivo depletion of microglia abolished rTMS-induced synaptic plasticity [18]. In our study, we showed that a 5 min rMS treatment of THP-1-derived macrophages did not significantly alter transcriptional expression levels of IL-6, IL-1β, and TNF-α. However, when THP-1-derived macrophages were pretreated with rMS prior to *S. aureus* infection, a decrease in the transcriptional expression levels of genes coding for IL-1β and TNF-β was observed. While the effects of rMS may vary depending on the specific cell type and variations in stimulatory parameters such as intensity, duration and frequency of treatment, it is reasonable to postulate that rMS may have a significant effect on the inflammatory processes of innate immune cells. Therefore, considering the diverse and still largely unexplored impact of rMS on cell–cell and cell–microenvironment interactions, we have decided to investigate the cellular and molecular responses induced by rMS in an in vitro model of *S. aureus*-infected macrophages.

We have previously demonstrated that the anti-inflammatory effect of Nrf2 signaling pathway in our in vitro model of *S. aureus*-infected THP-1-derived macrophages was obtained through repression of p38 and JNK MAPK phosphorylation. *S. aureus*-mediated activation of p38 and JNK MAPK signaling pathways was inhibited by the pretreatment of THP-1-derived macrophages with sulforaphane, a well-established Nrf2 activator. Repression of p38 and JNK MAPK phosphorylation led to a significant decrease in transcriptional expression levels of pro-inflammatory cytokines [5]. In the present study, we observed an increase in the phosphorylation of p38 and JNK MAPK in THP-1-derived macrophages infected with *S. aureus*. rMS treatment of macrophages specifically repressed *S. aureus*-induced p38 MAPK phosphorylation and decreased phosphorylated p38 nuclear localization, while showing no effect of rMS on *S. aureus*-mediated activation of JNK MAPK.

Interestingly, intracellular bacterial pathogens have evolved various strategies to escape degradation mechanisms. Early activation of the p38 MAPK signaling pathway is one of the strategies employed by *S. aureus* to survive in the host cell [26]. Indeed, activation of p38 MAPK by *S. aureus* has been found to block autophagy degradation and ensure bacterial survival within the host cell [27]. Specifically, it has been shown that p38 MAPK can activate the NF-κB, which in turn promotes the expression of p62.

rMS treatment of THP-1-derived macrophages resulted in the activation of the Nrf2 signaling pathway, accompanied by a decrease in Keap1 protein level. The lack of any change in oxidative stress observed in rMS-stimulated macrophages suggests that the activation of Nrf2 by rMS does not occur through the canonical mechanism. Instead, our data indicate that rMS activates Nrf2 via a non-canonical mechanism involving the upregulation of the autophagy-adaptor protein p62 and its subsequent phosphorylation. Phosphorylation of p62 enhances its binding affinity to Keap1, resulting in Nrf2 accumulation and facilitating its translocation to the nucleus [21]. The phosphorylated p62-Keap1 complex leads to autophagic degradation. Notably, Nrf2 also positively regulates the transcriptional expression of p62 due to the presence of ARE sequences in the p62 promoter, establishing a positive feedback loop [28].

The molecular mechanism responsible for the activation of p62 in response to rMS treatment in our in vitro cell model remains unclear. Previous studies have suggested that mammalian target of rapamycin complex 1 (mTORC1) may be involved in the phosphorylation of the Ser349 residue, which is located within the KIR domain of p62 and implicated in Keap1 affinity binding [21]. Therefore, it would be valuable to explore whether the modulation of p62 by rMS is dependent on mTORC1 activity. Furthermore, studies conducted on differentiated Neuro-2a cells have shown that high-frequency rMS treatment led to the activation of Ca^{2+}-calmodulin-dependent protein kinase-cAMP-response element binding protein (CREB) signaling pathway, resulting in increased brain-derived neurotrophic factor (BDNF) expression and synaptic plasticity. Moreover, rMS has been reported to reduce neuronal cell death and promote cell proliferation while inhibiting apoptosis through activation of the ERK and AKT-signaling pathways [29]. In addition, bone mesenchymal stromal cells stimulated with high-frequency rTMS demonstrated autophagy flux through the activation of the NMDAR-Ca^{2+}-ERK-mTOR signaling pathway [19].

Autophagy is a fundamental biological process against *S. aureus* and has been shown to play a crucial role in the control of *S. aureus* growth. Previous studies have described a recruitment of p62 to *S. aureus* in vitro in fibroblasts and epithelial cells [27]. Also, a recent report has revealed the interaction of p62 with *S. aureus* in vivo in neutrophils of zebrafish [30]. The disruption of p62 expression in neutrophils resulted in reduced zebrafish survival after *S. aureus* infection. Since both the p62 and p38 MAPK pathways are well-known players in autophagy, and here are involved in the bactericidal activity of rMS-treated macrophages, it would be interesting to investigate the potential role of autophagy in modulating macrophage function against pathogens via rMS. Overall, these findings indicate a complex interplay between Nrf2/Keap1/p62, *S. aureus*, and the p38 MAPK pathway in response to rMS treatment. Therefore, it would be interesting to investigate the potential impact of rMS on autophagy. While the effects of rMS have been widely acknowledged in clinical and preclinical models, it remains a challenge to apply rMS on systemic immune cells and elicit specific responses due to the limited area targeted by magnetic stimulation. One potential therapeutic strategy to consider would involve the promising developments in ex vivo-activated macrophage-based therapies [31].

5. Conclusions

In summary, we propose a molecular signaling pathway activated by rMS in THP-1-derived macrophages based on our findings (Figure 8). rMS treatment activates the Nrf2 signaling pathway through the Nrf2/Keap1/p62 pathway. The p62/Keap1 complex undergoes autophagy degradation, while unbound Nrf2 translocates to the nucleus and promotes the transcriptional upregulation of genes involved in detoxification, antioxidant defense, and anti-inflammatory mechanisms. In addition, rMS treatment inhibits the p38 MAPK signaling pathway, which is implicated in *S. aureus* intracellular survival and pro-inflammatory responses.

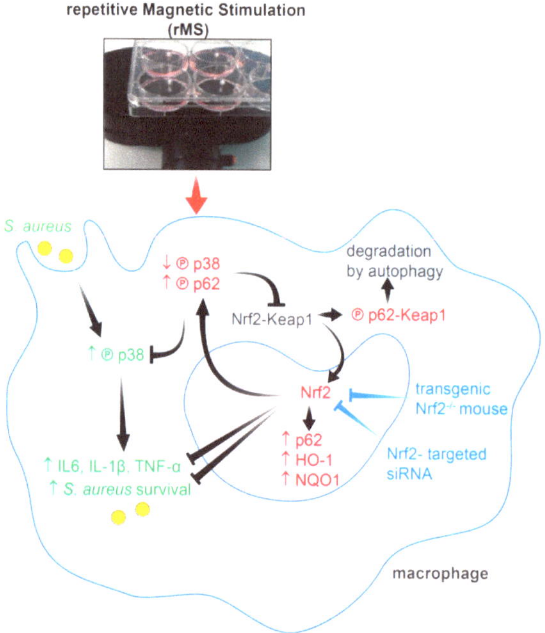

Figure 8. Schematic representation of the p62/Keap1/Nrf2 non-canonical pathway activated by rMS in THP-1-derived macrophages infected with *S. aureus*. THP-1-derived macrophages infected with *S. aureus* mediate an activation of the p38 MAPK pathway, which regulates the gene expression of pro-inflammatory cytokines such as IL-6, IL-1β, and TNF-α, and promote *S. aureus* intracellular survival. In THP-1-derived macrophages, treatment with rMS results in an increase in phosphorylated p62 at Ser349, which results in increased interaction with Keap1 and the release of Nrf2. The phosphorylated p62-Keap1 complex is identified for degradation via autophagy. Nrf2 is then free to translocate to the nucleus, interact with small Maf proteins, and bind antioxidant response element (ARE), leading to the upregulation of genes coding for antioxidant, detoxifying, and anti-inflammatory proteins, as well as coding for p62. Nrf2 silencing was obtained by utilizing specific siRNA in THP-1-derived macrophages or utilizing alveolar macrophages isolated from Nrf2 knockout mice.

Supplementary Materials: The following supporting information can be downloaded at: https://www.mdpi.com/article/10.3390/antiox12091695/s1, Figure S1: THP-1-derived macrophages were subjected to rMS protocol 5 Hz and 15 Hz and cells were lysed at the indicated time.

Author Contributions: Conceptualization, T.B.D. and M.B.; Formal analysis, T.B.D.; Funding acquisition, M.B.; Investigation, T.B.D.; Methodology, T.B.D., A.C., T.C. and J.A.; Supervision, T.B.D.; Validation, T.B.D. and M.B.; Visualization, T.B.D.; Writing—original draft, T.B.D.; Writing—review and editing, T.B.D. and M.B. All authors have read and agreed to the published version of the manuscript.

Funding: This study was supported by the National Institute of Health and Medical Research (INSERM) funding, the University of Versailles Saint-Quentin-en-Yvelines funding, and the Chancellerie des Universités de Paris Legs Poix funding.

Institutional Review Board Statement: Breeding of the C57BL/6J Nrf2 knockout (Nrf2$^{-/-}$) mouse strain was conducted according to National and European legislation and institutional guidelines for the care and use of laboratory animals approved by the Animal Ethics Committee CEEA47 PELVIPHARM (2Care animal facility, University of Versailles Saint-Quentin-en-Yvelines, UFR Santé Simone Veil) and the Ministry of Higher Education, Research and Innovation.

Informed Consent Statement: Not applicable.

Data Availability Statement: All the data are contained within the article and the Supplementary Figure S1.

Conflicts of Interest: The authors declare no conflict of interest.

Abbreviations

cDNA: complementary deoxyribonucleic acid; CFU: colony-forming unit; DAPI: 4′, 6-diamidino-2-phenylindole; DMSO: dimethyl sulfoxide; ERK: extracellular signal-regulated kinase; HO-1: Heme oxygenase-1; IL-1β: interleukin-1β; IL-6: interleukin-6; JNK: c-Jun N-terminal kinase; Keap1: Kelch-like ECH-associated protein 1; Maf: musculoaponeurotic fibrosarcoma; MOI: multiplicity of infection; MTT: 3-(4,5-dimethylthiazol-2-yl)-2,5-diphenyltetrazolium bromide; NQO1: NAD(P)H quinone dehydrogenase 1; Nrf2/NFE2L2: nuclear factor erythroid 2-related factor 2; p38 MAPK: p38 mitogen-activated protein kinase; p62/SQSTM1: sequestosome 1; PBS: phosphate-buffered saline; PCR: polymerase chain reaction; PFA: paraformaldehyde; RNA: ribonucleic acid; RIPA: radioimmunoprecipitation assay buffer; rMS: repetitive magnetic stimulation; ROS: reactive oxygen species; rTMS: repetitive transcranial magnetic stimulation; siRNA: small interfering RNA; THP-1: human leukemic cell line; TNF-α: tumor necrosis factor-α.

References

1. Lazarov, T.; Juarez-Carreno, S.; Cox, N.; Geissmann, F. Physiology and Diseases of Tissue-Resident Macrophages. *Nature* **2023**, *618*, 698–707. [CrossRef]
2. DeLeo, F.R.; Otto, M.; Kreiswirth, B.N.; Chambers, H.F. Community-Associated Meticillin-Resistant *Staphylococcus aureus*. *Lancet* **2010**, *375*, 1557–1568. [CrossRef]
3. Martin, F.J.; Parker, D.; Harfenist, B.S.; Soong, G.; Prince, A. Participation of Cd11c(+) Leukocytes in Methicillin-Resistant *Staphylococcus aureus* Clearance from the Lung. *Infect. Immun.* **2011**, *79*, 1898–1904. [CrossRef]
4. Nelson, R.E.; Hyun, D.; Jezek, A.; Samore, M.H. Mortality, Length of Stay, and Healthcare Costs Associated with Multidrug-Resistant Bacterial Infections among Elderly Hospitalized Patients in the United States. *Clin. Infect. Dis.* **2022**, *74*, 1070–1080. [CrossRef] [PubMed]
5. Deramaudt, T.B.; Ali, M.; Vinit, S.; Bonay, M. Sulforaphane Reduces Intracellular Survival of *Staphylococcus aureus* in Macrophages through Inhibition of Jnk and P38 Mapk-Induced Inflammation. *Int. J. Mol. Med.* **2020**, *45*, 1927–1941. [CrossRef] [PubMed]
6. Ellington, J.K.; Elhofy, A.; Bost, K.L.; Hudson, M.C. Involvement of Mitogen-Activated Protein Kinase Pathways in *Staphylococcus aureus* Invasion of Normal Osteoblasts. *Infect. Immun.* **2001**, *69*, 5235–5242. [CrossRef]
7. Gomez, M.I.; Lee, A.; Reddy, B.; Muir, A.; Soong, G.; Pitt, A.; Cheung, A.; Prince, A. *Staphylococcus aureus* Protein a Induces Airway Epithelial Inflammatory Responses by Activating Tnfr1. *Nat. Med.* **2004**, *10*, 842–848. [CrossRef]
8. Kujime, K.; Hashimoto, S.; Gon, Y.; Shimizu, K.; Horie, T. P38 Mitogen-Activated Protein Kinase and C-Jun-Nh2-Terminal Kinase Regulate Rantes Production by Influenza Virus-Infected Human Bronchial Epithelial Cells. *J. Immunol.* **2000**, *164*, 3222–3228. [CrossRef]
9. Bonay, M.; Roux, A.L.; Floquet, J.; Retory, Y.; Herrmann, J.L.; Lofaso, F.; Deramaudt, T.B. Caspase-Independent Apoptosis in Infected Macrophages Triggered by Sulforaphane Via Nrf2/P38 Signaling Pathways. *Cell Death Discov.* **2015**, *1*, 15022. [CrossRef]
10. Boutten, A.; Goven, D.; Artaud-Macari, E.; Boczkowski, J.; Bonay, M. Nrf2 Targeting: A Promising Therapeutic Strategy in Chronic Obstructive Pulmonary Disease. *Trends Mol. Med.* **2011**, *17*, 363–371. [CrossRef] [PubMed]
11. Silva-Islas, C.A.; Maldonado, P.D. Canonical and Non-Canonical Mechanisms of Nrf2 Activation. *Pharmacol. Res.* **2018**, *134*, 92–99. [CrossRef] [PubMed]
12. Komatsu, M.; Kurokawa, H.; Waguri, S.; Taguchi, K.; Kobayashi, A.; Ichimura, Y.; Sou, Y.S.; Ueno, I.; Sakamoto, A.; Tong, K.I.; et al. The Selective Autophagy Substrate P62 Activates the Stress Responsive Transcription Factor Nrf2 through Inactivation of Keap1. *Nat. Cell Biol.* **2010**, *12*, 213–223. [CrossRef] [PubMed]

13. Hast, B.E.; Goldfarb, D.; Mulvaney, K.M.; Hast, M.A.; Siesser, P.F.; Yan, F.; Hayes, D.N.; Major, M.B. Proteomic Analysis of Ubiquitin Ligase Keap1 Reveals Associated Proteins That Inhibit Nrf2 Ubiquitination. *Cancer Res.* **2013**, *73*, 2199–2210. [CrossRef] [PubMed]

14. Karapetian, R.N.; Evstafieva, A.G.; Abaeva, I.S.; Chichkova, N.V.; Filonov, G.S.; Rubtsov, Y.P.; Sukhacheva, E.A.; Melnikov, S.V.; Schneider, U.; Wanker, E.E.; et al. Nuclear Oncoprotein Prothymosin Alpha Is a Partner of Keap1: Implications for Expression of Oxidative Stress-Protecting Genes. *Mol. Cell. Biol.* **2005**, *25*, 1089–1099. [CrossRef] [PubMed]

15. Ma, J.; Cai, H.; Wu, T.; Sobhian, B.; Huo, Y.; Alcivar, A.; Mehta, M.; Cheung, K.L.; Ganesan, S.; Kong, A.N.; et al. Palb2 Interacts with Keap1 to Promote Nrf2 Nuclear Accumulation and Function. *Mol. Cell. Biol.* **2012**, *32*, 1506–1517. [CrossRef] [PubMed]

16. Vinhas, A.; Almeida, A.F.; Goncalves, A.I.; Rodrigues, M.T.; Gomes, M.E. Magnetic Stimulation Drives Macrophage Polarization in Cell to-Cell Communication with Il-1beta Primed Tendon Cells. *Int. J. Mol. Sci.* **2020**, *21*, 5441. [CrossRef]

17. Somaa, F.A.; de Graaf, T.A.; Sack, A.T. Transcranial Magnetic Stimulation in the Treatment of Neurological Diseases. *Front. Neurol.* **2022**, *13*, 793253. [CrossRef]

18. Eichler, A.; Kleidonas, D.; Turi, Z.; Fliegauf, M.; Kirsch, M.; Pfeifer, D.; Masuda, T.; Prinz, M.; Lenz, M.; Vlachos, A. Microglial Cytokines Mediate Plasticity Induced by 10 Hz Repetitive Magnetic Stimulation. *J. Neurosci.* **2023**, *43*, 3042–3060. [CrossRef]

19. Wang, X.; Zhou, X.; Bao, J.; Chen, Z.; Tang, J.; Gong, X.; Ni, J.; Fang, Q.; Liu, Y.; Su, M. High-Frequency Repetitive Transcranial Magnetic Stimulation Mediates Autophagy Flux in Human Bone Mesenchymal Stromal Cells Via Nmda Receptor-Ca(2+)-Extracellular Signal-Regulated Kinase-Mammalian Target of Rapamycin Signaling. *Front. Neurosci.* **2019**, *13*, 1225. [CrossRef]

20. Itoh, K.; Mochizuki, M.; Ishii, Y.; Ishii, T.; Shibata, T.; Kawamoto, Y.; Kelly, V.; Sekizawa, K.; Uchida, K.; Yamamoto, M. Transcription Factor Nrf2 Regulates Inflammation by Mediating the Effect of 15-Deoxy-Delta(12,14)-Prostaglandin J(2). *Mol. Cell. Biol.* **2004**, *24*, 36–45. [CrossRef]

21. Ichimura, Y.; Waguri, S.; Sou, Y.S.; Kageyama, S.; Hasegawa, J.; Ishimura, R.; Saito, T.; Yang, Y.; Kouno, T.; Fukutomi, T.; et al. Phosphorylation of P62 Activates the Keap1-Nrf2 Pathway During Selective Autophagy. *Mol. Cell* **2013**, *51*, 618–631. [CrossRef] [PubMed]

22. Fang, L.; Wu, H.M.; Ding, P.S.; Liu, R.Y. Tlr2 Mediates Phagocytosis and Autophagy through Jnk Signaling Pathway in *Staphylococcus aureus*-Stimulated Raw264.7 Cells. *Cell. Signal* **2014**, *26*, 806–814. [CrossRef] [PubMed]

23. Zong, X.; Li, Y.; Liu, C.; Qi, W.; Han, D.; Tucker, L.; Dong, Y.; Hu, S.; Yan, X.; Zhang, Q. Theta-Burst Transcranial Magnetic Stimulation Promotes Stroke Recovery by Vascular Protection and Neovascularization. *Theranostics* **2020**, *10*, 12090–12110. [CrossRef] [PubMed]

24. Elliott, J.P.; Smith, R.L.; Block, C.A. Time-Varying Magnetic Fields: Effects of Orientation on Chondrocyte Proliferation. *J. Orthop. Res.* **1988**, *6*, 259–264. [CrossRef]

25. Luo, J.; Feng, Y.; Li, M.; Yin, M.; Qin, F.; Hu, X. Repetitive Transcranial Magnetic Stimulation Improves Neurological Function and Promotes the Anti-Inflammatory Polarization of Microglia in Ischemic Rats. *Front. Cell. Neurosci.* **2022**, *16*, 878345. [CrossRef]

26. Miller, M.; Dreisbach, A.; Otto, A.; Becher, D.; Bernhardt, J.; Hecker, M.; Peppelenbosch, M.P.; van Dijl, J.M. Mapping of Interactions between Human Macrophages and *Staphylococcus aureus* Reveals an Involvement of Map Kinase Signaling in the Host Defense. *J. Proteome Res.* **2011**, *10*, 4018–4032. [CrossRef] [PubMed]

27. Neumann, Y.; Bruns, S.A.; Rohde, M.; Prajsnar, T.K.; Foster, S.J.; Schmitz, I. Intracellular *Staphylococcus aureus* Eludes Selective Autophagy by Activating a Host Cell Kinase. *Autophagy* **2016**, *12*, 2069–2084. [CrossRef]

28. Jain, A.; Lamark, T.; Sjottem, E.; Larsen, K.B.; Awuh, J.A.; Overvatn, A.; McMahon, M.; Hayes, J.D.; Johansen, T. P62/Sqstm1 Is a Target Gene for Transcription Factor Nrf2 and Creates a Positive Feedback Loop by Inducing Antioxidant Response Element-Driven Gene Transcription. *J. Biol. Chem.* **2010**, *285*, 22576–22591. [CrossRef]

29. Baek, A.; Kim, J.H.; Pyo, S.; Jung, J.H.; Park, E.J.; Kim, S.H.; Cho, S.R. The Differential Effects of Repetitive Magnetic Stimulation in an in Vitro Neuronal Model of Ischemia/Reperfusion Injury. *Front. Neurol.* **2018**, *9*, 50. [CrossRef]

30. Gibson, J.F.; Prajsnar, T.K.; Hill, C.J.; Tooke, A.K.; Serba, J.J.; Tonge, R.D.; Foster, S.J.; Grierson, A.J.; Ingham, P.W.; Renshaw, S.A.; et al. Neutrophils Use Selective Autophagy Receptor Sqstm1/P62 to Target *Staphylococcus aureus* for Degradation in Vivo in Zebrafish. *Autophagy* **2021**, *17*, 1448–1457. [CrossRef]

31. Mishra, A.K.; Malonia, S.K. Advancing Cellular Immunotherapy with Macrophages. *Life Sci.* **2023**, *328*, 121857. [CrossRef] [PubMed]

Review

NRF2, a Superstar of Ferroptosis

Ruihan Yan [†]**, Bingyi Lin** [†]**, Wenwei Jin, Ling Tang, Shuming Hu** *** and Rong Cai** ***

Department of Biochemistry & Molecular Cell Biology, Shanghai Jiao Tong University School of Medicine, Shanghai 200025, China; super_han@sjtu.edu.cn (R.Y.); linbingyi@sjtu.edu.cn (B.L.); ref-rain@sjtu.edu.cn (W.J.); tanglingyi@sjtu.edu.cn (L.T.)

* Correspondence: hu_shuming1999@sjtu.edu.cn (S.H.); rongcai@shsmu.edu.cn (R.C.)

[†] These authors contribute equally to this work.

Abstract: Ferroptosis is an iron-dependent and lipid peroxidation-driven cell death cascade, occurring when there is an imbalance of redox homeostasis in the cell. Nuclear factor erythroid 2-related factor 2 (NFE2L2, also known as NRF2) is key for cellular antioxidant responses, which promotes downstream genes transcription by binding to their antioxidant response elements (AREs). Numerous studies suggest that NRF2 assumes an extremely important role in the regulation of ferroptosis, for its various functions in iron, lipid, and amino acid metabolism, and so on. Many pathological states are relevant to ferroptosis. Abnormal suppression of ferroptosis is found in many cases of cancer, promoting their progression and metastasis. While during tissue damages, ferroptosis is recurrently promoted, resulting in a large number of cell deaths and even dysfunctions of the corresponding organs. Therefore, targeting NRF2-related signaling pathways, to induce or inhibit ferroptosis, has become a great potential therapy for combating cancers, as well as preventing neurodegenerative and ischemic diseases. In this review, a brief overview of the research process of ferroptosis over the past decade will be presented. In particular, the mechanisms of ferroptosis and a focus on the regulation of ferroptosis by NRF2 will be discussed. Finally, the review will briefly list some clinical applications of targeting the NRF2 signaling pathway in the treatment of diseases.

Keywords: NRF2; ferroptosis; antioxidant; metabolism

Citation: Yan, R.; Lin, B.; Jin, W.; Tang, L.; Hu, S.; Cai, R. NRF2, a Superstar of Ferroptosis. *Antioxidants* **2023**, *12*, 1739. https://doi.org/10.3390/antiox12091739

Academic Editor: Marcel Bonay

Received: 17 August 2023
Revised: 1 September 2023
Accepted: 3 September 2023
Published: 8 September 2023

1. Introduction

Ferroptosis was first observed in 2003 by Dixon, in whose observation, cancer cells with RAS mutations were selectively eliminated by erastin, which was newly discovered by Dolma and his colleague [1]. Subsequently, Yagoda [2] and Yang [3] in 2007 and 2008, respectively, found that iron chelators could inhibit this kind of cell death and discovered another compound, RSL3 (RAS synthetic lethal 3), which could activate it. In 2012, when studying the mechanism behind erastin killing cancer cells, Dixon et al. officially coined this mode of death as ferroptosis based on its unique features [4].

The morphological, biochemical, and genetic characteristics of ferroptosis are distinct from other types of cell death [4]. Morphologically, ferroptosis does not perform typical features of necrosis (such as plasma membrane breakdown and generalized swelling of the cytoplasm and organelles), neither does it have the properties of apoptosis (such as cellular and nuclear volume reduction, nuclear fragmentation, and formation of apoptotic bodies). The main recognizable ultrastructural characteristic of ferroptosis is the apparently altered mitochondrial morphology. Earlier studies showed that mitochondria in ferroptosis cells were significantly contracted along with increased membrane density [2,4]. Later studies illustrated that ferroptosis caused mitochondrial swelling, reduced cristae, and caused outer membrane rupture [5–8]. Biochemically, there are prominent features that can distinguish ferroptosis from others. For example, intracellular depletion of GSH (glutathione), decreased activity of GPX4 (glutathione peroxidase 4), and accumulation of lipid peroxides are regarded as essential factors to promote ferroptosis [3,5]. In genetics, ferroptosis is a biological process regulated by multiple genes, involving regulatory mechanisms in iron homeostasis, lipid peroxidation metabolism, amino acid metabolism, and so on.

Just as its name implies, NRF2 (nuclear factor erythroid 2-related factor 2, NFE2L2) originated from the study of erythropoiesis. In 1994, Moi et al. [9] identified NRF2 as one of the DNA-binding proteins that could bind to AP-1/NF-E2 (activating protein-1/nuclear factor erythroid 2) in the locus control region of β-globin, assisting NF-E2 to mediate globin gene expression. Subsequently, in 1997, pioneering work by Itoh and his colleague demonstrated that NRF2 acted an essential part in the transcriptional induction of phase II detoxification genes carrying ARE (antioxidant response elements), after which roles of NRF2 in erythrocyte generation were marginalized, with studies shifting their focus from erythropoiesis to toxicology [10].

Most importantly, so far, almost all genes that participate in ferroptosis are involved in the transcriptional regulation of NRF2, including glutathione regulation (such as System X_c^- and GPX4), iron regulation (such as FTH1 (ferritin heavy chain 1), FTL (ferritin light chain) and FPN1 (ferrous iron exporter ferroportin 1)), NADPH regeneration (such as G6PD (glucose-6-phosphate dehydrogenase), and ME1 (malic enzyme 1)), and so on [11–14], which are described in this review too. In addition, studies have suggested that NRF2 indirectly participated in the regulation of lipid metabolism via PPARγ (peroxisome proliferator-activated receptor-γ) whose abundance was associated with cell sensitivity to ferroptosis [6]. Hence, with abundant upstream and downstream effectors, targeting NRF2 signaling pathway can effectively affect ferroptosis. In other words, therapies that promote or inhibit ferroptosis through modulating NRF2 signaling pathway are feasible and are worth the time and effort required for their exploration.

2. Ferroptosis

The key process of ferroptosis is Fenton reaction—Fe^{2+} and O_2 (or ROS, reactive oxygen species) function as strong oxidants and peroxidize PL-PUFA (phospholipid with polyunsaturated fatty acyl tail) to PL-PUFA-OOH [15] in the catalysis of POR (cytochrome p450 oxidoreductases) [16] or LOXs (lipoxygenases) [17]. PL-PUFA-OOH then triggers the followed reactions of ferroptosis (Figure 1), though the mechanism of which is still not quite clear.

2.1. Processes That Promote Ferroptosis

Other than PE (phosphatidylethanolamines) [18], PUFAs (polyunsaturated fatty acids) might be the prime substrates for lipid peroxidation according to recent data [19]. PL-PUFAs are sensitive to peroxidation because of their C=C bond. PL-PUFAs are biosynthesized from acyl-CoA in multi-steps in the catalysis of ACC (acetyl coenzyme A carboxylase) [20,21], ACSL4 (acyl-CoA synthetase long-chain family member 4) [22], and LPCAT3 (lysophosphatidylcholine acyltransferase 3) [23]. PUFAs must be esterified into membrane PL-PUFAs so that they can be peroxidized to transmit the ferroptosis signal.

Energy stress (such as glucose starvation) inhibits lipogenesis and ferroptosis through activation of AMPK (adenosine-monophosphate-activated protein kinase), which inhibits the activity of ACC and biosynthesis of PUFAs [17]. E-cadherin-mediated cell–cell contacts activate Hippo signaling and thus inhibit nuclear translocation of NRF2 and the activity of transcriptional co-regulator YAP (Yes-associated protein), which promote transcription of ACSL4, TfR1 (transferrin receptor 1), and possibly other contributors of ferroptosis. Thus, closure of Hippo pathway and increased YAPs can make cells more sensitive to ferroptosis [24].

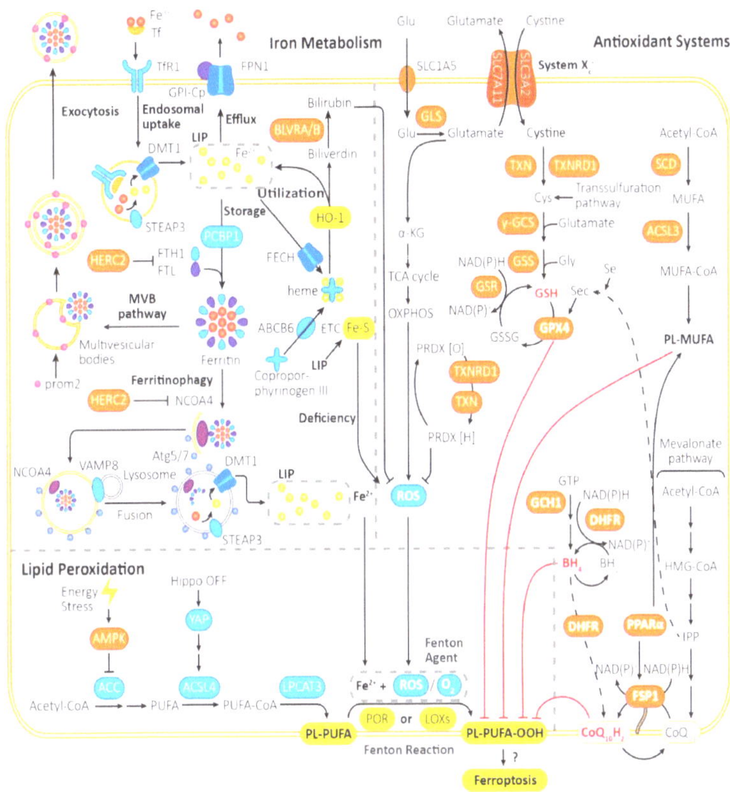

Figure 1. The mechanism of ferroptosis. Ferroptosis is driven by iron-dependent lipid peroxidation, the reaction of PL-PUFA with Fe^{2+} and ROS under the catalysis of POR or LOXs. Ferroptosis is regulated by lipid metabolism, iron metabolism, and intracellular antioxidant system. (1) Lipid metabolism: PL-PUFA, the prime substrate for lipid peroxidation, is synthesized from Acetyl-CoA in multiple steps, which are catalyzed by ACC, ACSL4, and LPCAT3. Then, it is prone to be peroxidized to PL-PUFA-OOH, which triggers the followed reactions of ferroptosis, though the initiation mechanism is still not quite clear. The peroxidation process is mainly driven by Fenton reaction, or mediated by POR or LOXs. By activating AMPK, energy stress can inhibit the activity of ACC, thus turning on an energy stress-protective program against ferroptosis. (2) Iron metabolism: The peroxidation of PL-PUFA is driven by the labile iron pool and iron-dependent enzymes. Fe^{3+} is imported into cells by Tf through TfR1, then being reduced by STEAP3 to Fe^{2+}, forming the labile iron pool. Excess Fe^{2+} can be expelled via FPN1, be stored in ferritin, which can be exported through MVB pathway or release Fe^{2+} through ferritinophagy, or be utilized to synthesize heme with coproporphyrinogen III through ABCB6 and FECH, or regenerated when heme is metabolized by HO-1. (3) Intracellular antioxidant system: There are mainly three pathways for removing PL-PUFA-OOHs. Cyst(e)ine/GSH/GPX4 axis: It is one of the most important pathways in ferroptosis inhibition. It requires uptake of cystine via system X_c^-, reduction of cystine to cysteine, and biosynthesis of GSH. GPX4-mediated reduction of PL-PUFA-OOH to PL-PUFA-OH, and regeneration of GSH from GSSG. FSP1/CoQ10 axis: It functions independently of GPX4. $CoQ_{10}H_2$ can remove lipid peroxides, and then consume NADPH to regenerate $CoQ_{10}H_2$ by FSP1. GCH1/BH4/DHFR system: BH4 can be synthesized from GTP under the catalysis of GCH1, being converted into BH2 and clearing lipid peroxidants, and then regenerated by DHFR. Furthermore, saturated fatty acid can be transformed into PL-MUFA in multi-steps, which is the competing substrate against PL-PUFA and can suppress ferroptosis. Furthermore, PRDX can reduce the content of ROS, which can be generated from PRDX[O] through NADPH and TRX.

2.1.1. LIP (Liable Iron Pool): Iron Drives Lipid Peroxidation with the Aid of ROS

Fe^{3+} formed from intestinal absorption or erythrocyte degradation can be captured by Tf and imported into cells through TfR1 [25], after which it is reduced to Fe^{2+} by STEAP3 (six transmembrane epithelial antigen of the prostate 3) [26] and imported via DMT1 (ferrous ion membrane transport protein) [27]. Fe^{2+} is maintained in the LIP, bound to low molecular weight compounds, including GSH (glutathione) [28].

The excess Fe^{2+} can be oxidized and then exported by FPN1 (ferrous iron exporter ferroportin 1) in the plasma membrane [29], utilized by mitochondria for heme synthesis, or oxidized and stored in ferritins by PCBP1 (poly-(rC)-binding protein 1) [30]. Heme synthesis in the mitochondria is regulated by ABCB6 (ATP-binding cassette subfamily B member 6) and FECH (ferrochelatase), which, respectively, transport coproporphyrinogen III and Fe^{2+} from the cytosol to the mitochondrial intermembrane space to form heme [31]. For heme degradation, HO-1 (heme-oxygenase 1) breaks down heme into biliverdin and liable Fe^{2+}. Biliverdin is further metabolized to a potent antioxidant called bilirubin by BLVRA/BLVRB (biliverdin reductase A and biliverdin reductase B) [32,33]. Thus, HO-1 may play a dual role in ferroptosis. Some studies suggested that HO-1 promotes ferroptosis [34,35], while others reported opposing results that knockout or knockdown of HO-1 exacerbates ferroptosis [36–38].

Ferritins can be exported through MVBs (prom2-mediated ferritin-containing multivesicular bodies) when necessary [39], while some can be degraded into liable Fe^{2+} through ferritinophagy triggered by NCOA4 (nuclear receptor coactivator 4) [40]. In ferritinophagy, NCOA4 recruits ferritins into autophagosome, and then VAMP8 (vesicle associated membrane protein 8) mediates autophagosome–lysosome fusion [40], after which newly freed Fe^{3+} can be reduced and transported into cytosol LIP by STEAP3 and DMT1 [41]. Except NCOA4, there are other ferritinophagy activation proteins, such as ULK1 (unc-51 like kinase 1), atg5 (recombinant autophagy-related protein 5), and atg7 (recombinant autophagy-related protein 7) [42].

2.1.2. Run-Out Mitochondrial Respiration Boosts ROS Generation

The out-of-control respiration of mitochondria is assumed to act as a promoting effect on ferroptosis because of the generation of more ROS [43]. Glutaminolysis is assumed to be an important factor for the run-out mitochondrial respiration. Glutamate is a precursor of α-KG (α-ketoglutaric acid), and hence, glutamate converted into α-KG accelerates the TCA (tricarboxylic acid) cycle and mitochondrial ROS generation (with hyperpolarization of inner mitochondrial membrane potential) [44].

2.2. Processes That Inhibit Ferroptosis

2.2.1. Cyst(e)ine/GSH/GPX4 System

Cysteine and glutamate are essential for GSH biosynthesis to protect against oxidation. Cystine is imported by system X_c^-, reduced into cysteine by TXN/TXNRD1 or through transsulfuration pathway [45] (reductive power from NAD(P)H), and utilized in GSH synthesis in the catalysis of GCL (glutamate-cysteine ligase, or γ-GCS, γ-glutamylcysteine synthetase) and GSS (glutathione synthetase). GPX4 (glutathione peroxidase 4) converts GSH into GSSG (oxidized glutathione) and reduces L-OOH (lipid peroxides) to the corresponding L-OH (lipid alcohols) [46]. TXN (thioredoxin) and TXNRD1 (thioredoxin reductase) contribute to the reduction of cystine to cysteine [47,48].

System X_c^- is an amino acid anti-porter in membrane, composed of two subunits, SLC7A11 (xCT) and SLC3A2, which imports cystine and exports glutamate in a ratio of 1:1 [4]. And SLC7A11 can be inhibited by p53 [49]. Since GPX4 is a selenoprotein, its biosynthesis relies on the co-translational incorporation of Sec (selenocysteine) [50].

2.2.2. FSP1/CoQ$_{10}$/NAD(P)H System

FSP1 (ferroptosis suppressor protein 1) can convert CoQ$_{10}$ (ubiquinone) to CoQ$_{10}$H$_2$ (ubiquinol) with NAD(P)H, adding to the antioxidation capacity [51]. The intermediate of CoQ$_{10}$, IPP (isopentenyl pyrophosphoric acid, or isopentenyl-PP), is as important as CoQ$_{10}$

for the maturation of selenocysteine tRNA for GPX4 [52]. FSP1 can be activated by PPARα (peroxisome proliferator-activated receptor-α) [53].

2.2.3. GCH1/BH$_4$/DHFR System

GCH1 (GTP cyclohydrolase 1) is the rate-limiting enzyme for BH$_4$ (tetrahydrobiopterin) synthesis. BH$_4$ can reduce lipid peroxidants and be converted into BH$_2$ (dihydrobiopterin), which can be regenerated from BH$_2$ with NAD(P)H by DHFR (dihydrofolate reductase). BH$_4$ can also promote the synthesis of CoQ$_{10}$ by DHFR [54].

2.2.4. Lipid Metabolism: PL-MUFA Inhibits Peroxidation of PL-PUFA

MUFA (monounsaturated fatty acid) biosynthesis requires SCD (stearoyl-CoA desaturase), which converts saturated fatty acids into MUFAs [55]. MUFA can be converted into MUFA-CoA by ACSL3 (acyl coenzyme A synthetase long-chain family member 3) and ultimately forms PL-MUFA (phospholipid with monounsaturated fatty acyl tail), the competing substrate against PL-PUFA [56]. Thus, the biosynthesis of PL-MUFA inhibits ferroptosis.

3. NRF2 Structure and Function

3.1. NRF2 Structure

NRF2 (nuclear factor erythroid 2-related factor 2, also named NFE2L2) contains 605 amino acids with seven highly conserved homology domains, named Neh1 (NRF2-ECH homology domain-1) to Neh7 (Figure 2). The Neh1 domain contains CNC/bZIP (Cap-N-Collar/basic leucine zipper) motifs, which can be bound with sMAF (small musculoaponeurotic fibrosarcoma) to form heterodimers that bind to AREs [57,58]. The amino terminal domain or carboxyl terminal domain is named Neh2 or Neh3, individually. The DLG and ETGE motifs on Neh2 domain bind with DC domain of Keap1 (kelch-like ECH-ssociated protein 1) homodimer [44,59], leading to the ubiquitination and degradation of NRF2 [60]. The Neh3 domain includes a VFLVPK motif, which is a critical binding site for CHD6 (chromodomain helicase DNA binding protein 6) [61,62]. Nch4 and Neh5 are acidic conserved sequences, which enhance NRF2 downstream genes expression via binding CBP (cAMP responsive element binding protein) [63,64]. Neh6, a serine-rich conserved sequence, could be phosphorylated by GSK-3β (glycogen synthase kinase 3β), leading to the degradation of NRF2 [63]. The Neh6 domain likely contains another phosphorylation-dependent motif, the DSAPGS motif, an extra binding site for β-TrCP (β-transducin repeat-containing protein) [65]. The Neh7 functions as a binding site for RXRα, capable of inhibiting NRF2 and its AREs [66,67].

3.2. NRF2 Regulation

The major approach to change NRF2 activity is regulating the stability of proteins, of which the Keap1-Cul3-Rbx1 axis is the most important regulator. In addition, NRF2 activity can also be regulated at the transcriptional and post-transcriptional levels.

As for the stability of proteins, Keap1 is considered to play a crucial role in the regulation of NRF2 stability. Therefore, NRF2 stability regulation is artificially divided into Keap1-dependent regulation and Keap1-independent regulation. As mentioned above, under normal circumstances, two molecules of Keap1 bind to the ETGE and DLG motifs on the Neh2 domain of NRF2 via its Kelch repeat domain [44,59]. As an adaptor protein for the Cul3 (Cullin3) E3 ubiquitin ligase complex, it can assemble with Cul3 and Rbx1 into a functional E3 ubiquitin ligase (Keap1-Cul3-E3), which leads to continuous ubiquitination and degradation of NRF2, ensuring low levels of intracellular NRF2 [59]. However, when cells are subjected to internal and external stimuli such as oxidative stress, Keap1 is inactivated due to modification of cysteine residues [68], competitive binding of p62 [69], and so on, causing NRF2 to dissociate from Keap1, and it becomes stable and subsequently transfers to the nucleus. Within the nucleus, NRF2 increases the transcription of antioxidant genes by interacting with other protein factors (such as sMaf) and binding to the ARE.

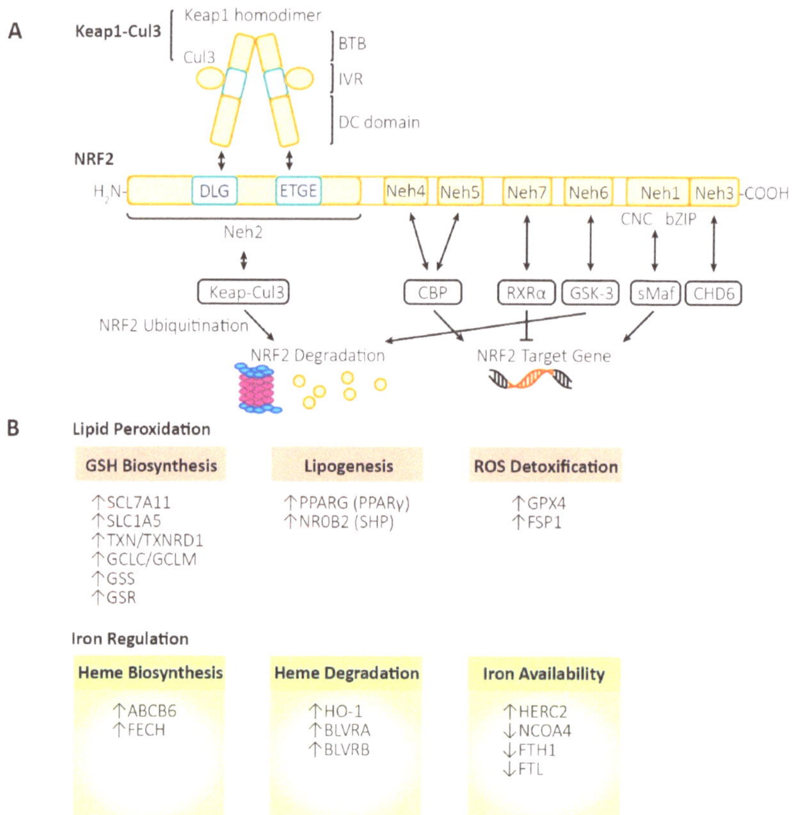

Figure 2. The structure and functions of NRF2. (**A**) Structure and regulation of NRF2. Keap1, kelch-like ECH associated protein 1; Cul3, cullin3 E3 ubiquitin ligase; BTB, broad complex, tramtrack, and bric-a-brac domain; IVR, intermediate region; DC domain, DGR (double glycine repeat) and CTR (carboxyl terminal region) domain; NRF2, nuclear factor erythroid 2-related factor 2; Neh1~6, NRF2-ECH homology domain-1~6; CNC, Cap-N-Collar motif; bZIP, basic leucine zipper motif; sMaf, small musculoaponeurotic fibrosarcoma; CHD6, chromodomain helicase DNA binding protein 6; CBP, cAMP responsive element binding protein; GSK-3, glycogen synthase kinase 3; RXRα, retinoid X receptor alpha. (**B**) Functions of NRF2. Lipid peroxidation (labeled in orange, referring to processes that inhibit ferroptosis): SLC7A11 (xCT), solute carrier family 7 member 11, a heteromeric, sodium-independent, highly specific cysteine-glutamate transport system; SLC1A5, solute carrier family 1 member 5, a sodium-dependent neutral amino acid transporter; TXN, thioredoxin; TXNRD1, thioredoxin reductase 1; GCLC/GCLM, glutamate-cysteine ligase catalytic/modifier subunit, comprising heterodimer γ-GCS (gamma-glutamylcysteine synthetase); GSS, glutathione synthetase; GSR, glutathione-disulfide reductase; PPARG (PPARγ), peroxisome proliferator-activated receptor gamma; NR0B2 (SHP), nuclear receptor subfamily 0 group B member 2 (small heterodimer partner); GPX4, glutathione peroxidase 4; FSP1 (AIFM2), ferroptosis-suppressor-protein 1 (apoptosis inducing factor mitochondria associated 2). Iron regulation (labeled in green, referring to processes that promote ferroptosis): ABCB6, ATP-binding cassette subfamily B member 6, a member of the heavy metal importer subfamily, playing a role in porphyrin transport; FECH, ferrochelatase; HO-1, heme oxygenase 1; BLVRA/BLVRB, biliverdin reductase A/B; HERC2, HECT, and RLD domain containing E3 ubiquitin protein ligase 2; NCOA4, nuclear receptor coactivator 4, androgen receptor coactivator; FTH1, ferritin heavy chain 1; FTL, ferritin light chain.

There are two main ways to regulate NRF2 stability independent of Keap1. One is the β-TrCP-Cul1-dependent pathway. Under the condition of homeostasis, NRF2 is phosphorylated by GSK-3β, after which it can be recognized by β-TrCP, resulting in the ubiquitination of NRF2 by the β-TrCP-Cul1 E3 ubiquitin ligase complex and later degradation by proteasome [63,70,71]. Under stress, the PI3K (phosphoinositide 3-kinase)/Akt pathway is activated, leading to the suppression of GSK-3β by phosphorylation, thus activating NRF2 [72]. The other is the Hrd1-dependent pathway. Hrd1 is an E3 ubiquitin ligase that binds directly to NRF2 and causes the ubiquitination and degradation of NRF2, negatively regulating NRF2 protein levels [73].

As for the regulation at the transcriptional level, Hrd1 is mainly involved in some transcriptional factors that can bind to specific sites in the promoter of NRF2 gene, such as AhR (aryl hydrocarbon receptor) [74], NF-κB (nuclear factor-κB) [75], NRF2 [76], and so on, which can increase the transcription of NRF2. Regulation of NRF2 also occurs at the post-transcriptional level. For example, NRF2 is regulated by a variety of miRNA genes, some of which (such as miR-144) can be locally complementary to the mRNA of NRF2 to silence the NRF2 gene [77]. In addition, in some tumor cells, the mRNA precursor of NRF2 can undergo alternative splicing, resulting in loss of the domain interacting with Keap1, so that Keap1 fails to bind to NRF2 normally, leading to NRF2 stabilization and a decrease in ferroptosis in cancer cells [78].

3.3. NRF2 Functions

3.3.1. Antioxidant-Dependent Function

(1). Glutathione metabolism

Two crucial enzymes in GSH biosynthesis are regulated by NRF2—GCL (composed of GCLC/GCLM, glutamate-cysteine ligase catalytic/modulatory subunits) and GSS [11,76,79–81]. SLC7A11, one of the two subunits of system X_c^-, is also regulated by NRF2 in GSH metabolism, which is responsible for transporting cystine into the cell, thereby increasing cysteine content and promoting the process of GSH generation [80,81].

With the assistance of GSH, GPX4 exerts the activity of reducing peroxides. GSH serves as a reducing agent that donates electrons to xenobiotics or electrophilic organisms, thus transforming into its oxidized form GSSG. Research has shown that GPX4 mainly acts on H_2O_2 or organic hydrogen peroxide compounds, as well as ONOO⁻ [82]. The resulting agent, GSSG, is subsequently reduced for the maintenance of intracellular GSH pool. This process is completed by GSR (glutathione-disulfide reductase) with NADPH [83]. According to reports, both GPX4 and GR are found being transcriptionally regulated by NRF2 [84,85].

(2). FSP1-CoQ$_{10}$ axis

FSP1 is a lipophilic antioxidant, which has been confirmed to be targeted by NRF2. The enzyme regenerates ubiquinol from ubiquinone, suppressing the formation of PL-PUFA-OOHs [19,86]. By regulating the expression level of FSP1, NRF2 maintains low levels of cellular peroxides.

(3). Components of the erythrotoxin family

Besides GSH, another system that resists oxidative stress is the erythrotoxin system. PRDX (peroxiredoxin) can reduce ROS level and is oxidized during this process. Oxidized PRDX is then reduced by TXN (thioredoxin), with the participation of cysteine oxidized by ROS. TXNRD1 (thioredoxin reductase 1) is used to rescue oxidized TXN with NADPH [87,88]. The expression of PRDX, TXN, and TXNRD1 mentioned above are all controlled by NRF2 at transcription level [89].

3.3.2. Antioxidant-Independent Function

NRF2 can exert its antioxidant-independent effects by orchestrating the genes expression in phase II detoxification processes [90]. It activates the transcription of genes encoding detoxification enzymes such as GSTs (glutathione S-transferases) and NQO1 (NAD(P)H:quinone oxidoreductase 1) [90]. These enzymes play pivotal roles in cellular detoxification, facilitating the conjugation and elimination of harmful substances and xenobiotics. In the realm of inflammation, NRF2 activation demonstrates anti-inflammatory properties by modulating the expression of genes involved in inflammatory signaling pathways [91]. In recent years, NRF2 has been found to play a critical role in cellular metabolism.

(1). Heme/iron metabolism

As an E3 ubiquitin ligase, HERC2 is responsible for the degradation of NCOA4 (ferritin cargo receptor mediating ferritinophagy) and FBXL5 (IRP1/2 regulator mediating ferritin synthesis) and thus downregulates liable iron level. HERC2 is upregulated by NRF2 to fight against ferroptosis [40]. FTL and FTH1, light and heavy chains of ferritin, respectively, are controlled by NRF2. FTL is the main stabilizer of ferritin, and FTH1 contains active sites of iron oxidase, through which Fe^{2+} is oxidized into Fe^{3+} and is stored in the central core. FPN1/SLC40A1 responsible for iron excretion from cells is also controlled by NRF2. In addition, HMOX1/HO-1, ABCB6, FECH, SLC48A1 (member A1 of lipolytic vector family 48), and BLVRA/B are also positively regulated by NRF2 [31,32,92–94]. Among them, SLC48A1 is responsible for transporting the released heme from the lysosome to the cytoplasm after the degradation of hemoglobin in aged red blood cells, reducing the degradation of heme.

(2). NADPH generation in glucose metabolism

NADPH generation occurs in many pathways such as the PPP (pentose phosphate pathway), NADK (NAD kinase)-catalyzed phosphorylation of NADH, and the transformation of isocitrate to α-KG by IDH (NADP-dependent isocitrate dehydrogenase). NRF2 directly regulates the transcription of various PPP enzymes, such as G6PD (glucose-6-phosphate dehydrogenase) and other oxidative PPP enzymes [95], comprising reversible and irreversible reactions [96]. NRF2 transcriptionally regulates ME1 (malic enzyme 1) and IDH1 (isocitrate dehydrogenase 1) as well, contributing to the generation of NADPH [95]. NADPH is not only crucial for the synthesis of fatty acids (such as PUFAs), cholesterol, and nucleotides [97], but also important in antioxidation axes such as GPX4, FSP1, and DHFR [19,98,99].

(3). Lipid metabolism

Lipid synthesis highly consumes NADPH and competes with cellular antioxidant systems. Studies have shown that activation of NRF2 mitigates the generation and desaturation of fatty acid [13,54] to increase NADPH content for detoxification in murine liver.

PPARγ (peroxisome proliferator-activated receptor gamma) and NR0B2 (nuclear receptor subfamily 0, group B, member 2) are nuclear receptors that play significant parts in the regulation and control of cholesterol metabolism, which may aid in anti-peroxidation. Both genes can be activated by NRF2 and thus play a crucial role in mitigating accumulation of lipid peroxides [93,100,101].

(4). Amino acid metabolism

NRF2 positively regulates xCT (SLC7A11, the subunit of system X_c^-), which imports cystine and exports glutamate [102]. NRF2 also regulates TXN and TXNRD1 at transcriptional level, both of which help cystine to be reduced to cysteine [47,48]. SLC1A5 (solute-linked carrier family A1 member 5), which can be activated by NRF2, plays an essential role in glutamine uptake, which contributes to GSH biosynthesis [103]. But some studies showed that NRF2-actived cells are in a state of glutamate deficiency due to elevated xCT, GCLC/GCLM, and GSS. As a result, less α-KG is generated from glutamate and TCA cycle is restricted, hence inhibiting generation of ROS by ETC (electron transport chain), the downstream process of TCA cycle [104].

These antioxidant-independent functions of NRF2 mentioned above illustrate its multifaceted role in maintaining cellular homeostasis and influencing disease pathogenesis. Elucidating the intricate mechanisms underlying these functions holds promise for the development of innovative therapeutic approaches by targeting NRF2 activation in various pathological conditions associated with inflammation, metabolic disorders, and tissue damages.

4. NRF2 in Ferroptosis

NRF2 is kept low by three E3 ubiquitin ligases (KEAP1-CUL3-RBX1, SCF/βTrCP, and synoviolin/Hrd1). In the presence of gene mutation, endogenously induced modification, competitive binding of other interacting particles, and inhibition of exogenous drugs, NRF2 will be activated and transferred into the nucleus to initiate the transcription of target genes (Table 1).

Table 1. Major roles of NRF2 in ferroptosis (via regulating downstream genes).

Symbol	Whole Name	Function	Upregulation or Downregulation	Main Factors for Inhibiting Ferroptosis
ABCB6	ATP-binding cassette subfamily B member 6	Heme synthesis (transport protoporphyrin to mitochondria)	↑upregulation	Decrease the capacity of LIP
FECH	Ferrochelatase	Heme synthesis (insert ferrous iron into protoporphyrin)	↑upregulation	Decrease the capacity of LIP
FTH1	Ferritin heavy chain 1	Iron storage (oxidize ferrous iron into ferric iron)	↓downregulation	Decrease the capacity of LIP
FTL	Ferritin light chain	Iron storage (stabilize ferritin)	↓downregulation	Decrease the capacity of LIP
FPN1	Ferroportin 1	Iron excretion	↑upregulation	Decrease the capacity of LIP
HO-1	Heme-oxygenase 1	Heme degradation (decompose heme into ferrous iron and biliverdin)	↑upregulation	Increase the capacity of LIP but generate antioxidants (HO-1 exerts dual effects on ferroptosis)
BLVRA/B	Biliverdin reductase A/B	Heme degradation (metabolize biliverdin into bilirubin)	↑upregulation	Generate antioxidant: bilirubin
HERC2	HECT and RLD domain containing E3 ubiquitin protein ligase 2	Inhibit ferritinophagy (NCOA4 and FBXL5 ubiquitination and degradation)	↑upregulation	Decrease the capacity of LIP
SLC7A11	Solute carrier family 7 member 11	GSH synthesis (import cystine and export glutamate)	↑upregulation	Generate antioxidant: GSH
SLC1A5	Solute carrier family 1 member 5	GSH synthesis (absorb glutamine)	↑upregulation	Generate antioxidant: GSH
TXN/TXNRD1	Thioredoxin/Thioredoxin reductase 1	GSH synthesis (reduce cystine to cysteine)	↑upregulation	Generate antioxidant: GSH
GCLC/GCLM	Glutamate-cysteine ligase catalytic/modifier subunits	GSH synthesis (connect glutamate and cysteine)	↑upregulation	Generate antioxidant: GSH
GSS	Glutathione synthetase	GSH synthesis (connect glycine and γ-glutamyl cysteine)	↑upregulation	Generate antioxidant: GSH
GSR	Glutathione-disulfide reductase	GSH synthesis (reduce GSSG to GSH)	↑upregulation	Generate antioxidant: GSH
GPX4	Glutathione peroxidase 4	ROS detoxification (reduce lipid peroxides by oxidizing GSH to GSSG)	↑upregulation	Decrease the accumulation of lipid peroxides
FSP1	Ferroptosis-suppressor-protein 1	Ubiquinol synthesis (convert ubiquinone to ubiquinol with NAD(P)H)	↑upregulation	Generate antioxidant: ubiquinol

Table 1. *Cont.*

Symbol	Whole Name	Function	Upregulation or Downregulation	Main Factors for Inhibiting Ferroptosis
PPARγ	Peroxisome proliferator-activated receptor gamma	ROS detoxification (mitigate lipid peroxidation as a nuclear receptor)	↑upregulation	Decrease the accumulation of lipid peroxides
NR0B2	Nuclear receptor subfamily 0, group B, member 2	ROS detoxification (mitigate lipid peroxidation as a nuclear receptor)	↑upregulation	Decrease the accumulation of lipid peroxides

4.1. Target NRF2 Signaling Pathway and Promote Ferroptosis in Cancer Therapy

Ferroptosis has emerged as a subject of significant interest within the cancer research field. Its distinct mechanistic and morphological attributes differentiate ferroptosis from other cell death modalities, thereby rendering it a promising avenue for cancer therapeutic exploration. The inhibition of ferroptosis observed across various cancer types, coupled with its dynamic role as a tumor suppressor during cancer progression, suggests that manipulating the regulation of ferroptosis holds promise as an interventional strategy in tumor treatment [105]. Consequently, small molecules that reprogram cancer cells to undergo ferroptosis are regarded as potent therapeutic agents for cancer therapy. Ferroptosis inducers can be classified into several different groups, each acting through different mechanisms.

4.1.1. Four Types of FIN (Ferroptosis Inducers)

(1). Class I FIN: system X_c^- inhibitors

System X_c^- inhibitors, including erastin, can block uptake of cystine and reduce cysteine content, leading to a significant reduction in GSH levels, lipid ROS-mediated damage, and finally ferroptosis [46]. Biochemical and metabolomic analyses have demonstrated a notable decrease in GSH levels following erastin treatment [106]. Studies have also observed decreased GPX4 activity in various cancer cells treated with erastin [107]. Moreover, AUR (auranofin) is able to trigger ferroptosis in liver by inhibiting the activity of TXNRD and facilitating lipid peroxidation when given a high dose while a lower dose might be relatively safe [108]. Similar to erastin, SAS (sulfasalazine) can inhibit system X_C^- and then trigger ferroptosis [109].

(2). Class II FIN: glutathione peroxidase 4 (GPX4) inhibitors

The conventional inducer of ferroptosis, RSL3 ((1S,3R)-RSL3), functions by covalently binding to the active site seleno-cysteine residue of the glutathione peroxidase 4 (GPX4) enzyme, irreversibly inhibiting its activity. This inhibition or loss of GPX4 activity results in the accumulation of lipid peroxides, which are considered lethal signals for ferroptosis. Different from RSL3, FIN56, derived from CIL56, can induce degradation of GPX4 with the participation of ACC and activate SQS which leads to CoQ_{10} depletion, rather than directly inhibit GPX4 activity [21]. Furthermore, exhaustion of GSH as well as the inactivation of GPX4 may play an important role in cisplatin-induced ferroptosis and apoptosis in A549 and HCT116 cells [110].

(3). Class III FIN: ubiquinone generation inhibitors

Statins, including fluvastatin, lovastatin, and simvastatin, are a class of pharmaceutical agents utilized for the reduction of blood cholesterol levels [111]. These drugs exert their cholesterol-lowering effects by inhibiting HMGCR (3-hydroxy-3-methylglutaryl-coenzyme A reductase), an enzyme responsible for the rate-limiting step in the mevalonate pathway, which is involved in cholesterol synthesis. By impeding the generation of isopentenyl pyrophosphate, an intermediate metabolite in the mevalonate pathway, statins can also hinder the biosynthesis of selenoproteins, including GPX4, and CoQ10, hence enhance ferroptosis [21]. The established clinical safety profile of statins, coupled with obesity being a significant risk factor for cancer, has led to the initiation of numerous clinical trials investigating the efficacy of statins as monotherapy or combination therapy in various types of tumors [112].

(4). Class IV FIN: lipid metabolism disruptors

Excessive iron levels induce the generation of reactive oxygen species (ROS), leading to an escalation in lipid peroxide levels. This lipid peroxide accumulation within cell membranes serves as a source of lipid ROS, which further contributes to the initiation of ferroptosis [2–4].

FINO2 (an endoperoxide containing 1,2-dioxolane) can directly oxidizes iron, leading to widespread lipid peroxidation via Fenton reaction and finally ferroptosis ultimately [113]. Lapatinib is able to attenuate the expression of ferroportin and ferritin and promotes the expression of transferrin, causing an increase in intracellular iron ion levels and subsequently ferroptosis [114]. Studies have shown that artemisinin derivatives induce ferroptosis in tumor cells by oxidizing iron, generating more ROS and fostering Fenton reaction [115].

4.1.2. Clinical Application of Targeting NRF2 and Ferroptosis in Cancer Therapy

NRF2 can fight against ferroptosis by regulating its downstream antioxidant genes, but this also means that it maintains the survival of tumor cells while protecting normal cells. In the past decade, many studies have revealed that the activation of NRF2 in cancer cells promotes cancer progression [116–118] and metastasis [119], and confers cancer cells resistance to chemotherapy and radiotherapy [120,121]. Based on the mechanism of ferroptosis inducers mentioned above, drugs have been designed and invented in order to induce ferroptosis by targeting NRF2 signaling pathway in cancer therapy, and some of which are feasible enough to be utilized and evaluated in clinical trials (Table 2). Several drugs are described in detail below, by considering some cancer types as examples.

Table 2. NRF2 inhibitors and activators.

Inhibitors

Compound	Model Organism	Mechanism	Diseases	References
Sorafenib	HT-1080 Calu-1 et al.	Inhibit NRF2/xCT signaling pathway	HCC	[109]
Metformin	HepG2, HUH-7, BALB/C nude mice	Inhibit NRF2 nuclear translocation	HCC resistant to sorafinib	[122]
APG	BEL-7402,BEL-7402/ADM	Enhance NRF2 degradation (upregulate miR-101 expression)	HCC	[123]
(1S,3R)-RSL3	BJeLR et al.	Inhibit NRF2/GPX4 signaling pathway	BC	[124]
Fascin	HS578T MDA-MB-231 et al.	Inhibit NRF2/xCT signaling pathway (induce xCT degradation)	BC	[125]
Isoorientin	A549	Inhibit SIRT6/NRF2/GPX4 signaling pathway	Lung cancer resistant to DDP	[126]
ZVI-NP	H1299, H460, A549	Enhance NRF2 degradation (by GSK3/β-TrCP)	Lung cancer	[127]
Wogonin	AsPC-1, PANC-1, HPDE6-C7	Inhibit NRF2/GPX4 signaling pathway	Pancreatic cancer	[128]
Trigonelline	Athymic BALB/C male nude mice	Silence NRF2 by RNA interference	Drug-resistant head and neck cancer	[37]
Sulfasalazine	OSC19, HSC-2 et al.	Inhibit NRF2/xCT signaling pathway	Lymphoma, head and neck squamous cell carcinoma, gastric tumor	[129,130]

Table 2. *Cont.*

Activators				
Compound	Model Organism	Mechanism	Diseases	References
Levistilide A	MDA-MB-231, MCF-7	Activate NRF2/HO-1 signaling pathway	BC	[131]
Tagitinin C	CRC cell lines	Enhance NRF2 nuclear translocation	Colorectal cancer	[132]
DMF	Mouse embryonic fibroblasts (MEFs)	Enhance NRF2 stabilization (modify key cysteine residues on Keap1)	Alzheimer's disease	[133]
SIRT1	Diabetic Sprague Dawley rat model, H9C2	Enhance NRF2 nuclear translocation	Myocardial ischemia-reperfusion injury	[134]
MiR-24-3p	C57BL/6 mice, H9C2	Inhibit Keap1 expression Activate NRF2/HO-1 signaling pathway	Myocardial ischemia-reperfusion injury	[135]
MiR-24	Diabetic Sprague Dawley rat model VSMCs	Activate NRF2/HO-1 signaling pathway	Diabetics (oxidative stress in VSMCs stimulated by high glucose)	[135]
Dexmedetomidine	Transient global cerebral ischemia rat model	Activate NRF2/HO-1 signaling pathway	Brain injury related to transient global cerebral ischemia	[136]
BCP	Cerebral ischemia/reperfusion SD rat model	Enhance NRF2 nuclear translocation	Ischemia brain injury	[137]
XJZ	C57BL/6 mice	Activate p62/Keap1/NRF2 pathway	Gastric mucosal injury	[138]
AS	Specific-pathogen-free (SPF) male Kunming (KM) mice	Enhance NRF2 nuclear translocation	Sepsis-induced lung injury	[139]
MiR-200a	Goto-Kakizaki (GK) rat model	Enhance NRF2 nuclear translocation (inhibit Keap1 translation)	Experimental diabetic nephric injury	[140]

(1). Liver cancer

HCC (Hepatocellular carcinoma) is the primary malignant tumor in the liver, accounting for 90% of all liver tumors, and is the fourth leading cause of cancer death worldwide [141,142]. Sorafenib has been in the market for more than ten years as a first-line targeted drug for patients with advanced liver cancer. As a multi-kinase inhibitor, sorafenib targets tumors mainly by inhibiting tumor angiogenesis. Years of studies on the cytotoxicity of sorafenib in vitro have found that sorafenib induces an iron-dependent cell death different from apoptosis by inhibiting SLC7A11 [109], an element in the NRF2 downstream signaling pathway, thus blocking the transmembrane transport of cysteine and suppressing the biosynthesis of GSH. Results show that such death can be protected by inhibitors of ferroptosis such as liproxstatin-1, Fer-1 (ferrostatin-1) or DFO (deferoxamine), indicating that sorafenib can induce ferroptosis of liver cancer cells [143]. However, many cancer cell types are known to have resistance to ROS-induced chemotherapy drugs, partly via increasing the expression of some of key antioxidant genes. For instance, sorafenib resistance (both primary and acquired), in which upregulation of NRF2-related signaling pathway in cancer cells plays a core role, is considered to be the main cause of poor prognosis in HCC patients. Currently, some novel treatment strategies to overcome drug resistance has been developed [144]. For instance, the combination therapy of metformin and sorafenib on HCC cells can induce ferroptosis through the p62-Keap1-NRF2 antioxidant signaling pathway. By preventing the translocation of NRF2, metformin reduces HO-1 expression, making HCC cells more sensitive to sorafenib and enhancing its anti-tumor effect [122]. Artesunate can also be used to sensitize sorafenib-induced ferroptosis by promoting lysosomal function [145].

Some nutraceuticals composed of active chemicals derived from natural sources are also capable of combating HCC. For example, APG (Apigenin, 4′,5,7-trihydroxyflavone), a natural bioflavonoid extensively existing in numerous fruits and vegetables, can achieve anti-tumor effect by altering the miR-101/NRF2-related pathway [123]. As a post-transcriptional regulator, miRNAs (microRNAs), a type of noncoding RNA with the length of 18–25 nucleotides, target the 3′-UTRs of mRNA, inhibiting its translation or causing its degradation [146]. As for the pharmacological mechanisms of APG, it is able to upregulate the expression of miR-101 [123], which implies that APG can attenuate the expression of NRF2 at the protein and mRNA levels and reduce NRF2 target genes expression, thus inducing ferroptosis of tumor cells.

(2).   Breast cancer

About 1.7 million people worldwide are diagnosed with breast cancer every year, and about 500,000 people die of breast cancer [147]. (1S,3R)-RSL3 can inhibit the peroxidase activity of GPX4 in BC (breast cancer) cells, causing the accumulation of peroxides and then ferroptosis [124]. It is worth mentioning that drug-resistant breast cancer cells often show stronger dependence on GPX4 and are more sensitive to GPX4 inhibitors [148]. Therefore, GPX4 can be a potential target to overcome drug resistance in breast cancer.

Fascin is able to enhance the vulnerability of breast cancer to erastin-induced ferroptosis by degrading xCT through ubiquitin-mediated proteasome degradation pathway [125]. In addition, siramesine, a lysosomotropic agent, and lapatinib, a potent double tyrosine kinase inhibitor of EGFR (epidermal growth factor receptor, ErbB-1) and ErbB-2 [114], can co-operatively induce ferroptosis-mediated cell death in BC cells [149,150]. Concerning the underlying mechanism, Ma et al. [150] demonstrated that the combination of siramesine and lapatinib attenuates the expression of ferroportin and ferritin and promotes the expression of transferrin, causing an increase in intracellular iron ion levels. Excess iron then participates in the Fenton reaction, raising the generation of ROS and leading to ferroptosis in MDA MB 231 and SKBR3 cells.

Moreover, LA (Levistilide A), an active compound extracted from Chuanxiong Rhizoma, is extensively utilized in cancer treatment in traditional oriental medicine. Previous studies have suggested that it is capable of promoting apoptosis in colon cancer cells through the ER (endoplasmic reticulum) stress pathway induced by ROS accumulation, exerting an anti-tumor effect [151]. In 2022, a study explored the underlying mechanism of LA-induced ferroptosis in vitro and eventually confirmed that LA treatment activated NRF2/HO-1 signaling pathway (upregulation of the expression of NRF2 and its downstream molecule HO-1), which enhanced ROS-induced ferroptosis in BC cells due to the dual effects of HO-1. Furthermore, LA also eliminate BC cells by significantly decreasing cell viability in a dose-dependent manner and disrupting the structure and function of mitochondria [131].

(3).   Lung cancer

Lung cancer is responsible for massive cancer-related deaths worldwide. In 2018, the global incidence of lung cancer took the lead among all kinds of cancers [152]. NSCLC (non-small-cell lung cancer) is the most common type of lung cancer, representing over 80% of all lung cancers. DDP (Cisplatin) is regarded as the first-in-class drug for advanced and non-targetable NSCLC. Studies have indicated that DDP exerts the ability of inducing both ferroptosis and apoptosis in the NSCLC cell lines A549 and HCT116 cells [110]. It is confirmed that the major part of intracellular DDP can bind to GSH, forming the Pt-GS complex [153]. Therefore, similar to erastin, via GSH depletion and the inactivation of GPXs, DDP induces ferroptosis of cancer cells. However, it is essential to mention that anti-tumor effects of the combined application of erastin and DDP surpasses that of single use of DDP [110].

However, drug resistance to DDP is a major barrier to lung cancer treatment as well. A recent study revealed that the combination of isoorientin (also known as homoorientin) and DDP treatment resulted in an apparent decrease in the viability of drug-resistant cells through in vitro and in vivo experiments. In terms of mechanism, isoorientin acts as a modulator that controls the SIRT6/NRF2/GPX4 signaling pathway (downregulation of the expression of SIRT6 protein, NRF2, and GPX4), thereby regulating ferroptosis and reversing drug resistance in lung cancer cells [126].

A new drug, ZVI-NP (zero-valent-iron nanoparticle), which has been widely developed to treat contaminated groundwater or wastewater [154], is now gradually coming into sight for its anti-tumor efficacy. Previously, ZVI-NP has been found to be preferentially converted to ferric ions in lysosomes of cancer cells rather than in normal cells, since cancer cells are more acidic in intra-organelle compartments. Sudden release of iron ions further induced a ROS surge in cancer cells, and led to ferroptosis in cancer cells [155]. In 2021, when exploring the potential role of ZVI-NP in regulating the tumor microenvironment in lung cancer models, a study identified a novel molecular mechanism in inducing ferroptosis. Mechanistically, ZVI-NP enhanced phosphate-dependent ubiquitination and degradation of NRF2, thereby triggering ferroptosis along with excessive oxidative stress and lipid peroxidation. Moreover, ZVI-NP impaired the self-renewal capacity of cancer cells and inhibited angiogenesis in endothelial cells [127], all of which profoundly opened up the potential of new advanced cancer therapies that reduced side effects and enhanced efficacy.

(4). Pancreatic cancer

Pancreatic cancer ranks seventh in cancer-related mortality according to a survey conducted by the American Cancer Society in 2021 [156]. Moreover, a study has indicated that pancreatic carcinoma will represent the second leading cause of cancer-related deaths in the United States by 2030 [157]. Wogonin, as a main element composing *Scutellaria baicalensis* [158], is found to exhibit strong anti-inflammatory activity and potential for treating tumors in vitro and in vivo. A recent study has shown that wogonin decreases the GSH content in cells by suppressing the NRF2-mediated GPX4 pathway, leading to ROS accumulation and the enhancement of lipid peroxidation in pancreatic cancer cells [128], which indicates that wogonin could be potentially used for the treatment of pancreatic carcinoma.

(5). Other cancers

AS (Artesunate), a water-soluble derivative of artemisinin, induces ferroptosis in head and neck cancer cells by depleting GSH and causing ROS accumulation. Furthermore, silence of NRF2 by RNA interference or inhibition of NRF2 by trigonelline can enhance the killing effect of AS on drug-resistant head and neck cancer cells in vivo and in vitro [37]. Sulfasalazine is an azo-bridged agent with anti-inflammatory activity, originally synthesized from the antibiotic sulfapyridine in 1940. It was confirmed to be a potent inhibitor of xCT in 2001 [129]. Through disturbing cystine absorption, sulfasalazine mitigates the biosynthesis of GSH, resulting in ferroptosis [4] of some types of cancer cells in vivo and in vitro [130], such as lymphoma [129]. Sulfasalazine is supposed to be non-toxic so that it can be utilized in combination with other drugs to fight against drug resistance and side effects [159]. Okazaki et al. [160] found that the combination of dyclonine and sulfasalazine restrained the growth of head and neck squamous cell carcinoma or gastric tumors with high expression of ALDH3A1, which contributes to resistance to sulfasalazine monotherapy. Tagitinin C, a sesquiterpene lactone isolated from Tithonia diversifolia, is widely found in the Asteraceae [161], with ample pharmacological effects, including anti-tumor, anti-virus, anti-fibrosis, anti-parasite, and cardio-protective activity [162–164]. Mechanistically, it induced ER stress and oxidative stress, which activated the nuclear translocation of NRF2, resulting in oxidative cell microenvironment, ferroptosis, and suppressed growth of colorectal cancer cells. As a downstream effector of NRF2, the expression of HO-1 increased significantly, which led to an increase in the LIP and promoted lipid peroxidation. Furthermore, tagitinine C exhibited synergistic anti-tumor effects together with the erastin [132].

In general, it is reasonable to believe that more and more treatment strategies with higher efficacy and prognosis will be found in the future. The application of drugs that target NRF2 signaling pathway and induce ferroptosis to kill tumors is both novel and promising.

4.2. Targeting of NRF2 Signaling Pathway and Inhibition of Ferroptosis in Disease Therapy
4.2.1. Diseases Associated with Ferroptosis

(1). Neurodegenerative diseases

Ferroptosis is presented in Alzheimer's disease, Parkinson's disease, and many other neurodegenerative diseases. Iron is the most abundant transition metal in the brain, participating in numerous metabolic reactions. Disorder of iron homeostasis in the brain can lead to some neurodegenerative diseases, such as Alzheimer's disease and Parkinson's disease [165–167]. Iron gradually accumulates in the brain with age, which changes the cerebral iron metabolism. In animal models of neurodegenerative diseases as well as aging populations, varying degrees of elevation of iron content in the brain have been observed [168,169]. Studies also show that chronic exposure to iron for mice causes dysfunction of membrane-transport protein, imbalance of intracellular iron homeostasis, and significant elevation of ROS and MDA (malondialdehyde), which ultimately lead to the dysfunction of nervous system [170]. Numerous studies reveal that disorder of iron metabolism has an influence on the normal folding of Aβ protein (amyloid β-protein) and phosphorylation of tau protein, resulting in oxidative stress and metal toxicity, which can cause DNA, lipid, and protein damage and ultimately lead to Alzheimer's disease [171–173]. Studies also indicate that cognitive function of Alzheimer's patients can be stabilized when applying antioxidants and iron chelators like α-lipoic acid, which blocks tau-induced iron overload, lipid peroxidation, and ferroptosis-related inflammation [174]. Parkinson disease is characterized by mitochondrial dysfunction, iron accumulation, copper depletion, and GSH exhaustion in the brain, some of which are also the most important features of ferroptosis [175–179]. Iron chelators (ferroptosis inhibitor) like deferiprone show therapeutic effect in Parkinson's patients in a randomized controlled trial [180].

(2). Cardiovascular diseases

Ferroptosis plays an influential role in cardiovascular diseases. Recent studies have found a significant decrease in GPX4 expression in the early and middle stages of myocardial infarction [181]. Human umbilical cord blood-derived mesenchymal stem cell exosomes may inhibit ferroptosis by targeting miR-23a-3p, thereby achieving a protective effect in alleviating acute myocardial infarction [182]. Ferroptosis also plays an important role in hypertension. Pulmonary hypertension can cause a decrease in GSH levels and an increase in iron, ROS, and lipid peroxides [183]. The cardiotoxicity of anthracyclines (e.g., doxorubicin) is also associated with ferroptosis. Studies have shown that doxorubicin can lead to heme degradation, causing mitochondrial dysfunction and a significant increase in the content of iron, ROS, and other markers of ferroptosis [35]. In DOX (doxorubicin)-induced cardiomyopathy, the administration of ferroptosis inhibitor Fer-1 (ferrostatin-1) can significantly reduce the mortality rate of mice, while the mortality rate of mice administered with inhibitors of apoptosis, necrosis, and autophagy is not significantly reduced [35]. Ferroptosis may also play a significant role in sepsis. In animal models of sepsis with cecum ligation and puncture, the levels of GSH and GPX4 decreased accompanied by elevated levels of iron and lipid peroxides [184]. Many other types of cardiovascular diseases, such as arrhythmias and sickle cell anemia, are also considered to be associated with ferroptosis [185,186].

(3). Organic injuries

Ferroptosis has been investigated in many organic injuries as well. For example, in mice with sepsis, multiple organic injuries caused by ferroptosis can be detected. Ferroptosis has also been proved to play an important role in LPS-induced acute lung injury, in which Fer-1 also displays effective therapeutic effect [187]. In various acute renal injuries, like rhabdomyolysis-induced acute kidney injury and folic acid-induced acute kidney injury, iron accumulation and ROS have been observed and can be suppressed by ferroptosis inhibitors [188–190]. Fe^{2+} produced by myoglobin metabolism directly induces lipid peroxidation in proximal tubular epithelial cells in rhabdomyolysis, which may be an important mechanism of rhabdomyolysis inducing acute kidney injury [189]. And the application of iron chelator deferoxamine is able to alleviate rhabdomyolysis-induced kidney injury in mice [191].

4.2.2. Drugs and Clinical Experiments

There have been many studies focusing on the role of NRF2 and showing that regulation of NRF2 and related signaling pathway can alleviate neurodegeneration, cardiovascular diseases, and many organic injuries.

(1). Neurodegenerative diseases

NRF2 has promising prospects as a novel target for the treatment of neurodegenerative diseases, since Alzheimer's disease is closely related to oxidative stress, mitochondrial dysfunction, and ferroptosis. The activation of NRF2 can upregulate the expression of xCT, enhance glutamate secretion, and restore GPX4 activity, thereby inhibiting ferroptosis. Therefore, drugs activating NRF2 could become a new approach to treat Alzheimer's disease. DMF (Tecfidera®) can be regarded as the most successful case reported in activating NRF2 [192,193]. The drug is originally used to treat psoriasis patients, but new research shows that after entering the body, it transformed so as to covalently modify key cysteine residues on KEAP1, thus blocking NRF2 ubiquitination and promoting NRF2 stabilization as well as subsequent activation of NRF2 target genes [133]. In addition, many drugs that can be used to treat Alzheimer's disease by activating NRF2, such as resveratrol, pyridoxine, and NBP (n-butylphthalide), have also entered clinical trials.

(2). Cardiovascular diseases

Targeting NRF2 has broad prospects in the treatment of cardiovascular diseases. SIRT1 (Sirtuin1) is a NAD-dependent deacetylase which plays an important role in a variety of cardiovascular diseases caused by diabetes. It can increase the localization of NRF2 in the nucleus, thereby alleviating myocardial ischemia-reperfusion injury in the diabetic Sprague Dawley rat model [134]. The effect of SIRT1 may be achieved by deacetylation of NRF2 and reduction of NRF2 ubiquitination [194]. At the same time, miRNAs can indirectly regulate NRF2 expression by modulating the upstream protein NRF2. MiR-24-3p that has been shown to modulate the Keap1/NRF2 pathway during myocardial ischemia-reperfusion [135]. In highly glucose-stimulated rat and mouse cardiomyocyte models, inhibition of miR-144, miR-155, and miR-503 activates NRF2 to attenuate cellular oxidative stress [195–197]. MiR-24 has been shown to activate the NRF2/HO-1 signaling pathway, playing a critical role in combating oxidative stress in VSMCs (vascular smooth muscle cells) stimulated by high glucose [198].

(3). Organic injuries

In brain injury related to transient global cerebral ischemia in rats, study shows that preischemia dexmedetomidine administration has neuroprotective effect by enhancing NRF2/HO-1 expression and reducing caspase-3 activity [136]. In ischemia brain injury, BCP (β-Caryophyllene) has protective effects, whose mechanism is related to the activation of the NRF2/HO-1 pathway [137]. Moreover, study shows that XJZ (Xiaojianzhong decoction), which consists of six Chinese herbal medicines and extracts, can attenuate ferroptosis mediated by oxidative stress in the gastric mucosal injury by upregulating the p62/Keap1/NRF2 pathway [138]. In sepsis-induced lung injury, AS (artesunate) upregulates the expression of antiferroptosis systems like NRF2 and HO-1, thus attenuating neutrophil infiltration and pathological damage in lung tissue [139]. Studies in the type II diabetic nephropathy rat model show that experimental diabetic nephric injury can be alleviated by targeting the Keap1/NRF2 pathway using MiR-200a [140].

Although drugs targeting NRF2 show therapeutic functions, and have a promising future, studies on these drugs are still limited to the experimental stage, and further clinical trials are needed for application. Multiple aspects, including safety, efficacy, and economy, should be taken into consideration before clinical application.

5. Conclusions and Future Perspectives

With a further understanding of the mechanism of ferroptosis, the criticality of NRF2 is demonstrated more obviously, whose rich value embodies in both antioxidant-dependent and antioxidant-independent functions. As a key regulator of the cellular antioxidant response, NRF2 is widely involved in iron, lipid, amino acid, glucose metabolisms, etc. Therefore, through pharmacological modulation of NRF2 and its downstream effectors, prevention and treatment of pathologies, such as numerous types of cancers, neurodegenerative diseases, cardiovascular diseases, and so on, can be achieved. Despite several unclear issues, such as the specific connection between lipid peroxidation and ferroptosis, there is no denying the fact that therapies targeting NRF2 signaling pathway exert extraordinary efficacy. This kind of approach can be used to treat various ferroptosis-related diseases and may have a brilliant therapeutic future.

Author Contributions: R.Y., B.L., W.J., L.T. and R.C. figured out the idea of writing this review. R.Y., B.L., W.J. and L.T. summarized the published results and drafted the manuscript. S.H. and R.C. revised the manuscript. All authors have read and agreed to the published version of the manuscript.

Funding: The review was funded by the Innovative Research Team of High-level Local Universities in Shanghai and supported by the National Natural Science Foundation of China (81872230 and 82173352 to R.C.).

Institutional Review Board Statement: Not applicable.

Informed Consent Statement: Not applicable.

Data Availability Statement: Not applicable.

Conflicts of Interest: The authors declare no conflict of interest.

References

1. Dolma, S.; Lessnick, S.L.; Hahn, W.C.; Stockwell, B.R. Identification of genotype-selective antitumor agents using synthetic lethal chemical screening in engineered human tumor cells. *Cancer Cell* **2003**, *3*, 285–296. [CrossRef] [PubMed]
2. Yagoda, N.; von Rechenberg, M.; Zaganjor, E.; Bauer, A.J.; Yang, W.S.; Fridman, D.J.; Wolpaw, A.J.; Smukste, I.; Peltier, J.M.; Boniface, J.J.; et al. RAS-RAF-MEK-dependent oxidative cell death involving voltage-dependent anion channels. *Nature* **2007**, *447*, 864–868. [CrossRef] [PubMed]
3. Yang, W.S.; Stockwell, B.R. Synthetic lethal screening identifies compounds activating iron-dependent, nonapoptotic cell death in oncogenic-RAS-harboring cancer cells. *Chem. Biol.* **2008**, *15*, 234–245. [CrossRef] [PubMed]
4. Dixon, S.J.; Lemberg, K.M.; Lamprecht, M.R.; Skouta, R.; Zaitsev, E.M.; Gleason, C.E.; Patel, D.N.; Bauer, A.J.; Cantley, A.M.; Yang, W.S.; et al. Ferroptosis: An iron-dependent form of nonapoptotic cell death. *Cell* **2012**, *149*, 1060–1072. [CrossRef] [PubMed]
5. Friedmann Angeli, J.P.; Schneider, M.; Proneth, B.; Tyurina, Y.Y.; Tyurin, V.A.; Hammond, V.J.; Herbach, N.; Aichler, M.; Walch, A.; Eggenhofer, E.; et al. Inactivation of the ferroptosis regulator Gpx4 triggers acute renal failure in mice. *Nat. Cell Biol.* **2014**, *16*, 1180–1191. [CrossRef]
6. Doll, S.; Proneth, B.; Tyurina, Y.Y.; Panzilius, E.; Kobayashi, S.; Ingold, I.; Irmler, M.; Beckers, J.; Aichler, M.; Walch, A.; et al. ACSL4 dictates ferroptosis sensitivity by shaping cellular lipid composition. *Nat. Chem. Biol.* **2017**, *13*, 91–98. [CrossRef]
7. Maiorino, M.; Conrad, M.; Ursini, F. GPx4, Lipid Peroxidation, and Cell Death: Discoveries, Rediscoveries, and Open Issues. *Antioxid. Redox Signal* **2018**, *29*, 61–74. [CrossRef]
8. Ingold, I.; Berndt, C.; Schmitt, S.; Doll, S.; Poschmann, G.; Buday, K.; Roveri, A.; Peng, X.; Porto Freitas, F.; Seibt, T.; et al. Selenium Utilization by GPX4 Is Required to Prevent Hydroperoxide-Induced Ferroptosis. *Cell* **2018**, *172*, 409–422.e21. [CrossRef]
9. Moi, P.; Chan, K.; Asunis, I.; Cao, A.; Kan, Y.W. Isolation of NF-E2-related factor 2 (Nrf2), a NF-E2-like basic leucine zipper transcriptional activator that binds to the tandem NF-E2/AP1 repeat of the beta-globin locus control region. *Proc. Natl. Acad. Sci. USA* **1994**, *91*, 9926–9930. [CrossRef]
10. Itoh, K.; Chiba, T.; Takahashi, S.; Ishii, T.; Igarashi, K.; Katoh, Y.; Oyake, T.; Hayashi, N.; Satoh, K.; Hatayama, I.; et al. An Nrf2/small Maf heterodimer mediates the induction of phase II detoxifying enzyme genes through antioxidant response elements. *Biochem. Biophys. Res. Commun.* **1997**, *236*, 313–322. [CrossRef]
11. Sasaki, H.; Sato, H.; Kuriyama-Matsumura, K.; Sato, K.; Maebara, K.; Wang, H.; Tamba, M.; Itoh, K.; Yamamoto, M.; Bannai, S. Electrophile response element-mediated induction of the cystine/glutamate exchange transporter gene expression. *J. Biol. Chem.* **2002**, *277*, 44765–44771. [CrossRef]
12. Lee, J.M.; Calkins, M.J.; Chan, K.; Kan, Y.W.; Johnson, J.A. Identification of the NF-E2-related factor-2-dependent genes conferring protection against oxidative stress in primary cortical astrocytes using oligonucleotide microarray analysis. *J. Biol. Chem.* **2003**, *278*, 12029–12038. [CrossRef]
13. Wu, K.C.; Cui, J.Y.; Klaassen, C.D. Beneficial role of Nrf2 in regulating NADPH generation and consumption. *Toxicol. Sci.* **2011**, *123*, 590–600. [CrossRef]
14. Kerins, M.J.; Ooi, A. The Roles of NRF2 in Modulating Cellular Iron Homeostasis. *Antioxid. Redox Signal* **2018**, *29*, 1756–1773. [CrossRef]
15. Porter, N.A.; Caldwell, S.E.; Mills, K.A. Mechanisms of free radical oxidation of unsaturated lipids. *Lipids* **1995**, *30*, 277–290. [CrossRef] [PubMed]
16. Zou, Y.; Li, H.; Graham, E.T.; Deik, A.A.; Eaton, J.K.; Wang, W.; Sandoval-Gomez, G.; Clish, C.B.; Doench, J.G.; Schreiber, S.L. Cytochrome P450 oxidoreductase contributes to phospholipid peroxidation in ferroptosis. *Nat. Chem. Biol.* **2020**, *16*, 302–309. [CrossRef] [PubMed]

17. Kuhn, H.; Banthiya, S.; van Leyen, K. Mammalian lipoxygenases and their biological relevance. *Biochim. Biophys. Acta* **2015**, *1851*, 308–330. [CrossRef] [PubMed]

18. Kagan, V.E.; Mao, G.; Qu, F.; Angeli, J.P.; Doll, S.; Croix, C.S.; Dar, H.H.; Liu, B.; Tyurin, V.A.; Ritov, V.B.; et al. Oxidized arachidonic and adrenic PEs navigate cells to ferroptosis. *Nat. Chem. Biol.* **2017**, *13*, 81–90. [CrossRef]

19. Doll, S.; Freitas, F.P.; Shah, R.; Aldrovandi, M.; da Silva, M.C.; Ingold, I.; Goya Grocin, A.; Xavier da Silva, T.N.; Panzilius, E.; Scheel, C.H.; et al. FSP1 is a glutathione-independent ferroptosis suppressor. *Nature* **2019**, *575*, 693–698. [CrossRef]

20. Lee, H.; Zandkarimi, F.; Zhang, Y.; Meena, J.K.; Kim, J.; Zhuang, L.; Tyagi, S.; Ma, L.; Westbrook, T.F.; Steinberg, G.R.; et al. Energy-stress-mediated AMPK activation inhibits ferroptosis. *Nat. Cell Biol.* **2020**, *22*, 225–234. [CrossRef]

21. Shimada, K.; Skouta, R.; Kaplan, A.; Yang, W.S.; Hayano, M.; Dixon, S.J.; Brown, L.M.; Valenzuela, C.A.; Wolpaw, A.J.; Stockwell, B.R. Global survey of cell death mechanisms reveals metabolic regulation of ferroptosis. *Nat. Chem. Biol.* **2016**, *12*, 497–503. [CrossRef] [PubMed]

22. Küch, E.M.; Vellaramkalayil, R.; Zhang, I.; Lehnen, D.; Brügger, B.; Sreemmel, W.; Ehehalt, R.; Poppelreuther, M.; Füllekrug, J. Differentially localized acyl-CoA synthetase 4 isoenzymes mediate the metabolic channeling of fatty acids towards phosphatidylinositol. *Biochim. Biophys. Acta* **2014**, *1841*, 227–239. [CrossRef] [PubMed]

23. Shindou, H.; Shimizu, T. Acyl-CoA:lysophospholipid acyltransferases. *J. Biol. Chem.* **2009**, *284*, 1–5. [CrossRef] [PubMed]

24. Wu, J.; Minikes, A.M.; Gao, M.; Bian, H.; Li, Y.; Stockwell, B.R.; Chen, Z.N.; Jiang, X. Intercellular interaction dictates cancer cell ferroptosis via NF2-YAP signalling. *Nature* **2019**, *572*, 402–406. [CrossRef] [PubMed]

25. Frazer, D.M.; Anderson, G.J. The regulation of iron transport. *Biofactors* **2014**, *40*, 206–214. [CrossRef]

26. Ohgami, R.S.; Campagna, D.R.; Greer, E.L.; Antiochos, B.; McDonald, A.; Chen, J.; Sharp, J.J.; Fujiwara, Y.; Barker, J.E.; Fleming, M.D. Identification of a ferrireductase required for efficient transferrin-dependent iron uptake in erythroid cells. *Nat. Genet.* **2005**, *37*, 1264–1269. [CrossRef]

27. Fleming, M.D.; Romano, M.A.; Su, M.A.; Garrick, L.M.; Garrick, M.D.; Andrews, N.C. Nramp2 is mutated in the anemic Belgrade (b) rat: Evidence of a role for Nramp2 in endosomal iron transport. *Proc. Natl. Acad. Sci. USA* **1998**, *95*, 1148–1153. [CrossRef]

28. Patel, S.J.; Frey, A.G.; Palenchar, D.J.; Achar, S.; Bullough, K.Z.; Vashisht, A.; Wohlschlegel, J.A.; Philpott, C.C. A PCBP1-BolA2 chaperone complex delivers iron for cytosolic [2Fe-2S] cluster assembly. *Nat. Chem. Biol.* **2019**, *15*, 872–881. [CrossRef]

29. Anderson, G.J.; Vulpe, C.D. Mammalian iron transport. *Cell Mol. Life Sci.* **2009**, *66*, 3241–3261. [CrossRef]

30. Bogdan, A.R.; Miyazawa, M.; Hashimoto, K.; Tsuji, Y. Regulators of Iron Homeostasis: New Players in Metabolism, Cell Death, and Disease. *Trends Biochem. Sci.* **2016**, *41*, 274–286. [CrossRef]

31. Campbell, M.R.; Karaca, M.; Adamski, K.N.; Chorley, B.N.; Wang, X.; Bell, D.A. Novel hematopoietic target genes in the NRF2-mediated transcriptional pathway. *Oxid. Med. Cell Longev.* **2013**, *2013*, 120305. [CrossRef] [PubMed]

32. Agyeman, A.S.; Chaerkady, R.; Shaw, P.G.; Davidson, N.E.; Visvanathan, K.; Pandey, A.; Kensler, T.W. Transcriptomic and proteomic profiling of KEAP1 disrupted and sulforaphane-treated human breast epithelial cells reveals common expression profiles. *Breast Cancer Res. Treat.* **2012**, *132*, 175–187. [CrossRef] [PubMed]

33. Hirotsu, Y.; Katsuoka, F.; Funayama, R.; Nagashima, T.; Nishida, Y.; Nakayama, K.; Engel, J.D.; Yamamoto, M. Nrf2-MafG heterodimers contribute globally to antioxidant and metabolic networks. *Nucleic Acids Res.* **2012**, *40*, 10228–10239. [CrossRef]

34. Hassannia, B.; Wiernicki, B.; Ingold, I.; Qu, F.; Van Herck, S.; Tyurina, Y.Y.; Bayır, H.; Abhari, B.A.; Angeli, J.P.F.; Choi, S.M.; et al. Nano-targeted induction of dual ferroptotic mechanisms eradicates high-risk neuroblastoma. *J. Clin. Investig.* **2018**, *128*, 3341–3355. [CrossRef] [PubMed]

35. Fang, X.; Wang, H.; Han, D.; Xie, E.; Yang, X.; Wei, J.; Gu, S.; Gao, F.; Zhu, N.; Yin, X.; et al. Ferroptosis as a target for protection against cardiomyopathy. *Proc. Natl. Acad. Sci. USA* **2019**, *116*, 2672–2680. [CrossRef] [PubMed]

36. Adedoyin, O.; Boddu, R.; Traylor, A.; Lever, J.M.; Bolisetty, S.; George, J.F.; Agarwal, A. Heme oxygenase-1 mitigates ferroptosis in renal proximal tubule cells. *Am. J. Physiol. Renal Physiol.* **2018**, *314*, F702–F714. [CrossRef]

37. Roh, J.L.; Kim, E.H.; Jang, H.; Shin, D. Nrf2 inhibition reverses the resistance of cisplatin-resistant head and neck cancer cells to artesunate-induced ferroptosis. *Redox Biol.* **2017**, *11*, 254–262. [CrossRef]

38. Sun, X.; Ou, Z.; Chen, R.; Niu, X.; Chen, D.; Kang, R.; Tang, D. Activation of the p62-Keap1-NRF2 pathway protects against ferroptosis in hepatocellular carcinoma cells. *Hepatology* **2016**, *63*, 173–184. [CrossRef]

39. Brown, C.W.; Amante, J.J.; Chhoy, P.; Elaimy, A.L.; Liu, H.; Zhu, L.J.; Baer, C.E.; Dixon, S.J.; Mercurio, A.M. Prominin2 Drives Ferroptosis Resistance by Stimulating Iron Export. *Dev. Cell* **2019**, *51*, 575–586.e4. [CrossRef]

40. Anandhan, A.; Dodson, M.; Shakya, A.; Chen, J.; Liu, P.; Wei, Y.; Tan, H.; Wang, Q.; Jiang, Z.; Yang, K.; et al. NRF2 controls iron homeostasis and ferroptosis through HERC2 and VAMP8. *Sci. Adv.* **2023**, *9*, eade9585. [CrossRef]

41. Anandhan, A.; Dodson, M.; Schmidlin, C.J.; Liu, P.; Zhang, D.D. Breakdown of an Ironclad Defense System: The Critical Role of NRF2 in Mediating Ferroptosis. *Cell Chem. Biol.* **2020**, *27*, 436–447. [CrossRef] [PubMed]

42. Pajares, M.; Jiménez-Moreno, N.; García-Yagüe, Á.J.; Escoll, M.; de Ceballos, M.L.; Van Leuven, F.; Rábano, A.; Yamamoto, M.; Rojo, A.I.; Cuadrado, A. Transcription factor NFE2L2/NRF2 is a regulator of macroautophagy genes. *Autophagy* **2016**, *12*, 1902–1916. [CrossRef]

43. Gao, M.; Yi, J.; Zhu, J.; Minikes, A.M.; Monian, P.; Thompson, C.B.; Jiang, X. Role of Mitochondria in Ferroptosis. *Mol. Cell* **2019**, *73*, 354–363.e3. [CrossRef] [PubMed]

44. Tong, K.I.; Katoh, Y.; Kusunoki, H.; Itoh, K.; Tanaka, T.; Yamamoto, M. Keap1 recruits Neh2 through binding to ETGE and DLG motifs: Characterization of the two-site molecular recognition model. *Mol. Cell Biol.* **2006**, *26*, 2887–2900. [CrossRef] [PubMed]

45. Hayano, M.; Yang, W.S.; Corn, C.K.; Pagano, N.C.; Stockwell, B.R. Loss of cysteinyl-tRNA synthetase (CARS) induces the transsulfuration pathway and inhibits ferroptosis induced by cystine deprivation. *Cell Death Differ.* **2016**, *23*, 270–278. [CrossRef]

46. Seibt, T.M.; Proneth, B.; Conrad, M. Role of GPX4 in ferroptosis and its pharmacological implication. *Free Radic. Biol. Med.* **2019**, *133*, 144–152. [CrossRef]

47. Hawkes, H.J.; Karlenius, T.C.; Tonissen, K.F. Regulation of the human thioredoxin gene promoter and its key substrates: A study of functional and putative regulatory elements. *Biochim. Biophys. Acta* **2014**, *1840*, 303–314. [CrossRef]

48. Malhotra, D.; Portales-Casamar, E.; Singh, A.; Srivastava, S.; Arenillas, D.; Happel, C.; Shyr, C.; Wakabayashi, N.; Kensler, T.W.; Wasserman, W.W.; et al. Global mapping of binding sites for Nrf2 identifies novel targets in cell survival response through ChIP-Seq profiling and network analysis. *Nucleic Acids Res.* **2010**, *38*, 5718–5734. [CrossRef]

49. Jiang, L.; Hickman, J.H.; Wang, S.J.; Gu, W. Dynamic roles of p53-mediated metabolic activities in ROS-induced stress responses. *Cell Cycle* **2015**, *14*, 2881–2885. [CrossRef]

50. Friedmann Angeli, J.P.; Conrad, M. Selenium and GPX4, a vital symbiosis. *Free Radic. Biol. Med.* **2018**, *127*, 153–159. [CrossRef]

51. Melo, A.M.; Bandeiras, T.M.; Teixeira, M. New insights into type II NAD(P)H:quinone oxidoreductases. *Microbiol. Mol. Biol. Rev.* **2004**, *68*, 603–616. [CrossRef]

52. Warner, G.J.; Berry, M.J.; Moustafa, M.E.; Carlson, B.A.; Hatfield, D.L.; Faust, J.R. Inhibition of selenoprotein synthesis by selenocysteine tRNA[Ser]Sec lacking isopentenyladenosine. *J. Biol. Chem.* **2000**, *275*, 28110–28119. [CrossRef]

53. Venkatesh, D.; O'Brien, N.A.; Zandkarimi, F.; Tong, D.R.; Stokes, M.E.; Dunn, D.E.; Kengmana, E.S.; Aron, A.T.; Klein, A.M.; Csuka, J.M.; et al. MDM2 and MDMX promote ferroptosis by PPARα-mediated lipid remodeling. *Genes. Dev.* **2020**, *34*, 526–543. [CrossRef]

54. Kraft, V.A.N.; Bezjian, C.T.; Pfeiffer, S.; Ringelstetter, L.; Müller, C.; Zandkarimi, F.; Merl-Pham, J.; Bao, X.; Anastasov, N.; Kössl, J.; et al. GTP Cyclohydrolase 1/Tetrahydrobiopterin Counteract Ferroptosis through Lipid Remodeling. *ACS Cent. Sci.* **2020**, *6*, 41–53. [CrossRef]

55. Paton, C.M.; Ntambi, J.M. Biochemical and physiological function of stearoyl-CoA desaturase. *Am. J. Physiol. Endocrinol. Metab.* **2009**, *297*, E28–E37. [CrossRef]

56. Magtanong, L.; Ko, P.J.; To, M.; Cao, J.Y.; Forcina, G.C.; Tarangelo, A.; Ward, C.C.; Cho, K.; Patti, G.J.; Nomura, D.K.; et al. Exogenous Monounsaturated Fatty Acids Promote a Ferroptosis-Resistant Cell State. *Cell Chem. Biol.* **2019**, *26*, 420–432.e9. [CrossRef]

57. Katoh, Y.; Itoh, K.; Yoshida, E.; Miyagishi, M.; Fukamizu, A.; Yamamoto, M. Two domains of Nrf2 cooperatively bind CBP, a CREB binding protein, and synergistically activate transcription. *Genes Cells* **2001**, *6*, 857–868. [CrossRef]

58. Katsuoka, F.; Motohashi, H.; Ishii, T.; Aburatani, H.; Engel, J.D.; Yamamoto, M. Genetic evidence that small Maf proteins are essential for the activation of antioxidant response element-dependent genes. *Mol. Cell. Biol.* **2005**, *25*, 8044–8051. [CrossRef]

59. Kobayashi, M.; Itoh, K.; Suzuki, T.; Osanai, H.; Nishikawa, K.; Katoh, Y.; Takagi, Y.; Yamamoto, M. Identification of the interactive interface and phylogenic conservation of the Nrf2-Keap1 system. *Genes Cells* **2002**, *7*, 807–820. [CrossRef]

60. McMahon, M.; Itoh, K.; Yamamoto, M.; Hayes, J.D. Keap1-dependent proteasomal degradation of transcription factor Nrf2 contributes to the negative regulation of antioxidant response element-driven gene expression. *J. Biol. Chem.* **2003**, *278*, 21592–21600. [CrossRef]

61. Nioi, P.; Nguyen, T.; Sherratt, P.J.; Pickett, C.B. The carboxy-terminal Neh3 domain of Nrf2 is required for transcriptional activation. *Mol. Cell. Biol.* **2005**, *25*, 10895–10906. [CrossRef] [PubMed]

62. Moore, S.; Berger, N.D.; Luijsterburg, M.S.; Piett, C.G.; Stanley, F.K.T.; Schräder, C.U.; Fang, S.; Chan, J.A.; Schriemer, D.C.; Nagel, Z.D.; et al. The CHD6 chromatin remodeler is an oxidative DNA damage response factor. *Nat. Commun.* **2019**, *10*, 241. [CrossRef]

63. Salazar, M.; Rojo, A.I.; Velasco, D.; de Sagarra, R.M.; Cuadrado, A. Glycogen synthase kinase-3beta inhibits the xenobiotic and antioxidant cell response by direct phosphorylation and nuclear exclusion of the transcription factor Nrf2. *J. Biol. Chem.* **2006**, *281*, 14841–14851. [CrossRef] [PubMed]

64. Nam, L.B.; Keum, Y.S. Binding partners of NRF2: Functions and regulatory mechanisms. *Arch. Biochem. Biophys.* **2019**, *678*, 108184. [CrossRef] [PubMed]

65. Cuadrado, A. Structural and functional characterization of Nrf2 degradation by glycogen synthase kinase 3/beta-TrCP. *Free Radic. Biol. Med.* **2015**, *88*, 147–157. [CrossRef]

66. Wang, X.J.; Hayes, J.D.; Henderson, C.J.; Wolf, C.R. Identification of retinoic acid as an inhibitor of transcription factor Nrf2 through activation of retinoic acid receptor alpha. *Proc. Natl. Acad. Sci. USA* **2007**, *104*, 19589–19594. [CrossRef]

67. Wang, H.; Liu, K.; Geng, M.; Gao, P.; Wu, X.; Hai, Y.; Li, Y.; Li, Y.; Luo, L.; Hayes, J.D.; et al. RXRα inhibits the NRF2-ARE signaling pathway through a direct interaction with the Neh7 domain of NRF2. *Cancer Res.* **2013**, *73*, 3097–3108. [CrossRef]

68. Levonen, A.L.; Landar, A.; Ramachandran, A.; Ceaser, E.K.; Dickinson, D.A.; Zanoni, G.; Morrow, J.D.; Darley-Usmar, V.M. Cellular mechanisms of redox cell signalling: Role of cysteine modification in controlling antioxidant defences in response to electrophilic lipid oxidation products. *Biochem. J.* **2004**, *378*, 373–382. [CrossRef]

69. Komatsu, M.; Kurokawa, H.; Waguri, S.; Taguchi, K.; Kobayashi, A.; Ichimura, Y.; Sou, Y.S.; Ueno, I.; Sakamoto, A.; Tong, K.I.; et al. The selective autophagy substrate p62 activates the stress responsive transcription factor Nrf2 through inactivation of Keap1. *Nat. Cell Biol.* **2010**, *12*, 213–223. [CrossRef]

70. Rada, P.; Rojo, A.I.; Chowdhry, S.; McMahon, M.; Hayes, J.D.; Cuadrado, A. SCF/{beta}-TrCP promotes glycogen synthase kinase 3-dependent degradation of the Nrf2 transcription factor in a Keap1-independent manner. *Mol. Cell Biol.* **2011**, *31*, 1121–1133. [CrossRef]

71. Rada, P.; Rojo, A.I.; Evrard-Todeschi, N.; Innamorato, N.G.; Cotte, A.; Jaworski, T.; Tobón-Velasco, J.C.; Devijver, H.; García-Mayoral, M.F.; Van Leuven, F.; et al. Structural and functional characterization of Nrf2 degradation by the glycogen synthase kinase 3/β-TrCP axis. *Mol. Cell Biol.* **2012**, *32*, 3486–3499. [CrossRef]

72. Martin, D.; Rojo, A.I.; Salinas, M.; Diaz, R.; Gallardo, G.; Alam, J.; De Galarreta, C.M.; Cuadrado, A. Regulation of heme oxygenase-1 expression through the phosphatidylinositol 3-kinase/Akt pathway and the Nrf2 transcription factor in response to the antioxidant phytochemical carnosol. *J. Biol. Chem.* **2004**, *279*, 8919–8929. [CrossRef] [PubMed]

73. Wu, T.; Zhao, F.; Gao, B.; Tan, C.; Yagishita, N.; Nakajima, T.; Wong, P.K.; Chapman, E.; Fang, D.; Zhang, D.D. Hrd1 suppresses Nrf2-mediated cellular protection during liver cirrhosis. *Genes. Dev.* **2014**, *28*, 708–722. [CrossRef]

74. Miao, W.; Hu, L.; Scrivens, P.J.; Batist, G. Transcriptional regulation of NF-E2 p45-related factor (NRF2) expression by the aryl hydrocarbon receptor-xenobiotic response element signaling pathway: Direct cross-talk between phase I and II drug-metabolizing enzymes. *J. Biol. Chem.* **2005**, *280*, 20340–20348. [CrossRef] [PubMed]

75. Rushworth, S.A.; Zaitseva, L.; Murray, M.Y.; Shah, N.M.; Bowles, K.M.; MacEwan, D.J. The high Nrf2 expression in human acute myeloid leukemia is driven by NF-κB and underlies its chemo-resistance. *Blood* **2012**, *120*, 5188–5198. [CrossRef]

76. Kwak, M.K.; Itoh, K.; Yamamoto, M.; Kensler, T.W. Enhanced expression of the transcription factor Nrf2 by cancer chemopreventive agents: Role of antioxidant response element-like sequences in the nrf2 promoter. *Mol. Cell Biol.* **2002**, *22*, 2883–2892. [CrossRef]

77. Sangokoya, C.; Telen, M.J.; Chi, J.T. microRNA miR-144 modulates oxidative stress tolerance and associates with anemia severity in sickle cell disease. *Blood* **2010**, *116*, 4338–4348. [CrossRef] [PubMed]

78. Goldstein, L.D.; Lee, J.; Gnad, F.; Klijn, C.; Schaub, A.; Reeder, J.; Daemen, A.; Bakalarski, C.E.; Holcomb, T.; Shames, D.S.; et al. Recurrent Loss of NFE2L2 Exon 2 Is a Mechanism for Nrf2 Pathway Activation in Human Cancers. *Cell Rep.* **2016**, *16*, 2605–2617. [CrossRef]

79. Ishii, T.; Itoh, K.; Takahashi, S.; Sato, H.; Yanagawa, T.; Katoh, Y.; Bannai, S.; Yamamoto, M. Transcription factor Nrf2 coordinately regulates a group of oxidative stress-inducible genes in macrophages. *J. Biol. Chem.* **2000**, *275*, 16023–16029. [CrossRef]

80. Yang, H.; Magilnick, N.; Lee, C.; Kalmaz, D.; Ou, X.; Chan, J.Y.; Lu, S.C. Nrf1 and Nrf2 regulate rat glutamate-cysteine ligase catalytic subunit transcription indirectly via NF-kappaB and AP-1. *Mol. Cell Biol.* **2005**, *25*, 5933–5946. [CrossRef]

81. Chan, J.Y.; Kwong, M. Impaired expression of glutathione synthetic enzyme genes in mice with targeted deletion of the Nrf2 basic-leucine zipper protein. *Biochim. Biophys. Acta* **2000**, *1517*, 19–26. [CrossRef]

82. Singh, A.; Rangasamy, T.; Thimmulappa, R.K.; Lee, H.; Osburn, W.O.; Brigelius-Flohé, R.; Kensler, T.W.; Yamamoto, M.; Biswal, S. Glutathione peroxidase 2, the major cigarette smoke-inducible isoform of GPX in lungs, is regulated by Nrf2. *Am. J. Respir. Cell Mol. Biol.* **2006**, *35*, 639–650. [CrossRef] [PubMed]

83. Jobbagy, S.; Vitturi, D.A.; Salvatore, S.R.; Turell, L.; Pires, M.F.; Kansanen, E.; Batthyany, C.; Lancaster, J.R., Jr.; Freeman, B.A.; Schopfer, F.J. Electrophiles modulate glutathione reductase activity via alkylation and upregulation of glutathione biosynthesis. *Redox Biol.* **2019**, *21*, 101050. [CrossRef] [PubMed]

84. Wu, S.; Lu, H.; Bai, Y. Nrf2 in cancers: A double-edged sword. *Cancer Med.* **2019**, *8*, 2252–2267. [CrossRef]

85. Amaral, J.H.; Rizzi, E.S.; Alves-Lopes, R.; Pinheiro, L.C.; Tostes, R.C.; Tanus-Santos, J.E. Antioxidant and antihypertensive responses to oral nitrite involves activation of the Nrf2 pathway. *Free Radic. Biol. Med.* **2019**, *141*, 261–268. [CrossRef]

86. Bersuker, K.; Hendricks, J.M.; Li, Z.; Magtanong, L.; Ford, B.; Tang, P.H.; Roberts, M.A.; Tong, B.; Maimone, T.J.; Zoncu, R.; et al. The CoQ oxidoreductase FSP1 acts parallel to GPX4 to inhibit ferroptosis. *Nature* **2019**, *575*, 688–692. [CrossRef]

87. Karlenius, T.C.; Tonissen, K.F. Thioredoxin and Cancer: A Role for Thioredoxin in all States of Tumor Oxygenation. *Cancers* **2010**, *2*, 209–232. [CrossRef]

88. Balsera, M.; Buchanan, B.B. Evolution of the thioredoxin system as a step enabling adaptation to oxidative stress. *Free Radic. Biol. Med.* **2019**, *140*, 28–35. [CrossRef]

89. Cuadrado, A.; Rojo, A.I.; Wells, G.; Hayes, J.D.; Cousin, S.P.; Rumsey, W.L.; Attucks, O.C.; Franklin, S.; Levonen, A.L.; Kensler, T.W.; et al. Therapeutic targeting of the NRF2 and KEAP1 partnership in chronic diseases. *Nat. Rev. Drug Discov.* **2019**, *18*, 295–317. [CrossRef]

90. Tonelli, C.; Chio, I.I.C.; Tuveson, D.A. Transcriptional Regulation by Nrf2. *Antioxid. Redox Signal* **2018**, *29*, 1727–1745. [CrossRef]

91. Saha, S.; Buttari, B.; Panieri, E.; Profumo, E.; Saso, L. An Overview of Nrf2 Signaling Pathway and Its Role in Inflammation. *Molecules* **2020**, *25*, 5474. [CrossRef] [PubMed]

92. Alam, J.; Stewart, D.; Touchard, C.; Boinapally, S.; Choi, A.M.; Cook, J.L. Nrf2, a Cap'n'Collar transcription factor, regulates induction of the heme oxygenase-1 gene. *J. Biol. Chem.* **1999**, *274*, 26071–26078. [CrossRef] [PubMed]

93. Chorley, B.N.; Campbell, M.R.; Wang, X.; Karaca, M.; Sambandan, D.; Bangura, F.; Xue, P.; Pi, J.; Kleeberger, S.R.; Bell, D.A. Identification of novel NRF2-regulated genes by ChIP-Seq: Influence on retinoid X receptor alpha. *Nucleic Acids Res.* **2012**, *40*, 7416–7429. [CrossRef] [PubMed]

94. Hübner, R.H.; Schwartz, J.D.; De Bishnu, P.; Ferris, B.; Omberg, L.; Mezey, J.G.; Hackett, N.R.; Crystal, R.G. Coordinate control of expression of Nrf2-modulated genes in the human small airway epithelium is highly responsive to cigarette smoking. *Mol. Med.* **2009**, *15*, 203–219. [CrossRef]

95. Mitsuishi, Y.; Taguchi, K.; Kawatani, Y.; Shibata, T.; Nukiwa, T.; Aburatani, H.; Yamamoto, M.; Motohashi, H. Nrf2 redirects glucose and glutamine into anabolic pathways in metabolic reprogramming. *Cancer Cell* **2012**, *22*, 66–79. [CrossRef]

96. Patra, K.C.; Hay, N. The pentose phosphate pathway and cancer. *Trends Biochem. Sci.* **2014**, *39*, 347–354. [CrossRef]

97. Ahn, C.S.; Metallo, C.M. Mitochondria as biosynthetic factories for cancer proliferation. *Cancer Metab.* **2015**, *3*, 1. [CrossRef]

98. Mandal, P.K.; Seiler, A.; Perisic, T.; Kölle, P.; Banjac Canak, A.; Förster, H.; Weiss, N.; Kremmer, E.; Lieberman, M.W.; Bannai, S.; et al. System x(c)- and thioredoxin reductase 1 cooperatively rescue glutathione deficiency. *J. Biol. Chem.* **2010**, *285*, 22244–22253. [CrossRef]

99. Soula, M.; Weber, R.A.; Zilka, O.; Alwaseem, H.; La, K.; Yen, F.; Molina, H.; Garcia-Bermudez, J.; Pratt, D.A.; Birsoy, K. Metabolic determinants of cancer cell sensitivity to canonical ferroptosis inducers. *Nat. Chem. Biol.* **2020**, *16*, 1351–1360. [CrossRef]

100. Cho, H.Y.; Gladwell, W.; Wang, X.; Chorley, B.; Bell, D.; Reddy, S.P.; Kleeberger, S.R. Nrf2-regulated PPAR{gamma} expression is critical to protection against acute lung injury in mice. *Am. J. Respir. Crit. Care Med.* **2010**, *182*, 170–182. [CrossRef]

101. Huang, J.; Tabbi-Anneni, I.; Gunda, V.; Wang, L. Transcription factor Nrf2 regulates SHP and lipogenic gene expression in hepatic lipid metabolism. *Am. J. Physiol. Gastrointest. Liver Physiol.* **2010**, *299*, G1211–G1221. [CrossRef] [PubMed]

102. Lewerenz, J.; Hewett, S.J.; Huang, Y.; Lambros, M.; Gout, P.W.; Kalivas, P.W.; Massie, A.; Smolders, I.; Methner, A.; Pergande, M.; et al. The cystine/glutamate antiporter system x(c)(-) in health and disease: From molecular mechanisms to novel therapeutic opportunities. *Antioxid. Redox Signal* **2013**, *18*, 522–555. [CrossRef] [PubMed]

103. He, F.; Antonucci, L.; Karin, M. NRF2 as a regulator of cell metabolism and inflammation in cancer. *Carcinogenesis* **2020**, *41*, 405–416. [CrossRef] [PubMed]

104. Romero, R.; Sayin, V.I.; Davidson, S.M.; Bauer, M.R.; Singh, S.X.; LeBoeuf, S.E.; Karakousi, T.R.; Ellis, D.C.; Bhutkar, A.; Sánchez-Rivera, F.J.; et al. Keap1 loss promotes Kras-driven lung cancer and results in dependence on glutaminolysis. *Nat. Med.* **2017**, *23*, 1362–1368. [CrossRef]

105. Wang, H.; Cheng, Y.; Mao, C.; Liu, S.; Xiao, D.; Huang, J.; Tao, Y. Emerging mechanisms and targeted therapy of ferroptosis in cancer. *Mol. Ther.* **2021**, *29*, 2185–2208. [CrossRef]

106. Skouta, R.; Dixon, S.J.; Wang, J.; Dunn, D.E.; Orman, M.; Shimada, K.; Rosenberg, P.A.; Lo, D.C.; Weinberg, J.M.; Linkermann, A.; et al. Ferrostatins inhibit oxidative lipid damage and cell death in diverse disease models. *J. Am. Chem. Soc.* **2014**, *136*, 4551–4556. [CrossRef]

107. Yu, Y.; Xie, Y.; Cao, L.; Yang, L.; Yang, M.; Lotze, M.T.; Zeh, H.J.; Kang, R.; Tang, D. The ferroptosis inducer erastin enhances sensitivity of acute myeloid leukemia cells to chemotherapeutic agents. *Mol. Cell Oncol.* **2015**, *2*, e1054549. [CrossRef]

108. Yang, L.; Wang, H.; Yang, X.; Wu, Q.; An, P.; Jin, X.; Liu, W.; Huang, X.; Li, Y.; Yan, S.; et al. Auranofin mitigates systemic iron overload and induces ferroptosis via distinct mechanisms. *Signal Transduct. Target. Ther.* **2020**, *5*, 138. [CrossRef]

109. Dixon, S.J.; Patel, D.N.; Welsch, M.; Skouta, R.; Lee, E.D.; Hayano, M.; Thomas, A.G.; Gleason, C.E.; Tatonetti, N.P.; Slusher, B.S.; et al. Pharmacological inhibition of cystine-glutamate exchange induces endoplasmic reticulum stress and ferroptosis. *eLife* **2014**, *3*, e02523. [CrossRef]

110. Guo, J.; Xu, B.; Han, Q.; Zhou, H.; Xia, Y.; Gong, C.; Dai, X.; Li, Z.; Wu, G. Ferroptosis: A Novel Anti-tumor Action for Cisplatin. *Cancer Res. Treat.* **2018**, *50*, 445–460. [CrossRef]

111. Notarnicola, M.; Altomare, D.F.; Correale, M.; Ruggieri, E.; D'Attoma, B.; Mastrosimini, A.; Guerra, V.; Caruso, M.G. Serum lipid profile in colorectal cancer patients with and without synchronous distant metastases. *Oncology* **2005**, *68*, 371–374. [CrossRef]

112. Chen, X.; Kang, R.; Kroemer, G.; Tang, D. Broadening horizons: The role of ferroptosis in cancer. *Nat. Rev. Clin. Oncol.* **2021**, *18*, 280–296. [CrossRef]

113. Dodson, M.; Castro-Portuguez, R.; Zhang, D.D. NRF2 plays a critical role in mitigating lipid peroxidation and ferroptosis. *Redox Biol.* **2019**, *23*, 101107. [CrossRef] [PubMed]

114. Wood, E.R.; Truesdale, A.T.; McDonald, O.B.; Yuan, D.; Hassell, A.; Dickerson, S.H.; Ellis, B.; Pennisi, C.; Horne, E.; Lackey, K.; et al. A unique structure for epidermal growth factor receptor bound to GW572016 (Lapatinib): Relationships among protein conformation, inhibitor off-rate, and receptor activity in tumor cells. *Cancer Res.* **2004**, *64*, 6652–6659. [CrossRef] [PubMed]

115. Haynes, R.K.; Cheu, K.W.; N'Da, D.; Coghi, P.; Monti, D. Considerations on the mechanism of action of artemisinin antimalarials: Part 1--the 'carbon radical' and 'heme' hypotheses. *Infect. Disord. Drug Targets* **2013**, *13*, 217–277. [CrossRef]

116. Satoh, H.; Moriguchi, T.; Takai, J.; Ebina, M.; Yamamoto, M. Nrf2 prevents initiation but accelerates progression through the Kras signaling pathway during lung carcinogenesis. *Cancer Res.* **2013**, *73*, 4158–4168. [CrossRef]

117. Tao, S.; Rojo de la Vega, M.; Chapman, E.; Ooi, A.; Zhang, D.D. The effects of NRF2 modulation on the initiation and progression of chemically and genetically induced lung cancer. *Mol. Carcinog.* **2018**, *57*, 182–192. [CrossRef]

118. DeNicola, G.M.; Karreth, F.A.; Humpton, T.J.; Gopinathan, A.; Wei, C.; Frese, K.; Mangal, D.; Yu, K.H.; Yeo, C.J.; Calhoun, E.S.; et al. Oncogene-induced Nrf2 transcription promotes ROS detoxification and tumorigenesis. *Nature* **2011**, *475*, 106–109. [CrossRef] [PubMed]

119. Wang, H.; Liu, X.; Long, M.; Huang, Y.; Zhang, L.; Zhang, R.; Zheng, Y.; Liao, X.; Wang, Y.; Liao, Q.; et al. NRF2 activation by antioxidant antidiabetic agents accelerates tumor metastasis. *Sci. Transl. Med.* **2016**, *8*, 334–351. [CrossRef] [PubMed]

120. Padmanabhan, B.; Tong, K.I.; Ohta, T.; Nakamura, Y.; Scharlock, M.; Ohtsuji, M.; Kang, M.I.; Kobayashi, A.; Yokoyama, S.; Yamamoto, M. Structural basis for defects of Keap1 activity provoked by its point mutations in lung cancer. *Mol. Cell* **2006**, *21*, 689–700. [CrossRef]

121. Singh, A.; Misra, V.; Thimmulappa, R.K.; Lee, H.; Ames, S.; Hoque, M.O.; Herman, J.G.; Baylin, S.B.; Sidransky, D.; Gabrielson, E.; et al. Dysfunctional KEAP1-NRF2 interaction in non-small-cell lung cancer. *PLoS Med.* **2006**, *3*, e420. [CrossRef]

122. Tang, K.; Chen, Q.; Liu, Y.; Wang, L.; Lu, W. Combination of Metformin and Sorafenib Induces Ferroptosis of Hepatocellular Carcinoma Through p62-Keap1-Nrf2 Pathway. *J. Cancer* **2022**, *13*, 3234–3243. [CrossRef] [PubMed]
123. Gao, A.M.; Zhang, X.Y.; Ke, Z.P. Apigenin sensitizes BEL-7402/ADM cells to doxorubicin through inhibiting miR-101/Nrf2 pathway. *Oncotarget* **2017**, *8*, 82085–82091. [CrossRef] [PubMed]
124. Yang, W.S.; SriRamaratnam, R.; Welsch, M.E.; Shimada, K.; Skouta, R.; Viswanathan, V.S.; Cheah, J.H.; Clemons, P.A.; Shamji, A.F.; Clish, C.B.; et al. Regulation of ferroptotic cancer cell death by GPX4. *Cell* **2014**, *156*, 317–331. [CrossRef] [PubMed]
125. Chen, C.; Xie, B.; Li, Z.; Chen, L.; Chen, Y.; Zhou, J.; Ju, S.; Zhou, Y.; Zhang, X.; Zhuo, W.; et al. Fascin enhances the vulnerability of breast cancer to erastin-induced ferroptosis. *Cell Death Dis.* **2022**, *13*, 150. [CrossRef] [PubMed]
126. Feng, S.; Li, Y.; Huang, H.; Huang, H.; Duan, Y.; Yuan, Z.; Zhu, W.; Mei, Z.; Luo, L.; Yan, P. Isoorientin reverses lung cancer drug resistance by promoting ferroptosis via the SIRT6/Nrf2/GPX4 signaling pathway. *Eur. J. Pharmacol.* **2023**, *954*, 175853. [CrossRef]
127. Hsieh, C.H.; Hsieh, H.C.; Shih, F.S.; Wang, P.W.; Yang, L.X.; Shieh, D.B.; Wang, Y.C. An innovative NRF2 nano-modulator induces lung cancer ferroptosis and elicits an immunostimulatory tumor microenvironment. *Theranostics* **2021**, *11*, 7072–7091. [CrossRef]
128. Liu, X.; Peng, X.; Cen, S.; Yang, C.; Ma, Z.; Shi, X. Wogonin induces ferroptosis in pancreatic cancer cells by inhibiting the Nrf2/GPX4 axis. *Front. Pharmacol.* **2023**, *14*, 1129662. [CrossRef]
129. Gout, P.W.; Buckley, A.R.; Simms, C.R.; Bruchovsky, N. Sulfasalazine, a potent suppressor of lymphoma growth by inhibition of the x(c)-cystine transporter: A new action for an old drug. *Leukemia* **2001**, *15*, 1633–1640. [CrossRef]
130. Wahl, C.; Liptay, S.; Adler, G.; Schmid, R.M. Sulfasalazine: A potent and specific inhibitor of nuclear factor kappa B. *J. Clin. Investig.* **1998**, *101*, 1163–1174. [CrossRef]
131. Jing, S.; Lu, Y.; Zhang, J.; Ren, Y.; Mo, Y.; Liu, D.; Duan, L.; Yuan, Z.; Wang, C.; Wang, Q. Levistilide a Induces Ferroptosis by Activating the Nrf2/HO-1 Signaling Pathway in Breast Cancer Cells. *Drug Des. Devel Ther.* **2022**, *16*, 2981–2993. [CrossRef] [PubMed]
132. Wei, R.; Zhao, Y.; Wang, J.; Yang, X.; Li, S.; Wang, Y.; Yang, X.; Fei, J.; Hao, X.; Zhao, Y.; et al. Tagitinin C induces ferroptosis through PERK-Nrf2-HO-1 signaling pathway in colorectal cancer cells. *Int. J. Biol. Sci.* **2021**, *17*, 2703–2717. [CrossRef] [PubMed]
133. Takaya, K.; Suzuki, T.; Motohashi, H.; Onodera, K.; Satomi, S.; Kensler, T.W.; Yamamoto, M. Validation of the multiple sensor mechanism of the Keap1-Nrf2 system. *Free Radic. Biol. Med.* **2012**, *53*, 817–827. [CrossRef] [PubMed]
134. Zhang, B.; Zhai, M.; Li, B.; Liu, Z.; Li, K.; Jiang, L.; Zhang, M.; Yi, W.; Yang, J.; Yi, D.; et al. Honokiol Ameliorates Myocardial Ischemia/Reperfusion Injury in Type 1 Diabetic Rats by Reducing Oxidative Stress and Apoptosis through Activating the SIRT1-Nrf2 Signaling Pathway. *Oxid. Med. Cell Longev.* **2018**, *2018*, 3159801. [CrossRef]
135. Xiao, X.; Lu, Z.; Lin, V.; May, A.; Shaw, D.H.; Wang, Z.; Che, B.; Tran, K.; Du, H.; Shaw, P.X. MicroRNA miR-24-3p Reduces Apoptosis and Regulates Keap1-Nrf2 Pathway in Mouse Cardiomyocytes Responding to Ischemia/Reperfusion Injury. *Oxid. Med. Cell Longev.* **2018**, *2018*, 7042105. [CrossRef]
136. Park, Y.H.; Park, H.P.; Kim, E.; Lee, H.; Hwang, J.W.; Jeon, Y.T.; Lim, Y.J. The antioxidant effect of preischemic dexmedetomidine in a rat model: Increased expression of Nrf2/HO-1 via the PKC pathway. *Braz. J. Anesthesiol.* **2023**, *73*, 177–185. [CrossRef]
137. Hu, Q.; Zuo, T.; Deng, L.; Chen, S.; Yu, W.; Liu, S.; Liu, J.; Wang, X.; Fan, X.; Dong, Z. β-Caryophyllene suppresses ferroptosis induced by cerebral ischemia reperfusion via activation of the NRF2/HO-1 signaling pathway in MCAO/R rats. *Phytomedicine* **2022**, *102*, 154112. [CrossRef]
138. Chen, J.; Zhang, J.; Chen, T.; Bao, S.; Li, J.; Wei, H.; Hu, X.; Liang, Y.; Liu, F.; Yan, S. Xiaojianzhong decoction attenuates gastric mucosal injury by activating the p62/Keap1/Nrf2 signaling pathway to inhibit ferroptosis. *Biomed. Pharmacother.* **2022**, *155*, 113631. [CrossRef]
139. Cao, T.H.; Jin, S.G.; Fei, D.S.; Kang, K.; Jiang, L.; Lian, Z.Y.; Pan, S.H.; Zhao, M.R.; Zhao, M.Y. Artesunate Protects Against Sepsis-Induced Lung Injury Via Heme Oxygenase-1 Modulation. *Inflammation* **2016**, *39*, 651–662. [CrossRef]
140. Civantos, E.; Bosch, E.; Ramirez, E.; Zhenyukh, O.; Egido, J.; Lorenzo, O.; Mas, S. Sitagliptin ameliorates oxidative stress in experimental diabetic nephropathy by diminishing the miR-200a/Keap-1/Nrf2 antioxidant pathway. *Diabetes Metab. Syndr. Obes.* **2017**, *10*, 207–222. [CrossRef]
141. Bray, F.; Ferlay, J.; Soerjomataram, I.; Siegel, R.L.; Torre, L.A.; Jemal, A. Global cancer statistics 2018: GLOBOCAN estimates of incidence and mortality worldwide for 36 cancers in 185 countries. *CA Cancer J. Clin.* **2018**, *68*, 394–424. [CrossRef]
142. EASL Clinical Practice Guidelines: Management of hepatocellular carcinoma. *J. Hepatol.* **2018**, *69*, 182–236. [CrossRef]
143. Feng, H.; Schorpp, K.; Jin, J.; Yozwiak, C.E.; Hoffstrom, B.G.; Decker, A.M.; Rajbhandari, P.; Stokes, M.E.; Bender, H.G.; Csuka, J.M.; et al. Transferrin Receptor Is a Specific Ferroptosis Marker. *Cell Rep.* **2020**, *30*, 3411–3423. [CrossRef] [PubMed]
144. Jin, Y.; Yang, R.; Ding, J.; Zhu, F.; Zhu, C.; Xu, Q.; Cai, J. KAT6A is associated with sorafenib resistance and contributes to progression of hepatocellular carcinoma by targeting YAP. *Biochem. Biophys. Res. Commun.* **2021**, *585*, 185–190. [CrossRef] [PubMed]
145. Li, Z.J.; Dai, H.Q.; Huang, X.W.; Feng, J.; Deng, J.H.; Wang, Z.X.; Yang, X.M.; Liu, Y.J.; Wu, Y.; Chen, P.H.; et al. Artesunate synergizes with sorafenib to induce ferroptosis in hepatocellular carcinoma. *Acta Pharmacol. Sin.* **2021**, *42*, 301–310. [CrossRef] [PubMed]
146. Bai, L.; Wang, H.; Wang, A.H.; Zhang, L.Y.; Bai, J. MicroRNA-532 and microRNA-3064 inhibit cell proliferation and invasion by acting as direct regulators of human telomerase reverse transcriptase in ovarian cancer. *PLoS ONE* **2017**, *12*, e0173912. [CrossRef]
147. Harbeck, N.; Gnant, M. Breast cancer. *Lancet* **2017**, *389*, 1134–1150. [CrossRef] [PubMed]
148. Hangauer, M.J.; Viswanathan, V.S.; Ryan, M.J.; Bole, D.; Eaton, J.K.; Matov, A.; Galeas, J.; Dhruv, H.D.; Berens, M.E.; Schreiber, S.L.; et al. Drug-tolerant persister cancer cells are vulnerable to GPX4 inhibition. *Nature* **2017**, *551*, 247–250. [CrossRef]

149. Ma, S.; Dielschneider, R.F.; Henson, E.S.; Xiao, W.; Choquette, T.R.; Blankstein, A.R.; Chen, Y.; Gibson, S.B. Ferroptosis and autophagy induced cell death occur independently after siramesine and lapatinib treatment in breast cancer cells. *PLoS ONE* **2017**, *12*, e0182921. [CrossRef]

150. Ma, S.; Henson, E.S.; Chen, Y.; Gibson, S.B. Ferroptosis is induced following siramesine and lapatinib treatment of breast cancer cells. *Cell Death Dis.* **2016**, *7*, e2307. [CrossRef]

151. Yang, Y.; Zhang, Y.; Wang, L.; Lee, S. Levistolide A Induces Apoptosis via ROS-Mediated ER Stress Pathway in Colon Cancer Cells. *Cell Physiol. Biochem.* **2017**, *42*, 929–938. [CrossRef] [PubMed]

152. Cao, M.; Chen, W. Epidemiology of lung cancer in China. *Thorac. Cancer* **2019**, *10*, 3–7. [CrossRef] [PubMed]

153. Min, Y.; Mao, C.Q.; Chen, S.; Ma, G.; Wang, J.; Liu, Y. Combating the drug resistance of cisplatin using a platinum prodrug based delivery system. *Angew. Chem. Int. Ed. Engl.* **2012**, *51*, 6742–6747. [CrossRef] [PubMed]

154. Zou, Y.; Wang, X.; Khan, A.; Wang, P.; Liu, Y.; Alsaedi, A.; Hayat, T.; Wang, X. Environmental Remediation and Application of Nanoscale Zero-Valent Iron and Its Composites for the Removal of Heavy Metal Ions: A Review. *Environ. Sci. Technol.* **2016**, *50*, 7290–7304. [CrossRef] [PubMed]

155. Yang, L.X.; Wu, Y.N.; Wang, P.W.; Huang, K.J.; Su, W.C.; Shieh, D.B. Silver-coated zero-valent iron nanoparticles enhance cancer therapy in mice through lysosome-dependent dual programed cell death pathways: Triggering simultaneous apoptosis and autophagy only in cancerous cells. *J. Mater. Chem. B* **2020**, *8*, 4122–4131. [CrossRef]

156. Sung, H.; Ferlay, J.; Siegel, R.L.; Laversanne, M.; Soerjomataram, I.; Jemal, A.; Bray, F. Global Cancer Statistics 2020: GLOBOCAN Estimates of Incidence and Mortality Worldwide for 36 Cancers in 185 Countries. *CA Cancer J. Clin.* **2021**, *71*, 209–249. [CrossRef]

157. Rahib, L.; Smith, B.D.; Aizenberg, R.; Rosenzweig, A.B.; Fleshman, J.M.; Matrisian, L.M. Projecting cancer incidence and deaths to 2030: The unexpected burden of thyroid, liver, and pancreas cancers in the United States. *Cancer Res* **2014**, *74*, 2913–2921. [CrossRef]

158. Liu, X.Q.; Jiang, L.; Li, Y.Y.; Huang, Y.B.; Hu, X.R.; Zhu, W.; Wang, X.; Wu, Y.G.; Meng, X.M.; Qi, X.M. Wogonin protects glomerular podocytes by targeting Bcl-2-mediated autophagy and apoptosis in diabetic kidney disease. *Acta Pharmacol. Sin.* **2022**, *43*, 96–110. [CrossRef]

159. Guan, J.; Lo, M.; Dockery, P.; Mahon, S.; Karp, C.M.; Buckley, A.R.; Lam, S.; Gout, P.W.; Wang, Y.Z. The xc-cystine/glutamate antiporter as a potential therapeutic target for small-cell lung cancer: Use of sulfasalazine. *Cancer Chemother. Pharmacol.* **2009**, *64*, 463–472. [CrossRef]

160. Okazaki, S.; Shintani, S.; Hirata, Y.; Suina, K.; Semba, T.; Yamasaki, J.; Umene, K.; Ishikawa, M.; Saya, H.; Nagano, O. Synthetic lethality of the ALDH3A1 inhibitor dyclonine and xCT inhibitors in glutathione deficiency-resistant cancer cells. *Oncotarget* **2018**, *9*, 33832–33843. [CrossRef]

161. Ranti, I.; Wahyuningsih, M.S.H.; Wirohadidjojo, Y.W. The antifibrotic effect of isolate tagitinin C from tithonia diversifolia (Hemsley) A. Gray on keloid fibroblast cell. *Pan Afr. Med. J.* **2018**, *30*, 264. [CrossRef] [PubMed]

162. Liao, M.H.; Lin, W.C.; Wen, H.C.; Pu, H.F. Tithonia diversifolia and its main active component tagitinin C induce survivin inhibition and G2/M arrest in human malignant glioblastoma cells. *Fitoterapia* **2011**, *82*, 331–341. [CrossRef] [PubMed]

163. Zhao, L.; Dong, J.; Hu, Z.; Li, S.; Su, X.; Zhang, J.; Yin, Y.; Xu, T.; Zhang, Z.; Chen, H. Anti-TMV activity and functional mechanisms of two sesquiterpenoids isolated from Tithonia diversifolia. *Pestic. Biochem. Physiol.* **2017**, *140*, 24–29. [CrossRef]

164. Gonçalves-Santos, E.; Vilas-Boas, D.F.; Diniz, L.F.; Veloso, M.P.; Mazzeti, A.L.; Rodrigues, M.R.; Oliveira, C.M.; Fernandes, V.H.C.; Novaes, R.D.; Chagas-Paula, D.A.; et al. Sesquiterpene lactone potentiates the immunomodulatory, antiparasitic and cardioprotective effects on anti-Trypanosoma cruzi specific chemotherapy. *Int. Immunopharmacol.* **2019**, *77*, 105961. [CrossRef]

165. Kasarskis, E.J.; Tandon, L.; Lovell, M.A.; Ehmann, W.D. Aluminum, calcium, and iron in the spinal cord of patients with sporadic amyotrophic lateral sclerosis using laser microprobe mass spectroscopy: A preliminary study. *J. Neurol. Sci.* **1995**, *130*, 203–208. [CrossRef] [PubMed]

166. Kaur, D.; Yantiri, F.; Rajagopalan, S.; Kumar, J.; Mo, J.Q.; Boonplueang, R.; Viswanath, V.; Jacobs, R.; Yang, L.; Beal, M.F.; et al. Genetic or pharmacological iron chelation prevents MPTP-induced neurotoxicity in vivo: A novel therapy for Parkinson's disease. *Neuron* **2003**, *37*, 899–909. [CrossRef] [PubMed]

167. Zhu, W.; Xie, W.; Pan, T.; Xu, P.; Fridkin, M.; Zheng, H.; Jankovic, J.; Youdim, M.B.; Le, W. Prevention and restoration of lactacystin-induced nigrostriatal dopamine neuron degeneration by novel brain-permeable iron chelators. *FASEB J.* **2007**, *21*, 3835–3844. [CrossRef]

168. Buijs, M.; Doan, N.T.; van Rooden, S.; Versluis, M.J.; van Lew, B.; Milles, J.; van der Grond, J.; van Buchem, M.A. In vivo assessment of iron content of the cerebral cortex in healthy aging using 7-Tesla T2*-weighted phase imaging. *Neurobiol. Aging* **2017**, *53*, 20–26. [CrossRef]

169. Lei, P.; Ayton, S.; Finkelstein, D.I.; Spoerri, L.; Ciccotosto, G.D.; Wright, D.K.; Wong, B.X.; Adlard, P.A.; Cherny, R.A.; Lam, L.Q.; et al. Tau deficiency induces parkinsonism with dementia by impairing APP-mediated iron export. *Nat. Med.* **2012**, *18*, 291–295. [CrossRef]

170. Li, L.B.; Chai, R.; Zhang, S.; Xu, S.F.; Zhang, Y.H.; Li, H.L.; Fan, Y.G.; Guo, C. Iron Exposure and the Cellular Mechanisms Linked to Neuron Degeneration in Adult Mice. *Cells* **2019**, *8*, 198. [CrossRef]

171. Connor, J.R.; Menzies, S.L.; St Martin, S.M.; Mufson, E.J. A histochemical study of iron, transferrin, and ferritin in Alzheimer's diseased brains. *J. Neurosci. Res.* **1992**, *31*, 75–83. [CrossRef] [PubMed]

172. Spotorno, N.; Acosta-Cabronero, J.; Stomrud, E.; Lampinen, B.; Strandberg, O.T.; van Westen, D.; Hansson, O. Relationship between cortical iron and tau aggregation in Alzheimer's disease. *Brain* **2020**, *143*, 1341–1349. [CrossRef] [PubMed]

173. Becerril-Ortega, J.; Bordji, K.; Fréret, T.; Rush, T.; Buisson, A. Iron overload accelerates neuronal amyloid-β production and cognitive impairment in transgenic mice model of Alzheimer's disease. *Neurobiol. Aging* **2014**, *35*, 2288–2301. [CrossRef] [PubMed]

174. Zhang, Y.H.; Wang, D.W.; Xu, S.F.; Zhang, S.; Fan, Y.G.; Yang, Y.Y.; Guo, S.Q.; Wang, S.; Guo, T.; Wang, Z.Y.; et al. α-Lipoic acid improves abnormal behavior by mitigation of oxidative stress, inflammation, ferroptosis, and tauopathy in P301S Tau transgenic mice. *Redox Biol.* **2018**, *14*, 535–548. [CrossRef]

175. Ayton, S.; Lei, P.; Duce, J.A.; Wong, B.X.; Sedjahtera, A.; Adlard, P.A.; Bush, A.I.; Finkelstein, D.I. Ceruloplasmin dysfunction and therapeutic potential for Parkinson disease. *Ann. Neurol.* **2013**, *73*, 554–559. [CrossRef]

176. Dexter, D.T.; Wells, F.R.; Lees, A.J.; Agid, F.; Agid, Y.; Jenner, P.; Marsden, C.D. Increased nigral iron content and alterations in other metal ions occurring in brain in Parkinson's disease. *J. Neurochem.* **1989**, *52*, 1830–1836. [CrossRef]

177. Davies, K.M.; Hare, D.J.; Cottam, V.; Chen, N.; Hilgers, L.; Halliday, G.; Mercer, J.F.; Double, K.L. Localization of copper and copper transporters in the human brain. *Metallomics* **2013**, *5*, 43–51. [CrossRef]

178. Pearce, R.K.; Owen, A.; Daniel, S.; Jenner, P.; Marsden, C.D. Alterations in the distribution of glutathione in the substantia nigra in Parkinson's disease. *J. Neural. Transm.* **1997**, *104*, 661–677. [CrossRef]

179. Bender, A.; Krishnan, K.J.; Morris, C.M.; Taylor, G.A.; Reeve, A.K.; Perry, R.H.; Jaros, E.; Hersheson, J.S.; Betts, J.; Klopstock, T.; et al. High levels of mitochondrial DNA deletions in substantia nigra neurons in aging and Parkinson disease. *Nat. Genet.* **2006**, *38*, 515–517. [CrossRef]

180. Devos, D.; Moreau, C.; Devedjian, J.C.; Kluza, J.; Petrault, M.; Laloux, C.; Jonneaux, A.; Ryckewaert, G.; Garçon, G.; Rouaix, N.; et al. Targeting chelatable iron as a therapeutic modality in Parkinson's disease. *Antioxid. Redox Signal* **2014**, *21*, 195–210. [CrossRef]

181. Park, T.J.; Park, J.H.; Lee, G.S.; Lee, J.Y.; Shin, J.H.; Kim, M.W.; Kim, Y.S.; Kim, J.Y.; Oh, K.J.; Han, B.S.; et al. Quantitative proteomic analyses reveal that GPX4 downregulation during myocardial infarction contributes to ferroptosis in cardiomyocytes. *Cell Death Dis.* **2019**, *10*, 835. [CrossRef] [PubMed]

182. Song, Y.; Wang, B.; Zhu, X.; Hu, J.; Sun, J.; Xuan, J.; Ge, Z. Human umbilical cord blood-derived MSCs exosome attenuate myocardial injury by inhibiting ferroptosis in acute myocardial infarction mice. *Cell Biol. Toxicol.* **2021**, *37*, 51–64. [CrossRef] [PubMed]

183. Jin, R.; Yang, R.; Cui, C.; Zhang, H.; Cai, J.; Geng, B.; Chen, Z. Ferroptosis due to Cystathionine γ Lyase/Hydrogen Sulfide Downregulation Under High Hydrostatic Pressure Exacerbates VSMC Dysfunction. *Front. Cell Dev. Biol.* **2022**, *10*, 829316. [CrossRef] [PubMed]

184. Wang, C.; Yuan, W.; Hu, A.; Lin, J.; Xia, Z.; Yang, C.F.; Li, Y.; Zhang, Z. Dexmedetomidine alleviated sepsis-induced myocardial ferroptosis and septic heart injury. *Mol. Med. Rep.* **2020**, *22*, 175–184. [CrossRef] [PubMed]

185. Dai, C.; Kong, B.; Qin, T.; Xiao, Z.; Fang, J.; Gong, Y.; Zhu, J.; Liu, Q.; Fu, H.; Meng, H.; et al. Inhibition of ferroptosis reduces susceptibility to frequent excessive alcohol consumption-induced atrial fibrillation. *Toxicology* **2022**, *465*, 153055. [CrossRef]

186. Menon, A.V.; Liu, J.; Tsai, H.P.; Zeng, L.; Yang, S.; Asnani, A.; Kim, J. Excess heme upregulates heme oxygenase 1 and promotes cardiac ferroptosis in mice with sickle cell disease. *Blood* **2022**, *139*, 936–941. [CrossRef]

187. Liu, P.; Feng, Y.; Li, H.; Chen, X.; Wang, G.; Xu, S.; Li, Y.; Zhao, L. Ferrostatin-1 alleviates lipopolysaccharide-induced acute lung injury via inhibiting ferroptosis. *Cell Mol. Biol. Lett.* **2020**, *25*, 10. [CrossRef]

188. Liang, N.N.; Zhao, Y.; Guo, Y.Y.; Zhang, Z.H.; Gao, L.; Yu, D.X.; Xu, D.X.; Xu, S. Mitochondria-derived reactive oxygen species are involved in renal cell ferroptosis during lipopolysaccharide-induced acute kidney injury. *Int. Immunopharmacol.* **2022**, *107*, 108687. [CrossRef]

189. Zager, R.A.; Foerder, C.A. Effects of inorganic iron and myoglobin on in vitro proximal tubular lipid peroxidation and cytotoxicity. *J. Clin. Investig.* **1992**, *89*, 989–995. [CrossRef]

190. Martin-Sanchez, D.; Ruiz-Andres, O.; Poveda, J.; Carrasco, S.; Cannata-Ortiz, P.; Sanchez-Niño, M.D.; Ruiz Ortega, M.; Egido, J.; Linkermann, A.; Ortiz, A.; et al. Ferroptosis, but Not Necroptosis, Is Important in Nephrotoxic Folic Acid-Induced AKI. *J. Am. Soc. Nephrol.* **2017**, *28*, 218–229. [CrossRef]

191. Paller, M.S. Hemoglobin- and myoglobin-induced acute renal failure in rats: Role of iron in nephrotoxicity. *Am. J. Physiol.* **1988**, *255*, F539–F544. [CrossRef] [PubMed]

192. Havrdova, E.; Giovannoni, G.; Gold, R.; Fox, R.J.; Kappos, L.; Phillips, J.T.; Okwuokenye, M.; Marantz, J.L. Effect of delayed-release dimethyl fumarate on no evidence of disease activity in relapsing-remitting multiple sclerosis: Integrated analysis of the phase III DEFINE and CONFIRM studies. *Eur. J. Neurol.* **2017**, *24*, 726–733. [CrossRef] [PubMed]

193. Cuadrado, A. NRF2 in neurodegenerative diseases. *Curr. Opin. Toxicol.* **2016**, *1*, 46–53. [CrossRef]

194. Huang, K.; Gao, X.; Wei, W. The crosstalk between Sirt1 and Keap1/Nrf2/ARE anti-oxidative pathway forms a positive feedback loop to inhibit FN and TGF-β1 expressions in rat glomerular mesangial cells. *Exp. Cell Res.* **2017**, *361*, 63–72. [CrossRef]

195. Miao, Y.; Wan, Q.; Liu, X.; Wang, Y.; Luo, Y.; Liu, D.; Lin, N.; Zhou, H.; Zhong, J. miR-503 Is Involved in the Protective Effect of Phase II Enzyme Inducer (CPDT) in Diabetic Cardiomyopathy via Nrf2/ARE Signaling Pathway. *Biomed. Res. Int.* **2017**, *2017*, 9167450. [CrossRef]

196. Li, Y.; Duan, J.Z.; He, Q.; Wang, C.Q. miR-155 modulates high glucose-induced cardiac fibrosis via the Nrf2/HO-1 signaling pathway. *Mol. Med. Rep.* **2020**, *22*, 4003–4016. [CrossRef]

197. Yu, M.; Liu, Y.; Zhang, B.; Shi, Y.; Cui, L.; Zhao, X. Inhibiting microRNA-144 abates oxidative stress and reduces apoptosis in hearts of streptozotocin-induced diabetic mice. *Cardiovasc. Pathol.* **2015**, *24*, 375–381. [CrossRef]
198. Zhang, J.; Cai, W.; Fan, Z.; Yang, C.; Wang, W.; Xiong, M.; Ma, C.; Yang, J. MicroRNA-24 inhibits the oxidative stress induced by vascular injury by activating the Nrf2/Ho-1 signaling pathway. *Atherosclerosis* **2019**, *290*, 9–18. [CrossRef]

Article

Inhibitory Effects of *Ehretia tinifolia* Extract on the Excessive Oxidative and Inflammatory Responses in Lipopolysaccharide-Stimulated Mouse Kupffer Cells

Jae Sung Lim [1,†], Sung Ho Lee [1,†], Hyosuk Yun [2,†], Da Young Lee [1], Namki Cho [1], Guijae Yoo [3], Jeong Uk Choi [1], Kwang Youl Lee [1], Tran The Bach [4], Su-Jin Park [5,*] and Young-Chang Cho [1,*]

[1] College of Pharmacy and Research Institute of Pharmaceutical Sciences, Chonnam National University, 77 Yongbong-ro, Gwangju 61186, Republic of Korea; dr.jslim7542@gmail.com (J.S.L.); puzim23@gmail.com (S.H.L.); dlekdud0914@naver.com (D.Y.L.); cnamki@jnu.ac.kr (N.C.); cju0667@jnu.ac.kr (J.U.C.); kwanglee@jnu.ac.kr (K.Y.L.)

[2] Department of Chemistry, Chonnam National University, Gwangju 61186, Republic of Korea; hslov2aron@hanmail.net

[3] Korea Food Research Institute, 245, Nongsaengmyeong-ro, Iseo-myeon, Wanju-Gun 55365, Republic of Korea; gjyoo@kfri.re.kr

[4] Institute of Ecology and Biological Resources, Vietnam Academy of Science and Technology (VAST), 18 Hoang Quoc Viet, Cau Giay, Ha Noi 122000, Vietnam; tranthebach@yahoo.com

[5] Functional Biomaterial Research Center, Korea Research Institute of Bioscience and Biotechnology, 181 Ipsin-gil, Jeongeup-si 56212, Republic of Korea

* Correspondence: sjpark@kribb.re.kr (S.-J.P.); yccho@jnu.ac.kr (Y.-C.C.)

† These authors contributed equally to this work.

Citation: Lim, J.S.; Lee, S.H.; Yun, H.; Lee, D.Y.; Cho, N.; Yoo, G.; Choi, J.U.; Lee, K.Y.; Bach, T.T.; Park, S.-J.; et al. Inhibitory Effects of *Ehretia tinifolia* Extract on the Excessive Oxidative and Inflammatory Responses in Lipopolysaccharide-Stimulated Mouse Kupffer Cells. *Antioxidants* **2023**, *12*, 1792. https://doi.org/10.3390/antiox12101792

Academic Editor: Young-Sam Keum

Received: 1 September 2023
Revised: 19 September 2023
Accepted: 20 September 2023
Published: 22 September 2023

Abstract: *Ehretia tinifolia* (*E. tinifolia*) L., an evergreen tree with substantial biological activity, including antioxidant and anti-inflammatory effects, has been used in many herbal and traditional medicines. To elucidate its antioxidant and anti-inflammatory activity and the underlying mechanisms, we applied a methanol extract of *E. tinifolia* (ETME) to lipopolysaccharide (LPS)-stimulated mouse immortalized Kupffer cells. ETME suppressed the LPS-induced increase in nitric oxide, a mediator for oxidative stress and inflammation, and restored LPS-mediated depletion of total glutathione level by stabilizing antioxidative nuclear factor erythroid 2-related factor 2 (Nrf2) and the subsequent increase in heme oxygenase-1 levels. Furthermore, ETME inhibited the LPS-induced production of pro-inflammatory cytokines, including tumor necrosis factor-α, interleukin (IL)-1β, and IL-6. The inhibitory effects of ETME on pro-inflammatory responses were regulated by ETME-mediated dephosphorylation of mitogen-activated protein kinases (MAPKs: p38, p44/p42, and stress-associated protein kinase/c-Jun N-terminal kinase) and inhibition of nuclear localization of nuclear factor kappa B (NF-κB). These results suggest that ETME is a possible candidate for protecting Kupffer cells from LPS-mediated oxidative stress and excessive inflammatory responses by activating antioxidant Nrf2/HO-1 and inhibiting pro-inflammatory NF-κB and MAPKs, respectively.

Keywords: *Ehretia tinifolia*; anti-inflammatory; antioxidant; MAPK; NF-κB; Nrf2

1. Introduction

Inflammation is a physiological response by the immune system to harmful stimuli, playing an important role in the body's defense mechanisms. However, uncontrolled inflammation can contribute to the development of various diseases including autoimmune diseases, metabolic syndrome, neurodegenerative diseases, cancers, and cardiovascular diseases [1]. Although non-steroidal anti-inflammatory drugs (NSAIDs) are mostly used for the treatment of patients suffering from pain and inflammatory disorders, NSAIDs have severe adverse effects including the gastrointestinal toxicities, cardiovascular risks, renal injuries, hypertension, and hepatotoxicity [2]. Therefore, there is a need to discover better anti-inflammatory drugs.

Natural products have massive structural and chemical diversity that cannot be matched by any synthetic libraries of small molecules. They continue to encourage novel discoveries in the fields of chemistry, biology, and medicine. Moreover, natural products are evolutionarily optimized as drug-like molecules, making them the best sources of drugs and drug leads [3]. Given their extensive history of use, natural products have shown promise in the treatment of inflammatory disorders and even certain types of cancers [4–6]. The extracts of whole plants [7], fruit [8], leaves [9], and plant-based formulations [10] have considerable antioxidant and anti-inflammatory effects [11,12]. Harnessing the therapeutic properties of natural products could lead to the development of safer and more effective anti-inflammatory therapies.

Ehretia tinifolia (*E. tinifolia*) L. is a plant species in the family Boraginaceae. It is also known as a 'pinguica' and a tree with small, round, fragrant, and sweet yellow fruit widely used as food in Mexico and the United States [13,14]. It has been used as a traditional medicine for the treatment of urinary track disorder by reducing uric acid. Also, the bark of this plant has been used for wound healing, and the flowers and leaves have been used to treat bloody vomiting [15–18]. Recently, it was reported that the fruit of *E. tinifolia* has antioxidant effects [19], which are associated with the phenol content of its polar organic extracts [20]. Recent examination of the chemical composition and biological activity of the fruit revealed rosmarinic acid as one of the main active components [21]. This hydroxylated compound occurs in many herbal plants and has been extensively studied for its antioxidant and anti-inflammatory activities [22,23]. However, other parts such as leaf, flower, and branch, except for the fruit of *E. tinifolia*, have not been studied yet.

To elucidate the antioxidant and anti-inflammatory effects of *E. tinifolia*, we tested the effects of a methanol extract of the leaf, flower, and branch of *E. tinifolia* (ETME) on lipopolysaccharide (LPS)-stimulated immortalized mouse Kupffer cells (ImKCs). Kupffer cells, macrophages in the liver derived from monocytes in the bloodstream, play an important role in clearing foreign substances, including immune complexes, bacterial components, and endotoxins, from the portal circulation [24]. They are activated by numerous molecules, including bacterial endotoxins such as LPS. When activated, they secrete various pro-inflammatory cytokines, including tumor necrosis factor (TNF)-α and several types of interleukins (ILs), which act as inflammatory cytokines, eliminating aberrant toxic or foreign substances and initiating healing [25].

Our findings reveal that ETME exerts anti-inflammatory effects by inhibiting the mitogen-activated protein kinase (MAPK) and nuclear factor kappa-light-chain-enhancer of activated B cells (NF-κB) signaling pathways and antioxidant activity by inducing the nuclear factor erythroid 2-related factor 2 (Nrf2)/heme oxygenase 1 (HO-1) signaling pathway. This fruit has the potential to treat liver diseases such as liver fibrosis.

2. Materials and Methods

2.1. Methanol Extraction of E. tinifolia

E. tinifolia plant material (whole leaves, flowers, and branches of the plant) was collected in the Da Chais community, Lac Duong district, Lam Dong province, Vietnam, by Dr. Tran The Bach (Institute of Ecology and Biological Resources, Hanoi, Vietnam), who also verified the identification. Voucher specimens (KRIB41338 and VK4876) were deposited in the herbarium of the Korea Research Institute of Bioscience and Biotechnology (Daejeon, Republic of Korea). A total of 50 g of the plant material was dried in the shade, powdered, added to 1 L of high-performance liquid chromatography (HPLC) grade methanol, and extracted via 30 cycles of ultrasonication (40 kHz, 1500 W, 15 min per cycle, with a 120 min rest between cycles) at room temperature, using an ultrasonic extractor (SDN-900H; SD-ULTRASONIC Co., Ltd., Seoul, Republic of Korea). This mixture was then filtered and dried at 40 °C under reduced pressure to obtain ETME.

2.2. Ultra-HPLC-Quadrupole Time-of-Flight Mass Spectrometry (UPLC-Q-TOF-MS) Analysis

The dried sample extracts (1 mg/mL) were dissolved in aqueous methanol for UPLC-Q-TOF-MS analysis using an LTQ IT mass spectrometer (Thermo Fisher Scientific, Waltham, MA, USA) equipped with an electrospray interface, RS column compartment, and RS pump (Dionex Corporation, Sunnyvale, CA, USA). Chromatographic separation was performed on a Thermo Fisher Scientific Syncronis C18 HPLC column (100 × 2.1 mm internal diameter; 1.7 μm particle size) with an injection volume of 2 μL. The mobile phase comprised water (solvent A) with 0.1% formic acid (*v/v*) and acetonitrile (solvent B) with 0.1% formic acid (*v/v*) at a flow rate of 0.4 mL/min; the column temperature was 35 °C. The solvent gradient was 5% B for 1 min, increased to 100% B for 20 min, maintained for 2.5 min, decreased to 5% B for 1 min, and maintained at 10% B for the final 3 min. The total run time was 25 min. The mass spectra and photodiode array range in the negative mode were tuned for m/z 50–1200 and 200–600 nm, respectively. The collision voltage was 4 V, and the source voltage was ±1 kV.

2.3. Radical Scavenging (DPPH) Assay

High-concentration stocks of ETME (100 mg/mL), quercetin (100 mg/mL), and rosmarinic acid (40 mg/mL) were prepared in DMSO solution, and then ETME and rosmarinic acid were prepared into each treatment concentration [ETME: 10, 20, 40, 80, and 160 μg/mL, rosmarinic acid: 5 μg/mL (13.88 μM), 10 μg/mL (27.75 μM), and 20 μg/mL (55.51 μM)] by half serial dilution using 100% methanol. And the quercetin was diluted in 100% methanol [10 μg/mL (33.09 μM) and 30 μg/mL (99.26 μM)]. Then, 100 μL of each diluent was transferred into 96-well plates. Afterwards, 100 μL of 0.2 mM DPPH solution was added and incubated for 30 min in a 37 °C incubator. After incubation, absorbance was measured at 517 nm using a Synergy H1 Hybrid Microplate Reader. The DPPH inhibition was expressed as % using the OD value of the experimental group and the value of 100% methanol solution (as a negative control). Quercetin (Sigma-Aldrich, St. Louis, MO, USA) was used as a positive control. The calculation formula of DPPH inhibition is as follows:

$$\text{DPPH inhibition (\%)} = (\text{OD}_{517\text{nm Sample}} - \text{OD}_{517\text{ Methanol}})/\text{OD}_{517\text{ Methanol}} \times 100$$

2.4. Cell Culture

LPS-stimulated ImKCs (#SCC119, Sigma-Aldrich, St. Louis, MO, USA) were cultured in Dulbecco's modified Eagle's medium (DMEM; #LM0001-05; WELGENE Inc., Gyeongsan, Republic of Korea) supplemented with 10% fetal bovine serum (16000044; Gibco, Waltham, MA, USA), 100 IU/mL penicillin, and 100 μg/mL streptomycin (#30-002-cl; Corning Inc., Corning, NY, USA) at 37 °C in a CO_2 incubator. Dried extract of ETME was dissolved in 100% DMSO (stock concentration of ETME: 160 mg/mL). When cells were treated with ETME, it was diluted in the cell culture medium and the concentration of DMSO was equally controlled at 0.1%, a non-toxic concentration in cells, in all experimental groups.

2.5. Cell Viability Assay

The ImKCs were seeded into 96-well plates (4×10^4/well). The cells were cultured with various concentrations (10, 20, 40, 80, or 160 μg/mL) of ETME for 24 h. Cell viability was measured using an EZ-Cytox cell viability assay kit (#EZ-1000; DoGenBio, Seoul, Republic of Korea). Briefly, the cells were incubated with the EZ-Cytox solution (containing a water-soluble tetrazolium salt) for 2 h at 37 °C. Cell viability was determined by quantifying the metabolic conversion of the tetrazolium salt to formazan dye. The absorbance of the supernatant was measured at 450 nm using a microplate reader (Synergy HTX; BioTek Instruments, Inc., Winooski, VT, USA). Cell viability was calculated using the following equation:

$$\text{Cell viability (\%)} = (\text{OD}_{450\text{ Experimental group}} - \text{OD}_{450\text{ Background control}})/(\text{OD}_{450\text{ Untreated group}} - \text{OD}_{450\text{ Background control}}) \times 100$$

where $OD_{420\ Background\ control}$ represented the optical density of EZ-Cytox solution with the medium without a cell.

2.6. Lactate Dehydrogenase (LDH) Release Assay

Release assays for lactate dehydrogenase (LDH, an indicator of cell damage) were performed as previously described [6]. Cytotoxicity was determined using an LDH assay kit (#DG-LDH500; DoGenBio) according to the manufacturer's protocol. The cells were seeded in triplicate into 12-well plates (reaching 60% confluence on day 0), then pretreated with ETME (10, 20, 40, 80, or 160 µg/mL) for 2 days. Thereafter, the cells were treated with LPS (1 µg/mL) for 24 h in a CO_2 incubator. The culture supernatants were collected, and 10 µL of the supernatant was transferred into 96-well plates, followed by incubation with 100 µL of LDH solution at room temperature in the dark for 30 min. Optical density (OD) was measured at 450 nm using a Synergy HTX microplate reader (BioTek Instruments). Cytotoxicity was calculated as previously described [6] using the following equation:

$$\text{Cytotoxicity (\%)} = [(OD_{450\ Experimental\ control} - OD_{450\ Background\ control}) - (OD_{450\ LDH\_min\ control} - OD_{450\ Background\ control})]/[(OD_{450\ LDH\_max\ control} - OD_{450\ Background\ control}) - (OD_{450\ LDH\_min\ control} - OD_{450\ Background\ control})] \times 100$$

where $OD_{450\ Background\ control}$ represented the optical density of LDH in the complete medium, $OD_{450\ LHD\_max\ control}$ represented the maximum amount of LDH released by lysis from the cells, and $OD_{450\ LDH\_min\ control}$ represented the minimum amount of LDH released by cells that had died naturally. For the "Volume" control, we added lysis solution to the complete medium.

2.7. Glutathione (GSH) Assay

A GSH assay kit (#703002; Cayman Chemical Company, Ann Arbor, MI, USA) was used to determine the amount of GSH in the cells, according to the manufacturer's guidelines. The ImKCs were seeded into 6-well plates (1.0×10^6 cells/well). Cells were pretreated with various concentrations of ETME (20, 40, 80, or 160 µg/mL) for 2 h, then stimulated with LPS (1 µg/mL) for 10 min. Thereafter, the cells were washed with cold phosphate-buffered saline (PBS) and collected by centrifugation. The cell pellet was sonicated in 50 mM cold MES buffer (from the GSH assay kit) and centrifuged at $10,000 \times g$ for 15 min at 4 °C. Next, the supernatant was collected in a new tube, and 1 M 2-vinylpyridine (#13229-2; Sigma-Aldrich) was added at a volume one-tenth that of the supernatant. The samples and standards were transferred into 96-well plates, and the assay cocktail mixture (from the GSH assay kit) was added to each sample at four times the volume of the sample. The plate was then incubated on an orbital shaker for 25 min in the dark, and absorbance was measured using a Synergy HTX microplate reader (BioTek Instruments) at 410 nm. We calculated GSH as follows. A linear regression equation ($Y = \beta X + \alpha$; Y: absorbance, β: slope, X: sample's concentration, α: intercept) with the concentration of the standard reagent (glutathione) in each experimental analysis as the X-axis and the absorbance (or optical density, OD) value as the Y-axis was calculated, and linear regression (r), slope, and Y-intercept were obtained. The obtained OD value in each experiment was substituted into a linear regression equation to calculate each experimental value. And its average value was used as the titer of the experimental group. When using a diluted sample, the sample's dilution factor was reflected in the calculation. Moreover, if the linear regression (r) was 0.990 or more, the linear regression equation was valid.

2.8. Measurement of Nitric Oxide (NO) Production

The ImKCs were seeded into 12-well plates (4.0×10^5 cells). The cells were pretreated with various concentrations of ETME (10, 20, 40, 80, or 160 µg/mL) for 2 h, then stimulated with LPS (1 µg/mL) for 24 h. Thereafter, the cell supernatants (100 µL) were transferred into new 96-well plates, and 100 µL Griess reagent (1% sulfanilamide, 0.1% N-1-naphthylethylenediamine dihydrochloride, and 2.5% phosphoric acid) was added.

Absorbance at 540 nm was measured using a Synergy HTX microplate reader (BioTek Instruments). We calculated NO as follows. A linear regression equation ($Y = \beta X + \alpha$; Y: absorbance, β: slope, X: sample's concentration, α: intercept) with the concentration of the standard reagent (nitrite) in each experimental analysis as the X-axis and the absorbance (or optical density, OD) value as the Y-axis was calculated, and linear regression (r), slope, and Y-intercept were obtained. The obtained OD value in each experiment was substituted into a linear regression equation to calculate each experimental value. And its average value was used as the titer of the experimental group. When using a diluted sample, the sample's dilution factor was reflected in the calculation. Moreover, if the linear regression (r) was 0.990 or more, the linear regression equation was valid.

2.9. Western Blotting

The cells were pretreated with various concentrations of ETME (20, 40, 80, or 160 µg/mL) for 2 h, then stimulated with LPS (1 µg/mL) for 15 min or 24 h. The cells were lysed with rapid immunoprecipitation assay (RIPA) buffer (#RC2002-050-00; Biosesang, Seongnam, Republic of Korea) containing a protease inhibitor cocktail and phosphatase inhibitor cocktails II and III (#P8340, #P5726, and #P0044, respectively; Sigma-Aldrich). The whole-cell lysate was denatured in 5× sodium dodecyl sulfate (SDS) sample buffer at 95 °C for 10 min, separated by SDS-polyacrylamide gel electrophoresis, then transferred onto nitrocellulose membranes. To block nonspecific binding, the membranes were incubated in 5% nonfat dry milk (#SKI500, LPS Solution, Daejeon, Republic of Korea) in Tris-buffered saline and Tween-20 (25 mM Tris-HCl pH 8.0, 125 mM NaCl, and 0.1% Tween-20) for 1 h at room temperature. The membranes were incubated with anti-inducible nitric oxide synthase (anti-iNOS; #610332; BD Biosciences, San Diego, CA, USA), anti-β-actin, anti-p38, anti-p44/42 (#sc-47778, #sc-7972, and #sc-514302, respectively; Santa Cruz Biotechnology Inc., Dallas, TX, USA), anti-stress-associated protein kinase/c-Jun N-terminal kinase (anti-SAPK/JNK), anti-phospho-SAPK/JNK, anti-phospho-p38, and anti-phospho-p44/42 (#9252, #9251, #9211, and #9101, respectively; Cell Signaling Technology, Danvers, MA, USA) antibodies at 4 °C overnight, then incubated with horseradish peroxidase-conjugated (HRP) secondary antibodies (Cell Signaling Technology) for 1 h at room temperature. Western blotting substrate for enhanced chemiluminescence (#BWP0200; Biomax, Seoul, Republic of Korea) was used to detect HRP-conjugated secondary antibodies. Protein expression was analyzed using a ChemiDoc imaging system (Amersham Imager 680; GE Healthcare, Chicago, IL, USA) and quantified using ImageJ (National Institutes of Health, Bethesda, MD, USA). The quantification graph of protein expression was basically calculated based on the obtained values from each protein band. First, the protein bands were quantified with the specific length and height of each protein band under the same conditions in quantitative software. Then, the value of target proteins was normalized by the value of internal control (β-actin, GAPDH, and Lamin B1). Finally, the normalized values were graphed and expressed as protein expression. In the case of the phospho-form, proteins were normalized by the values of total protein based on the values of β-actin, and the normalized values were graphed.

2.10. Enzyme-Linked Immunosorbent Assay (ELISA)

The ImKCs were seeded into 12-well plates (4.0×10^5 cells/well). The cells were pretreated with various concentrations of ETME (20, 40, 80, or 160 µg/mL) for 2 h, then stimulated with LPS (1 µg/mL) for 24 h. The expression of the indicated cytokines in cell supernatants was measured using an ELISA kit. Purified anti-IL-6 (#554400; BD Pharmingen, San Diego, CA, USA), anti-IL-1β, and anti-TNF-α (#14-7012-85 and #14-7423-85, respectively; Thermo Fisher Scientific Inc., Waltham, MA, USA) antibodies were coated onto 96-well plates and incubated overnight at 4 °C. The plates were washed three times with 0.05% Tween-20 in PBS, then incubated with 1% bovine serum albumin (BSA) in PBS for 1 h at room temperature. The supernatants and standard solutions were incubated for 2 h at room temperature and washed three times. Thereafter, the plate was incubated with detection antibodies for 1 h at room temperature and washed three times, then incubated with streptavidin-conjugated alkaline phosphatase (AKP; #554065; BD Pharmingen) solution for 30 min at room temperature, and washed five times. Finally, the plate was incubated in the dark with a substrate buffer (pH 9.8), comprising 10% diethanolamine, 0.1% $MgCl_2 \cdot 6H_2O$, 0.2% NaN_3 (#3032-4400, #5503-44, and #7530-4105, respectively; Daejung Chemicals & Metals Co., Siheung, Republic of Korea), and 4-nitrophenyl phosphate (#N2765; Sigma-Aldrich). Subsequently, 1 N NaOH was added to stop the reaction. Absorbance at 450 nm was measured using a Synergy HTX microplate reader (BioTek Instruments). We calculated IL-6, IL-1β, and TNF-α as follows. A linear regression equation ($Y = \beta X + \alpha$; Y: absorbance, β: slope, X: sample's concentration, α: intercept) with the concentration of the standard reagents (IL-6, IL-1β, and TNF-α) in each experimental analysis as the X-axis and the absorbance (or optical density, OD) value as the Y-axis was calculated, and linear regression (r), slope, and Y-intercept were obtained. The obtained OD value in each experiment was substituted into a linear regression equation to calculate each experimental value. And its average value was used as the titer of the experimental group. When using a diluted sample, the sample's dilution factor was reflected in the calculation. Moreover, if the linear regression (r) was 0.990 or more, the linear regression equation was valid.

2.11. Fractionation of Nuclear and Cytoplasm Proteins

Nuclear and cytoplasmic fractionation was performed according to the manufacturer's protocol using the NE-PER Nuclear and Cytoplasmic Extraction Reagent Kit (#78833; Thermo Fisher Scientific Inc.). Briefly, ImKCs were seeded into 6-well plates (2.0×10^6 cells/well), pretreated with 160 µg/mL of ETME for 2 h, then stimulated with LPS (1 µg/mL) for 10 min. The cells were washed twice with ice-cold PBS and lysed with 200 µL cytoplasmic lysis buffer (from the fractionation kit) on ice for 10 min. The lysates were centrifuged at maximum speed for 10 min at 4 °C, and the supernatants were collected to obtain the cytoplasmic fraction. Next, the pellet was resuspended in 100 µL nuclear extraction buffer (from the fractionation kit) on ice for 40 min and vortexed every 10 min. After centrifugation for 10 min at 4 °C, the supernatant was collected to obtain the nuclear fraction. Western blotting was performed using anti-GAPDH (a cytoplasmic marker), anti-NF-κB p65 (#sc-365062 and #sc-8008, respectively; Santa Cruz Biotechnology Inc.), anti-Lamin B1 (a nuclear marker), and anti-inhibitor of κB (IκB) (#9242 and #9242, respectively; Cell Signaling Technology) antibodies.

2.12. Immunofluorescence Staining

The ImKCs were seeded into 24-well plates (2.0×10^5 cells/well) with coverslips. The cells were pretreated with 160 µg/mL ETME for 2 h, then stimulated with LPS (1 µg/mL) for 10 min. The cells were fixed with a 4% paraformaldehyde solution (#PC2031-050-00; Biosesang) and permeabilized with 0.1% Triton X-100 (#T8787; Sigma-Aldrich) for 10 min at room temperature. Next, the cells were blocked with PBS containing 1% BSA at room temperature for 1 h, then incubated with an anti-NF-κB antibody at 4 °C overnight. Thereafter, the cells were incubated with secondary antibodies in the dark for 1 h. Finally, the cells were fixed onto glass slides using a mounting solution (#S36936; Thermo Fisher Scientific Inc.), and fluorescent images were captured using a confocal microscope (Nikon AX R; Nikon Instruments, Tokyo, Japan).

2.13. Statistical Analysis

The data are presented as the mean $\pm$ standard error of the mean (SEM) and were analyzed using Prism v8.0 (GraphPad Inc., San Diego, CA, USA). The results were analyzed using the nonparametric Mann–Whitney U test, with $p < 0.05$ considered statistically significant. All experiments were performed in triplicates.

3. Results

3.1. Rosmarinic Acid Is a Major Consituent of ETME

Previous studies revealed that rosmarinic acid is the major component of *E. tinifolia* through a phytochemical assay-guided fractionation [21] and it has antioxidant and anti-inflammatory properties [22,23]. Before investigating the effects of ETME on oxidative stress and inflammation, an evaluation of the major components in ETME that exhibit antioxidant and anti-inflammatory effects was performed. The UPLC-Q-TOF-MS and the total ion chromatogram (TIC) of the MeOH extract are shown in Figure 1. The analysis revealed the presence of 22 phytochemicals belonging to various subclasses such as oligosaccharides, flavonoids, phenolic acids, and lignans, as detailed in Table 1. One of the most abundant compounds in ETME observed by the mass analysis was the predominant precursor ion at m/z 359, which can be attributed to rosmarinic acid as confirmed by a standard product retention time (retention time of 11.30 min; Supplementary Figure S1). Compound methyl rosmarinate was identified at m/z 374, as previously reported in *E. tinifolia* ethyl acetate portion [26], and was known for its antioxidant and antifungal activities [22,27]. For the identification of the remaining compounds, the m/z values of the molecular ion $[M-H]^-$ and $[M + HCOO]^-$ were compared with the Waters Unifi Software Traditional Medicine Library (Waters Corporation, Milford, MA, USA) and corroborated from the previously reported literature [28–42]. Among these compounds, schisantherin A, kaempferol-3,7-diglucoside, procyanidin A2, and tocopherol exert anti-inflammatory and antioxidant activity and promote reproductive functions [43–46], which explain the potent antioxidant and anti-inflammatory activity observed in *E. tinifolia*. Through UPLC-Q-TOF-MS spectral analysis, it is evident that rosmarinic acid comprises a larger proportion relative to other components. Therefore, it can be inferred that rosmarinic acid predominantly exerts its influence on antioxidant and anti-inflammatory effects. Herein, an additional experiment was performed to assess ETME's antioxidant properties by measuring rosmarinic acid-mediated antioxidant activity. Rosmarinic acid showed a dose-dependent increase in DPPH inhibition and showed a high degree of inhibition at 20 µg/mL similar to those of quercetin 10 µg/mL, a well-known antioxidant molecule, indicating that ETME might exhibit antioxidant properties (Supplementary Figure S2). These data prompted us to study the antioxidant and anti-inflammatory effects and underlying regulatory mechanisms of action of ETME in ImKCs.

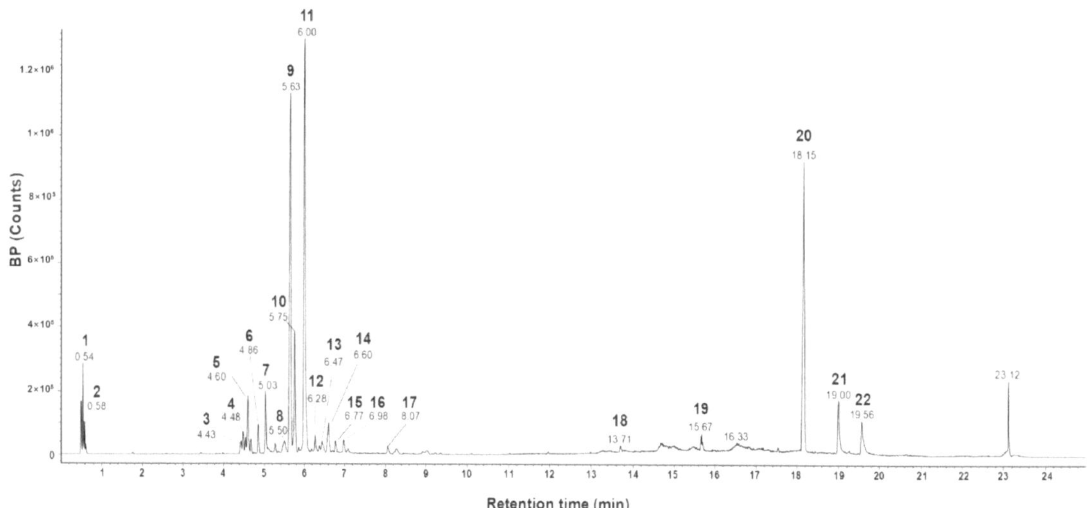

Figure 1. UPLC-Q-TOF-MS spectrometry analysis of ETME.

Table 1. Tentative identification of constituents in the ETME using UPLC-Q-TOF-MS spectrometry analysis. TML: traditional medicine library.

#	Observed RT (min)	Neutral Mass (Da)	Observed Neutral Mass (Da)	Observed (m/z)	Adducts	Component Name
1	0.54	504.16903	504.1691	503.1619	−H, +HCOO	Raffinose [28]
2	0.58	150.05282	150.0527	195.0509	+HCOO, −H	Pentose (TML)
3	4.43	242.09429	242.0938	241.0865	−H	Flavanthrinin [29]
4	4.48	536.20463	536.2039	581.2022	+HCOO	Schisantherin A [30]
5	4.60	610.15338	610.1547	609.1475	−H	Kaempferol-3,7-diglucoside [31]
6	4.86	594.15847	594.1591	593.1519	−H	Genistein-7,4′-di-O-β-D-glucoside [32]
7	5.03	286.08412	286.0831	285.0759	−H	Phyllodulcin [33]
8	5.50	314.07904	314.0787	313.0714	−H	Ermanin [34]
9	5.63	360.08452	360.0848	359.0775	−H	Rosmarinic acid [35]
10	5.75	312.06339	312.0631	357.0613	+HCOO, −H	2-Acetyl emodin (TML)
11	6.00	286.06887	286.0688	321.0399	+Cl	Uralenneoside (TML)
12	6.28	340.0583	340.0584	339.0511	−H	Versicolorin B [36]
13	6.47	374.10017	374.1001	373.0928	−H	Methyl rosmarinate [26]
14	6.60	354.07395	354.0741	353.0668	−H	5-Methoxysterigmatocystn [38]
15	6.77	336.027	336.0268	335.0195	−H	Rufescidride [39]
16	6.99	518.32435	518.3231	517.3159	−H	2-Hydroxyesculentic acid (TML)
17	8.07	576.12678	576.124	575.1167	−H	Procyanidin A2 [40]
18	13.71	256.24023	256.2403	255.233	−H	Methyl pentadecanote (TML)
19	15.67	510.17373	510.1744	555.1726	+HCOO	Globularinin (TML)
20	18.15	498.37091	498.3713	497.3641	−H, +HCOO	Acetyl-β-boswellic acid [41]
21	19.00	206.16707	206.1668	205.1595	−H	Longicamphenylone (TML)
22	19.57	416.36543	416.3664	461.3646	+HCOO	γ-Tocopherol [42]

3.2. ETME Exerted Antioxidant Properties under Non-Cytotoxic Concentrations

To verify non-cytotoxic concentrations of ETME on ImKCs' cell viability, a cell viability assay based on tetrazolium conversion and an LDH assay were carried out in the absence or presence of LPS. Both assays revealed that ETME showed no significant cytotoxicity in ImKCs at various concentrations of up to 160 μg/mL (Figure 2a,b). Therefore, subsequent experiments were conducted at ETME concentrations of up to 160 μg/mL. Next, to examine whether ETME has antioxidant properties, DPPH assay was conducted with ETME and quercetin. ETME exhibited a dose-dependent increase in DPPH inhibition and showed a similar effect at 160 μg/mL with that of quercetin, indicating that ETME has antioxidant properties (Figure 2c).

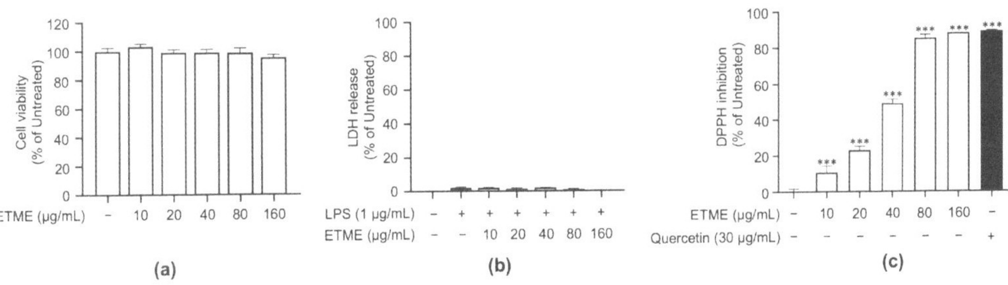

Figure 2. Effects of ETME on cell viability of ImKCs and free radical scavenging activity. (**a**) ImKCs were treated with various concentrations of ETME for 24 h. Cell viability was then measured using the EZ-Cytox reagent and compared with that of the untreated group. (**b**) The ImKCs were treated with LPS (1 μg/mL) in the presence of ETME (10, 20, 40, 80, or 160 μg/mL) for 24 h. The supernatants were collected and analyzed using the lactate dehydrogenase (LDH) assay kit. Cells treated with lysis buffer were used as a positive control (100% LDH release). (**c**) The antioxidant effect of ETME was analyzed by DPPH assay. Quercetin was used as a positive control. The data are the mean ± standard error of the mean (SEM) of three independent experiments. Differences between groups were analyzed using the Mann–Whitney U test. ETME: *E. tinifolia* methanol extract; DPPH: 2,2-diphenyl-1-picryhydrazyl. *** $p < 0.001$ vs. untreated group.

3.3. ETME Rescued GSH Levels and Inhibited NO Production

Based on the radical scavenging effect of ETME, the antioxidant property of ETME was evaluated by measuring total reduced GSH levels in LPS-treated ImKCs. As shown in Figure 3a, the LPS-mediated increase in oxidative stress was detected by the decrease in total reduced GSH levels in LPS-treated ImKCs. The LPS-mediated decrease in total reduced GSH levels was recovered by ETME treatment (Figure 3a), indicating that ETME has antioxidant properties in ImKCs. The production of NO, a well-known mediator for oxidative stress and inflammation [47], was significantly induced in LPS-treated ImKCs. ETME suppressed LPS-induced NO production in a dose-dependent manner (Figure 3b). The LPS-stimulated ImKCs express pro-inflammatory enzymes, such as iNOS, which play an important role in NO production [48]. Consistent with NO production, LPS-induced iNOS expression was significantly suppressed by ETME in ImKCs (Figure 3c), implying that ETME alleviates NO production in LPS-stimulated ImKCs through the transcriptional inhibition of iNOS.

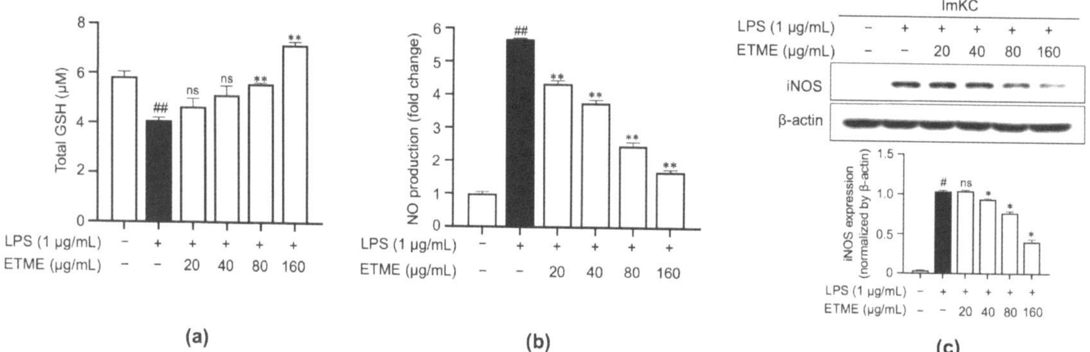

Figure 3. Effects of ETME on total glutathione (GSH), nitric oxide (NO) production, and inducible nitric oxide synthase (iNOS) expression in ImKCs. The ImKCs were treated with LPS (1 µg/mL) at the indicated concentration of ETME (20, 40, 80, or 160 µg/mL) for 24 h. (**a**) ImKCs were pretreated with ETME (20, 40, 80, or 160 µg/mL) for 2 h, then stimulated with LPS (1 µg/mL) for 10 min. Total GSH was measured using a GSH assay kit. (**b**) NO production in the culture supernatant was measured using a Griess assay. NO secretion was calculated using a standard curve of nitrite standard-solution concentrations. (**c**) Inducible nitric oxide synthase (iNOS) expression was detected by Western blotting. β-actin was used as a loading control. Protein expression was normalized to that of β-actin. The data are the mean ± SEM of three independent experiments. Differences between groups were analyzed using the Mann–Whitney U test. # $p < 0.05$, ## $p < 0.01$ vs. LPS-untreated group; * $p < 0.05$, ** $p < 0.01$ vs. LPS-treated group; ns: not significant.

3.4. ETME Inhibited Pro-Inflammatory Cytokine Production

To investigate whether ETME inhibits the secretion of pro-inflammatory cytokines, including TNF-α, IL-1β, and IL-6, in ImKCs, supernatants were collected following the treatment of ETME in the presence of LPS and subjected to an ELISA. While the secretion of pro-inflammatory cytokines was markedly induced in LPS-stimulated ImKCs, the ETME treatment inhibited secretory levels of those cytokines in a dose-dependent manner (Figure 4), revealing the anti-inflammatory effects of ETME in LPS-stimulated ImKCs.

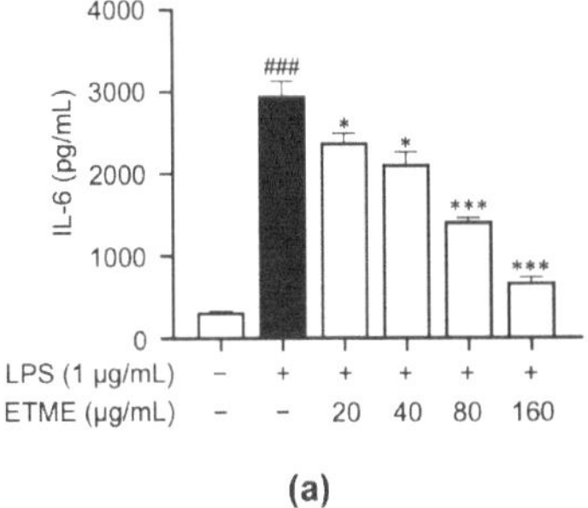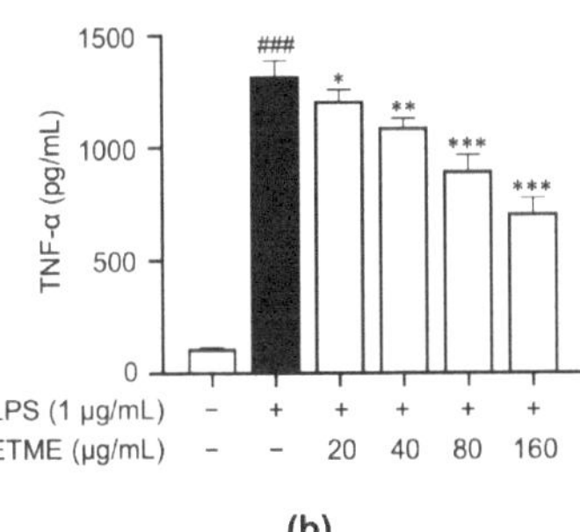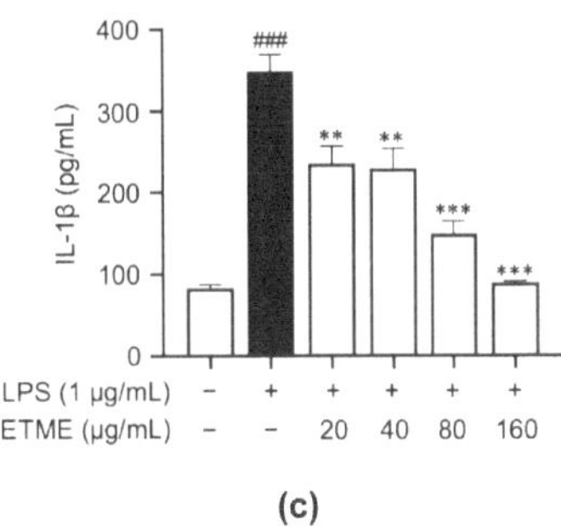

Figure 4. Inhibitory effects of ETME on pro-inflammatory cytokine production. ImKCs were treated with LPS in the presence of ETME (0, 20, 40, 80, or 160 μg/mL). (**a–c**) After stimulation for 24 h, the culture supernatants were collected and analyzed for interleukin (IL)-6, tumor necrosis factor (TNF)-α, and IL-1β production via an enzyme-linked immunosorbent assay (ELISA). The data presented are the mean ± SEM of three independent experiments. Differences between groups were analyzed using the Mann–Whitney U test. ### $p < 0.001$ vs. LPS-untreated group; * $p < 0.05$, ** $p < 0.01$, *** $p < 0.001$ vs. LPS-treated group.

3.5. ETME Exerted Anti-Inflammatory Activity by Activating HO-1

The results shown in Figures 2–4 indicate that ETME alleviates LPS-induced oxidative stress and pro-inflammatory responses. To clarify whether ETME regulates oxidative stress through the activation of antioxidant signaling molecules, Sn protoporphyrin (SnPP), a well-known HO-1 inhibitor [49], was co-treated with ETME in LPS-stimulated ImKCs. The ETME-mediated inhibition of NO production was significantly attenuated by SnPP treatment (Figure 5a). These results revealed that the ETME-mediated inhibition of NO is regulated by the ETME-induced increase in antioxidant expression. To further investigate whether the increase in antioxidant HO-1 expression is involved in the anti-inflammatory effect of ETME, the secretory levels of pro-inflammatory cytokines were measured in the same supernatants. SnPP treatment partially abolished the inhibitory effects of ETME on pro-inflammatory cytokine production (Figure 5b), suggesting that the ETME-mediated increase in HO-1 expression leads to its antioxidant and anti-inflammatory effects in LPS-stimulated ImKCs.

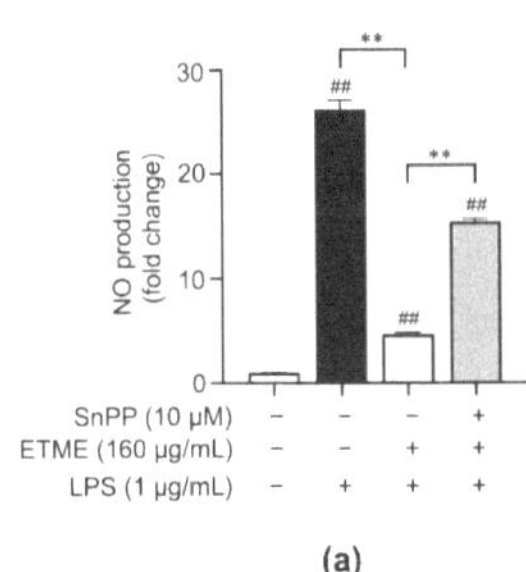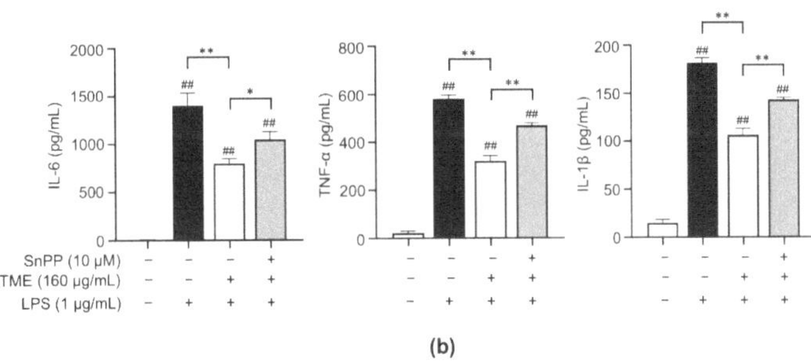

Figure 5. The protective action of ETME on NO production and pro-inflammatory cytokine production is heme oxygenase-1 (HO-1)-dependent. ImKCs were treated with the indicated combinations of LPS (1 μg/mL), ETME (160 μg/mL), and Sn protoporphyrin (SnPP, 10 μM) for 24 h. (**a**) NO production in the culture supernatant was measured using a Griess assay. NO secretion was calculated using a standard curve of nitrite standard-solution concentration. (**b**) The culture supernatants were collected and analyzed for IL-6, TNF-α, and IL-1β production via an ELISA. The data presented are the mean ± SEM of three independent experiments. Differences between groups were analyzed using the Mann–Whitney U test. ## $p < 0.01$ vs. LPS-untreated group; * $p < 0.05$, ** $p < 0.01$ between paired groups.

3.6. ETME Increased HO-1 and Nrf2 Expression Thereby Promoting Its Antioxidant Activity

Nrf2 is a transcription factor responsible for regulating the expression of genes, such as HO-1, involved in cellular redox balance and protective antioxidant systems [50]. Nrf2 pathway activation is thus a possible explanation for HO-1 induction. As shown in Figure 6a, ETME significantly induced expression levels of Nrf2 at high concentrations (80 and 160 µg/mL) compared to LPS-stimulated ImKCs. Further investigation was conducted to evaluate Nrf2 accumulation in the nucleus. Immunofluorescence data showed that ETME increased nuclear Nrf2 expression (Figure 6b). These results indicate that ETME exhibits antioxidant activity via the nuclear accumulation of Nrf2 protein and subsequent increase in HO-1 expression.

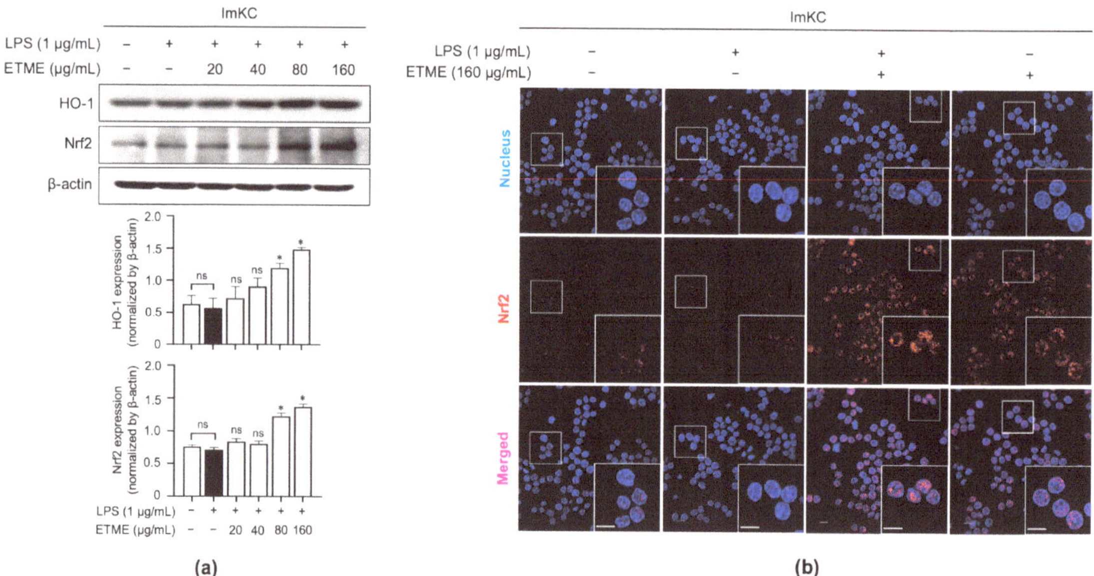

Figure 6. Effects of ETME on nuclear factor erythroid 2-related factor 2 (Nrf2) nuclear localization and HO-1 expression. (**a**) ImKCs were treated with LPS (1 µg/mL) in the presence of ETME (0, 24, 40, 80, or 160 µg/mL) for 24 h, after which total protein was extracted. Western blotting was used to detect HO-1 and Nrf2 expression, with β-actin as a loading control. Protein expression was normalized to that of each loading control. The data presented are the mean ± SEM of three independent experiments. (**b**) The nuclear translocation of Nrf2 (red) in cells was analyzed by confocal microscopy. Nuclei were stained with 4′,6-diamidino-2-phenylindole (blue). Scale bar, 10 µm. Differences between groups were analyzed using the Mann–Whitney U test. * $p < 0.05$ vs. LPS-treated group; ns: not significant.

3.7. ETME Inhibited NF-κB/p65 Nuclear Translocation

The NF-κB signaling pathway, a major regulator of inflammatory responses in macrophages upon LPS stimulation [51,52], is activated primarily by the nuclear localization of NF-κB/p65 after IκB degradation [52]. Total proteins were fractionated into cytosolic and nucleus fractions and subjected to immunoblotting analysis to detect IκB and NF-κB/p65. In the cytosolic fraction, LPS stimulation induced IκB degradation, whereas ETME reversed the LPS-induced IκB degradation (Figure 7a). Considering NF-κB/p65 levels in both fractions, ETME inhibited the LPS-induced translocation of NF-κB/p65 into the nucleus from the cytosol (Figure 7a). The ETME-mediated inhibition of LPS-induced NF-κB/p65 translocation into the nucleus was further detected by immunofluorescence imaging (Figure 7b). These results suggest that ETME exerts anti-inflammatory effects by inhibiting NF-κB signaling activation.

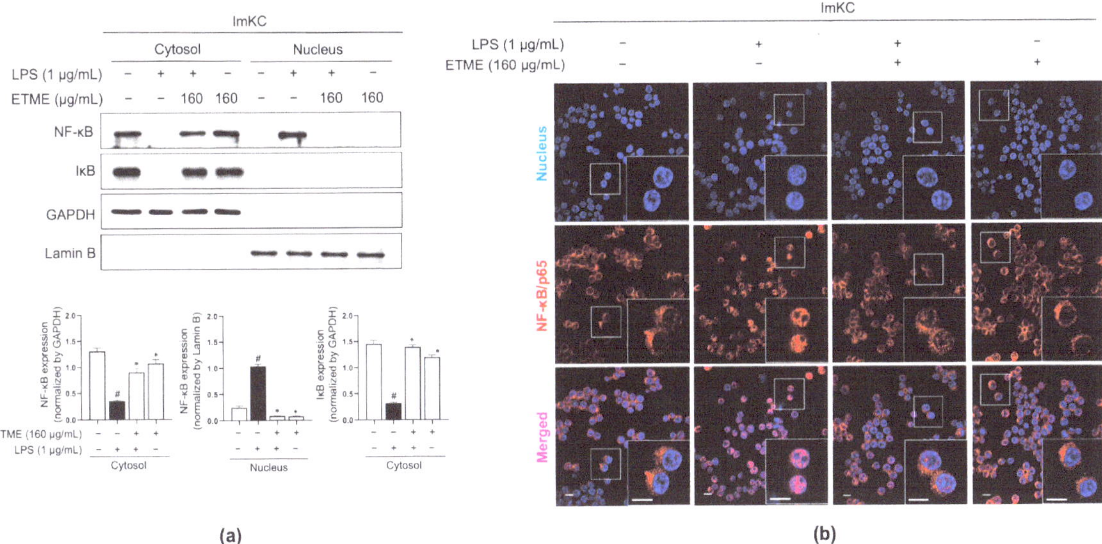

Figure 7. Inhibitory effects of ETME on nuclear factor kappa-light-chain-enhancer of activated B cells (NF-κB)/p65 nuclear translocation. ImKCs were pretreated with ETME (160 μg/mL) for 2 h, then stimulated with LPS (1 μg/mL) for 15 min. (**a**) NF-κB/p65 and inhibitor of κB (IκB) expression in cytosolic and nuclear extracted-protein samples was detected by Western blotting. Glyceraldehyde 3-phosphate dehydrogenase (GAPDH) was used as a cytosolic loading control and Lamin B as a nuclear loading control. Protein expression was normalized to that of each loading control. The data presented are the mean ± SEM of three independent experiments. Differences between groups were analyzed using the Mann–Whitney U test. # $p < 0.05$ vs. LPS-untreated group; * $p < 0.05$ vs. LPS-treated group. (**b**) The nuclear translocation of NF-κB/p65 (red) in cells was analyzed by confocal microscopy. Nuclei were stained with 4′,6-diamidino-2-phenylindole (blue). Scale bar, 10 μm.

3.8. ETME Inhibited MAPK Phospohrylation

MAPK pathways are also major inflammatory signaling pathways, which are activated and characterized by LPS-mediated phosphorylation [53,54]. The engagement of MAPK signaling pathways was investigated by measuring the ETME-mediated decrease in MAPK phosphorylation in LPS-treated ImKCs. The phosphorylation of MAPKs (p38, p44/42, and JNK) was markedly reduced by ETME treatment in a dose-dependent manner (Figure 8), suggesting that the anti-inflammatory effect is mediated by the suppression of MAPK activation in LPS-stimulated ImKCs.

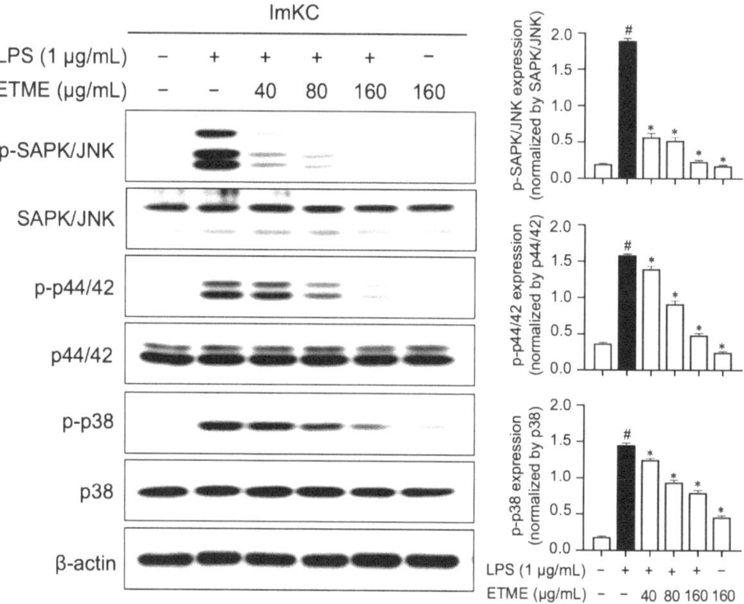

Figure 8. Inhibitory effects of ETME on mitogen-activated protein kinase (MAPK) signaling. The ImKCs were pretreated with ETME (40, 80, or 160 µg/mL) for 2 h, then stimulated with LPS (1 µg/mL) for 15 min. The expression of proteins associated with the MAPK signaling pathway (p38, p44/42, and SAPK/JNK) was detected by Western blotting, with β-actin as a loading control. The expression of phosphorylated (p-) protein (p-p38, p-p44/42, and p-SAPK/JNK) was normalized to that of the total-form protein (p38, p44/42, and SAPK/JNK). The data presented are the mean ± SEM of three independent experiments. Differences between groups were analyzed using the Mann–Whitney U test. $\# p < 0.05$ vs. LPS-untreated group; $^{*} p < 0.05$ vs. LPS-treated group.

4. Discussion

NO plays an important role in liver physiology and pathophysiology [55,56]. It is produced as a by-product when L-arginine is oxidized to citrulline by the action of three isomorphs: neuronal NOS (nNOS), iNOS, and endothelial NOS (eNOS). nNOS and eNOS are constitutively expressed. Although iNOS is not expressed under resting conditions, it is induced by immunological stimuli such as LPS [57]. Given the oxidative stress and pro-inflammatory nature of iNOS, and the beneficial role of iNOS inhibition in liver fibrosis, iNOS has been considered a potential therapeutic target for various diseases, including liver fibrosis [58,59], septic shock [60], and asthma [61]. Therefore, the inhibitory effect of ETME on NO production via the inhibition of LPS-induced iNOS expression in ImKCs suggests that ETME may be a therapeutic candidate for oxidative stress and inflammatory diseases in the liver, including liver fibrosis.

HO-1, an inducible enzyme, plays a cytoprotective role against oxidative stress by removing cytotoxic free heme and generating antioxidants [62,63]. Recent studies have revealed that it also exhibits anti-inflammatory activities against numerous inflammatory diseases. For example, rare HO-1 deficiencies in humans and animal models are characterized by high levels of chronic inflammation and increased sensitivity to oxidative stress [64–67]. Therefore, the immunomodulatory functions of HO-1 provide a promising therapeutic target for treating a broad range of oxidative and inflammatory diseases [62,63,68,69]. Here, ETME increased HO-1 expression in LPS-stimulated ImKCs (Figure 6a), and SnPP, an HO-1 inhibitor, significantly restored the ETME-induced reduction in NO and pro-inflammatory cytokines (Figure 5). These results suggest that ETME's anti-inflammatory effects are closely related to its antioxidative properties, implying that ETME could be a valuable medication for treating various diseases via its cooperative regulatory effects on inflammation and oxidative stress.

The Keap1-Nrf2 pathway is a protective response to oxidative stress in macrophages [70]. Under homeostatic conditions, Keap1 tightly associates with Nrf2 and acts as a member of E3 ubiquitin ligase, which tightly regulates ubiquitination and proteasomal degradation of Nrf2 [71]. In response to oxidative stress, Nrf2 escapes ubiquitination by dissociating from Keap1, accumulating within the cell, translocating to the nucleus, and promoting its antioxidant transcription program [72]. Immunoblotting and immunofluorescence data revealed that Nrf2 accumulation in the nucleus was enhanced by the ETME treatment in the absence or presence of LPS (Figure 6). Cellular Nrf2 expression could be induced by enhancing its transcription and stabilizing it through the inhibition of its poly-ubiquitination. To clarify this, ML385 and brusatol, Nrf2 inhibitors that inhibit Nrf2 transcription [73] and stimulate its polyubiquitination [74], respectively, were treated and compared to the ETME-treated group. However, both ML385 and brusatol did not alleviate ETME-mediated NO inhibition in LPS-stimulated macrophages, indicating that the ETME-mediated accumulation of Nrf2 in the nucleus is not regulated by its transcription and polyubiquitination. Further study using other inhibitors, such as ascorbic acid, an electrophilic modifier of Keap1, or trigonelline, an inhibitor of nuclear translocation of Nrf2, could clarify the mechanism for the ETME-mediated nuclear increase in Nrf2 and subsequent antioxidant responses in macrophages.

The pharmacological activities of plant extracts are closely related to the pharmacological activities of various phytochemicals contained in the extracts. In this study, UPLC-Q-TOF-MS spectrometry analysis of ETME revealed that rosmarinic acid (component 9), uralenneoside (component 11), and acetyl-β-boswellic acid (component 20) were identified as the major components (Figure 1 and Table 1). Among these components, rosmarinic acid was found to exhibit both antioxidant [22] and anti-inflammatory effects [75]. In detail, rosmarinic acid elicits neuroprotection in ischemic stroke via Nrf2/HO-1 pathway [76] and ameliorates acute liver damage and fibrogenesis accompanied by enhanced Nrf2/HO-1 expression [77]. Rosmarinic acid is also known to inhibit LPS-induced NO production and iNOS in RAW264.7 macrophage via suppressing NF-κB signaling [78]. Also, rosmarinic acid attenuates the LPS-stimulated proinflammatory mediators such as TNF-α, IL-8, and iNOS through the inhibition of three MAPKs and NF-κB signaling in vascular smooth muscle cell [79]. Based on the high similarity between our results and previous reports on rosmarinic acid, it seems that the antioxidant and anti-inflammatory effects of ETME are mainly attributed to rosmarinic acid. Meanwhile, the known effects of these two components (uralenneoside and acetyl-β-boswellic acid) are insufficient to account for the antioxidant and anti-inflammatory pharmacological effects of ETME. Acetyl-β-boswellic acid was reported to have only limited anti-inflammatory property through the inhibition of 5-lipoxygenase and MAPKs [80,81], and no reports are available for the antioxidant and anti-inflammatory effects of uralenneoside. Since the effect of an extract is due to several constituents, additional studies to elucidate the antioxidant and anti-inflammatory properties of uralenneoside and acetyl-β-boswellic acid are required to clarify ETME-mediated effects.

5. Conclusions

This study elucidates the antioxidant and anti-inflammatory effects of a methanol extract of *E. tinifolia* and its underlying mechanisms of action. ETME significantly restored total GSH levels and suppressed NO and pro-inflammatory cytokine production. These inhibitions were mediated by the activation of the antioxidant Nrf2/HO-1 pathway and the inhibition of NF-κB translocation and MAPK phosphorylation. These results provide evidence supporting the traditional pharmacological efficacy reported for ETME, suggesting its potential application as a candidate capable of alleviating pathological conditions mediated by inflammation or oxidative stress.

Supplementary Materials: The supporting information can be downloaded at: https://www.mdpi.com/article/10.3390/antiox12101792/s1. Figure S1: HPLC chromatogram of Rosmarinic acid and ETME. Figure S2: Radical scavenging effect of rosmarinic acid.

Author Contributions: Conceptualization, J.S.L., Y.-C.C., S.H.L. and K.Y.L.; validation, H.Y., N.C. and G.Y.; methodology, J.S.L., J.U.C. and D.Y.L.; formal analysis, J.S.L., S.H.L., H.Y., D.Y.L., N.C., G.Y. and J.U.C.; resources, T.T.B.; writing—original draft preparation, J.S.L., S.H.L. and H.Y.; writing—review and editing, K.Y.L., Y.-C.C. and S.-J.P.; supervision, Y.-C.C. and S.-J.P.; project administration, J.S.L. and Y.-C.C.; funding acquisition, J.S.L., Y.-C.C., S.-J.P. and G.Y. All authors have read and agreed to the published version of the manuscript.

Funding: This study was financially supported by the National Research Foundation of Korea (NRF) grant funded by the ICT (NRF-2020R1C1C1007261) and the Ministry of Education (NRF-2022R1I1A1A01056975), the Korea Research Institute of Bioscience and Biotechnology (KRIBB) Research Initiative Program (KGM5242322), and the Main Research Program E0210300 of the Korea Food Research Institute (KFRI) funded by the Ministry of Science.

Institutional Review Board Statement: Not applicable.

Informed Consent Statement: Not applicable.

Data Availability Statement: The data presented in this study are available in the article and Supplementary Material.

Conflicts of Interest: The authors declare no conflict of interest.

References

1. Chen, L.; Deng, H.; Cui, H.; Fang, J.; Zuo, Z.; Deng, J.; Li, Y.; Wang, X.; Zhao, L. Inflammatory responses and inflammation-associated diseases in organs. *Oncotarget* **2018**, *9*, 7204–7218. [CrossRef]
2. Bindu, S.; Mazumder, S.; Bandyopadhyay, U. Non-steroidal anti-inflammatory drugs (NSAIDs) and organ damage: A current perspective. *Biochem. Pharmacol.* **2020**, *180*, 114147. [CrossRef] [PubMed]
3. Newman, D.J.; Cragg, G.M. Natural products as sources of new drugs over the 30 years from 1981 to 2010. *J. Nat. Prod.* **2012**, *75*, 311–335. [CrossRef] [PubMed]
4. Atanasov, A.G.; Zotchev, S.B.; Dirsch, V.M.; International Natural Product Sciences, T.; Supuran, C.T. Natural products in drug discovery: Advances and opportunities. *Nat. Rev. Drug Discov.* **2021**, *20*, 200–216. [CrossRef] [PubMed]
5. Lim, J.S.; Lee, S.H.; Lee, S.R.; Lim, H.-J.; Roh, Y.-S.; Won, E.J.; Cho, N.; Chun, C.; Cho, Y.-C. Inhibitory effects of Aucklandia lappa decne. Extract on inflammatory and oxidative responses in LPS-treated macrophages. *Molecules* **2020**, *25*, 1336. [CrossRef] [PubMed]
6. Lee, S.H.; Cho, Y.-C.; Lim, J.S. Costunolide, a sesquiterpene lactone, suppresses skin cancer via induction of apoptosis and blockage of cell proliferation. *Int. J. Mol. Sci.* **2021**, *22*, 2075. [CrossRef]
7. Hammer, K.D.; Hillwig, M.L.; Solco, A.K.; Dixon, P.M.; Delate, K.; Murphy, P.A.; Wurtele, E.S.; Birt, D.F. Inhibition of prostaglandin E_2 production by anti-inflammatory hypericum perforatum extracts and constituents in RAW264.7 Mouse Macrophage Cells. *J. Agric. Food Chem.* **2007**, *55*, 7323–7331. [CrossRef]
8. Li, C.; Wang, M.H. Anti-inflammatory effect of the water fraction from hawthorn fruit on LPS-stimulated RAW 264.7 cells. *Nutr. Res. Pract.* **2011**, *5*, 101–106. [CrossRef]
9. Qnais, E.Y.; Abu-Dieyeh, M.; Abdulla, F.A.; Abdalla, S.S. The antinociceptive and anti-inflammatory effects of Salvia officinalis leaf aqueous and butanol extracts. *Pharm. Biol.* **2010**, *48*, 1149–1156. [CrossRef]
10. Guardia, T.; Rotelli, A.E.; Juarez, A.O.; Pelzer, L.E. Anti-inflammatory properties of plant flavonoids. Effects of rutin, quercetin and hesperidin on adjuvant arthritis in rat. *Il Farmaco* **2001**, *56*, 683–687. [CrossRef]
11. Azab, A.; Nassar, A.; Azab, A.N. Anti-Inflammatory Activity of Natural Products. *Molecules* **2016**, *21*, 1321. [CrossRef]
12. Shen, N.; Wang, T.; Gan, Q.; Liu, S.; Wang, L.; Jin, B. Plant flavonoids: Classification, distribution, biosynthesis, and antioxidant activity. *Food Chem.* **2022**, *383*, 132531. [CrossRef] [PubMed]
13. Hadjichambis, A.; Paraskeva-Hadjichambi, D.; Della, A.; Giusti, M.E.; De Pasquale, C.; Lenzarini, C.; Censorii, E.; Gonzales-Tejero, M.R.; Sanchez-Rojas, C.P.; Ramiro-Gutierrez, J.M.; et al. Wild and semi-domesticated food plant consumption in seven circum-Mediterranean areas. *Int. J. Food Sci. Nutr.* **2008**, *59*, 383–414. [CrossRef] [PubMed]
14. Miller, J.S. A Revision of the New World Species of Ehretia (Boraginaceae). *Ann. Mo. Bot. Gard.* **1989**, *76*, 1050–1076. [CrossRef]
15. Emes Boronda, M.; Ochurte Espinoza, C.; Castañeda Silva, G.; Peralta González, B. *Flora medicinal Indígena de México: Treinta y Cinco Monografías del Atlas de las Plantas de la Medicina Tradicional Mexicana*; Instituto Nacional Indigenista: México, DF, Mexico, 1994; 3 Volumes.
16. Martínez, M. Las plantas medicinales de México. In *Las plantas medicinales de México*; Ediciones Botas: México, Mexico, 1991; p. 656.

17. Lentz, D.L.; Dickau, R. *Seeds of the Central America and Southern Mexico: The Economic Species*; New York Botanical Garden: New York, NY, USA, 2005.

18. Liogier, H. *Plantas medicinales de Puerto Rico y del Caribe.*; Iberoamericana de Ediciones: San Juan, Puerto Rico, 1990; Volume 566.

19. Dzib-Guerra, W.D.; Escalante-Erosa, F.; Garcia-Sosa, K.; Derbre, S.; Blanchard, P.; Richomme, P.; Pena-Rodriguez, L.M. Anti-Advanced Glycation End-product and Free Radical Scavenging Activity of Plants from the Yucatecan Flora. *Pharmacogn. Res.* **2016**, *8*, 276–280. [CrossRef]

20. Pío-León, J.F.; Díaz-Camacho, S.P.; López, M.G.; Montes-Avila, J.; López-Angulo, G.; Delgado-Vargas, F. Physicochemical, nutritional, and antioxidant characteristics of the fruit of Ehretia tinifolia. *Rev. Mex. Biodivers.* **2012**, *83*, 273–280. [CrossRef]

21. Monroy-García, I.N.; Carranza-Torres, I.E.; Carranza-Rosales, P.; Oyón-Ardoiz, M.; García-Estévez, I.; Ayala-Zavala, J.F.; Morán-Martínez, J.; Viveros-Valdez, E. Phenolic Profiles and Biological Activities of Extracts from Edible Wild Fruits Ehretia tinifolia and Sideroxylon lanuginosum. *Foods* **2021**, *10*, 2710. [CrossRef]

22. Adomako-Bonsu, A.G.; Chan, S.L.; Pratten, M.; Fry, J.R. Antioxidant activity of rosmarinic acid and its principal metabolites in chemical and cellular systems: Importance of physico-chemical characteristics. *Toxicol. Vitr.* **2017**, *40*, 248–255. [CrossRef]

23. Rocha, J.; Eduardo-Figueira, M.; Barateiro, A.; Fernandes, A.; Brites, D.; Bronze, R.; Duarte, C.M.; Serra, A.T.; Pinto, R.; Freitas, M.; et al. Anti-inflammatory effect of rosmarinic acid and an extract of Rosmarinus officinalis in rat models of local and systemic inflammation. *Basic Clin. Pharmacol. Toxicol.* **2015**, *116*, 398–413. [CrossRef]

24. Krenkel, O.; Tacke, F. Liver macrophages in tissue homeostasis and disease. *Nat. Rev. Immunol.* **2017**, *17*, 306–321. [CrossRef]

25. Su, G.L. Lipopolysaccharides in liver injury: Molecular mechanisms of Kupffer cell activation. *Am. J. Physiol. Gastrointest. Liver Physiol.* **2002**, *283*, G256–G265. [CrossRef] [PubMed]

26. Li, L.; Xu, L.J.; He, Z.D.; Yang, Q.Q.; Peng, Y.; Xiao, P.G. Chemical study on ethyl acetate portion of Ehretia thyrsiflora, boraginaceae species of Kudingcha. *Zhongguo Zhong Yao Za Zhi* **2008**, *33*, 2121–2123. [PubMed]

27. Niculae, M.; Hanganu, D.; Oniga, I.; Benedec, D.; Ielciu, I.; Giupana, R.; Sandru, C.D.; Ciocârlan, N.; Spinu, M. Phytochemical Profile and Antimicrobial Potential of Extracts Obtained from Thymus marschallianus Willd. *Molecules* **2019**, *24*, 3101. [CrossRef]

28. Koley, S.; Chu, K.L.; Gill, S.S.; Allen, D.K. An efficient LC-MS method for isomer separation and detection of sugars, phosphorylated sugars, and organic acids. *J. Exp. Bot.* **2021**, *73*, 2938–2952. [CrossRef] [PubMed]

29. Shi, M.-Z.; Yu, Y.-L.; Zhu, S.-C.; Cao, J.; Ye, L.-H. Nontargeted metabonomics-assisted two-dimensional ion mobility mass spectrometry point imaging to identify plant teas. *LWT* **2022**, *167*, 113852. [CrossRef]

30. Yang, J.-M.; Ip, S.-P.P.; Yeung, H.-K.J.; Che, C.-T. HPLC-MS analysis of Schisandra lignans and their metabolites in Caco-2 cell monolayer and rat everted gut sac models and in rat plasma. *Acta Pharm. Sin. B* **2011**, *1*, 46–55. [CrossRef]

31. Le Gall, G.; DuPont, M.S.; Mellon, F.A.; Davis, A.L.; Collins, G.J.; Verhoeyen, M.E.; Colquhoun, I.J. Characterization and Content of Flavonoid Glycosides in Genetically Modified Tomato (*Lycopersicon esculentum*) Fruits. *J. Agric. Food Chem.* **2003**, *51*, 2438–2446. [CrossRef]

32. Abdallah, H.M.; Al-Abd, A.M.; Asaad, G.F.; Abdel-Naim, A.B.; El-halawany, A.M. Isolation of Antiosteoporotic Compounds from Seeds of Sophora japonica. *PLoS ONE* **2014**, *9*, e98559. [CrossRef]

33. Jung, C.H.; Kim, Y.; Kim, M.S.; Lee, S.; Yoo, S.H. The establishment of efficient bioconversion, extraction, and isolation processes for the production of phyllodulcin, a potential high intensity sweetener, from sweet hydrangea leaves (*Hydrangea macrophylla Thunbergii*). *Phytochem. Anal.* **2016**, *27*, 140–147. [CrossRef]

34. Zakaria, Z.A.; Balan, T.; Azemi, A.; Omar, M.; Mohtarrudin, N.; Ahmad, Z.; Abdullah, M.; Mohd Desa, M.N.; Teh, L.K.; Salleh, M.Z. Mechanism(s) of action underlying the gastroprotective effect of ethyl acetate fraction obtained from the crude methanolic leaves extract of Muntingia calabura. *BMC Complement. Altern. Med.* **2016**, *16*, 78. [CrossRef]

35. Damašius, J.; Venskutonis, R.; Kaškonienė, V.; Maruska, A. Fast Screening of the Main Phenolic Acids with Antioxidant Properties in Common Spices Using On-Line HPLC/UV/DPPH Radical Scavenging Assay. *Anal. Methods* **2014**, *6*, 2774–2779. [CrossRef]

36. Jakšić, D.; Puel, O.; Canlet, C.; Kopjar, N.; Kosalec, I.; Klarić, M. Cytotoxicity and genotoxicity of versicolorins and 5-methoxysterigmatocystin in A549 cells. *Arch. Toxicol.* **2012**, *86*, 1583–1591. [CrossRef] [PubMed]

37. Avasthi, A.S.; Bhatnagar, M.; Sarkar, N.; Kitchlu, S.; Ghosal, S. Bioassay guided screening, optimization and characterization of antioxidant compounds from high altitude wild edible plants of Ladakh. *J. Food Sci. Technol.* **2016**, *53*, 3244–3252. [CrossRef] [PubMed]

38. Cao, T.Q.; Liu, Z.; Dong, L.; Lee, H.; Ko, W.; Vinh, L.B.; Tuan, N.Q.; Kim, Y.C.; Sohn, J.H.; Yim, J.H.; et al. Identification of Potential Anti-Neuroinflammatory Inhibitors from Antarctic Fungal Strain *Aspergillus* sp. SF-7402 via Regulating the NF-κB Signaling Pathway in Microglia. *Molecules* **2022**, *27*, 2851. [CrossRef] [PubMed]

39. Silva, S.; Da, S.; Lopes, A.; De, M.; Agra, M.D.F.; Vasconcelos, E.; Da-Cunha, L.; Barbosa Filho, J.; Silva, M.; Braz-Filho, R. A new arylnaphthalene type lignan from *Cordia rufescens* A. DC. (*Boraginaceae*). *Arkivoc* **2004**, *2004*, 54–58. [CrossRef]

40. De Taeye, C.; Caullet, G.; Eyamo Evina, V.J.; Collin, S. Procyanidin A2 and Its Degradation Products in Raw, Fermented, and Roasted Cocoa. *J. Agric. Food Chem.* **2017**, *65*, 1715–1723. [CrossRef] [PubMed]

41. Katragunta, K.; Siva, B.; Kondepudi, N.; Vadaparthi, P.R.R.; Rama Rao, N.; Tiwari, A.K.; Suresh Babu, K. Estimation of boswellic acids in herbal formulations containing *Boswellia serrata* extract and comprehensive characterization of secondary metabolites using UPLC-Q-Tof-MS(e). *J. Pharm. Anal.* **2019**, *9*, 414–422. [CrossRef]

42. Habib, H.; Finno, C.J.; Gennity, I.; Favro, G.; Hales, E.; Puschner, B.; Moeller, B.C. Simultaneous quantification of vitamin E and vitamin E metabolites in equine plasma and serum using LC-MS/MS. *J. Vet. Diagn. Investig.* **2021**, *33*, 506–515. [CrossRef]

43. Ci, X.; Ren, R.; Xu, K.; Li, H.; Yu, Q.; Song, Y.; Wang, D.; Li, R.; Deng, X. Schisantherin A exhibits anti-inflammatory properties by down-regulating NF-kappaB and MAPK signaling pathways in lipopolysaccharide-treated RAW 264.7 cells. *Inflammation* **2010**, *33*, 126–136. [CrossRef]

44. Fang, Y.; Wang, H.; Xia, X.; Yang, L.; He, J. Kaempferol 3-O-(2G-glucosylrutinoside)-7-O-glucoside isolated from the flowers of *Hosta plantaginea* exerts anti-inflammatory activity via suppression of NF-κB, MAPKs and Akt pathways in RAW 264.7 cells. *Biomed. Pharmacother.* **2022**, *153*, 113295. [CrossRef]

45. Wang, Q.Q.; Gao, H.; Yuan, R.; Han, S.; Li, X.X.; Tang, M.; Dong, B.; Li, J.X.; Zhao, L.C.; Feng, J.; et al. Procyanidin A2, a polyphenolic compound, exerts anti-inflammatory and anti-oxidative activity in lipopolysaccharide-stimulated RAW264.7 cells. *PLoS ONE* **2020**, *15*, e0237017. [CrossRef]

46. Frankel, E.N. The antioxidant and nutritional effects of tocopherols, ascorbic acid and beta-carotene in relation to processing of edible oils. *Bibl. Nutr. Dieta* **1989**, *43*, 297–312. [CrossRef]

47. Papi, S.; Ahmadizar, F.; Hasanvand, A. The role of nitric oxide in inflammation and oxidative stress. *Immunopathol. Persa* **2019**, *5*, e08. [CrossRef]

48. Xue, Q.; Yan, Y.; Zhang, R.; Xiong, H. Regulation of iNOS on Immune Cells and Its Role in Diseases. *Int. J. Mol. Sci.* **2018**, *19*, 3805. [CrossRef]

49. Yang, G.; Nguyen, X.; Ou, J.; Rekulapelli, P.; Stevenson, D.K.; Dennery, P.A. Unique effects of zinc protoporphyrin on HO-1 induction and apoptosis. *Blood* **2001**, *97*, 1306–1313. [CrossRef] [PubMed]

50. Loboda, A.; Damulewicz, M.; Pyza, E.; Jozkowicz, A.; Dulak, J. Role of Nrf2/HO-1 system in development, oxidative stress response and diseases: An evolutionarily conserved mechanism. *Cell. Mol. Life Sci.* **2016**, *73*, 3221–3247. [CrossRef]

51. Sharif, O.; Bolshakov, V.N.; Raines, S.; Newham, P.; Perkins, N.D. Transcriptional profiling of the LPS induced NF-kappaB response in macrophages. *BMC Immunol.* **2007**, *8*, 1. [CrossRef]

52. Hobbs, S.; Reynoso, M.; Geddis, A.V.; Mitrophanov, A.Y.; Matheny, R.W., Jr. LPS-stimulated NF-kappaB p65 dynamic response marks the initiation of TNF expression and transition to IL-10 expression in RAW 264.7 macrophages. *Physiol. Rep.* **2018**, *6*, e13914. [CrossRef]

53. Moens, U.; Kostenko, S.; Sveinbjornsson, B. The Role of Mitogen-Activated Protein Kinase-Activated Protein Kinases (MAP-KAPKs) in Inflammation. *Genes* **2013**, *4*, 101–133. [CrossRef] [PubMed]

54. Chan, E.D.; Riches, D.W. IFN-gamma + LPS induction of iNOS is modulated by ERK, JNK/SAPK, and p38(mapk) in a mouse macrophage cell line. *Am. J. Physiol. Cell Physiol.* **2001**, *280*, C441–C450. [CrossRef] [PubMed]

55. Clemens, M.G. Nitric oxide in liver injury. *Hepatology* **1999**, *30*, 1–5. [CrossRef] [PubMed]

56. Milbourne, E.A.; Bygrave, F.L. Does nitric oxide play a role in liver function? *Cell. Signal.* **1995**, *7*, 313–318. [CrossRef] [PubMed]

57. MacMicking, J.; Xie, Q.W.; Nathan, C. Nitric oxide and macrophage function. *Annu. Rev. Immunol.* **1997**, *15*, 323–350. [CrossRef] [PubMed]

58. Anavi, S.; Eisenberg-Bord, M.; Hahn-Obercyger, M.; Genin, O.; Pines, M.; Tirosh, O. The role of iNOS in cholesterol-induced liver fibrosis. *Lab. Investig.* **2015**, *95*, 914–924. [CrossRef] [PubMed]

59. Cohen-Naftaly, M.; Friedman, S.L. Current status of novel antifibrotic therapies in patients with chronic liver disease. *Therap. Adv. Gastroenterol.* **2011**, *4*, 391–417. [CrossRef]

60. Stahl, W.; Matejovic, M.; Radermacher, P. Inhibition of nitric oxide synthase during sepsis: Revival because of isoform selectivity? *Shock* **2010**, *34*, 321–322. [CrossRef]

61. Nathan, C. Is iNOS beginning to smoke? *Cell* **2011**, *147*, 257–258. [CrossRef]

62. Immenschuh, S.; Ramadori, G. Gene regulation of heme oxygenase-1 as a therapeutic target. *Biochem. Pharmacol.* **2000**, *60*, 1121–1128. [CrossRef]

63. Campbell, N.K.; Fitzgerald, H.K.; Dunne, A. Regulation of inflammation by the antioxidant haem oxygenase 1. *Nat. Rev. Immunol.* **2021**, *21*, 411–425. [CrossRef]

64. Kapturczak, M.H.; Wasserfall, C.; Brusko, T.; Campbell-Thompson, M.; Ellis, T.M.; Atkinson, M.A.; Agarwal, A. Heme oxygenase-1 modulates early inflammatory responses: Evidence from the heme oxygenase-1-deficient mouse. *Am. J. Pathol.* **2004**, *165*, 1045–1053. [CrossRef]

65. Yachie, A.; Niida, Y.; Wada, T.; Igarashi, N.; Kaneda, H.; Toma, T.; Ohta, K.; Kasahara, Y.; Koizumi, S. Oxidative stress causes enhanced endothelial cell injury in human heme oxygenase-1 deficiency. *J. Clin. Investig.* **1999**, *103*, 129–135. [CrossRef] [PubMed]

66. Poss, K.D.; Tonegawa, S. Reduced stress defense in heme oxygenase 1-deficient cells. *Proc. Natl. Acad. Sci. USA* **1997**, *94*, 10925–10930. [CrossRef] [PubMed]

67. Radhakrishnan, N.; Yadav, S.P.; Sachdeva, A.; Pruthi, P.K.; Sawhney, S.; Piplani, T.; Wada, T.; Yachie, A. Human heme oxygenase-1 deficiency presenting with hemolysis, nephritis, and asplenia. *J. Pediatr. Hematol. Oncol.* **2011**, *33*, 74–78. [CrossRef] [PubMed]

68. Paine, A.; Eiz-Vesper, B.; Blasczyk, R.; Immenschuh, S. Signaling to heme oxygenase-1 and its anti-inflammatory therapeutic potential. *Biochem. Pharmacol.* **2010**, *80*, 1895–1903. [CrossRef] [PubMed]

69. Rahman, I.; Biswas, S.K.; Kirkham, P.A. Regulation of inflammation and redox signaling by dietary polyphenols. *Biochem. Pharmacol.* **2006**, *72*, 1439–1452. [CrossRef]

70. Wang, P.; Geng, J.; Gao, J.; Zhao, H.; Li, J.; Shi, Y.; Yang, B.; Xiao, C.; Linghu, Y.; Sun, X.; et al. Macrophage achieves self-protection against oxidative stress-induced ageing through the Mst-Nrf2 axis. *Nat. Commun.* **2019**, *10*, 755. [CrossRef]

71. Villeneuve, N.F.; Lau, A.; Zhang, D.D. Regulation of the Nrf2-Keap1 antioxidant response by the ubiquitin proteasome system: An insight into cullin-ring ubiquitin ligases. *Antioxid. Redox Signal.* **2010**, *13*, 1699–1712. [CrossRef]

72. Kobayashi, M.; Yamamoto, M. Nrf2-Keap1 regulation of cellular defense mechanisms against electrophiles and reactive oxygen species. *Adv. Enzyme Regul.* **2006**, *46*, 113–140. [CrossRef]

73. Singh, A.; Venkannagari, S.; Oh, K.H.; Zhang, Y.Q.; Rohde, J.M.; Liu, L.; Nimmagadda, S.; Sudini, K.; Brimacombe, K.R.; Gajghate, S.; et al. Small Molecule Inhibitor of NRF2 Selectively Intervenes Therapeutic Resistance in KEAP1-Deficient NSCLC Tumors. *ACS Chem. Biol.* **2016**, *11*, 3214–3225. [CrossRef]

74. Ren, D.; Villeneuve, N.F.; Jiang, T.; Wu, T.; Lau, A.; Toppin, H.A.; Zhang, D.D. Brusatol enhances the efficacy of chemotherapy by inhibiting the Nrf2-mediated defense mechanism. *Proc. Natl. Acad. Sci. USA* **2011**, *108*, 1433–1438. [CrossRef]

75. Luo, C.; Zou, L.; Sun, H.; Peng, J.; Gao, C.; Bao, L.; Ji, R.; Jin, Y.; Sun, S. A Review of the Anti-Inflammatory Effects of Rosmarinic Acid on Inflammatory Diseases. *Front. Pharmacol.* **2020**, *11*, 153. [CrossRef] [PubMed]

76. Cui, H.Y.; Zhang, X.J.; Yang, Y.; Zhang, C.; Zhu, C.H.; Miao, J.Y.; Chen, R. Rosmarinic acid elicits neuroprotection in ischemic stroke via Nrf2 and heme oxygenase 1 signaling. *Neural Regen. Res.* **2018**, *13*, 2119–2128. [CrossRef] [PubMed]

77. Domitrovic, R.; Skoda, M.; Vasiljev Marchesi, V.; Cvijanovic, O.; Pernjak Pugel, E.; Stefan, M.B. Rosmarinic acid ameliorates acute liver damage and fibrogenesis in carbon tetrachloride-intoxicated mice. *Food Chem. Toxicol.* **2013**, *51*, 370–378. [CrossRef] [PubMed]

78. Qiao, S.; Li, W.; Tsubouchi, R.; Haneda, M.; Murakami, K.; Takeuchi, F.; Nisimoto, Y.; Yoshino, M. Rosmarinic acid inhibits the formation of reactive oxygen and nitrogen species in RAW264.7 macrophages. *Free Radic. Res.* **2005**, *39*, 995–1003. [CrossRef]

79. Chen, C.P.; Lin, Y.C.; Peng, Y.H.; Chen, H.M.; Lin, J.T.; Kao, S.H. Rosmarinic Acid Attenuates the Lipopolysaccharide-Provoked Inflammatory Response of Vascular Smooth Muscle Cell via Inhibition of MAPK/NF-kappaB Cascade. *Pharmaceuticals* **2022**, *15*, 437. [CrossRef]

80. Siddiqui, M.Z. Boswellia serrata, a potential antiinflammatory agent: An overview. *Indian J. Pharm. Sci.* **2011**, *73*, 255–261. [CrossRef]

81. Zhang, P.Y.; Yu, B.; Men, W.J.; Bai, R.Y.; Chen, M.Y.; Wang, Z.X.; Zeng, T.; Zhou, K. Acetyl-alpha-boswellic acid and Acetyl-beta-boswellic acid protects against caerulein-induced pancreatitis via down-regulating MAPKs in mice. *Int. Immunopharmacol.* **2020**, *86*, 106682. [CrossRef]

MDPI AG

Grosspeteranlage 5

4052 Basel

Switzerland

Tel.: +41 61 683 77 34

Antioxidants Editorial Office

E-mail: antioxidants@mdpi.com

www.mdpi.com/journal/antioxidants